Statistical
Methods
for Psychology

SIXTH EDITION

Statistical Methods for Psychology

David C. Howell
University of Vermont

THOMSON ™

WADSWORTH

Australia • Brazil • Canada • Mexico • Singapore • Spain
United Kingdom • United States

THOMSON
™
WADSWORTH

Publisher: *Vicki Knight*
Assistant Editor: *Jennifer Keever*
Editorial Assistant: *Sheila Walsh*
Technology Project Manager: *Adrian Paz*
Vice President, Director of Marketing: *Caroline Croley*
Marketing Assistant: *Natasha Coats*
Senior Marketing Communications Manager: *Kelley McAllister*
Project Manager, Editorial Production: *Karol Jurado*
Creative Director: *Rob Hugel*
Senior Art Director: *Vernon Boes*
Senior Print Buyer: *Barbara Britton*

Senior Permissions Editor: *Joohee Lee*
Production Service: *Sara Dovre Wudali/Interactive Composition Corporation*
Text Designer: *Roy Neuhaus*
Copy Editor: *Robin Gold*
Cover Designer: *Roy Neuhaus*
Cover Image: *Masato Tokiwa/Photonica/Getty Images*
Cover Printer: *Coral Graphic Services, Inc.*
Compositor: *Interactive Composition Corporation*
Printer: *Quebecor World/Dubuque*

Library of Congress Control Number: 2006920846

Student Edition: ISBN 978-0-495-01287-0
Student Edition: ISBN 0-495-01287-4

519, 5024

Thomson Higher Education
10 Davis Drive
Belmont, CA 94002-3098
USA

For more information about our products, contact us at:
Thomson Learning Academic Resource Center
1-800-423-0563

For permission to use material from this text or product, submit a request online at **http://www.thomsonrights.com**.

Any additional questions about permissions can be submitted by e-mail to **thomsonrights@thomson.com**.

To Donna

Brief Contents

Contents

CHAPTER 16   Analyses of Variance and Covariance as General Linear Models   555

CHAPTER 17   Log-Linear Analysis   605

CHAPTER 18   Resampling and Nonparametric Approaches to Data   635

Preface

This sixth edition of *Statistical Methods for Psychology,* like the previous editions, surveys statistical techniques commonly used in the behavioral and social sciences, especially psychology and education. Although it is designed for students at the intermediate level or above, it does not assume that students have had either a previous course in statistics or a course in mathematics beyond high-school algebra. Those students who have had an introductory course will find that the early material provides a welcome review. The book is suitable for either a one-term or a full-year course, and I have used it successfully for both. Because I have found that students, and faculty, frequently refer back to the book from which they originally learned statistics when they have a statistical problem, I have included material that will make the book a useful reference for future use. The instructor who wants to omit this material will have no difficulty doing so. I have cut back on that material, however, to include only what is likely to be useful. The idea of including every interesting idea had led to a book that was beginning to be daunting.

My intention in writing this book was to explain the material at an intuitive level. This should not be taken to mean that the material is "watered down," but only that the emphasis is on conceptual understanding. The student who can successfully derive the sampling distribution of *t,* for example, may not have any understanding of how that distribution is to be used. With respect to this example, my aim has been to concentrate on the meaning of a sampling distribution and to show the role it plays in the general theory of hypothesis testing. In my opinion, this approach allows students to gain a better understanding, than would a more technical approach, of the way a particular test works and of the interrelationships among tests.

Changes in the Sixth Edition

This sixth edition contains several new, or expanded, features that make the book more appealing to the student and more relevant to the actual process of methodology and data analysis:

- I have responded to the issue faced by the American Psychological Association's committee on null hypothesis testing, and have included additional material on effect size and magnitude of effect. The coverage in this edition goes well beyond the coverage in previous editions and should serve as a thorough introduction to the material.

• In chapters in which students are expected to draw conclusions from data, I have included a section on how they would go about writing up the results. Students often find that a difficult thing to do.

• I have included a discussion of a proposal put forth by Jones and Tukey (2000) in which they reconceived of hypothesis testing in ways that I find very helpful. However, I have retained the more traditional approach because students will be expected to be familiar with it.

• In the previous edition, I largely converted to using definitional formulae rather than computational ones. Thus in the analysis of variance, for example, the total sum of squares is expressed in terms of squared deviations around the mean, rather than as the sum of squared observations minus the grand total squared over *N*. I continued that approach here and omitted some of the material that seemed to focus more on calculation than on understanding. (I did retain calculational material when it would be important to someone coming back to the book as a reference.)

• There is a greater emphasis on computer solutions, especially in later chapters. I have emphasized SPSS more than other packages because that is the one most students use. However, I have retained some printouts from other packages.

• Data for all examples and problems are available on the web, except for those that can more quickly be entered by hand. We have done away with the CD-ROM and put everything on the web (www.uvm.edu/~dhowell/methods/) where it is more accessible to users.

• I have spent a substantial amount of time pulling together material for instructors and students and placing it on web pages on the Internet. Users can readily get at additional (and complex) examples, discussion of topics that aren't covered in the text, data, other sources on the Internet, demonstrations that would be suitable for class or for a lab, and so on. All of this is easily available to anyone with a web browser (such as Firefox or Internet Explorer) and with an Internet connection. I will continue to add to this material and will encourage people to use it and critique it. I have also included in the book links to websites that provide Java applets that perform interesting and useful statistical calculations. The address of my own, more general, website is http://www.uvm.edu/~dhowell/StatPages/ StatHomePage.html (capitalization in this address is critical).

• In the last edition, I added a major section to Chapter 18 on nonparametric statistics to account for the recent work on resampling statistics and bootstrapping. I have retained that material here and plan to expand on it even more through my website. Computer solutions to these problems are now readily available, and affordable, and there has been a surge of interest in these procedures. They are particularly useful for helping students to understand the logic behind hypothesis testing and to actually be involved in designing a test. I have written a simple Windows program that will carry out a number of resampling and bootstrapping procedures, and this is freely available over the Internet.

This edition shares with its predecessors two underlying themes that are more or less independent of the statistical hypothesis tests that are the main content of the book:

• The first theme is the importance of looking at the data before jumping in with a hypothesis test. With this in mind, I discuss, in detail, plotting data, looking for outliers, and checking assumptions. (Graphical displays are used extensively.) I try to do this with each data set as soon as I present it, even though the data set may be intended as an example of a sophisticated statistical technique. As examples, see pages 311 and 497.

- The second theme is the importance of the relationship between the statistical test to be employed and the theoretical questions being posed by the experiment. To emphasize this relationship, I use real examples in an attempt to make the student understand the purpose behind the experiment and the predictions made by the theory. For this reason, I sometimes use one major example as the focus for an entire section, or even a whole chapter. For example, interesting data on the moon illusion from a well-known study by Kaufman and Rock (1962) are used in several forms of the *t* test (pages 179–182), and most of Chapter 12 is organized around an important study of morphine addiction by Siegel (1975). Chapter 17 on log-linear models, which has been extensively revised in this edition, is built around Pugh's study of the "blame-the-victim" strategy in prosecutions for rape. Each of these examples should have direct relevance for students. The increased emphasis on effect sizes in this edition helps drive home the point that one must think carefully about one's data and research questions.

Although no one would be likely to call this book controversial, I have felt it important to express opinions on a number of controversial issues. After all, the controversies within statistics are part of what makes it an interesting discipline. For example, I have argued that the underlying measurement scale is not as important as some have suggested, and I have argued for a particular way of treating the analyses of variance with unequal group sizes (unless there is a compelling reason to do otherwise). I do not expect every instructor to agree with me, and actually, I hope that some will not. This offers the opportunity to give students opposing views and help them understand the issues. It seems to me that it is unfair and frustrating to the student to present a large number of multiple comparison procedures (which I do), and then to walk away and leave that student with no recommendation about which procedure is best for his or her problem.

In addition to the answer section at the back of the book, there is a solutions manual for the students, with extensive worked solutions to odd-numbered exercises. This manual can be purchased from the publisher and is available at the afore-mentioned website. The data sets for all examples and exercises that were previously on a CD-ROM are now available on the Internet at www.uvm.edu/~dhowell/methods/. In addition, a separate instructor's manual with worked out solutions to all problems is available from the publisher.

Acknowledgments

I would like to thank the following reviewers who read the manuscript and provided valuable feedback: Jamison Fargo, Utah State University; Susan Cashin, University of Wisconsin–Milwaukee; Bill Frederickson, University of Central Oklahoma; Angus MacDonald, University of Minnesota; Carl Scott, University of St. Thomas–Houston; William Smith, California State University–Fullerton; and Karl Wuensch, East Carolina University, who has provided valuable guidance over many editions. In the fifth edition, I received helpful comments and suggestions from Kenneth J. Berry, Colorado State University; Tim Bockes, Nazareth College; Richard Lehman, Franklin and Marshall College; Tim Robinson, Virginia Tech; Paul R. Shirley, University of California–Irvine; Mathew Spackman, Brigham Young University; Mary Uley, Lindenwood University; and Christy Witt, Louisiana State University. Their influence is still evident in this edition.

The publishing staff was exceptionally helpful throughout, and I would like to thank Dory Schaeffer, Marketing Manager; Vernon Boes, Creative Director; Adrian Paz, Technology Project Manager; and Karol Jurado, Production Project Manager. Sara Dovre Wudali and Robin Gold were a pleasure to work with on the production of the final book.

David C. Howell
University of Vermont

About the Author

Professor Howell is Emeritus Professor at the University of Vermont. After gaining his Ph.D. from Tulane University in 1967, he was associated with the University of Vermont until retiring as chair of the Department of Psychology in 2002. He also spent two separate years as visiting professor at two universities in the United Kingdom.

Professor Howell is the author of several books and many journal articles and book chapters. He continues to write in his retirement and was most recently the co-editor, with Brian Everitt, of *The Encyclopedia of Statistics in Behavioral Sciences,* published by Wiley. He has recently authored a number of chapters in various books on research design and statistics.

Professor Howell now lives in Colorado where he enjoys the winter snow and is an avid skier and hiker.

CHAPTER 1

Basic Concepts

Objectives

To examine the kinds of problems presented in this book and the issues involved in selecting a statistical procedure.

Contents

STRESS IS SOMETHING that we are all forced to deal with throughout life. It arises in our daily interactions with those around us, in our interactions with the environment, in the face of an impending exam, and, for many students, in the realization that they are required to take a statistics course. Although most of us learn to respond and adapt to stress, the learning process is often slow and painful. This rather grim preamble may not sound like a great way to introduce a course on statistics, but it leads to a description of a practical research project, which in turn illustrates a number of important statistical concepts. I was involved in a very similar project a number of years ago, so this example is far from hypothetical.

A group of educators has put together a course designed to teach high-school students how to manage stress. They need an outside investigator, however, who can tell them how well the course is working and, in particular, whether students who take the course have fewer problems with stress, and handle it better, than do students who have not taken the course. For the moment we will assume that we are charged with the task of designing an evaluation of their program. The experiment that we design will not be complete, but it will illustrate some of the issues involved in designing and analyzing experiments and some of the statistical concepts with which you must be familiar.

1.1   Important Terms

Although the program in stress management was designed for high-school students, it clearly would be impossible to apply it to the population of all high-school students in the country. First, there are far too many such students. Moreover, it makes no sense to apply a program to everyone until we know whether it is a useful program. Instead of dealing with the entire population of high-school students, we will draw a sample of students from that population and apply the program to them. But we will not draw just any old sample. We would like to draw a **random sample.** To do this, we would follow a particular set of procedures to ensure that each and every element of the population has an equal chance of being selected.[1] Having drawn our sample of students, we will **randomly assign** half the subjects to a group that will receive the stress-management program and half to a group that will not receive the program.

random sample

randomly assign

This description has already brought out several concepts that need further elaboration; namely, a population, a sample, a random sample, and random assignment. A **population** is the entire collection of events (students' scores, people's incomes, rats' running speeds, etc.) in which you are interested. Thus, if you are interested in the stress-management scores of all high-school students in the United States, then the collection of all high-school students' stress scores would form a population—in this case, a population of many millions of elements. If, on the other hand, you were interested in the stress scores of high-school seniors only in Fairfax, Vermont (a town of fewer than 4,000 inhabitants), the population would consist of only about 100 elements.

population

The point is that a population can range from a relatively small set of numbers, which can be collected easily, to a large but finite set of numbers, which would be impractical to collect in their entirety, to an infinite set of numbers, such as the set of all possible cartoon drawings that students could theoretically produce, which would be impossible to collect. Unfortunately for us, the populations we are interested in are usually very large. The practical consequence is that we seldom if ever measure entire populations. Instead, we are

---

[1] Although random samples are the ideal, they are often totally impractical because of constraints that life imposes on us. This does not stop our doing research, but it does place limits on the conclusions that we can draw from our samples. In practice, random assignment, to be discussed next, is far more important.

sample

forced to draw only a **sample** of observations from that population and to use that sample to infer something about the characteristics of the population.

Assuming that the sample is truly random, we not only can estimate certain characteristics of the population, but also can have a very good idea of how accurate our estimates are. To the extent that the sample is not random, our estimates may or may not be meaningful, because the sample may or may not accurately reflect the entire population. Studies based

external validity

on nonrandom samples may lack **external validity.** A sample drawn from a small town in Nebraska would not produce a valid estimate of the percentage of the population of the United States that is Hispanic—nor would a sample drawn solely from the American Southwest. On the other hand, a sample from a small town in Nebraska might give us a reasonable estimate of the reaction time of people to stimuli presented suddenly.

Before going on, let us clear up one point that tends to confuse many people. The problem is that one person's sample might be another person's population. For example, if I were to conduct a study on the effectiveness of this book as a teaching instrument, one class's scores on an examination might be considered by me to be a sample, albeit a nonrandom one, of the population of scores of all students using, or potentially using, this book. The class instructor, on the other hand, is probably not terribly concerned about this book, but instead cares only about his or her own students. He or she would regard the same set of scores as a population. In turn, someone interested in the teaching of statistics might regard my population (everyone using my book) as a very nonrandom sample from a larger population (everyone using any textbook in statistics). Thus, the definition of a population depends on what you are interested in studying.

The fact that I have used nonrandom samples here to make a point should not lead you to think that randomness is not important. On the contrary, it is the cornerstone of most statistical procedures. As a matter of fact, the relevant population could be defined as the collection of elements from which samples have been randomly drawn. But this leads to a major theoretical problem. In our stress study, for example, it is highly unlikely that we would seriously consider drawing a truly random sample of U.S. high-school students and administering the stress-management program to them. How then are we going to take advantage of methods and procedures based on the assumption of random sampling? This question has no clear answer other than to realize the limitations of our methodology and to draw our conclusions accordingly. To the extent that we think that our sample is not representative of U.S. high-school students, we must limit our interpretation of the results. To the extent to which the sample is representative of the population, our estimates have validity.

random assignment

internal validity

While we are dealing with random selection of subjects, we must consider the related concept of random assignment. Whereas random selection concerns the *source* of our data and is important for generalizing the results of our study to the whole population, **random assignment** of subjects (once selected) to treatment groups is fundamental to the integrity, or **internal validity,** of our experiment. It helps ensure that our results mean what we think they mean (which, for example, would not be the case if we put all our timid subjects in one group and our more self-confident subjects in the other). In actual practice, random assignment is usually far more important than random sampling.

variable

Having dealt with the selection of subjects and their assignment to treatment groups, it is time to consider the data that will result. Because we want to study the ability of subjects to deal with stress, and because the response to stress is a function of many variables, a critical aspect of planning the study involves selecting the variables to be studied. A **variable** is a property of an object or event that can take on different values. For example, hair color is a variable because it is a property of an object (hair) and can take on different values (brown, yellow, red, gray, etc.). With respect to our evaluation, such things as self-confidence, social support, gender, degree of personal control, and treatment group are all relevant variables, and each element of the population takes on *one* value of each variable. We can further

discrete variables

continuous variables

discriminate between **discrete variables,** such as gender or high-school class, which take on only a limited number of values, and **continuous variables,** such as age and self-esteem score, which can assume, at least in theory, any value between the lowest and highest points on the scale.[2] As you will see, this distinction plays an important role in the way we treat data.

Closely related to the distinction between discrete and continuous variables is the distinction between measurement and categorical data. By **measurement data** (sometimes called **quantitative data**), we mean the results of any sort of measurement—for example, grades on a test, people's weights, scores on a scale of self-esteem, and so on. In all cases, some sort of instrument (in its broadest sense) has been used to measure something.

measurement data

quantitative data

categorical data

frequency data

qualitative data

On the other hand, **categorical data** (also known as **frequency data** or **qualitative data**) are illustrated in such statements as, "There are 34 females and 26 males in our study" or "Fifteen people were classed as 'highly anxious,' 33 as 'neutral,' and 12 as 'low anxious.'" Here we are categorizing things, and our data consist of frequencies for each category (hence the name categorical data). Several hundred subjects might be involved in our study, but the results (data) would consist of only two or three numbers—the number of subjects falling in each category. In contrast, if instead of sorting people with respect to high, medium, and low anxiety, we had assigned them each a score based on some more-or-less continuous scale of anxiety, we would be dealing with measurement data, and the data would consist of scores for each subject on that variable. Note that in both situations the variable is labeled *anxiety*. As with most distinctions, the one between measurement and categorical data can be pushed too far. The distinction is useful, however, and the answer to the question of whether a variable is a measurement or a categorical one is almost always clear in practice.

independent variables

dependent variables

In statistics, we also dichotomize the concept of a variable in an additional way. We speak of **independent variables** (those that are manipulated by the experimenter) and **dependent variables** (those that are not under the experimenter's control—the data). In our example, group membership is an independent variable because we control it. We decide what the treatments will be and who will receive each treatment. The data—such as self-esteem scores, age, personal control, and so on—are the dependent variables. Basically, the study is about the independent variables, and the results of the study (the data) are the dependent variables. Independent variables may be either quantitative or qualitative, whereas dependent variables are generally, but certainly not always, quantitative.[3]

1.2 Descriptive and Inferential Statistics

Returning to our intervention program for stress, once we have chosen the variables to be measured and the schools have administered the program to the students, we are left with a collection of raw data—the scores. There are two primary divisions of the field of statistics that are concerned with the use we make of these data.

descriptive statistics

Whenever our purpose is merely to describe a set of data, we are employing **descriptive statistics.** For example, one of the first things that we would want to do with our data is to graph them, to calculate means (averages) and other measures, and to look for extreme scores or oddly shaped distributions of scores. These procedures are called descriptive

---

[2] Actually, a continuous variable is one in which *any* value between the extremes of the scale (e.g., 32.485687 . . .) is possible. In practice, however, we treat a variable as continuous whenever it can take on many different values and we treat it as discrete whenever it can take on only a few different values.

[3] Many people have difficulty remembering which is the dependent variable and which is the independent variable. Notice that both "dependent" and "data" start with a "d."

statistics because they are primarily aimed at describing the data. The field of descriptive statistics was once looked down on as a rather uninteresting field populated primarily by those who drew distorted-looking graphs for such publications as *Time* magazine. Twenty-five years ago, however, John Tukey developed what he called exploratory statistics, or **exploratory data analysis (EDA).** He showed the necessity of paying close attention to the data and examining them in detail before invoking more technically involved procedures. In this book, we will spend some time considering exploratory methods and Tukey's general orientation toward data.

exploratory data analysis (EDA)

After we have described our data in detail and are satisfied that we understand what the numbers have to say on a superficial level, we will be particularly interested in what is called **inferential statistics.** In fact, most of this book will deal with inferential statistics. In designing our experiment, we acknowledged that it was not possible to measure the entire population, and therefore we drew samples from that population. Our basic questions, however, deal with the population itself. We might want to ask, for example, about the average self-esteem score for an entire population of students who could have taken our program, even though all that we really have is the average score for a sample of students who actually went through the program.

inferential statistics

A measure, such as the average self-esteem score, that refers to an entire population is called a **parameter.** That same measure, when it is calculated from a sample of data that we have collected, is called a **statistic.** Parameters are the *real* entities of interest, and the corresponding statistics are *guesses* at reality. Although most of what we will do in this book deals with sample statistics (or guesses, if you prefer), keep in mind that the reality of interest is the corresponding population parameter. We want to *infer* something about the characteristics of the population (parameters) from what we know about the characteristics of the sample (statistics). In a similar vein, we are particularly interested in knowing whether the average self-esteem score of a population of students potentially enrolled in our program is higher, or lower, than the average self-esteem score of students who might not be enrolled. Again we are dealing with the area of inferential statistics because we are inferring characteristics of populations from characteristics of samples.

parameter

statistic

1.3 Measurement Scales

The topic of measurement scales is one that some writers think is crucial and others think is irrelevant. Although I tend to side with the latter group, it is important that you have some familiarity with the general issue. (You do not have to agree with something to think that it is worth studying. After all, evangelists claim to know a great deal about sin, though they hardly advocate it.) An additional benefit of this discussion is that you will begin to realize that statistics as a subject is not merely a cut-and-dried set of facts but, rather, is a set of facts put together with a variety of interpretations and opinions.

Probably the foremost leader of those who see measurement scales as crucial to the choice of statistical procedures was S. S. Stevens.[4] Zumbo and Zimmerman (2000) have discussed measurement scales at considerable length and remind us that Stevens's system has to be seen in its historical context. In the 1940s and 1950s, Stevens was attempting to defend psychological research against those in the "hard sciences" who had a restricted view of scientific measurement. He was trying to make psychology "respectable." Stevens spent much of his very distinguished professional career developing measurement scales for

---

[4] Chapter 1 in Stevens's *Handbook of Experimental Psychology* (1951) is an excellent reference for anyone wanting to examine the substantial mathematical issues underlying this position.

the field of psychophysics and made important contributions. However, outside of that field there has been little effort in psychology to develop the kinds of scales that Stevens pursued, nor has there been much real interest. The criticisms that so threatened Stevens have largely evaporated, and with them much of the belief that measurement scales critically influence the statistical procedures that are appropriate.

Nominal Scales

nominal scales

In a sense, **nominal scales** are not really scales at all; they do not scale items along any dimension, but rather label them. Variables such as gender and political-party affiliation are nominal variables. Such categorical data are usually measured on a nominal scale, because we merely assign category labels (e.g., male or female; Republican, Democrat, or Independent) to observations. A numerical example of a nominal scale is the set of numbers assigned to football players. Frequently, these numbers have no meaning other than that they are convenient labels to distinguish the players from one another. Letters or pictures of animals could just as easily be used.

Ordinal Scales

ordinal scale

The simplest true scale is an **ordinal scale,** which orders people, objects, or events along some continuum. An excellent example of such a scale is the ranks in the Navy. A commander is lower in prestige than a captain, who in turn is lower than a rear admiral. However, there is no reason to think that the *difference* in prestige between a commander and a captain is the same as that between a captain and a rear admiral. An example from psychology would be the Holmes and Rahe (1967) scale of life stress. Using this scale, you count (sometimes with differential weightings) the number of changes (marriage, moving, new job, etc.) that have occurred during the past 6 months of a person's life. Someone who has a score of 20 is presumed to have experienced more stress than someone with a score of 15, and the latter in turn is presumed to have experienced more stress than someone with a score of 10. Thus, people are ordered, in terms of stress, by the number of changes in their recent lives. This is an example of an ordinal scale, because nothing is implied about the differences between points on the scale. We do not assume, for example, that the difference between 10 and 15 points represents the same difference in stress as the difference between 15 and 20 points. Distinctions of that sort must be left to interval scales.

Interval Scales

interval scale

With an **interval scale,** we have a measurement scale in which we can legitimately speak of differences between scale points. A common example is the Fahrenheit scale of temperature, where a 10-point difference has the same meaning anywhere along the scale. Thus, the difference in temperature between 10° F and 20° F is the same as the difference between 80° F and 90° F. Notice that this scale also satisfies the properties of the two preceding ones. What we do not have with an interval scale, however, is the ability to speak meaningfully about ratios. Thus, we cannot say, for example, that 40° F is half as hot as 80° F, or twice as hot as 20° F. We have to use ratio scales for that purpose. (In this regard, it is worth noting that when we perform perfectly legitimate conversions from one interval scale to another—for example, from the Fahrenheit to the Celsius scale of temperature— we do not even keep the same ratios. Thus, the ratio between 40° and 80° on a Fahrenheit scale is different from the ratio between 4.4° and 26.7° on a Celsius scale, although the temperatures are comparable. This highlights the arbitrary nature of ratios when dealing with interval scales.)

Ratio Scales

ratio scale

A **ratio scale** is one that has a *true* zero point. Notice that the zero point must be a true zero point and not an arbitrary one, such as 0° F or even 0° C. (A true zero point is the point corresponding to the absence of the thing being measured. Since 0° F and 0° C do not represent the absence of temperature or molecular motion, they are not true zero points.) Examples of ratio scales are the common physical ones of length, volume, time, and so on. With these scales, we not only have the properties of the preceding scales but we also can speak about ratios. We can say that in physical terms 10 seconds is twice as long as 5 seconds, that 100 lb is one-third as heavy as 300 lb, and so on.

You might think that the kind of scale with which we are working would be obvious. Unfortunately, especially with the kinds of measures we collect in the behavioral sciences, this is rarely the case. Consider for a moment the situation in which an anxiety questionnaire is administered to a group of high-school students. If you were foolish enough, you might argue that this is a ratio scale of anxiety. You would maintain that a person who scored 0 had no anxiety at all and that a score of 80 reflected twice as much anxiety as did a score of 40. Although most people would find this position ridiculous, with certain questionnaires you might be able to build a reasonable case. Someone else might argue that it is an interval scale and that, although the zero point was somewhat arbitrary (the student receiving a 0 was somewhat anxious but your questions failed to detect it), equal differences in scores represent equal differences in anxiety. A more reasonable stance might be to say that the scores represent an ordinal scale: A 95 reflects more anxiety than an 85, which in turn reflects more than a 75, but equal differences in scores do not reflect equal differences in anxiety. For an excellent and readable discussion of measurement scales, see Hays (1981, pp. 59–65).

As an example of a form of measurement that has a scale that depends on its use, consider the temperature of a house. We generally speak of Fahrenheit temperature as an interval scale. We have just used it as an example of one, and there is no doubt that, to a physicist, the difference between 62° F and 64° F is exactly the same as the difference between 92° F and 94° F. If we are measuring temperature as an index of *comfort,* rather than as an index of molecular activity, however, the same numbers no longer form an interval scale. To a person sitting in a room at 62° F, a jump to 64° F would be distinctly noticeable (and welcome). The same cannot be said about the difference between room temperatures of 92° F and 94° F. This points up the important fact that *it is the underlying variable that we are measuring (e.g., comfort), not the numbers themselves, that is important in defining the scale.* As a scale of *comfort,* degrees Fahrenheit do not form an interval scale—they don't even form a nominal scale because comfort would increase with temperature to a point and would then start to decrease.

There usually is no unanimous agreement concerning the measurement scale employed, so the individual user of statistical procedures must decide which scale best fits the data. All that can be asked of the user is that he or she think about the problem carefully before coming to a decision, rather than simply assume that the standard answer is necessarily the best answer.

The Role of Measurement Scales

I stated earlier that writers disagree about the importance assigned to measurement scales. Some authors have ignored the problem totally, whereas others have organized whole textbooks around the different scales. A reasonable view is that the central issue is the absolute necessity of separating in our minds the numbers we collect from the objects or events to which they refer. Such an argument was made for the example of room temperature, where the scale (interval or ordinal) depended on whether we were interested in measuring some physical attribute of temperature or its effect on people (i.e., comfort). A difference of 2° F is the same, *physically,* anywhere on the scale, but a difference of 2° F when a room is already warm may not *feel* as large as does a difference of 2° F when a room is relatively cool. In other words, we have an interval scale of the physical units but no more than an ordinal scale of comfort (again, up to a point).

Because statistical tests use numbers without considering the objects or events to which those numbers refer, we may carry out any of the standard mathematical operations (addition, multiplication, etc.) regardless of the nature of the underlying scale. An excellent, entertaining, and highly recommended paper on this point is one by Lord (1953), entitled "On the Statistical Treatment of Football Numbers," in which he argues that these numbers can be treated in any way you like because, "The numbers do not remember where they came from" (p. 751).

The problem arises when it is time to interpret the results of some form of statistical manipulation. At that point, we must ask whether the statistical results are related in any meaningful way to the objects or events in question. Here we are no longer dealing with a statistical issue, but with a methodological one. No *statistical* procedure can tell us whether the fact that one group received higher scores than another on an anxiety questionnaire reveals anything about group differences in underlying anxiety levels. Moreover, to be satisfied because the questionnaire provides a ratio scale of anxiety *scores* (a score of 50 is twice as large as a score of 25) is to lose sight of the fact that we set out to measure anxiety, which may not increase in an orderly way with increases in scores. Our statistical tests can apply only to the numbers that we obtain, and the validity of statements about the objects or events that we think we are measuring hinges primarily on our knowledge of those objects or events, not on the measurement scale. We do our best to ensure that our measures relate as closely as possible to what we want to measure, but our results are ultimately only the numbers we obtain and our faith in the relationship between those numbers and the underlying objects or events.[5]

From the preceding discussion, the apparent conclusion—and the one accepted in this book—is that the underlying measurement scale is not crucial in our choice of statistical techniques. A certain amount of common sense is required in interpreting the results of these statistical manipulations. Only a fool would conclude that a painting that was judged as excellent by one person and contemptible by another ought therefore to be classified as mediocre.

1.4 Using Computers

In the not-too-distant past, most statistical analyses were done on desktop or hand calculators, and textbooks were written accordingly. Methods have changed, however, and most calculations are now done by computers.

This book attempts to deal with the increased availability of computers by incorporating them into the discussion. The level of computer involvement increases substantially as the book proceeds and as computations become more laborious. For the simpler procedures, the calculational formulae are important in defining the concept. For example, the formula for a standard deviation or a *t* test defines and makes meaningful what a standard deviation or a

---

[5] As Cohen (1965) has pointed out, "Thurstone once said that in psychology we measure men by their shadows. Indeed, in clinical psychology we often measure men by their shadows while they are dancing in a ballroom illuminated by the reflections of an old-fashioned revolving polyhedral mirror" (p. 102).

t test actually is. In those cases, hand calculation is emphasized even though examples of computer solutions are also given. Later in the book, when we discuss multiple regression or log-linear models, for example, the formulae become less informative. The formula for deriving regression coefficients with five predictors, or the formula for estimating expected frequencies in a complex log-linear model, would not reasonably be expected to add to your understanding of such statistics. In those situations, we will rely almost exclusively on computer solutions.

At present, many statistical software packages are available to the typical researcher or student conducting statistical analyses. Whereas a few years ago we would have broken those into mainframe packages and microcomputer packages, we no longer make that distinction. All the major mainframe packages are now available for microcomputers, doing virtually the same analyses. The most important large statistical packages, which will carry out nearly every analysis that statisticians have invented, are Minitab®, SAS®, SPSS™, and SYSTAT. These are highly reliable and relatively easy-to-use packages, and one or more of them is generally available in any college or university computer center. Many examples of their use are scattered throughout this book. Each has its own set of supporters (my preference may become obvious as we go along), but they are all excellent. Choosing among them hinges on subtle differences.

In speaking about statistical packages, we should mention the widely available spreadsheets such as Excel. These programs are capable of performing a number of statistical calculations, and they produce reasonably good graphics and are an excellent way of carrying out hand calculations. They force you to go about your calculations logically, while retaining all intermediate steps for later examination. Statisticians often rightly criticize such programs for the accuracy of their results with very large samples or with samples of unusual data, but they are extremely useful for small to medium-sized problems. They also have the advantage that most people have one or more of them installed on their personal computers.

1.5   The Plan of the Book

Our original example, the examination of the effects of a program of stress management, offers an opportunity to illustrate the book's organization. In the process of running the study, we will be collecting data on many variables. One of the first things we will do with these data is to plot them, to look at the distribution for each variable, to calculate means and standard deviations, and so on. These techniques will be discussed in Chapter 2.

Following an exploratory analysis of the data, we will apply several inferential procedures. For example, we will want to compare the mean score on a scale of coping skills for a group who received stress-management training with the mean score for a group who did not receive such training. Techniques for making these kinds of comparisons will be discussed in Chapters 7, 11, 12, 13, 14, 16, and 18, depending on the complexity of our experiment, the number of groups to be compared, and the degree to which we are willing to make certain assumptions about our data.

We might also want to ask questions dealing with the relationships between variables rather than the differences among groups. For example, we might like to know whether a person's level of behavior problems is related to his score on self-esteem, or whether a person's coping scores can be predicted from variables such as her self-esteem and social support. Techniques for asking these kinds of questions will be considered in Chapters 9, 10, 15, and 17, depending on the type of data we have and the number of variables involved.

Most students (and courses) never seem to make it all the way through any book. In this case, that would mean skipping Chapter 18 on nonparametric analyses. I think that would

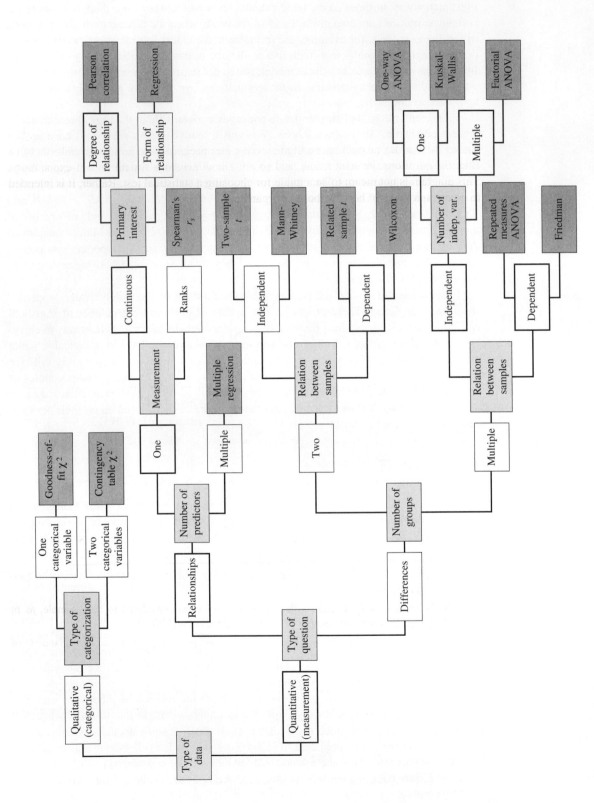

Figure 1.1 Decision tree

be unfortunate because that chapter focuses on some of the newer, and important, work on bootstrapping and resampling methods. These methods have become much more popular with the drastic increases in computing power, and they make considerable intuitive sense. I would recommend that you at least skim that chapter early on, and go back to it for the relevant material as you work through the rest of the book. You do not need an extensive background to understand what is there, and reading it will give you a real step up on analyses that you will see in the literature. (I believe that it will also give you a much better understanding of the parametric analyses in the remainder of the book.)

Figure 1.1 on the facing page provides an organizational scheme that distinguishes among the various procedures on the basis of a number of dimensions, such as the type of data, the questions we want to ask, and so on. The dimensions should be self-explanatory. This diagram is not meant to be a guide for choosing a statistical test. Rather, it is intended to give you a sense of how the book is organized.

Key Terms

Random sample (1.1)

Randomly assign (1.1)

Population (1.1)

Sample (1.1)

External validity (1.1)

Random assignment (1.1)

Internal validity (1.1)

Variable (1.1)

Discrete variables (1.1)

Continuous variables (1.1)

Measurement data (1.1)

Quantitative data (1.1)

Categorical data (1.1)

Frequency data (1.1)

Qualitative data (1.1)

Independent variables (1.1)

Dependent variables (1.1)

Descriptive statistics (1.2)

Exploratory data analysis (EDA) (1.2)

Inferential statistics (1.2)

Parameter (1.2)

Statistic (1.2)

Nominal scales (1.3)

Ordinal scale (1.3)

Interval scale (1.3)

Ratio scale (1.3)

Exercises

1.1   Under what conditions would the entire student body of your college or university be considered a population?

1.2   Under what conditions would the entire student body of your college or university be considered a sample?

1.3   If the student body of your college or university were considered to be a sample, as in Exercise 1.2, would this sample be random or nonrandom? Why?

1.4   Why would choosing names from a local telephone book not produce a random sample of the residents of that city? Who would be underrepresented and who would be overrepresented?

1.5   Give two examples of independent variables and two examples of dependent variables.

1.6   Write a sentence describing an experiment in terms of an independent and a dependent variable.

1.7   Give three examples of continuous variables.

1.8   Give three examples of discrete variables.

1.9   Give an example of a study in which we are interested in estimating the average score of a population.

1.10   Give an example of a study in which we do not care about the actual numerical value of a population average, but want to know whether the average of one population is greater than the average of a different population.

1.11   Give three examples of categorical data.

1.12   Give three examples of measurement data.

1.13   Give an example in which the thing we are studying could be either a measurement or a categorical variable.

1.14   Give one example of each kind of measurement scale.

1.15   Give an example of a variable that could be said to be measured on a ratio scale for some purposes and on an interval or ordinal scale for other purposes.

1.16   We trained rats to run a straight-alley maze by providing positive reinforcement with food. On trial 12, a rat lay down and went to sleep halfway through the maze. What does this say about the measurement scale when speed is used as an index of learning?

1.17   What does Exercise 1.16 say about speed used as an index of motivation?

1.18   Give two examples of studies in which our primary interest is in looking at relationships between variables.

1.19   Give two examples of studies in which our primary interest is in looking at differences among groups.

Discussion Questions

1.20   The *Chicago Tribune* of July 21, 1995, reported on a study by a fourth-grade student named Beth Peres. In the process of collecting evidence in support of her campaign for a higher allowance, she polled her classmates on what they received for an allowance. She was surprised to discover that the 11 girls who responded reported an average allowance of $2.63 per week, whereas the 7 boys reported an average of $3.18, 21% more than for the girls. At the same time, boys had to do fewer chores to earn their allowance than did girls. The story had considerable national prominence and raised the question of whether the income disparity for adult women relative to adult men may actually have its start very early in life.

   a.   What are the dependent and independent variables in this study, and how are they measured?

   b.   What kind of a sample are we dealing with here?

   c.   How could the characteristics of the sample influence the results she obtained?

   d.   How might Beth go about "random sampling"? How would she go about "random assignment"?

   e.   If random assignment is not possible in this study, does that have negative implications for the validity of the study?

   f.   What are some of the variables that might influence the outcome of this study separate from any true population differences between boys' and girls' incomes?

   g.   Distinguish clearly between the descriptive and inferential statistical features of this example.

1.21   The *Journal of Public Health* published data on the relationship between smoking and health (see Landwehr & Watkins [1987]). They reported the cigarette consumption per adult for 21 mostly Western and developed countries, along with the coronary heart disease rate for each country. The data clearly show that coronary heart disease is highest in those countries with the highest cigarette consumption.

   a.   Why might the sampling in this study have been limited to Western and developed countries?

   b.   How would you characterize the two variables in terms of what we have labeled "scales of measurement"?

c. If our goal is to study the health effects of smoking, how do these data relate to that over-all question?

d. What other variables might need to be considered in such a study?

e. It has been reported that tobacco companies are making a massive advertising effort in Asia. At present, only 7% of Chinese women smoke (compared with 61% of Chinese men). How would a health psychologist go about studying the health effects of likely changes in the incidence of smoking among Chinese women?

CHAPTER 2

Describing and Exploring Data

Objectives

To show how data can be reduced to a more interpretable form by using graphical representation and measures of central tendency and dispersion.

Contents

A COLLECTION OF RAW DATA, taken by itself, is no more exciting or informative than junk mail before Election Day. Whether you have neatly arranged the data in rows on a data collection form or scribbled them on the back of an out-of-date announcement you tore from the bulletin board, a collection of numbers is still just a collection of numbers. To be interpretable, they first must be organized in some sort of logical order. The following actual experiment illustrates some of these steps.

How do human beings process information that is stored in their short-term memory? If I asked you to tell me whether the number "6" was included as one of a set of five digits that you just saw presented on a screen, do you use *sequential processing* to search your short-term memory of the screen and say "Nope, it wasn't the first digit; nope, it wasn't the second," and so on? Or do you use *parallel processing* to compare the digit "6" with your memory of all the previous digits at the same time? The latter approach would be faster and more efficient, but human beings don't always do things in the fastest and most efficient manner. How do *you* think that you do it? How do you search back through your memory and identify the person who just walked in as Jennifer? Do you compare her one at a time with all the women her age whom you have met, or do you make comparisons in parallel? (This second example uses long-term memory rather than short-term memory, but the questions are analogous.)

In 1966, Sternberg ran a simple, famous, and important study that examined how people recall data from short-term memory. On a screen in front of the subject, he briefly presented a *comparison* set of one, three, or five digits. Shortly after each presentation he flashed a single test digit on the screen and required the subject to push one button (the positive button) if the test digit had been included in the comparison set or another button (the negative button) if the test digit had not been part of the comparison set. For example, the two stimuli might look like this:

Comparison	2	7	4	8	1
Test			5		

(Remember, the two sets of stimuli were presented sequentially, not simultaneously, so only one of those lines was visible at a time.) The numeral "5" was not part of the comparison set, so the subject should have responded by pressing the negative button. Sternberg measured the time, in 100ths of a second, that the subject took to respond. This process was repeated over many randomly organized trials. Because Sternberg was interested in how people process information, he was interested in how reaction times varied as a function of the number of digits in the comparison set and as a function of whether the test digit was a positive or negative instance for that set. (If you make comparisons sequentially, the time to make a decision should increase as the number of digits in the comparison set increases. If you make comparisons in parallel, the number of digits in the comparison set shouldn't matter.)

Although Sternberg's goal was to compare data for the different conditions, we can gain an immediate impression of our data by taking the full set of reaction times, regardless of the stimulus condition. The data in Table 2.1 were collected in an experiment similar to Sternberg's but with only one subject—myself. No correction of responses was allowed, and the data presented here come only from correct trials.

Table 2.1   Reaction time data from number identification experiment

Comparison Stimuli*	Reaction Times, in 100ths of a Second																
	40	41	47	38	40	37	38	47	45	61	54	67	49	43	52	39	46
1Y	47	45	43	39	49	50	44	53	46	64	51	40	41	44	48	50	42
	90	51	55	60	47	45	41	42	72	36	43	94	45	51	46	52	
	52	45	74	56	53	59	43	46	51	40	48	47	57	54	44	56	47
1N	62	44	53	48	50	58	52	57	66	49	59	56	71	76	54	71	104
	44	67	45	79	46	57	58	47	73	67	46	57	52	61	72	104	
	73	83	55	59	51	65	61	64	63	86	42	65	62	62	51	62	72
3Y	55	58	46	67	56	52	46	62	51	51	61	60	75	53	59	56	50
	43	58	67	52	56	80	53	72	62	59	47	62	53	52	46	60	
	73	47	63	63	56	66	72	58	60	69	74	51	49	69	51	60	52
3N	72	58	74	59	63	60	66	59	61	50	67	63	61	80	63	60	64
	64	57	59	58	59	60	62	63	67	78	61	52	51	56	95	54	
	39	65	53	46	78	60	71	58	87	77	62	94	81	46	49	62	55
5Y	59	88	56	77	67	79	54	83	75	67	60	65	62	62	62	60	58
	67	48	51	67	98	64	57	67	55	55	66	60	57	54	78	69	
	66	53	61	74	76	69	82	56	66	63	69	76	71	65	67	67	55
5N	65	58	64	65	81	69	69	63	68	70	80	68	63	74	61	85	125
	59	61	74	76	62	83	58	72	65	61	95	58	64	66	66	72	

*Y = Yes, test stimulus was included; N = No, it was not included 1, 3, and 5 refer to the number of digits in the comparison stimuli

2.1   Plotting Data

As you can see, there are simply too many numbers in Table 2.1 for us to be able to interpret them at a glance. One of the simplest methods to reorganize data to make them more intelligible is to plot them in some sort of graphical form. There are several common ways in which data can be represented graphically. Some of these methods are frequency distributions, histograms, and stem-and-leaf displays, which we will discuss in turn.

Frequency Distributions

frequency distribution

As a first step, we can make a **frequency distribution** of the data as a way of organizing them in some sort of logical order. For our example, we would count the number of times that each possible reaction time occurred. For example, the subject responded in 50/100 of a second 5 times and in 51/100 of a second 12 times. On one occasion he became flustered and took 1.25 seconds (125/100 of a second) to respond. The frequency distribution for these data is presented in Table 2.2, which reports how often each reaction time occurred,

Table 2.2   Frequency distribution of reaction times

Reaction Time, in 100ths of a Second	Frequency	Reaction Time, in 100ths of a Second	Frequency
36	1	71	4
37	1	72	8
38	2	73	3
39	3	74	6
40	4	75	2
41	3	76	4
42	3	77	2
43	5	78	3
44	5	79	2
45	6	80	3
46	11	81	2
47	9	82	1
48	4	83	3
49	5	84	0
50	5	85	1
51	12	86	1
52	10	87	1
53	8	88	1
54	6	89	0
55	7	90	1
56	10	91	0
57	7	92	0
58	12	93	0
59	11	94	2
60	12	95	2
61	11	96	0
62	14	97	0
63	10	98	1
64	7	99	0
65	8	. . .	. . .
66	8	. . .	. . .
67	14	104	2
68	2	. . .	. . .
69	7	125	1
70	1		

and in Figure 2.1, on which the data are plotted so they can be seen graphically. (The term "frequency distribution" is often used to describe both Table 2.2 and Figure 2.1.)

From the distribution shown in Table 2.2 and graphically in Figure 2.1, we can see a wide distribution of reaction times, with times as low as 36/100 of a second and as high as 125/100 of a second. The data tend to cluster around about 60/100, with most of the scores between 40/100 and 90/100. This tendency was not apparent from the unorganized data shown in Table 2.1.

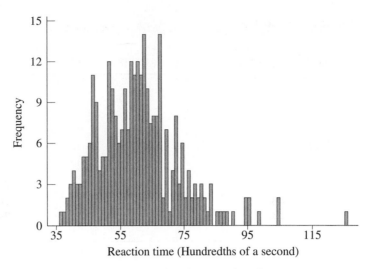

Figure 2.1   Plot of reaction times against frequency

2.2 Histograms

In the preceding discussion of frequency distributions, Figure 2.1 plotted the frequencies of the individual values of reaction time. But when we are dealing with a variable, such as this one, that has many different values, each individual value often occurs with low frequency, and there is often substantial fluctuation of the frequencies in adjacent intervals. Notice, for example, that there are fourteen 67s, but only two 68s. In situations such as this, it makes

histogram
more sense to group adjacent values together into a **histogram.**[1] Doing so obscures some of the random "noise" that is not likely to be meaningful, but preserves important trends in the data. We might, for example, group the data into blocks of 5/100 of a second, combining the frequencies for all outcomes between 35 and 39, between 40 and 44, and so on. An example of such a distribution is shown in Table 2.3.

In Table 2.3, I have reported the upper and lower boundaries of the intervals as whole integers, for the simple reason that it makes the table easier to read. However, you should realize that the true limits of the interval (known as the real lower limit and the real upper limit) are decimal values that fall halfway between the top of one interval and the bottom of

real lower limit
the next. The **real lower limit** of an interval is the smallest value that would be classed as falling into the interval. Similarly, an interval's **real upper limit** is the largest value that

real upper limit
would be classed as being in the interval. For example, had we recorded reaction times to the nearest thousandth of a second, rather than to the nearest hundredth, the interval 35–39 would include all values between 34.5 and 39.5 because values falling between those points would be rounded up or down into that interval. (People often become terribly worried

---

[1] Different people seem to mean different things when they talk about a "histogram." Some use it for the distribution of the data regardless of whether or not categories have been combined (they would call Figure 2.1 a histogram), and others reserve it for the case where adjacent categories are combined. You can probably tell by now that I am not a stickler for such distinctions, and I will use "histogram" and "frequency distribution" more or less interchangeably. I hope that is not confusing to anyone.

Table 2.3   Grouped frequency distribution

Interval	Midpoint	Frequency	Cumulative Frequency	Interval	Midpoint	Frequency	Cumulative Frequency
35–39	37	7	7	85–89	87	4	291
40–44	42	20	27	90–94	92	3	294
45–49	47	35	62	95–99	97	3	297
50–54	52	41	103	100–104	102	2	299
55–59	57	47	150	105–109	107	0	299
60–64	62	54	204	110–114	112	0	299
65–69	67	39	243	115–119	117	0	299
70–74	72	22	265	120–124	122	0	299
75–79	77	13	278	125–129	127	1	300
80–84	82	9	287				

about what we would do if a person had a score of exactly 39.50000000 and therefore sat right on the breakpoint between two intervals. Don't worry about it. First, it doesn't happen very often. Second, you can always flip a coin. Third, there are many more important things to worry about. Just make up an arbitrary rule of what you will do in those situations, and then stick to it. This is one of those non-issues that make people think the study of statistics is confusing, boring, or both.)

midpoints

The **midpoints** listed in Table 2.3 are the averages of the upper and lower limits and are presented for convenience. When we plot the data, we often plot the points as if they all fell at the midpoints of their respective intervals.

Table 2.3 also lists the frequencies with which scores fell in each interval. For example, there were seven reaction times between 35/100 and 39/100 of a second. The distribution in Table 2.3 is shown as a histogram in Figure 2.2.

People often ask about the optimal number of intervals to use when grouping data. Although there is no right answer to this question, somewhere around 10 intervals is usually reasonable.[2] In this example I used 19 intervals because the numbers naturally broke that way and because I had a lot of observations. In general and when practical, it is best to use natural breaks in the number system (e.g., 0–9, 10–19, . . . or 100–119, 120–139) rather than to break up the range into exactly 10 arbitrarily defined intervals. However, if another kind of limit makes the data more interpretable, then use those limits. Remember that you are trying to make the data meaningful—don't try to follow a rigid set of rules made up by someone who has never seen your problem.

Notice in Figure 2.2 that the reaction time data are generally centered on 50–70 hundredths of a second, that the distribution rises and falls fairly regularly, and that the distribution trails off to the right. We would expect such times to trail off to the right (referred to as being positively skewed) because there is some limit on how *quickly* the person can respond, but really no limit on how *slowly* he can respond. Notice also the extreme value of 125 hundredths. This value is called an **outlier** because it is widely separated from the rest of the data. Outliers frequently represent errors in recording data, but in this particular case it was just a trial in which the subject couldn't make up his mind which button to push.

outlier

---

[2] One interesting scheme for choosing an optimal number of intervals is to set it equal to the integer closest to $\sqrt{N}$, where N is the number of observations. Applying that suggestion here would leave us with $\sqrt{N} = \sqrt{300} = 17.32 = 17$ intervals, which is close to the 19 that I actually used.

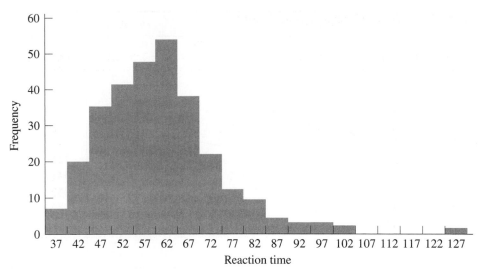

Figure 2.2 Grouped distribution of reaction times

2.3 Stem-and-Leaf Displays

stem-and-leaf display

exploratory data analysis (EDA)

leading digits

most significant digits

stem

trailing digits

less significant digits

leaves

Although histograms and frequency distributions are commonly used methods of presenting data, each has its drawbacks. Because histograms portray grouped data, they lose the actual numerical values of the individual scores in each interval. Frequency distributions, on the other hand, retain the values of the individual observations, but they can be difficult to use when they do not summarize the data sufficiently. An alternative approach that avoids both of these criticisms is the **stem-and-leaf display.**

John Tukey (1977), as part of his general approach to data analysis, known as **exploratory data analysis (EDA),** developed a variety of methods for displaying data in visually meaningful ways. One of the simplest of these methods is a stem-and-leaf display, which you will see presented by most major statistical software packages. I can't start with the reaction time data here, because that would require a slightly more sophisticated display due to the large number of observations. Instead, I'll use a hypothetical set of data in which we record the amount of time (in minutes per week) that each of 100 students spends playing electronic games. Some of the raw data are given in Figure 2.3. On the left side of the figure is a portion of the data (data from students who spend between 40 and 80 minutes per week playing games) and on the right is the complete stem-and-leaf display that results.

From the raw data in Figure 2.3, you can see that there are several scores in the 40s, another bunch in the 50s, two in the 60s, and some in the 70s. We refer to the tens' digits—here 4, 5, 6, and 7—as the **leading digits** (sometimes called the **most significant digits**) for these scores. These leading digits form the **stem,** or vertical axis, of our display. Within the set of 14 scores that were in the 40s, you can see that there was one 40, two 41s, one 42, two 43s, one 44, no 45s, three 46s, one 47, one 48, and two 49s. The units' digits 0, 1, 2, 3, and so on, are called the **trailing** (or **less significant**) **digits.** They form the **leaves**—the horizontal elements—of our display.[3]

---

[3] It is not always true that the tens' digits form the stem and the units' digits the leaves. For example, if the data ranged from 100 to 1000, the hundreds' digits would form the stem, the tens' digits the leaves, and we would ignore the units' digits.

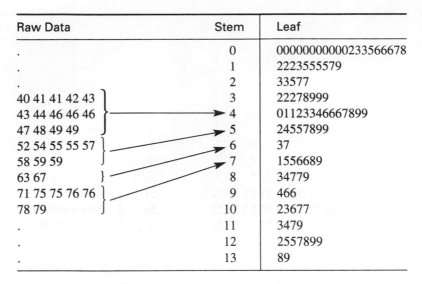

Raw Data	Stem	Leaf
·	0	00000000000233566678
·	1	2223555579
·	2	33577
40 41 41 42 43	3	22278999
43 44 46 46 46 →	4	01123346667899
47 48 49 49 →	5	24557899
52 54 55 55 57 →	6	37
58 59 59 →	7	1556689
63 67 →	8	34779
71 75 75 76 76	9	466
78 79	10	23677
·	11	3479
·	12	2557899
·	13	89

Figure 2.3   Stem-and-leaf display of electronic game data

On the right side of Figure 2.3 you can see that next to the stem entry of 4 you have one 0, two 1s, a 2, two 3s, a 4, three 6s, a 7, an 8, and two 9s. These leaf values correspond to the units' digits in the raw data. Similarly, note how the leaves opposite the stem value of 5 correspond to the units' digits of all responses in the 50s. From the stem-and-leaf display you could completely regenerate the raw data that went into that display. For example, you can tell that 11 students spent zero minutes playing electronic games, one student spent two minutes, two students spent three minutes, and so on. Moreover, the shape of the display looks just like a sideways histogram, giving you all of the benefits of that method of graphing data as well.

One apparent drawback of this simple stem-and-leaf display is that for some data sets it will lead to a grouping that is too coarse for our purposes. In fact, that is why I needed to use hypothetical data for this introductory example. When I tried to use the reaction time data, I found that the stem for 50 (i.e., 5) had 88 leaves opposite it, which was a little silly. Not to worry; Tukey was there before us and figured out a clever way around this problem.

If the problem is that we are trying to lump together everything between 50 and 59, perhaps what we should be doing is breaking that interval into smaller intervals. We *could* try using the intervals 50–54, 55–59, and so on. But then we couldn't just use 5 as the stem, because it would not distinguish between the two intervals. Tukey suggested using "5*" to represent 50–54, and "5." to represent 55–59. But that won't solve our problem here because the categories still are too coarse. So Tukey suggested an alternative scheme where "5*" represents 50–51, "5t" represents 52–53, "5f" represents 54–55, "5s" represents 56–57, and "5." represents 58–59. (Can you guess why he used those particular letters? *Hint:* "Two" and "three" both start with "t.") If we apply this scheme to the data on reaction times, we obtain the results shown in Figure 2.4. In deciding on the number of stems to use, the problem is similar to selecting the number of categories in a histogram. Again, you want to do something that makes sense and that conveys information in a meaningful way. The one restriction is that the stems should be the same width. You would not let one stem be 50–54, and another 60–69.

Raw Data	Stem	Leaf
36 37 38 38 39 39 39 40	3s	67
40 40 40 41 41 41 42 42	3.	88999
42 43 43 43 43 43 44 44	4*	0000111
44 44 44 45 45 45 45 45	4t	22233333
45 46 46 46 46 46 46 46	4f	44444555555
46 46 46 46 47 47 47 47	4s	6666666666677777777777
47 47 47 47 47 48 48 48	4.	888899999
48 49 49 49 49 49 50 50	5*	000001111111111111
50 50 50 51 51 51 51 51	5t	2222222222333333333
51 51 51 51 51 51 51 52	5f	4444445555555
52 52 52 52 52 52 52 52	5s	666666666677777777
52 53 53 53 53 53 53 53	5.	888888888888899999999999
53 54 54 54 54 54 54 55	6*	0000000000000011111111111
55 55 55 55 55 55	6t	2222222222222223333333333
. . .	6f	444444455555555
	6s	66666666777777777777777
	6.	889999999
	7*	01111
	7t	22222222333
	7f	44444455
	7s	666677
	7.	88899
	8*	00011
	8t	2333
	8f	5
	8s	67
	8.	8
	9*	0
	9t	
	9f	4455
	9s	
	9.	8
	High	104; 104; 125

Figure 2.4   Stem-and-leaf display for reaction time data

Notice that in Figure 2.4 I did not list the extreme values as I did the others. I used the word *High* in place of the stem and then inserted the actual values. I did this to highlight the presence of extreme values, as well as to conserve space.

Stem-and-leaf displays can be particularly useful for comparing two different distributions. Such a comparison is accomplished by plotting the two distributions on opposite sides of the stem. Figure 2.5 shows the actual distribution of numerical grades of males and females in a course I taught on experimental methods that included a substantial statistics component. These are actual data. Notice the use of stems such as 6* (for 60–64), and 6. (for 65–69). In addition, notice the code at the bottom of the table that indicates how entries translate to raw scores. This particular code says that |4*|,1 represents 41, not 4.1 or 410. Finally, notice that the figure nicely illustrates the difference in performance between the male students and the female students.

Male	Stem	Female
	3*	
6	3.	
	4*	1
	4.	
	5*	
	5.	
2	6*	03
	6.	568
32200	7*	0144
88888766666655	7.	555556666788899
4432221000	8*	0000011112222334444
7666666555	8.	556666666666667788888899
422	9*	000000000133
	9.	56
Code |4*|1 = 41		

Figure 2.5 Grades (in percent) for an actual course in experimental methods, plotted separately by gender

2.4 Alternative Methods of Plotting Data

The previous sections dealt with only a few ways of plotting data. There is an almost unlimited number of other ways that data can be presented, some of which are quite ingenious and informative. A few examples are shown in Figures 2.6, 2.7, and 2.8. These examples were chosen because they illustrate how displays can be used to reveal interesting features of data.

Two comments are in order about how we plot data. First, the point of representing data graphically is to communicate to an audience. If there is a better way to communicate, then use it. Rules of graphical presentation are intended as guides to clearer presentation, not as prescriptive rules that may never be broken. This point was made earlier in the discussion about the number of intervals that should be used for a histogram, but it goes beyond histograms. So the first "rule" is this: *If it aids understanding, do it; if it doesn't, don't.*

The second rule is to keep things simple. Generally, the worst graphics are those that include irrelevant features that only add to the confusion. Tufte (1983) calls such material "chart junk," and you should avoid it. Perhaps the worst sin, in the opinion of many, is plotting something in three dimensions that could be better plotted in two. There are legitimate reasons for three-dimensional plots (Figure 2.7 is one), but three dimensions are more likely to confuse the issue than to clarify it. Unfortunately, many graphics packages encourage the addition of unnecessary, and confusing dimensions. Graphics should look utilitarian, neat, and orderly; they should rarely look "pretty."

Figure 2.6 is the distribution, by age and gender, of the populations of Mexico, Spain, the United States, and Sweden. This figure clearly portrays differences between countries in their age distributions. (Compare Mexico and Sweden, for example.) By having males and females plotted back to back, we can also see the effects of gender differences in life expectancy. The older age groups in three countries contain more females than males. In Mexico, it appears that men begin to outnumber women in their early twenties. This type of distribution was common in the past when many women died in childbirth, and we might start looking there for an explanation.

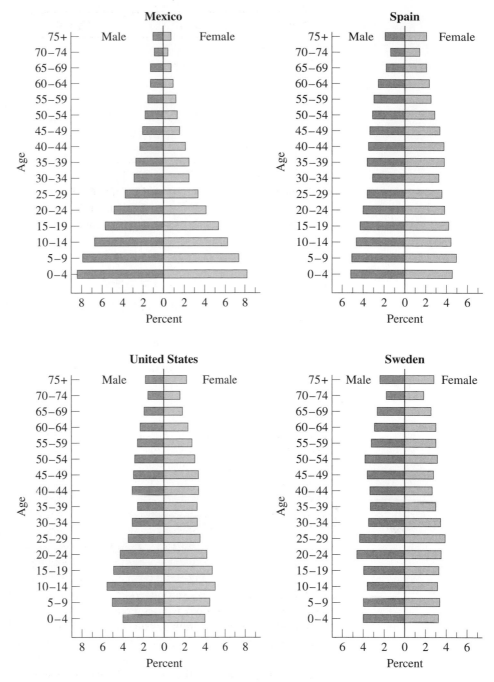

From *Social Indicators: 1976*. U.S. Department of Commerce. U.S. Government Printing Office, 1977.

Figure 2.6   Population, by gender and age, for selected countries: 1970

Figure 2.7 is a comparison of the causes of death of Vermont residents in 1900 and 1981. This figure violates the guidelines about using three-dimensional figures, but it does that to promote clarity, not to be cute. Notice that the various causes have been ordered from bottom to top in decreasing order of magnitude. From this figure it is immediately apparent that, whereas almost one-third of the deaths in 1900 were attributable to a high rate of infant

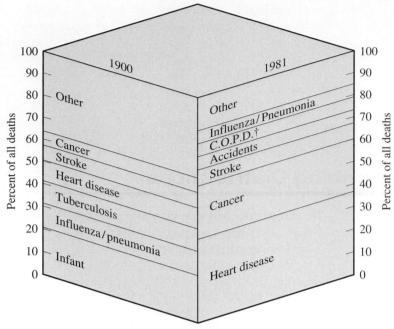

† Chronic obstructive pulmonary disease

Figure 2.7   Major causes of death in Vermont, 1900 and 1981

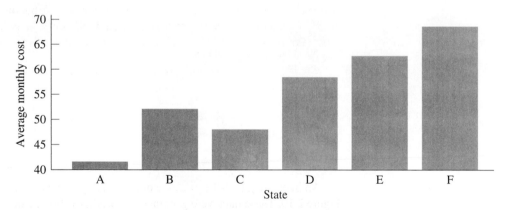

Figure 2.8   Average monthly electric bills in New England

mortality and to tuberculosis, neither of those sources contributed noticeably to death rates in 1981. On the other hand, cancer and heart disease, which together accounted for 60% of all deaths in 1981, played a much-reduced role in 1900, accounting for less than 15% of all deaths. Some of these differences could be the result of changes in diagnostic ability, but others are very real.

Figure 2.8 is included primarily as an example of how data are sometimes presented in an accurate but misleading manner. In an attempt to illustrate in a limited space the fact that consumers in one state pay less than consumers in other states for the same amount of electricity (750 kilowatt hours), a local utility company (which prefers to remain nameless) published a graph similar to the one shown in Figure 2.8. Although the *data* would be accurately portrayed in such a figure, the *relationships* among those data would not. For example, because the values on the ordinate start at $40, it appears that customers in State C pay about seven times as much (their bar is about seven times as tall) as do customers in State A. In fact, the charges were actually $41.02 and $47.11, with State C customers paying 15%, not 700%, more. When you are presenting data in a graph such as this, it is important to start the ordinate at 0 or, if that is not practical, at least far enough from the smallest value to avoid giving an inaccurate visual message.

More examples of distorted representation of data can be found in the excellent and entertaining paper by Wainer (1984).

2.5 Describing Distributions

symmetric

bimodal

unimodal

modality

negatively skewed

positively skewed

skewness

The distributions of scores illustrated in Figures 2.1 and 2.2 were more or less regularly shaped distributions, rising to a maximum and then dropping away smoothly—although even those figures were not completely symmetric. However not all distributions are peaked in the center and fall off evenly to the sides (see the stem-and-leaf display in Figure 2.3), and it is important to understand the terms used to describe different distributions. Consider the two distributions shown in Figure 2.9(a) and (b). These plots are of data that were computer generated to come from populations with specific shapes. These plots, and the other four in Figure 2.9, are based on samples of 1,000 observations, and the slight irregularities are just random variability. Both of the distributions in Figure 2.9(a) and (b) are called **symmetric** because they have the same shape on both sides of the center.

The distribution shown in Figure 2.9(a) came from what we will later refer to as a normal distribution. The distribution in Figure 2.9(b) is referred to as **bimodal,** because it has two peaks. The term *bimodal* is used to refer to any distribution that has two predominant peaks, whether or not those peaks are of exactly the same height. If a distribution has only one major peak, it is called **unimodal.** The term used to refer to the number of major peaks in a distribution is **modality.**

Next consider Figure 2.9(c) and (d). These two distributions obviously are not symmetric. The distribution in Figure 2.9(c) has a tail going out to the left, whereas that in Figure 2.9(d) has a tail going out to the right. We say that the former is **negatively skewed** and the latter **positively skewed.** (*Hint:* To help you remember which is which, notice that negatively skewed distributions point to the negative, or small, numbers, and that positively skewed distributions point to the positive end of the scale.) There are statistical measures of the degree of asymmetry, or **skewness,** but they are not commonly used in the social sciences.

An interesting real-life example of a positively skewed bimodal distribution is shown in Figure 2.10. These data were generated by Bradley (1963), who instructed subjects to press a button as quickly as possible whenever a small light came on. Most of the data points are smoothly distributed between roughly 7 and 17 hundredths of a second, but a small but noticeable cluster of points lies between 30 and 70 hundredths, trailing off to the right. This second cluster of points was obtained primarily from trials on which the subject missed the button on the first try. Their inclusion in the data significantly affects the distribution's shape. An experimenter who had such a collection of data might seriously consider treating times greater than some maximum separately, on the grounds that those times were more a reflection of the *accuracy* of a psychomotor response than a measure of the *speed* of that response. Alternatively, it might be possible to apply a transformation to the data to reduce skewness. Transformations will be discussed shortly.

It is important to consider the difference between Bradley's data, shown in Figure 2.10, and the data that I generated, shown in Figures 2.1 and 2.2. Both distributions are positively skewed, but my data generally show longer reaction times without the second cluster of

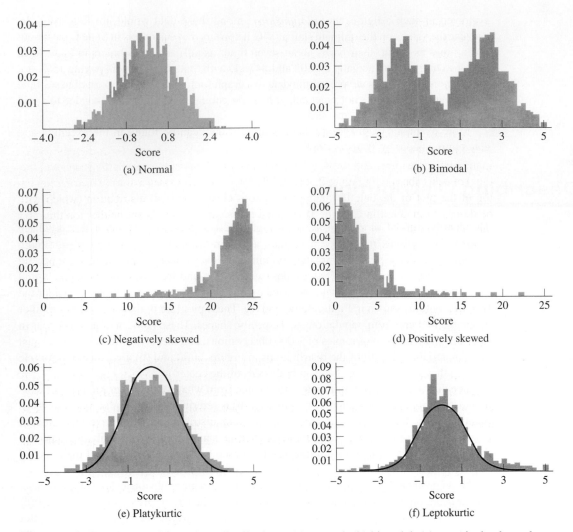

Figure 2.9 Shapes of frequency distributions: (a) normal, (b) bimodal, (c) negatively skewed, (d) positively skewed, (e) platykurtic, and (f) leptokurtic

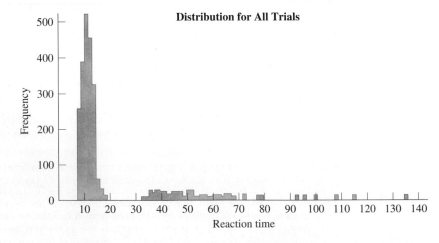

Figure 2.10 Frequency distribution of Bradley's reaction time data

points. One difference was that I was making a decision on *which* button to press, whereas Bradley's subjects only had to press a single button whenever the light came on. In addition, the program I was using to present stimuli recorded data only from *correct* responses, not from errors. There was no chance to correct and hence nothing equivalent to missing the button on the first try and having to press it again. I point out these differences to illustrate that differences in the way in which data are collected can have noticeable effects on the kinds of data we see.

kurtosis

The last characteristic of a distribution that we will examine is kurtosis. **Kurtosis** has a specific mathematical definition, but basically it refers to the relative concentration of scores in the center, the upper and lower ends (tails), and the shoulders (between the center and the tails) of a distribution. In Figure 2.9(e) and (f) I have superimposed a normal distribution on top of the plot of the data to make comparisons clear. A normal distribution (which will be described in detail in Chapter 3) is called **mesokurtic.** Its tails are neither too thin nor too thick, and there are neither too many nor too few scores concentrated in the center. If you start with a normal distribution and move scores from both the center and the tails into the shoulders, the curve becomes flatter and is called **platykurtic.** This is clearly seen in Figure 2.9(e), where the central portion of the distribution is much too flat. If, on the other hand, you moved scores from the shoulders into both the center and the tails, the curve becomes more peaked with thicker tails. Such a curve is called **leptokurtic,** and an example is Figure 2.9(f). Notice in this distribution that there are too many scores in the center *and* too many scores in the tails.[4]

mesokurtic

platykurtic

leptokurtic

Skewness and kurtosis, although not commonly used measures in the social sciences, are convenient verbal labels used to describe distributions. As an educated person, you should know what a positively skewed distribution is, even though it is unlikely that you will ever want to compute a numerical index of skewness. You should be aware, however, that these measures are important to statisticians, who might rightly be annoyed by the cavalier way I seem to be dismissing them. I downplay these measures because they are less useful to a user of statistics than to the professional.

It is important to recognize that quite large samples of data are needed before we can have a good idea about the shape of a distribution, especially its kurtosis. With sample sizes of around 30, the best we can reasonably expect to see is whether the data tend to pile up in the tails of the distribution or are markedly skewed in one direction or another.

2.6 Using Computer Programs to Display Data

Many statistics texts assume that simple data analyses will be carried out by hand with the help of a standard calculator. This is probably a good approach to teaching, but today computer programs carry out most real analyses. Thus, it is important for you to be familiar with reading and interpreting the results on computer printouts. For that reason, most chapters in this book will include samples of computer solutions for examples previously analyzed by hand. These solutions will be obtained by using several different programs, to allow you to see alternative approaches. In the early chapters, we will focus on a commonly

---

[4] I would like to thank Karl Wuensch of East Carolina University for his helpful suggestions on understanding skewness and kurtosis. His ideas are reflected here, although I'm not sure that he would be satisfied by my statements on kurtosis. Karl has spent a lot of time thinking about kurtosis and made a good point recently when he stated in an electronic mail discussion, "I don't think my students really suffer much from not understanding kurtosis well, so I don't make a big deal out of it." You should have a general sense of what kurtosis is, but you should focus your attention on other, more important issues. Except in the extreme, most people, including statisticians, are unlikely to be able to look at a distribution and tell whether it is platykurtic or leptokurtic without further calculations.

available program called Minitab because it is one of the easiest general-purpose programs to use and many students have access to it. Along with the printout you will see screen shots showing the menu selections required to produce the output. In later chapters, we will move toward SPSS, which is probably the most widely used statistical package. SAS will also make occasional appearances. If your instructor uses one statistical program and I demonstrate something using a different program, all is not lost. I think you will be surprised how much similarity there is between programs, both in steps you go through and the resulting output.

Exhibit 2.1 shows a portion of the Minitab menu system and the spreadsheet that holds the data for the electronic games data shown in Figure 2.3. The **Graph** menu has been expanded, and you can see where to choose a histogram and a stem-and-leaf display produced by Minitab. The full set of observations was reconstructed from Figure 2.3.

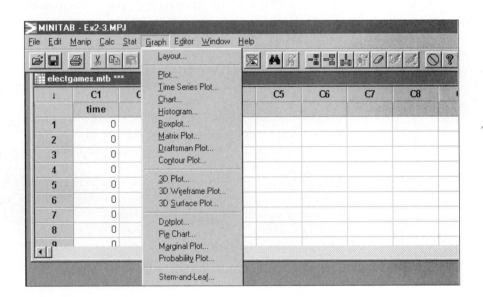

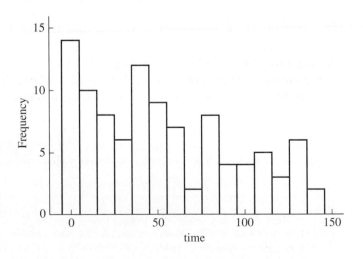

(*continued*)

Exhibit 2.1 Minitab analysis of electronic game data

20	0	00000000000233566678
30	1	2223555579
35	2	33577
43	3	22278999
(14)	4	01123346667899
43	5	24557899
35	6	37
33	7	1556689
26	8	34779
21	9	466
18	10	23677
13	11	3479
9	12	2557899
2	13	89

Exhibit 2.1 *(Continued)*

depth

The stem-and-leaf display generated by Minitab is equivalent to ours, although an additional column on the left contains cumulative frequencies (often referred to as **depth**) running inward from each end. For example, 20 people had scores less than or equal to 8, the largest entry for the first stem. In addition, 10 people had scores between 12 and 19, meaning a total of 30 people had scores less than or equal to 19. Similarly, we find that 43 people had scores less than or equal to 39. You should be able to find each of these values in the table. The number in parentheses is the frequency (noncumulative) for the interval that contains the middle value. In this case, 14 people fell in the interval containing the middle value. You can see that the distribution is positively skewed because the cumulative frequencies pile up much more quickly when we go from low scores toward the center than when we go from high scores toward the center. This is even more apparent in the stem-and-leaf display itself.

So far in our discussion almost no mention has been made of the numbers themselves. We have seen how data can be organized and presented in the form of distributions, and we have discussed a number of ways in which distributions can be characterized: symmetry or its lack (skewness), kurtosis, and modality. As useful as this information might be in certain situations, it is inadequate in others. We still do not know the average speed of a simple decision reaction time nor how alike or dissimilar are the reaction times for individual trials. To obtain this knowledge, we must reduce the data to a set of measures that carry the information we need. The questions to be asked refer to the location, or central tendency, and to the dispersion, or variability, of the distributions along the underlying scale. Measures of these characteristics will be considered in Sections 2.8 and 2.9. But before going to those sections we need to set up a notational system that we can use in that discussion.

2.7 Notation

Any discussion of statistical techniques requires a notational system for expressing mathematical operations. You might be surprised to learn that no standard notational system has been adopted. Although several attempts to formulate a general policy have been made, the fact remains that no two textbooks use exactly the same notation.

The notational systems commonly used range from the very complex to the very simple. The more complex systems gain precision at the expense of easy intelligibility, whereas the simpler systems gain intelligibility at the expense of precision. Because the loss of precision is usually minor when compared with the gain in comprehension, in this book we will adopt an extremely simple system of notation.

Notation of Variables

The general rule is that an uppercase letter, often X or Y, will represent a variable as a whole. The letter and a subscript will then represent an individual value of that variable. Suppose for example that we have the following five scores on the length of time (in seconds) that third-grade children can hold their breath: [45, 42, 35, 23, 52]. This set of scores will be referred to as X. The first number of this set (45) can be referred to as X_1, the second (42) as X_2, and so on. When we want to refer to a single score without specifying which one, we will refer to X_i, where i can take on any value between 1 and 5. In practice, the use of subscripts is often a distraction, and they are generally omitted if no confusion will result.

Summation Notation

sigma

One of the most common symbols in statistics is the uppercase Greek letter **sigma** $\left(\sum\right)$, which is the standard notation for summation. It is readily translated as "add up, or sum, what follows." Thus, $\sum X_i$ is read "sum the X_is." To be perfectly correct, the notation for summing all N values of X is $\sum_{i=1}^{N} X_i$, which translates to "sum all of the X_is from $i = 1$ to $i = N$." In practice, we seldom need to specify what is to be done this precisely, and in most cases all subscripts are dropped and the notation for the sum of the X_i is simply $\sum X$.

Several extensions of the simple case of $\sum X$ must be noted and thoroughly understood. One of these is $\sum X^2$, which is read as "sum the squared values of X" (i.e., $45^2 + 42^2 + 35^2 + 23^2 + 52^2 = 8{,}247$). (Note that this is quite different from $\left(\sum X\right)^2$, which tells us to sum the Xs and then square the result. This would equal $(45 + 42 + 35 + 23 + 52)^2 = (197)^2 = 38{,}809) = \left(\sum X\right)^2$. *The general rule, which always applies, is to perform operations within parentheses before performing operations outside parentheses.* Thus, for $\left(\sum X\right)^2$, we sum the values of X and *then* we square the result, as opposed to $\sum X^2$, for which we square the Xs before we sum.

Another common expression, when data are available on two variables (X and Y), is $\sum XY$, which means "sum the products of the corresponding values of X and Y." The use of these and other terms will be illustrated in the following example.

Imagine a simple experiment in which we record the anxiety scores (X) of five students and also record the number of days during the last semester that they missed a test because they were absent from school (Y). The data and simple summation operations on them are illustrated in Table 2.4. Some of these operations have been discussed already, and others will be discussed in the next few chapters.

Double Subscripts

A common notational device is to use two or more subscripts to specify exactly which value of X you have in mind. Suppose, for example, that we were given the data shown in Table 2.5. If we want to specify the entry in the ith row and jth column, we will denote this as X_{ij}. Thus, the score on the third trial of Day 2 is $X_{2,3} = 13$. Some notational systems use $\sum_{i=1}^{2} \sum_{j=1}^{5} X_{ij}$, which translates as "sum the X_{ij}s where i takes on values 1 and 2 and j takes on all values from 1 to 5." You need to be aware of this system of notation because some other textbooks use it. In this book, however, the simpler, but less precise, $\sum X$ is used where possible, with $\sum X_{ij}$ used only when absolutely necessary, and $\sum \sum X_{ij}$ never appearing.

Table 2.4   Illustration of operations involving summation notation

	Anxiety Score (X)	Tests Missed (Y)	X^2	Y^2	$X - Y$	XY
	10	3	100	9	7	30
	15	4	225	16	11	60
	12	1	144	1	11	12
	9	1	81	1	8	9
	10	3	100	9	7	30
Sum	56	12	650	36	44	141

$$\sum X = (10 + 15 + 12 + 9 + 10) = 56$$
$$\sum Y = (3 + 4 + 1 + 1 + 3) = 12$$
$$\sum X^2 = (10^2 + 15^2 + 12^2 + 9^2 + 10^2) = 650$$
$$\sum Y^2 = (3^2 + 4^2 + 1^2 + 1^2 + 3^2) = 36$$
$$\sum (X - Y) = (7 + 11 + 11 + 8 + 7) = 44$$
$$\sum XY = (10)(3) + (15)(4) + (12)(1) + (9)(1) + (10)(3) = 141$$
$$\left(\sum X\right)^2 = 56^2 = 3136$$
$$\left(\sum Y\right)^2 = 12^2 = 144$$
$$\left(\sum (X - Y)\right)^2 = 44^2 = 1936$$
$$\left(\sum X\right)\left(\sum Y\right) = (56)(12) = 672$$

Table 2.5   Hypothetical data illustrating notation

		Trial					
		1	2	3	4	5	Total
Day	1	8	7	6	9	12	42
	2	10	11	13	15	14	63
	Total	18	18	19	24	26	105

You must thoroughly understand notation if you are to learn even the most elementary statistical techniques. You should study Table 2.4 until you fully understand all the procedures involved.

2.8   Measures of Central Tendency

We have seen how to display data in ways that allow us to begin to draw some conclusions about what the data have to say. Plotting data shows the general shape of the distribution and gives a visual sense of the general magnitude of the numbers involved. In this section you will see several statistics that can be used to represent the "center" of the distribution. These statistics are called measures of central tendency. In the next section we will go a step further and look at measures that deal with how the observations are dispersed around that central tendency, but first we must address how we identify the center of the distribution.

measures of central tendency

measures of location

The phrase **measures of central tendency,** or sometimes **measures of location,** refers to the set of measures that reflect where on the scale the distribution is centered. These measures differ in how much use they make of the data, particularly of extreme values, but they are all trying to tell us something about where the center of the distribution lies. The three major measures of central tendency are the mode, which is based on only a few data points; the median, which ignores most of the data; and the mean, which is calculated from all of the data. We will discuss these in turn, beginning with the mode, which is the least used (and often the least useful) measure.

The Mode

mode (Mo)

The **mode (Mo)** can be defined simply as the most common score, that is, the score obtained from the largest number of subjects. Thus, the mode is that value of X that corresponds to the highest point on the distribution. If two *adjacent* times occur with equal (and greatest) frequency, a common convention is to take an average of the two values and call that the mode. If, on the other hand, two *nonadjacent* reaction times occur with equal (or nearly equal) frequency, we say that the distribution is bimodal and would most likely report both modes. For example, the distribution of time spent playing electronic games is roughly bimodal (see Figure 2.3), with peaks at the intervals of 0–9 minutes and 40–49 minutes. (You might argue that it is trimodal, with another peak at 120+ minutes, but that is a catchall interval for "all other values," so it does not make much sense to think of it as a modal value.)

The reaction time data present us with a very minor problem. There are actually two modal values (at 62 and 67, with 14 scores falling at each of these times). These points are too close to really call the distribution "bimodal," but too far apart to feel good about averaging them. My preference would be either to report both values or to simply say that the modal value is "in the 60s," and leave it at that. (We rarely use the mode in a very precise way, so there isn't much to be lost by being a bit imprecise in reporting it.) When many data points are examined in an ungrouped fashion, there is probably so much unimportant variability from one value to another that it doesn't make much sense to worry about the mode or to be very concerned with bimodality. When the data have been grouped into a reasonably stable looking plot, as in Figure 2.2, the mode is considerably more meaningful.

The Median

median (Mdn)

The **median (Mdn)** is the score that corresponds to the point at or below which 50% of the scores fall when the data are arranged in numerical order. By this definition, the median is also called the 50th percentile.[5] For example, consider the numbers (5, 8, 3, 7, 15). If the numbers are arranged in numerical order (3, 5, 7, 8, 15), the middle score would be 7, and it would be called the median. Suppose, however, that there were an even number of scores, for example (5, 11, 3, 7, 15, 14). Rearranging, we get (3, 5, 7, 11, 14, 15), and no score has 50% of the values below it. That point actually falls between the 7 and the 11. In such a case the average (9) of the two middle scores (7 and 11) is commonly taken as the median.[6]

---

[5] A specific percentile is defined as the point on a scale at or below which a specified percentage of scores fall.

[6] The definition of the median is another one of those things about which statisticians love to argue. The definition given here, in which the median is defined as a *point* on a distribution of numbers, is the one most critics prefer. It is also in line with the statement that the median is the 50th percentile. On the other hand, there are many who are perfectly happy to say that the median is either the middle *number* in an ordered series (if N is odd) or the average of the two middle *numbers* (if N is even). Reading these arguments is a bit like going to a faculty meeting when there is nothing terribly important on the agenda. The less important the issue, the more there is to say about it.

median location

A term that we will need shortly is the **median location.** The median location of N numbers is defined as follows:

$$\text{Median location} = \frac{N+1}{2}$$

Thus, for five numbers the median location $= (5+1)/2 = 3$, which simply means that the median is the third number in an ordered series. For 12 numbers, the median location $= (12+1)/2 = 6.5$; the median falls between, and is the average of, the sixth and seventh numbers.

For the data on reaction times in Table 2.2, the median location $= (300+1)/2 = 150.5$. When the data are arranged in order, the 150th time is 59 and the 151st time is 60; thus the median is $(59+60)/2 = 59.5$ hundredths of a second. You can calculate this for yourself from Table 2.2. For the electronic games data there are 100 scores, and the median location is 50.5. We can tell from the stem-and-leaf display in Figure 2.3 that the 50th score is 44 and the 51st score is 46. The median would be 45, which is the average of these two values.

The Mean

mean

The most common measure of central tendency, and one that really needs little explanation, is the mean, or what people generally have in mind when they use the word average. The **mean** $(\overline{X})$ is the sum of the scores divided by the number of scores and is usually designated $\overline{X}$ (read "X bar").[7] It is defined (using the summation notation given on page 32) as follows:

$$\overline{X} = \frac{\sum X}{N}$$

where $\sum X$ is the sum of all values of X, and N is the number of X values. As an illustration, the mean of the numbers 3, 5, 12, and 5 is

$$\frac{3+5+12+5}{4} = \frac{25}{4} = 6.25$$

For the reaction time data in Table 2.2, the sum of the observations is 18,078. When we divide that number by $N = 300$, we get $18,078/300 = 60.26$. Notice that this answer agrees well with the median, which we found to be 59.5. The mean and the median will be close whenever the distribution is nearly symmetric (as defined on page 27). It also agrees well with the modal interval (60–64).

We could calculate the mean for the electronics game data by obtaining the raw data values from the stem-and-leaf display, summing those values, and dividing by 100. For that example, the mean would be $5204/100 = 52.04$. Later in this chapter you will see how to use Minitab to save yourself considerable work in calculating the mean for large data sets.

Relative Advantages and Disadvantages of the Mode, the Median, and the Mean

Only when the distribution is symmetric will the mean and the median be equal, and only when the distribution is symmetric and unimodal will all three measures be the same. In all other cases—including almost all situations with which we will deal—some measure of

---

[7] The American Psychological Association would like us to use M for the mean instead of $\overline{X}$, but I have used $\overline{X}$ for so many years that it would offend my delicate sensibilities to give it up. The rest of the statistical world generally agrees with me on this, so we will use $\overline{X}$ throughout.

central tendency must be chosen. A set of rules governing when to use a particular measure of central tendency would be convenient. However, there are no such rules. Nevertheless it is possible to make intelligent choices among the three measures.

The Mode

The mode is the most commonly occurring score. By definition, then, it is a score that actually occurred, whereas the mean and sometimes the median may be values that never appear in the data. The mode also has the obvious advantage of representing the largest number of people. Someone who is running a small store would do well to concentrate on the mode. If 80% of your customers want the giant economy family size and 20% want the teeny-weeny, single-person size, it wouldn't seem particularly wise to aim for some other measure of location and stock only the regular size.

Related to these two advantages is that, by definition, the probability that an observation drawn at random (X_i) will be equal to the mode is greater than the probability that it will be equal to any other specific score. Finally, the mode has the advantage of being applicable to nominal data, which, if you think about it, is not true of the median or the mean.

The mode has its disadvantages, however. We have already seen that the mode depends on how we group our data. Another disadvantage is that it may not be particularly representative of the entire collection of numbers. This disadvantage is illustrated in the electronic game data (see Figure 2.3), in which the modal interval equals 0–9, which probably reflects the fact that a large number of people do not play video games (difficult as that may be to believe). Using that interval as the mode would be to ignore all those people who do play.

The Median

The major advantage of the median, which it shares with the mode, is that it is unaffected by extreme scores. The medians of both (5, 8, 9, 15, 16) and (0, 8, 9, 15, 206) are 9. Many experimenters find this characteristic to be useful in studies in which extreme scores occasionally occur but have no particular significance. For example, the average trained rat can run down a short runway in approximately 1 to 2 seconds. Every once in a while this same rat will inexplicably stop halfway down, scratch himself, poke his nose at the photocells, and lie down to sleep. In that instance it is of no practical significance whether he takes 30 seconds or 10 minutes to get to the other end of the runway. It may even depend on when the experimenter gives up and pokes him with a pencil. If we ran a rat through three trials on a given day and his times were (1.2, 1.3, and 20 seconds), that would have the same meaning to us—in terms of what it tells us about the rat's knowledge of the task—as if his times were (1.2, 1.3, and 136.4 seconds). In both cases the median would be 1.3; however, his daily *mean* would be quite different in the two cases (7.5 versus 46.3 seconds). This problem frequently induces experimenters to work with the median rather than the mean time per day.

The median has another point in its favor, when contrasted with the mean, which those writers who get excited over scales of measurement like to point out. The calculation of the median does not require any assumptions about the interval properties of the scale. With the numbers (5, 8, and 11), the object represented by the number 8 is in the middle, no matter how close or distant it is from objects represented by 5 and 11. When we say that the *mean* is 8, however, we, or our readers, may be making the implicit assumption that the underlying distance between objects 5 and 8 is the same as the underlying distance between objects 8 and 11. Whether or not this assumption is reasonable is up to the experimenter to determine. I prefer to work on the principle that if it is an absurdly unreasonable assumption, the experimenter will realize that and take appropriate steps. If it is not absurdly unreasonable, then its practical effect on the results most likely will be negligible. (This problem of scales of measurement was discussed in more detail earlier.)

A major disadvantage of the median is that it does not enter readily into equations and is thus more difficult to work with than the mean. It is also not as stable from sample to sample as the mean, as we will see shortly.

The Mean

Of the three principal measures of central tendency, the mean is by far the most common. It would not be too much of an exaggeration to say that for many people statistics is nearly synonymous with the study of the mean.

As we have already seen, certain disadvantages are associated with the mean: It is influenced by extreme scores, its value may not actually exist in the data, and its interpretation in terms of the underlying variable being measured requires at least some faith in the interval properties of the data. You might be inclined to politely suggest that if the mean has all the disadvantages I have just ascribed to it, then maybe it should be quietly forgotten and allowed to slip into oblivion along with statistics like the "critical ratio," a statistical concept that hasn't been heard of for years. The mean, however, is made of sterner stuff.

The mean has several important advantages that far outweigh its disadvantages. Probably the most important of these from a historical point of view (though not necessarily from your point of view) is that the mean can be manipulated algebraically. In other words, we can use the mean in an equation and manipulate it through the normal rules of algebra, specifically because we can write an equation that defines the mean. Because you cannot write a standard equation for the mode or the median, you have no real way of manipulating those statistics using standard algebra. Whatever the mean's faults, this accounts in large part for its widespread application. The second important advantage of the mean is that it has several desirable properties with respect to its use as an estimate of the population mean. In particular, if we drew many samples from some population, the sample means that resulted would be more stable (less variable) estimates of the central tendency of that population than would the sample medians or modes. The fact that the sample mean is generally a better estimate of the population mean than is the mode or the median is a major reason that it is so widely used.

Obtaining Measures of Central Tendency Using Minitab

For small sets of data it is perfectly reasonable to compute measures of central tendency by hand. With larger sample sizes or data sets with many variables, however, it is much simpler to let a computer program do the work. Minitab is ideally suited to this purpose because it is easy to use, versatile, and widely available. (For that matter, you can do a good job of calculating the mean with Excel, although statisticians often get on their high horses about the numerical properties of the Excel subroutines for weird samples that we hardly ever see.)

Suppose that as part of a large study on teaching effectiveness, we asked each of 15 students in class to record the number of different annoying mannerisms exhibited by their instructor (e.g., dropping chalk, losing chalk, arranging and rearranging lecture notes, pacing, alternately standing and sitting, and all those other activities the counting of which makes the lecture pass more quickly). These data are illustrated in Exhibit 2.2 along with the menus required to produce the three common measures of central tendency. We can obtain the mean and the median directly, but to get the mode we need to produce a histogram (or a stem-and-leaf display) and then look for the most frequently appearing interval.

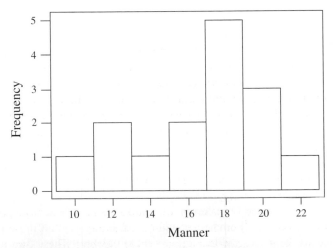

Descriptive Statistics

Variable	N	Mean	Median	TrMean	StDev	SE Mean
Manner	15	16.400	17.000	16.385	3.562	0.920

Variable	Minimum	Maximum	Q1	Q3
Manner	10.000	23.000	14.000	19.000

Exhibit 2.2   Minitab menu choices and output for measures of central tendency on the number of annoying mannerisms

From Exhibit 2.2 you can see that the mean (16.4), the median (17), and the mode (18) are all about the same, and that the distribution is fairly smooth but slightly negatively skewed. We can also see from the histogram that there is considerable disagreement among students concerning the number of annoying mannerisms exhibited by the instructor. This dispersion on either side of the mean is discussed in the next section.

2.9 Measures of Variability

In the previous section we considered several measures related to the center of a distribution. However, an average value for the distribution (whether it be the mode, the median, or the mean) fails to give the whole story. We need some additional measure (or measures) to indicate the degree to which individual observations are clustered about or, equivalently, deviate from that average value. The average may reflect the general location of most of the scores, or the scores may be distributed over a wide range of values, and the "average" may not be very representative of the full set of observations. Everyone has had experience with examinations on which all students received approximately the same grade and with those on which the scores ranged from excellent to dreadful. Measures referring to the differences between these two situations are what we have in mind when we speak of **dispersion,** or variability, around the median, the mode, or any other point. In general, we will refer specifically to dispersion around the mean.

dispersion

As an example of a situation in which we might expect differences in variability from one group to another, consider the case in which two sections of Computer Science 1 are given the same exam. One section is told that the exam will count for 20% of their final grade, and the other section is told that it will count for 90%. It is probably reasonable to expect that the two sections will have roughly the same mean. However, because "pressure" has a facilitative effect on the performance of some people and a disruptive effect on the performance of others, we might expect more variability in the section for whom the exam counts for 90% of their final grade. Do these hypotheses correspond with your experience? If not, what would you expect to happen? Does your expectation lead to differences in means, variances, both, or neither?

A second example was recommended by Weaver (1999) and is based on something with which I'm sure you are all familiar—the standard growth chart for infants. Such a chart appears in Figure 2.11, in the bottom half of the chart, where you can see the normal range of girls' weights between birth and 36 months. The bold line labeled "50" through the center represents the mean weight at each age. The two lines on each side represent the limits within which we expect the middle half of the distribution to fall, the next two lines as you go each way from the center enclose the middle 80% and the middle 90% of children, respectively. From this figure it is easy to see the increase in dispersion as children increase in age. The weights of most newborns lie within 1 pound of the mean, whereas the weights of 3-year-olds are spread out over about 5 pounds on each side of the mean. Obviously the mean is increasing too, though we are more concerned here with dispersion.

For our final illustration we will take some interesting data collected by Langlois and Roggman (1990) on the perceived attractiveness of faces. Think for a moment about some of the faces you consider attractive. Do they tend to have unusual features (e.g., prominent noses or unusual eyebrows), or are the features rather ordinary? Langlois and Roggman were interested in investigating what makes faces attractive. Toward that end, they presented students with computer-generated pictures of faces. Some of these pictures had been created by averaging together snapshots of four different people to create a composite. We will label

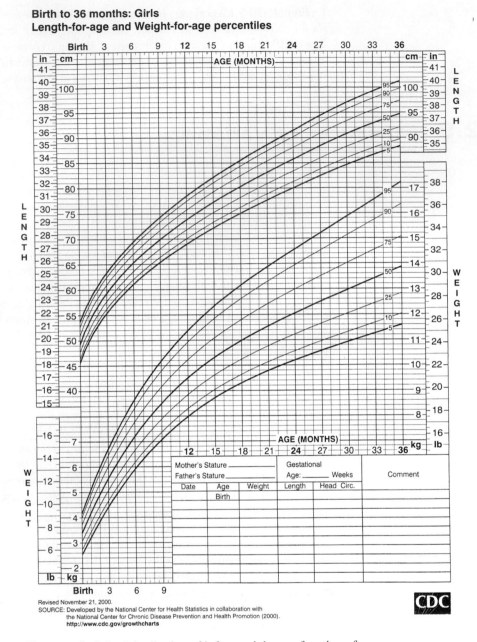

Birth to 36 months: Girls
Length-for-age and Weight-for-age percentiles

Figure 2.11   Distribution of infant weight as a function of age

these photographs Set 4. Other pictures (Set 32) were created by averaging across snapshots of 32 different people. As you might suspect, when you average across four people, there is still room for individuality in the composite. For example, some composites show thin faces, while others show round ones. However, averaging across 32 people usually gives results that are very "average." Noses are neither too long nor too short, ears don't stick out too far nor sit too close to the head, and so on. Students were asked to examine the resulting pictures and rate each one on a 5-point scale of attractiveness. The authors were primarily interested in determining whether the *mean* rating of the faces in Set 4 was less than the

Table 2.6   Rated attractiveness of composite faces

Set 4		Set 32	
Picture	Composite of 4 Faces	Picture	Composite of 32 Faces
1	1.20	21	3.13
2	1.82	22	3.17
3	1.93	23	3.19
4	2.04	24	3.19
5	2.30	25	3.20
6	2.33	26	3.20
7	2.34	27	3.22
8	2.47	28	3.23
9	2.51	29	3.25
10	2.55	30	3.26
11	2.64	31	3.27
12	2.76	32	3.29
13	2.77	33	3.29
14	2.90	34	3.30
15	2.91	35	3.31
16	3.20	36	3.31
17	3.22	37	3.34
18	3.39	38	3.34
19	3.59	39	3.36
20	4.02	40	3.38
	Mean = 2.64		Mean = 3.26

mean rating of the faces in Set 32. It was, suggesting that faces with distinctive characteristics are judged as less attractive than more ordinary faces. In this section, however, we are more interested in the degree of *similarity* in the ratings of faces. We suspect that composites of 32 faces would be more homogeneous, and thus would be rated more similarly, than would composites of four faces.

The data are shown in Table 2.6.[8] From the table you can see that Langlois and Roggman correctly predicted that Set 32 faces would be rated as more attractive than Set 4 faces. (The means were 3.26 and 2.64, respectively.) But notice also that the ratings for the composites of 32 faces are considerably more homogeneous than the ratings of the composites of four faces. Figure 2.12 plots these sets of data as standard histograms.

Even though it is apparent from Figure 2.12 that there is greater variability in the rating of composites of 4 photographs than in the rating of composites of 32 photographs, some sort of measure is needed to reflect this difference in variability. A number of measures could be used, and they will be discussed in turn, starting with the simplest.

---

[8] These data are not the actual numbers that Langlois and Roggman collected, but they have been generated to have exactly the same mean and standard deviation as the original data. Langlois and Roggman used six composite photographs per set. I have used 20 photographs per set to make the data more applicable to my purposes in this chapter. The conclusions that you would draw from these data, however, are exactly the same as the conclusions you would draw from theirs.

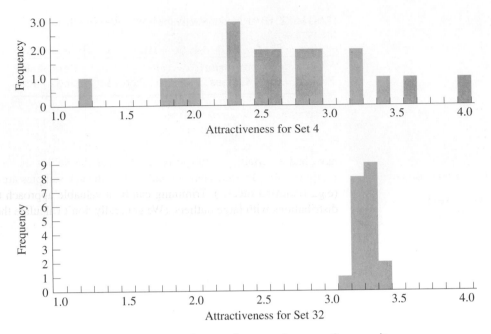

Figure 2.12 Distribution of scores for attractiveness of composite

Range

range

The **range** is a measure of distance, namely the distance from the lowest to the highest score. For our data, the range for Set 4 is $(4.02 - 1.20) = 2.82$ units; for Set 32 it is $(3.38 - 3.13) = 0.25$ unit. The range is an exceedingly common measure and is illustrated in everyday life by such statements as "The price of red peppers fluctuates over a 3-dollar range from \$.99 to \$3.99 per pound." The range suffers, however, from a total reliance on extreme values, or, if the values are *unusually* extreme, on outliers. As a result, the range may give a distorted picture of the variability.

Interquartile Range and Other Range Statistics

interquartile range

first quartile

third quartile

second quartile

The **interquartile range** represents an attempt to circumvent the problem of the range's heavy dependence on extreme scores. An interquartile range is obtained by discarding the upper 25% and the lower 25% of the distribution and taking the range of what remains. The point that cuts off the lowest 25% of the distribution is called the **first quartile,** and is usually denoted as Q_1. Similarly the point that cuts off the upper 25% of the distribution is called the **third quartile** and is denoted Q_3. (The median is the **second quartile, Q_2.**) The difference between the first and third quartiles $(Q_3 - Q_1)$ is the interquartile range. We can calculate the interquartile range for the data on attractiveness of faces by omitting the lowest five scores and the highest five scores and determining the range of the remainder. In this case, the interquartile range for Set 4 would be 0.58 and the interquartile range for Set 32 would be only 0.11.

The interquartile range plays an important role in a useful graphical method known as a boxplot. This method will be discussed in Section 2.10.

The interquartile range suffers from problems that are just the opposite of those found with the range. Specifically, the interquartile range discards too much of the data. If we want to know whether one set of photographs is judged more variable than another, it may not make much sense to toss out those scores that are most extreme and thus vary the most from the mean.

There is nothing sacred about eliminating the upper and lower 25% of the distribution before calculating the range. Actually, we could eliminate any percentage we wanted, as long as we could justify that number to ourselves and to others. What we really want to do is eliminate those scores that are likely to be errors or attributable to unusual events without eliminating the variability that we seek to study.

trimmed samples

trimmed statistics

trimmed means

Winsorized

An old technique that has received renewed attention in recent years (Wilcox, 2005) involves trimming the extreme values from samples before taking the mean. Samples that have had a certain percentage (e.g., 10%) of the values in each tail removed are called **trimmed samples,** and statistics calculated on such samples are called **trimmed statistics** (e.g., **trimmed means**). Trimming can be a valuable approach to skewed distributions or distributions with large outliers. (We generally don't calculate the "trimmed standard deviation." Instead we normally create a **Winsorized** sample by dropping the lowest 10% (for example) of the scores and replacing them by the smallest score that remains, then dropping the highest 10% and *replacing* those by the highest score which remains, and then computing the standard deviation on the modified data.) Trimmed and Winsorized samples have begun to play a more important role in statistical practice.

The Average Deviation

At first glance it would seem that if we want to measure how scores are dispersed around the mean (i.e., deviate from the mean), the most logical thing to do would be to obtain all the deviations (i.e., $X_i - \overline{X}$) and average them. You might reasonably think that the more widely the scores are dispersed, the greater the deviations and therefore the greater the average of the deviations. However, common sense has led you astray here. If you calculate the deviations from the mean, some scores will be above the mean and have a positive deviation, whereas others will be below the mean and have negative deviations. In the end, the positive and negative deviations will balance each other out and the sum of the deviations will be zero. This will not get us very far.

The Mean Absolute Deviation

mean absolute deviation (m.a.d.)

If you think about the difficulty in trying to get something useful out of the average of the deviations, you might well be led to suggest that we could solve the whole problem by taking the absolute values of the deviations. (The absolute value of a number is the value of that number with any minus signs removed. The absolute value is indicated by vertical bars around the number, e.g., $|-3| = 3$.) The suggestion to use absolute values makes sense because we want to know *how much* scores deviate from the mean without regard to whether they are above or below it. The measure suggested here is a perfectly legitimate one and even has a name: the **mean absolute deviation (m.a.d.)**. The sum of the absolute deviations is divided by N (the number of scores) to yield an average (mean) deviation: m.a.d. For all its simplicity and intuitive appeal, the mean absolute deviation has not played an important role in statistical methods. Much more useful measures, the variance and the standard deviation, are normally used instead.

The Variance

sample variance

population variance

The measure that we will consider in this section, the **sample variance** (s^2), represents a different approach to the problem of the deviations themselves averaging to zero. (When we are referring to the **population variance,** we use σ^2 (lowercase sigma) as the symbol.)

In the case of the variance, we take advantage of the fact that the square of a negative number is positive. Thus, we sum the *squared* deviations rather than the absolute deviations. Because we want an average, we next divide that sum by some function of N, the number of scores. Although you might reasonably expect that we would divide by N, we actually divide by $(N - 1)$. We use $(N - 1)$ as a divisor for the *sample variance* because, as we will see shortly, it leaves us with a sample variance that is a better estimate of the corresponding population variance. (The population variance is calculated by dividing the sum of the squared deviations, for each value in the population, by N rather than $[N - 1]$. However, we only rarely calculate a population variance; we almost always estimate it from a sample variance.)

If it is important to specify more precisely the variable to which s^2 refers, we can subscript it with a letter representing the variable. Thus, if we denote the data in Set 4 as X, the variance could be denoted as s_X^2. (You could refer to $s_{\text{Set 4}}^2$, but long subscripts are usually awkward. In general, we label variables with simple letters like X and Y.)

For our example, we can calculate the sample variances of Set 4 and Set 32 as follows:[9]

Set 4(X)

$$s_X^2 = \frac{\sum(X - \overline{X})^2}{N - 1}$$

$$= \frac{(1.20 - 2.64)^2 + (1.82 - 2.64)^2 + \cdots + (4.02 - 2.64)^2}{20 - 1}$$

$$= \frac{8.1569}{19} = 0.4293$$

Set 32(Y)

$$s_Y^2 = \frac{\sum(Y - \overline{Y})^2}{N - 1}$$

$$= \frac{(3.13 - 3.26)^2 + (3.17 - 3.26)^2 + \cdots + (3.38 - 3.26)^2}{20 - 1}$$

$$= \frac{0.0903}{19} = 0.0048$$

From these calculations we see that the difference in variances reflects the differences we see in the distributions.

Although the variance is an exceptionally important concept and one of the most commonly used statistics, it does not have the direct intuitive interpretation we would like. Because it is based on *squared* deviations, the result is in squared units. Thus, Set 4 has a mean attractiveness rating of 2.64 and a variance of 0.4293 *squared* unit. But squared units are awkward things to talk about and have little meaning with respect to the data. Fortunately, the solution to this problem is simple: Take the square root of the variance.

The Standard Deviation

standard deviation

The **standard deviation** (s or σ) is defined as the positive square root of the variance and, for a sample, is symbolized as s (with a subscript identifying the variable if necessary) or,

---

[9] In these calculations and others throughout the book, my answers may differ slightly from those that you obtain for the same data. If so, the difference is most likely caused by rounding. If you repeat my calculations and arrive at a similar, though different, answer, that is sufficient.

occasionally, as SD.[10] (The notation σ is used in reference to a *population* standard deviation.) The following formula defines the sample standard deviation:

$$s_X = \sqrt{\frac{\sum(X - \overline{X})^2}{N - 1}}$$

For our example,

$$s_X = \sqrt{s_X^2} = \sqrt{0.4293} = 0.6552$$

$$s_Y = \sqrt{s_Y^2} = \sqrt{0.0048} = 0.0689$$

For convenience, I will round these answers to 0.66 and 0.07, respectively.

If you look at the formula for the standard deviation, you will see that the standard deviation, like the mean absolute deviation, is basically a measure of the average of the deviations of each score from the mean. Granted, these deviations have been squared, summed, and so on, but at heart they are still deviations. And even though we have divided by $(N - 1)$ instead of N, we still have obtained something very much like a mean or an "average" of these deviations. Thus, we can say without too much distortion that attractiveness ratings for Set 4 deviated, on the average, 0.66 unit from the mean, whereas attractiveness ratings for Set 32 deviated, on the average, only 0.07 unit from the mean. This way of thinking about the standard deviation as a sort of average deviation goes a long way toward giving it meaning without doing serious injustice to the concept.

These results tell us two interesting things about attractiveness. If you were a subject in this experiment, the fact that computer averaging of many faces produces similar composites would be reflected in the fact that your ratings of Set 32 would not show much variability—all those images are judged to be pretty much alike. Second, the fact that those ratings have a higher mean than the ratings of faces in Set 4 reveals that averaging over many faces produces composites that seem more attractive. Does this conform to your everyday experience? I, for one, would have expected that faces judged attractive would be those with distinctive features, but I would have been wrong. Go back and think again about those faces you class as attractive. Are they really distinctive? If so, do you have an additional hypothesis to explain the findings?

We can also look at the standard deviation in terms of how many scores fall no more than a standard deviation above or below the mean. For a wide variety of reasonably symmetric and mound-shaped distributions, we can say that approximately two-thirds of the observations lie within one standard deviation of the mean (for a normal distribution, which will be discussed in Chapter 3, it is almost exactly two-thirds). Although there certainly are exceptions, especially for badly skewed distributions, this rule is still useful. If I told you that for traditional jobs the mean starting salary for college graduates this year is expected to be $28,000 with a standard deviation of $4,000, you probably would not be far off to conclude that about two-thirds of graduates who take these jobs will earn between $24,000 and $32,000. In addition, most (e.g., 95%) fall within 2 standard deviations.

Computational Formulae for the Variance and the Standard Deviation

The previous expressions for the variance and the standard deviation, although perfectly correct, are incredibly unwieldy for any reasonable amount of data. They are also prone to rounding errors because they usually involve squaring fractional deviations. They are excellent definitional formulae, but we will now consider a more practical set of calculational

---

[10] The American Psychological Association prefers to abbreviate the standard deviation as "SD," but everyone else uses "*s*."

formulae. These formulae are algebraically equivalent to the ones we have seen, so they will give the same answers but with much less effort.

The definitional formula for the sample variance was given as

$$s_X^2 = \frac{\sum(X - \overline{X})^2}{N - 1}$$

A more practical computational formula is

$$s_X^2 = \frac{\sum X^2 - \frac{(\sum X)^2}{N}}{N - 1}$$

Similarly, for the sample standard deviation

$$s_X = \sqrt{\frac{\sum(X - \overline{X})^2}{N - 1}}$$

$$= \sqrt{\frac{\sum X^2 - \frac{(\sum X)^2}{N}}{N - 1}}$$

Recently people whose opinions I respect have suggested that I should remove such formulae as these from the book because people rarely calculate variances by hand anymore. Although that is true, and I only wave my hands at them in my own courses, many people still believe it is important to be able to do the calculation. More important, perhaps, is the fact that we will see these formulae again in different disguises, and it helps to understand what is going on if you recognize them for what they are. However, I agree with those critics of more complex formulae, and in those cases I have restructured recent editions of the text around definitional formulae.

Applying the computational formula for the sample variance for Set 4, we obtain

$$s_X^2 = \frac{\sum X^2 - \frac{(\sum X)^2}{N}}{N - 1}$$

$$= \frac{1.20^2 + 1.82^2 + \cdots + 4.02^2 - \frac{52.89^2}{20}}{19}$$

$$= \frac{148.0241 - \frac{52.89^2}{20}}{19} = 0.4293$$

Note that the answer we obtained here is exactly the same as the answer we obtained by the definitional formula. Note also, as pointed out earlier, that $\sum X^2 = 148.0241$ is quite different from $\left(\sum X\right)^2 = 52.89^2 = 2797.35$. I leave the calculation of the variance for Set 32 to you.

You might be somewhat reassured to learn that the level of mathematics required for the previous calculations is about as much as you will need anywhere in this book—not because I am watering down the material, but because an understanding of most applied statistics does not require much in the way of advanced mathematics. (I told you that you learned it all in high school.)

The Influence of Extreme Values on the Variance and Standard Deviation

The variance and standard deviation are very sensitive to extreme scores. To put this differently, extreme scores play a disproportionate role in determining the variance. Consider a set of data that range from roughly 0 to 10, with a mean of 5. From the definitional formula for the variance, you will see that a score of 5 (the mean) contributes nothing to the variance, because the deviation score is 0. A score of 6 contributes $1/(N-1)$ to s^2, since $(X - \overline{X})^2 = (6-5)^2 = 1$. A score of 10, however, contributes $25/(N-1)$ units to s^2, since $(10-5)^2 = 25$. Thus, although 6 and 10 deviate from the mean by 1 and 5 units, respectively, their relative contributions to the variance are 1 and 25. This is what we mean when we say that large deviations are disproportionately represented. You might keep this in mind the next time you use a measuring instrument that is "OK because it is unreliable only at the extremes." It is just those extremes that may have the greatest effect on the interpretation of the data.

The Coefficient of Variation

One of the most common things we do in statistics is to compare the means of two or more groups, or even two or more variables. Comparing the variability of those groups or variables, however, is also a legitimate and worthwhile activity. Suppose, for example, that we have two competing tests for assessing long-term memory.

One of the tests typically produces data with a mean of 15 and a standard deviation of 3.5. The second, quite different, test produces data with a mean of 75 and a standard deviation of 10.5. All other things being equal, which test is better for assessing long-term memory? We might be inclined to argue that the second test is better, in that we want a measure on which there is enough variability that we are able to study differences among people, and the second test has the larger standard deviation. However, keep in mind that the two tests also differ substantially in their means, and this difference must be considered.

If you think for a moment about the fact that the standard deviation is based on deviations from the mean, it seems logical that a value could more easily deviate substantially from a large mean than from a small one. For example, if you rate teaching effectiveness on a 7-point scale with a mean of 3, it would be impossible to have a deviation greater than 4. On the other hand, on a 70-point scale with a mean of 30, deviations of 10 or 20 would be common. Somehow we need to account for the greater opportunity for large deviations in the second case when we compare the variability of our two measures. In other words, when we look at the standard deviation, we must keep in mind the magnitude of the mean as well.

The simplest way to compare standard deviations on measures that have quite different means is simply to scale the standard deviation by the magnitude of the mean. That is what **coefficient of variation (CV)** we do with the **coefficient of variation** (CV).[11] We will define that coefficient as simply the standard deviation divided by the mean:

$$\text{CV} = \frac{\text{Standard deviation}}{\text{Mean}} = \frac{s_X}{\overline{X}} \times 100$$

(We multiply by 100 to express the result as a percentage.) To return to our memory-task example, for the first measure, $\text{CV} = (3.5/15) \times 100 = 23.3$. Here the standard deviation is approximately 23% of the mean. For the second measure, $\text{CV} = (10.5/75) \times 100 = 14$. In this case, the coefficient of variation for the second measure is about half as large as for the first. If I could be convinced that the larger coefficient of variation in the first measure was not attributable simply to sloppy measurement, I would be inclined to choose the first measure over the second.

To take a second example, Katz, Lautenschlager, Blackburn, and Harris (1990) asked students to answer a set of multiple-choice questions from the Scholastic Aptitude Test[12] (SAT). One group read the relevant passage and answered the questions. Another group answered the

---

[11] I want to thank Andrew Gilpin (personal communication, 1990) for reminding me of the usefulness of the coefficient of variation. It is a meaningful statistic that is often overlooked.

[12] The test is now known simply as the SAT, or, more recently, the SAT-I.

questions without having read the passage on which they were based—sort of like taking a multiple-choice test on Mongolian history without having taken the course. The data follow:

	Read Passage	Did Not Read Passage
Mean	69.6	46.6
SD	10.6	6.8
CV	15.2	14.6

The ratio of the two standard deviations is $10.6/6.8 = 1.56$, meaning that the Read group had a standard deviation that was more than 50% larger than that of the Did Not Read group. On the other hand, the coefficients of variation are virtually the same for the two groups, suggesting that any difference in variability between the groups can be explained by the higher scores in the first group. (Incidentally, chance performance would have produced a mean of 20 with a standard deviation of 4. Even without reading the passage, students score well above chance levels just by intelligent guessing.)

In using the coefficient of variation, it is important to keep in mind the nature of the variable that you are measuring. If its scale is arbitrary, you might not want to put too much faith in the coefficient. But perhaps you don't want to put too much faith in the variance either. This is a place where a little common sense is particularly useful.

The Mean and Variance as Estimators

I pointed out in Chapter 1 that we generally calculate measures such as the mean and variance to use as *estimates* of the corresponding values in the populations. Characteristics of samples are called statistics and are designated by Roman letters (e.g., $\overline{X}$). Characteristics of populations are called parameters and are designated by Greek letters. Thus, the population mean is symbolized by μ (mu). In general, then, we use statistics as estimates of parameters.

If the purpose of obtaining a statistic is to use it as an estimator of a parameter, then it should come as no surprise that our choice of a statistic (and even how we define it) is based partly on how well that statistic functions as an estimator of the parameter in question. Actually, the mean is usually preferred over other measures of central tendency because of its performance as an estimator of μ. The variance (s^2) is defined as it is, with ($N - 1$) in the denominator, specifically because of the advantages that accrue when s^2 is used to estimate the population variance (σ^2).

Four properties of estimators are of particular interest to statisticians and heavily influence the choice of the statistics we compute. These properties are those of sufficiency, unbiasedness, efficiency, and resistance. They are discussed here simply to give you a feel for why some measures of central tendency and variability are regarded as more important than others. It is not critical that you have a thorough understanding of estimation and related concepts, but you should have a general appreciation of the issues involved.

Sufficiency

sufficient statistic

A statistic is a **sufficient statistic** if it contains (makes use of) all the information in a sample. You might think this is pretty obvious because it certainly seems reasonable to base your estimates on all the data. The mean does exactly that. The mode, however, uses only the most common observations, ignoring all others, and the median uses only the middle one, again ignoring the values of other observations. Similarly, the range, as a measure of dispersion, uses only the two most extreme (and thus most unrepresentative) scores. Here you see one of the reasons that we emphasize the mean as our measure of central tendency.

Unbiasedness

Suppose we have a population for which we somehow know the mean (μ), say, the heights of all basketball players in the NBA. If we were to draw one sample from that population and calculate the sample mean ($\overline{X}_1$), we would expect $\overline{X}_1$ to be reasonably close to μ, particularly if N is large, because it is an estimator of μ. So if the average height in this population is 7.0' ($\mu = 7.0'$), we would expect a sample of, say, 10 players to have an average height of approximately 7.0' as well, although it probably would not be exactly equal to 7.0'. (We can write $\overline{X}_1 \approx 7$, where the symbol $\approx$ means "approximately equal.") Now suppose we draw another sample and obtain its mean ($\overline{X}_2$). (The subscript is used to differentiate the means of successive samples. Thus, the mean of the 43rd sample, if we drew that many, would be denoted by $\overline{X}_{43}$.) This mean would probably also be reasonably close to μ, but we would not expect it to be exactly equal to μ or to $\overline{X}_1$. If we were to keep up this procedure and draw sample means ad infinitum, we would find that *the average of the sample means would be precisely equal to* μ. Thus, we say that the **expected value** (i.e., the long-range

expected value

average of many, many samples) of the sample mean is equal to μ, the population mean that it is estimating. An estimator whose expected value equals the parameter to be estimated is

unbiased estimator

called an **unbiased estimator** and that is a very important property for a statistic to possess. Both the sample mean and the sample variance are unbiased estimators of their corresponding parameters. (This is why we used ($N - 1$) as the denominator of the formula for the sample variance.) By and large, unbiased estimators are like unbiased people—they are nicer to work with than biased ones.

Efficiency

efficiency

Estimators are also characterized in terms of **efficiency.** Suppose that a population is symmetric: Thus, the values of the population mean and median are equal. Now suppose that we want to estimate the mean of this population (or, alternatively, its median). If we drew many samples and calculated their means, we would find that the means ($\overline{X}$) clustered relatively closely around μ. The medians of the same samples, however, would cluster more loosely around μ. This is so even though the median is also an unbiased estimator in this situation because the expected value of the median in this case would also equal μ. The fact that the sample means cluster more closely around μ than do the sample medians indicates that the mean is more *efficient* as an estimator. (In fact, it is the most efficient estimator of μ.) Because the mean is more likely to be closer to μ (i.e., a more accurate estimate) than the median, it is a better statistic to use to estimate μ.

Although it should be obvious that efficiency is a relative term (a statistic is more or less efficient than some other statistic), statements that such and such a statistic is "efficient" should really be taken to mean that the statistic is more efficient than all other statistics as an estimate of the parameter in question. Both the sample mean, as an estimate of μ, and the sample variance, as an estimate of σ^2, are efficient estimators in that sense. The fact that both the mean and the variance are unbiased and efficient is the major reason that they play such an important role in statistics. These two statistics will form the basis for most of the procedures discussed in the remainder of this book.

Resistance

The last property of an estimator to be considered concerns the degree to which the estimator is influenced by the presence of outliers. Recall that the median is relatively uninfluenced by outliers, whereas the mean can drastically change with the inclusion of one or two extreme scores. In a very real sense we can say that the median "resists" the influence of

resistance   these outliers, whereas the mean does not. This property is called the **resistance** of the estimator. In recent years, considerably more attention has been placed on developing resistant
estimators—such as the trimmed mean discussed earlier. These should soon start filtering
down to the level of everyday data analysis, though they aren't there yet.

The Sample Variance as an Estimator of the Population Variance

The sample variance offers an excellent example of what was said in the discussion of
unbiasedness. You may recall that I earlier sneaked in the divisor of $N - 1$ instead of N for
the calculation of the variance and standard deviation. Now is the time to explain why. (You
may be perfectly willing to take the statement that we divide by $N - 1$ on faith, but I get a lot
of questions about it, so I guess you will just have to read the explanation—or skip it.)

There are a number of ways to explain why sample variances require $N - 1$ as the
denominator. Perhaps the simplest is phrased in terms of what has been said about the sample variance (s^2) as an unbiased estimate of the population variance (σ^2). Assume for the
moment that we have an infinite number of samples (each containing N observations) from
one population and that we know the population variance. Suppose further that we are
foolish enough to calculate sample variances as

$$\frac{\sum (X - \overline{X})^2}{N}$$

(Note the denominator.) If we take the average of these sample variances, we find

$$\text{Average } \frac{\sum (X - \overline{X})^2}{N} = E\left[\frac{\sum (X - \overline{X})^2}{N} \right] = \frac{(N - 1)\sigma^2}{N}$$

where $E[\ ]$ is read as "the expected value of (whatever is in brackets)." Thus the average
value of $\sum (X - \overline{X})^2/N$ is not σ^2. It is a biased estimator.

Degrees of Freedom

**degrees of
freedom**

The foregoing discussion is very much like saying that we divide by $N - 1$ *because it works*.
But *why* does it work? To explain this, we must first consider **degrees of freedom** (*df*).
Assume that you have in front of you the three numbers 6, 8, and 10. Their mean is 8. You are
now informed that you may change any of these numbers, as long as the mean is kept constant at 8. How many numbers are you free to vary? If you think that you are free to vary all
three numbers, you are wrong. If you change all three of them in some haphazard fashion, the
mean almost certainly will no longer equal 8. Only two of the numbers can be *freely* changed
if the mean is to remain constant. For example, if you change the 6 to a 7 and the 10 to a 13,
the remaining number is determined; it must be 4 if the mean is to be 8. If you had 50 numbers and were given the same instructions, you would be free to vary only 49 of them; the
50th would be determined.

Now let us go back to the formulae for the population and sample variances and see why
we lost one degree of freedom in calculating the sample variances.

$$\sigma^2 = \frac{\sum (X - \mu)^2}{N} \qquad s^2 = \frac{\sum (X - \overline{X})^2}{N - 1}$$

In the case of σ^2, μ is known and does not have to be estimated from the data. Thus, no
df are lost and the denominator is N. In the case of s^2, however, μ is not known and must be
estimated from the sample mean ($\overline{X}$). Once you have estimated μ from $\overline{X}$, you have fixed it

for purposes of estimating variability. Thus, you lose that degree of freedom that we discussed, and you have only $N - 1$ *df* left ($N - 1$ scores free to vary). We lose this one degree of freedom *whenever* we estimate a mean. It follows that the denominator (the number of scores on which our estimate is based) should reflect this restriction. It represents the number of independent pieces of data.

2.10   Boxplots: Graphical Representations of Dispersions and Extreme Scores

boxplot

box-and-whisker plot

Earlier you saw how stem-and-leaf displays represent data in several meaningful ways at the same time. Such displays combine data into something very much like a histogram, while retaining the individual values of the observations. In addition to the stem-and-leaf display, John Tukey has developed other ways of looking at data, one of which gives greater prominence to the dispersion of the data. This method is known as a **boxplot,** or, sometimes, **box-and-whisker plot.**

The data and the accompanying stem-and-leaf display in Table 2.7 were taken from normal- and low-birthweight infants participating in a study of infant development at the University of Vermont and represent preliminary data on the length of hospitalization of 38 normal-birthweight infants. Data on three infants are missing for this particular variable and are represented by an asterisk (*). (Asterisks are included to emphasize that we should not just ignore missing data.) Because the data vary from 1 to 10, with two exceptions, all the leaves are zero. The zeros really just fill in space to produce a histogram-like distribution. Examination of the data as plotted in the stem-and-leaf display reveals that the distribution is positively skewed with a median stay of 3 days. Near the bottom of the stem you will see the entry HI and the values 20 and 33. These are extreme values, or outliers, and are set off in this way to highlight their existence. Whether they are large enough to make us suspicious is one of the questions a boxplot is designed to address. The last line of the stem-and-leaf display indicates the number of missing observations.

Table 2.7   Data and stem-and-leaf display on length of hospitalization for full-term newborn infants (in days)

Data			Stem-and-Leaf	
2	1	7	1	000
1	33	2	2	000000000
2	3	4	3	00000000000
3	*	4	4	0000000
3	3	10	5	00
9	2	5	6	0
4	3	3	7	0
20	6	2	8	
4	5	2	9	0
1	*	*	10	0
3	3	4	HI	20, 33
2	3	4	Missing = 3	
3	2	3		
2	4			

Tukey originally defined boxplots in terms of special measures that he devised. Most people now draw boxplots using more traditional measures, and I am adopting that approach in this edition.

We defined the median location of a set of N scores as $(N + 1)/2$. When the median location is a whole number, as it will be when N is odd, then the median is simply the value that occupies that location in an ordered arrangement of data. When the median location is a fractional number (i.e., when N is even), the median is the average of the two values on each side of that location. For the data in Table 2.7, the median location is $(38 + 1)/2 = 19.5$, and the median is 3. To construct a boxplot, we are also going to take the first and third quartiles, defined earlier. The easiest way to do this is to define the **quartile location,** which is defined as

quartile location

$$\text{Quartile location} = \frac{\text{Median location} + 1}{2}$$

If the median location is a fractional value, the fraction should be dropped from the numerator when you compute the quartile location. The quartile location is to the quartiles what the median location is to the median. It tells us where, in an ordered series, the quartile values[13] are to be found. For the data on hospital stay, the quartile location is $(19 + 1)/2 = 10$. Thus, the quartiles are going to be the tenth scores from the bottom and from the top. These values are 2 and 4, respectively. For data sets without tied scores, or for large samples, the quartiles will bracket the middle 50% of the scores.

To complete the concepts required for understanding boxplots, we need to consider three more terms: the interquartile range, inner fences, and adjacent values. As we saw earlier, the interquartile range is simply the range between the first and third quartiles. For our data, the interquartile range $4 - 2 = 2$. An **inner fence** is a point that falls 1.5 times the interquartile range below or above the appropriate quartile. Because the interquartile range is 2 for our data, the inner fence is $2 \times 1.5 = 3$ points farther out than the quartiles. Because our quartiles are the values 2 and 4, the inner fences will be at $2 - 3 = -1$ and $4 + 3 = 7$. **Adjacent values** are those actual values in the data that are no more extreme (no farther from the median) than the inner fences. Because the smallest value we have is 1, that is the closest value to the lower inner fence and is the lower adjacent value. The upper inner fence is 7, and because we have a 7 in our data, that will be the higher adjacent value. The calculations for all the terms we have just defined are shown in Table 2.8.

inner fence

adjacent values

Inner fences and adjacent values can cause some confusion. Think of a herd of cows scattered around a field. The fence around the field represents the inner fence of the boxplot. The cows closest to but still inside the fence are the adjacent values. Don't worry about the cows that have escaped outside the fence and are wandering around on the road. They are not involved in the calculations at this point. (They will be the outliers.)

Now we are ready to draw the boxplot. First, we draw and label a scale that covers the whole range of the obtained values. This has been done at the bottom of Table 2.8. We then draw a rectangular box from Q_1 to Q_3, with a vertical line representing the location of the median. Next we draw lines (**whiskers**) from the quartiles out to the adjacent values. Finally we plot the locations of all points that are more extreme than the adjacent values.

whiskers

From Table 2.8 we can see several important things. First, the central portion of the distribution is reasonably symmetric. This is indicated by the fact that the median lies in the center of the box and was apparent from the stem-and-leaf display. We can also see that the distribution is positively skewed, because the whisker on the right is substantially longer than the one on the left. This also was apparent from the stem-and-leaf display, although not so clearly. Finally, we see that we have four outliers, where an outlier is defined here as

---

[13] Tukey referred to the quartiles in this situation as "hinges," but little is lost by thinking of them as the quartiles.

Table 2.8   Calculation and boxplots for data from Table 2.7

Median location $= (N + 1)/2 = (38 + 1)/2 = 19.5$
Median $= 3$
Quartile location $=$ (median location† $+ 1)/2 = (19 + 1)/2 = 10$
$Q_1 = $ 10th lowest score $= 2$
$Q_3 = $ 10th highest score $= 4$
Interquartile range $= 4 - 2 = 2$
Interquartile range $* 1.5 = 2 * 1.5 = 3$
Lower inner fence $= Q_1 - 1.5$(interquartile range) $= 2 - 3 = -1$
Upper inner fence $= Q_3 + 1.5$(interquartile range) $= 4 + 3 = 7$
Lower adjacent value $=$ smallest value $\geq$ lower fence $= 1$
Upper adjacent value $=$ largest value $\leq$ upper fence $= 7$

0	5	10	15	20	25	30	35

$\dagger$ Drop any fractional values.

any value more extreme than the whiskers (and therefore more extreme than the adjacent values). The stem-and-leaf display did not show the position of the outliers nearly so graphically as does the boxplot.

Outliers deserve special attention. An outlier could represent an error in measurement, in data recording, or in data entry, or it could represent a legitimate value that just happens to be extreme. For example, our data represent length of hospitalization, and a full-term infant might have been born with a physical defect that required extended hospitalization. Because these are actual data, it was possible to go back to hospital records and look more closely at the four extreme cases. On examination, it turned out that the two most extreme scores were attributable to errors in data entry and were readily correctable. The other two extreme scores were caused by physical problems of the infants. Here a decision was required by the project director as to whether the problems were sufficiently severe to cause the infants to be dropped from the study (both were retained as subjects). The two corrected values were 3 and 5 instead of 33 and 20, respectively, and a new boxplot for the corrected data is shown in Table 2.9. This boxplot is identical to the one shown in Table 2.8 except for the spacing and the two largest values. (You should verify for yourself that the corrected data set would indeed yield this boxplot.)

Table 2.9   Boxplot for corrected data from Table 2.8

0	2	4	6	8	10

From what has been said, it should be evident that boxplots are extremely useful tools for examining data with respect to dispersion. I find them particularly useful for screening data for errors and for highlighting potential problems before subsequent analyses are carried out. Boxplots are presented often in the remainder of this book as visual guides to the data.

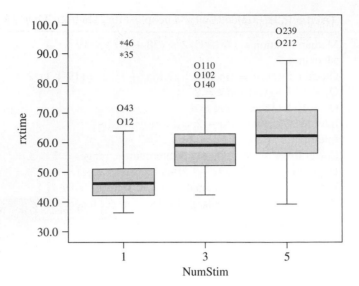

Figure 2.13 Boxplot of reaction times as a function of number of stimuli in the original set of stimuli

A word of warning: Different statistical computer programs may vary in the ways they define the various elements in boxplots. (See Frigge, Hoaglin, and Iglewicz [1989] for an extensive discussion of this issue.) You may find two different programs that produce slightly different boxplots for the same set of data. They may even identify different outliers. However, boxplots are normally used as informal heuristic devices, and subtle differences in definition are rarely, if ever, a problem. I mention the potential discrepancies here simply to explain why analyses that you do on the data in this book may come up with slightly different results if you use different computer programs.

Many people think that the real usefulness of boxplots comes when we want to compare several groups. We will use the example with which we started this chapter, where we have recorded the reaction times of response to the question of whether a specific digit was presented in a previous slide, as a function of the number of stimuli on that slide. The boxplot in Figure 2.13, produced by SPSS, shows the reaction times for those cases in which the stimulus was actually present, broken down by the number of stimuli in the original. The outliers are indicated by their identification number. The most obvious conclusion from this figure is that as the number of stimuli in the original increases, reaction times also increase. We can also see that the distributions are reasonably symmetric (the boxes are roughly centered on the medians), and there are a few outliers, all of which are long reaction times.

2.11 Obtaining Measures of Dispersion Using Minitab

Earlier we saw how to use Minitab to calculate measures of central tendency. We can also use Minitab to calculate measures of dispersion, as shown in Exhibit 2.3, which is based on our previous data on annoying mannerisms of instructors (Exhibit 2.2). Note that the simple **Stat/Descriptive statistics/Display descriptive statistics** menu command (not shown here) supplies nearly all the statistics we need. The **Graph/Boxplot** commands generate the necessary graphic. (Notice that the dependent variable Manner is entered as Y (measurement) and that nothing is entered as X (category). If we had collected data on male and female instructors, we could have plotted their results separately by entering Sex as X.)

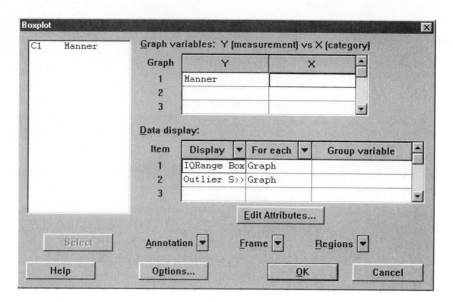

Descriptive Statistics: depvar

Variable	N	Mean	Median	TrMcan	StDev	SE Mean
depvar	15	16.400	17.000	16.385	3.562	0.920

Variable	Minimum	Maximum	Q1	Q3
depvar	10.000	23.000	14.000	19.000

Boxplot of depvar

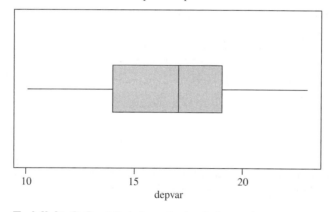

Exhibit 2.3   Minitab analysis of electronic game data

2.12 Percentiles, Quartiles, and Deciles

quartiles

A distribution has many properties besides its location and dispersion. We saw one of these briefly when we considered boxplots, where **quartiles** were defined as those values that divide the distribution into fourths. Thus, the first quartile cuts off the lowest 25%, the second quartile cuts off the lowest 50%, and the third quartile cuts off the lowest 75%. (Note that the second quartile is also the median.) These percentiles were shown clearly on the growth chart in Figure 2.11. If we want to examine finer gradations of the distribution, we can look at **deciles**, which divide the distribution into tenths, with the first decile cutting off the lowest

deciles

percentiles

10%, the second decile cutting off the lowest 20%, and so on. Finally, most of you have had experience with **percentiles,** which are values that divide the distribution into hundredths. Thus, the 81st percentile is that point on the distribution below which 81% of the scores lie.

quantiles

fractiles

Quartiles, deciles, and percentiles are the three most common examples of a general class of statistics known by the generic name of **quantiles,** or, sometimes, **fractiles.** We will not have much to say about quantiles in this book, but they are usually covered extensively in more introductory texts (e.g., Howell, 2004). They also play an important role in many of the techniques of exploratory data analysis advocated by Tukey.

2.13 The Effect of Linear Transformations on Data

linear transformations

Frequently, we want to transform data in some way. For instance, we may want to convert feet into inches, inches into centimeters, degrees Fahrenheit into degrees Celsius, test grades based on 79 questions to grades based on a 100-point scale, four- to five-digit incomes into one- to two-digit incomes, and so on. Fortunately, all of these transformations fall within a set called **linear transformations,** in which we multiply each X by some constant (possibly 1) and add a constant (possibly 0):

$$X_{new} = bX_{old} + a$$

where a and b are our constants. (Transformations that use exponents, logarithms, trigonometric functions, etc., are classed as nonlinear transformations.) An example of a linear transformation is the formula for converting degrees Celsius to degrees Fahrenheit:

$$F = 9/5(C) + 32.$$

As long as we content ourselves with linear transformations, a set of simple rules defines the mean and variance of the observations on the new scale in terms of their means and variances on the old one:

1. Adding (or subtracting) a constant to (or from) a set of data adds (or subtracts) that same constant to (or from) the mean:

    For $X_{new} = X_{old} \pm a$: $\overline{X}_{new} = \overline{X}_{old} \pm a$

2. Multiplying (or dividing) a set of data by a constant multiplies (or divides) the mean by the same constant:

    For $X_{new} = bX_{old}$: $\overline{X}_{new} = b\overline{X}_{old}$

    For $X_{new} = X_{old}/b$: $\overline{X}_{new} = \overline{X}_{old}/b$

3. Adding or subtracting a constant to (or from) a set of scores leaves the variance and standard deviation unchanged:

    For $X_{new} = X_{old} \pm a$: $s^2_{new} = s^2_{old}$

4. Multiplying (or dividing) a set of scores by a constant multiplies (or divides) the variance by the square of the constant and the standard deviation by the constant:

    For $X_{new} = bX_{old}$: $s^2_{new} = b^2 s^2_{old}$ and $s_{new} = b s_{old}$

    For $X_{new} = X_{old}/b$: $s^2_{new} = s^2_{old}/b^2$ and $s_{new} = s_{old}/b$

The following example illustrates these rules. In each case, the constant used is 3.

Addition of a constant:

Old Data	$\overline{X}$	s^2	s	New Data	$\overline{X}$	s^2	s
4, 8, 12	8	16	4	7, 11, 15	11	16	4

Multiplication by a constant:

Old Data	$\overline{X}$	s^2	s	New Data	$\overline{X}$	s^2	s
4, 8, 12	8	16	4	12, 24, 36	24	144	12

Reflection as a Transformation

reflection

A very common and useful transformation concerns reversing the order of a scale. For example, assume that we asked subjects to indicate on a 5-point scale the degree to which they agree or disagree with each of several items. To prevent the subjects from simply checking the same point on the scale all the way down the page without thinking, we phrase half of our questions in the positive direction and half in the negative direction. Thus, given a 5-point scale where 5 represents "strongly agree" and 1 represents "strongly disagree," a 4 on "I hate movies" would be comparable to a 2 on "I love plays." If we want the scores to be comparable, we need to rescore the negative items (for example), converting a 5 to a 1, a 4 to a 2, and so on. This procedure is called **reflection** and is quite simply accomplished by a linear transformation. We merely write $X_{new} = 6 - X_{old}$. The constant (6) is just the largest value on the scale plus 1. It should be evident that when we reflect a scale, we also reflect its mean but have no effect on its variance or standard deviation. This is true by Rule 3 in the preceding list.

Standardization

deviation scores

centering

standard scores

standardization

One common linear transformation often employed to rescale data involves subtracting the mean from each observation. Such transformed observations are called **deviation scores,** and the transformation itself is often referred to as **centering** because we are centering the mean at 0. Centering is most often used in regression, which is discussed later in the book. An even more common transformation involves creating deviation scores and then dividing the deviation scores by the standard deviation. Such scores are called **standard scores,** and the process is referred to as **standardization.** Basically, standardized scores are simply transformed observations that are measured in standard deviation units. Thus, for example, a standardized score of 0.75 is a score that is 0.75 standard deviation above the mean; a standardized score of −0.43 is a score that is 0.43 standard deviation below the mean. I will have much more to say about standardized scores when we consider the normal distribution in Chapter 3. I mention them here specifically to show that we can compute standardized scores regardless of whether or not we have a normal distribution (defined in Chapter 3). People often think of standardized scores as being normally distributed, but there is absolutely no requirement that they be. Standardization is a simple linear transformation of the raw data, and, as such, does not alter the shape of the distribution.

Nonlinear Transformations

Whereas linear transformations are usually used to convert the data to a more meaningful format—such as expressing them on a scale from 0 to 100, putting them in standardized form, and so on, nonlinear transformations are usually invoked to change the shape of a distribution. As we saw, linear transformations do not change the underlying shape of a distribution. Nonlinear transformations, on the other hand, can make a skewed distribution look more symmetric and can reduce the effects of outliers.

Some nonlinear transformations are quite common. If we had an intervention program where we scored each participant before (X) and after (Y) the intervention, one logical measure would be a person's performance after the intervention (Y). Another would be the

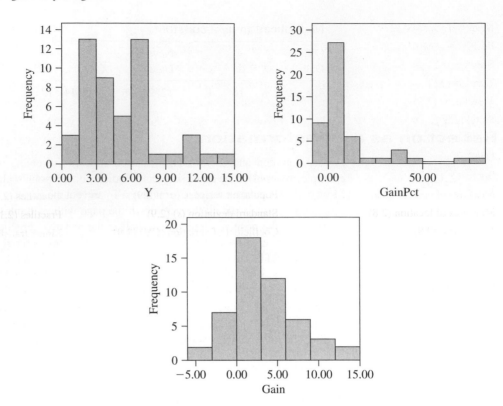

Figure 2.14   Alternative measures of the effect of an intervention

gain in performance from pre- to post-intervention, as measured by $(Y - X)$. A third alternative would be to record the change as a function of the original score. This would be $(Y - X))/Y$. We might use this measure because we assume that how much a person's score increases is related to how poorly they did originally. Figure 2.14 portrays the histograms for these three measures based on hypothetical data.

From Figure 2.14 you can see that the three alternative measures, the second two of which are nonlinear transformations of X and Y, have quite different distributions. In this case, as you will see later in the book, the use of gain scores[14] would probably be preferable because of the symmetric nature of the distribution. Later in this book you will see how to use other nonlinear transformations (e.g., square root and logarithmic) to make the shape of the distribution more symmetrical.

Key Terms

Frequency distribution (2.1)	Stem-and-leaf display (2.3)	Least significant digits (2.3)
Histogram (2.2)	Exploratory data analysis (EDA) (2.3)	Leaves (2.3)
Real lower limit (2.2)	Leading digits (2.3)	Symmetric (2.5)
Real upper limit (2.2)	Most significant digits (2.3)	Bimodal (2.5)
Midpoints (2.2)	Stem (2.3)	Unimodal (2.5)
Outlier (2.2)	Trailing digits (2.3)	Modality (2.5)

---

[14] The use of gain scores is not without its detractors, however, for reasons having little or nothing to do with the shape of the distribution.

Exercises

Many of the following exercises can be solved using either computer software or pencil and paper. The choice is up to you or your instructor. Any software package should be able to work these problems. Some of the exercises refer to an actual research study (Howell & Huessy, 1981, 1985), which is described at the beginning of Appendix Data Set. These data are available from my Web site (http://www.uvm.edu/~dhowell/methods/DataFiles/Add.dat).

2.1 Any of you who have listened to children tell stories will recognize that children differ from adults in that they tend to recall stories as a sequence of actions rather than as an overall plot. Their descriptions of a movie are filled with the phrase "and then . . ." An experimenter with supreme patience asked 50 children to tell her about a given movie. Among other variables, she counted the number of "and then . . ." statements, which is the dependent variable. The data follow:

18	15	22	19	18	17	18	20	17	12	16	16	17	21	23	18	20	21	20
20	15	18	17	19	20	23	22	10	17	19	19	21	20	18	18	24	11	19
31	16	17	15	19	20	18	18	40	18	19	16							

a. Plot an ungrouped frequency distribution for these data.

b. What is the general shape of the distribution?

2.2 Create a histogram for the data in Exercise 2.1 using a reasonable number of intervals.

2.3 What difficulty would you encounter in making a stem-and-leaf display of the data in Exercise 2.1?

2.4 As part of the study described in Exercise 2.1, the experimenter obtained the same kind of data for 50 adults. The data follow:

10	12	5	8	13	10	12	8	7	11	11	10	9	9	11	15	12	17	14	10	9
8	15	16	10	14	7	16	9	1	4	11	12	7	9	10	3	11	14	8	12	5	10
9	7	11	14	10	15	9														

    a. What can you tell just by looking at these numbers? Do children and adults seem to recall stories in the same way?

    b. Plot an ungrouped frequency distribution for these data using the same scale on the axes as you used for the children's data in Exercise 2.1.

    c. Overlay the frequency distribution from part (b) on the one from Exercise 2.1.

2.5   Use a back-to-back stem-and-leaf display (see Figure 2.5) to compare the data from Exercises 2.1 and 2.4.

2.6   Make a cumulative frequency distribution for the data in Exercise 2.1.

2.7   Make a cumulative frequency distribution for the data in Exercise 2.4.

2.8   Create a positively skewed set of data and plot it.

2.9   Create a bimodal set of data that represents some actual phenomenon and plot it.

2.10 In my undergraduate research methods course, women generally do a bit better than men. This year I had the grades shown in the following Minitab boxplots. What might you conclude from these boxplots?

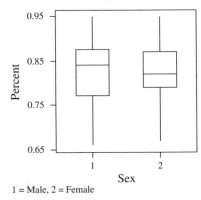

1 = Male, 2 = Female

2.11 In Exercise 2.10, what would be the first and third quartiles for males and females?

2.12 The following stem-and-leaf displays show the individual grades separately for males and females. From these results, what would you conclude about any differences between males and females?

Stem-and-leaf of Percent Sex = 1 (Male) N = 29 Leaf Unit = 0.010			Stem-and-leaf of Percent Sex = 2 (Female) N = 78 Leaf Unit = 0.010		
3	6	677	2	6	77
3	6		3	6	8
3	7		6	7	000
5	7	33	10	7	2233
7	7	45	15	7	45555
7	7		15	7	
10	7	999	22	7	8899999
12	8	01	34	8	011111111111
14	8	22	(8)	8	22222233
(4)	8	4455	36	8	445555555
11	8	6677	27	8	666777777
7	8	8	18	8	888889999
6	9		9	9	00001
6	9	23	4	9	333
4	9	4445	1	9	5

2.13 What would you predict to be the shape of the distribution of the number of movies attended per month for the next 200 people you meet?

2.14 Draw a histogram for the data for GPA in Appendix Data Set using reasonable intervals.

2.15 Create a stem-and-leaf display for the ADDSC score in Appendix Data Set.

2.16 What three interesting facts about the populations of Mexico and Spain can be seen in Figure 2.6?

2.17 In a hypothetical experiment, researchers rated 10 Europeans and 10 North Americans on a 12-point scale of musicality. The data for the Europeans were [10 8 9 5 10 11 7 8 2 7]. Using X for this variable,

    a.   What are X_3, X_5, and X_8?

    b.   Calculate $\sum X$.

    c.   Write the summation notation from part (b) in its most complex form.

2.18 The data for the North Americans in Exercise 2.17 were [9 9 5 3 8 4 6 6 5 2]. Using Y for this variable,

    a.   What are Y_1 and Y_{10}?

    b.   Calculate $\sum Y$.

2.19 Using the data from Exercise 2.17,

    a.   Calculate $\left(\sum X\right)^2$ and $\sum X^2$.

    b.   Calculate $\sum X/N$, where $N =$ the number of scores.

    c.   What do you call what you calculated in part (b)?

2.20 Using the data from Exercise 2.18,

    a.   Calculate $\left(\sum Y\right)^2$ and $\sum Y^2$.

    b.   Calculate $\dfrac{\sum Y^2 - \dfrac{\left(\sum Y\right)^2}{N}}{N-1}$

    c.   Calculate the square root of the answer for part (b).

    d.   What are the units of measurement for parts (b) and (c)?

2.21 Using the data from Exercises 2.17 and 2.18, record the two data sets side by side in columns, name the columns $X + Y$, and treat the data as paired.

    a.   Calculate $\sum XY$.

    b.   Calculate $\sum X \sum Y$.

    c.   Calculate $\dfrac{\sum XY - \dfrac{\sum X \sum Y}{N}}{N-1}$

    (You will come across these calculations again in Chapter 9.)

2.22 Use the data from Exercises 2.17 and 2.18 to show that

    a. $\sum (X + Y) = \sum X + \sum Y$.

    b. $\sum XY \neq \sum X \sum Y$.

    c. $\sum CX = C \sum X$. (where C represents any arbitrary constant)

    d. $\sum X^2 \neq \left(\sum X\right)^2$.

2.23 In Table 2.1 (p. 17), the reaction time data are broken down separately by the number of digits in the comparison stimulus. Create three stem-and-leaf displays, one for each set of data, and place them side-by-side. (Ignore the distinction between positive and negative instances.) What kinds of differences do you see among the reaction times under the three conditions?

2.24 Sternberg ran his original study (the one that is replicated in Table 2.1) to investigate whether people process information simultaneously or sequentially. He reasoned that if they process information simultaneously, they would compare the test stimulus against all digits in the comparison stimulus at the same time, and the time to decide whether a digit was part of the comparison set would not depend on how many digits were in the comparison. If people process information sequentially, the time to come to a decision would increase with the number of digits in the comparison. Which hypothesis do you think the figures you drew in Exercise 2.23 support?

2.25   In addition to comparing the three distributions of reaction times, as in Exercise 2.23, how else could you use the data from Table 2.1 to investigate how people process information?

2.26   One frequent assumption in statistical analyses is that observations are independent of one another. (Knowing one response tells you nothing about the magnitude of another response.) How would you characterize the reaction time data in Table 2.1, just based on what you know about how they were collected? (A lack of independence would not invalidate anything we have done with these data in this chapter.)

2.27   The following figure is adapted from a paper by Cohen, Kaplan, Cunnick, Manuck, and Rabin (1992), which examined the immune response of nonhuman primates raised in stable and unstable social groups. In each group, animals were classed as high or low in affiliation, measured by the amount of time they spent in close physical proximity to other animals. Higher scores on the immunity measure represent greater immunity to disease. How would you interpret these results?

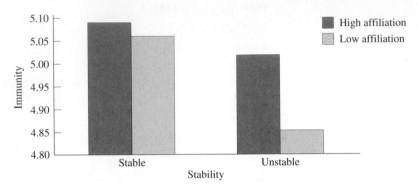

2.28   Rogers and Prentice-Dunn (1981) had subjects deliver shock to their fellow subjects as part of a biofeedback study. They recorded the amount of shock that the subjects delivered to white participants and black participants when the subjects had and had not been insulted by the experimenter. Their results are shown in the accompanying figure. Interpret these results.

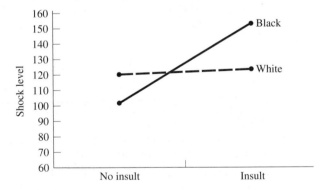

2.29   The following data represent U.S. college enrollments by census categories as measured in 1982 and 1991. Plot the data in a form that represents the changing ethnic distribution of college students in the United States. (The data entries are in thousands.)

Ethnic Group	1982	1991
White	9,997	10,990
Black	1,101	1,335
Native American	88	114
Hispanic	519	867
Asian	351	637
Foreign	331	416

2.30   The following data represent the number of AIDS cases in the United States among people aged 13–29. Plot these data to show the trend over time. (The data are in thousands of cases.)

Year	Cases
1981–1982	196
1983	457
1984	960
1985	1685
1986	2815
1987	4385
1988	6383
1989	6780
1990	5483

2.31   The following data represent the total number of households, the number of households headed by women, and family size from 1960 to 1990. Present these data in such a way to reveal any changes in U.S. demographics. What do the data suggest about how a social scientist might look at the problems facing the United States? (Households are given in thousands.)

Year	Total Households	Households Headed by Females	Family Size
1960	52,799	4,507	3.33
1970	63,401	5,591	3.14
1975	71,120	7,242	2.94
1980	80,776	8,705	2.76
1985	86,789	10,129	2.69
1987	89,479	10,445	2.66
1988	91,066	10,608	2.64
1989	92,830	10,890	2.62
1990	92,347	10,890	2.63

2.32   Make up a set of data for which the mean is greater than the median.

2.33   Make up a positively skewed set of data. Does the mean fall above or below the median?

2.34   Make up a unimodal set of data for which the mean and median are equal but are different from the mode.

2.35   A group of 15 rats running a straight-alley maze required the following number of trials to perform at a predetermined criterion level:

Trials required to reach criterion: 18   19   20   21   22   23   24
Number of rats (frequency):   1   0   4   3   3   3   1

Calculate the mean and median of the required number of trials for this group.

2.36   Given the following set of data, demonstrate that subtracting a constant (e.g., 5) from every score reduces all measures of central tendency by that constant: 8 7 12 14 3 7.

2.37   Given the following set of data, show that multiplying each score by a constant multiplies all measures of central tendency by that constant: 8 3 5 5 6 2.

2.38   Create a sample of 10 numbers that has a mean of 8.6. How does this illustrate the point we discussed about degrees of freedom?

2.39   The accompanying output was produced by SPSS applied to the data on ADDSC and GPA described in Appendix Data Set. How do these answers on measures of central tendency compare to what you would predict from the answers to Exercises 2.14 and 2.15?

Descriptive Statistics

	ADDSC	GPA	Valid N (listwise)
N	88	88	88
Minimum	26	1	
Maximum	85	4	
Mean	52.60	2.46	
Std. Deviation	12.42	.86	
Variance	154.311	.742	

Descriptive Statistics for ADDSC and GPA

2.40   In one or two sentences, describe what the following graphic has to say about the grade point averages for the students in our sample.

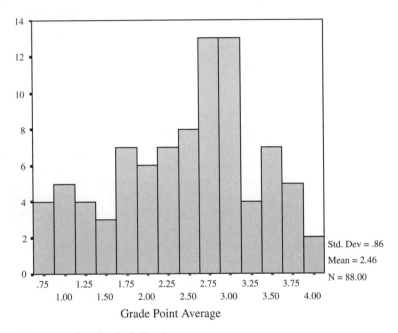

Histogram for Grade Point Average

2.41   Calculate the range, variance, and standard deviation for the data in Exercise 2.1.

2.42   Calculate the range, variance, and standard deviation for the data in Exercise 2.4.

2.43   Compare the answers to Exercises 2.41 and 2.42. Is the standard deviation for children substantially greater than that for adults?

2.44   In Exercise 2.1, what percentage of the scores fall within plus or minus two standard deviations from the mean?

2.45   In Exercise 2.4, what percentage of the scores fall within plus or minus two standard deviations from the mean?

2.46   Given the following set of data, demonstrate that adding a constant to, or subtracting a constant from, each score does not change the standard deviation. (What happens to the mean when a constant is added or subtracted?) [5 4 2 3 4 9 5].

2.47   Given the data in Exercise 2.46, show that multiplying or dividing by a constant multiplies or divides the standard deviation by that constant. How is this related to what happens to the mean under similar conditions?

2.48   Using the results demonstrated in Exercises 2.46 and 2.47, transform the following set of data to a new set that has a standard deviation of 1.00: [5 8 3 8 6 9 9 7].

2.49   Use your answers to Exercises 2.46 and 2.47 to modify your answer to Exercise 2.48 such that the new set of data has a mean of 0 and a standard deviation of 1.00. (Note: The solution of Exercises 2.48 and 2.49 will be elaborated further in Chapter 3.)

2.50   Create a boxplot for the data in Exercise 2.1.

2.51   Create a boxplot for the data in Exercise 2.4.

2.52   Create a boxplot for the variable ADDSC in Appendix Data Set.

2.53   Compute the coefficient of variation to compare the variability in usage of "and then . . ." statements by children and adults in Exercises 2.1 and 2.4.

2.54   For the data in Appendix Data Set, the GPA has a mean of 2.456 and a standard deviation of 0.8614. Compute the coefficient of variation as defined in this chapter.

2.55   The data set named BadCancr.dat (at www.uvm.edu/~dhowell/methods/DataSets/BadCancer.dat) has been deliberately corrupted by entering errors into a perfectly good data set (named Cancer.dat). The purpose of this corruption was to give you experience in detecting and correcting the kinds of errors that appear almost every time we attempt to use a newly entered data set. Every error in here is one that I and almost everyone I know have come across countless times. Some of them are so extreme that most statistical packages will not run until they are corrected. Others are logical errors that will allow the program to run, producing meaningless results. (No college student is likely to be 10 years old or receive a score of 15 on a 10-point quiz.) The variables in this set are described in the Appendix Computer Exercises for the file Cancer.dat. That description tells where each variable should be found and the range of its legitimate values. You can use any statistical package available to read the data. Standard error messages will identify some of the problems, visual inspection will identify others, and computing descriptive statistics or plotting the data will help identify the rest. In some cases, the appropriate correction will be obvious. In other cases, you will just have to delete the offending values. When you have cleaned the data, use your program to compute a final set of descriptive statistics on each of the variables. This problem will take a fair amount of time. I have found that it is best to have students work in pairs.

2.56   Compute the 10% trimmed mean for the data in Table 2.6, Set 32.

2.57   Compute the 10% Winsorized standard deviation for the data in Table 2.6, Set 32.

2.58   Draw a boxplot to illustrate the difference between reaction times to positive and negative instances in reaction time for the data in Table 2.1. (These data can be found at www.uvm.edu/~dhowell/methods/DataSets/RxTime.dat)

2.59   Under what conditions will a transformation alter the shape of a distribution?

Discussion Question

2.60   In the exercises in Chapter 1, we considered the study by a fourth-grade girl who examined the average allowance of her classmates. You may recall that 7 boys reported an average allowance of $3.18, and 11 girls reported an average allowance of $2.63. These data raise some interesting statistical issues. Without in any way diminishing the value of what the fourth-grade student did, let's look at the data more closely. The article in the paper reported that the highest allowance for a boy was $10, whereas the highest for a girl was $9. It also reported that the girls' two lowest allowances were $0.50 and $0.51, but the lowest reported allowance for a boy was $3.00.

   a.   Create a set of data for boys and girls that would produce these results. (No, I did not make an error in reporting the results that were given.)

   b.   What is the most appropriate measure of central tendency to report in this situation?

   c.   What does the available information suggest to you about the distribution of allowances for the two genders? What would the means be if we trimmed extreme allowances from each group?

CHAPTER 3

The Normal Distribution

Objectives

To develop the concept of the normal distribution and show how it can be used to draw inferences about observations.

Contents

normal distribution

FROM WHAT HAS BEEN SAID in the preceding chapters, it is apparent that we are going to be very much concerned with distributions—distributions of data, hypothetical distributions of populations, and sampling distributions. Of all the possible forms that distributions can take, the class known as the **normal distribution** is by far the most important for our purposes.

Before elaborating on the normal distribution, however, it is worth a short digression to explain just why we are so interested in distributions in general. The critical factor is that there is an important link between distributions and probabilities. If we know something about the distribution of events (or of sample statistics), we know something about the probability that one of those events (or statistics) is likely to occur. To see the issue in its simplest form, take the lowly pie chart. (This is the only time you will see a pie chart in this book, because I find it very difficult to compare little slices of pie in different orientations to see which one is larger. There are much better ways to present data. However, the pie chart serves a useful purpose here.)

The pie chart shown in Figure 3.1 is taken from a U.S. Department of Justice report on probation and parole.[1] It shows the status of all individuals convicted of a criminal offense. From this figure you can see that 9% were in jail, 19% were in prison, 61% were on probation, and the remaining 11% were on parole. You can also see that the percentages in each category are directly reflected in the percentage of the area of the pie that each wedge occupies. The area taken up by each segment is directly proportional to the percentage of individuals in that segment. Moreover, if we declare that the total area of the pie is 1.00 unit, then the area of each segment is equal to the proportion of observations falling in that segment.

It is easy to go from speaking about areas to speaking about probabilities. The concept of probability will be elaborated in Chapter 5, but even without a precise definition of probability we can make an important point about areas of a pie chart. For now, simply think of probability in its common everyday usage, referring to the likelihood that some event will occur. From this perspective, it is logical to conclude that, because 19% of those convicted of a federal crime are currently in prison, if we were to randomly draw the name of one person from a list of convicted individuals, the probability is .19 that the individual would be in prison. To put this in slightly different terms, if 19% of the area of the pie is allocated to prison, then the probability that a person would fall in that segment is .19.

This pie chart also allows us to explore the addition of areas. It should be clear that if 19% are in prison and 9% are in jail, then $19 + 9 = 28\%$ are incarcerated. In other words, we can find the percentage of individuals in one of several categories just by adding the percentages for each category. The same thing holds for areas, in the sense that we can find the percentage of incarcerated individuals by adding the areas devoted to prison and to jail. And finally, if we can find percentages by adding areas, we can also find probabilities by adding areas. Thus, the probability of being incarcerated is the probability of being in one of the two segments associated with incarceration, which we can get by summing the two areas (or their associated probabilities).

There are other ways to present data besides pie charts. Two of the simplest are a histogram (discussed in Chapter 2) and its closely related cousin, the bar chart. Figure 3.2 is a redrawing of Figure 3.1 in the form of a bar chart. Although this figure does not contain any new information, it has two advantages over the pie chart. First, it is easier to compare categories, because the only thing we need to look at is the height of the bar, rather than trying to compare the lengths of two different arcs in different orientations. The second advantage is that the bar chart is visually more like the common distributions we will deal with, in that the various levels or categories are spread out along the horizontal dimension, and the

---

[1] The figures given here were for 1982. For 1993, the figures were Jail = 9%, Prison = 19%, Probation = 58%, and Parole = 14%. (For the 2003 data, see http://www.ojp.usdoj.gov/bjs/pub/pdf/ppus03.pdf.) What does this tell us about changes in how we deal with convictions? A search of the Internet under the words "correctional supervision" reveals all sorts of interesting statistics, which seem to vary with the politics of the person reporting them.

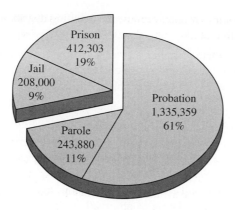

Figure 3.1   Pie chart showing persons under correctional supervision, by type of supervision, on December 31, 1982

NOTE: The prison data are from the U.S. Department of Justice, Bureau of Justice Statistics, *Prisoners in 1982,* Bulletin NCJ-87933 (Washington, D.C.: U.S. Government Printing Office, 1983). The jail data are from the U.S. Department of Justice, Bureau of Justice Statistics, *Jail Inmates 1982,* Bulletin NCJ-87161 (Washington, D.C.: U.S. Department of Justice, February 1983). The parole data and the probation data are from the annual Uniform Parole Reports and National Probation Reports surveys.

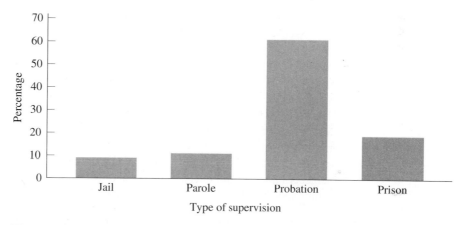

Figure 3.2   Bar chart showing persons under correctional supervision, by type of supervision

percentages (or frequencies) in each category are shown along the vertical dimension. (However, in a bar chart the values on the *X*-axis can form a nominal scale, as they do here. This is not true in a histogram.) Here again, you can see that the various areas of the distribution are related to probabilities. Further, you can see that we can meaningfully sum areas in exactly the same way that we did in the pie chart. When we move to more common distributions, particularly the normal distribution, the principles of areas, percentages, probabilities, and the addition of areas or probabilities carry over almost without change.

3.1 The Normal Distribution

Now we'll move closer to the normal distribution. I stated earlier that the normal distribution is one of the most important distributions we will encounter. There are several reasons for this:

1. Many of the dependent variables that we see are commonly assumed to be normally distributed in the population. That is to say, we frequently assume that if we were to obtain

the whole population of observations, the resulting distribution would closely resemble the normal distribution.

2. If we can assume that a variable is at least approximately normally distributed, then the techniques that are discussed in this chapter allow us to make a number of inferences (either exact or approximate) about values of that variable.

3. The theoretical distribution of the hypothetical set of sample means obtained by drawing an infinite number of samples from a specified population can be shown to be approximately normal under a wide variety of conditions. Such a distribution is called the sampling distribution of the mean and is discussed and used extensively throughout the remainder of this book.

4. Most of the statistical procedures we will employ have, somewhere in their derivation, an assumption that the population of observations (or of measurement errors) is normally distributed.

To introduce the normal distribution, we will look at one additional data set that is approximately normal (and would be even closer to normal if we had more observations). The data we are going to look at were collected using the Achenbach Youth Self-Report form (Achenbach, 1991b), a frequently used measure of behavior problems that produces scores on a number of different dimensions. We are going to focus on the dimension of Total Behavior Problems, which represents the total number of behavior problems reported by the child (weighted by the severity of the problem). (Examples of Behavior Problem categories are "Argues," "Impulsive," "Shows off," and "Teases.") Figure 3.3 is a histogram of data from 289 junior-high-school students. A higher score represents more behavior problems. You can see that this distribution has a center very near 50 and is fairly symmetrically distributed on each side of that value, with the scores ranging between about 25 and 75. The standard deviation of this distribution is approximately 10. The distribution is not perfectly even—it has some bumps and valleys—but overall it is fairly smooth, rising in the center and falling off at the ends. (The actual mean and standard deviation for this particular sample are 49.1 and 10.56, respectively.)

One thing that you might note from this distribution is that if you add the frequencies of subjects falling in the intervals 52–54 and 54–56, you will find that 54 students obtained scores between 52 and 56. Because there are 289 observations in this sample, $54/289 = 19\%$

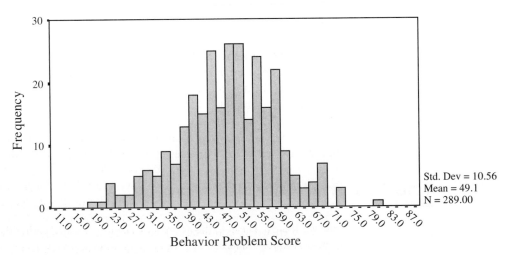

Figure 3.3 Histogram showing distribution of total behavior problem scores

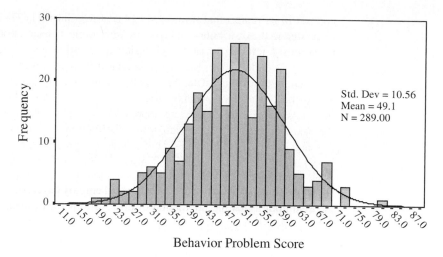

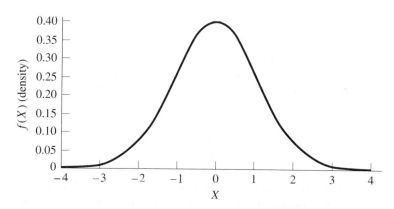

Figure 3.4   A characteristic normal distribution representing the distribution of behavior problem scores

Figure 3.5   A characteristic normal distribution with values of X on the abscissa and density on the ordinate

of the observations fell in this interval. This illustrates the comments made earlier on the addition of areas.

If we take this same set of data and fit a smooth curve to it, we will have a distribution that closely resembles the normal distribution. This distribution is just a stylized version that leaves out the bumps and valleys. Figure 3.4 shows the data from Figure 3.3 plotted with a smoothed curve (actually a normal distribution) fit to the data. The smooth distribution shown in Figure 3.4 is a characteristic normal distribution. It is a symmetric, unimodal distribution, frequently referred to as "bell shaped," and has limits of $\pm\infty$. The **abscissa,** or horizontal axis, represents different possible values of X, whereas the **ordinate,** or vertical axis, is referred to as the density and is related to (but not the same as) the frequency or probability of occurrence of X. The concept of density is discussed in further detail in the Chapter 5.

We often discuss the normal distribution by showing a generic kind of distribution with X on the abscissa and density on the ordinate. Such a distribution is shown in Figure 3.5.

The normal distribution has a long history. It was originally investigated by Abraham De Moivre (1667–1754), who was interested in its use to describe the results of games of chance (gambling). The distribution was defined precisely by Pierre-Simon Laplace (1749–1827)

abscissa

ordinate

and put in its more usual form by Carl Friedrich Gauss (1777–1855), both of whom were interested in the distribution of errors in astronomical observations. The normal distribution is variously referred to as the Gaussian distribution and as the "normal law of error." Adolphe Quételet (1796–1874), a Belgian astronomer, was the first to apply the distribution to social and biological data. Apparently having nothing better to do with his time, he collected chest measurements of Scottish soldiers and heights of French soldiers. He found that both sets of measurements were approximately normally distributed. Quételet interpreted the data to indicate that the mean of this distribution was the ideal at which nature was aiming, and observations to each side of the mean represented error (a deviation from nature's ideal). (For 5'8" males like myself, it is somehow comforting to think of all those bigger guys as nature's mistakes.) Although we no longer think of the mean as nature's ideal, this is a useful way to conceptualize variability around the mean. Actually, we still use the word *error* to refer to deviations from the mean. Francis Galton (1822–1911) carried Quételet's ideas further and gave the normal distribution a central role in psychological theory, especially the theory of mental abilities. Some would insist that Galton was *too* successful in this endeavor, and we tend to assume that measures are normally distributed even when they are not. I won't argue the issue here.

Mathematically, the normal distribution is defined as

$$f(X) = \frac{1}{\sigma\sqrt{2\pi}}(e)^{-(X-\mu)^2/2\sigma^2}$$

where π and e are constants ($\pi = 3.1416$ and $e = 2.7183$), and μ and σ are the mean and the standard deviation, respectively, of the distribution. If μ and σ are known, the ordinate, $f(X)$, for any value of X can be obtained simply by substituting the appropriate values for μ, σ, and X and solving the equation. This is not nearly as difficult as it looks, but in practice you are unlikely ever to have to make the calculations. The cumulative form of this distribution is tabled, and we can simply read the information we need from the table.

Those of you who have had a course in calculus may recognize that the area under the curve between any two values of X (say, X_1 and X_2), and thus the probability that a randomly drawn score will fall within that interval, can be found by integrating the function over the range from X_1 to X_2. Those of you who have not had such a course can take comfort from the fact that tables are readily available in which this work has already been done for us or by use of which we can easily do the work ourselves. Such a table appears in Appendix z (page 694).

You might be excused at this point for wondering why anyone would want to table such a distribution in the first place. Just because a distribution is common (or at least commonly assumed) doesn't automatically suggest a reason for having an appendix that tells all about it. The reason is quite simple. By using Appendix z, we can readily calculate the probability that a score drawn at random from the population will have a value lying between any two specified points (X_1 and X_2). Thus, by using statistical tables we can make probability statements in answer to a variety of questions. You will see examples of such questions in the rest of this chapter. They will also appear in many other chapters throughout the book.

3.2 The Standard Normal Distribution

A problem arises when we try to table the normal distribution because the distribution depends on the values of the mean and the standard deviation (μ and σ) of the distribution. To do the job right, we would have to make up a different table for every possible combination of the values of μ and σ, which certainly is not practical. The solution to this problem

**standard normal
distribution**

is quite simple. What we actually have in the table is what is called the **standard normal distribution,** which has a mean of 0 and a standard deviation of 1. Such a distribution is often designated as $N(0,1)$, where N refers to the fact that it is normal, 0 is the value of μ, and 1 is the value of σ^2. ($N(\mu, \sigma^2)$ is the more general expression.) Given the standard normal distribution in the appendix and a set of rules for transforming any normal distribution to standard form and vice versa, we can use Appendix z to find the areas under any normal distribution.

Consider the distribution shown in Figure 3.6, with a mean of 50 and a standard deviation of 10 (variance of 100). It represents the distribution of *an entire population* of Total Behavior Problem scores from the Achenbach Youth Self-Report form, of which the data in Figures 3.3 and 3.4 are a sample. If we knew something about the areas under the curve in Figure 3.6, we could say something about the probability of various values of Behavior Problem scores and could identify, for example, those scores that are so high that they are obtained by only 5% or 10% of the population. You might wonder why we would want to do this, but it is often important in diagnosis to be able to separate extreme scores from more typical scores.

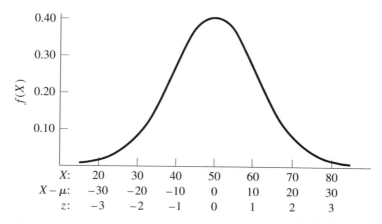

Figure 3.6   A normal distribution with various transformations on the abscissa

The only tables of the normal distribution that are readily available are those of the *standard* normal distribution. Therefore, before we can answer questions about the probability that an individual will get a score above some particular value, we must first transform the distribution in Figure 3.6 (or at least specific points along it) to a standard normal distribution. That is, we want to be able to say that a score of X_i from a normal distribution with a mean of 50 and a variance of 100—often denoted $N(50,100)$—is comparable to a score of z_i from a distribution with a mean of 0 and a variance, and standard deviation, of 1— denoted $N(0,1)$. Then anything that is true of z_i is also true of X_i, and z and X are compara-

pivotal statistic

ble variables. (Statisticians sometimes call z a **pivotal statistic** because its distribution does not depend on the values of μ and σ^2.)

From Exercise 2.36 we know that subtracting a constant from each score in a set of scores reduces the mean of the set by that constant. Thus, if we subtract 50 (the mean) from all the values for X, the new mean will be $50 - 50 = 0$. (More generally, the distribution of $[X - \mu]$ has a mean of 0.) The effect of this transformation is shown in the second set of values for the abscissa in Figure 3.6. We are halfway there because we now have the mean down to 0, although the standard deviation (σ) is still 10. We also know from Exercise 2.37 that if we multiply or divide all values of a variable by a constant (e.g., 10), we multiply or divide the standard deviation by that constant. Thus, if we divide all scores by 10, the standard deviation will now be $10/10 = 1$, which is just what we wanted. We will call this transformed distribution z and define it, on the basis of what we have done, as

$$z = \frac{X - \mu}{\sigma}$$

For our particular case, where $\mu = 50$ and $\sigma = 10$,

$$z = \frac{X - \mu}{\sigma} = \frac{X - 50}{10}$$

The third set of values (labeled z) for the abscissa in Figure 3.6 shows the effect of this transformation. Note that aside from a linear transformation of the numerical values, the data have not been changed in any way. The distribution has the same shape and the observations continue to stand in the same relation to each other as they did before the transformation. It should not come as a great surprise that changing the unit of measurement does not change the shape of the distribution or the relative standing of observations. Whether we measure the quantity of alcohol that people consume per week in ounces or in milliliters really makes no difference in the relative standing of people. It just changes the numerical values on the abscissa. (The town drunk is still the town drunk, even if now his liquor is measured in milliliters.) It is important to realize exactly what converting X to z has accomplished. A score that used to be 60 is now 1. That is, a score that used to be one standard deviation (10 points) above the mean remains one standard deviation above the mean, but now is given a new value of 1. A score of 45, which was 0.5 standard deviation *below* the mean, now is given the value of -0.5, and so on. In other words, a z score represents the number of standard deviations that X_i is above or below the mean—a positive z score being above the mean and a negative z score being below the mean.

z scores

The equation for z is completely general. We can transform any distribution to a distribution of z **scores** simply by applying this equation. Keep in mind, however, the point that was just made. The *shape* of the distribution is unaffected by a linear transformation. That means that *if the distribution was not normal before it was transformed, it will not be normal afterward.* Some people believe that they can "normalize" (in the sense of producing a normal distribution) their data by transforming them to z. It just won't work.

You can see what happens when you draw random samples from a population that is normal by going to http://www.anu.edu.au/nceph/surfstat/surfstat-home/surfstat.html and clicking on "Hotlist for Java Applets." Just click on the histogram, and it will present another histogram that you can modify in various ways. By repeatedly clicking "start" without clearing, you can add cases to the sample. It is useful to see how the distribution approaches a normal distribution as the number of observations increases. (And how "non-normal" a distribution with a small sample size can look.)

3.3 Using the Tables of the Standard Normal Distribution

As already mentioned, the standard normal distribution is extensively tabled. Such a table can be found in Appendix z, part of which is reproduced in Table 3.1.[2] To see how we can make use of this table, consider the normal distribution represented in Figure 3.7. This

---

[2] If you prefer electronic tables, many small Java programs are available on the Internet. One good source is http://calculators.stat.ucla.edu/cdf/normal/normalcalc.php, which allows you to calculate all sorts of things. (The UCLA homepage (www.stat.ucla.edu) is an excellent source of statistical information.) An online video displaying properties of the normal distribution is available at http://huizen.dds.nl/~berrie/normal.html. One of my favorite programs for calculating z probabilities is at http://psych.colorado.edu/~mcclella/java/zcalc.html.

Table 3.1   The normal distribution (abbreviated version of Appendix z)

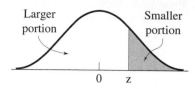

z	Mean to z	Larger Portion	Smaller Portion	z	Mean to z	Larger Portion	Smaller Portion
0.00	0.0000	0.5000	0.5000	0.45	0.1736	0.6736	0.3264
0.01	0.0040	0.5040	0.4960	0.46	0.1772	0.6772	0.3228
0.02	0.0080	0.5080	0.4920	0.47	0.1808	0.6808	0.3192
0.03	0.0120	0.5120	0.4880	0.48	0.1844	0.6844	0.3156
0.04	0.0160	0.5160	0.4840	0.49	0.1879	0.6879	0.3121
0.05	0.0199	0.5199	0.4801	0.50	0.1915	0.6915	0.3085
. . .	. . .	. . .	. . .	. . .	. . .	. . .	. . .
0.97	0.3340	0.8340	0.1660	1.42	0.4222	0.9222	0.0778
0.98	0.3365	0.8365	0.1635	1.43	0.4236	0.9236	0.0764
0.99	0.3389	0.8389	0.1611	1.44	0.4251	0.9251	0.0749
1.00	0.3413	0.8413	0.1587	1.45	0.4265	0.9265	0.0735
1.01	0.3438	0.8438	0.1562	1.46	0.4279	0.9279	0.0721
1.02	0.3461	0.8461	0.1539	1.47	0.4292	0.9292	0.0708
1.03	0.3485	0.8485	0.1515	1.48	0.4306	0.9306	0.0694
1.04	0.3508	0.8508	0.1492	1.49	0.4319	0.9319	0.0681
1.05	0.3531	0.8531	0.1469	1.50	0.4332	0.9332	0.0668
. . .	. . .	. . .	. . .	. . .	. . .	. . .	. . .
1.95	0.4744	0.9744	0.0256	2.40	0.4918	0.9918	0.0082
1.96	0.4750	0.9750	0.0250	2.41	0.4920	0.9920	0.0080
1.97	0.4756	0.9756	0.0244	2.42	0.4922	0.9922	0.0078
1.98	0.4761	0.9761	0.0239	2.43	0.4925	0.9925	0.0075
1.99	0.4767	0.9767	0.0233	2.44	0.4927	0.9927	0.0073
2.00	0.4772	0.9772	0.0228	2.45	0.4929	0.9929	0.0071
2.01	0.4778	0.9778	0.0222	2.46	0.4931	0.9931	0.0069
2.02	0.4783	0.9783	0.0217	2.47	0.4932	0.9932	0.0068
2.03	0.4788	0.9788	0.0212	2.48	0.4934	0.9934	0.0066
2.04	0.4793	0.9793	0.0207	2.49	0.4936	0.9936	0.0064
2.05	0.4798	0.9798	0.0202	2.50	0.4938	0.9938	0.0062

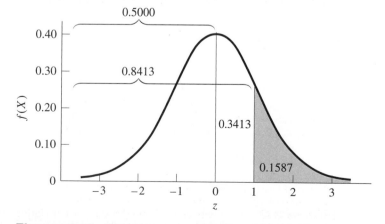

Figure 3.7   Illustrative areas under the normal distribution

might represent the standardized distribution of the Behavior Problem scores as seen in Figure 3.6. Suppose we want to know how much of the area under the curve is above one standard deviation from the mean, if the total area under the curve is taken to be 1.00. (Remember that we care about areas because they translate directly to probabilities.) We already have seen that z scores represent standard deviations from the mean, and thus, we know that we want to find the area above $z = 1$.

Only the positive half of the normal distribution is tabled. Because the distribution is symmetric, any information given about a positive value of z applies equally to the corresponding negative value of z. (The table in Appendix z also contains a column labeled "y". This is just the height (density) of the curve corresponding to that value of z. I have not included it here to save space and because it is rarely used.) From Table 3.1 (or Appendix z), we find the row corresponding to $z = 1.00$. Reading across that row, we can see that the area from the *mean to $z = 1$* is 0.3413, the area in the *larger portion* is 0.8413, and the area in the *smaller portion* is 0.1587. (If you visualize the distribution being divided into the segment below $z = 1$ [the unshaded part of Figure 3.7] and the segment above $z = 1$ [the shaded part], the meanings of the terms *larger portion* and *smaller portion* become obvious.) Thus, the answer to our original question is 0.1587. Because we already have equated the terms *area* and *probability*, we now can say that if we sample a child at random from the population of children, and if Behavior Problem scores are normally distributed, then the probability that the child will score more than one standard deviation above the mean of the population (i.e., above 60) is .1587. Because the distribution is symmetric, we also know that the probability that a child will score more than one standard deviation *below* the mean of the population is also .1587.

Now suppose that we want the probability that the child will be more than one standard deviation (10 points) from the mean *in either direction*. This is a simple matter of the summation of areas. Because we know that the normal distribution is symmetric, then the area below $z = -1$ will be the same as the area above $z = +1$. This is why the table does not contain negative values of z—they are not needed. We already know that the areas in which we are interested are each 0.1587. Then the total area outside $z = \pm 1$ must be $0.1587 + 0.1587 = 0.3174$. The converse is also true. If the area outside $z = \pm 1$ is 0.3174, then the area between $z = +1$ and $z = -1$ is equal to $1 - 0.3174 = 0.6826$. Thus, the probability that a child will score between 40 and 60 is .6826.

To extend this procedure, consider the situation in which we want to know the probability that a score will be between 30 and 40. A little arithmetic will show that this is simply the probability of falling between 1.0 standard deviation below the mean and 2.0 standard deviations below the mean. This situation is diagrammed in Figure 3.8. (*Hint:* It is always

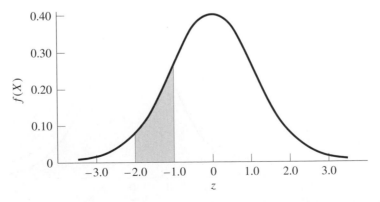

Figure 3.8 Area between 1.0 and 2.0 standard deviations below the mean

wise to draw simple diagrams such as Figure 3.8. They eliminate many errors and make clear the area or areas for which you are looking.)

From Appendix z, we know that the area from the mean to $z = -2.0$ is 0.4772 and from the mean to $z = -1.0$ is 0.3413. The difference is these two areas must represent the area between $z = -2.0$ and $z = -1.0$. This area is $0.4772 - 0.3413 = 0.1359$. Thus, the probability that Behavior Problem scores drawn at random from a normally distributed population will be between 30 and 40 is .1359.

3.4 Setting Probable Limits on an Observation

For a final example, consider the situation in which we want to identify limits within which we have some specified degree of confidence that a child sampled at random will fall. In other words, we want to make a statement of the form, "If I draw a child at random from this population, 95% of the time her score will lie between _____ and _____." In Figure 3.9, you can see the limits we want—the limits that include 95% of the scores in the population.

If we are looking for the limits within which 95% of the scores fall, we also are looking for the limits beyond which the remaining 5% of the scores fall. To rule out this remaining 5%, we want to find that value of z that cuts off 2.5% at each end, or "tail," of the distribution. (We do not need to use symmetric limits, but we typically do because they usually make the most sense and produce the shortest interval.) From Appendix z, we see that these values are $z = \pm 1.96$. Thus, we can say that 95% of the time a child's score sampled at random will fall between 1.96 standard deviations above the mean and 1.96 standard deviations below the mean.

Because we generally want to express our answers in terms of raw Behavior Problem scores, rather than z scores, we must do a little more work. To obtain the raw score limits, we simply work the formula for z backward, solving for X instead of z. Thus, if we want to state the limits encompassing 95% of the population, we want to find those scores that are 1.96 standard deviations above and below the mean of the population. This can be written as

$$z = \frac{X - \mu}{\sigma}$$

$$\pm 1.96 = \frac{X - \mu}{\sigma}$$

$$X - \mu = \pm 1.96\sigma$$

$$X = \mu \pm 1.96\sigma$$

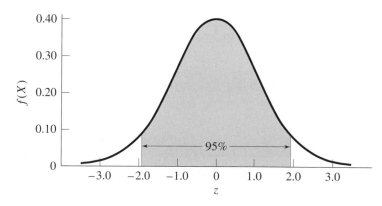

Figure 3.9   Values of z that enclose 95% of the behavior problem scores

where the values of X corresponding to $(\mu + 1.96\sigma)$ and $(\mu - 1.96\sigma)$ represent the limits we seek. For our example the limits will be

$$\text{Limits} = 50 \pm (1.96)(10) = 50 \pm 19.6 = 30.4 \text{ and } 69.6.$$

So the probability is .95 that a child's score (X) chosen at random would be between 30.4 and 69.6. We may not be very interested in low scores, because they don't represent problems. But anyone with a score of 69.6 or higher is a problem to someone. Only 2.5% of children score at least that high.

What we have just discussed is closely related to, but not quite the same as, what we will later consider under the heading of confidence limits. The major difference is that here we knew the population mean and were trying to estimate where a single observation (X) would fall. When we discuss confidence limits, we will have a sample mean (or some other statistic) and will want to set limits that have a probability of .95 of bracketing the population mean (or some other relevant parameter). You do not need to know anything at all about confidence limits at this point. I simply mention the issue to forestall any confusion in the future.

3.5 Measures Related to z

standard scores

We already have seen that the z formula given earlier can be used to convert a distribution with any mean and variance to a distribution with a mean of 0 and a standard deviation (and variance) of 1. We frequently refer to such transformed scores as **standard scores.** There also are other transformational scoring systems with particular properties, some of which people use every day without realizing what they are.

A good example of such a scoring system is the common IQ. The raw scores from an IQ test are routinely transformed to a distribution with a mean of 100 and a standard deviation of 15 (or 16 in the case of the Binet). Knowing this, you can readily convert an individual's IQ (e.g., 120) to his or her position in terms of standard deviations above or below the mean (i.e., you can calculate the z score). Because IQ scores are more or less normally distributed, you can then convert z into a percentage measure by use of Appendix z. (In this example, a score of 120 has approximately 91% of the scores below it. This is known as the 91st **percentile.**)

percentile

Another common example is a nationally administered examination, such as the SAT. The raw scores are transformed by the producer of the test and reported as coming from a distribution with a mean of 500 and a standard deviation of 100 (at least that was the case when the tests were first developed). Such a scoring system is easy to devise. We start by converting raw scores to z scores (using the obtained raw score mean and standard deviation). We then convert the z scores to the particular scoring system we have in mind. Thus, New score = New SD * (z) + New mean, where z represents the z score corresponding to the individual's raw score. For the SAT, New score = $100(z) + 500$. Scoring systems such as the one used on the Achenbach Youth Self-Report checklist, which have a mean set at 50 and a standard deviation set at 10, are called **T scores** (the T is always capitalized). These tests are useful in psychological measurement because they have a common frame of reference. For example, people become used to seeing a cutoff score of 63 as identifying the highest 10% of the subjects.

T scores

Key Terms

Normal distribution (Introduction)	Standard normal distribution (3.2)	Standard scores (3.5)
Abscissa (3.1)	Pivotal statistic (3.2)	Percentile (3.5)
Ordinate (3.1)	z score (3.2)	T scores (3.5)

Exercises

3.1   Assume that the following data represent a population with $\mu = 4$ and $\sigma = 1.63$: $X =$ [1 2 2 3 3 3 4 4 4 4 5 5 5 6 6 7]

a.   Plot the distribution as given.

b.   Convert the distribution in part (a) to a distribution of $X - \mu$.

c.   Go the next step and convert the distribution in part (b) to a distribution of z.

3.2   Using the distribution in Exercise 3.1, calculate z scores for $X = 2.5$, 6.2, and 9. Interpret these results.

3.3   Suppose we want to study the errors found in the performance of a simple task. We ask a large number of judges to report the number of people seen entering a major department store in one morning. Some judges will miss some people, and some will count others twice, so we don't expect everyone to agree. Suppose we find that the mean number of shoppers reported is 975 with a standard deviation of 15. Assume that the distribution of counts is normal.

a.   What percentage of the counts will lie between 960 and 990?

b.   What percentage of the counts will lie below 975?

c.   What percentage of the counts will lie below 990?

3.4   Using the example from Exercise 3.3:

a.   What two values of X (the count) would encompass the middle 50% of the results?

b.   75% of the counts would be less than _____ .

c.   95% of the counts would be between _____ and _____ .

3.5   The person in charge of the project in Exercise 3.3 counted only 950 shoppers entering the store. Is this a reasonable answer if he was counting conscientiously? Why or why not?

3.6   A set of reading scores for fourth-grade children has a mean of 25 and a standard deviation of 5. A set of scores for ninth-grade children has a mean of 30 and a standard deviation of 10. Assume that the distributions are normal.

a.   Draw a rough sketch of these data, putting both groups in the same figure.

b.   What percentage of the fourth graders score better than the average ninth grader?

c.   What percentage of the ninth graders score worse than the average fourth grader? (We will come back to the idea behind these calculations when we study power in Chapter 8.)

3.7   Under what conditions would the answers to parts (b) and (c) of Exercise 3.6 be equal?

3.8   A certain diagnostic test is indicative of problems only if a child scores in the lowest 10% of those taking the test (the 10th percentile). If the mean score is 150 with a standard deviation of 30, what would be the diagnostically meaningful cutoff?

3.9   A dean must distribute salary raises to her faculty for next year. She has decided that the mean raise is to be $2,000, the standard deviation of raises is to be $400, and the distribution is to be normal.

a.   The most productive 10% of the faculty will have a raise equal to or greater than $_____ .

b.   The 5% of the faculty who have done nothing useful in years will receive no more than $_____ each.

3.10   We have sent everyone in a large introductory course out to check whether people use seat belts. Each student has been told to look at 100 cars and count the number of people wearing seat belts. The number found by any given student is considered that student's score. The mean score for the class is 44, with a standard deviation of 7.

a.   Diagram this distribution, assuming that the counts are normally distributed.

b. A student who has done very little work all year has reported finding 62 seat belt users out of 100. Do we have reason to suspect that the student just made up a number rather than actually counting?

3.11 A number of years ago a friend of mine produced a diagnostic test of language problems. A score on her scale is obtained simply by counting the number of language constructions (e.g., plural, negative, passive) that the child produces correctly in response to specific prompts from the person administering the test. The test had a mean of 48 and a standard deviation of 7. Parents had trouble understanding the meaning of a score on this scale, and my friend wanted to convert the scores to a mean of 80 and a standard deviation of 10 (to make them more like the kinds of grades parents are used to). How could she have gone about her task?

3.12 Unfortunately, the whole world is not built on the principle of a normal distribution. In the preceding example, the real distribution is badly skewed because most children do not have language problems and therefore produce all constructions correctly.

a. Diagram how the distribution might look.

b. How would you go about finding the cutoff for the bottom 10% if the distribution is not normal?

3.13 In October 1981, the mean and the standard deviation on the Graduate Record Exam (GRE) for all people taking the exam were 489 and 126, respectively. What percentage of students would you expect to have a score of 600 or less? (This is called the percentile rank of 600.)

3.14 In Exercise 3.13, what score would be equal to or greater than 75% of the scores on the exam? (This score is called the 75th percentile.)

3.15 For all seniors and nonenrolled college graduates taking the GRE in October 1981, the mean and the standard deviation were 507 and 118, respectively. How does this change the answers to Exercises 3.13 and 3.14?

3.16 What does the answer to Exercise 3.15 suggest about the importance of reference groups?

3.17 What is the 75th percentile for GPA described in Appendix Data Set? (This is the point below which 75% of the observations are expected to fall.)

3.18 Assuming that the Behavior Problem scores discussed in this chapter come from a population with a mean of 50 and a standard deviation of 10, what would be a diagnostically meaningful cutoff if you wanted to identify those children who score in the highest 2% of the population?

3.19 In Section 3.5, I said that T scores are designed to have a mean of 50 and a standard deviation of 10 and that the Achenbach Youth Self-Report measure produces T scores. The data in Figure 3.3 do not have a mean and standard deviation of exactly 50 and 10. Why do you suppose that this is so?

Discussion Questions

3.20 If you go back to the reaction time data presented as a frequency distribution in Table 2.2 and Figure 2.1, you will see that they are not normally distributed. For these data, the mean is 60.26 and the standard deviation is 13.01. By simple counting, you can calculate exactly what percentage of the sample lies above or below ± 1.0, 1.5, 2.0, 2.5, and 3.0 standard deviations from the mean. You can also calculate, from tables of the normal distribution, what percentage of scores would lie above or below those cutoffs if the distribution were perfectly normal. Calculate these values and then consider the implications of using a normal distribution to approximate a distribution that is not normal. Where are you off by quite a bit, and where are you reasonably accurate? Consider this from the point of view of both one-tailed and two-tailed areas. How would your answers change if the sample had been very much larger or very much smaller?

3.21 The data plotted in the following figure represent the distribution of salaries paid to new full-time assistant professors in U.S. doctoral departments of psychology in 1999–2000. The data are available on the web site (www.uvm.edu/~dhowell/methods/) as Ex3-21.dat. Although the data are obviously skewed to the right, what would you expect to happen if you treated these data as if they were normally distributed? What explanation could you hypothesize to account for the extreme values?

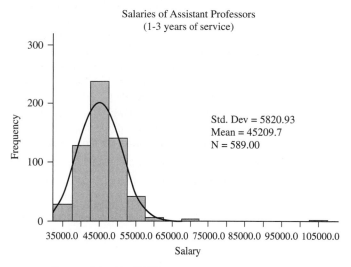

Salaries of Assistant Professors
(1-3 years of service)

Cases weighted by FREQ

3.22   The data file named sat.dat on the web site contains data on SAT scores for all 50 states as well as the amount of money spent on education and the percentage of students taking the SAT in that state. (The data are described under "general data sets" at www.uvm.edu/~ dhowell/methods/DataFiles/DataSets.html.) Draw a histogram of the combined SAT scores. Is this distribution normal? The variable *adjcomb* is the combined score adjusted for the percentage of students in that state who took the exam. What can you tell about this variable? How does its distribution differ from that for the unadjusted scores?

CHAPTER 4

Sampling Distributions and Hypothesis Testing

Objectives

To lay the groundwork for the procedures discussed in this book by examining the general theory of hypothesis testing and describing specific concepts as they apply to all hypothesis tests.

Contents

IN CHAPTER 2, we examined a number of different statistics and saw how they might be used to describe a set of data or to represent the frequency of the occurrence of some event. Although the description of the data is important and fundamental to any analysis, it is not sufficient to answer many of the most interesting problems we encounter. In a typical experiment, we might treat one group of people in a special way and want to see whether their scores differ from the scores of people in general. Or we might offer a treatment to one group but not to a control group and want to compare the means of the two groups on some variable. Descriptive statistics will not tell us, for example, whether the difference between a sample mean and a hypothetical population mean, or the difference between two obtained sample means, is small enough to be explained by chance alone or whether it represents a true difference that might be attributable to the effect of our experimental treatment(s).

Statisticians frequently use phrases such as "variability due to chance" or "sampling error" and assume that you know what they mean. Probably you do, but if you do not, you are headed for confusion in the remainder of this book unless we spend a minute clarifying the meaning of these terms. We will begin with a simple example.

In Chapter 3, we considered the distribution of Total Behavior Problem scores from the Achenbach Youth Self-Report form. Total Behavior Problem scores are normally distributed in the population (i.e., the complete population of such scores is approximately normally distributed) with a population mean (μ) of 50 and a population standard deviation (σ) of 10. We know that different children show different levels of problem behaviors and therefore have different scores. We also know that if we took a sample of children, their sample mean would probably not equal exactly 50. One sample of children might have a mean of 49, but a second sample might have a mean of 52.3. The actual sample means would depend on the particular children who happened to be included in the sample. This expected variability from sample to sample is what is meant when we speak of "variability due to chance." The phrase refers to the fact that statistics (in this case, means) obtained from samples naturally vary from one sample to another.

sampling error   Along the same lines, the term **sampling error** often is used in this context as a synonym for variability due to chance. It indicates that the numerical value of a sample statistic probably will be in error (i.e., will deviate from the parameter it is estimating) as a result of the particular observations that happened to be included in the sample. In this context, "error" does not imply carelessness or mistakes. In the case of behavior problems, one random sample might just happen to include an unusually obnoxious child, whereas another sample might happen to include an unusual number of relatively well-behaved children.

4.1 Two Simple Examples Involving Course Evaluations and Rude Motorists

One example that we will investigate when we discuss correlation and regression looks at the relationship between how students evaluate a course and the grade they expect to receive in that course. Many faculty feel strongly about this topic because even the best instructors turn to the semiannual course evaluation forms with some trepidation—perhaps the same amount of trepidation with which many students open their grade report form. Some faculty think that a course is good or bad independently of how well a student feels he or she will do in terms of a grade. Others feel that a student who seldom came to class and who will do poorly as a result will also unfairly rate the course as poor. Finally, there are those who argue that students who do well and experience success take something away from the course other than just a grade and that those students will generally rate the course highly. But the relationship between course ratings and student performance is an empirical question and, as such, can be answered by looking at relevant data. Suppose that in a random sample of 50 courses we find a general trend for students in a course in which they expect to do well to rate the course highly, and for students to rate courses in which they expect to do poorly as low in overall quality. How do we tell whether this trend in our small data set is representative of a trend among students in general or just an odd result that would disappear if we ran the study again? (For your own interest, make your prediction of what kind of results we will find. We will return to this issue later.)

A second example comes from a study by Doob and Gross (1968), who investigated the influence of perceived social status. They found that if an old, beat-up (low-status) car failed to start when a traffic light turned green, 84% of the time the driver of the second car in line honked the horn. However, when the stopped car was an expensive, high-status car, the following driver only honked 50% of the time. These results could be explained in one of two ways:

1. The difference between 84% in one sample and 50% in a second sample is attributable to sampling error (random variability among samples); therefore, we cannot conclude that perceived social status influences horn-honking behavior.

2. The difference between 84% and 50% is large and reliable. The difference is not attributable to sampling error; therefore, we conclude that people are less likely to honk at drivers of high-status cars.

Although the statistical calculations required to answer this question are different from those used to answer the one about course evaluations (because the first deals with relationships and the second deals with proportions), the underlying logic is fundamentally the same.

hypothesis testing These examples of course evaluations and horn honking are two kinds of questions that fall under the heading of **hypothesis testing.** This chapter is intended to present the theory of hypothesis testing in as general a way as possible, without going into the specific techniques or properties of any particular test. I will focus largely on the situation involving differences instead of the situation involving relationships, but the logic is basically the same. (You will see additional material on examining relationships in Chapter 9.) I am very deliberately glossing over details of computation because my purpose is to explore the concepts of hypothesis testing without involving anything but the simplest technical details.

We need to be explicit about what the problem is here. The reason for having hypothesis testing in the first place is that data are ambiguous. Suppose that we want to decide whether larger classes receive lower student ratings. We all know that some large classes are terrific, and others are really dreadful. Similarly, there are both good and bad small classes. So if we collect data on large classes, for example, the mean of several large classes will depend to some extent on which large courses just happen to be included in our sample. If we reran our data collection with a new random sample of large classes, that mean would almost certainly be different. A similar situation applies for small classes. When we find a difference between the means of samples of large and small classes, we know that the difference would come out slightly differently if we collected new data. So a difference between the means is ambiguous. Is it greater than zero because large classes are worse than small ones, or because of the particular samples we happened to pick? Well, if the difference is quite large, it probably reflects differences between small and large classes. If it is quite small, it probably reflects just random noise. But how large is "large" and how small is "small"? That is the problem we are beginning to explore, and that is the subject of this chapter.

If we are going to look at either of the two examples laid out earlier, or at a third one to follow, we need to find some way of deciding whether we are looking at a small chance fluctuation between the horn-honking rates for low- and high-status cars or a difference that is sufficiently large for us to believe that people are much less likely to honk at those they consider higher in status. If the differences are small enough to attribute to chance variability, we may well not worry about them further. On the other hand, if we can rule out chance as the source of the difference, we probably need to look further. This decision about chance is what we mean by hypothesis testing.

4.2 Sampling Distributions

In addition to course evaluations and horn honking, we will add a third example, which is one to which we can all relate. It involves those annoying people who spend what seems to us an unreasonable amount of time vacating the parking space we are waiting for. Ruback and Juieng (1997) ran a simple study in which they divided drivers into two groups of 100 participants each—those who had someone waiting for their space and those who did not. Ruback and Juieng then recorded the amount of time that it took the driver to leave the parking space. For those drivers who had no one waiting, it took an average of 32.15 seconds to leave the space. For those who did have someone waiting, it took an average of 39.03 seconds. For each of these groups, the standard deviation of waiting times was 14.6 seconds. Notice that a driver took 6.88 seconds longer to leave a space when someone was waiting for it. (If you think about it, 6.88 seconds is a long time if you are the person doing the waiting.)

There are two possible explanations here. First, it is entirely possible that having someone waiting doesn't make any difference in how long it takes to leave a space, and that normally drivers who have no one waiting for them take, on average, the same length of time as do drivers who have someone waiting. In that case, the difference that we found is just a result of the particular samples we happened to obtain. What we are saying here is that if we had whole populations of drivers in each of the two conditions, the populations' means (μ_{nowait} and μ_{wait}) would be identical and any difference we find in our samples is sampling error. The alternative explanation is that the population means really are different and that people actually do take longer to leave a space when there is someone waiting for it. If the sample means had come out to be 32.15 and 32.18, you and I would probably side with the first explanation—or at least not be willing to reject it. If the means had come out to be 32.15 and 59.03, we would probably be likely to side with the second explanation—having someone waiting actually makes a difference. But the difference we found is actually somewhere in between, and we need to decide which explanation is more reasonable.

sampling distributions

We want to answer the question "Is the obtained difference too great to be attributable to chance?" To do this, we have to use what are called **sampling distributions,** which tell us specifically what degree of sample-to-sample variability we can expect by chance as a function of sampling error.

The most basic concept underlying all statistical tests is the sampling distribution of a statistic. It is fair to say that if we did not have sampling distributions, we would not have any statistical tests. Roughly speaking, sampling distributions tell us what values we might (or might not) expect to obtain for a particular statistic under a set of predefined conditions (e.g., what the sample differences between our two samples might be expected to be *if* the true means of the populations from which those samples came are equal?) In addition, the standard deviation of that distribution of differences between sample means (known as the "standard error" of the distribution) reflects the variability that we would expect to find in the values of that statistic (differences between means) over repeated trials. Sampling distributions provide the opportunity to evaluate the likelihood (given the value of a sample statistic) that such predefined conditions actually exist.

Basically, the sampling distribution of a statistic can be thought of as the distribution of values obtained for that statistic over repeated sampling (i.e., running the experiment, or drawing samples, an unlimited number of times). Sampling distributions are almost always derived mathematically, but it is easier to understand what they represent if we consider how they could, in theory, be derived empirically with a simple sampling experiment.

sampling distribution of the differences between means

We will take as an illustration the **sampling distribution of the differences between means** because it relates directly to our example of waiting times in parking lots. The sampling distribution of differences between means is the distribution of differences between means of an infinite number of random samples drawn under certain specified conditions (e.g., under the condition that the true means of our populations are equal). Suppose we have two populations with known means and standard deviations (Here we will suppose that the two population means are 35 and the standard deviation is 15, though what the values are is not critical to the logic of our argument.) Further suppose that we draw a very large number (theoretically an infinite number) of pairs of random samples from these populations, each sample consisting of 100 scores. For each sample we will calculate its sample mean and then the difference between the two means in that draw. When we finish drawing all the pairs of samples, we will plot the distribution of these differences. Such a distribution would be a sampling distribution of the difference between means and might look like the one presented in Figure 4.1. The center of this distribution is at 0.0, because we expect that, on average, differences between sample means will be 0.0. (The individual means themselves will be roughly 35.) We can see from this figure that differences between sample means of approximately −1.5 and 1.5, for example, are quite likely to occur when we sample from identical populations. We also can see that it is extremely unlikely that we would draw samples from these populations that differ by 4.5 or more. Knowing the kinds of values to expect for the difference of means of samples drawn from these populations allows us to turn the question around and ask whether an obtained sample mean difference can be taken as evidence in favor of the hypothesis that we actually are sampling from identical populations—or populations with the same mean.

Notice here that the most common event we would find in drawing pairs of samples is that the means don't differ. ($\mu_{nowait} - \mu_{wait}$) = 0. (That is the mode [and the mean] of that distribution.) It is also fairly common to find differences of 1.5 or 2, though it is rare to find differences of 4.5.

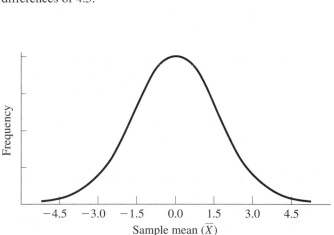

Figure 4.1   Distribution of difference between means, each based on 100 scores

4.3 Theory of Hypothesis Testing

Preamble

One of the major ongoing discussions in statistics in the behavioral sciences relates to hypothesis testing. The logic and theory of hypothesis testing has been debated for at least 75 years, but recently that debate has intensified considerably. The exchanges on this topic have not always been constructive (referring to your opponent's position as "bone-headedly misguided," "a perversion of the scientific method," or "ridiculous" usually does not win them to your cause), but some real and positive changes have come as a result. The changes are sufficiently important that much of this chapter, and major parts of the rest of the book, have been rewritten to accommodate them.

The arguments about the role of hypothesis testing concern several issues. First, and most fundamental, is hypothesis testing a sensible procedure? I think that it is, and whether it is or isn't, the logic involved is related to so much of what we do, and is so central to what you will see in the experimental literature, that you have to understand it whether you approve of it or not. Second, what logic will we use for hypothesis testing? The dominant logic has been an amalgam of positions put forth by R. A. Fisher and by Neyman and Pearson, dating from the 1920s and 1930s. (This amalgam is one to which both Fisher and Neyman and Pearson would express deep reservations, but it has grown to be employed by many, particularly in the behavioral sciences.) We will discuss that approach first, but follow it by more recent conceptualizations that lead to roughly the same point, but do so in what many feel is a more logical and rational process. Third, and perhaps most importantly, what do we need to consider *in addition to* traditional hypothesis testing? Running a statistical test and declaring a difference to be statistically significant at "$p < .05$" is no longer sufficient. A hypothesis test can only suggest whether a relationship is reliable or it is not, or that a difference between two groups is likely to be the result of chance, or that it probably is not. In addition to running a hypothesis test, we need to tell our readers something about the difference itself, about confidence limits on that difference, and about the power of our test. This will involve a change in emphasis from earlier editions, and will affect how I describe results in the rest of the book. I think the basic conclusion is that simple hypothesis testing, no matter how you do it, is important, but it is not enough. If the debate has done nothing else, getting us to that point has been very important. You can see that we have a lot to cover, but once you understand the positions and the proposals, you will have a better grasp of the issues than most people in your field.

The American Psychological Association recently put together a task force to look at the general issue of hypothesis tests, and its report is now available (Wilkinson, 1999; see also http://www.loyola.edu/library/ref/articles/Wilkinson.pdf). Further discussion of this issue was included in an excellent paper by Nickerson (2000). These two documents do a very effective job of summarizing current thinking in the field. These recommendations have influenced the coverage of material in this book, and you will see more frequent references to confidence limits and effect size measures than you would have seen in previous editions.

The Traditional Approach to Hypothesis Testing

For the next several pages, we will consider the traditional treatment of hypothesis testing. This is the treatment that you will find in almost any statistics text and is something that you need to fully understand. The concepts here are central to what we mean by hypothesis testing, no matter who is speaking about it.

We have just been discussing sampling distributions, which lie at the heart of the treatment of research data. We do not go around obtaining sampling distributions, either mathematically or empirically, simply because they are interesting to look at. We have important reasons for doing so. The usual reason is that we want to test some hypothesis. Let's go back to the sampling distribution of differences in mean times that it takes people to leave a parking space. We want to test the hypothesis that the obtained difference between sample means could reasonably have arisen had we drawn our samples from populations with the same mean. This is another way of saying that we want to know whether the mean departure time when someone is waiting is different from the mean departure time when there is no one waiting. One way we can test such a hypothesis is to have some idea of the probability of obtaining a difference in sample means as extreme as 6.88 seconds *if* we actually sampled observations from populations with the same mean. The answer to this question is precisely what a sampling distribution is designed to provide.

Suppose we obtained (constructed) the sampling distribution plotted in Figure 4.1. Suppose further, for the sake of argument, that our sample mean difference was 1.5 seconds and that we then determined from the sampling distribution that the probability of a sample mean difference as high as 1.5 seconds is .16. (How we determine this probability is not important here.) Our reasoning could then go as follows: "If we did in fact sample from populations with the same mean, the probability of obtaining a sample mean difference as high as 1.5 seconds is .16—that is not a terribly high probability, but it certainly isn't a low probability event. Because a sample mean difference at least as great as 1.5 is often obtained from populations with equal means, we have no reason to doubt that our two samples came from such populations."

Alternatively, suppose we obtained a sample mean difference of 10 seconds and calculated from the sampling distribution that the probability of a sample mean difference as large as 10, when the population means are equal, was only .0008. Our argument could then go like this: "*If* we did obtain our samples from populations with equal means, the probability of obtaining a sample mean difference as large as 10 is only .0008—an unlikely event. Because a sample mean difference that large is unlikely to be obtained from such populations, we can reasonably conclude that these samples probably came from populations with different means."

It is important to realize the steps in this example because the logic is typical of most tests of hypotheses. The actual test consisted of several stages:

research hypothesis

1. We wanted to test the hypothesis, often called the **research hypothesis,** that people backing out of a parking space take longer when someone is waiting.

2. We obtained random samples of behaviors under the two conditions.

null hypothesis

3. We set up the hypothesis (called the **null hypothesis,** H_0) that the samples were drawn from populations with the same means. This hypothesis states that leaving times do not depend on whether someone is waiting.

4. We then obtained the sampling distribution of the differences between means under the assumption that H_0 (the null hypothesis) is true (i.e., we obtained the sampling distribution of the differences between means when the population means are equal.)

5. Given the sampling distribution, we calculated the probability of a mean difference *at least as large* as the one we actually obtained between the means of our two samples.

6. On the basis of that probability, we made a decision: either to reject or fail to reject H_0. Because H_0 states the means of the populations are equal, rejection of H_0 represents a belief that they are unequal, although the actual value of the difference in population means remains unspecified.

The preceding discussion is oversimplified in the sense that in practice we also would need to consider (either directly or by estimation) the value of σ^2, the population variance(s), and N, the sample size(s). But again, those are specifics we can deal with when the time comes. The logic of the approach is representative of the logic of most, if not all, statistical tests.

1. Begin with a research hypothesis.
2. Set up the null hypothesis.
3. Construct the sampling distribution of the particular statistic on the assumption that H_0 is true.
4. Collect some data.
5. Compare the sample statistic to that distribution.
6. Reject or retain H_0, depending on the probability, under H_0, of a sample statistic as extreme as the one we have obtained.

The First Stumbling Block

I probably slipped something past you there, and you need to at least notice. This is one of the very important issues that motivates the fight over hypothesis testing, and it is something that you need to understand even if you can't do much about it. What I imagine that you would like to know is "what is the probability that the null hypothesis (drivers don't take longer when people are waiting) is true *given* the data we obtained?" But that is not what I gave you, and it is not what I am going to give you in the future. I gave you the answer to a different question, which is "what is the probability that I would have obtained these data *given* that the null hypothesis is true?" I don't know how to give you an answer to the question you would like to answer—not because I am a terrible statistician, but because the answer is much too difficult in most situations. However, the answer that I did give you is still useful—and is used all the time. When the police ticket a driver for drunken driving because he can't drive in a straight line and can't speak coherently, they are saying, "*if he were sober, he would not behave this way. Because he behaves this way, we will conclude that he is not sober.*" This logic remains central to most approaches to hypothesis testing.

4.4 The Null Hypothesis

As we have seen, the concept of the null hypothesis plays a crucial role in the testing of hypotheses. People frequently are puzzled by the fact that we set up a hypothesis that is directly counter to what we hope to show. For example, if we hope to demonstrate the research hypothesis that college students do not come from a population with a mean self-confidence score of 100, we immediately set up the null hypothesis that they do. Or if we hope to demonstrate the validity of a research hypothesis that the means (μ_1 and μ_2) of the populations from which two samples are drawn are different, we state the null hypothesis that the population means are the same (or, equivalently, $\mu_1 - \mu_2 = 0$). (The term "null hypothesis" is most easily seen in this second example, in which it refers to the hypothesis that the difference between the two population means is zero, or *null*—some people call this the *nil null,* but that complicates the issue too much). We use the null hypothesis for several reasons. The philosophical argument, put forth by Fisher when he first introduced the concept, is that we can never prove something to be true, but we can prove something to be false. Observing 3,000 people with two arms does not prove the statement, "Everyone has two arms." However, finding one person with three arms does disprove the original statement beyond any shadow of a doubt. Although one might argue with Fisher's basic position—and many people have—the null hypothesis retains its dominant place in statistics.

A second and more practical reason for employing the null hypothesis is that it provides us with the starting point for any statistical test. Consider the case in which you want to show that the mean self-confidence score of college students is greater than 100. Suppose further that you were granted the privilege of proving the truth of some hypothesis. What hypothesis are you going to test? Should you test the hypothesis that $\mu = 101$, or maybe the hypothesis that $\mu = 112$, or how about $\mu = 113$? The point is that in almost all research in the behavioral sciences we do not have a *specific* **alternative** (research) **hypothesis** in mind, but without one we cannot construct the sampling distribution we need. (This was one of the arguments raised against the original approach of Neyman and Pearson because they often spoke as if there were a specific alternative hypothesis to be tested, rather than just the diffuse negation of the null.) However, if we start off by assuming H_0: $\mu = 100$, we can immediately set about obtaining the sampling distribution for $\mu = 100$ and then, if our data are convincing, reject that hypothesis and conclude that the mean score of college students is greater than 100, which is what we wanted to show in the first place.

alternative hypothesis

Statistical Conclusions

When the data differ markedly from what we would expect if the null hypothesis were true, we simply reject the null hypothesis and there is no particular disagreement about what our conclusions mean—we conclude that the null hypothesis is false. (This is not to suggest that we still don't need to tell our readers more about what we have found.) The interpretation is murkier and more problematic, however, when the data do not lead us to reject the null hypothesis. How are we to interpret a nonrejection? Shall we say that we have "proved" the null hypothesis to be true? Or shall we claim that we can "accept" the null, or that we shall "retain" it, or that we shall "withhold judgment"?

The problem of how to interpret a nonrejected null hypothesis has plagued students in statistics courses for more than 50 years, and it will probably continue to do so (but see Section 4.10). The idea that if something is not false then it must be true is too deeply ingrained in common sense to be dismissed lightly.

The one thing on which all statisticians agree is that we can never claim to have "proved" the null hypothesis. As was pointed out, the fact that the next 3,000 people we meet all have two arms certainly does not prove the null hypothesis that all people have two arms. In fact, we know that many perfectly normal people have fewer than two arms. Failure to reject the null hypothesis often means that we have not collected enough data.

The issue is easier to understand if we use a concrete example. Wagner, Compas, and Howell (1988) conducted a study to evaluate the effectiveness of a program for teaching high-school students to deal with stress. If this study found that students who participate in such a program had significantly fewer stress-related problems than did students in a control group who did not have the program, then we could, without much debate, conclude that the program was effective. However, if the groups did not differ at some predetermined level of statistical significance, what could we conclude?

We know we cannot conclude from a nonsignificant difference that we have proved that the mean of a population of scores of treatment subjects is the same as the mean of a population of scores of control subjects. The two treatments may lead to subtle differences that we were not able to identify conclusively with our relatively small sample of observations.

Fisher's position was that a nonsignificant result is an inconclusive result. For Fisher, the choice was between rejecting a null hypothesis and suspending judgment. He would have argued that a failure to find a significant difference between conditions could result from the fact that the students who participated in the program handled stress only *slightly* better than did control subjects, or that they handled it only slightly less well, or that there was no difference between the groups. For Fisher, a failure to reject H_0 merely means that our data are insufficient to allow us to choose among these three alternatives; therefore, we must suspend judgment. You will see this position return when we shortly discuss a proposal by Jones and Tukey (2000).

A slightly different approach was taken by Neyman and Pearson (1933), who took a much more pragmatic view of the results of an experiment. In our example, Neyman and Pearson would be concerned with the problem faced by the school board, who must decide whether to continue spending money on this stress-management program that we are providing for them. The school board would probably not be impressed if we told them that our study was inconclusive and then asked them to give us money to continue operating the program until we had sufficient data to state confidently whether or not the program was beneficial (or harmful). In the Neyman–Pearson position, one either rejects or *accepts* the null hypothesis. But when we say that we "accept" a null hypothesis, however, we do not mean that we take it to be proven as true. We simply mean that we will *act as if* it is true, at least until we have more adequate data. Whereas given a nonsignificant result, the ideal school board from Fisher's point of view would continue to support the program until we finally were able to make up our minds, the school board with a Neyman–Pearson perspective would conclude that the available evidence is not sufficient to defend continuing to fund the program, and they would cut off our funding.

This discussion of the Neyman–Pearson position has been much oversimplified, but it contains the central issue of their point of view. The debate between Fisher on the one hand and Neyman and Pearson on the other was a lively (and rarely civil) one, and present practice contains elements of both viewpoints. Most statisticians prefer to use phrases such as "retain the null hypothesis" and "fail to reject the null hypothesis" because these make clear the tentative nature of a nonrejection. These phrases have a certain Fisherian ring to them. On the other hand, the important emphasis on Type II errors (failing to reject a *false* null hypothesis), which we will discuss in Section 4.7, is clearly an essential feature of the Neyman–Pearson school. If you are going to choose between two alternatives (accept or reject), then you have to be concerned with the probability of falsely accepting as well as that of falsely rejecting the null hypothesis. Fisher would never accept a null hypothesis in the first place, so he did not need to worry much about the probability of accepting a false one.[1] We will return to this whole question in Section 4.10, where we will consider an alternative approach, after we have developed several other points. First, however, we need to consider some basic information about hypothesis testing so as to have a vocabulary and an example with which to go further into hypothesis testing. This information is central to any discussion of hypothesis testing under any of the models that have been proposed.

4.5 Test Statistics and Their Sampling Distributions

We have been discussing the sampling distribution of the mean, but the discussion would have been essentially the same had we dealt instead with the median, the variance, the range, the correlation coefficient (as in our course evaluation example), proportions (as in our horn-honking example), or any other statistic you care to consider. (Technically, the

---

[1] Excellent discussions of the differences between the theories of Fisher on the one hand, and Neyman and Pearson on the other can be found in Chapter 4 of Gigerenzer, Swijtink, Porter, Daston, Beatty, & Krüger (1989), Lehman (1993), and Oakes (1990). The central issues involve the concept of probability, the idea of an infinite population or infinite resampling, and the choice of a critical value, among other things. The controversy is far from a simple one.

shapes of these distributions would be different, but I am deliberately ignoring such issues in this chapter.) The statistics just mentioned usually are referred to as **sample statistics** because they describe characteristics of samples. There is a whole different class of statistics called **test statistics,** which are associated with specific statistical procedures and which have their own sampling distributions. Test statistics are statistics such as t, F, and χ^2, which you may have run across in the past. (If you are not familiar with them, don't worry—we will consider them separately in later chapters.) This is not the place to go into a detailed explanation of any test statistics. I put this chapter where it is because I don't want readers to think that they are supposed to worry about technical issues. This chapter is the place, however, to point out that the sampling distributions for test statistics are obtained and used in essentially the same way as the sampling distribution of the mean.

As an illustration, consider the sampling distribution of the statistic t, which will be discussed in Chapter 7. For those who have never heard of the t test, it is sufficient to say that the t test is often used, among other things, to determine whether two samples were drawn from populations with the same means. Let μ_1 and μ_2 represent the means of the populations from which the two samples were drawn. The null hypothesis is the hypothesis that the two population means are equal, in other words, H_0: $\mu_1 = \mu_2$ (or $\mu_1 - \mu_2 = 0$). If we were extremely patient, we could empirically obtain the sampling distribution of t when H_0 is true by drawing an infinite number of pairs of samples, all from two identical populations, calculating t for each pair of samples (by methods to be discussed later), and plotting the resulting values of t. In that case, H_0 must be true because we forced it to be true by drawing the samples from identical populations. The resulting distribution is the sampling distribution of t when H_0 is true. If we later had two samples that produced a particular value of t, we would test the null hypothesis by comparing our sample t to the sampling distribution of t. We would reject the null hypothesis if our obtained t did not look like the kinds of t values that the sampling distribution told us to expect when the null hypothesis is true.

I could rewrite the preceding paragraph, substituting χ^2, or F, or any other test statistic in place of t, with only minor changes dealing with how the statistic is calculated. Thus, you can see that all sampling distributions can be obtained in basically the same way (calculate and plot an infinite number of statistics by sampling from identical populations).

4.6 Using the Normal Distribution to Test Hypotheses

Much of the discussion so far has dealt with statistical procedures that you do not yet know how to use. I did this deliberately to emphasize the point that the logic and the calculations behind a test are two separate issues. However, we now can use what you already know about the normal distribution to test some simple hypotheses. In the process, we can deal with several fundamental issues that are more easily seen by use of a concrete example.

An important use of the normal distribution is to test hypotheses, either about individual observations or about sample statistics such as the mean. In this chapter, we will deal with individual observations, leaving the question of testing sample statistics until later chapters. Note, however, that in the usual case we test hypotheses about sample statistics such as the mean rather than about individual observations. I am starting with an example of an individual observation because the explanation is somewhat clearer. Because we are dealing with only single observations, the sampling distribution invoked here will be the distribution of individual scores (rather than the distribution of means or differences between means). The basic logic is the same, and we are using an example of individual scores only because it simplifies the explanation and is something with which you have had experience.

Psychologists who study neurological functioning have a battery of tests at their disposal. A common test is simple finger tapping speed, which is useful for diagnosing hidden brain damage. (For example, people with brain damage to the dorsal lateral frontal lobes are especially slow in the speed of finger tapping, but are often unaware of their loss of behavioral competency.) For a simple example, assume we know that the mean rate of finger tapping of normal healthy adults is 100 taps in 20 seconds, with a standard deviation of 20, and that tapping speeds are normally distributed in the population. We already know that the tapping rate is slower among people with dorsal lateral frontal lobe damage. Suppose that an individual has just been sent to us who taps at a rate of 70 taps in 20 seconds. Is his score sufficiently below the mean for us to assume that he did not come from a population of neurologically healthy people? This situation is diagrammed in Figure 4.2, in which the arrow indicates the location of our piece of data (the person's score).

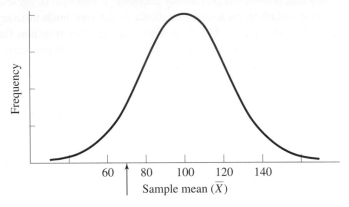

Figure 4.2   Location of a person's tapping score on a distribution of scores of neurologically healthy people

The logic of the solution to this problem is the same as the logic of hypothesis testing in general. We begin by assuming that the individual's score *does* come from the population of healthy scores. This is the null hypothesis (H_0). If H_0 is true, we automatically know the mean and the standard deviation of the population from which he was supposedly drawn (100 and 20, respectively). With this information, we are in a position to calculate the probability that a score *as low as* his would be obtained from this population. If the probability is very low, we can reject H_0 and conclude that he did not come from the healthy population. Conversely, if the probability is not particularly low, then the data represent a reasonable result under H_0, and we would have no reason to doubt its validity and thus no reason to doubt that the person is healthy. Keep in mind that we are not interested in the probability of a score *equal* to 70 (which, because the distribution is continuous, would be infinitely small) but, rather, in the probability that the score would be at least as low as (i.e., less than or equal to) 70.

The individual had a score of 70. We want to know the probability of obtaining a score *at least as low as* 70 if H_0 is true. We already know how to find this—it is the area below 70 in Figure 4.2. All we have to do is convert 70 to a z score and then refer to Appendix z (page 694).

$$z = \frac{X - \mu}{\sigma} = \frac{70 - 100}{20} = \frac{-30}{20} = -1.5$$

From Appendix z, we can see that the probability of a z score of -1.5 or below is .0668. (Locate $z = 1.50$ in the table and then read across to the column headed "Smaller Portion.")

decision-making

At this point, we have to become involved in the **decision-making** aspects of hypothesis testing. We must decide whether an event with a probability of .0668 is sufficiently unlikely to cause us to reject H_0. Here we will fall back on arbitrary conventions that have been established over the years. The rationale for these conventions will become clearer as we go along, but for the time being keep in mind that they are merely conventions. One convention calls for rejecting H_0 if the probability under H_0 is less than or equal to .05 ($p \leq .05$), and another convention—one that is more conservative with respect to the probability of rejecting H_0—calls for rejecting H_0 whenever the probability under H_0 is less than or equal to .01. These values of .05 and .01 are often referred to as the **rejection level,** or the **significance level,** of the test. (When we say that a difference is statistically significant at the .05 level, we mean that a difference that large would occur less than 5% of the time if the null were true.) Whenever the probability obtained under H_0 is less than or equal to our predetermined significance level, we will reject H_0. Another way of stating this is to say that any outcome whose probability under H_0 is less than or equal to the significance level falls in the **rejection region** because such an outcome leads us to reject H_0.

rejection level (significance level)

rejection region

For the purpose of setting a standard level of rejection for this book, we will use the .05 level of statistical significance, keeping in mind that some people would consider this level to be too lenient.[2] For our particular example, we have obtained a probability value of $p = .0668$, which obviously is greater than .05. Because we have specified that we will not reject H_0 unless the probability of the data under H_0 is less than .05, we must conclude that we have no reason to decide that the person did not come from a population of healthy people.

More specifically, we conclude that a finger-tapping rate of 70 reasonably could have come from a population of scores with a mean equal to 100 and a standard deviation equal to 20. It is important to note that we have not shown that this person is healthy, but only that we have insufficient reason to believe that he is not. It may be that he is just acquiring the disease and therefore is not quite as different from normal as is usual for his condition. Or maybe he has the disease at an advanced stage but just happens to be an unusually fast tapper. This is an example of the fact that we can never say that we have proved the null hypothesis. We can conclude only that this person does not tap sufficiently slowly for an illness, if any, to be statistically detectable.

4.7 Type I and Type II Errors

Whenever we reach a decision with a statistical test, there is always a chance that our decision is the wrong one. Although this is true of almost all decisions, statistical or otherwise, the statistician has one point in her favor that other decision makers normally lack. She not only makes a decision by some rational process, but she can also specify the conditional probabilities of a decision's being in error. In everyday life, we make decisions with only subjective feelings about what is probably the right choice. The statistician, however, can state quite precisely the probability that she would make an erroneously rejection of H_0 if it were true. This ability to specify the probability of erroneously rejecting a true H_0 follows directly from the logic of hypothesis testing.

---

[2] The particular view of hypothesis testing described here is the classical one that a null hypothesis is rejected if the probability of obtained the data when the null hypothesis is true is less than the predefined significance level, and not rejected if that probability is greater than the significance level. Currently, a substantial body of opinion holds that such cut-and-dried rules are inappropriate and that more attention should be paid to the probability value itself. In other words, the classical approach (using a .05 rejection level) would declare $p = .051$ and $p = .150$ to be (equally) "statistically nonsignificant" and $p = .048$ and $p = .0003$ to be (equally) "statistically significant." The alternative view would think of $p = .051$ as "nearly significant" and $p = .0003$ as "very significant." Although this view has much to recommend it, especially given current trends to move away from only reporting statistical significance of results, it will not be wholeheartedly adopted here. Most computer programs do print out exact probability levels, and those values, when interpreted judiciously, can be useful. The difficulty comes in defining what is meant by "interpreted judiciously."

Consider the finger-tapping example, this time ignoring the score of the individual sent to us. The situation is diagrammed in Figure 4.3, in which the distribution is the distribution of scores from healthy subjects, and the shaded portion represents the lowest 5% of the distribution. The actual score that cuts off the lowest 5% is called the **critical value.** Critical values are those values of X (the variable) that describe the boundary or boundaries of the rejection region(s). For this particular example, the critical value is 67.

critical value

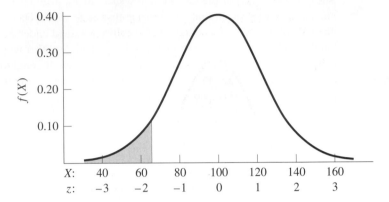

Figure 4.3 Lowest 5% of scores from clinically healthy people

If we have a decision rule that says to reject H_0 whenever an outcome falls in the lowest 5% of the distribution, we will reject H_0 whenever an individual's score falls in the shaded area; that is, whenever a score as low as his has a probability of .05 or less of coming from the population of healthy scores. Yet by the very nature of our procedure, 5% of the scores from perfectly healthy people will themselves fall in the shaded portion. Thus, if we actually have sampled a person who is healthy, we stand a 5% chance of his score being in the shaded tail of the distribution, causing us erroneously to reject the null hypothesis. This kind of error (rejecting H_0 when it is actually true) is called a **Type I error,** and its conditional probability (the probability of rejecting the null hypothesis given that it is true) is designated as α **(alpha)**, the size of the rejection region. In the future, whenever we represent a probability by α, we will be referring to the probability of a Type I error.

Type I error

α (alpha)

Keep in mind the "conditional" nature of the probability of a Type I error. I know that sounds like jargon, but what it means is that you should be sure you understand that when we speak of a Type I error we mean the probability of rejecting H_0 *given that it is true*. We are not saying that we will reject H_0 on 5% of the hypotheses we test. We would hope to run experiments on important and meaningful variables and, therefore, to reject H_0 often. But when we speak of a Type I error, we are speaking only about rejecting H_0 in those situations in which the null hypothesis happens to be true.

You might feel that a 5% chance of making an error is too great a risk to take and suggest that we make our criterion much more stringent, by rejecting, for example, only the lowest 1% of the distribution. This procedure is perfectly legitimate, but realize that the more stringent you make your criterion, the more likely you are to make another kind of error—failing to reject H_0 when it is false and H_1 is true. This type of error is called a **Type II error,** and its probability is symbolized by β **(beta)**.

Type II error

β (beta)

The major difficulty of Type II errors stems from the fact that if H_0 is false, we almost never know what the true distribution (the distribution under H_1) would look like for the population from which our data came. We know only the distribution of scores under H_0. Put in the present context, we know the distribution of scores from healthy people but not from nonhealthy people. It may be that people suffering from some neurological disease tap, on average, considerably more slowly than healthy people, or it may be that they tap, on

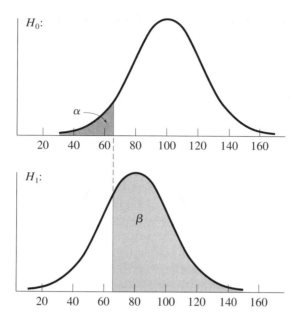

Figure 4.4   Areas corresponding to α and β for tapping speed example

average, only a little more slowly. This situation is illustrated in Figure 4.4, in which the distribution labeled H_0 represents the distribution of scores from healthy people (the set of observations expected under the null hypothesis), and the distribution labeled H_1 represents our hypothetical distribution of nonhealthy scores (the distribution under H_1). Remember that the curve H_1 is only hypothetical. We really do not know the location of the nonhealthy distribution, other than that it is lower (slower speeds) than the distribution of H_0. (I have arbitrarily drawn that distribution with a mean of 80 and a standard deviation of 20.)

The darkly shaded portion in the top half of Figure 4.4 represents the rejection region. Any observation falling in that area (i.e., to the left of about 67) would lead to rejection of the null hypothesis. If the null hypothesis is true, we know that our observation will fall in this area 5% of the time. Thus, we will make a Type I error 5% of the time.

The lightly shaded portion in the bottom half of Figure 4.4 represents the probability (β) of a Type II error. This is the situation of a person who was actually drawn from the nonhealthy population but whose score was not sufficiently low to cause us to reject H_0.

In the particular situation illustrated in Figure 4.4, we can actually calculate β by using the normal distribution to calculate the probability of obtaining a score *greater than* 67 (the critical value) if $\mu = 80$ and $\sigma = 20$. The actual calculation is not important for your understanding of β; because this chapter was designed specifically to avoid calculation, I will simply state that this probability (i.e., the area labeled β) is .74. Thus, for this example, 74% of the time when we have a person who is actually nonhealthy (i.e., H_1 is actually true), we will make a Type II error by failing to reject H_0 when it is false (as medical diagnosticians, we leave a lot to be desired).

From Figure 4.4, you can see that if we were to reduce the level of α (the probability of a Type I error) from .05 to .01 by moving the rejection region to the left, it would reduce the probability of Type I errors but would increase the probability of Type II errors. Setting α at .01 would mean that $\beta = .908$. You can see that there is room for debate about what level of significance to use. The decision rests primarily on your opinion concerning the relative importance of Type I and Type II errors for the kind of study you are conducting. If it were important to avoid Type I errors (such as telling someone that he has a disease when he

Table 4.1   Possible outcomes of the decision-making process

	True State of the World	
Decision	H_0 True	H_0 False
Reject H_0	Type I error $p = \alpha$	Correct decision $p = 1 - \beta$ = Power
Don't reject H_0	Correct decision $p = 1 - \alpha$	Type II error $p = \beta$

does not), then you would set a stringent (i.e., small) level of α. If, on the other hand, you want to avoid Type II errors (telling someone to go home and take an aspirin when in fact he needs immediate treatment), you might set a fairly high level of α. (Setting $\alpha = .20$ in this example would reduce β to .44.) Unfortunately, in practice most people choose an arbitrary level of α, such as .05 or .01, and simply ignore β. In many cases, this may be all you can do. (You will probably use the alpha level that your instructor recommends.) In other cases, however, there is much more you can do, as you will see in Chapter 8.

I should stress again that Figure 4.4 is purely hypothetical. I was able to draw the figure only because I arbitrarily decided that speeds of nonhealthy people were normally distributed with a mean of 80 and a standard deviation of 20. The calculated answers would be different if I had chosen to draw it with a mean of 70 or a standard deviation of 10. In most everyday situations, we do not know the mean and the variance of that distribution and can make only educated guesses, thus providing only crude estimates of β. In practice, we can select a value of μ under H_1 that represents the *minimum* difference we would like to be able to detect because larger differences will have even smaller βs.

From this discussion of Type I and Type II errors, we can summarize the decision-making process with a simple table. Table 4.1 presents the four possible outcomes of an experiment. The items in this table should be self-explanatory, but there is one concept—power—that we have not yet discussed. The **power** of a test is the probability of rejecting H_0 when it is actually false. Because the probability of *failing* to reject a false H_0 is β, then power must equal $1 - \beta$. Those who want to know more about power and its calculation will find power covered in Chapter 8.

power

4.8   One- and Two-Tailed Tests

The preceding discussion brings us to a consideration of one- and two-tailed tests. In our tapping example, we knew that nonhealthy subjects tapped more slowly than healthy subjects; therefore, we decided to reject H_0 only if a subject tapped too slowly. However, suppose our subject had tapped 180 times in 20 seconds. Although this is an exceedingly unlikely event to observe from a healthy subject, it did not fall in the rejection region, which consisted *solely* of low rates. As a result, we find ourselves in the position of not rejecting H_0 in the face of a piece of data that is very unlikely, but not in the direction expected.

The question then arises as to how we can protect ourselves against this type of situation (if protection is thought necessary). The answer is to specify before we run the experiment that we are going to reject a given percentage (say 5%) of the *extreme* outcomes, both those that are extremely high and those that are extremely low. But if we reject the lowest 5% and the highest 5%, then we would reject H_0 a total of 10% of the time when it is actually true, that is, $\alpha = .10$. We are rarely willing to work with α as high as .10 and prefer to see it set no higher than .05. The way to accomplish this is to reject the lowest 2.5% and the highest 2.5%, making a total of 5%.

**one-tailed
(directional) test**

**two-tailed
(nondirectional)
test**

The situation in which we reject H_0 for only the lowest (or only the highest) tapping speeds is referred to as a **one-tailed,** or **directional, test.** We make a prediction of the direction in which the individual will differ from the mean, and our rejection region is located in only one tail of the distribution. (That makes sense when we know that brain damage is only associated with slow tapping speeds.) When we reject extremes in both tails, we have what is called a **two-tailed,** or **nondirectional, test.** It is important to keep in mind that although we gain something with a two-tailed test (the ability to reject the null hypothesis for extreme scores in either direction), we also lose something. A score that would fall in the 5% rejection region of a one-tailed test may not fall in the rejection region of the corresponding two-tailed test because now we reject only 2.5% in each tail.

In the finger-tapping example, the decision between a one- and a two-tailed test might seem reasonably clear-cut. We know that people with a given disease tap more slowly; therefore, we care only about rejecting H_0 for low scores—high scores have no diagnostic importance. In other situations, however, we do not know which tail of the distribution is important (or both are), and we need to guard against extremes in either tail. The situation might arise when we are considering a campaign to persuade children not to start smoking. We might find that the campaign leads to a decrease in the incidence of smoking. Or, we might find that campaigns run by adults to persuade children not to smoke simply make smoking more attractive and exciting, leading to an increase is the number of children smoking. In either case, we would want to reject H_0.

In general, two-tailed tests are far more common than one-tailed tests for several reasons. The investigator may have no idea what the data will look like and therefore has to be prepared for any eventuality. Although this situation is rare, it does occur in some exploratory work.

Another common reason for preferring two-tailed tests is that the investigators are reasonably sure the data will come out one way but want to cover themselves in the event that they are wrong. This type of situation arises more often than you might think. (Carefully formed hypotheses have an annoying habit of being phrased in the wrong direction, for reasons that seem so obvious after the event.) The smoking example is a case in point, where there is some evidence that poorly contrived antismoking campaigns actually do more harm than good. A frequent question that arises when the data may come out the other way around is, "Why not plan to run a one-tailed test and then, if the data come out the other way, just change the test to a two-tailed test?" This kind of question comes from people who have no intention of being devious but who just do not fully understand the logic of hypothesis testing. If you start an experiment with the extreme 5% of the left-hand tail as your rejection region and then turn around and reject any outcome that happens to fall in the extreme 2.5% of the right-hand tail, you are working at the 7.5% level. In that situation, you will reject 5% of the outcomes in one direction (assuming that the data fall in the desired tail), and you are willing also to reject 2.5% of the outcomes in the other direction (when the data are in the unexpected direction). There is no denying that $5\% + 2.5\% = 7.5\%$. To put it another way, would you be willing to flip a coin for an ice cream cone if I have chosen "heads" but also reserve the right to switch to "tails" after I see how the coin lands? Or would you think it fair of me to shout, "Two out of three!" when the coin toss comes up in your favor? You would object to both of these strategies, and you should. For the same reason, the choice between a one-tailed test and a two-tailed one is made *before* the data are collected. It is also one of the reasons that two-tailed tests are usually chosen.

Although the preceding discussion argues in favor of two-tailed tests, and although in this book we generally confine ourselves to such procedures, there are no hard-and-fast rules. The final decision depends on what you already know about the relative severity of different kinds of errors. It is important to remember that with respect to a given tail of a distribution, the difference between a one-tailed test and a two-tailed test is that the latter just uses a different cutoff. A two-tailed test at $\alpha = .05$ is more liberal than a one-tailed test at $\alpha = .01$.[3]

If you have a sound grasp of the logic of testing hypotheses by use of sampling distributions, the remainder of this course will be relatively simple. For any new statistic you encounter, you will need to ask only two basic questions:

1. How and with which assumptions is the statistic calculated?
2. What does the statistic's sampling distribution look like under H_0?

If you know the answers to these two questions, you can accomplish your test by calculating the test statistic for the data at hand and comparing the statistic to the sampling distribution. Because the relevant sampling distributions are tabled in the appendices, all you really need to know is which test is appropriate for a particular situation and how to calculate its test statistic. (Of course, there is more to statistics than just hypothesis testing, so perhaps I'm doing a bit of overselling here. There is a great deal to understanding the field of statistics beyond how to calculate, and evaluate, a specific statistical test. Calculation is the easy part, especially with modern computer software.)

4.9 What Does It Mean to Reject the Null Hypothesis?

conditional
probabilities

One of the common problems that even well-trained researchers have with the null hypothesis is the confusion over what rejection really means. I mentioned this earlier when I discussed the fact that we calculate the probability of the data given that the null is true, rather than the probability of the null being true given the data. Suppose that we test a null hypothesis about the difference between two population means and reject it at $p = .045$. There is a temptation to say that such a result means that the probability of the null being true is .045. But that is *not* what this probability means. What we have shown is that *if the null hypothesis were true,* the probability of obtaining a difference between means as great as the difference we found is only .045. That is quite different from saying that the probability that the null is true is .045. What we are doing here is confusing the probability of the hypothesis given the data, and the probability of the data given the hypothesis. These are called **conditional probabilities,** and will be discussed in Chapter 5. The probability of .045 that we have here is the probability of the data given that H_0 is true, written $p(D \mid H_0)$— the vertical line is read "given." It is not the probability that H_0 is true given the data, written $p(H_0 \mid D)$. The best discussion of this issue that I have read is in an excellent paper by Nickerson (2000). Let me illustrate my major point with an example.

---

[3] One of the reviewers of an earlier edition of this book made the case for two-tailed tests even more strongly: "It is my (minority) belief that what an investigator *expects to be true* has absolutely no bearing *whatsoever* on the issue of one- versus two-tailed tests. Nature couldn't care less what psychologists' theories predict, and will often show patterns/trends in the opposite direction. Since our goal is to know the truth (not to prove we are astute at predicting), our tests must always allow for testing *both* directions. I say *always* do two-tailed tests, and if you are worried about β, jack the sample size up a bit to offset the loss in power" (D. Bradley, personal communication, 1983). I am personally inclined toward this point of view. Nature is notoriously fickle, or else we are notoriously inept at prediction. On the other hand, a second reviewer takes exception to this position. While acknowledging that Bradley's point is well considered, Rodgers, engaging in a bit of hyperbole, argues, "To generate a theory about how the world works that implies an expected direction of an effect, but then to hedge one's bet by putting some (up to ½) of the rejection region in the tail other than that predicted by the theory, strikes me as both scientifically dumb and slightly unethical. . . . Theory generation and theory testing are much closer to the proper goal of science than truth searching, and running one-tailed tests is quite consistent with those goals" (J. Rodgers, personal communication, 1986). Neither Bradley nor I would accept the judgment of being "scientifically dumb and slightly unethical," but I presented the two positions in juxtaposition because doing so gives you a flavor of the debate. Obviously there is room for disagreement on this issue.

Suppose that I create a computer-generated example where I know for a fact that the data for one sample came from a population with a mean of 54.28, and the data for a second sample came from a population with a mean of 54.25. (It is very easy to use a program like SPSS to generate such samples.) Here I *know for a fact* that the null hypothesis is false. In other words, the probability that the null hypothesis is true is .00—that is, $p(H_0) = .00$. However, if I have two small samples I might happen to get a result such as 54.26 and 54.36, and that result would have a very high probability of occurring even in the situation where the null hypothesis is true and both means were, say, 54.28. Thus, the probability of the data given a true null hypothesis might be .75, for example, and yet we know that the probability that the null is really true is exactly .00. (Using probability terminology, we can write $p(H_0) = .00$ and $p(D \mid H_0) = .75$). Alternatively, assume that I created a situation where I know that the null is true. For example, I set up populations where both means are 54.00. It is easy to imagine getting samples with means of 53 and 54.5. If the null is really true, the probability of getting means this difference may be .33, for example. Thus, the probability that the null is true is fixed, by me, at 1.00, yet the probability of the data when the null is true is .33. (Using probability terminology again, we can write $p(H_0) = 1.00$ and $p(D \mid H_0) = .33$.) Notice that in both of these cases there is a serious discrepancy between the probability of the null being true and the probability of the data given the null. You will see several instances like this throughout the book whenever I sample data from known populations. Never confuse the probability value associated with a test of significance with the probability that the null hypothesis is true. They are very different things.

4.10   An Alternative View of Hypothesis Testing

What I have presented so far about hypothesis testing is the traditional approach. It is found in virtually every statistics text, and you need to be very familiar with it. However, there has recently been an interest in different ways of looking at hypothesis testing, and a new approach proposed by Lyle Jones and John Tukey (2000) avoids some of the problems of the traditional approach.

We will begin with an example comparing two population means that is developed further in Chapter 7. Adams, Wright, and Lohr (1996) showed a group of homophobic heterosexual males and a group of nonhomophobic heterosexual males a videotape of sexually explicit erotic homosexual images, and recorded the resulting level of sexual arousal in the participants. The researchers were interested in seeing whether there was a difference in sexual arousal between the two categories of viewers. (Notice that I didn't say which group they expected to come out with the higher mean, just that there would be a difference.)

The traditional hypothesis testing approach would set up the null hypothesis that $\mu_h = \mu_n$, where μ_h is the population mean for homophobic males, and μ_n is the population mean for nonhomophobic males. The traditional alternative (two-tailed) hypothesis is that $\mu_h \neq \mu_v$. Many people have pointed out that the null hypothesis in such a situation is never going to be true. It is not reasonable to believe that if we had a population of all homophobic males their mean would be exactly equal to the mean of the population all nonhomophobic males to an unlimited number of decimal places. Whatever the means are, they will certainly differ by *at least* some trivial amount. So we know before we begin that the null hypothesis is false, and we might ask ourselves why we are testing the null in the first place. (Many people have asked that question.)

Jones and Tukey (2000) and Harris (2005) have argued that we really have three possible hypotheses or conclusions we could draw—Jones and Tukey speak primarily of "conclusions." One is that $\mu_h < \mu_n$, another is that $\mu_h > \mu_n$, and the third is that $\mu_h = \mu_n$. This third hypothesis is the traditional null hypothesis, and we have just said that it is never going to be true when means are carried to enough decimal places. These three hypotheses lead to three courses of action. If we test the first ($\mu_h < \mu_n$) and reject it, we conclude that homophobic males are more aroused than nonhomophobic males. If we test the second ($\mu_h > \mu_n$) and reject it, we conclude that homophobic males are less aroused than nonhomophobic males. If we cannot reject either of those hypotheses, we conclude that we have insufficient evidence to make a choice—the population means are almost certainly different, but we don't know which is the larger.

The difference between this approach and the traditional one may seem minor, but it is important. In the first place, when Jones and Tukey tell us something, we should definitely listen. These are not two guys who just got out of graduate school—they are two very highly respected statisticians. (If there were a Nobel Prize in statistics, John Tukey would have won it.) In the second place, this approach acknowledges that the null is never strictly true, but that sometimes the data do not allow us to draw conclusions about which mean is larger. So, instead of relying on fuzzy phrases like "fail to reject the null hypothesis" or "retain the null hypothesis," we simply do away with the whole idea of a null hypothesis and just conclude, "we can't decide whether μ_h is greater than μ_n, or is less than μ_n." In the third place, this looks as if we are running two one-tailed tests, but with an important difference. In a traditional one-tailed test, we must specify *in advance* which tail we are testing. If the result falls in the extreme of that tail, we reject the null and declare that $\mu_h < \mu_n$, for example. If the result does not fall in that tail, we must not reject the null, no matter how extreme it is in the other tail. But that is not what Jones and Tukey are suggesting. They do not require you to specify the direction of the difference before you begin.

Jones and Tukey are suggesting that we do not specify a tail in advance, but that we collect our data and determine whether the result is extreme in either tail. If it is extreme in the lower tail, we conclude that $\mu_h < \mu_n$. If it is extreme in the upper tail, we conclude that $\mu_h > \mu_n$. And if neither of those conditions applies, we declare that the data are insufficient to make a choice. (Notice that I didn't once use the word "reject" in the last few sentences. I said "conclude." The difference is subtle, but I think that it is important.)

But Jones and Tukey go a bit further and alter the significance level. First, we know that the probability that the null is true is .00. (In other words, $p(\mu_h = \mu_n) = 0$.) The difference may be trivially small, but there is a difference nonetheless. We cannot make an error by not rejecting the null because saying that we don't have enough evidence is not the same as incorrectly rejecting a hypothesis. As Jones and Tukey wrote,

> With this formulation, a conclusion is in error only when it is "a reversal," when it asserts one direction while the (unknown) truth is in the other direction. Asserting that the direction is not yet established may constitute a wasted opportunity, but it is not an error. We want to control the rate of error, the reversal rate, while minimizing wasted opportunity, that is, while minimizing indefinite results. (p. 412)

So one of two things is true—either $\mu_h > \mu_n$ or $\mu_h < \mu_n$. If $\mu_h > \mu_n$ is actually true, meaning that homophobic males are more aroused by homosexual videos, then the only error we can make is to erroneously conclude the reverse—that $\mu_h < \mu_n$. And the probability of that error is, at most, .025 if we were to use the traditional two-tailed test with 2.5% of the area in each tail. If, on the other hand, $\mu_h < \mu_n$, the only error we can make is to conclude that $\mu_h > \mu_n$, the probability of which is also at most .025. Thus, if we use the traditional cutoffs of a two-tailed test, the probability of a Type I error is at most .025. Jones and Tukey go on to suggest that we could use the cutoffs corresponding to 5% in each tail (the traditional two-tailed test at $\alpha = .10$) and still have only a 5% chance of making a Type I error. Although this is true, I think that you will find that many traditionally trained colleagues, including journal reviewers, will start getting a bit "squirrelly" at this point, and you might not want to push your luck.

I wouldn't be surprised if at this point students are throwing up their hands with one of two objections. First would be the claim that we are just "splitting hairs." My answer to that is "no, we're not." These issues have been hotly debated in the literature, with some people arguing that we abandon hypothesis testing altogether (Hunter, 1997). The Jones–Tukey formulations make sense of hypothesis testing and increase statistical power if you follow all their suggestions. (I believe that they would prefer the phrase "drawing conclusions" to "hypothesis testing.") Second, students could very well be asking why I spent many pages laying out the traditional approach and then another page or two saying why it is all wrong. I tried to answer that at the beginning—the traditional approach is so ingrained in what we do that you cannot possibly get by without understanding it. It will lie behind most of the studies you read, and your colleagues will expect that you understand it. That there is an alternative, and better, approach does not release you from the need to understand the traditional approach. And unless you change α levels, as Jones and Tukey recommend, you will be doing almost the same things but coming to more sensible conclusions.

4.11 Effect Size

Earlier in the chapter I mentioned that there was a movement afoot to go beyond simple significance testing to report some measure of the size of an effect. In fact, some professional journals are already insisting on it. I will expand on this topic in some detail later, but it is worth noting here that I have already sneaked a measure of effect size past you, and I'll bet that nobody noticed. When writing about waiting for parking spaces to open up, I pointed out that Ruback and Juieng (1997) found a difference of 6.88 seconds, which is not trivial when you are the one doing the waiting. I could have gone a step further and pointed out that, because the standard deviation of waiting times was 14.6 seconds, we are seeing a difference of nearly half a standard deviation. Expressing the difference between waiting times in terms of the actual number of seconds or as being "more than half a standard deviation" provides a measure of how large the effect was—and a very reputable measure. There is much more to be said about effect sizes, but at least this gives you some idea of what we are talking about.

I should say one more thing on this topic. One of the difficulties in understanding the debates about hypothesis testing is that for years statisticians have been very sloppy in selecting their terminology. Thus, for example, in rejecting the null hypothesis, it is very common for researchers to report that they have found a "significant difference." Most readers could be excused for taking this to mean that the study has found an "important difference," but that is not at all what is meant. When statisticians and researchers say "significant," that is shorthand for "statistically significant." It merely means that the difference, even if trivial, is not due to chance. The recent emphasis on effect sizes is intended to go beyond statements about chance, and tell the reader something, though perhaps not much, about "importance." I will try in this book to insert the word "statistical" before "significant," when that is what I mean, but I can't promise to always remember.

4.12 A Final Worked Example

A number of years ago the mean on the verbal section of the Graduate Record Exam (GRE) was 489 with a standard deviation of 126. The statistics were based on all students taking the exam in that year, most of whom were native speakers of English. Suppose we have an application from an individual with a Chinese name who scored particularly low (e.g., 220). If this individual were a native speaker of English, that score would be sufficiently low for us to question his suitability for graduate school unless the rest of the documentation is considerably better. If, however, this student were not a native speaker of English, we would probably disregard the low score entirely, on the grounds that it is a poor reflection of his abilities.

I will stick with the traditional approach to hypothesis testing in what follows, though you should be able to see the difference between this and the Jones and Tukey approach. We have two possible choices here, namely that the individual is or is not a native speaker of English. If he is a native speaker, we know the mean and the standard deviation of the population from which his score was sampled: 489 and 126, respectively. If he is not a native speaker, we have no idea what the mean and the standard deviation are for the population from which his score was sampled. To help us to draw a reasonable conclusion about this person's status, we will set up the null hypothesis that this individual is a native speaker, or, more precisely, he was drawn from a population with a mean of 489; $H_0 : \mu = 489$. We will identify H_1 with the hypothesis that the individual is not a native speaker ($\mu \neq 489$). (Note that Jones and Tukey would [simultaneously] test $H_1: \mu < 489$ and $H_2: \mu > 489$, and would associate the null hypothesis with the conclusion that we don't have sufficient data to make a decision.)

For the traditional approach we now need to choose between a one-tailed and a two-tailed test. In this particular case, we will choose a one-tailed test on the grounds that the GRE is given in English, and it is difficult to imagine that a population of nonnative speakers would have a mean higher than the mean of native speakers of English on a test that is given in English. (*Note:* This does not mean that non-English speakers may not, singly or as a population, outscore English speakers on a fairly administered test. It just means that they are unlikely to do so, especially as a group, when both groups take the test in English.) Because we have chosen a one-tailed test, we have set up the alternative hypothesis as $H_1: \mu < 489$.

Before we can apply our statistical procedures to the data at hand, we must make one additional decision. We have to decide on a level of significance for our test. In this case, I have chosen to run the test at the 5% level, instead of at the 1% level, because I am using $\alpha = .05$ as a standard for this book and also because I am more worried about a Type II error than I am about a Type I error. If I make a Type I error and erroneously conclude that the student is not a native speaker when in fact he is, it is very likely that the rest of his credentials will exclude him from further consideration anyway. If I make a Type II error and do not identify him as a nonnative speaker, I am doing him a real injustice.

Next, we need to calculate the probability of a student receiving a score *at least as low as* 220 when $H_0 : \mu = 489$ is true. We first calculate the z score corresponding to a raw score of 220:

$$z = \frac{X - \mu}{\sigma} = \frac{(220 - 489)}{126} = \frac{-269}{126} = -2.13$$

We then go to tables of z to calculate the probability that we would obtain a z value less than or equal to -2.13. From Appendix z, we find that this probability is .017. Because this probability is less than the 5% significance level we chose to work with, we will reject the null hypothesis on the grounds that it is too unlikely that we would obtain a score as low as 220 if we had sampled an observation from a population of native speakers of English who had taken the GRE. Instead, we will conclude that we have an observation from an individual who is not a native speaker of English.

It is important to note that in rejecting the null hypothesis, we could have made a Type I error. We know that if we do sample speakers of English, 1.7% of them will score this low. It is possible that our applicant was a native speaker who just did poorly. All we are saying is that such an event is sufficiently unlikely that we will place our bets with the alternative hypothesis.

4.13 Back to Course Evaluations and Rude Motorists

We started this chapter with a discussion of the relationship between how students evaluate a course and the grade they expect to receive in that course. Our second example looked at the probability of motorists honking their horns at low- and high-status cars that did not move when a traffic light changed to green. As you will see in Chapter 9, the first example uses a correlation coefficient to represent the degree of relationship. The second example simply compares two proportions. Both examples can be dealt with using the techniques discussed in this chapter. In the first case, if there were no relationship between the grades and ratings, we would expect that the true correlation in the population of students is .00. We simply set up the null hypothesis that the population correlation is .00 and then ask about the probability that a sample of observations would produce a correlation as large as the one we obtained. In the second case, we set up the null hypothesis that there is no difference between the proportion of motorists *in the population* who honk at low- and high-status cars. Then we calculate the probability of obtaining a difference in sample proportions as large as the one we obtained (in our case, .34) if the null hypothesis is true. I do not expect you to be able to run these tests now, but you should have a general sense of the way we will set up the problem when we do learn to run them.

Key Terms

Sampling error (Introduction)

Hypothesis testing (4.1)

Sampling distributions (4.2)

Sampling distribution of the differences between means (4.2)

Research hypothesis (4.3)

Null hypothesis (H_0) (4.3)

Alternative hypothesis (H_1) (4.4)

Sample statistics (4.5)

Test statistics (4.5)

Decision making (4.6)

Rejection level (significance level) (4.6)

Rejection region (4.6)

Critical value (4.7)

Type I error (4.7)

α (alpha) (4.7)

Type II error (4.7)

β (beta) (4.7)

Power (4.7)

One-tailed test (directional test) (4.8)

Two-tailed test (nondirectional test) (4.8)

Conditional probabilities (4.9)

Exercises

4.1 Suppose I told you that last night's NHL hockey game resulted in a score of 26–13. You would probably decide that I had misread the paper and was discussing something other than a hockey score. In effect, you have just tested and rejected a null hypothesis.

    a. What was the null hypothesis?

    b. Outline the hypothesis-testing procedure that you have just applied.

4.2 For the past year, I have spent about $4.00 a day for lunch, give or take a quarter or so.

    a. Draw a rough sketch of this distribution of daily expenditures.

    b. If, without looking at the bill, I paid for my lunch with a $5 bill and received $.75 in change, should I worry that I was overcharged?

    c. Explain the logic involved in your answer to part (b).

4.3 What would be a Type I error in Exercise 4.2?

4.4 What would be a Type II error in Exercise 4.2?

4.5 Using the example in Exercise 4.2, describe what we mean by the rejection region and the critical value.

4.6   Why might I want to adopt a one-tailed test in Exercise 4.2, and which tail should I choose? What would happen if I chose the wrong tail?

4.7   A recently admitted class of graduate students at a large state university has a mean Graduate Record Exam (GRE) verbal score of 650 with a standard deviation of 50. (The scores are reasonably normally distributed.) One student, whose mother just happens to be on the board of trustees, was admitted with a GRE score of 490. Should the local newspaper editor, who loves scandals, write a scathing editorial about favoritism?

4.8   Why is such a small standard deviation reasonable in Exercise 4.7?

4.9   Why might (or might not) the GRE scores be normally distributed for the restricted sample (admitted students) in Exercise 4.7?

4.10   Imagine that you have just invented a statistical test called the Mode Test to test whether the mode of a population is some value (e.g., 100). The statistic (M) is calculated as

$$M = \frac{\text{Sample mode}}{\text{Sample range}}$$

Describe how you could obtain the sampling distribution of M. (*Note:* This is a purely fictitious statistic as far as I am aware.)

4.11   In Exercise 4.10, what would we call M in the terminology of this chapter?

4.12   Describe a situation in daily life in which we routinely test hypotheses without realizing it.

4.13   In Exercise 4.7, what would be the alternative hypothesis (H_1)?

4.14   Define "sampling error."

4.15   What is the difference between a "distribution" and a "sampling distribution"?

4.16   How would decreasing α affect the probabilities given in Table 4.1?

4.17   Give two examples of research hypotheses, and state the corresponding null hypotheses.

4.18   For the distribution in Figure 4.4, I said that the probability of a Type II error (β) is .74. Show how this probability was obtained.

4.19   Rerun the calculations in Exercise 4.18 for $\alpha = .01$.

4.20   In the example in Section 4.12, how would the test have differed if we had chosen to run a two-tailed test?

4.21   Describe the steps you would go through to develop the example given in this chapter about the course evaluations. In other words, how might you go about determining whether there truly is a relationship between grades and course evaluations?

4.22   Describe the steps you would go through to test the hypothesis that motorists are ruder to fellow drivers who drive low-status cars than to those who drive high-status cars.

Discussion Questions

4.23   In Chapter 1, we discussed a study of allowances for fourth-grade children. We considered that study again in the exercises for Chapter 2, where you generated data that might have been found in such a study.

   a.   Consider how you would go about testing the research hypothesis that boys receive more allowance than girls do. What would be the null hypothesis?

   b.   Would you use a one- or a two-tailed test?

   c.   What results might lead you to reject the null hypothesis and what might lead you to retain it?

   d.   What single thing might you do to make this study more convincing?

4.24   Simon and Bruce (1991), in demonstrating a different approach to statistics called "Resampling statistics,"[4] tested the null hypothesis that the mean price of liquor (in 1961) for the 16 "monopoly" states, where the state owned the liquor stores, was different from the mean price in the 26 "private" states, where liquor stores were privately owned. (The means were $4.35 and $4.84, respectively, giving you some hint at the effects of inflation.) For technical reasons, several states don't conform to this scheme and could not be analyzed.

a. What is the null hypothesis that we are really testing?

b. What label would you apply to $4.35 and $4.84?

c. If these are the only states that qualify for our consideration, why are we testing a null hypothesis in the first place?

d. Can you think of a situation where it does make sense to test a null hypothesis here?

4.25 Discuss the different ways that the traditional approach to hypothesis testing and the Jones and Tukey approach would address the question(s) inherent in the example of waiting times for a parking space.

4.26 What effect might the suggestion that experimenters report effect sizes have on the conclusions we draw from future research studies in psychology?

---

[4] The home page containing information on this approach is available at http://www.resample.com/. I will discuss resampling statistics at some length in Chapter 18.

CHAPTER 5

Basic Concepts of Probability

Objectives

To develop the concept of probability, present some basic rules for manipulating probabilities, outline the basic ideas behind Bayes' theorem, and introduce the binomial distribution and its role in hypothesis testing.

Contents

In CHAPTER 3, we began to make use of the concept of probability. For example, we saw that about 19% of children have Behavior Problem scores between 52 and 56 and thus concluded that if we chose a child at random, the probability that he or she would score between 52 and 56 is .19. When we begin concentrating on inferential statistics in Chapter 6, we will rely heavily on statements of probability. There, we will be making statements of the form, "If this hypothesis were correct, the probability is only .015 that we would have obtained a result as extreme as the one we actually obtained." If we are to rely on statements of probability, it is important to understand what we mean by probability and to understand a few basic rules for computing and manipulating probabilities. That is the purpose of this chapter.

The material covered in this chapter has been selected for two reasons. First, it is directly applicable to an understanding of the material presented in the remainder of the book. Second, it is intended to allow you to make simple calculations of probabilities that are likely to be useful to you. Material that does not satisfy either of these qualifications has been deliberately omitted. For example, we will not consider such things as the probability of drawing the queen of hearts, given that 14 cards, including the four of hearts, have already been drawn. Nor will we consider the probability that your desk light will burn out in the next 25 hours of use, given that it has already lasted 250 hours. The student who is interested in those topics is encouraged to take a course in probability theory, in which such material can be covered in depth.

5.1 Probability

analytic view

The concept of probability can be viewed in several different ways. There is no general agreement about what we mean by the word *probability*. The oldest and perhaps the most common definition of a probability is what is called the **analytic view.** Let's take an example that I have used through 10 editions of two books. (It once was a true example, but alas I have improved my health habits over the years.) I have a bag of caramels hidden in the drawer of my desk (hidden because I have learned not to trust my colleagues). The bag contains 85 of the light caramels, which I like, and 15 of the dark ones, which I save for candy-grubbing colleagues. Being hungry, I reach into the bag and grab a caramel at random. What is the probability that I will pull out a light-colored caramel? Most of you could answer this without knowing anything more about probability. Because 85 of the 100 caramels are light, and because I am sampling at random, the probability (p) of drawing a light caramel is $85/100 = .85$. This example illustrates one definition of probability:

> If an event can occur in A ways and can fail to occur in B ways, and if all possible ways are equally likely (e.g., each caramel in the bag has an equal chance of being drawn), then the probability of its occurrence is $A/(A + B)$, and the probability of its failing to occur is $B/(A + B)$.

Because there are 85 ways of drawing a light caramel (one for each of the 85 light caramels) and 15 ways of drawing a dark caramel, $A = 85$, $B = 15$, and $p(A) = 85/(85 + 15) = .85$.

frequentist view

An alternative view of probability is the **frequentist view.** Suppose that we keep drawing caramels from the bag, noting the color on each draw. In conducting this sampling study, we **sample with replacement,** meaning that each caramel is replaced before the next one is

sample with replacement

drawn. If we made a very large number of draws, we would find that (very nearly) 85% of the draws would result in a light caramel. Thus, we might define probability as the limit[1] of the relative frequency of occurrence of the desired event that we approach as the number of draws increases.

subjective probability

Yet a third concept of probability is advocated by a number of theorists. That is the concept of **subjective probability.** By this definition, probability represents an individual's subjective belief in the likelihood of the occurrence of an event. For example, the statement, "I think that tomorrow will be a good day," is a subjective statement of degree of belief, which probably has very little to do with the long-range relative frequency of the occurrence of good days, and in fact may have no mathematical basis whatsoever. This is not to say that such a view of probability has no legitimate claim for our attention. Subjective probabilities play an extremely important role in human decision-making and govern all aspects of our behavior. Just think of the number of decisions you make based on subjective beliefs in the likelihood of certain outcomes. You order pasta for dinner because it is probably better than the mystery meat special; you plan to go skiing tomorrow because the weather forecaster says that there is an 80% chance of snow overnight; you bet your money on a horse because you think that the odds of its winning are better than the 6:1 odds the bookies are offering. We will shortly discuss what is called Bayes' theorem, which is essential to the use of subjective probabilities. Statistical decisions as we will make them here generally will be stated relative to frequentist or analytical approaches, even so, the *interpretation* of those probabilities has a strong subjective component.

Although the particular definition that you or I prefer may be important to each of us, any of the definitions will lead to essentially the same result for hypothesis testing, the discussion of which runs through the rest of the book. (It should be said that those who favor subjective probabilities often disagree with the general hypothesis-testing orientation.) Actually, most people use the different approaches interchangeably. When we say that the probability of losing at Russian roulette is 1/6, we are referring to the fact that one of the gun's six cylinders has a bullet in it. When we buy a particular car because *Consumer Reports* says it has a good repair record, we are responding to the fact that a high proportion of these cars have been relatively trouble-free. When we say that the probability of the Yankees winning the pennant is high, we are stating our subjective belief in the likelihood of that event. But when we reject some hypothesis because there is a very low probability that the actual data would have been obtained if the hypothesis had been true, it may not be important which view of probability we hold.

5.2   Basic Terminology and Rules

event

The basic bit of data for a probability theorist is called an **event**. Statisticians use the word *event* to cover just about anything. An event can be the occurrence of a king when we deal from a deck of cards, a score of 36 on a scale of likability, a classification of "female" for the next person appointed to the Supreme Court, or the mean of a sample. Whenever you speak of the probability of something, the "something" is called an event. When we are dealing with a process as simple as flipping a coin, the event is the outcome of that flip— either heads or tails. When we draw caramels out of a bag, the possible events are light and

---

[1] The word *limit* refers to the fact that as we sample more and more caramels, the proportion of light will get closer and closer to some value. After 100 draws, the proportion might be .83; after 1,000 draws it might be .852; after 10,000 draws it might be .8496, and so on. Notice that the answer is coming closer and closer to $p = .8500000 \ldots$. The value that is being approached is called the *limit*.

dark. When we speak of a grade in a course, the possible events are the letters A, B, C, D, and F.

independent events

Two events are said to be **independent events** when the occurrence or nonoccurrence of one has no effect on the occurrence or nonoccurrence of the other. The voting behaviors of two randomly chosen subjects normally would be assumed to be independent, especially with a secret ballot, because how one person votes could not be expected to influence how the other will vote. However, the voting behaviors of two members of the same family probably would not be independent events, because those people share many of the same beliefs and attitudes. This would be true even if those two people were careful not to let the other see their ballots.

mutually exclusive

exhaustive

Two events are said to be **mutually exclusive** if the occurrence of one event precludes the occurrence of the other. For example, the standard college classes of first year, sophomore, junior, and senior are mutually exclusive because one person cannot be a member of more than one class. A set of events is said to be **exhaustive** if it includes all possible outcomes. Thus, the four college classes in the previous example are exhaustive with respect to full-time undergraduates, who have to fall in one or another of those categories—if only to please the registrar's office. At the same time, they are not exhaustive with respect to total university enrollments, which include graduate students, medical students, nonmatriculated students, hangers-on, and so forth.

As you already know, or could deduce from our definitions of probability, probabilities range between .00 and 1.00. If some event has a probability of 1.00, then it *must* occur. (Very few things have a probability of 1.00, including the probability that I will be able to keep typing until I reach the end of this paragraph.) If some event has a probability of .00, it is certain *not* to occur. The closer the probability comes to either extreme, the more likely or unlikely is the occurrence of the event.

Basic Laws of Probability

Two important theorems are central to any discussion of probability. (If my use of the word *theorems* makes you nervous, substitute the word *rules*.) They are often referred to as the additive and multiplicative rules.

The Additive Rule

To illustrate the additive rule, we will complicate the caramel example by eating many of the light candies and replacing them with wooden cubes. We now have 30 light caramels, 15 dark caramels, and 55 rather tasteless wooden cubes. Given these frequencies, we know from the analytic definition of probability that $p(\text{light}) = 30/100 = .30$, $p(\text{dark}) = 15/100 = .15$, and $p(\text{wooden}) = 55/100 = .55$. But what is the probability that I will draw

additive law of probability

a caramel, either light or dark, instead of a piece of wood? Here we need the **additive law of probability.**

> Given a set of mutually exclusive events, the probability of the occurrence of one event or another is equal to the sum of their separate probabilities.

Thus, $p(\text{light or dark}) = p(\text{light}) + p(\text{dark}) = .30 + .15 = .45$. Notice that we have imposed the restriction that the events must be mutually exclusive, meaning that the occurrence of one event precludes the occurrence of the other. If a caramel is light, it can't be dark. This requirement is important. About one-half of the population of this country is female, and about one-half of the population has traditionally feminine names. But the probability that a person chosen at random will be female *or* will have a feminine name is obviously not .50 + .50 = 1.00. Here the two events are *not* mutually exclusive. However, the probability that a girl born in Vermont in 1987 was named Ashley or Sarah, the two most common girls' names in that year, equals $p(\text{Ashley}) + p(\text{Sarah}) = .044 + .032 = .076$. Here the names are mutually exclusive because you can't have both Ashley *and* Sarah as your first name (unless your parents got carried away and combined the two with a hyphen).

The Multiplicative Rule

multiplicative law of probability

Let's continue with the bag of caramels in which $p(\text{light}) = .30$, $p(\text{dark}) = .15$, and $p(\text{wooden}) = .55$. Suppose I draw two caramels, replacing the first before drawing the second. What is the probability that I will draw a light caramel on the first trial *and* a light one on the second? Here we need to invoke the **multiplicative law of probability.**

> The probability of the joint occurrence of two or more independent events is the product of their individual probabilities.

Thus $p(\text{light, light}) = p(\text{light}) \times p(\text{light}) = .30 \times .30 = .09$. Similarly, the probability of a light caramel followed by a dark one is $p(\text{light, dark}) = p(\text{light}) \times p(\text{dark}) = .30 \times .15 = .045$. Notice that we have restricted ourselves to independent events, meaning the occurrence of one event can have no effect on the occurrence or nonoccurrence of the other. Because gender and name are not independent, it would be wrong to state that $p(\text{female with feminine name}) = .50 \times .50 = .25$. However, it most likely would be correct to state that $p(\text{female, born in January}) = .50 \times 1/12 = .50 \times .083 = .042$, because I know of no data to suggest that gender is dependent on birth month. (If month and gender were related, my calculation would be wrong.)

In Chapter 6, we will use the multiplicative law to answer questions about the independence of two variables. An example from that chapter will help illustrate a specific use of this law. In a study to be discussed in Chapter 6, Geller, Witmer, and Orebaugh (1976) wanted to test the hypothesis that what someone did with a supermarket flyer depended on whether the flyer contained a request not to litter. Geller et al. distributed flyers with and without this message and at the end of the day searched the store to find where the flyers had been left. Testing their hypothesis involves, in part, calculating the probability that a flyer would contain a message about littering *and* would be found in a trash can. We need to calculate what this probability would be if the two events (contains message about littering and flyer in trash) are independent, as would be the case if the message had no effect. *If* we assume that these two events are independent, the multiplicative law tells us that $p(\text{message, trash}) = p(\text{message}) \times p(\text{trash})$. In their study, 49% of the flyers contained a message, so the probability that a flyer chosen at random would contain the message is .49. Similarly, 6.8% of the flyers were later found in the trash, giving $p(\text{trash}) = .068$. Therefore, if the two events are independent, $p(\text{message, trash}) = .49 \times .068 = .033$. (In fact, 4.5% of the flyers with messages were found in the trash, which is a bit higher than we would expect if the ultimate disposal of the flyers were independent of the message. If this difference is reliable, what does this suggest to you about the effectiveness of the message?)

Finally, we can take a simple example that illustrates both the additive and the multiplicative laws. What is the probability that over two trials (sampling with replacement) I will draw one light caramel and one dark one, *ignoring the order in which they are drawn?* First, we use the multiplicative rule to calculate

$$p(\text{light, dark}) = .30 \times .15 = .045$$

$$p(\text{dark, light}) = .15 \times .30 = .045$$

Because these two outcomes satisfy our requirement (and because they are the only ones that do), we now need to know the probability that one or the other of these outcomes will occur. Here we apply the additive rule:

$$p(\text{light, dark}) + p(\text{dark, light}) = .045 + .045 = .09$$

Thus, the probability of obtaining one caramel of each color over two draws is .09—that is, it will occur a little less than one-tenth of the time.

Students sometimes get confused over the additive and multiplicative laws because they almost sound the same when you hear them quickly. One useful idea is to realize the difference between the situations in which the rules apply. In those situations in which you use the additive rule, you know that you are going to have *one* outcome. A caramel that you draw may be dark or light, but there is only going to be one of them. In the multiplicative case, we are speaking about at least *two* outcomes (e.g., the probability that we will get one dark caramel *and* one light one). For single outcomes, we add probabilities; for multiple independent outcomes we multiply them.

Sampling with Replacement

Why do I keep referring to "sampling with replacement?" The answer goes back to the issue of independence. Considering the example with light and dark caramels, we had 85 light caramels and 15 dark ones in the bag. On the first trial, the probability of a dark caramel is $15/100 = .15$. If I put that caramel back before I draw again, there will still be an 85:15 split, and the probability of a dark caramel on the next draw will still be $15/100 = .15$. But if I did not replace the caramel, the probability of a dark caramel on Trial 2 would depend on the result of Trial 1. If I had drawn a dark caramel on Trial 1, there will be 14 dark ones and 85 light ones remaining, and $p(\text{dark}) = 14/99 = .1414$. If I had drawn a light caramel on Trial 1, for Trial 2 $p(\text{dark}) = 15/99 = .1515$. So when I sample with replacement, $p(\text{dark})$ stays the same from trial to trial, whereas when I sample without replacement the probability keeps changing. To take an extreme example, if I sample without replacement, what is the probability of exactly 16 dark caramels out of 60 draws? The answer, of course, is .00, because there are only 15 caramels and it is impossible to draw 16 of them. Sampling with replacement, however, would produce a possible result, though the probability would only be .0077.

Joint and Conditional Probabilities

Two types of probabilities play an important role in discussions of probability: joint probabilities and conditional probabilities.

joint probability

A **joint probability** is defined simply as the probability of the co-occurrence of two or more events. For example, in Geller's study of supermarket flyers, the probability that a flyer would *both* contain a message about littering *and* be found in the trash is a joint probability, as is the probability that a flyer would both contain a message about littering and be found stuffed down behind the Raisin Bran. Given two events, their joint probability is denoted as $p(A, B)$, just as we have used $p(\text{light, dark})$ or $p(\text{message, trash})$. If those two events are independent, then the probability of their joint occurrence can be found by using the multiplicative law, as we have just seen. If they are *not* independent, the probability of their joint occurrence is more complicated to compute and will differ from what it would be if the events were independent. We won't compute that probability here.

conditional probability

A **conditional probability** is the probability that one event will occur *given* that some other event has occurred. The probability that a person will contract AIDS given that he or she is an intravenous drug user is a conditional probability. The probability that an advertising flyer will be thrown in the trash given that it contains a message about littering is another example. A third example is a phrase that occurs repeatedly throughout this book: "If the null hypothesis is true, the probability of obtaining a result such as this is . . ." Here I have substituted the word *if* for *given,* but the meaning is the same.

With two events, A and B, the conditional probability of A given B is denoted, by use of a vertical bar, as $p(A \mid B)$, for example, $p(\text{AIDS} \mid \text{drug user})$ or $p(\text{trash} \mid \text{message})$.

We often assume, with some justification, that parenthood breeds responsibility. People who have spent years acting in careless and irrational ways somehow seem to turn into different people once they become parents, changing many of their old behavior patterns. (Just wait a few years.) Suppose that a radio station sampled 100 people, 20 of whom had children. They found that 30 of the people sampled used seat belts, and that 15 of those people had children. The results are shown in Table 5.1.

Table 5.1   The relationship between parenthood and seat belt use

Parenthood	Wear Seat belt	Do Not Wear Seat belt	Total
Children	15	5	20
No children	15	65	80
Total	30	70	100

The information in Table 5.1 allows us to calculate the simple, joint, and conditional probabilities. The simple probability that a person sampled at random will use a seat belt is $30/100 = .30$. The joint probability that a person will have children *and* will wear a seat belt is $15/100 = .15$. The conditional probability of a person using a seat belt given that he or she has children is $15/20 = .75$. Do not confuse joint and conditional probabilities. As you can see, they are quite different. You might wonder why I didn't calculate the joint probability here by multiplying the appropriate simple probabilities. The use of the multiplicative law requires that parenthood and seat belt use be independent. In this example they are not, because the data show that whether people use seat belts depends very much on whether or not they have children. (If I had assumed independence, I would have predicted the joint probability to be $.30 \times .20 = .06$, which is less than half the size of the actual obtained value.)

To take another example, the probability that you have been drinking alcoholic beverages and that you have an accident is a joint probability. This probability is not very high, because relatively few people are drinking at any one time and relatively few people have accidents. However, the probability that you have an accident given that you have been drinking, or, in reverse, the probability that you have been drinking given that you have an accident, are both much higher. At night, the conditional probability of $p(\text{drinking} \mid \text{accident})$ approaches .50 because nearly half of all automobile accidents at night in the United States involve alcohol. I don't know the conditional probability of $p(\text{accident} \mid \text{drinking})$, but I do know that it is much higher than the **unconditional probability** of an accident, that is, $p(\text{accident})$.

unconditional probability

5.3   Discrete versus Continuous Variables

In Chapter 1, I made a distinction between discrete and continuous variables. As mathematicians view things, a discrete variable is one that can take on a countable number of different values, whereas a continuous variable is one that can take on an infinite number of different values. For example, the number of people attending a specific movie theater tonight is a discrete variable because we literally can count the number of people entering the theater, and there is no such thing as a fractional person. However, the distance between two people in a study of personal space is a continuous variable because the distance could be 2', or 2.8', or 2.8173754814'. Although the distinction given here is technically correct, common usage is somewhat different.

In practice, when we speak of a discrete variable, we *usually* mean a variable that takes on one of a relatively small number of possible values (e.g., a five-point scale of socioeconomic status). A variable that can take on one of many possible values is generally treated as a continuous variable if the values represent at least an ordinal scale. Thus, we usually treat an IQ score as a continuous variable, even though we recognize that IQ scores come in whole units and we will not find someone with an IQ of 105.317. In Chapter 3, I referred to the Achenbach Total Behavior Problem score as normally distributed, even though I know that it can only take on positive integers, whereas a normal distribution can take on all values between ±∞. I treat those scores as normal because they are close enough to normal that my results will be reasonably accurate.

The distinction between discrete and continuous variables is reintroduced here because the *distributions* of the two kinds of variables are treated somewhat differently in probability theory. With discrete variables, we can speak of the probability of a specific outcome. With continuous variables, on the other hand, we need to speak of the probability of obtaining a value that falls within a specific *interval*.

5.4   Probability Distributions for Discrete Variables

An interesting example of a discrete probability distribution is seen in Figure 5.1. The data plotted in this figure come from a study by Campbell, Converse, and Rodgers (1976), in which they asked 2164 respondents to rate on a 1–5 scale the importance they attach to various aspects of their lives (1 = extremely important, 5 = not at all important). Figure 5.1 presents the distribution of responses for several of these aspects. The possible values of X (the rating) are presented on the abscissa (X-axis), and the relative frequency (or probability) of people choosing that response is plotted on the ordinate (Y-axis). From the figure, you can see that the distributions of responses to questions concerning health, friends, and savings are quite different. The probability that a person chosen at random will consider his or her health to be extremely important is .70, whereas the probability that the same person will consider a large bank account to be extremely important is only .16. (So much for the stereotypic American Dream.) Campbell et al. collected their data in the mid-1970s. Would you expect to find similar results today? How might they differ?

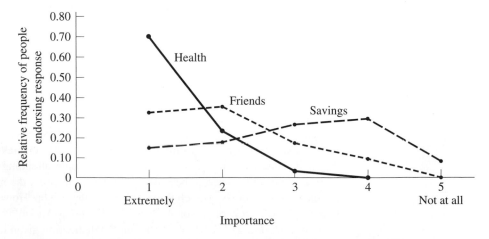

Figure 5.1   Distributions of importance ratings of three aspects of life

5.5 Probability Distributions for Continuous Variables

When we move from discrete to continuous probability distributions, things become more complicated. We dealt with a continuous distribution when we considered the normal distribution in Chapter 3. You may recall that in that chapter we labeled the ordinate of the distribution "density." We also spoke in terms of intervals rather than in terms of specific outcomes. Now we need to elaborate somewhat on those points.

Figure 5.2 shows the approximate distribution of the age at which children first learn to walk (based on data from Hindley, Filliozat, Klackenberg, Nicolet-Meister, & Sand, 1966). The mean is approximately 14 months, the standard deviation is approximately three months, and the distribution is positively skewed. You will notice that in this figure the ordinate is labeled "density," whereas in Figure 5.1 it was labeled "relative frequency."

density   **Density** is not synonymous with probability, and it is probably best thought of as merely the height of the curve at different values of *X*. At the same time, the fact that the curve is higher near 14 months than it is near 12 months tells us that children are more likely to walk at around 14 months than at about one year. The reason for changing the label on the ordinate is that we now are dealing with a continuous distribution rather than a discrete one. If you think about it for a moment, you will realize that although the highest point of the curve is at 14 months, the probability that a child picked at random will first walk at *exactly* 14 months (i.e., 14.00000000 months) is infinitely small—statisticians would argue that it is in fact 0. Similarly, the probability of first walking at 14.00000001 months also is infinitely small. This suggests that it does not make any sense to speak of the probability of any *specific* outcome. On the other hand, we know that many children start walking at *approximately* 14 months, and it does make considerable sense to speak of the probability of obtaining a score that falls within some specified *interval*. For example, we might be interested in the probability that an infant will start walking at 14 months plus or minus one-half month. Such an interval is shown in Figure 5.3. If we arbitrarily define the total area under the curve to be 1.00, then the shaded area in Figure 5.3 between points *a* and *b* will be equal to the probability that an infant chosen at random will begin walking at this time. Those of you who have had calculus will probably recognize that if we knew the form of the equation that describes this distribution (i.e., if we knew the equation for the curve), we would simply need to integrate the function over the interval from *a* to *b*. For those of you who have not had calculus, it is sufficient to know that the distributions with which we will work are adequately approximated by other distributions that have already been tabled. In this book we will never integrate functions, but we will often refer to tables of distributions. You have

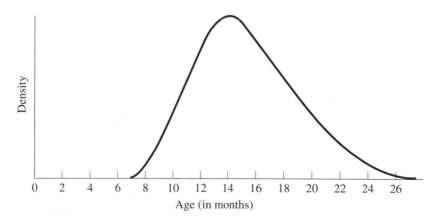

Figure 5.2   Age at which a child first walks unaided

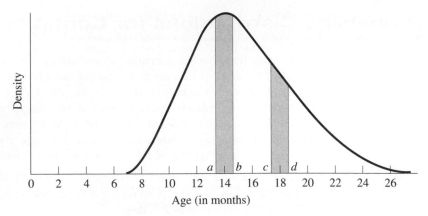

Figure 5.3 Probability of first walking during four-week intervals centered on 14 and 18 months

already had experience with this procedure with regard to the normal distribution in Chapter 3.

We have just considered the area of Figure 5.3 between *a* and *b*, which is centered on the mean. However, the same things could be said for any interval. In Figure 5.3 you can also see the area that corresponds to the period that is one-half month on either side of 18 months (denoted as the shaded area between *c* and *d*). Although there is not enough information in this example for us to calculate actual probabilities, it should be clear by inspection of Figure 5.3 that the one-month interval around 14 months has a higher probability (greater shaded area) than the one-month interval around 18 months.

A good way to get a feel for areas under a curve is to take a piece of transparent graph paper and lay it on top of the figure (or use a regular sheet of graph paper and hold the two up to a light). If you count the number of squares that fall within a specified interval and divide by the total number of squares under the whole curve, you will approximate the probability that a randomly drawn score will fall within that interval. The smaller the size of the individual squares on the graph paper, the more accurate the approximation will be.

5.6 Permutations and Combinations

We will set continuous distributions aside for now. We will concentrate on two discrete distributions (the binomial and the multinomial) that can be used to develop the chi-square test in Chapter 6. First, we must consider the concepts of permutations and combinations, which are required for a discussion of those distributions.

combinatorics

The special branch of mathematics dealing with the number of ways in which objects can be put together (e.g., the number of different ways of forming a three-person committee with five people available) is known as **combinatorics.** Although not many instances in this book require knowledge of combinatorics, there are enough of them to make it necessary to briefly define the concepts of permutations and combinations and to give formulae for their calculation.

Permutations

We will start with a simple example that is easily expanded into a more useful and relevant one. Assume that four people have entered a lottery for ice-cream cones. The names are placed in a hat and drawn. The person whose name is drawn first wins a double-scoop cone, the second wins a single-scoop cone, the third wins just the cone, and the fourth wins nothing. Assume that the people are named Aaron, Barbara, Cathy, and David, abbreviated A, B, C, and D. The following orders in which the names are drawn are all possible.

A	B	C	D		B	A	C	D		C	A	B	D		D	A	B	C
A	B	D	C		B	A	D	C		C	A	D	B		D	A	C	B
A	C	B	D		B	C	A	D		C	B	A	D		D	B	A	C
A	C	D	B		B	C	D	A		C	B	D	A		D	B	C	A
A	D	B	C		B	D	A	C		C	D	A	B		D	C	A	B
A	D	C	B		B	D	C	A		C	D	B	A		D	C	B	A

permutation

Each of these 24 orders presents a unique arrangement (called a **permutation**) of the four names taken four at a time. If we represent the number of permutations (arrangements) of N things taken r at a time as P_r^N, then

$$P_r^N = \frac{N!}{(N-r)!}$$

N factorial

where the symbol $N!$ is read N **factorial** and represents the product of all integers from N to 1. [In other words, $N! = N(N-1)(N-2)(N-3)\cdots(1)$. By definition, $0! = 1$].

For our example of drawing four names for four entrants,

$$P_4^4 = \frac{4!}{(4-4)!} = \frac{4!}{0!} = \frac{4 \cdot 3 \cdot 2 \cdot 1}{1} = 24$$

which agrees with the number of listed permutations.

Now, few people would get very excited about winning a cone without any ice cream in it, so let's eliminate that prize. Then out of the four people, only two will win on any drawing. The order in which those two winners are drawn is still important, however, because the first person whose name is drawn wins a larger cone. In this case, we have four names but are drawing only two out of the hat (because the other two are both losers). Thus, we want to know the number of permutations of four names taken two at a time, $\left(P_2^4\right)$. We can easily write down these permutations and count them:

A	B		B	A		C	A		D	A
A	C		B	C		C	B		D	B
A	D		B	D		C	D		D	C

Or we can calculate the number of permutations directly:

$$P_2^4 = \frac{4!}{(4-2)!} = \frac{4 \cdot 3 \cdot 2 \cdot 1}{2} = 12$$

Here there are 12 possible orderings of winners, and the ordering makes an important difference—it determines not only who wins, but also which winner receives the larger cone.

Now we will take a more useful example involving permutations. Suppose we are designing an experiment studying physical attractiveness judged from slides. We are concerned that the order of presentation of the slides is important. Given that we have six slides to present, in how many different ways can these be arranged? This again is a question of permutations because the ordering of the slides is important. More specifically, we want to know the permutations of six slides taken six at a time. Or, suppose that we have six slides, but any given subject is going to see only three. Now how many orders can be used? This is a question about the permutations of six slides taken three at a time.

For the first problem, in which subjects are presented with all six slides, we have

$$P_6^6 = \frac{6!}{(6-6)!} = \frac{6!}{0!} = \frac{6 \cdot 5 \cdot 4 \cdot 3 \cdot 2 \cdot 1}{1} = 720$$

Thus, there are 720 different ways of arranging six slides. If we want to present all possible arrangements to each participant, we are going to need 720 trials, or some multiple of that. That is a lot of trials. For the second problem, where we have six slides but show only three to any one subject, we have

$$P_3^6 = \frac{6!}{(6-3)!} = \frac{6!}{3!} = \frac{6 \cdot 5 \cdot 4 \cdot 3 \cdot 2 \cdot 1}{6} = 120$$

If we want to present all possible arrangements to each subject, we need 120 trials, a result that may still be sufficiently large to lead us to modify our design. This is one reason we often use random orderings rather than try to present all possible orderings.

Combinations

combinations

To return to the ice-cream lottery, suppose we now decide that we will award only single-dip cones to the two winners. We will still draw the names of two winners out of a hat, *but we will no longer care which of the two names was drawn first*—the result AB is for all practical purposes the same as the result BA because in each case Aaron and Barbara win a cone. When the order in which names are drawn is no longer important, we are no longer interested in permutations. Instead, we are now interested in what are called **combinations.** We want to know the number of possible combinations of winning names, but not the order in which they were drawn.

We can enumerate these combinations as

A	B	B	C
A	C	B	D
A	D	C	D

There are six of them. In other words, out of four people, we could compile six different sets of winners. (If you look back to the previous enumeration of permutations of winners, you will see that we have just combined outcomes containing the same names.)

Normally, we do not want to enumerate all possible combinations just to find out how many of them there are. To calculate the number of *combinations* of N things taken r at a time C_r^N, we will define

$$C_r^N = \frac{N!}{r!(N-r)!}$$

For our example,

$$C_2^4 = \frac{4!}{2!(4-2)!} = \frac{4 \cdot 3 \cdot 2 \cdot 1}{2 \cdot 1 \cdot 2 \cdot 1} = 6$$

Let's return to the example involving slides to be presented to subjects. When we were dealing with permutations, we worried about the way in which each set of slides was arranged; that is, we worried about all possible orderings. Suppose we no longer care about the order of the slides within sets, but we need to know how many different sets of slides we could form if we had six slides but took only three at a time. This is a question of combinations.

For six slides taken three at a time, we have

$$C_3^6 = \frac{6!}{3!(6-3)!} = \frac{6 \cdot 5 \cdot 4 \cdot 3 \cdot 2 \cdot 1}{3 \cdot 2 \cdot 1 \cdot 3 \cdot 2 \cdot 1} = 20$$

If we wanted every subject to get a different set of three slides but did not care about the order within a set, we would need 20 subjects.

Later in the book we will discuss procedures, called *permutation tests,* in which we imagine that the data we have are all the data we could collect, but we want to imagine what the sample means would likely be if the N scores fell into our two different experimental groups (of n_1 and n_2 scores) *purely at random.* To solve that problem, we could calculate the number of different ways the observations could be assigned to groups, which is just the number of combinations of N things taken n_1 and n_2 at a time. (Please don't ask why it's called a permutation test if we are dealing with combinations—I haven't figured that out yet.) Knowing the number of different ways that data could have occurred at random, we will calculate the percentage of those outcomes that would have produced differences in means at least as extreme as the difference we found. That would be the probability of the data given H_0: true, often written $p(D|H_0)$. I mention this here only to give you an illustration of when we would want to know how to calculate permutations and combinations.

5.7   Bayes' Theorem

Bayes' theorem

We have one more basic element of probability theory to cover before we go on to use those basics in particular applications. This section is new to this edition, not because **Bayes' theorem** is new (it was developed by Thomas Bayes and first read before the Royal Society in London in 1764—3 years after Bayes' death), but because it is becoming important that people in the behavioral sciences know what the theorem is about, even if they forget the details of how to use it. (You can always look up the details.)

Bayes' theorem tells us how to accumulate information to revise estimates of probabilities. By "accumulate information" I mean a process in which you continually revise a probability estimate as more information comes in. Suppose that I tell you that Fred was murdered and ask you for your personal (subjective) probability that Willard committed the crime. You think he is certainly capable of it and not a very nice person, so you say $p = .15$. Then I say that Willard was seen near the crime that night, and you raise your probability to $p = .20$. Then I say that Willard owns the right type of gun, and you might raise your probability to $p = .25$. Then I say that a fairly reliable witness says Willard was at a baseball game with him at the time, and you drop your probability to $p = .10$. And so on. This is a process of accumulating information to come up with a probability that some event occurred. For those interested in Bayesian statistics, probabilities are usually subjective or personal probabilities, meaning that they are a statement of personal belief, rather than having a frequentist or analytic basis as defined at the beginning of the chapter. Bayes' theorem will work perfectly well with any kind of probability, but it is most often seen with subjective probabilities.

Let's take a simple example. Suppose that I asked you to give me your estimate of the probability that I am writing this section in April. You have no idea what month it is, so you would probably say that the probability is $1/12 = .083$, if we ignore the fact that some months have more days than others. (The probability that it is not April is $11/12 = .917 = 1 - .083$.) Now I tell you that I can look out my office window and see that it snowed in the mountains in the last few days. Although that is not definitive information, it certainly should be helpful. (You probably would be inclined to doubt that it is July.) What you want to do is to revise your earlier estimate ($p = .083$) on the basis of this new information. That is what Bayes' theorem allows you to do.

prior probability

First, define $p(A)$ as the **prior probability** that it is April. We call it a prior probability because it is the probability you estimate *before* I tell you anything about snow. Define $p(NA)$ as the probability that it is not April, which we have specified as .917. We will define

posterior
probability

$p(A|S)$ as the **posterior probability** that it is April given that you know that it snowed in the last few days. This is the probability that we ultimately want to estimate. We call it a posterior probability because it is your revised probability *after* receiving information about snow. Last, we need to consider the probability that it would snow last night if the month really is April ($p(S|A)$ and the probability it would snow last night if this is not April (NA), which is represented by $p(S|NA)$. You could either look these probabilities up in meteorological tables, or you could just take a reasonable guess—making them subjective probabilities. I will choose to guess based on what I know about the annual snow patterns where I live. I will guess that the probability of having snow in the last few days *given that it is April* is .20, and the probability that it snowed *given that it is not April* is .10 (remember all those summer months when it doesn't snow).[2]

So now I have the following probabilities

$$p(A) = 1/12 = .083 \qquad p(NA) = 11/12 = .917$$
$$p(S|A) = .20 \qquad p(S|NA) = .10$$

What we now want is the probability of it being April, *given the data about snow*. Bayes' theorem tells us that this probability is given by

$$p(A|S) = \frac{p(S|A)p(A)}{p(S|A)p(A) + p(S|NA)p(NA)}$$

Substituting what we already know, we have

$$p(A|S) = \frac{p(S|A)p(A)}{p(S|A)p(A) + p(S|NA)p(NA)}$$

$$= \frac{(.20)(.083)}{(.20)(.083) + (.10)(.917)}$$

$$= \frac{.0166}{.0166 + .0917} = \frac{.0166}{.1083} = .1533$$

When you didn't know anything about snow, your best estimate of the probability was .083. Once you know that it snowed, you were able to revise that probability to .153, nearly doubling it. This it the kind of task that Bayes' theorem was designed to solve. It allows you to accumulate information and update your estimates.

A lot of work in human decision making has been based on applications of Bayes' theorem. Much of it focuses on comparing what people *should* say in a situation, with what they *actually* say, for the purpose of characterizing how people really make decisions. A famous problem was posed to decision makers by Tversky and Kahneman (1980). This problem involved deciding which cab company was involved in an accident. We are told that there was an accident involving one of the two cab companies (Green Cab and Blue Cab) in the city, but we are not told which one it was. We know that 85% of the cabs in a given city are Green, and 15% are Blue. The prior probabilities then, based on the percentage of Green and Blue cabs, are .85 and .15. If that were all you knew and were then told that someone was just run over by a cab, your best estimate would be that the probability that it was a Green cab is .85. Then a witness comes along who thinks that it was a

---

[2] The probability of snow in April may look high to you, but I live in the Colorado Rockies, and those mountains are at 10,000 to 12,000 feet. (So why do I give my affiliation as the University of Vermont? Because I retired from there and am now "Professor Emeritus." They give me the title, and I give them the credit. Fair trade.)

Blue cab. You might think that was conclusive, but identifying colors at night is not a fool-proof task, and the insurance company tested our informant and found that he was able to identify colors at night with only 80% accuracy. Thus, if you show him a Blue cab, the probability that he will correctly say Blue is .80, and the probability that he will incorrectly say Green is .20. (Similarly, if the cab is Green.) So our conditional probability that the cab was a Blue cab, given that he said it was Blue is .80, and the conditional probability that it was Green, given that he said it was Blue is .20. This information is sufficient to allow you to calculate the posterior probability that the cab was a Blue cab given that the witness said it was blue.

In the following formula, let B stand for the event that it was a Blue cab, and let b stand for the event that the witness called it blue. Similarly for G and g.

$$p(B|b) = \frac{p(b|B)p(B)}{p(b|B)p(B) + p(g|B)p(G)}$$

$$= \frac{(.80)(.15)}{(.80)(.15) + (.20)(.85)}$$

$$= \frac{.12}{.12 + .17} = \frac{.12}{.29} = .414$$

Most of the participants in Tversky and Kahneman's experiment guessed that the probability that it was the Blue Cab was around .80, when in fact the correct answer is approximately .41. Thus, Kahneman and Tversky concluded that judges place too much weight on the witness' testimony, and not enough weight on the prior probabilities. Here is a situation where the discrepancy between what judges say and what they should say gives us clues about the strategies that judges use and where they go wrong.

A Generic Formula

The formulae given previously were framed in terms of the specific example under discussion. It may be helpful to have a more generic formula that you can adapt to your own purposes. Suppose that we are asking about the probability that some hypothesis (H) is true, given certain data (D). For our examples, H represented "the month is April" or "it was the Blue Cab company." The D represent "it snowed" or "the witness reported that the cab was blue." The symbol $\overline{H}$ is read "not H" and stands for the case where the hypothesis is false. Then

$$p(H|D) = \frac{p(D|H)p(H)}{p(D|H)p(H) + p(D|\overline{H})p(\overline{H})}$$

Back to the Hypothesis Testing

In Chapter 4, we discussed hypothesis testing and different approaches to it. Bayes' theorem has an important contribution to make to that discussion, although I am only going to touch on the issue here. (I want you to understand the nature of the argument, but it is not reasonable to expect you to go much beyond that.) Recall that I said that in some ways a hypothesis test is not really designed to answer the question we would ideally like to answer. We want to collect some data and then ask about the probability that the null hypothesis is true given the data. But instead, our statistical procedures tell us the probability that we would obtain those data given that the null hypothesis (H_0) is true. In other words, we want

$p(H_0|D)$ when what we really have is $p(D|H_0)$. Many people have pointed out that we could have the answer we seek if we simply apply Bayes' theorem,

$$p(H_0|D) = \frac{p(D|H_0)p(H_0)}{p(D|H_0)p(H_0) + p(D|H_1)p(H_1)}$$

where H_0 stands for the null hypothesis, H_1 stands for the alternative hypothesis, and D stands for the data.

The problem here is that we don't know most of the necessary probabilities. We could *estimate* those probabilities, but those would only be estimates. It is one thing to be able to calculate that the probability of April is .083 because April is one of 12 months in the year. But it is quite a different thing to be able to estimate the probability that the null hypothesis is true. Using the example of waiting times in parking lots, you and I might have quite different prior probability estimates that people leave a parking space at the same speed whether or not there is someone waiting. In addition, our statistical test is designed to give us $p(D|H_0)$, which is helpful. But where do we obtain $p(D|H_1)$ from if we don't have a specific alternative hypothesis is mind (other than the negation of the null)? It was one thing to estimate it when we had something concrete (like all months but April), but considerably more difficult when the alternative is that people leave *more slowly* when someone is waiting if we don't know *how much* more slowly. The probabilities would differ dramatically if we think of "5 seconds more slowly" compared with "25 seconds more slowly." That these probabilities we need are hard, or impossible, to determine has stood in the way of developing this as a general approach to hypothesis testing—though many have tried. (One approach is to chose a variety of reasonable estimates, and note how the results hold up under those different estimates. If most believable estimates lead to the same conclusion, that tells us something useful.)

Bayesian statistics

I don't mean to suggest that the application of Bayes' theorem (known as **Bayesian statistics**) is hopeless—it certainly is not. Many people are very interested in that approach, though its use is mostly restricted to situations where the null and alternative hypotheses are sharply defined, such as H_0: $\mu = 0$ and H_1: $\mu = 3$. But I have never seen clearly specified alternative hypotheses in the behavioral sciences.

5.8   The Binomial Distribution

binomial distribution

We now have all the information on probabilities and combinations that we need for understanding one of the most common probability distributions—the **binomial distribution.** This distribution will be discussed briefly, and you will see how it can be used to test simple hypotheses. I don't think that I can write a chapter on probability without discussing the binomial distribution, but many students and instructors would be more than happy if I skipped this topic. There certainly are many applications for it (the sign test to be discussed shortly is one example), but I would easily forgive you for not wanting to memorize the necessary formulae—you can always look them up.

Bernoulli trial

The binomial distribution deals with situations in which each of a number of independent trials results in one of two mutually exclusive outcomes. Such a trial is called a **Bernoulli trial** (after a famous mathematician of the same name). The most common example of a Bernoulli trial is flipping a coin, and the binomial distribution could be used to give us the probability of, for example, 3 heads out of 5 tosses of a coin. Most people don't get excited by the prospect of flipping coins, so think of calculating the probability that 20 out of your 30 cancer patients will survive a diagnosis of lung cancer if the probability of survival for any one of them is .70.

The binomial distribution is an example of a discrete, rather than a continuous, distribution because one can flip coins and obtain 3 heads or 4 heads, but not, for example, 3.897 heads. Similarly one can have 21 survivors or 22 survivors, but not anything in between.

Mathematically, the binomial distribution is defined as

$$p(X) = C_X^N p^X q^{(N-X)} = \frac{N!}{X!(N-X)!} p^X q^{(N-X)}$$

where

$p(X)$ = The probability of X successes

N = The number of trials

p = The probability of a success on any one trial

$q = (1 - p)$ = The probability of a failure on any one trial

C_X^N = The number of combinations of N things taken X at a time

The notation for combinations has been changed from r to X because the symbol X is used to refer to data. Whether we call something r or X is arbitrary; the choice is made for convenience or intelligibility.

success

failure

The words **success** and **failure** are used as arbitrary labels for the two alternative outcomes. If we are talking about cancer, the meaning is obvious. If we are talking about whether a driver will turn left or right at a fork, the designation is arbitrary. We will require that the trials be independent of one another, meaning that the result of trial$_i$ has no influence on trial$_j$.

To illustrate the application of this formula, suppose we are interested in studying the art of wine tasting and the relationship between quality and price. As part of our study, we ask a judge to taste two glasses of wine and pick the one that she thinks is the more expensive one. This task is repeated 10 times, each time with a different pair of wines. Assume for the moment that our wine taster really does not know the first thing about wines (or that quality and price are completely unrelated—which is sometimes the case). Assuming that there are no extraneous factors to bias the judge's decision (such as a tendency to choose the darker-colored wine), then on each trial the probability of her being correct (i.e., correctly identifying the more expensive wine) is .50 because there are only two wines to choose from. Now suppose we want to know the probability that our judge will somehow manage to make 9 (X) correct choices out of 10 (N) trials when the null hypothesis ($p = .50$) is true. The probability of being correct on any one trial is denoted p and equals .50, whereas the probability of being incorrect on any one trial is denoted q and also equals .50. Then we have

$$p(X) = \frac{N!}{X!(N-X)!} p^X q^{(N-X)}$$

$$p(9) = \frac{10!}{9!1!}(.50^9)(.50^1)$$

But $10! = 10 \cdot 9 \cdot 8 \cdot \cdots \cdot 2 \cdot 1 = 10 \cdot 9!$ so

$$p(9) = \frac{10 \cdot \cancel{9}!}{\cancel{9}!1!}(.50^9)(.50^1)$$

$$= 10(.001953)(.50) = .0098$$

Thus, the probability of making 9 correct choices out of 10 trials with $p = .50$ is remote, occurring approximately 1 time out of every 100 tasting sessions.

As a second example, the probability of 6 correct choices out of 10 trials is the probability of any one such outcome (p^6q^4) times the number of possible 6:4 outcomes C_6^{10}). Thus,

$$p(6) = \frac{N!}{X!(N-X)!}p^X q^{(N-X)}$$

$$= \frac{10!}{6!4!}(.5)^6(.5)^4$$

$$= \frac{10 \cdot 9 \cdot 8 \cdot 7 \cdot \cancel{6!}}{\cancel{6!}4 \cdot 3 \cdot 2 \cdot 1}(.5)^{10}$$

$$= \frac{5040}{24}(.00098)$$

$$= .2058$$

Plotting Binomial Distributions

You will notice that the probability of six correct choices is greater than the probability of nine of them. This is what we would expect because we are assuming that our judge is operating at random and would be right about as often as she is wrong. If we were to calculate the probabilities for each outcome between 0 and 10 correct out of 10, we would find the results shown in Table 5.2. Observe from this table that the sum of those probabilities is 1, reflecting the fact that all possible outcomes have been considered.

Now that we have calculated the probabilities of the individual outcomes, we can plot the distribution of the results, as has been done in Figure 5.4. Although this distribution resembles many of the distributions we have seen, it differs from them in two important ways. First, notice that the ordinate has been labeled "probability" instead of "frequency." This is because Figure 5.4 is not a frequency distribution at all but, rather, is a probability distribution. This distinction is important. With frequency, or relative frequency, distributions, we were plotting the obtained outcomes of some experiment—that is, we were plotting real data. Here we are not plotting real data; instead, we are plotting the probability that some event or another will occur.

To reiterate a point made earlier, the fact that the ordinate (Y-axis) represents probabilities instead of densities (as in the normal distribution) reflects the fact that the

Table 5.2   Binomial distribution for $p = .50, N = 10$

Number Correct	Probability
0	.001
1	.010
2	.044
3	.117
4	.205
5	.246
6	.205
7	.117
8	.044
9	.010
10	.001
	1.000

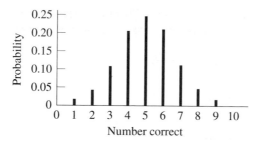

Figure 5.4   Binomial distribution when $N = 10$ and $p = .50$

binomial distribution deals with discrete rather than continuous outcomes. With a continuous distribution such as the normal distribution, the probability of any specified individual outcome is near 0. (The probability that you weight 158.214567 pounds is vanishingly small.) With a discrete distribution, however, the data fall into one or another of relatively few categories, and probabilities for individual events can be obtained easily. In other words, with discrete distributions we deal with the probability of individual events, whereas with continuous distributions we deal with the probability of intervals of events.

The second way this distribution differs from many others we have discussed is that although it is a sampling distribution, it is obtained mathematically rather than empirically. The values on the abscissa represent statistics (the number of successes as obtained in a given experiment) rather than individual observations or events. We have already discussed sampling distributions in Chapter 4, and what we said there applies directly to what we will consider in this chapter.

The Mean and Variance of a Binomial Distribution

In Chapter 2, we saw that it is possible to describe a distribution in many ways—we can discuss its mean, its standard deviation, its skewness, and so on. From Figure 5.4 we can see that the distribution for the outcomes for our judge is symmetric. This will always be the case for $p = q = .50$, but not for other values of p and q. Furthermore, the mean and standard deviation of any binomial distribution are easily calculated. They are always:

$$\text{Mean} = Np$$
$$\text{Variance} = Npq$$
$$\text{Standard deviation} = \sqrt{Npq}$$

For example, Figure 5.4 shows the binomial distribution when $N = 10$ and $p = .50$. The mean of this distribution is $10(.5) = 5$ and the standard deviation is $\sqrt{10(.5)(.5)} = \sqrt{2.5} = 1.58$.

We will see shortly that being able to specify the mean and standard deviation of any binomial distribution is exceptionally useful when it comes to testing hypotheses. First, however, it is necessary to point out two more considerations.

In the wine-tasting example, we dealt with a judge who was choosing at random ($p = q = .50$). Had we chosen to use a different judge—one who had a higher or lower probability of being correct on any one trial—the arithmetic would have been the same but the results would have been different. For purposes of illustration, three distributions obtained with different values of p are plotted in Figure 5.5.

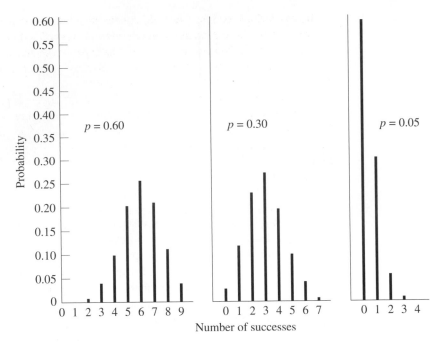

Figure 5.5 Binomial distributions for $N = 10$ and $p = .60, .30,$ and $.05$

For the distribution on the left of Figure 5.5, the judge knows a bit about wine and can be expected to choose the more expensive wine at slightly greater than chance levels with a probability of .60 of being correct on any given trial. The distribution in the middle represents the results expected from a judge who has a probability of only .30 of being correct on each trial. The distribution on the right represents the behavior of a judge with a nearly unerring ability to choose the *wrong wine*. On each trial, this judge had a probability of only .05 of being correct. From these three distributions, you can see that, for a given number of trials, as p and q depart more and more from .50, the distributions become more and more skewed although the mean and standard deviation are still Np and $\sqrt{Npq}$, respectively. Moreover, it is important to point out (although it is not shown in Figure 5.5, in which N is always 10) that as the number of trials increases, the distribution approaches normal, regardless of the values of p and q. As a rule of thumb, as long as both Np and Nq are greater than about 5, the distribution is close enough to normal that our estimates won't be far in error if we treat it as normal.

5.9 Using the Binomial Distribution to Test Hypotheses

Many of the situations for which the binomial distribution is useful in testing hypotheses are handled equally well by the chi-square test, discussed in Chapter 6. For that reason, this discussion will be limited to those cases for which the binomial distribution is uniquely useful.

In the previous sections, we dealt with the situation in which a person was judging wines, and we saw how to calculate the distribution of possible outcomes and their probabilities over $N = 10$ trials. Now suppose we turn the question around and ask whether the available data from a set of wine-tasting trials can be taken as evidence that our judge really can identify expensive wines at better than chance levels.

For example, suppose we had our judge make eight comparisons of an inexpensive wine with an expensive one, and the judge has been correct on seven out of eight trials. Do these data indicate that she is operating at a better than chance level? Put another way, are we likely to have seven out of eight correct choices if the judge is really operating by blind guessing?

Following the procedure outlined in Chapter 4, we can begin by stating our research hypothesis as the hypothesis that the judge knows a good wine when she tastes it (at least that is presumably what we set out to demonstrate). In other words, the *research* hypothesis (H_1) is that her performance is at better than chance levels ($p > .50$). (We have chosen a one-tailed test merely to simplify the example; in general, we would prefer to use a two-tailed test.) The *null* hypothesis is that the judge's behavior does not differ from chance ($H_0: p = .50$). The sampling distribution of the number of correct choices out of eight trials, given that the null hypothesis is true, is provided by the binomial distribution with $p = .50$. Rather than calculate the probability of each of the possible number of correct choices (as we did in Figure 5.5, for example), all we need to do is calculate the probability of seven correct choices and the probability of eight correct choices because we want to know the probability of our judge doing *at least* as well as she did if she were choosing randomly.

Letting N represent the number of trials (eight) and X represent the number of correct trials, the probability of seven correct trials out of eight is given by

$$p(X) = C_X^N p^X q^{(N-X)}$$

$$p(7) = C_7^8 p^7 q^1$$

$$= \frac{8!}{7!1!}(.5)^7(.5)^1 = 8(.0078)(.5) = 8(.0039) = .0312$$

Thus, the probability of making seven correct choices out of eight by chance is .0312. But we know that we test null hypotheses by asking questions of the form, "What is the probability of *at least* this many correct choices if H_0 is true?" In other words, we need to sum $p(7)$ and $p(8)$:

$$p(8) = C_8^8 p^8 q^0 = 1(.0039)(1) = .0039$$

Then

$$p(7) = .0312$$
$$+ p(8) = \underline{.0039}$$
$$p(7 \text{ or } 8) = .0351$$

Here we see that the probability of at least seven correct choices is approximately .035. Earlier, we said that we will reject H_0 whenever the probability of a Type I error (α) is less than or equal to .05. We have just determined that the probability of making at least seven correct choices out of eight is only .035 if H_0 is true (i.e., if $p = .50$), so we will reject

H_0 and conclude that our judge is performing at better than chance levels. In other words, her performance is better than we would expect if she were just guessing.[3]

The Sign Test

sign test

Another example of the use of the binomial to test hypotheses is one of the simplest tests we have: the **sign test.** Although the sign test is very simple, it is also very useful in a variety of settings. Suppose we hypothesize that when people know each other they tend to be more accepting of individual differences. As a test of this hypothesis, we asked a group of first-year male students matriculating at a small college to rate 12 target subjects (also male) on physical appearance (higher scores represent greater attractiveness). At the end of the first semester, when students have come to know one another, we again ask them to rate those same 12 targets. Assume we obtain the data in Table 5.3, where each entry is the median rating that person (target) received when judged by participants in the experiment on a 30-point scale.

Table 5.3   Median ratings of physical appearance at the beginning and end of the semester

Target	1	2	3	4	5	6	7	8	9	10	11	12
Beginning	12	21	10	8	14	18	25	7	16	13	20	15
End	15	22	16	14	17	16	24	8	19	14	28	18
Gain	3	1	6	6	3	−2	−1	1	3	1	8	3

The gain score in this table was computed by subtracting the score obtained at the beginning of the semester from the one obtained at the end of the semester. For example, the first target was rated 3 points higher at the end of the semester than at the beginning. Notice that in 10 of the 12 cases the score at the end of the semester was higher than at the beginning. In other words, the sign was positive. (The sign test gets its name from the fact that we look at the sign, but not the magnitude, of the difference.)

Consider the null hypothesis in this example. If familiarity does not affect ratings of physical appearance, we would not expect a systematic change in ratings (assuming that no other variables are involved). Ignoring tied scores, which we don't have anyway, we would expect that by chance about half the ratings would increase and half the ratings would decrease over the course of the semester. Thus, under H_0, $p(\text{higher}) = p(\text{lower}) = .50$. The binomial can now be used to compute the probability of obtaining at least 10 out of 12 improvements if H_0 is true:

$$p(10) = \frac{12!}{10!2!}(.5)^{10}(.5)^2 = .0161$$

$$p(11) = \frac{12!}{11!1!}(.5)^{11}(.5)^1 = .0029$$

$$p(12) = \frac{12!}{12!0!}(.5)^{12}(.5)^0 = .0002$$

---

[3] One problem with discrete distributions is that rarely is there a set of outcomes with a probability of exactly .05. In our particular example with 7 correct guesses, we rejected the null because $p = .035$. If we had found 6 correct choices the probability would have been .133, and we would have failed to reject the null. There is no possible outcome with a tail area of exactly .05. So we are faced with the choice of a case where the critical value is either too conservative or two liberal. One proposal that has been seriously considered is to use what is called the "mid-p" value, which takes one half of the probability of the observed outcome, plus all the probabilities of more extreme outcomes. For a discussion of this approach, see Berger (2005).

From these calculations we see that the probability of at least 10 improvements = .0161+ .0029 + .0002 = .0192 if the null hypothesis is true and ratings are unaffected by familiarity. Because this probability is less than our traditional cutoff of .05, we will reject H_0 and conclude that ratings of appearance have increased during the semester. (Although variables other than familiarity could explain this difference, at the very least our test has shown that there is a significant difference to be explained.)

5.10 The Multinomial Distribution

multinomial distribution

The binomial distribution we have just examined is a special case of a more general distribution, the **multinomial distribution.** In binomial distributions, we deal with events that can have only one of two outcomes—a coin could land heads or tails, a wine could be judged as more expensive or less expensive, and so on. In many situations, however, an event can have more than two possible outcomes—a roll of a die has six possible outcomes; a maze might present three choices (right, left, and center); political opinions could be classified as For, Against, or Undecided. In these situations, we must invoke the more general multinomial distribution.

If we define the probability of each of k events (categories) as $p_1, p_2, \ldots, p_k$ and want to calculate the probability of exactly X_1 outcomes of event$_1$, X_2 outcomes of event$_2$, ..., X_k outcomes of event$_k$, this probability is given by

$$p(X_1, X_2, \ldots, X_k) = \frac{N!}{X_1! X_2! \cdots X_k!} p_1^{X_1} p_2^{X_2} \cdots p_k^{X_k}$$

where N has the same meaning as in the binomial. Note that when $k = 2$ this is actually the binomial distribution, where $p_2 = 1 - p_1$ and $X_2 = N - X_1$.

As a brief illustration, suppose we had a die with two black sides, three red sides, and one white side. If we roll this die, the probability of a black side coming up is 2/6 = .333, the probability of a red is 3/6 = .500, and the probability of a white is 1/6 = .167. If we roll the die 10 times, what is the probability of obtaining exactly four blacks, five reds, and one white? This probability is given as

$$p(4, 5, 1) = \frac{10!}{4! 5! 1!} (.333)^4 (.500)^5 (.167)^1$$
$$= 1260 (.333)^4 (.500)^5 (.167)^1 = 1260 (.000064)$$
$$= .081$$

At this point, this is all we will say about the multinomial. It will appear again in Chapter 6, when we discuss chi-square, and forms the basis for some of the other tests you are likely to run into in the future.

Key Terms

Analytic view (5.1)

Frequentist view (5.1)

Sample with replacement (5.1)

Subjective probability (5.1)

Event (5.2)

Independent (5.2)

Mutually exclusive (5.2)

Exhaustive (5.2)

Additive law of probability (5.2)

Multiplicative law of probability (5.2)

Joint probability (5.2)

Conditional probability (5.2)

Unconditional probability (5.2)

Density (5.5)

Combinatorics (5.6)

Exercises

5.1   Give an example of an analytic, a relative-frequency, and a subjective view of probability.

5.2   Assume that you have bought a ticket for the local fire department lottery and that your brother has bought two tickets. You have just read that 1,000 tickets have been sold.

   a.   What is the probability that you will win the grand prize?

   b.   What is the probability that your brother will win?

   c.   What is the probability that you or your brother will win?

5.3   Assume the same situation as in Exercise 5.2, except that a total of only 10 tickets were sold and there are two prizes.

   a.   Given that you don't win first prize, what is the probability that you will win second prize? (The first prize-winning ticket is not put back in the hopper.)

   b.   What is the probability that your brother will win first prize and you will win second prize?

   c.   What is the probability that you will win first prize and your brother will win second prize?

   d.   What is the probability that the two of you will win the first and second prizes?

5.4   Which parts of Exercise 5.3 deal with joint probabilities?

5.5   Which parts of Exercise 5.3 deal with conditional probabilities?

5.6   Make up a simple example of a situation in which you are interested in joint probabilities.

5.7   Make up a simple example of a situation in which you are interested in conditional probabilities.

5.8   In some homes, a mother's behavior seems to be independent of her baby's, and vice versa. If the mother looks at her child a total of 2 hours each day, and the baby looks at the mother a total of 3 hours each day, and if they really do behave independently, what is the probability that they will look at each other at the same time?

5.9   In Exercise 5.8, assume that both the mother and child are asleep from 8:00 p.m. to 7:00 a.m. What would the probability be now?

5.10   In the example dealing with what happens to supermarket flyers, we found that the probability that a flyer carrying a "do not litter" message would end up in the trash if what people do with flyers is independent of the message that is on them, was .033. I also said that 4.5% of those messages actually ended up in the trash. What does this tell you about the effectiveness of messages?

5.11   Give an example of a common continuous distribution for which we have some real interest in the probability that an observation will fall within some specified interval.

5.12   Give an example of a continuous variable that we routinely treat as if it were discrete.

5.13   Give two examples of discrete variables.

5.14   A graduate-admissions committee has finally come to realize that it cannot make valid distinctions among the top applicants. This year, the committee rated all 300 applicants and randomly chose 10 from those in the top 20%. What is the probability that any particular applicant will be admitted (assuming you have no knowledge of her or his rating)?

5.15   With respect to Exercise 5.14,

   a.   What is the conditional probability that a person will be admitted given that she has the highest faculty rating among the 300 students?

   b.   What is the conditional probability given that she has the lowest rating?

5.16   Using the file ADD.dat (from www.uvm.edu/~dhowell/methods/),

   a.   What is the probability that a person drawn at random will have an ADDSC score greater than 50 if the scores are normally distributed with a mean of 52.6 and a standard deviation of 12.4?

   b.   What percentage of the sample actually exceeded 50?

5.17   Using the file ADD.dat,

   a.   What is the probability that a male will have an ADDSC score greater than 50 if the scores are normally distributed with a mean of 54.3 and a standard deviation of 12.9?

   b.   What percentage of the male sample actually exceeded 50?

5.18   Using the file ADD.dat, what is the empirical probability that a person will drop out of school given that he or she has an ADDSC score of at least 60? Here we do not need to assume normality.

5.19   How might you use conditional probabilities to determine if an ADDSC cutoff score in ADD.dat of 66 is predictive of whether or not a person will drop out of school?

5.20   Using the file ADD.dat scores, compare the conditional probability of dropping out of school given an ADDSC score of at least 60, which you computed in Exercise 5.18, with the unconditional probability that a person will drop out of school regardless of his or her ADDSC score.

5.21   In a 5-choice task, subjects are asked to choose the stimulus that the experimenter has arbitrarily determined to be correct; the 10 subjects can only guess on the first trial. Plot the sampling distribution of the number of correct choices on trial 1.

5.22   Refer to Exercise 5.21. What would you conclude if 6 of 10 subjects were correct on trial 2?

5.23   Refer to Exercise 5.21. What is the minimum number of correct choices on a trial necessary for you to conclude that the subjects as a group are no longer performing at chance levels?

5.24   People who sell cars are often accused of treating male and female customers differently. Make up a series of statements to illustrate simple, joint, and conditional probabilities with respect to such behavior. How might we begin to determine if those accusations are true?

5.25   Assume you are a member of a local human rights organization. How might you use what you know about probability to examine discrimination in housing?

5.26   In a study of human cognition, we want to look at recall of different classes of words (nouns, verbs, adjectives, and adverbs). Each subject will see one of each. We are afraid that there may be a sequence effect, however, and want to have different subjects see the different classes in a different order. How many subjects will we need if we are to have one subject per order?

5.27   Refer to Exercise 5.26. Assume we have just discovered that, because of time constraints, each subject can see only two of the four classes. The rest of the experiment will remain the same, however. Now how many subjects do we need? (*Warning:* Do not actually try to run an experiment like this unless you are sure you know how you will analyze the data.)

5.28   In a learning task, a subject is presented with five buttons. He must learn to press a certain three of them in a predetermined order. What chance does that subject have of pressing correctly on the first trial?

5.29 An ice-cream shop has six different flavors of ice cream, and you can order any combination of any number of them (but only one scoop of each flavor). How many different ice-cream cone combinations could they truthfully advertise? (We do not care if the Oreo Mint is above or below the Raspberry-Pistachio. Each cone must have at least one scoop of ice cream—an empty cone doesn't count.)

5.30 We are designing a study in which six external electrodes will be implanted in a rat's brain. The six-channel amplifier in our recording apparatus blew two channels when the research assistant took it home to run her stereo. How many different ways can we record from the brain? (It makes no difference what signal goes on which channel.)

5.31 In a study of knowledge of current events, we give a 20-item true–false test to a class of college seniors. One of the not-so-alert students gets 11 answers right. Do we have any reason to believe that he has done anything other than guess?

5.32 Earlier in this chapter I stated that the probability of drawing 16 dark caramels out of 60 draws, with replacement, was .0077. Reproduce that result. (*Warning:* your calculator will be computing some very large numbers, which may lead to substantial rounding error. The value of .0077 is what my calculator produced.)

5.33 This question is not an easy one, and requires putting together material in Chapters 3, 4, and 5. Suppose we make up a driving test that we have good reason to believe should be passed by 60% of all drivers. We administer it to 30 drivers, and 22 pass it. Is the result sufficiently large to cause us to reject H_0 ($p = .60$)? This problem is too unwieldy to be approached by solving the binomial for $x = 22, 23, \ldots, 30$. But you do know the mean and variance of the binomial, and something about its shape. With the aid of a diagram of what the distribution would look like, you should be able to solve the problem.

5.34 Make up a simple experiment for which a sign test would be appropriate.

a. Create reasonable data and run the test.

b. Draw the appropriate conclusion.

Discussion Questions

5.35 The "law of averages," or the "gambler's fallacy," is the oft-quoted belief that if random events have come out one way for a number of trials, they are "due" to come out the other way on one of the next few trials. (For example, it is the (mistaken) belief that if a fair coin has come up heads on 18 out of the last 20 trials, it has a better than 50:50 chance of coming up tails on the next trial to balance things out.) The gambler's fallacy is just that, a fallacy—coins have an even worse memory of their past performance than I do. Ann Watkins, in the spring 1995 edition of *Chance* magazine, reported a number of instances of people operating as if the "law of averages" were true. One of the examples that Watkins gave was a letter to Dear Abby in which the writer complained that she and her husband had just had their eighth child and eighth girl. She criticized fate and said that even her doctor had told her that the law of averages was in her favor 100 to 1. Watkins also cited another example in which the writer noted that fewer English than American men were fat, but the English must be fatter to keep the averages the same. And, finally, she quotes a really remarkable application of this (non-) law in reference to Marlon Brando: "Brando has had so many lovers, it would only be surprising if they were all of one gender; the law of averages alone would make him bisexual." (*Los Angeles Times,* 18 September 1994, Book Reviews, p. 13) What is wrong with each of these examples? What underlying belief system would seem to lie behind such a law? How might you explain to the woman who wrote to Dear Abby that she really wasn't owed a boy to "make up" for all those girls?

5.36 At age 40, 1% of women can be expected to have breast cancer. Of those women with breast cancer, 80% will have positive mammographies. In addition, 9.6% of women who do not have breast cancer will have a positive mammography. If a woman in this age group tests positive for breast cancer, what is the probability that she actually has it? Use Bayes' theorem to solve this problem. (*Hint*: Letting BC stand for "breast cancer," we have $p(BC) = .01$, $p(+|BC) = .80$, and $p(+|\overline{BC}) = .096$. You want to solve for $p(BC|+)$.)

5.37   The answer that you found in 5.36 is probably much lower than the answer that you expected knowing that 80% of women with breast cancer have positive mammographies. Why is it so low?

5.38   What would happen to the answer to Exercise 5.36 if we were able to refine our test so that only 5% of women without breast cancer test positive? (In other words, we reduce the rate of false positives.)

CHAPTER 6

Categorical Data and Chi-Square

Objectives

To present the chi-square test as a procedure for testing hypotheses when the data are categorical and to examine other measures that clarify the meaning of our results.

Contents

IN SAINT-EXUPÉRY'S, *The Little Prince,* the narrator, remarking that he believes the prince came from an asteroid known as B-612, explains his attention to such a trivial detail as the precise number of the asteroid with the following comment:

> Grown-ups love figures. When you tell them you have made a new friend, they never ask you any questions about essential matters. They never say to you, "What does his voice sound like? What games does he love best? Does he collect butterflies?" Instead they demand: "How old is he? How many brothers has he? How much does he weigh? How much does his father make?" Only from these figures do they think they have learned anything about him.[1]

In some ways, the first chapters of this book have concentrated on dealing with the kinds of numbers Saint-Exupéry's grown-ups like so much. This chapter will be devoted to the analysis of largely nonnumerical data.

In Chapter 1, I drew a distinction between measurement data (sometimes called quantitative data) and categorical data (sometimes called frequency data). When we deal with measurement data, each observation represents a score along some continuum, and the most common statistics are the mean and the standard deviation. When we deal with categorical data, on the other hand, the data consist of the frequencies of observations that fall into each of two or more categories ("Does your friend have a gravelly voice or a high-pitched voice?" or "Is he a collector of butterflies, coins, or baseball cards?").

In Chapter 5, we examined the use of the binomial distribution to test simple hypotheses. In those cases, we were limited to situations in which an individual event had one of only two possible outcomes, and we merely asked whether, over repeated trials, one outcome occurred (statistically) significantly more often than the other.

In this chapter, we will expand the kinds of situations that we can evaluate. We will deal with the case in which a single event can have two *or more* possible outcomes, and then with the case in which we have two variables and we want to test null hypotheses concerning their independence. For both of these situations, the appropriate statistical test will be the chi-square (χ^2) test.

chi-square (χ^2)

The term **chi-square** (χ^2) has two distinct meanings in statistics, which leads to some confusion. In one meaning, it is used to refer to a particular mathematical distribution that exists in and of itself without any necessary referent in the outside world. In the second meaning, it is used to refer to a statistical test that has a resulting test statistic distributed in approximately the same way as the χ^2 distribution. When you hear someone refer to chi-square, they usually have this second meaning in mind. (The test itself was developed by Karl Pearson [1900] and is often referred to as **Pearson's chi-square** to distinguish it from other tests that also produce a χ^2 statistic—for example, Friedman's test, discussed in Chapter 18, and the likelihood ratio tests discussed at the end of this chapter and in Chapter 17.) You need to be familiar with both meanings of the term, however, if you are to use the test correctly and intelligently and if you are to understand many of the other statistical procedures that follow.

Pearson's chi-square

6.1 The Chi-Square Distribution

chi-square (χ^2) distribution

The **chi-square (χ^2) distribution** is the distribution defined by

$$f(\chi^2) = \frac{1}{2^{\frac{k}{2}}\Gamma(k/2)}\chi^{2[(k/2)-1]}e^{\frac{-(\chi^2)}{2}}$$

---

[1] Antoine de Saint-Exupéry, *The Little Prince,* trans. Katherine Woods (New York: Harcourt Brace, 1943), pp. 15–16.

This is a rather messy-looking function and most readers will be pleased to know that they will not have to work with it in any arithmetic sense. We do need to consider some of its features, however, to understand what the distribution of χ^2 is all about. The first thing that should be mentioned, if only in the interest of satisfying healthy curiosity, is that the term $\Gamma(k/2)$ in the denominator, called a **gamma function,** is related to what we normally mean by *factorial*. In fact, when the argument of gamma $(k/2)$ is an integer, then $\Gamma(k/2) = [(k/2) - 1]!$. We need gamma functions in part because arguments are not always integers. Mathematical statisticians have a lot to say about gamma, but we'll stop here.

gamma function

A second and more important feature of this equation is that the distribution has only one parameter (k). Everything else is either a constant or else the value of χ^2 for which we want to find the ordinate $[f(\chi^2)]$. Whereas the normal distribution was a two-parameter function, with μ and σ as parameters, χ^2 is a one-parameter function with k as the only parameter. When we move from the mathematical to the statistical world, k will become our degrees of freedom. (We often signify the degrees of freedom by subscripting χ^2. Thus, χ_3^2 is read "chi-square with three degrees of freedom." Alternatively, some authors write it as $\chi^2(3)$.)

Figure 6.1 shows the plots for several different χ^2 distributions, each representing a different value of k. From this figure, we can see that the distribution changes markedly with changes in k, becoming more symmetric as k increases. It is also apparent that the mean and variance of each χ^2 distribution increase with increasing values of k and are directly related to k. It can be shown that in all cases

$$\text{Mean} = k$$
$$\text{Variance} = 2k$$

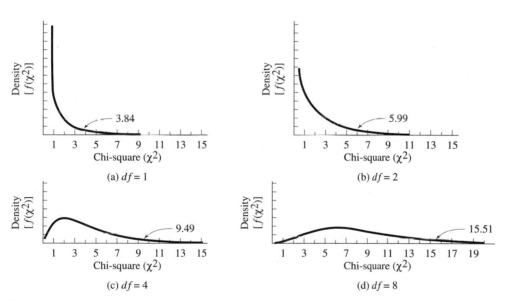

(a) $df = 1$ (b) $df = 2$

(c) $df = 4$ (d) $df = 8$

Figure 6.1   Chi-square distributions for $df = 1, 2, 4,$ and 8 (arrows indicate critical values at alpha $= .05$)

6.2 The Chi-Square Goodness-of-Fit Test—One-Way Classification

chi-square test

We now turn to what is commonly referred to as the **chi-square test,** which is based on the χ^2 distribution. We will first examine the test as it is applied to one-dimensional tables and then as applied to two-dimensional tables (contingency tables).

The following example is based on one of the most famous experiments in animal learning, conducted by Tolman, Ritchie, and Kalish (1946). At the time of the original study, Tolman was engaged in a theoretical debate with Clark Hull and the latter's students on whether a rat in a maze learns a discrete set of motor responses (Hull) or forms some sort of cognitive map of the maze and responds on the basis of that map (Tolman). At issue was the fundamental question of whether animals learn by stimulus-response conceptions or whether there is room for a cognitive interpretation of animal behavior. (To put this in less academic language, "Do animals think?" Though that doesn't seem like such a radical question now, I assure you that it was a very radical question in the 1940s.) The statistical test in question is called a **goodness-of-fit test** because it asks whether there is a "good fit" between the data (observed frequencies) and the theory (expected frequencies).

goodness-of-fit test

In a simple and ingenious experiment, Tolman and his colleagues first taught a rat to run down a starting alley of a maze into a large circular area. From the circular area, another alley exited straight across from the entrance but then turned and ended up in a goal box, which was actually to the right of the circular area. After the rats had learned the task ("go to the circular area and exit straight across"), Tolman changed the task by making the original exit alley a dead end and by adding several new alleys, one of which pointed in the direction of the original goal box. Thus, the rat had several choices, one of which included the original alley and one of which included a new alley that pointed directly toward the goal. The maze is shown in Figure 6.2, with the original exit alley drawn with solid lines and the new alleys drawn with dotted lines. If Hull was correct, the rat would learn a stimulus-response sequence during the first part of the experiment and would therefore continue to make the same set of responses, thus entering the now dead-end alley. If Tolman was right and the rat learned a cognitive map of the situation, then the rat would enter the alley on the *right* because it knew that the food was "over there to the right." As Tolman was the one who published the study, you can probably guess how it came out—the rats chose the alley on the right more often than the others. But we still need some way of testing whether the preference for the alley on the right was the result of chance (the rats entered the alleys at random) or whether the data support a general preference for the right alley. Do the data represent a "good fit" to a random choice model? Tolman certainly hoped not because he wanted to show that they had learned something.

It certainly looks as if animals were choosing Alley D much more than the others, which is what Tolman expected, but how can we be sure?

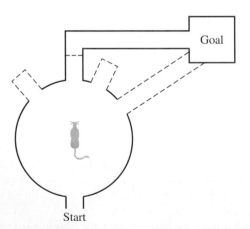

	Alley Chosen			
	A	B	C	D
Observed	4	5	8	15
Expected	8	8	8	8

Figure 6.2 Schematic diagram of Tolman's maze. The alleys are labeled A through D from left to right.

**observed
frequencies**

**expected
frequencies**

The most common and important formula for χ^2 involves a comparison of observed and expected frequencies. The **observed frequencies,** as the name suggests, are the frequencies you actually observed in the data—the numbers in the table in Figure 6.2. The **expected frequencies** are the frequencies you would expect *if the null hypothesis were true.* We want to test the null hypothesis that rats enter alleys at random. In this case, we have 32 rats, each making independent choices. (If we used the same four rats 8 times, we would probably have strong reservations about this assumption of independence.) We have four alleys, so if the rats are responding at random, rather than on the basis of what they have learned about the maze, we would expect that one-quarter of them would enter each alley. That means that we would expect frequencies of 8 for each alley. Instead, we got frequencies of 4, 5, 8, and 15. The standard formula for the chi-square test looks at the difference between these observed and expected frequencies.

$$\chi^2 = \sum \frac{(O - E)^2}{E}$$

This formula should make a certain amount of intuitive sense. Start with the numerator. If the null hypothesis is true, the observed and expected frequencies (O and E) would be reasonably close together and the numerator would be small, even after it is squared. Moreover, how large the difference between O and E would be ought to depend on how large a number we expected. If we were talking about 1,000 animals entering each alley, an $O - E$ difference of 5 would be trivial. But if we expected 8 animals to enter each alley, an $O - E$ difference of 5 would be substantial. To keep the squared size of the difference in perspective relative to the number of observations we expect, we divide the former by the later. Finally, we sum all of the alleys to combine these relative differences. (If you wonder why we square the numerator, work out what would happen with these, or any other data, if we did not.)

First, I will go ahead and calculate the χ^2 statistic for these data using the observed and expected frequencies given in the table.

$$\chi^2 = \sum \frac{(O - E)^2}{E}$$
$$= \frac{(4 - 8)^2}{8} + \frac{(5 - 8)^2}{8} + \frac{(8 - 8)^2}{8} + \frac{(15 - 8)^2}{8}$$
$$= 9.25$$

The Tabled Chi-Square Distribution

**tabled
distribution of χ^2**

Now that we have obtained a value of χ^2, we must refer it to the χ^2 distribution to determine the probability of a value of χ^2 at least this extreme if the null hypothesis of a chance distribution were true. We can do this through the use of the standard tabled distribution of χ^2.

The **tabled distribution of χ^2,** like that of most other statistics, differs in a very important way from the *tabled* standard normal distribution that we saw in Chapter 3. We will use a simple illustration. Consider the distribution of χ^2 for 1 *df* shown in Figure 6.1. Although it is certainly true that we could construct a table of exactly the same form as that for the standard normal distribution, allowing us to determine what percentage of the values are greater than any arbitrary value of χ^2, this would be tremendously time-consuming and wasteful. We would have to make up a new table for every reasonable number of degrees of freedom. It is not uncommon to want as many as 30 *df*, which would require 30 separate tables, each the size of Appendix *z*. Such a procedure would be particularly wasteful because most users would need only a small fraction of each of these tables. If we want to reject H_0 at the .05 level, all that we really care about is whether our value of χ^2 is greater or less than the value of χ^2 that cuts off the upper 5% of the distribution. Thus, for our particular purposes, all we need to know is the 5% cutoff point for each *df*. Other people might want the 2.5% cutoff, 1% cutoff, and so on, but it is hard to imagine wanting the 17% cutoff, for example. Thus, tables of χ^2 such as the one given in Appendix χ^2, part of which is reproduced in Table 6.1, are designed to supply only those values that might be of general interest.

Table 6.1 Upper percentage points of the χ^2 distribution

df	.995	.990	.975	.950	.900	.750	.500	.250	.100	.050	.025	.010	.005
1	0.00	0.00	0.00	0.00	0.02	0.10	0.45	1.32	2.71	3.84	5.02	6.63	7.88
2	0.01	0.02	0.05	0.10	0.21	0.58	1.39	2.77	4.61	5.99	7.38	9.21	10.60
3	0.07	0.11	0.22	0.35	0.58	1.21	2.37	4.11	6.25	**7.82**	9.35	11.35	12.84
4	0.21	0.30	0.48	0.71	1.06	1.92	3.36	5.39	7.78	9.49	11.14	13.28	14.86
5	0.41	0.55	0.83	1.15	1.61	2.67	4.35	6.63	9.24	11.07	12.83	15.09	16.75
6	0.68	0.87	1.24	1.64	2.20	3.45	5.35	7.84	10.64	12.59	14.45	16.81	18.55
7	0.99	1.24	1.69	2.17	2.83	4.25	6.35	9.04	12.02	14.07	16.01	18.48	20.28
8	1.34	1.65	2.18	2.73	3.49	5.07	7.34	10.22	13.36	15.51	17.54	20.09	21.96
9	1.73	2.09	2.70	3.33	4.17	5.90	8.34	11.39	14.68	16.92	19.02	21.66	23.59
...	...	...	...	...	...	...	...	...	...	...	...	...	...

Look for a moment at Table 6.1. Down the leftmost column you will find the degrees of freedom. In each of the other columns, you will find the critical values of χ^2 cutting off the percentage of the distribution labeled at the top of that column. Thus, for example, you will see that for 3 *df* a χ^2 of 7.82 cuts off the upper 5% of the distribution. (Note the boldfaced entry in Table 6.1.)

Returning to our example, we have found a value of $\chi^2 = 9.25$ on 3 *df*. We have already seen that, with 3 *df*, a χ^2 of 7.82 cuts off the upper 5% of the distribution. Because our obtained value $(\chi^2_{obt}) = 9.25$ is greater than $\chi^2_{.05} = 7.82$, we reject the null hypothesis and conclude that the obtained frequencies differed from those expected under the null hypothesis by more than could be attributed to chance.[2] In other words, Tolman's rats were not behaving randomly—they look as if they knew what they were doing.

6.3 Two Classification Variables: Contingency Table Analysis

In the previous example, we considered the case in which data are categorized along only one dimension (classification variable). Often, however, data are categorized with respect to two (or more) variables, and we are interested in asking whether those variables are independent of one another. To put this in the reverse, we often are interested in asking whether the distribution of one variable is *contingent* on a second variable. In this situation, we will construct a **contingency table** showing the distribution of one variable at each level of the other. An excellent example is offered by a study by Pugh (1983) on the "blaming the victim" phenomenon in prosecutions for rape.

contingency table

Pugh conducted a thorough and complex study examining how juries come to decisions in rape cases. He examined a number of variables, but we will collapse two of them

---

[2] Notice that here the subscript for χ^2 (i.e., obt and .05) do not refer to the degrees of freedom, but designate either the obtained value of $\chi^2 [\chi^2_{obt}]$ or the value of χ^2 that cuts off the largest 5% of the distribution $[\chi^2_{.05}]$. When we want to designate both the degrees of freedom and the level of alpha we write something like $\chi^2_{.05}(1) = 3.84$.

and simply look at his data about (1) whether the defendant was found innocent or guilty, and (2) whether the defense alleged that the victim was somehow partially at fault for the rape. Pugh's actual data are presented in Table 6.2 in the form of such a contingency table.

Table 6.2   Pugh's data on decisions in rape cases

| | Verdict | | |
Fault	Guilty	Not Guilty	Total
Low	153 (127.559)	24 (49.441)	177
High	105 (130.441)	76 (50.559)	181
Total	258	100	358

For the moment, ignore the numbers in parentheses. This table shows some evidence that jurors assign guilt partly on the basis of the perceived faults of the victim. Notice that when the victim was seen as low in fault, approximately 86% (153/177) of the time the defendant was found guilty. On the other hand, when the victim was seen as high in fault, the defendant was found guilty only 58% (105/181) of the time.

Expected Frequencies for Contingency Tables

marginal totals

cell

row total

column totals

The expected frequencies in a contingency table represent those frequencies that we would expect if the two variables forming the table (here, guilt and victim blame) were independent. For a contingency table, the expected frequency for a given cell is obtained by multiplying together the totals for the row and column in which the cell is located and dividing by the total sample size (N). (These totals are known as **marginal totals,** because they sit at the margins of the table.) If E_{ij} is the expected frequency for the **cell** in row i and column j, R_i and C_j are the corresponding **row** and **column totals,** and N is the total number of observations, we have the following formula[3]:

$$E_{ij} = \frac{R_i C_j}{N}$$

For our example

$$E_{11} = \frac{177 \times 258}{358} = 127.559$$

$$E_{12} = \frac{177 \times 100}{358} = 49.441$$

$$E_{21} = \frac{181 \times 258}{358} = 130.441$$

$$E_{22} = \frac{181 \times 100}{358} = 50.559$$

These values are shown in parentheses in Table 6.2.

---

[3] This formula for the expected values is derived directly from the formula for the probability of the joint occurrence of two *independent* events given in Chapter 5 on probability. For this reason, the expected values that result are those that would be expected if H_0 were true and the variables were independent. A large discrepancy in the fit between expected and observed would reflect a large departure from independence, which is what we want to test.

Calculation of Chi-Square

Now that we have the observed and expected frequencies in each cell, the calculation of χ^2 is straightforward. We simply use the same formula that we have been using all along, although we sum our calculations over all cells in the table.

$$\chi^2 = \sum \frac{(O - E)^2}{E}$$
$$= \frac{(153 - 127.559)^2}{127.559} + \frac{(24 - 49.441)^2}{49.441} + \frac{(105 - 130.441)^2}{130.441} + \frac{(76 - 50.559)^2}{50.559}$$
$$= 35.93$$

Degrees of Freedom

Before we can compare our value of χ^2 to the value in Appendix χ^2, we must know the degrees of freedom. For the analysis of contingency tables, the degrees of freedom are given by

$$df = (R - 1)(C - 1)$$

where

$R =$ the number of rows in the table

and

$C =$ the number of columns in the table

For our example we have $R = 2$ and $C = 2$; therefore, we have $(2 - 1)(2 - 1) = 1$ df. It may seem strange to have only 1 df when we have four cells, but you can see that once you know the row and column totals, you need to know only one cell frequency to be able to determine the rest.

Evaluation of χ^2

With 1 df, the critical value of χ^2, as found in Appendix χ^2, is 3.84. Because our value of 35.93 exceeds the critical value, we will reject the null hypothesis that the variables are independent of each other. In this case, we will conclude that whether a defendant is found guilty depends in part on whether the victim is portrayed by the defending lawyer as being at fault for the rape. How do these results fit with how you think you would judge the case?

Correcting for Continuity

Yates's correction for continuity

Many books advocate that for simple 2×2 tables such as Table 6.2, we should employ what is called **Yates's correction for continuity,** especially when the expected frequencies are small. (The correction merely involves reducing the absolute value of each numerator by 0.5 units before squaring.) There is an extensive literature debating the pros and cons of Yates's correction, with firmly held views on both sides. However, the common availability of Fisher's Exact Test, to be discussed next, makes Yates's correction superfluous.

Fisher's Exact Test

Fisher introduced what is called Fisher's Exact Test in 1934 at a meeting of the Royal Statistical Society. (Good [2001] has pointed out that one of the speakers who followed Fisher referred to Fisher's presentation as "the braying of the Golden Ass." Statistical debates at that time were far from boring, and no doubt Fisher had something equally kind to say about his critic.)

Without going into details, Fisher's proposal was to take all possible 2×2 tables that could be formed from the fixed set of marginal totals. He then determined the proportion of those tables whose results are as extreme, or more so, than the table we obtained from our data. If this proportion is less than α, we reject the null hypothesis that the two variables are independent and conclude that there is a statistically significant relationship between the two variables that make up our contingency table. (This is classed as a *conditional test* because it is conditioned on the marginal totals actually obtained, instead of all possible marginal totals given the total sample size.) I am assuming that you will do the calculations using statistical software rather than by hand.

fixed marginals

Fisher's Exact Test has been controversial since he proposed it. One problem concerns the fact that it is a conditional test (conditional on the **fixed marginals**). Some have argued that if you repeated the experiment exactly, you would likely find different marginal totals and have asked why those additional tables should not be included in the calculation. Making the test unconditional on the marginals complicates the calculations considerably. This may sound like an easy debate to resolve, but if you read the extensive literature surrounding fixed and random marginals, you will find that it is a difficult debate to follow and you will probably come away thoroughly confused. (An excellent discussion of some of the issues can be found in Agresti (2002), pages 95–96.)

Fisher's Exact Test also leads to controversy because of the issue of one-tailed versus two-tailed tests and what outcomes would constitute a "more extreme" result in the opposite tail. Instead of going into how to determine what is a more extreme outcome, I will avoid that complication by simply telling you to decide in advance whether you want a one- or a two-tailed test, and then to report the values given by standard statistical software. (I haven't given you any calculational formula for Fisher's Exact Test because I cannot imagine that you would ever do the calculations by hand.) Virtually all common statistical software prints out Fisher's Exact Test results along with Pearson's chi-square and related test statistics.

Fisher's Exact Test versus Pearson's Chi-Square

We now have at least two statistical tests for 2×2 contingency tables, and will soon have a third—which one should we use? Probably the most common solution is to go with Pearson's chi-square; perhaps because "that is what we have always done." In previous editions of this book I recommended against Fisher's Exact Test, primarily because of the conditional nature of it. However, in recent years there has been an important growth of interest in permutation and randomization tests, of which Fisher's Exact Test is an example. (This approach is discussed extensively in Chapter 18.) I am extremely impressed with the logic and simplicity of such tests and have come to side with Fisher's Exact Test. In most cases, the conclusion you will draw will be the same for the two approaches, though this is not always the case. When we come to tables larger than 2×2, Fisher's approach does not apply, without modification, and there we almost always use the Pearson chi-square. (But see Howell & Gordon, 1976.)

6.4 Chi-Square for Larger Contingency Tables

The Pugh example involved two variables (Verdict and Fault), each of which had two levels. We referred to this design as a 2×2 contingency table; it is a special case of the more general $R \times C$ designs, where, again, R and C represent the number of rows and columns.

Table 6.3   Data from Geller, Witmer, and Orebaugh (1976)
(expected frequencies in parentheses)

	Location			
Instructions	Trash can	Litter	Removed	
Control	41 (61.66)	385 (343.98)	477 (497.36)	903
Message	80 (59.34)	290 (331.02)	499 (478.64)	869
	121	675	976	1772

As an example of a larger contingency table, consider the study by Geller, Witmer, and Orebaugh (1976) mentioned in Chapter 5. These authors were studying littering behavior and were interested, among other things, in whether a message about not littering would be effective if placed on the handbills that are often given out in supermarkets advertising the daily specials. To oversimplify a fairly complex study, two of Geller's conditions involved passing out handbills in a supermarket. Under one condition (Control), the handbills contained only a listing of the daily specials. In the other condition (Message), the handbills also included the notation, "Please don't litter. Please dispose of this properly." At the end of the day, Geller and his students searched the store for handbills. They recorded the number that were found in trash cans; the number that were left in shopping carts, on the floor, and various places where they didn't belong (denoted litter); and the number that could not be found and were apparently removed from the premises. The data obtained under the two conditions are shown in Table 6.3 and are taken from a larger table reported by Geller et al. Expected frequencies are shown in parentheses and were obtained exactly as they were in the previous example [$E_{ij} = (R_i)(C_j)/N$].

The calculation of χ^2 is carried out just as it was earlier:

$$\chi^2 = \sum \frac{(O - E)^2}{E}$$
$$= \frac{(41 - 61.66)^2}{61.66} + \frac{(385 - 343.98)^2}{343.98} + \cdots + \frac{(499 - 478.64)^2}{478.64}$$
$$= 25.79$$

There are two *df* for Table 6.3 because $(R - 1)(C - 1) = (2 - 1)(3 - 1) = 2$. The critical value of $\chi^2_{.05} = 5.99$. Our value of 25.79 is larger than 5.99, so we are led to reject H_0 and to conclude that the location in which the handbills were left depended to some extent on the instructions given. In other words, Instructions and Location are not independent. From the data, it is evident that when subjects were asked not to litter, a higher percentage of handbills were thrown in the trash can or taken out of the store, and fewer were left lying in shopping carts or on floors and shelves.

As we have seen, the chi-square test can be applied to two-dimensional tables of any size (and, in some situations, tables of more dimensions). The calculations are always the same. The problem with larger tables, however, is one of interpretation. If a 2 × 2 chi-square is statistically significant, it is usually pretty obvious what the results mean. We just have to look at the contingency table. But with larger tables, it is not always clear. In the Geller et al. (1976) example, was chi-square significant because of a disparate distribution in the "litter" column, or the "trash" column, or in all three columns? There are statistical techniques to help tease this apart, but they are not common. Often larger contingency tables are collapsed back to 2×2 tables for ease of interpretation. We will see a similar kind of issue raised when we consider odds ratios shortly.

Computer Analyses

Chi-square statistics can be produced by computer programs in two different ways. Suppose that we had a data file containing Pugh's data on convictions for rape. One column (Fault) would contain a 1 if that defendant's lawyer had tried to assign high blame to the victim or a 0 if he assigned low blame. A second column (Guilt) would contain a 1 if the defendant was found guilty, and a 0 if not. (Alternatively, we could code the Fault variable as "Little" or "Much" depending on whether the victim was assigned little or much fault by the attorney. We could also code Guilt as "Guilty" and "Not Guilty." There would be 358 lines of data, one for each case. We could then ask SPSS (or almost any other program) to cross tabulate Fault against Guilt. This analysis is presented in Exhibit 6.1.

Exhibit 6.1 contains several statistics we have not yet discussed. In Exhibit 6.1b, the likelihood ratio test is one that we shall take up shortly and is simply another approach to calculating chi-square. The three statistics in Exhibit 6.1c (phi, Cramér's V, and the contingency coefficient) will also be discussed later in this chapter, as will the odds ratio shown in Exhibit 6.1d. Each of these four statistics is an attempt to assess the size of the effect.

If you didn't already have a data file for Pugh's data, you would probably not be eager to create a file of 358 lines just to calculate a simple chi-square. Fortunately, there is an alternative approach that is much quicker. Basically, we create one line of data for each

Fault * Guilt Crosstabulation

Count

		Guilt		Total
		Guilty	Not Guilty	
Fault	Little	153	24	177
	Much	105	76	181
Total		258	100	358

Exhibit 6.1a   Cross tabulation of Fault versus Guilt From Pugh's data on conviction for rape

Chi-Square Tests

	Value	df	Asymp. Sig. (2-sided)	Exact Sig. (2-sided)	Exact Sig. (1-sided)
Pearson Chi-Square	35.930[b]	1	.000		
Continuity Correction[a]	34.532	1	.000		
Likelihood Ratio	37.351	1	.000		
Fisher's Exact Test				.000	.000
N of Valid Cases	358				

[a] Computed only for a 2×2 table

[b] 0 cells (.0%) have expected count less than 5. The minimum expected count is 49.44.

Exhibit 6.1b   Test statistics for analysis of Pugh's data

Symmetric Measures

		Value	Approx. Sig.
Nominal by Nominal	Phi	.317	.000
	Cramer's V	.317	.000
	Contingency Coefficient	.302	.000
N of Valid Cases		358	

Exhibit 6.1c Measures of association for Pugh's data

Risk Estimate

	Value	95% Confidence Interval	
		Lower	Upper
Odds Ratio for Fault (Little / Much)	4.614	2.738	7.776
For cohort Guilt = Guilty	1.490	1.299	1.709
For cohort Guilt = NotGuilty	.323	.214	.486
N of Valid Cases	358		

Exhibit 6.1d Risk estimates on Pugh's data

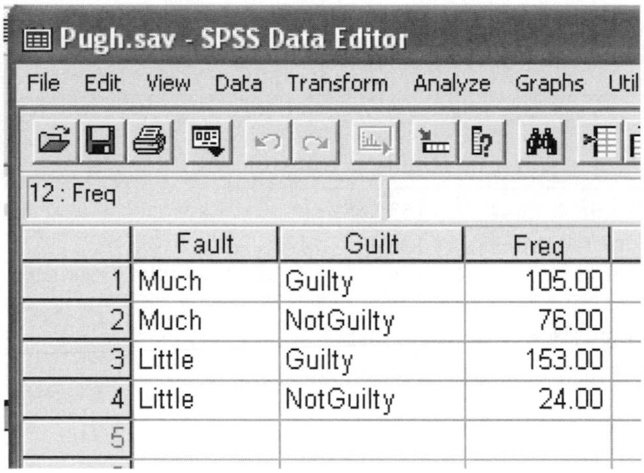

Exhibit 6.2 SPSS data file for analysis of Pugh's experiment

possible cell in the table, and then add a column (here labeled *Freq*) that reports how many observations fell in that cell. A screen shot of such a table is shown in Exhibit 6.2.

Once we have entered the frequencies, simply go to **Data/Weight cases** menu and instruct SPSS to weight each combination of Fault and Guilt by the Freq variable. Similar commands can be carried out in most software. The rest of the calculations can then be carried out just as we did earlier.

Exhibit 6.1b contains the printout of the test statistics for testing the null hypothesis of independence between Fault and Guilt. You can see that we obtained the same value of χ^2 (35.93) that we obtained earlier by hand. The next entry is the value of χ^2 with a continuity correction, as we discussed earlier. I suggest ignoring this. Fisher's Exact Test follows, and here it leads to the same conclusion as Pearson's chi-square. (You will not find Fisher's Exact Test printed out with larger tables because it was not designed for them nor will you see a chi-square value printed in column 2 because the test does not produce one.)

Small Expected Frequencies

small expected frequency

One of the most important requirements for using the Pearson chi-square test concerns the size of the expected frequencies. We have already met this requirement briefly in discussing corrections for continuity. Before defining more precisely what we mean by *small,* we should examine why a **small expected frequency** causes so much trouble.

For a given sample size, there are often a limited number of different contingency tables that you could obtain and, thus, a limited number of different values of chi-square. If only a few different values of χ^2_{obt} are possible, then the χ^2 distribution, which is continuous, cannot provide a reasonable approximation to the distribution of our statistic. We cannot closely fit a discrete distribution having relatively few values with a continuous one. Those cases that result in only a few possible values of χ^2_{obt}, however, are those with small expected frequencies in one or more cells. (This is directly analogous to the fact that if you flip a coin three times, there are only four possible values for the number of heads, and the resulting sampling distribution certainly cannot be approximated by the normal distribution.)

We have seen that difficulties arise when we have small expected frequencies, but the question of how small is small remains. Those conventions that do exist are conflicting and have only minimal claims to preference over one another. Probably the most common is to require that all expected frequencies should be at least five. This is a conservative position and I don't feel overly guilty when I violate it. Bradley and colleagues (1979) ran a computer-based sampling study. They used tables ranging in size from 2×2 to 4×4 and found that for those applications likely to arise in practice, the actual percentage of Type I errors rarely exceeds .06, even for *total* samples sizes as small as 20, unless the row or column marginal totals are drastically skewed. Camilli and Hopkins (1979) demonstrated that even with quite small expected frequencies, the test produces few Type I errors in the 2×2 case as long as the total sample size is greater than or equal to eight, but they, and Overall (1980), point to the extremely low power to reject a false H_0 that such tests possess. With small sample sizes, power is more likely to be a problem than are inflated Type I error rates.

One major advantage of Fisher's Exact Test is that it is not based on the χ^2 distribution and, thus, is not affected by a lack of continuity. One of the strongest arguments for that test is that it applies well to cases with small expected frequencies.

6.5   Chi-Square for Ordinal Data

Chi-square is an important statistic for analyzing categorical data, but it can sometimes fall short of what we need. If you apply chi-square to a contingency table, and then rearrange one or more rows or columns and calculate chi-square again, you will arrive at exactly the same answer. That is as it should be because chi-square does not take the ordering of the rows or columns into account.

But what do you do if the order of the rows or columns does make a difference? How can you take that ordinal information and make it part of your analysis? An interesting example of just such a situation was provided in a query that I received from Jennifer Mahon at the University of Leicester, in England.

Ms. Mahon collected data on the treatment for eating disorders. She was interested in how likely participants were to remain in treatment or drop out, and she wanted to examine this relative to the number of traumatic events they had experienced in childhood. Her general hypothesis was that participants who had experienced more traumatic events during childhood would be more likely to drop out of treatment. Notice that her hypothesis treats the number of traumatic events as an ordered variable, which is something that chi-square ignores. There is a solution to this problem, but it is more appropriately covered after we have talked about correlations. I will come back to this problem in Chapter 10 and show you one approach. (Many of you could skip now to Chapter 10, Section 10.4, and be able to follow the discussion.) I mention it here because it comes up most often when discussing χ^2.

6.6 Summary of the Assumptions of Chi-Square

assumptions of χ^2

Because of the widespread misuse of chi-square still prevalent in the literature, it is important to pull together in one place the underlying **assumptions of** χ^2. For a thorough discussion of the misuse of χ^2, see the paper by Lewis and Burke (1949) and the subsequent rejoinders to that paper. These articles are not yet out of date, although it has been more than 50 years since they were written. A somewhat more recent discussion of many of the issues Lewis and Burke (1949) raised can be found in Delucchi (1983).

The Assumption of Independence

At the beginning of this chapter, we assumed that *observations* were independent of one another. The word *independence* has been used in two different ways in this chapter, and it is important to keep these two uses separate. A basic assumption of χ^2 deals with the independence of *observations* and is the assumption, for example, that one participant's choice among brands of coffee has no effect on another participant's choice. This is what we are referring to when we speak of an assumption of independence. We also spoke of the independence of *variables* when we discussed contingency tables. In this case, independence is what is being tested, whereas in the former use of the word, it is an assumption. So, we want the *observations* to be independent and we are testing the independence of *variables*.

It is not uncommon to find cases in which the assumption of independence of observations is violated, usually by having the same participant respond more than once. A typical illustration of the violation of the independence assumption occurred when a former student categorized the level of activity of each of five animals on each of four days. When he was finished, he had a table similar to this:

	Activity		
High	Medium	Low	Total
10	7	3	20

This table looks legitimate until you realize that there were only five animals, and thus, each animal was contributing four tally marks toward the cell entries. If an animal exhibited high activity on Day 1, it is likely to have exhibited high activity on other days. The observations are not independent, and we can make a better-than-chance prediction of one score knowing another score. This kind of error is easy to make, but it is an error nevertheless. The best guard against it is to make certain that the total of all observations (N) equals precisely the number of participants in the experiment.

Inclusion of Nonoccurrences

Although the requirement that nonoccurrences be included has not yet been mentioned specifically, it is inherent in the derivation. It is probably best explained by an example.

Suppose that out of 20 students from rural areas, 17 were in favor of having daylight savings time (DST) all year. Out of 20 students from urban areas, only 11 were in favor of DST on a permanent basis. We want to determine if significantly more rural students than urban students are in favor of DST. One *erroneous* method of testing this would be to set up the following data table on the number of students favoring DST:

	Rural	Urban	Total
Observed	17	11	28
Expected	14	14	28

nonoccurrences

We could then compute $\chi^2 = 1.29$ and fail to reject H_0. This data table, however, does not take into account the *negative* responses, which Lewis and Burke (1949) call **nonoccurrences.** In other words, it does not include the numbers of rural and urban students *opposed* to DST. However, the derivation of chi-square assumes that we have included both those opposed to DST and those in favor of it. So we need a table such as this one:

	Rural	Urban	
Yes	17	11	28
No	3	9	12
	20	20	40

Now $\chi^2 = 4.29$, which is significant at $\alpha = .05$, resulting in an entirely different interpretation of the results.

Perhaps a more dramatic way to see why we need to include nonoccurrences can be shown by assuming that 17 out of *2,000* rural students and 11 out of 20 urban students preferred DST. Consider how much different the interpretation of the two tables would be. Certainly, our analysis must reflect the difference between the two data sets, which would not be the case if we failed to include nonoccurrences.

Failure to consider the nonoccurrences invalidates the test and reduces the value of χ^2, leaving you less likely to reject H_0. Again, you must be sure that the total (N) equals the number of participants in the study.

6.7   One- and Two-Tailed Tests

People are often confused about whether chi-square is a one- or a two-tailed test. This confusion results from the fact that there are different ways of defining what we mean by a one- or a two-tailed test. If we think of the sampling distribution of χ^2, we can argue that χ^2 is a one-tailed test because we reject H_0 only when our value of χ^2 lies in the extreme right tail of the distribution. On the other hand, if we think of the underlying data on which our obtained χ^2 is based, we could argue that we have a two-tailed test. If, for example, we were using chi-square to test the fairness of a coin, we would reject H_0 if it produced too many heads *or* if it produced too many tails because either event would lead to a large value of χ^2.

The preceding discussion is not intended to start an argument about semantics (it does not really matter whether you think of the test as one-tailed or two); rather, it is intended to point out one weakness of the chi-square test, so that you can take this into account. The weakness is that the test, *as normally applied,* is nondirectional. To take a simple example, consider the situation in which you want to show that increasing amounts of quinine added to an animal's food make it less appealing. You take 90 rats and offer them a choice of three bowls of food that differ in the amount of quinine that has been added. You then count the number of animals selecting each bowl of food. Suppose the data are

Amount of Quinine		
Small	Medium	Large
39	30	21

The computed value of χ^2 is 5.4, which, on 2 df, is not significant at $p < .05$.

The important fact about the data is that any of the six possible configurations of the same frequencies (such as 21, 30, 39) would produce the same value of χ^2, and you receive no credit for the fact that the configuration you obtained is precisely the one that you predicted. Thus, you have made a *multi-tailed* test when you actually have a specific prediction of the direction in which the totals will be ordered. I referred to this problem a few pages back when discussing a problem Jennifer Mahon raised. A solution to this problem will be given in Chapter 10 (Section 10.4), where I discuss creating a correlational measure of the relationship between the two variables.

6.8 Likelihood Ratio Tests

likelihood ratios

An alternative approach to analyzing categorical data is based on **likelihood ratios.** For large sample sizes, the two tests are equivalent, though for small sample size the standard Pearson chi-square is thought to be better approximated by the exact chi-square distribution than is the likelihood ratio chi-square (Agresti, 1990). Likelihood ratio tests are heavily used in log-linear models for analyzing contingency tables because of their additive properties. Log-linear models will be discussed in Chapter 17. Such models are particularly important when we want to analyze multidimensional contingency tables. Such models are being used more and more, and you should be exposed at least minimally to such methods.

Without going into detail, the general idea of a likelihood ratio can be described quite simply. Suppose we collect data and calculate the probability or likelihood of the data occurring given that the null hypothesis is true. We also calculate the likelihood that the data would occur under some alternative hypothesis (the hypothesis for which the data are most probable). If the data are much more likely for some alternative hypothesis than for H_0, we would be inclined to reject H_0. However, if the data are almost as likely under H_0 as they are for some other alternative, we would be inclined to retain H_0. Thus, the likelihood ratio (the ratio of these two likelihoods) forms a basis for evaluating the null hypothesis.

Using likelihood ratios, it is possible to devise tests, frequently referred to as "maximum likelihood χ^2," for analyzing both one-dimensional arrays and contingency tables. For the development of these tests, see Mood (1950) or Mood and Graybill (1963).

For the one-dimensional goodness-of-fit case,

$$\chi^2_{(C-1)} = 2 \sum O_i \ln \left(\frac{O_i}{E_i} \right)$$

where O_i and E_i are the observed and expected frequencies for each cell and "ln" denotes the natural logarithm (logarithm to the base e). This value of χ^2 can be evaluated using the standard table of χ^2 on $C - 1$ degrees of freedom.

For analyzing contingency tables, we can use essentially the same formula,

$$\chi^2_{(R-1)(C-1)} = 2 \sum O_{ij} \ln \left(\frac{O_{ij}}{E_{ij}} \right)$$

where O_{ij} and E_{ij} are the observed and expected frequencies in each cell. The expected frequencies are obtained just as they were for the standard Pearson chi-square test. This statistic is evaluated with respect to the χ^2 distribution on $(R - 1)(C - 1)$ degrees of freedom.

As an illustration of the use of the likelihood ratio test for contingency tables, consider the data found in the Pugh (1983) study. The cell and marginal frequencies follow:

	Verdict		
Fault	Guilty	Not Guilty	
Low	153	24	177
High	105	76	181
	258	100	358

$$\chi^2 = 2 \sum O_{ij} \ln \left(\frac{O_{ij}}{E_{ij}} \right)$$

$$= 2 \left[153 \ln \left(\frac{153}{127.559} \right) + 24 \ln \left(\frac{24}{49.441} \right) + 105 \ln \left(\frac{105}{130.441} \right) + 76 \ln \left(\frac{76}{50.559} \right) \right]$$

$$= 2[153(0.1819) + 24(-0.7227) + 105(-0.2170) + 76(0.4076)]$$

$$= 2[18.6785] = 37.36$$

This answer agrees with the likelihood ratio statistic found in Exhibit 6.16. It is a χ^2 on 1 df, and because it exceeds $\chi^2_{.05}(1) = 3.84$, it will lead to rejection of H_0. The decision of the juror depends in part on how the victim is portrayed.

6.9   Effect Sizes

The fact that a relationship is "statistically significant" doesn't tell us very much about whether it is of practical significance. The fact that two independent variables are not statistically independent does not mean that the lack of independence is important or worthy of our attention. In fact, if you allow the sample size to grow large enough, almost any two variables would likely show a statistically significant lack of independence.

What we need, then, are ways to go beyond a simple test of significance to present one or more statistics that reflect the size of the effect we are looking at. There are two different types of measures designed to represent the size of an effect. One type, called the **d-family** by Rosenthal (1994), is based on one or more measures of the *differences* between groups or levels of the independent variable. For example, as we will see in a moment, the probability of being found guilty of rape is about 30% higher for dependents in Pugh's Low Fault condition than for those in the High Fault condition. The other type of measure, called the **r-family,** represents some sort of correlation coefficient between the two independent variables. We will discuss correlation thoroughly in Chapter 9, but I will discuss these measures here because they are appropriate at this time. Measures in the r-family are often called **"measures of association."**

d-family

r-family

measures of
association

An Example

prospective
study

An important study of the beneficial effects of small daily doses of aspirin on reducing heart attacks in men was reported in 1988. More than 22,000 physicians were administered aspirin or a placebo, and the incidence of later heart attacks was recorded. The data follow in Table 6.4. Notice that this design is a **prospective study** because the treatments (aspirin versus no aspirin) were applied and then future outcome was determined. (A

Table 6.4   The effect of aspirin on the incidence of heart attacks

	Outcome		
	Heart Attack	No Heart Attack	
Aspirin	104	10,933	11,037
Placebo	189	10,845	11,034
	293	21,778	22,071

retrospective study retrospective study would select people who had, or had not, experienced a heart attack and then look backward in time to see whether they had been in the habit of taking aspirin in the past.)

For these data, $\chi^2 = 25.014$ on one degree of freedom, which is statistically significant at $\alpha = .05$, indicating that there is a relationship between whether or not one takes aspirin daily and whether one later has a heart attack.[4]

d-family: Risks and Odds

Two important concepts with categorical data, especially for 2×2 tables, are the concepts of risks and odds. These concepts are closely related, and often confused, but they are basically very simple.

For the aspirin data, 0.94% (104/11, 037) of people in the aspirin group and 1.71% (189/11, 034) of those in the control group suffered a heart attack during the study. (Unless you are a middle-aged male worrying about your health, the numbers look rather small. But they are important.) These two statistics are commonly referred to as **risk** estimates because they describe the risk that someone with, or without, aspirin will suffer a heart attack. Risk measures offer a useful way of looking at the size of an effect.

risk

The **risk difference** is simply the difference between the two proportions. In our example, the difference is $1.71\% - 0.94\% = 0.77\%$. Thus, there is about three-quarters of a percentage point difference between the two conditions. Put another way, the difference in risk between a male taking aspirin and one not taking aspirin is about three-quarters of 1%. This may not appear to be very large, but keep in mind that we are talking about heart attacks, which are serious events.

risk difference

One problem with a risk difference is that its magnitude depends on the overall level of risk. Heart attacks are quite low risk events, so we would not expect a huge difference between the two conditions. (In contrast, when we looked at Pugh's data on convictions for rape, where the probability of being convicted was quite high, there was a lot of room for the two conditions to differ, and we saw a 30 percentage point difference. Does that mean that Pugh's study found a much larger effect size? Well, it depends—it certainly did with respect to risk difference.)

risk ratio

relative risk

Another way to compare the risks is to form a **risk ratio,** also called **relative risk,** that is just the ratio of the two risks. For the heart attack data, the risk ratio is

$$RR = Risk_{\text{no apsirin}}/Risk_{\text{aspirin}} = 1.71\%/0.94\% = 1.819$$

---

[4] It is important to note that, although taking aspirin daily is associated with a lower rate of heart attack, more recent data have shown that there are important negative side effects. Current literature suggests that Omega-3 fish oil is at least as effective with fewer side effects.

Thus, the risk of having a heart attack if you do not take aspirin is 1.8 times higher than if you do take aspirin. That strikes me as quite a difference.

We must consider a third measure of effect size, and that is the odds ratio. At first glance, odds and odds ratios look like risk and risk ratios, and they are often confused, even by people who know better. (In a previous edition, I referred to odds, but described them as risks, much to my chagrin.) Recall that we defined the risk of a heart attack in the aspirin group as the number having a heart attack divided by the *total number of people in that group.* (e.g., $104/11,037 = 0.0094 = 0.94\%$.) The **odds** of having a heart attack for a member of the aspirin group is the number having a heart attack divided by the number *not having a heart attack.* (e.g., $104/10,933 = 0.0095.$) The difference (though very slight) comes in what we use as the denominator—risk uses the total sample size and is thus the proportion of people in that condition who experience a heart attack. Odds uses as a denominator the number not having a heart attack and is thus the ratio of the number having an attack versus the number not having an attack. Because the denominators are so much alike in this example, the results are almost indistinguishable. That is certainly not always the case. In Pugh's example, the risk of being convicted of rape in the low fault condition are $153/177 = 0.864$ (86% of the cases are convicted), whereas the odds of being convicted in the low fault condition are $153/24 = 6.375$ (the odds of being convicted are 6.4 times the odds of being found innocent).

Just as we can form a risk ratio by dividing the two risks, we can form an **odds ratio** by dividing the two odds. For the aspirin example, the odds of heart attack given that you did not take aspirin were $189/10,845 = 0.017$. The odds of a heart attack given that you did take aspirin were $104/10,933 = 0.010$. The odds ratio is simply the ratio of these two odds and is

$$OR = \frac{Odds \mid No\, Aspirin}{Odds \mid Aspirin} = \frac{0.0174}{0.0095} = 1.83$$

Thus, the odds of a heart attack without aspirin are 1.83 times higher than the odds of a heart attack with aspirin.[5]

Why do we have to complicate things by having both odds ratios and risk ratios because they often look very much alike? That is a very good question, and it has some good answers. Risk is something that I think most of us understand. When we say the risk of having a heart attack in the No Aspirin condition is 0.0171, we are saying that 1.7% of the participants in that condition had a heart attack, and that is pretty straightforward. When we say that the odds of a heart attack in that condition are 0.0174, we are saying that the chances of having a heart attack are 1.7% of the chances of not having a heart attack. That may be a popular way of setting bets on race horses, but it leaves me dissatisfied. So why have an odds ratio in the first place?

The odds ratio has at least two things in its favor. In the first place, it can be calculated in situations in which a true risk ratio cannot be. In a retrospective study, where we find a group of people with heart attacks and another group of people without heart attacks, and look back to see if they took aspirin, we can't really calculate *risk*. Risk is future oriented. If we give 1,000 people aspirin and withhold it from 1,000 others, we can look at these people 10 years down the road and calculate the risk (and risk ratio) of heart attacks. But if

odds

odds ratio

---

[5] In computing an odds ratio, there is no rule about which odds go in the numerator and which in the denominator. It depends on convenience. Where reasonable, I prefer to put the larger value in the numerator to make the ratio come out greater than 1.0, simply because I find it easier to talk about that way. If we reversed them in this example, we would find OR = 0.546, and conclude that your odds of having a heart attack in the aspirin condition are about half of what they are in the No Aspirin condition. That is simply the inverse of the original OR $(0.546 = 1/1.83)$.

we take 1,000 people with (and without) heart attacks and look backward, we can't really calculate risk because we have sampled heart attack patients at far greater than their normal rate in the population (50% of our sample has had a heart attack, but certainly 50% of the population does not suffer from heart attacks). But we can always calculate odds ratios. And, when we are talking about low probability events, such as having a heart attack, the odds ratio is usually a very good estimate of what the risk ratio would be.[6] The odds ratio is equally valid for prospective, retrospective, and cross-sectional sampling designs. That is important.

A second important advantage of the odds ratio is that taking the natural log of the odds ratio [ln(OR)] gives us a statistic that is extremely useful in a variety of situations. Two of these are logistic regression and log-linear models, both of which are discussed later in the book. I don't expect most people to be excited by the fact that a logarithmic transformation of the odds ratio has interesting statistical properties, but that is a very important point nonetheless.

r-family: Phi and Cramér's V

The measures that we have discussed are sometimes called *d*-family measures because they focus on comparing differences between conditions—either by calculating the difference directly or by using ratios of risks or odds. An older, and more traditional set of measures, sometimes called "measures of association," but now frequently called "*r*-family measures" looks at the correlation between two variables. We won't come to correlation until Chapter 9, but I would expect that you already know enough about correlation to understand what follows.

There are a great many measures of association, and I have no intention of discussing most of them. One of the nicest discussions of these can be found in Nie, Hull, Jenkins, Steinbrenner, and Bent (1970). (If your instructor is very old—like me—he or she probably remembers it fondly as the old "maroon SPSS manual." It is such a classic that it is very likely to be available in your university library or through interlibrary loan.)

Phi (ϕ)

phi (ϕ)

In the case of 2×2 tables, a correlation coefficient that we will consider in Chapter 10 serves as a good measure of association. This coefficient is called **phi** (ϕ)**,** and it represents the correlation between two variables, each of which is a dichotomy (a dichotomy is a variable that takes on one of two distinct values.). If we coded Aspirin as 1 or 2, for Yes and No, and coded Heart Attack as 1 for Yes and 2 for No, and then correlated the two variables (see Chapters 9 and 10), the result would be phi. (It doesn't even matter what two numbers we use as values for coding, as long as one condition always gets one value and the other always gets a different, but consistent, value.)

An easier way to calculate ϕ for these data is by the relation

$$\phi = \sqrt{\frac{\chi^2}{N}}$$

For the aspirin data in Table 6.4, $\chi^2 = 25.014$, $\phi = \sqrt{25.014/22,071} = .034$. That does not appear to be a very large correlation, but we are speaking about a major life-threatening event, and even a small correlation can be meaningful.

---

[6] The odds ratio can be defined as $OR = RR(\frac{1-p_2}{1-p_1})$, where $OR =$ odds ratio, $RR =$ relative risk, p_1 is the population proportion of heart attacks in one group, and p_2 is the population proportion of heart attacks in the other group. When these two proportions are close to 0, numerator and denominator nearly cancel each other and $OR = RR$.

Cramér's V

The difficulty with phi is that it applies only to 2×2 tables and, therefore, is not of any use with larger contingency tables. Cramér (1946) proposed a way around this problem by defining

$$V = \sqrt{\frac{\chi^2}{N(k-1)}}$$

where N is the sample size and k is defined as the smaller of R and C.

Cramér's V

Cramér's V can be seen as a simple extension of ϕ. Note that when $k = 2$, it *is* ϕ. Its usefulness applies to larger tables. We can calculate Cramér's V for the data on littering in Table 6.3 as follows:

$$V = \sqrt{\frac{\chi^2}{N(k-1)}} = \sqrt{\frac{25.79}{(1772)(1)}} = .121$$

The problem with V is that it is hard to give it a simple intuitive interpretation when there are more than two categories and they do not fall on an ordered dimension. There is a fairly technical explanation, but I am not going into it here, and I doubt that it would be very enlightening at this point.

I am not happy with the r-family of measures simply because I don't think that they have a meaningful interpretation in most situations. It is one thing to use a d-family measure like the odds ratio and declare that the odds of having a heart attack if you don't take aspirin are 1.83 times higher than the odds of having a heart attack if you do not take aspirin. I think that most people can understand what that statement means. But to use an r-family measure, such as phi, and say that the correlation between aspirin intake and heart attack is .034 doesn't seem to be telling them anything useful. (And squaring it and saying that aspirin usage accounts for 0.1% of the variance in heart attacks is even less helpful.) I would suggest that you stay away from the older r-family measures unless you really have a good reason to use them.

Effect Sizes for Larger Tables

Measures like odds ratios are most easily understood with 2×2 tables because it is clear what the odds represent. Things are very much messier with larger tables. We will see this distinction between two levels and multiple levels in several places in this book. If you think clearly about what it is you want to convey to your audience, I suspect that you will generally find that you really want to compare only two things. For example, in the littering study, you might want to compare the number of flyers littering the floor with the number of flyers that took themselves off some place—such as the trash or out of the store. I would suggest that after computing the overall chi-square for the 2×3 table, you simply recompile your contingency table into "Litter" and "Non-litter" and treat it as a 2×2. That is really what you probably want. And if that is the case, risk ratios and odds ratios will do very nicely. (When we come to the analysis of variance in Chapter 11, which looks a million miles away from contingency tables, you will see that frequently the questions we most care about also come down to comparing two groups or sets of groups.)

6.10   Measures of Agreement

We should discuss one more measure. It is not really a measure of effect size, like the previous measures, but it is an important statistic when you want to ask about the agreement between judges.

Kappa (κ)—A Measure of Agreement

kappa (κ)

An important statistic that is not based on chi-square but that does use contingency tables is **kappa (κ),** commonly known as Cohen's kappa (Cohen, 1960). This statistic measures interjudge agreement and is often used when we want to examine the reliability of ratings.

Suppose we asked a judge with considerable clinical experience to interview 30 adolescents and classify them as exhibiting (1) no behavior problems, (2) internalizing behavior problems (e.g., withdrawn), and (3) externalizing behavior problems (e.g., acting out). Anyone reviewing our work would be concerned with the reliability of our measure—how do we know that this judge was doing any better than flipping a coin? As a check, we ask a second judge to go through the same process and rate the same adolescents. We then set up a contingency table showing the agreements and disagreements between the two judges. Suppose the data are those shown in Table 6.5.

Table 6.5 Classification of behavior problems by two judges

| Judge II | Judge I | | | Total |
	No Problem	Internalizing	Externalizing	
No Problem	15 (10.67)	2	3	20
Internalizing	1	3 (1.20)	2	6
Externalizing	0	1	3 (1.07)	4
Total	16	6	8	30

Ignore the values in parentheses for the moment. In this table, Judge I classified 16 adolescents as exhibiting no problems, as shown by the total in column 1. Of those 16, Judge II agreed that 15 had no problems, but also classed 1 of them as exhibiting internalizing problems and 0 as exhibiting externalizing problems. The entries on the diagonal (15, 3, 3) represent agreement between the two judges, whereas the off-diagonal entries represent disagreement.

percentage of agreement

A simple (but unwise) approach to these data is to calculate the **percentage of agreement.** For this statistic, all we need to say is that out of 30 total cases, there were 21 cases (15 + 3 + 3) where the judges agreed. Then $21/30 = 0.70 = 70\%$ agreement. This measure has problems, however. Most adolescents in our sample exhibit no behavior problems, and both judges are (correctly) biased toward a classification of No Problem and away from the other classifications. The probability of No Problem for Judge I would be estimated as $16/30 = .53$. The probability of No Problem for Judge II would be estimated as $20/30 = .67$. If the two judges operated by pulling their diagnoses out of the air, the probability that they would both classify the same case as No Problem is $.53 \times .67 = .36$, which for 30 judgments would mean that $.36 \times 30 = 10.67$ agreements on No Problem alone, purely by chance.

Cohen (1960) proposed a chance-corrected measure of agreement known as kappa. To calculate kappa, we first need to calculate the expected frequencies for each of the diagonal cells assuming that judgments are independent. We calculate these the same way we calculate the standard chi-square test. For example, the expected frequency of both judges assigning a classification of No Problem, assuming that they are operating at random, is $(20 \times 16)/30 = 10.67$. For Internalizing, it is $(6 \times 6)/30 = 1.2$, and for Externalizing, it is $(4 \times 8)/30 = 1.07$. These values are shown in parentheses in the table.

We will now define kappa as

$$\kappa = \frac{\sum f_O - \sum f_E}{N - \sum f_E}$$

where f_O represents the observed frequencies on the diagonal and f_E represents the expected frequencies on the diagonal. Thus

$$\sum f_O = 15 + 3 + 3 = 21$$

and

$$\sum f_E = 10.67 + 1.20 + 1.07 = 12.94.$$

Then

$$\kappa = \frac{21 - 12.94}{30 - 12.94} = \frac{8.06}{17.06} = .47$$

Notice that this coefficient is considerably lower than the 70% agreement figure that we just calculated. Instead of 70% agreement, we have 47% agreement after correcting for chance.

If you examine the formula for kappa, you can see the correction that is being applied. In the numerator we subtract, from the number of agreements, the number of agreements that we would expect merely by chance. In the denominator, we reduce the total number of judgments by that same amount. We then form a ratio of the two chance-corrected values.

Cohen and others have developed statistical tests for the significance of kappa. However, its significance is rarely the issue. If kappa is low enough for us to even question its significance, the lack of agreement among our judges is a serious problem.

6.11   Writing Up the Results

We will take as our example Pugh's study of rape convictions (1983). If you were writing up these results, you would probably want to say something like the following:

> In examining the question of whether a defense lawyer's attempt to place blame on the victim of rape would influence a jury's decision in a rape case, jury participants were presented with a situation in which the victim was characterized by the defense as either partly responsible for the rape or not responsible. The jurors were then asked to make a judgment about whether the defendant was guilty or not guilty of the crime. When the victim was portrayed as low in fault, 86% of the time the defendant was judged to be guilty. When the victim was portrayed as high in fault, the defendant was judged guilty only 58% of the time. A chi-square test of the relationship between Fault and Guilt produced $\chi^2(1) = 35.93$, which is statistically significant at $p < .05$. This is associated with an odds ratio of 4.61, indicating that the odds of being found guilty of rape are more than 4.5 times higher in the condition in which the victim is portrayed as not bearing fault for the rape. The odds ratio would indicate that we are speaking of a meaningful difference between the two conditions.

Key Terms

Chi-square (χ^2) (Introduction)

Pearson's chi-square (Introduction)

Chi-square distribution (χ^2) (6.1)

Gamma function (6.1)

Chi-square test (6.2)

Goodness-of-fit test (6.2)

Observed frequencies (6.2)

Expected frequencies (6.2)

Tabled distribution of χ^2 (6.2)

Contingency table (6.3)

Marginal totals (6.3)

Cell (6.3)

Row total (6.3)

Column total (6.3)

Yates's correction for continuity (6.4)

Fixed marginals (6.4)

Small expected frequency (6.4)

Assumptions of χ^2 (6.6)

Nonoccurrences (6.6)

Likelihood ratios (6.8)

d-family (6.9)

Exercises

6.1   The chairperson of a psychology department suspects that some of her faculty members are more popular with students than are others. There are three sections of introductory psychology, taught at 10:00 a.m., 11:00 a.m., and 12:00 p.m. by Professors Anderson, Klatsky, and Kamm. The number of students who enroll for each is

Professor Anderson	Professor Klatsky	Professor Kamm
32	25	10

State the null hypothesis, run the appropriate chi-square test, and interpret the results.

6.2   From the point of view of designing a valid experiment (as opposed to the arithmetic of calculation), there is an important difference between Exercise 6.1 and the examples used in this chapter. The data in Exercise 6.1 will not really answer the question the chairperson wants answered. What is the problem, and how could the experiment be improved?

6.3   You have a theory that if you ask subjects to sort one-sentence characteristics of people (e.g., "I eat too fast") into five piles ranging from "not at all like me" to "very much like me," the percentage of items placed in each of the five piles will be approximately 10, 20, 40, 20, and 10. You have one of your friend's children sort 50 statements, and you obtain the following data: [8, 10, 20, 8, 4]. Do these data support your hypothesis?

6.4   To what population does the answer to Exercise 6.3 generalize? (*Hint:* From what population of observations might these observations be thought to be randomly sampled?)

6.5   In a classic study by Clark and Clark (1939), African American children were shown black dolls and white dolls and were asked to select the one with which they wanted to play. Of 252 children, 169 chose the white doll and 83 chose the black doll. What can we conclude about the behavior of these children?

6.6   Thirty years after the Clark and Clark study, Hraba and Grant (1970) repeated the study referred to in Exercise 6.5. The studies were not exactly equivalent, but the results were interesting. Hraba and Grant found that of 89 African American children, 28 chose the white doll and 61 chose the black doll. Run the appropriate chi-square test on their data and interpret the results.

6.7   Combine the data from Exercises 6.5 and 6.6 into a two-way contingency table and run the appropriate test. How does the question that the two-way classification addresses differ from the questions addressed by Exercises 6.5 and 6.6?

6.8   We know that smoking has all sorts of ill effects on people; among other things, there is evidence that it affects fertility. Weinberg and Gladen (1986) examined the effects of smoking and the ease with which women become pregnant. The researchers asked 586 women who had planned pregnancies how many menstrual cycles it had taken for them to become pregnant after discontinuing contraception. Weinberg and Gladen also sorted the women into whether they were smokers or nonsmokers. The data follow.

	1 cycle	2 cycles	3+ cycles	Total
Smokers	29	16	55	100
Nonsmokers	198	107	181	486
Total	227	123	236	586

Does smoking affect the ease with which women become pregnant? (I do not recommend smoking as a birth control device, regardless of your answer.)

6.9 In discussing the correction for continuity, we referred to the idea of fixed marginals, meaning that a replication of the study would produce the same row and column totals. Give an example of a study in which

a. No marginal totals are fixed.

b. One set of marginal totals is fixed.

c. Both sets of marginal totals (row and column) could reasonably be considered to be fixed. (This is a hard one.)

6.10 Howell and Huessy (1981) used a rating scale to classify children in a second-grade class as showing or not showing behavior commonly associated with attention deficit disorder (ADD). The researchers then classified these same children again when they were in fourth and fifth grades. When the children reached the end of the ninth grade, the researchers examined school records and noted which children were enrolled in remedial English. In the following data, all children who were ever classified as exhibiting behavior associated with ADD have been combined into one group (labeled ADD):

	Remedial English	Nonremedial English	
Normal	22	187	209
ADD	19	74	93
	41	261	302

Does behavior during elementary school affect class assignment during high school?

6.11 Use the data in Exercise 6.10 to demonstrate how chi-square varies as a function of sample size.

a. Double each cell entry and recompute chi-square.

b. What does your answer to (a) say about the role of the sample size in hypothesis testing?

6.12 In Exercise 6.10, children were classified as those who never showed ADD behavior and those who showed ADD behavior at least once in the second, fourth, or fifth grade. If we do not collapse across categories, we obtain the following data:

	Never	2nd	4th	2nd & 4th	5th	2nd & 5th	4th & 5th	2nd, 4th, & 5th
Remedial	22	2	1	3	2	4	3	4
Nonrem.	187	17	11	9	16	7	8	6

a. Run the chi-square test.

b. What would you conclude, ignoring the small expected frequencies?

c. How comfortable do you feel with these small expected frequencies? If you are not comfortable, how might you handle the problem?

6.13 In 2000, the State of Vermont legislature approved a bill authorizing "civil unions" between gay or lesbian partners. This was a very contentious debate with very serious issues raised by both sides. How the vote split along gender lines may tell us something important about the different ways in which males and females looked at this issue. The data follow. What would you conclude from these data?

	Vote		
	Yes	No	Total
Women	35	9	44
Men	60	41	101
Total	95	50	145

6.14 Stress has long been known to influence physical health. Visintainer, Volpicelli, and Seligman (1982) investigated the hypothesis that rats given 60 trials of inescapable shock would be less likely later to reject an implanted tumor than would rats that had received 60 trials of escapable shock or 60 no-shock trials. The researchers obtained the following data:

	Inescapable Shock	Escapable Shock	No Shock	
Reject	8	19	18	45
No Reject	22	11	15	48
	30	30	33	93

What could Visintainer et al. conclude from the results?

6.15 Darley and Latané (1968) asked subjects to participate in a discussion carried on over an intercom. Aside from the experimenter to whom they were speaking, subjects thought that there were zero, one, or four other people (bystanders) also listening over intercoms. Partway through the discussion, the experimenter feigned serious illness and asked for help. Darley and Latané noted how often the subject sought help for the experimenter as a function of the number of supposed bystanders. The data follow:

		Sought Assistance		
		Yes	No	
Number of	0	11	2	13
	1	16	10	26
Bystanders	4	4	9	13
		31	21	52

What could Darley and Latané conclude from the results?

6.16 In a study similar to the one in Exercise 6.15, Latané and Dabbs (1975) had a confederate enter an elevator and then "accidentally" drop a handful of pencils. They then noted whether bystanders helped pick them up. The data tabulate helping behavior by the gender of the bystander:

	Gender of Bystander		
	Female	Male	
Help	300	370	670
No Help	1003	950	1953
	1303	1320	2623

What could Latané and Dabbs conclude from the data? (Note that when we collapse over gender, only about one-quarter of the bystanders helped. That is not relevant to the question, but it is an interesting finding that could easily be missed by routine computer-based analyses.)

6.17 In a study of eating disorders in adolescents, Gross (1985) asked each of her subjects whether they would prefer to gain weight, lose weight, or maintain their present weight. (*Note:* Only 12% of the girls in Gross's sample were actually more than 15% above their normative weight—a common cutoff for a label of "overweight.") When she broke down the data for girls by race (African American versus white), she obtained the following results (other races have been omitted because of small sample sizes):

	Reducers	Maintainers	Gainers	
White	352	152	31	535
African American	47	28	24	99
	399	180	55	634

a. What conclusions can you draw from these data?

b. Ignoring race, what conclusion can you draw about adolescent girls' attitudes toward their own weight?

6.18 Use the likelihood ratio approach to analyze the data in Exercise 6.10.

6.19 Use the likelihood ratio approach to analyze the data in Exercise 6.12.

6.20 Would it be possible to calculate a one-way chi-square test on the data in row 2 of the table in Exercise 6.12? What hypothesis would you be testing if you did that? How would that hypothesis differ from the one you tested in Exercise 6.12?

6.21 Suppose we asked a group of 40 subjects whether they liked Monday Night Football, made them watch a game, and then asked them again. We would record the data as follows:

	Pro	Con	
Before	30	10	40
After	15	25	40
	45	35	80

Would chi-square calculated on such a table be appropriate? Why or why not?

6.22 As an alternative approach to the data in Exercise 6.21, you might find that after watching the game 20 people switched from Pro to Con and 5 people switched from Con to Pro. Thus, you can run a one-way chi-square test on the $20 + 5 = 25$ subjects who changed their opinion. (This is a test suggested by McNemar (1969) and is often referred to as McNemar's test.)

a. Run the test.

b. Explain how this tests the null hypothesis that you wanted to test.

6.23 From the SPSS printout in Exhibit 6.3

a. Verify the answer to Exercise 6.17a.

b. Interpret the row and column percentages.

c. What are the values labeled "Asymp. Sig."?

d. Interpret the coefficients.

RACE*GOAL Crosstabulation

			Goal			Total
			Gain	Lose	Maintain	
RACE	African-Amer	Count	24	47	28	99
		Expected Count	8.6	62.3	28.1	99.0
		% within RACE	24.2%	47.5%	28.3%	100.0%
		% within GOAL	43.6%	11.8%	15.6%	15.6%
		% of Total	3.8%	7.4%	4.4%	15.6%
	White	Count	31	352	152	535
		Expected Count	46.4	336.7	151.9	535.0
		% within RACE	5.8%	65.8%	28.4%	100.0%
		% within GOAL	56.4%	88.2%	84.4%	84.4%
		% of Total	4.9%	55.5%	24.0%	84.4%
Total		Count	55	399	180	634
		Expected Count	55.0	399.0	180.0	634.0
		% within RACE	8.7%	62.9%	28.4%	100.0%
		% within GOAL	100.0%	100.0%	100.0%	100.0%
		% of Total	8.7%	62.9%	28.4%	100.0%

Exhibit 6.3 Continued

Chi-Square Tests

	Value	df	Asymp. Sig. (2-sided)
Pearson Chi-Square	37.229[a]	2	.000
Likelihood Ratio	29.104	2	.000
N of Valid Cases	634		

[a] 0 cells (.0%) have expected count less than 5. The minimum expected count is 8.59.

Symmetric Measures

		Value	Approx. Sig.
Nominal by Nominal	Phi	.242	.000
	Cramer's V	.242	.000
	Contingency Coefficient	.236	.000
N of Valid Cases		634	

Exhibit 6.3   Continued

6.24   A more complete set of data on heart attacks and aspirin, from which Table 6.4 was taken, follows. Here we distinguish not just between Heart Attacks and No Heart Attacks, but also between Fatal and NonFatal attacks.

Myocardial Infarction

	Fatal Attack	NonFatal Attack	No Attack	Total
Placebo	18	171	10,845	11,034
Aspirin	5	99	10,933	11,037
Total	23	270	21,778	22,071

a.   Calculate both Pearson's chi-square and the likelihood ratio chi-square table. Interpret the results.

b.   Using only the data for the first two columns (those subjects with heart attacks), calculate both Pearson's chi-square and the likelihood ratio chi-square and interpret your results.

c.   Combine the Fatal and NonFatal heart attack columns and compare the combined column against the No Attack column, using both Pearson's and likelihood ratio chi-squares. Interpret these results.

d.   Sum the Pearson chi-squares in (b) and (c) and then the likelihood ratio chi-squares in (b) and (c), and compare each of these results with the results in (a). What do they tell you about the partitioning of chi-square?

e.   What do these results tell you about the relationship between aspirin and heart attacks?

6.25   For the results in Exercise 6.24, calculate and interpret

a.   Cramér's V

b.   Useful odds ratios

6.26   Compute the odds ratio for the data in Exercise 6.10. What do these values mean?

6.27   Compute the odds ratios for the data in Exercise 6.13. What do these ratios add to your understanding of the phenomena being studied?

6.28 Compute the odds in favor of seeking assistance for each of the groups in Exercise 6.15. Interpret the results.

6.29 Dabbs and Morris (1990) examined archival data from military records to study the relationship between high testosterone levels and antisocial behavior in males. Of 4016 men in the Normal Testosterone group, 10.0% had a record of adult delinquency. Of 446 men in the High Testosterone group, 22.6% had a record of adult delinquency. Is this relationship significant?

6.30 What is the odds ratio in Exercise 6.29? How would you interpret it?

6.31 In the study described in Exercise 6.29, 11.5% of the Normal Testosterone group and 17.9% of the High Testosterone group had a history of childhood delinquency.

   a. Is there a significant relationship between these two variables?

   b. Interpret this relationship.

   c. How does this result expand on what we already know from Exercise 6.29?

6.32 In a study examining the effects of individualized care of youths with severe emotional problems, Burchard and Schaefer (1990, personal communication) proposed to have caregivers rate the presence or absence of specific behaviors for each of 40 adolescents on a given day. To check for rater reliability, the researchers asked two raters to rate each adolescent. The following hypothetical data represent reasonable results for the behavior of "extreme verbal abuse."

	Rater A		
Rater B	Presence	Absence	
Presence	12	2	14
Absence	1	25	26
	13	27	40

   a. What is the percentage of agreement for these raters?

   b. What is Cohen's kappa?

   c. Why is kappa noticeably less than the percentage of agreement?

   d. Modify the raw data, keeping N at 40, so that the two statistics move even farther apart. How did you do this?

6.33 Many school children receive instruction on child abuse around the "good touch-bad touch" model, with the hope that such a program will reduce sexual abuse. Gibson and Leitenberg (2000) collected data from 818 college students, and recorded whether they had ever received such training and whether they had subsequently been abused. Of the 500 students who had received training, 43 reported that they had subsequently been abused. Of the 318 who had not received training, 50 reported subsequent abuse.

   a. Do these data present a convincing case for the efficacy of the sexual abuse prevention program?

   b. What is the odds ratio for these data, and what does it tell you?

Computer Exercises

6.34 In a data set named Mireault.dat and described in Appendix: Computer Data Sets, Mireault (1990) collected data from college students on the effects of the death of a parent. Leaving the critical variables aside for a moment, let's look at the distribution of students. The data set contains information on the gender of the students and the college (within the university) in which they were enrolled.

   a. Use any statistical package to tabulate Gender against College.

   b. What is the chi-square test on the hypothesis that College enrollment is independent of Gender?

   c. Interpret the results.

6.35 When we look at the variables in Mireault's data, we will want to be sure that there are no systematic differences of which we are ignorant. For example, if we found that the gender of the parent who died was an important variable in explaining some outcome variable, we would not like to later discover that the gender of the parent who died was in some way related to the gender of the subject, and that the effects of the two variables were confounded.

  a. Run a chi-square test on these two variables.

  b. Interpret the results.

  c. What would it mean to our interpretation of the relationship between gender of the parent and some other variable (e.g., subject's level of depression) if the gender of the parent is itself related to the gender of the subject?

6.36 Zuckerman, Hodgins, Zuckerman, and Rosenthal (1993) surveyed more than 500 people and asked a number of questions on statistical issues. In one question a reviewer warned a researcher that she had a high probability of a Type I error because she had a small sample size. The researcher disagreed. Subjects were asked, "Was the researcher correct?" The proportions of respondents, partitioned among students, assistant professors, associate professors, and full professors, who sided with the researcher and the total number of respondents in each category were as follows:

	Students	Assistant Professors	Associate Professors	Full Professors
Proportion	.59	.34	.43	.51
Sample size	17	175	134	182

(*Note*: These data mean that 59% of the 17 students who responded sided with the researcher. When you calculate the actual obtained frequencies, round to the nearest whole person.)

  a. Would you agree with the reviewer or with the researcher? Why?

  b. What is the error in logic of the person you disagreed with in (a)?

  c. How would you set up this problem to be suitable for a chi-square test?

  d. What do these data tell you about differences among groups of respondents?

6.37 The Zuckerman et al. paper referred to in the previous question hypothesized that faculty were less accurate than students because they have a tendency to give negative responses to such questions. ("There must be a trick.") How would you design a study to test such a hypothesis?

Discussion Questions

6.38 Hout, Duncan, and Sobel (1987) reported data on the relative sexual satisfaction of married couples. They asked each member of 91 married couples to rate the degree to which they agreed with "Sex is fun for me and my partner" on a four-point scale ranging from "never or occasionally" to "almost always." The data appear here:

Husband's Rating	Wife's Rating				TOTAL
	Never	Fairly Often	Very Often	Almost Always	
Never	7	7	2	3	19
Fairly Often	2	8	3	7	20
Very Often	1	5	4	9	19
Almost Always	2	8	9	14	33
TOTAL	12	28	18	33	91

a. How would you go about analyzing these data? Remember that you want to know more than just whether or not the two ratings are independent. Presumably you would like to show that as one spouse's ratings go up, so do the other's, and vice versa.

b. Use both Pearson's chi-square and the likelihood ratio chi-square.

c. What does Cramér's V offer?

d. What about odds ratios?

e. What about kappa?

f. Finally, what if you combined the Never and Fairly Often categories and the Very Often and Almost Always categories? Would the results be clearer, and under what conditions might this make sense?

6.39 In the previous question, we were concerned with whether husbands and wives rate their degree of sexual fun congruently (i.e., to the same degree). But suppose that women have different cut points on an underlying scale of "fun." For example, maybe women's idea of Fairly Often or Almost Always is higher than men's. (Maybe men would rate "a couple of times a month" as "Very Often" whereas women would rate "a couple of times a month" as "Fairly Often.") How would this affect your conclusions? Would it represent an underlying incongruency between males and females?

6.40 Use SPSS or another statistical package to calculate Fisher's Exact Test for the data in Exercise 6.13. How does it compare to the probability associated with Pearson's chi-square?

CHAPTER 7

Hypothesis Tests Applied to Means

Objectives

To introduce the *t* test as a procedure for testing hypotheses with measurement data, and to show how it can be used with several different designs. To describe ways of estimating the magnitude of any differences that do appear.

Contents

IN CHAPTERS 5 AND 6, we considered tests dealing with frequency (categorical) data. In those situations, the results of any experiment can usually be represented by a few subtotals—the frequency of occurrence of each category of response. In this and subsequent chapters, we will deal with a different type of data, which I have previously termed *measurement* or *quantitative data.*

In analyzing measurement data, our interest can focus either on differences between groups of subjects or on the relationship between two or more variables. The question of relationships between variables will be postponed until Chapters 9, 10, 15, and 16. This chapter will be concerned with the question of differences, and the statistic we will be most interested in will be the sample mean.

Low-birthweight (LBW) infants (who are often premature) are considered to be at risk for a variety of developmental difficulties. As part of an example we will return to later, suppose we took 25 LBW infants in an experimental group and 31 LBW infants in a control group, provided training to the parents of those in the experimental group on how to recognize the needs of LBW infants, and, when these children were 2 years old, obtained a measure of cognitive ability. Suppose that we found that the LBW infants in the experimental group had a mean score of 117.2, whereas those in the control group had a mean score of 106.7. Is the observed mean difference sufficient evidence for us to conclude that 2-year-old LBW children in the experimental group score higher, on average, than do 2-year-old LBW control children? We will answer this particular question later; I mention the problem here to illustrate the kind of question we will discuss in this chapter.

7.1 Sampling Distribution of the Mean

sampling distribution of the mean

central limit theorem

As you should recall from Chapter 4, the sampling distribution of a statistic is the distribution of values we would expect to obtain for that statistic if we drew an infinite number of samples from the population in question and calculated the statistic on each sample. Because we are concerned in this chapter with sample *means,* we need to know something about the **sampling distribution of the mean.** Fortunately, all the important information about the sampling distribution of the mean can be summed up in one very important theorem: the central limit theorem. The **central limit theorem** is a factual statement about the distribution of means. In an extended form, it states,

> Given a population with mean μ and variance σ^2, the sampling distribution of the mean (the distribution of sample means) will have a mean equal to μ (i.e., $\mu_{\overline{X}} = \mu$), a variance ($\sigma^2_{\overline{X}}$) equal to σ^2/n, and a standard deviation ($\sigma_{\overline{X}}$) equal to $\sigma/\sqrt{n}$. The distribution will approach the normal distribution as n, the *sample size,* increases.[1]

This is one of the most important theorems in statistics. Beyond telling us what the mean and variance of the sampling distribution of the mean must be for any given sample size, the theorem states that as n increases, the shape of this sampling distribution approaches normal, *whatever* the shape of the parent population. The importance of these facts will become clear shortly.

The rate at which the sampling distribution of the mean approaches normal as n increases is a function of the shape of the parent population. If the population is itself normal,

---

[1] The central limit theorem can be found stated in a variety of forms. The simplest form merely says that the sampling distribution of the mean approaches normal as n increases. The more extended form given here includes all the important information about the sampling distribution of the mean.

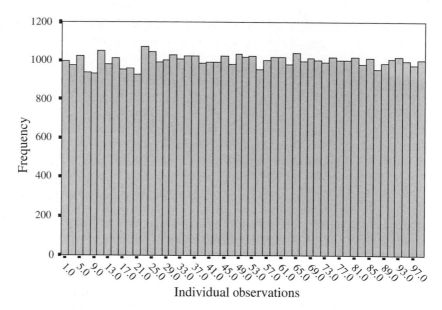

Figure 7.1   50,000 observations from a uniform distribution

the sampling distribution of the mean will be normal regardless of n. If the population is symmetric but nonnormal, the sampling distribution of the mean will be nearly normal even for small sample sizes, especially if the population is unimodal. If the population is markedly skewed, sample sizes of 30 or more may be required before the means closely approximate a normal distribution.

To illustrate the central limit theorem, suppose we have an infinitely large population of random numbers evenly distributed between 0 and 100. This population will have what is called a **uniform distribution**—every value between 0 and 100 will be equally likely. The distribution of 50,000 observations drawn from this population is shown in Figure 7.1. You can see that the distribution is very flat, as would be expected. For uniform distributions, the mean (μ) is known to be equal to one-half of the range (50), the standard deviation (σ) is known to be equal to 28.87 (the range divided by the square root of 12), and the variance (σ^2) is thus 833.33.

Now suppose we drew 5,000 samples of size 5 ($n = 5$) from this population and plotted the resulting sample *means*. Such sampling can be easily accomplished with a simple computer program; the results of just such a procedure are presented in Figure 7.2a, with a normal distribution superimposed. It is apparent that the distribution of means, although not exactly normal, is at least peaked in the center and trails off toward the extremes. (Actually, the superimposed normal distribution fits the data quite well.) The mean and standard deviation of this distribution are shown, and they are extremely close to $\mu = 50$ and $\sigma_{\overline{X}} = \sigma/\sqrt{n} = 28.87/\sqrt{5} = 12.91$. Any discrepancy between the actual values and those predicted by the central limit theorem is attributable to rounding error and to the fact that we did not draw an infinite number of samples.

Now suppose we repeated the entire procedure, only this time drawing 5,000 samples of 30 observations each. The results for these samples are plotted in Figure 7.2b. Here you see that just as the central limit theorem predicted, the distribution is approximately normal, the mean is again at $\mu = 50$, and the standard deviation has been reduced to approximately $28.87/\sqrt{30} = 5.27$.

uniform distribution

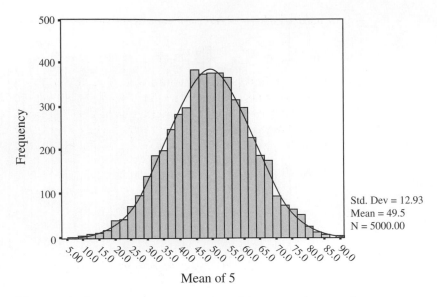

Figure 7.2a Sampling distribution of the mean when $n = 5$

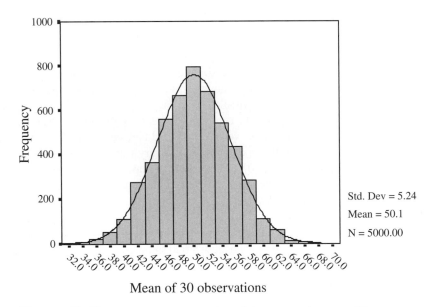

Figure 7.2b Sampling distribution of the mean when $n = 30$

7.2 Testing Hypotheses About Means—σ Known

From the central limit theorem, we know all the important characteristics of the sampling distribution of the mean. (We know its shape, its mean, and its standard deviation.) On the basis of this information, we are in a position to begin testing hypotheses about means. But first it might be well to go back to something we discussed with respect to the normal distribution. In Chapter 4, we saw that we could test a hypothesis about the population from which a single score (in that case, a finger-tapping score) was drawn by calculating

$$z = \frac{X - \mu}{\sigma}$$

and then, if the population is normally distributed, by obtaining the probability of a value of z as low as the one obtained by using the tables of the standard normal distribution. We ran a one-tailed test on the null hypothesis that the tapping rate (70) of a single individual was drawn at random from a normally distributed population of healthy subjects' tapping rates with a mean of 100 and a standard deviation of 20. We did this by calculating

$$z = \frac{X - \mu}{\sigma}$$
$$= \frac{70 - 100}{20} = \frac{-30}{20}$$
$$= -1.5$$

and then using Appendix z to find the area below $z = -1.5$.[2] This value is 0.0668. Thus, approximately 7% of the time, we would expect a score as low as this if we were sampling from a healthy population. This probability was not less than our preselected significance level of $\alpha = .05$, so we could not reject the null hypothesis. The tapping rate for the person we examined was not an unusual rate for healthy individuals. Although in this example we were testing a hypothesis about a single observation, the same logic applies to testing hypotheses about sample means. The only difference is that instead of comparing an observation with a distribution of observations, we will compare a mean with a distribution of means (the sampling distribution of the mean).

In most situations in which we test a hypothesis about a population mean, we don't have any knowledge about the variance of that population. (This is the main reason we have t tests, which are the main focus of this chapter.) However, in a limited number of situations we do know σ. A discussion of testing a hypothesis when σ is known provides a good transition from what we already know about the normal distribution to what we want to know about t tests. An example of behavior problem scores on the Achenbach Child Behavior Checklist (CBCL) (Achenbach, 1991a) is a useful example for this purpose because we know both the mean and the standard deviation for the population of Total Behavior Problems scores ($\mu = 50$ and $\sigma = 10$). Assume that a random sample of five children under stress had a mean score of 56.0. We want to test the null hypothesis that these five children are a random sample from a population of normal children (i.e., normal with respect to their general level of behavior problems). In other words, we want to test $H_0: \mu = 50$ against the alternative $H_1: \mu \neq 50$.

Because we know the mean and standard deviation of the population of general behavior problem scores, we can use the central limit theorem to obtain the sampling distribution when the null hypothesis is true. The central limit theorem states that if we obtain the sampling distribution of the mean from this population, it will have a mean of $\mu = 50$, a variance of $\sigma^2/n = 10^2/5 = 100/5 = 20$, and a standard deviation (usually referred to as the **standard error**)[3] of $\sigma/\sqrt{n} = 4.47$. This distribution is diagrammed in Figure 7.3. The arrow in Figure 7.3 represents the location of the sample mean.

standard error

---

[2] Recall that the normal distribution is symmetric, and thus there are no entries for negative values of z. The "smaller portion" for $z = -1.5$ is the same as the "smaller portion" for $z = +1.5$.

[3] The standard deviation of any sampling distribution is normally referred to as the *standard error* of that distribution. Thus, the standard deviation of means is called the standard error of the mean (symbolized by $\sigma_{\overline{X}}$), whereas the standard deviation of differences between means, which will be discussed shortly, is called the standard error of differences between means and is symbolized by $\sigma_{\overline{X}_1 - \overline{X}_2}$. Minor changes in terminology, such as calling a standard deviation a standard error, are not really designed to confuse students, though they probably have that effect.

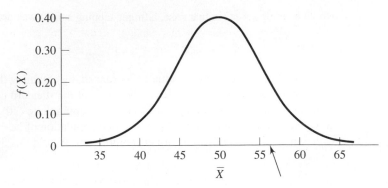

Figure 7.3   Sampling distribution of the mean for $n = 5$ drawn from a population with $\mu = 50$ and $\sigma = 10$

Because we know that the sampling distribution is normally distributed with a mean of 50 and a standard error of 4.47, we can find areas under the distribution by referring to tables of the standard normal distribution. Thus, for example, because two standard errors is $2(4.47) = 8.94$, the area to the right of $\overline{X} = 58.94$ is simply the area under the normal distribution greater than two standard deviations above the mean.

For our particular situation, we first need to know the probability of a sample mean greater than or equal to 56, and thus, we need to find the area above $\overline{X} = 56$. We can calculate this in the same way we did with individual observations, with only a minor change in the formula for z:

$$z = \frac{X - \mu}{\sigma} \quad \text{becomes} \quad z = \frac{\overline{X} - \mu}{\sigma_{\overline{X}}}$$

which can also be written as

$$z = \frac{\overline{X} - \mu}{\dfrac{\sigma}{\sqrt{n}}}$$

For our data, this becomes

$$z = \frac{56 - 50}{4.47} = \frac{6}{4.47} = 1.34$$

Notice that the equation for z used here is in the same form as our earlier formula for z. The only differences are that X has been replaced by $\overline{X}$ and σ has been replaced by $\sigma_{\overline{X}}$. These differences occur because we are now dealing with a distribution of means, and thus, the data points are now means, and the standard deviation in question is now the standard error of the mean (the standard deviation of means). The formula for z continues to represent (1) a point on a distribution, minus (2) the mean of that distribution, all divided by (3) the standard deviation of the distribution. Now rather than being concerned specifically with the distribution of $\overline{X}$, we have re-expressed the sample mean in terms of z scores and can now answer the question with regard to the standard normal distribution.

From Appendix z, we find that the probability of a z as large as 1.34 is .0901. Because we want a two-tailed test of H_0, we need to double the probability to obtain the probability of a deviation as large as 1.34 standard errors *in either direction* from the mean. This is $2(.0901) = .1802$. Thus, with a two-tailed test (that stressed children have a mean behavior problem score that is different in either direction from that of normal children) at the

.05 level of significance, we would not reject H_0 because the obtained probability is greater than .05. We would conclude that we have no evidence that stressed children show more or fewer behavior problems than other children.

7.3 Testing a Sample Mean When σ Is Unknown—The One-Sample *t* Test

The preceding example was chosen deliberately from among a fairly limited number of situations in which the population standard deviation (σ) is known. In the general case, we rarely know the value of σ and usually have to estimate it by way of the *sample* standard deviation (*s*). When we replace σ with *s* in the formula, however, the nature of the test changes. We can no longer declare the answer to be a *z* score and evaluate it using tables of *z*. Instead, we will denote the answer as *t* and evaluate it using tables of *t*, which are different from tables of *z*. The reasoning behind the switch from *z* to *t* is really rather simple. The basic problem that requires this change to *t* is related to the sampling distribution of the sample variance.

The Sampling Distribution of s^2

Because the *t* test uses s^2 as an estimate of σ^2, it is important that we first look at the sampling distribution of s^2. This sampling distribution gives us some insight into the problems we are going to encounter. We saw in Chapter 2 that s^2 is an *unbiased* estimate of σ^2, meaning that with repeated sampling, the average value of s^2 will equal σ^2. Although an unbiased estimator is a nice thing, it is not everything. The problem is that the shape of the sampling distribution of s^2 is positively skewed, especially for small samples. I drew 50,000 samples of $n = 5$ from a population with $\mu = 5$ and $\sigma^2 = 50$. I calculated the variance for each sample, and have plotted those 50,000 variances in Figure 7.4. Notice that the mean of this

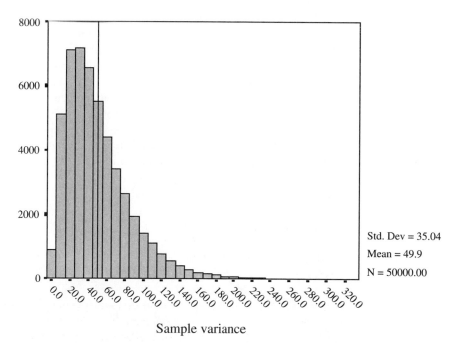

Std. Dev = 35.04
Mean = 49.9
N = 50000.00

Sample variance

Figure 7.4 Sampling distribution of the sample variance

distribution is almost exactly 50, reflecting the unbiased nature of s^2 as an estimate of σ^2. However, the distribution is very positively skewed. Because of the skewness of this distribution, an individual value of s^2 is more likely to underestimate σ^2 than to overestimate it, especially for small samples. Also because of this skewness, the resulting value of t is likely to be larger than the value of z that we would have obtained had σ been known and used.

The t Statistic

We are going to take the formula that we just developed for z,

$$z = \frac{\overline{X} - \mu}{\sigma_{\overline{X}}} = \frac{\overline{X} - \mu}{\dfrac{\sigma}{\sqrt{n}}} = \frac{\overline{X} - \mu}{\sqrt{\dfrac{\sigma^2}{n}}}$$

and substitute s for σ to give

$$t = \frac{\overline{X} - \mu}{s_{\overline{X}}} = \frac{\overline{X} - \mu}{\dfrac{s}{\sqrt{n}}} = \frac{\overline{X} - \mu}{\sqrt{\dfrac{s^2}{n}}}$$

We know that for any particular sample, s^2 is more likely than not to be smaller than the appropriate value of σ^2, so we can see that the t formula is more likely than not to produce a larger answer (in absolute terms) than we would have obtained if we had solved for t using the true but unknown value of σ^2 itself. (You can see this in Figure 7.4, where more than half of the observations fall to the left of σ^2.) As a result, it would not be fair to treat the answer as a z score and use the table of z. To do so would give us too many "significant" results—that is, we would make more than 5% Type I errors. (For example, when we were calculating z, we rejected H_0 at the .05 level of significance whenever z exceeded ± 1.96. If we create a situation in which H_0 is true, repeatedly draw samples of $n = 5$, and use s^2 in place of σ^2, we will obtain a value of ± 1.96 or greater more than 10% of the time. The t cutoff in this case is 2.776.)

The solution to our problem was supplied in 1908 by William Gosset, who worked for the Guinness Brewing Company and wrote under the pseudonym of Student, supposedly because the brewery would not allow him to publish under his own name. Gosset showed that if the data are sampled from a normal distribution, using s^2 in place of σ^2 would lead to

**Student's
t distribution**

a particular sampling distribution, now generally known as **Student's t distribution.** As a result of Gosset's work, all we have to do is substitute s^2, denote the answer as t, and evaluate t with respect to its own distribution, much as we evaluated z with respect to the normal distribution. The t distribution is tabled in Appendix t, and examples of the actual distribution of t for various sample sizes are shown graphically in Figure 7.5.

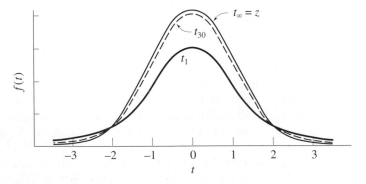

Figure 7.5 t distribution for 1, 30, and ∞ degrees of freedom

As you can see from Figure 7.5, the distribution of *t* varies as a function of the degrees of freedom, which for the moment we will define as one less than the number of observations in the sample. As $n \Rightarrow \infty$, $p(s^2 < \sigma^2) \Rightarrow p(s^2 > \sigma^2)$. (The symbol $\Rightarrow$ is read "approaches.") The skewness of the sampling distribution of s^2 disappears as the number of degrees of freedom increases, so the tendency for *s* to underestimate σ will also disappear. Thus, for an infinitely large number of degrees of freedom, *t* will be normally distributed and equivalent to *z*.

The test of one sample mean against a known population mean, which we have just performed, is based on the assumption that the sample was drawn from a normally distributed population. This assumption is required primarily because Gosset derived the *t* distribution assuming that the mean and variance are independent, which they are with a normal distribution. In practice, however, our *t* statistic can reasonably be compared with the *t* distribution whenever the sample size is sufficiently large to produce a normal sampling distribution of the mean. Most people would suggest that an *n* of 25 or 30 is "sufficiently large" for most situations, and for many situations it can be considerably smaller than that.

On the other hand, Wuensch (1993, personal communication) has argued convincingly that, at least with *very* skewed distributions, the fact that *n* is large enough to lead to a sampling distribution of the mean that appears to be normal does not guarantee that the resulting sampling distribution of *t* follows Student's *t* distribution. The derivation of *t* makes assumptions both about the distribution of means (which is under the control of the central limit theorem), and the variance, which is not controlled by that theorem.

Degrees of Freedom

I have mentioned that the *t* distribution is a function of the degrees of freedom (*df*). For the one-sample case, $df = n - 1$, the one degree of freedom has been lost because we used the sample mean in calculating s^2. To be more precise, we obtained the variance (s^2) by calculating the deviations of the observations from their own mean ($X - \overline{X}$), rather than from the population mean ($X - \mu$). Because the sum of the deviations about the mean $\left[\sum (X - \overline{X}) \right]$ is always zero, only $n - 1$ of the deviations are free to vary (the *n*th deviation is determined if the sum of the deviations is to be zero).

Psychomotor Abilities of Low-Birthweight Infants

An example drawn from an actual study of low-birthweight (LBW) infants will be useful at this point because that same general study can illustrate both this particular *t* test and other *t* tests to be discussed later in the chapter. Nurcombe et al. (1984) reported on an intervention program for the mothers of LBW infants. These infants present special problems for their parents because they are (superficially) unresponsive and unpredictable, in addition to being at risk for physical and developmental problems. The intervention program was designed to make mothers more aware of their infants' signals and more responsive to their needs, with the expectation that this would decrease later developmental difficulties often encountered with LBW infants. The study included three groups of infants: an LBW experimental group, an LBW control group, and a normal-birthweight (NBW) group. Mothers of infants in the last two groups did not receive the intervention treatment.

One of the dependent variables used in this study was the Psychomotor Development Index (PDI) of the Bayley Scales of Infant Development. This scale was first administered to all infants in the study when they were 6 months old. Because we would not expect to see

Table 7.1   Data for LBW infants on Psychomotor Development Index (PDI)

Raw Data			
96	120	112	100
125	96	86	124
89	104	116	89
127	89	89	124
102	104	120	102
112	92	92	102
120	124	83	116
108	96	108	96
92	108	108	95
120	86	92	100
104	100	120	120
89	92	102	98
92	98	100	108
89	117	112	126

Stem-and-Leaf Display

Stem	Leaf
8*	3
8.	6 6 9 9 9 9 9 9
9*	2 2 2 2 2 2
9.	5 6 6 6 6 8 8
10*	0 0 0 0 2 2 2 2 4 4 4
10.	8 8 8 8 8
11*	2 2 2
11.	6 6 7
12*	0 0 0 0 0 0 4 4 4
12.	5 6 7

Mean = 104.125
S.D. = 12.584
$N = 56$

Boxplot

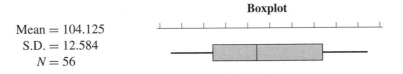

differences in psychomotor development between the two LBW groups as early as 6 months, it makes some sense to combine the data from the two groups and ask whether LBW infants in general are significantly different from the normative population mean of 100 usually found with this index.

The data for the LBW infants on the PDI are presented in Table 7.1. Included in this figure are a stem-and-leaf display and a boxplot. These two displays are important for examining the general nature of the distribution of the data and for searching for the presence of outliers.

From the stem-and-leaf display, we can see that the data, although not exactly normally distributed, at least are not badly skewed. Given our sample size (56), it is reasonable to assume that the sampling distribution of the mean would be reasonably normal. One interesting and unexpected finding that is apparent from the stem-and-leaf display is the prevalence of certain scores. For example, there are five scores of 108, but no other scores between 104 and 112. Similarly, there are six scores of 120, but no other scores between 117 and 124. Notice also that, with the exception of six scores of 89, there is a relative absence of odd numbers. A complete analysis of the data requires that we at least notice these oddities and try to track down their source. It would be worthwhile to examine the scoring process to see whether there is a reason why scores often tended to fall in bunches. It is probably an artifact of how raw scores are converted to scale scores, but it is worth checking. (Actually, if you check the scoring manual, you will find that these peculiarities are to be expected.) The fact that Tukey's exploratory data analysis (EDA) procedures lead us to notice these peculiarities is one of the great virtues of these methods. Finally, from the boxplot, we can see that there are no serious outliers we need to worry about, which makes our task noticeably easier.

From the data in Table 7.1, we can see that the mean PDI score for our LBW infants is 104.125. The norms for the PDI indicate that the population mean should be 100. Given the data, a reasonable first question concerns whether the mean of our LBW sample departs significantly from a population mean of 100. The t test is designed to answer this question.

From our formula for t and from the data, we have

$$t = \frac{\overline{X} - \mu}{s_{\overline{X}}} = \frac{\overline{X} - \mu}{\dfrac{s}{\sqrt{n}}}$$

$$= \frac{104.125 - 100}{\dfrac{12.584}{\sqrt{56}}} = \frac{4.125}{1.682}$$

$$= 2.45$$

This value will be a member of the t distribution on $56 - 1 = 55$ *df* if the null hypothesis is true—that is, if the data were sampled from a population with $\mu = 100$.

A t value of 2.45 in and of itself is not particularly meaningful unless we can evaluate it against the sampling distribution of t. For this purpose, the critical values of t are presented in Appendix t. This table differs in form from the table of the normal distribution (z) because instead of giving the area above and below each specific value of t, which would require too much space, the table instead gives those values of t that cut off particular critical areas—for example, the .05 and .01 levels of significance. We saw a similar situation with respect to the χ^2 distribution. Also, in contrast to z, a different t distribution is defined for each possible number of degrees of freedom. We want to work at the two-tailed .05 level, so we will want to know the value of t that cuts off $5/2 = 2.5\%$ in each tail. These critical values are generally denoted $t_{\alpha/2}$ or, in this case, $t_{.025}$. From the table of the t distribution in Appendix t, an abbreviated version of which is shown in Table 7.2, we find that the critical value of $t_{.025}$ (rounding to 50 *df* for purposes of the table) = 2.009. (This is sometimes written as $t_{.025}(50) = 2.009$ to indicate the degrees of freedom.) Because the obtained value of t, written t_{obt}, is greater than $t_{.025}$, we will reject H_0 at $\alpha = .05$, two-tailed, that our sample came from a population of observations with $\mu = 100$. Instead, we will conclude that our sample of LBW children differed from the general population of children on the PDI. In fact, their mean was statistically significantly *above* the normative population mean. This points out the advantage of using two-tailed tests because we would have expected this group to score below the normative mean. (This might also suggest that we check our scoring procedures to make sure we are not systematically overscoring our subjects. Actually, however, a number of other studies using the PDI have reported similarly high means.)

The Moon Illusion

It will be useful to consider a second example, this one taken from a classic paper by Kaufman and Rock (1962) on the moon illusion.[4] The moon illusion has fascinated psychologists for years and refers to the fact that when we see the moon near the horizon, it appears to be considerably larger than when we see it high in the sky. Kaufman and Rock concluded that this illusion could be explained on the basis of the greater *apparent*

---

[4] A more recent paper on this topic by Lloyd Kaufman and his son James Kaufman was published in the January 2000 issue of the *Proceedings of the National Academy of Sciences*.

Table 7.2   Percentage points of the *t* distribution

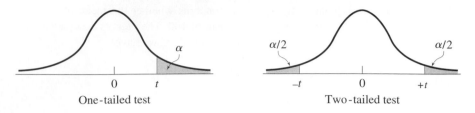

One-tailed test   Two-tailed test

	Level of Significance for One-Tailed Test								
	.25	.20	.15	.10	.05	.025	.01	.005	.0005

	Level of Significance for Two-Tailed Test								
df	.50	.40	.30	.20	.10	.05	.02	.01	.001
1	1.000	1.376	1.963	3.078	6.314	12.706	31.821	63.657	636.62
2	0.816	1.061	1.386	1.886	2.920	4.303	6.965	9.925	31.599
3	0.765	0.978	1.250	1.638	2.353	3.182	4.541	5.841	12.924
4	0.741	0.941	1.190	1.533	2.132	2.776	3.747	4.604	8.610
5	0.727	0.920	1.156	1.476	2.015	2.571	3.365	4.032	6.869
6	0.718	0.906	1.134	1.440	1.943	2.447	3.143	3.707	5.959
7	0.711	0.896	1.119	1.415	1.895	2.365	2.998	3.499	5.408
8	0.706	0.889	1.108	1.397	1.860	2.306	2.896	3.355	5.041
9	0.703	0.883	1.100	1.383	1.833	2.262	2.821	3.250	4.781
10	0.700	0.879	1.093	1.372	1.812	2.228	2.764	3.169	4.587
. . .	. . .	. . .	. . .	. . .	. . .	. . .	. . .	. . .	. . .
30	0.683	0.854	1.055	1.310	1.697	2.042	2.457	2.750	3.646
40	0.681	0.851	1.050	1.303	1.684	2.021	2.423	2.704	3.551
50	0.679	0.849	1.047	1.299	1.676	**2.009**	2.403	2.678	3.496
100	0.677	0.845	1.042	1.290	1.660	1.984	2.364	2.626	3.390
∞	0.674	0.842	1.036	1.282	1.645	1.960	2.326	2.576	3.291

SOURCE: The entries in this table were computed by the author.

distance of the moon when it is at the horizon. As part of a very complete series of experiments, the authors initially sought to estimate the moon illusion by asking subjects to adjust a variable "moon" that appeared to be on the horizon to match the size of a standard "moon" that appeared at its zenith, or vice versa. (In these measurements, they used an artifical moon created with special apparatus.) One of the first questions we might ask is whether there really is a moon illusion—that is, whether a larger setting is required to match a horizon moon or a zenith moon. The following data for 10 subjects are taken from Kaufman and Rock's paper and present the ratio of the diameter of the variable and standard moons. A ratio of 1.00 would indicate no illusion, whereas a ratio other than 1.00 would represent an illusion. (For example, a ratio of 1.50 would mean that the horizon moon appeared to have a diameter 1.50 times the diameter of the zenith moon.) Evidence in support of an illusion would require that we reject $H_0: \mu = 1.00$ in favor of $H_0: \mu \neq 1.00$.

Obtained ratio: 1.73   1.06   2.03   1.40   0.95
                   1.13   1.41   1.73   1.63   1.56

For these data, $n = 10$, $\overline{X} = 1.463$, and $s = 0.341$. A *t* test on H_0: $\mu = 1.00$ is given by

$$t = \frac{\overline{X} - \mu}{s_{\overline{X}}} = \frac{\overline{X} - \mu}{\dfrac{s}{\sqrt{n}}}$$

$$= \frac{1.463 - 1.000}{\dfrac{0.341}{\sqrt{10}}} = \frac{0.463}{0.108}$$

$$= 4.29$$

From Appendix *t*, with $10 - 1 = 9$ *df* for a two-tailed test at $\alpha = .05$, the critical value of $t_{.025}(9) = \pm 2.262$. The obtained value of *t* was 4.29. Because $4.29 > 2.262$, we can reject H_0 at $\alpha = .05$ and conclude that the true mean ratio under these conditions is not equal to 1.00. In fact, it is greater than 1.00, which is what we would expect on the basis of our experience. (It is always comforting to see science confirm what we have all known since childhood, but in this case, the results also indicate that Kaufman and Rock's experimental apparatus performed as it should.)

Confidence Interval on μ

point estimate

confidence limits

confidence interval

Confidence intervals are a useful way to convey the meaning of an experimental result that goes beyond the simple hypothesis test. The data on the moon illusion offer an excellent example of a case in which we are particularly interested in estimating the true value of μ—in this case, the true ratio of the perceived size of the horizon moon to the perceived size of the zenith moon. The sample mean $(\overline{X})$, as you already know, is an unbiased estimate of μ. When we have one specific estimate of a parameter, we call this a **point estimate.** There are also interval estimates, which are attempts to set limits that have a high probability of encompassing the true (population) value of the mean (the mean [μ] of a whole population of observations). What we want here are **confidence limits** on μ. These limits enclose what is called a **confidence interval.**[5] In Chapter 3, we saw how to set "probable limits" on an observation. A similar line of reasoning will apply here, where we attempt to set confidence limits on a parameter.

If we want to set limits that are likely to include μ given the data at hand, what we really want is to ask how large, or small, the true value of μ could be without causing us to reject H_0 if we ran a *t* test on the obtained sample mean. In other words, if μ were quite small (or quite large), we would have been unlikely to obtain the sample data. But for a whole range of values for μ, we would expect data like those we obtained. We want to calculate what those values of μ are.

An easy way to see what we are doing is to start with the formula for *t* for the one-sample case:

$$t = \frac{\overline{X} - \mu}{s_{\overline{X}}} = \frac{\overline{X} - \mu}{\dfrac{s}{\sqrt{n}}}$$

From the moon illusion data we know $\overline{X} = 1.463$, $s = 0.341$, $n = 10$. We also know that the critical two-tailed value for *t* at $\alpha = .05$ is $t_{.025}(9) = \pm 2.262$. We will substitute

---

[5] We often speak of "confidence limits" and "confidence interval" as if they were synonymous. The pretty much are, except that the limits are the end points of the interval. Don't be confused when you see them used interchangeably.

these values in the formula for t, but this time we will solve for the μ associated with this value of t.

$$t = \frac{\overline{X} - \mu}{\frac{s}{\sqrt{n}}} \qquad \pm 2.262 = \frac{1.463 - \mu}{\frac{0.341}{\sqrt{10}}} = \frac{1.463 - \mu}{0.108}$$

Rearranging to solve for μ, we have

$$\mu = \pm 2.262(0.108) + 1.463 = \pm 0.244 + 1.463$$

Using the $+0.244$ and -0.244 separately to obtain the upper and lower limits for μ, we have

$$\mu_{upper} = +0.244 + 1.463 = 1.707$$
$$\mu_{lower} = -0.244 + 1.463 = 1.219$$

and thus, we can write the 95% confidence limits as 1.219 and 1.707 and the confidence interval as

$$CI_{.95} = 1.219 \le \mu \le 1.707$$

Testing a null hypothesis about any value of μ outside these limits would lead to rejection of H_0, and testing a null hypothesis about any value of μ inside those limits would not lead to rejection. The general expression is

$$CI_{1-\alpha} = \overline{X} \pm t_{\alpha/2}(s_{\overline{X}}) = \overline{X} \pm t_{\alpha/2}\frac{s}{\sqrt{n}}$$

We have a 95% confidence interval because we used the two-tailed critical value of t at $\alpha = .05$. For the 99% limits we would take $t_{.01/2} = t_{.005} = \pm 3.250$. Then the 99% confidence interval is

$$CI_{.99} = \overline{X} \pm t_{.01/2}(s_{\overline{X}}) = 1.463 \pm 3.250(0.108) = 1.12 \le \mu \le 1.814$$

We can now say that the probability is .95 that intervals calculated as we have calculated the 95% interval earlier include the true mean ratio for the moon illusion. It is very tempting to say that the probability is .95 that the interval 1.219 to 1.707 includes the true mean ratio for the moon illusion, and the probability is .99 that the interval 1.112 to 1.814 includes μ. However, most statisticians would object to the statement of a confidence limit expressed in this way. They would argue that *before the experiment is run* and the calculations are made, an interval *of the form,*

$$\overline{X} \pm t_{.025}(s_{\overline{X}})$$

has a probability of .95 of encompassing μ. However, μ is a fixed (though unknown) quantity, and once the data are in, the specific interval 1.219 to 1.707 either includes the value of μ ($p = 1.00$) or it does not ($p = .00$). Put in slightly different form,

$$\overline{X} \pm t_{.025}(s_{\overline{X}})$$

is a random variable (it will vary from one experiment to the next), but the specific interval 1.219 to 1.707 is not a random variable and therefore does not have a probability associated with it. (Good [1999] has made the point that we place our confidence in the *method,* rather than in the *interval.*) Many would maintain that it is perfectly reasonable to say that my confidence is .95 that if you were to tell me the true value of μ, it would be found to lie between 1.219 and 1.707. But there are many people just lying in wait for you to say that the *probability* is .95 that μ lies between 1.219 and 1.707. When you do, they will pounce!

Note that neither the 95% nor the 99% confidence intervals that I computed includes the value of 1.00, which represents no illusion. We already knew this for the 95% confidence

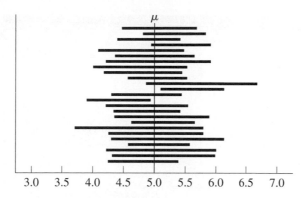

Figure 7.6 Confidence intervals computed on 25 samples from a population with $\mu = 5$

interval because we had rejected that null hypothesis when we ran our *t* test at that significance level.

I should add another way of looking at the interpretation of confidence limits. Statements of the form $p(1.219 < \mu < 1.707) = .95$ are not interpreted in the usual way. (Actually, I probably shouldn't use *p* in that equation.) The parameter μ is not a variable—it does not jump around from experiment to experiment. Rather, μ is a constant, and the interval is what varies from experiment to experiment. Thus, we can think of the parameter as a stake and the experimenter, in computing confidence limits, as tossing rings at it. Ninety-five percent of the time, a ring of specified width will encircle the parameter; 5% of the time, it will miss. A confidence statement is a statement of the probability that the ring has been on target; it is not a statement of the probability that the target (parameter) landed in the ring.

A graphic demonstration of confidence limits is shown in Figure 7.6. To generate this figure, I drew 25 samples of $n = 4$ from a population with a mean (μ) of 5. For every sample, a 95% confidence limit on μ was calculated and plotted. For example, the limits produced from the first sample (the top horizontal line) were approximately 4.46 and 5.72, whereas those for the second sample were 4.83 and 5.80. In this case, we know that the value of μ equals 5, so I have drawn a vertical line at that point. Notice that the limits for samples 12 and 14 do not include $\mu = 5$. We would expect that 95% confidence limits would encompass μ 95 times out of 100. Therefore, 2 misses out of 25 seems reasonable. Notice also that the confidence intervals vary in width. This variability is because the width of an interval is a function of the standard deviation of the sample, and some samples have larger standard deviations than others.

Using Minitab to Run One-Sample *t* Tests

With a large data set, it is often convenient to use a program such as Minitab to compute *t* values. Exhibit 7.1 shows how Minitab can be used to obtain a one-sample *t* test and confidence limits for the moon-illusion data. To get both the *t* test and the confidence limits, you have to specify separate analyses by clicking on different radio buttons. These buttons are shown in the first part of Exhibit 7.1. Notice that Minitab's results agree, within rounding error, with those we obtained by hand. Notice also that Minitab computes the exact

***p* level**   probability of a Type I error (the ***p* level**), rather than comparing *t* with a tabled value. Thus, whereas we concluded that the probability of a Type I error was *less than* .05, Minitab reveals that the actual probability is .0020. Most computer programs operate in this way.

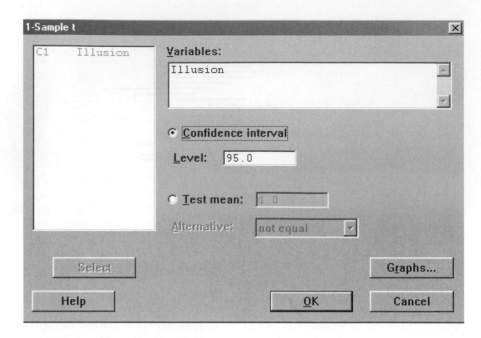

T Confidence Intervals

Variable	N	Mean	StDev	SE Mean	95.0 % CI
Illusion	10	1.463	0.341	0.108	(1.219, 1.707)

T-Test of the Mean

Test of mu = 1.000 vs mu not = 1.000

Variable	N	Mean	StDev	SE Mean	T	P
Illusion	10	1.463	0.341	0.108	4.30	0.0020

Exhibit 7.1 Minitab for one-sample *t*-test and confidence limits

7.4 Hypothesis Tests Applied to Means—Two Matched Samples

matched samples

repeated measures

related samples

matched-sample *t* test

In Section 7.3, we considered the situation in which we had one sample mean ($\overline{X}$) and wanted to test to see whether it was reasonable to believe that such a sample mean would have occurred if we had been sampling from a population with some specified mean (often denoted μ_0). Another way of phrasing this is to say that we were testing to determine whether the mean of the population from which we sampled (call it μ_1) was equal to some particular value given by the null hypothesis (μ_0). In this section, we will consider the case in which we have two **matched samples** (often called **repeated measures,** when the same subjects respond on two occasions, or **related samples,** correlated samples, paired samples, or dependent samples) and want to perform a test on the difference between their two means. In this case, we want what is sometimes called the **matched-sample *t* test.**

Table 7.3   Data from Everitt on weight gain

ID	1	2	3	4	5	6	7	8	9	10
Before	83.8	83.3	86.0	82.5	86.7	79.6	76.9	94.2	73.4	80.5
After	95.2	94.3	91.5	91.9	100.3	76.7	76.8	101.6	94.9	75.2
Diff	11.4	11.0	5.5	9.4	13.6	−2.9	−0.1	7.4	21.5	−5.3

ID	11	12	13	14	15	16	17	Mean	St. Dev
Before	81.6	82.1	77.6	83.5	89.9	86.0	87.3	83.23	5.02
After	77.8	95.5	90.7	92.5	93.8	91.7	98.0	90.49	8.48
Diff	−3.8	13.4	13.1	9.0	3.9	5.7	10.7	7.26	7.16

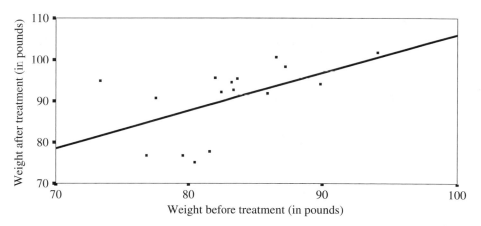

Figure 7.7   Relationship of weight before and after family therapy, for a group of 17 anorexic girls

Treatment of Anorexia

Everitt, in Hand, Daly, Lunn, McConway, and Ostrowski (1994), reported on family therapy as a treatment for anorexia. There were 17 girls in this experiment, and they were weighed before and after treatment. The weights of the girls, in pounds,[6] are given in Table 7.3. The row of difference scores was obtained by subtracting the Before score from the After score, so that a negative difference represents weight *loss,* and a positive difference represents a *gain.*

One of the first things we should probably do, although it takes us away from *t* tests for a moment, is to plot the relationship between Before Treatment and After Treatment weights, looking to see if there is, in fact, a relationship, and how linear that relationship is. Such a plot is given in Figure 7.7. Notice that the relationship is basically linear, with a slope quite near 1.0. Such a slope suggests that how much the girl weighed at the beginning of

---

[6] Everitt reported that these weights were in kilograms, but if so he has a collection of anorexic young girls whose mean weight is about 185 pounds, and that just doesn't sound reasonable. The example is completely unaffected by the units in which we record weight.

therapy did not seriously influence how much weight she gained or lost by the end of therapy. (We will discuss regression lines and slopes further in Chapter 9.)

The primary question we want is ask is whether subjects gained weight as a function of the therapy sessions. We have an experimental problem here because it is possible that weight gain resulted merely from the passage of time, and that therapy had nothing to do with it. However, I know from other data in that experiment that a group that did not receive therapy did not gain weight over the same period, which strongly suggests that the simple passage of time was not an important variable. If you were to calculate the weight of these girls before and after therapy, the means would be 83.23 and 90.49 lbs, respectively, which translates to a gain of a little over 7 pounds. However, we still need to test to see whether this difference is likely to represent a true difference in population means, or a chance difference. By this, I mean that we need to test the null hypothesis that the mean *in the population* of Before scores is equal to the mean *in the population* of After scores. In other words, we are testing $H_0: \mu_A = \mu_B$.

Difference Scores

difference scores

gain scores

Although it would seem obvious to view the data as representing two samples of scores, one set obtained before the therapy program and one after, it is also possible, and very profitable, to transform the data into one set of scores—the set of differences between X_1 and X_2 for each subject. These differences are called **difference scores,** or **gain scores,** and are shown in the row labeled "Diff" in Table 7.3. They represent the degree of weight gain between one measurement session and the next—presumably as a result of our intervention. If the therapy program had *no* effect (i.e., if H_0 is true), the average weight would not change from session to session. By chance, some participants would happen to have a higher weight on X_2 than on X_1, and some would have a lower weight, but *on the average* there would be no difference.

If we now think of our data as being the set of difference scores, the null hypothesis becomes the hypothesis that the mean of a population of difference scores (denoted μ_D) equals 0. Because it can be shown that $\mu_D = \mu_1 - \mu_2$, we can write $H_0: \mu_D = \mu_1 - \mu_2 = 0$. But now we can see that we are testing a hypothesis using *one* sample of data (the sample of difference scores), and we already know how to do that.

The *t* Statistic

We are now at precisely the same place we were in the previous section when we had a sample of data and a null hypothesis ($\mu = 0$). The only difference is that in this case the data are difference scores, and the mean and the standard deviation are based on the differences. Recall that *t* was defined as the difference between a sample mean and a population mean, divided by the standard error of the mean. Then we have

$$t = \frac{\overline{D} - 0}{s_{\overline{D}}} = \frac{\overline{D} - 0}{\frac{s_D}{\sqrt{N}}}$$

where and $\overline{D}$ and s_D are the mean and the standard deviation of the difference scores and N is the number of difference scores (i.e., the number of *pairs,* not the number of raw scores). From Table 7.3, we see that the mean difference score was 7.26, and the standard deviation of the differences was 7.16. For our data

$$t = \frac{\overline{D} - 0}{s_{\overline{D}}} = \frac{\overline{D} - 0}{\frac{s_D}{\sqrt{N}}} = \frac{7.26 - 0}{\frac{7.16}{\sqrt{17}}} = \frac{7.26}{1.74} = 4.18$$

Degrees of Freedom

The degrees of freedom for the matched-sample case are exactly the same as they were for the one-sample case. Because we are working with the difference scores, N will be equal to the number of differences (or the number of *pairs* of observations, or the number of *independent* observations—all of which amount to the same thing). Because the variance of these difference scores (s_D^2) is used as an estimate of the variance of a population of difference scores (σ_D^2) and because this sample variance is obtained using the sample mean ($\overline{D}$), we will lose one *df* to the mean and have $N - 1$ *df*. In other words, *df* = number of *pairs* minus 1.

We have 17 difference scores in this example, so we will have 16 degrees of freedom. From Appendix *t*, we find that for a two-tailed test at the .05 level of significance, $t_{.05}(16) = \pm 2.12$. Our obtained value of $t(4.18)$ exceeds 2.12, so we will reject H_0 and conclude that the difference scores were not sampled from a population of difference scores where $\mu_D = 0$. In practical terms, this means that the subjects weighed significantly more after the intervention program than before it. Although we would like to think that this means that the program was successful, keep in mind the possibility that this could just be normal growth. The fact remains, however, that for whatever reason, the weights were sufficiently higher on the second occasion to allow us to reject $H_0: \mu_D = \mu_1 - \mu_2 = 0$.

The Moon Illusion Revisited

As a second example, we will return to the work by Kaufman and Rock (1962) on the moon illusion. An important hypothesis about the source of the moon illusion was put forth by Holway and Boring (1940), who suggested that the illusion was because the observer looked straight at the moon with eyes level when it was on the horizon, whereas when the moon was at its zenith, the observer had to elevate his eyes as well as his head. Holway and Boring proposed that this difference in the elevation of the eyes was the cause of the illusion. Kaufman and Rock thought differently. To test Holway and Boring's hypothesis, Kaufman and Rock devised an apparatus that allowed them to present two artificial moons (one at the horizon and one at the zenith) and to control whether the subjects elevated their eyes to see the zenith moon. In one case, the subject was forced to put his head in such a position as to be able to see the zenith moon with eyes level. In the other case, the subject was forced to see the zenith moon with eyes raised. (The horizon moon was always viewed with eyes level.) In both cases, the dependent variable was the ratio of the perceived size of the horizon moon to the perceived size of the zenith moon (a ratio of 1.00 would represent no illusion). If Holway and Boring were correct, there should have been a greater illusion (larger ratio) in the eyes-elevated condition than in the eyes-level condition, although the moon was always perceived to be in the same place, the zenith. The actual data for this experiment are given in Table 7.4.

In this example, we want to test the *null* hypothesis that the means are equal under the two viewing conditions. Because we are dealing with related observations (each subject served under both conditions), we will work with the difference scores and test $H_0: \mu_D = 0$. Using a two-tailed test at $\alpha = .05$, the alternative hypothesis is $H_1: \mu_D \neq 0$.

From the formula for a *t* test on related samples, we have

$$t = \frac{\overline{D} - 0}{s_{\overline{D}}} = \frac{\overline{D} - 0}{\dfrac{s_D}{\sqrt{n}}}$$

$$= \frac{0.019 - 0}{\dfrac{0.137}{\sqrt{10}}} = \frac{0.019}{0.043}$$

$$= 0.44$$

Table 7.4 Magnitude of the moon illusion when zenith moon is viewed with eyes level and with eyes elevated

Observer	Eyes Elevated	Eyes Level	Difference (D)
1	1.65	1.73	−0.08
2	1.00	1.06	−0.06
3	2.03	2.03	0.00
4	1.25	1.40	−0.15
5	1.05	0.95	0.10
6	1.02	1.13	−0.11
7	1.67	1.41	0.26
8	1.86	1.73	0.13
9	1.56	1.63	−0.07
10	1.73	1.56	0.17

$$\overline{D} = 0.019$$
$$s_D = 0.137$$
$$s_{\overline{D}} = 0.043$$

From Appendix t, we find that $t_{.025}(9) = \pm 2.262$. Because $t_{\text{obt}} = 0.44$ is less than 2.262, we will fail to reject H_0 and will decide that we have no evidence to suggest that the illusion is affected by the elevation of the eyes.[7] (These data also include a second test of Holway and Boring's hypothesis because they would have predicted that there would not be an illusion if subjects viewed the zenith moon with eyes level. On the contrary, the data reveal a considerable illusion under this condition. A test of the significance of the illusion with eyes level can be obtained by the methods discussed in the Section 7.3, and the illusion is in fact statistically significant.)

Confidence Limits on Matched Samples

We can calculate confidence limits on matched samples in the same way we did for the one-sample case because in matched samples the data come down to a single column of difference scores. Returning to Everitt's data on anorexia, we have

$$t = \frac{\overline{D} - 0}{s_{\overline{D}}}$$

and thus

$$\text{CI}_{.95} = \overline{D} \pm t_{.05/2}(s_{\overline{D}}) = \overline{D} \pm t_{.025}\frac{s_D}{\sqrt{n}}$$

$$\text{CI}_{.95} = 7.26 \pm 2.12(1.74)$$

$$\text{CI}_{.95} = 7.26 \pm 3.69$$

$$= 3.57 \leq \mu \leq 10.95$$

---

[7] A glance at Appendix t will reveal that a t less than 1.96 (the critical value for z) will never be significant at $\alpha = .05$, regardless of the number of degrees of freedom. Moreover, unless you have at least 50 degrees of freedom, t values less than 2.00 will not be significant, often making it unnecessary for you even to bother looking at the table of t.

Notice that this confidence interval does not include $\mu_D = 0.0$, which is consistent with the fact that we rejected the null hypothesis.

Effect Size

In Chapter 6, we looked at effect size measures as a way of understanding the magnitude of the effect that we see in an experiment—rather than simply the statistical significance. When we are looking at the difference between two related measures we can, and should, also compute effect sizes. In this case, there is a slight complication as we will see shortly.

d-Family of Measures

A number of different effect sizes measures are often recommended, and for a complete coverage of this topic I suggest the reference by Kline (2004). As I did in Chapter 6, I am going to distinguish between measures based on differences between groups (the *d*-family) and measures based on correlations between variables (the *r*-family). However, in this chapter I am not going to discuss the *r*-family measures, partly because I find them less informative and partly because they are more easily and logically discussed in Chapter 11 when we come to the analysis of variance.

There is considerable confusion in the naming of measures, and for clarification on that score I refer the reader to Kline (2004). Here I will use the more common approach, which Kline points out is not quite technically correct, and refer to my measure as **Cohen's *d*.** Measures proposed by Hedges and by Glass are very similar and are often named almost interchangeably.

Cohen's *d*

The data on treatment of anorexia offer a good example of a situation in which it is relatively easy to report on the difference in ways that people will understand. All of us step onto a scale occasionally, and we have some general idea what it means to gain or lose 5 or 10 pounds. So for Everitt's data, we could simply report that the difference was significant ($t = 4.18$, $p < .05$) and that girls gained an average of 7.26 pounds. For girls who started out weighing, on average, 83 pounds, that is a substantial gain. In fact, it might make sense to convert pounds gained to a percentage and say that the girls increased their weight by $7.26/83.23 = 9\%$.

An alternative measure would be to report the gain in standard deviation units. This idea goes back to Cohen, who originally formulated the problem in terms of a statistic (*d*), where

$$d = \frac{\mu_1 - \mu_2}{\sigma}$$

In this equation, the numerator is the difference between two population means, and the denominator is the standard deviation of either population. In our case, we can modify that slightly to let the numerator be the mean gain ($\mu_{\text{After}} - \mu_{\text{Before}}$), and let the denominator be the population standard deviation *of the pretreatment weights*. To put this in terms of statistics, rather than parameters, we can substitute sample means and standard deviations instead of population values. This leaves us with

$$\hat{d} = \frac{\overline{X}_1 - \overline{X}_2}{s_{X_1}} = \frac{90.49 - 83.23}{5.02} = \frac{7.26}{5.02} = 1.45$$

I have put a "hat" over the *d* to indicate that we are calculating an estimate of *d*, and I have put the standard deviation of the pretreatment scores in the denominator. Our estimate tells us that, on average, the girls involved in family therapy gained nearly one and a half standard deviations of pretreatment weights over the course of therapy.

In this particular example, I find it easier to deal with the mean weight gain, rather than *d*, simply because I know something meaningful about weight. However, if this experiment had measured the girls' self-esteem, rather than weight, I would not know what to think if you said that they gained 7.26 self-esteem points because that scale means nothing to me. I would be impressed, however, if you said that they gained nearly one and a half standard deviation units in self-esteem.

The issue is not quite as simple as I have made it out to be because there are alternative ways of approaching the problem. One way would be to use the average of the pre- and post-score standard deviations, rather than just the standard deviation of the pre-scores. However, when we are measuring gain, it makes sense to me to measure it in the metric of the original weights. You may come across situations where you would think that it makes more sense to use the average standard deviation. In addition, it would be perfectly possible to use the standard deviation of the difference scores in the denominator for d. Kline (2004) discusses this approach and concludes, "If our natural reference for thinking about scores on (some) measure is their original standard deviation, it makes most sense to report standardized mean change (using that standard deviation)." However, the important point here is to keep in mind that such decisions often depend on substantive considerations in the particular research field, and no one measure is uniformly best.

Confidence Limits on d

Just as we were able to establish confidence limits on our estimate of the population mean (μ), we can establish confidence limits on d. However, it is not a simple process to do so, and I refer the reader to Kline (2004) or Cumming and Finch (2001). The latter provide a very inexpensive computer program to make these calculations.

Matched Samples

In many, but certainly not all, situations in which we will use the matched-sample t test, we will have two sets of data from the same subjects. For example, we might ask each of 20 people to rate their level of anxiety before and after donating blood. Or we might record ratings of level of disability made using two different scoring systems for each of 20 disabled individuals in an attempt to see whether one scoring system leads to generally lower assessments than does the other. In both examples, we would have 20 sets of numbers, two numbers for each person, and would expect these two sets of numbers to be related (or, in the terminology we will later adopt, to be correlated). Consider the blood-donation example. People differ widely in level of anxiety. Some seem to be anxious all of the time no matter what happens, and others just take things as they come and do not worry about anything. Thus, there should be a relationship between an individual's anxiety level before donating blood and her anxiety level after donating blood. In other words, if we know what a person's anxiety score was before donation, we can make a reasonable guess what it was after donation. Similarly, some people are severely disabled whereas others are only mildly disabled. If we know that a particular person received a high assessment using one scoring system, it is likely that he also received a relatively high assessment using the other system. The relationship between data sets does not have to be perfect—it probably never will be. The fact that we can make better-than-chance predictions is sufficient to classify two sets of data as matched or related.

In the two preceding examples, I chose situations in which each person in the study contributed two scores. Although this is the most common way of obtaining related samples, it is not the only way. For example, a study of marital relationships might involve asking husbands and wives to rate their satisfaction with their marriage, with the goal of testing to see whether wives are, on average, more or less satisfied than husbands. (You will see an example of just such a study in the exercises for this chapter.) Here, each individual would contribute only one score, but the couple *as a unit* would contribute a pair of scores. It is reasonable to assume that if the husband is very dissatisfied with the marriage, his wife is probably also dissatisfied, and vice versa, thus causing their scores to be related.

Many experimental designs involve related samples. They all have one thing in common, and that is that knowing one member of a pair of scores tells you something—maybe not much, but something—about the other member. Whenever this is the case, we say that the samples are matched.

Missing Data

Ideally, with matched samples we have a score on each variable for each case or pair of cases. If a subject participates in the pretest, she also participates in the posttest. If one member of a couple provides data, so does the other member. When we are finished collecting data, we have a complete set of paired scores. Unfortunately, experiments do not usually work out as cleanly as we would like.

Suppose, for example, that we want to compare scores on a checklist of children's behavior problems completed by mothers and fathers, with the expectation that mothers are more sensitive to their children's problems than are fathers and, thus, will produce higher scores. Most of the time both parents will complete the form. But there might be 10 cases where the mother sent in her form but the father did not, and 5 cases where we have a form from the father but not from the mother. The normal procedure in this situation is to eliminate the 15 pairs of parents where we do not have complete data, and then run a matched-sample t test on the data that remain. This is the way almost everyone would analyze the data. There is an alternative, however, that allows us to use all the data if we are willing to assume that data are missing at random and not systematically. (By this, I mean that we have to assume that we are not more likely to be missing Dad's data when the child is reported by Mom to have very few problems, nor are we less likely to be missing Dad's data for a very behaviorally disordered child.)

Bohj (1978) proposed an ingenious test in which you basically compute a matched-sample t for those cases in which both scores are present, then compute an additional independent group t (to be discussed next) between the scores of mothers without fathers and fathers without mothers, and finally combine the two t statistics. This combined t can then be evaluated against special tables. These tables are available in Wilcox (1986), and approximations to critical values of this combined statistic are discussed briefly in Wilcox (1987a). This test is sufficiently awkward that you would not use it simply because you are missing two or three observations. But it can be extremely useful when many pieces of data are missing. For a more extensive discussion, see Wilcox (1987b).

Using Computer Software for *t* Tests on Matched Samples

The use of almost any computer software to analyze matched samples can involve nothing more than using a compute command to create a variable that is the difference between the two scores we are comparing. We then run a simple one-sample t test to test the null hypothesis that those difference scores came from a population with a mean of 0. Alternatively, some software, such as SPSS, allows you to specify that you want a t on two related samples, and then to specify the two variables that represent those samples. This is very similar to what we have already done, so I will not repeat that here.

Writing Up the Results of a Dependent *t*

Suppose that we want to write up the results of Everitt's study of family therapy for anorexia. We would want to be sure to include the relevant sample statistics ($\overline{X}$, s^2, and N), as well as the test of statistical significance. But we would also want to include confidence limits on the mean weight gain following therapy, and our effect size estimate (d). We might write,

> Everitt ran a study on the effect of family therapy on weight gain in girls suffering from anorexia. He collected weight data on 17 girls before therapy, provided family therapy to the girls and their families, and then collected data on the girls' weight at the end of therapy.
>
> The mean weight gain for the $N = 17$ girls was 7.26 pounds, with a standard deviation of 7.16. A two-tailed t-test on weight gain was statistically significant ($t(16) = 4.18$, $p < .05$), revealing that on average the girls did gain weight over the course of therapy. A 95% confidence interval on mean weight gain was 3.57–10.95, which is a notable weight gain even at the low end of the interval. Cohen's $d = 1.45$, indicating that the girls' weight gain was nearly 1.5 standard deviations relative to their original pretest weights. It would appear that family therapy has made an important contribution to the treatment of anorexia in this experiment.

7.5   Hypothesis Tests Applied to Means—Two Independent Samples

One of the most common uses of the t test involves testing the difference between the means of two independent groups. We might want to compare the mean number of trials needed to reach criterion on a simple visual discrimination task for two groups of rats—one raised under normal conditions and one raised under conditions of sensory deprivation. Or we might want to compare the mean levels of retention of a group of college students asked to recall active declarative sentences and a group asked to recall passive negative sentences. Or, we might place subjects in a situation in which another person needed help; we could compare the latency of helping behavior when subjects were tested alone and when they were tested in groups.

In conducting any experiment with two independent groups, we would most likely find that the two sample means differed by some amount. The important question, however, is whether this difference is sufficiently large to justify the conclusion that the two samples were drawn from different populations—that is, using the example of helping behavior, is the mean of the population of latencies from singly tested subjects different from the mean of the population of latencies from group-tested subjects? Before we consider a specific example, however, we will need to examine the sampling distribution of differences between means and the t test that results from it.

Distribution of Differences Between Means

When we are interested in testing for a difference between the mean of one population (μ_1) and the mean of a second population (μ_2), we will be testing a null hypothesis of the form H_0: $\mu_1 - \mu_2 = 0$ or, equivalently, $\mu_1 = \mu_2$. Because the test of this null hypothesis involves the difference between independent sample means, it is important that we digress for a moment and examine the **sampling distribution of differences between means.** Suppose that we have two populations labeled X_1 and X_2 with means μ_1 and μ_2 and variances σ_1^2 and σ_2^2. We now draw pairs of samples of size n_1 from population X_1 and of size n_2 from population X_2, and record the means and the difference between the means for each pair of samples. Because we are sampling independently from each population, the sample means will be independent. (Means are paired only in the trivial and presumably irrelevant sense of being drawn at the same time.) The results of an infinite number of replications of this procedure

sampling distribution of differences between means

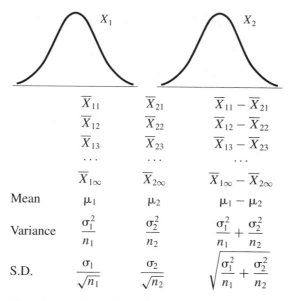

Figure 7.8 Schematic set of means and mean differences when sampling from two populations

are presented schematically in Figure 7.8. In the lower portion of this figure, the first two columns represent the sampling distributions of $\overline{X}_1$ and $\overline{X}_2$, and the third column represents the sampling distribution of mean differences $(\overline{X}_1 - \overline{X}_2)$. We are most interested in this third column because we are concerned with testing differences between means. The mean of this distribution can be shown to equal $\mu_1 - \mu_2$. The variance of this distribution of differences is given by what is commonly called the **variance sum law,** a limited form of which states

variance sum law

> The variance of a sum or difference of two *independent* variables is equal to the sum of their variances.[8]

We know from the central limit theorem that the variance of the distribution of $\overline{X}_1$ is σ_1^2/n_1 and the variance of the distribution of $\overline{X}_2$ is σ_2^2/n_2. Because the variables (sample means) are independent, the variance of the difference of these two variables is the sum of their variances. Thus

$$\sigma_{\overline{X}_1 - \overline{X}_2}^2 = \sigma_{\overline{X}_1}^2 + \sigma_{\overline{X}_2}^2 = \frac{\sigma_1^2}{n_1} + \frac{\sigma_2^2}{n_2}$$

Having found the mean and the variance of a set of differences between means, we know most of what we need to know. The general form of the sampling distribution of mean differences is presented in Figure 7.9.

The final point to be made about this distribution concerns its shape. An important theorem in statistics states that the sum or difference of two independent normally distributed variables is itself normally distributed. Because Figure 7.9 represents the difference between two sampling distributions of the mean, and because we know that the sampling distribution of means is at least approximately normal for reasonable sample sizes, the distribution in Figure 7.9 must itself be at least approximately normal.

---

[8] The complete form of the law omits the restriction that the variables must be independent and states that the variance of their sum or difference is $\sigma_{\overline{X}_1 \pm \overline{X}_2}^2 = \sigma_1^2 + \sigma_2^2 \pm 2\rho\sigma_1\sigma_2$ where the notation $\pm$ is interpreted as plus when we are speaking of their sum and as minus when we are speaking of their difference. The term ρ (rho) in this equation is the correlation between the two variables (to be discussed in Chapter 9) and is equal to zero when the variables are independent. (The fact that $\rho \neq 0$ when the variables are not independent was what forced us to treat the related sample case separately.)

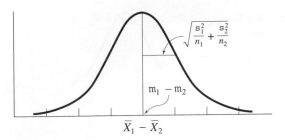

Figure 7.9   Sampling distribution of mean differences

The *t* Statistic

standard error of differences between means

Given the information we now have about the sampling distribution of mean differences, we can proceed to develop the appropriate test procedure. Assume for the moment that knowledge of the population variances (σ_i^2) is not a problem. We have earlier defined *z* as a statistic (a point on the distribution) minus the mean of the distribution, divided by the standard error of the distribution. Our statistic in the present case is ($\overline{X}_1 - \overline{X}_2$), the observed difference between the sample means. The mean of the sampling distribution is ($\mu_1 - \mu_2$), and, as we saw, the **standard error of differences between means**[9] is

$$\sigma_{\overline{X}_1 - \overline{X}_2} = \sqrt{\sigma_{\overline{X}_1}^2 + \sigma_{\overline{X}_2}^2} = \sqrt{\frac{\sigma_1^2}{n_1} + \frac{\sigma_2^2}{n_2}}$$

Thus, we can write

$$z = \frac{(\overline{X}_1 - \overline{X}_2) - (\mu_1 - \mu_2)}{\sigma_{\overline{X}_1 - \overline{X}_2}}$$

$$= \frac{(\overline{X}_1 - \overline{X}_2) - (\mu_1 - \mu_2)}{\sqrt{\dfrac{\sigma_1^2}{n_1} + \dfrac{\sigma_2^2}{n_2}}}$$

The critical value for $\alpha = .05$ is $z = \pm 1.96$ (two-tailed), as it was for the one-sample tests discussed earlier.

The preceding formula is not particularly useful except for the purpose of showing the origin of the appropriate *t* test because we rarely know the necessary population variances. (Such knowledge is so rare that it is not even worth imagining cases in which we would have it, although a few do exist.) We can circumvent this problem just as we did in the one-sample case, by using the sample variances as estimates of the population variances. This, for the same reasons discussed earlier for the one-sample *t*, means that the result will be distributed as *t* rather than *z*.

$$t = \frac{(\overline{X}_1 - \overline{X}_2) - (\mu_1 - \mu_2)}{s_{\overline{X}_1 - \overline{X}_2}}$$

$$= \frac{(\overline{X}_1 - \overline{X}_2) - (\mu_1 - \mu_2)}{\sqrt{\dfrac{s_1^2}{n_1} + \dfrac{s_2^2}{n_2}}}$$

---

[9] Remember that the standard deviation of any sampling distribution is called the standard error of that distribution.

The null hypothesis is generally the hypothesis that $\mu_1 - \mu_2 = 0$, so we will drop that term from the equation and write

$$t = \frac{(\overline{X}_1 - \overline{X}_2)}{s_{\overline{X}_1 - \overline{X}_2}} = \frac{(\overline{X}_1 - \overline{X}_2)}{\sqrt{\dfrac{s_1^2}{n_1} + \dfrac{s_2^2}{n_2}}}$$

Pooling Variances

Although the equation for t that we have just developed is appropriate when the sample sizes are equal, it requires some modification when the sample sizes are unequal. This modification is designed to improve the estimate of the population variance. One of the assumptions required in the use of t for two independent samples is that $\sigma_1^2 = \sigma_2^2$ (i.e., the samples come from populations with equal variances, regardless of the truth or falsity of H_0). The assumption is required regardless of whether n_1 and n_2 are equal. Such an assumption is often reasonable. We frequently begin an experiment with two groups of subjects who are equivalent and then do something to one (or both) group(s) that will raise or lower the scores by an amount equal to the effect of the experimental treatment. In such a case, it often makes sense to assume that the variances will remain unaffected. (Recall that adding or subtracting a constant—here, the treatment effect—to or from a set of scores has no effect on its variance.) Because the population variances are assumed to be equal, this common variance can be represented by the symbol σ^2, without a subscript.

In our data, we have two estimates of σ^2, namely s_1^2 and s_2^2. It seems appropriate to obtain some sort of an average of s_1^2 and s_2^2 on the grounds that this average should be a better estimate of σ^2 than either of the two separate estimates. We do not want to take the simple arithmetic mean, however, because doing so would give equal weight to the two estimates, even if one were based on considerably more observations. What we want is a **weighted average,** in which the sample variances are weighted by their degrees of freedom ($n_i - 1$). If we call this new estimate s_p^2 then

weighted average

$$s_p^2 = \frac{(n_1 - 1)s_1^2 + (n_2 - 1)s_2^2}{n_1 + n_2 - 2}$$

The numerator represents the sum of the variances, each weighted by their degrees of freedom, and the denominator represents the sum of the weights or, equivalently, the degrees of freedom for s_p^2.

The weighted average of the two sample variances is usually referred to as a **pooled variance estimate** (a rather inelegant name, but reasonably descriptive). Having defined the pooled estimate (s_p^2), we can now write

pooled variance estimate

$$t = \frac{(\overline{X}_1 - \overline{X}_2)}{s_{\overline{X}_1 - \overline{X}_2}} = \frac{(\overline{X}_1 - \overline{X}_2)}{\sqrt{\dfrac{s_p^2}{n_1} + \dfrac{s_p^2}{n_2}}} = \frac{(\overline{X}_1 - \overline{X}_2)}{\sqrt{s_p^2 \left(\dfrac{1}{n_1} + \dfrac{1}{n_2} \right)}}$$

Notice that both this formula for t and the one we have just been using involve dividing the difference between the sample means by an estimate of the standard error of the difference between means. The only difference concerns how this standard error is estimated. When the sample sizes are equal, it makes absolutely no difference whether or not you pool variances; the answer will be the same. When the sample sizes are unequal, however, pooling can make quite a difference.

Degrees of Freedom

Two sample variances (s_1^2 and s_2^2) have gone into calculating t. Each of these variances is based on squared deviations about their corresponding sample means, and therefore, each sample variance has $n_i - 1$ df. Across the two samples, therefore, we will have $(n_1 - 1) + (n_2 - 1) = (n_1 + n_2 - 2)$ df. Thus, the t for two independent samples will be based on $n_1 + n_2 - 2$ degrees of freedom.

Homophobia and Sexual Arousal

Adams, Wright, & Lohr (1996) were interested in some basic psychoanalytic theories that homophobia may be unconsciously related to the anxiety of being or becoming homosexual. They administered the Index of Homophobia to 64 heterosexual males and classed them as homophobic or nonhomophobic on the basis of their score. The researchers then exposed homophobic and nonhomophobic heterosexual men to videotapes of sexually explicit erotic stimuli portraying heterosexual and homosexual behavior, and recorded their level of sexual arousal. Adams et al. reasoned that if homophobia were unconsciously related to anxiety about one's own sexuality, homophobic individuals would show greater arousal to the homosexual videos than would nonhomophobic individuals.

In this example, we will examine only the data from the homosexual video. (There were no group differences for the heterosexual and lesbian videos.) The data in Table 7.5 were created to have the same means and pooled variance as the data that Adams et al. collected, so our conclusions will be the same as theirs.[10] The dependent variable is the degree of arousal at the end of the 4-minute video, with larger values indicating greater arousal.

Table 7.5 Data from Adams et al. on level of sexual arousal in homophobic and nonhomophobic heterosexual males

Homophobic						Nonhomophobic					
39.1	38.0	14.9	20.7	19.5	32.2	24.0	17.0	35.8	18.0	−1.7	11.1
11.0	20.7	26.4	35.7	26.4	28.8	10.1	16.1	−0.7	14.1	25.9	23.0
33.4	13.7	46.1	13.7	23.0	20.7	20.0	14.1	−1.7	19.0	20.0	30.9
19.5	11.4	24.1	17.2	38.0	10.3	30.9	22.0	6.2	27.9	14.1	33.8
35.7	41.5	18.4	36.8	54.1	11.4	26.9	5.2	13.1	19.0	−15.5	
8.7	23.0	14.3	5.3	6.3							

Mean	24.00		Mean	16.50	
Variance	148.87		Variance	139.16	
n	35		n	29	

Before we consider any statistical test, and ideally even before the data are collected, we must specify several features of the test. First, we must specify the null and alternative hypotheses:

$H_0: \mu_1 = \mu_2$

$H_1: \mu_1 \neq \mu_2$

The alternative hypothesis is bidirectional (we will reject H_0 if $\mu_1 < \mu_2$ or if $\mu_1 > \mu_2$, and thus we will use a two-tailed test. For the sake of consistency with other examples in this book, we will let $\alpha = .05$. It is important to keep in mind, however, that there is nothing particularly sacred about any of these decisions. (Think about how Jones and Tukey [2000]

---

[10] I actually added 12 points to each mean, largely to avoid many negative scores, but it doesn't change the results or the calculations in the slightest.

would have written this paragraph. Where would they have differed from what is here, and why might their approach be clearer?)

Given the null hypothesis as stated, we can now calculate t:

$$t = \frac{\overline{X}_1 - \overline{X}_2}{s_{\overline{X}_1 - \overline{X}_2}} = \frac{\overline{X}_1 - \overline{X}_2}{\sqrt{\dfrac{s_p^2}{n_1} + \dfrac{s_p^2}{n_2}}} = \frac{\overline{X}_1 - \overline{X}_2}{\sqrt{s_p^2\left(\dfrac{1}{n_1} + \dfrac{1}{n_2}\right)}}$$

Because we are testing H_0, $\mu_1 - \mu_2 = 0$, the $\mu_1 - \mu_2$ term has been dropped from the equation. We should pool our sample variances because they are so similar that we do not have to worry about heterogeneity of variance. Doing so we obtain

$$s_p^2 = \frac{(n_1 - 1)s_1^2 + (n_2 - 1)s_2^2}{n_1 + n_2 - 2}$$

$$= \frac{34(148.87) + 28(139.16)}{35 + 29 - 2} = 144.48$$

Notice that the pooled variance is slightly closer in value to s_1^2 than to s_2^2 because of the greater weight given s_1^2 in the formula. Then

$$t = \frac{\overline{X}_1 - \overline{X}_2}{\sqrt{\dfrac{s_p^2}{n_1} + \dfrac{s_p^2}{n_2}}} = \frac{(24.00 - 16.50)}{\sqrt{\dfrac{144.48}{35} + \dfrac{144.48}{29}}} = \frac{7.50}{\sqrt{9.11}} = 2.48$$

For this example, we have $n_1 - 1 = 34$ df for the homophobic group and $n_2 - 1 = 28$ df for the nonhomophobic group, making a total of $n_1 - 1 + n_2 - 1 = 62$ df. From the sampling distribution of t in Appendix t, $t_{.025}(62) \cong \pm 2.003$ (with linear interpolation). Because the value of t_{obt} far exceeds $t_{\alpha/2}$, we will reject H_0 (at $\alpha = .05$) and conclude that there is a difference between the means of the populations from which our observations were drawn. In other words, we will conclude (statistically) that $\mu_1 \neq \mu_2$ and (practically) that $\mu_1 > \mu_2$. In terms of the experimental variables, homophobic subjects show greater arousal to a homosexual video than do nonhomophobic subjects. (How would the conclusions of Jones and Tukey (2000) compare with the one given here?)

Confidence Limits on $\mu_1 - \mu_2$

In addition to testing a null hypothesis about population means (i.e., testing $H_0: \mu_1 - \mu_2 = 0$), it is useful to set confidence limits and effect sizes on the difference between μ_1 and μ_2. The logic for setting confidence limits is exactly the same as it was for the one-sample case. The calculations are also exactly the same except that we use the *difference* between the means and the standard error of *differences* between means in place of the mean and the standard error of the mean. Thus, for the 95% confidence limits on $\mu_1 - \mu_2$, we have

$$\text{CI}_{.95} = (\overline{X}_1 - \overline{X}_2) \pm t_{.05}\, s_{\overline{X}_1 - \overline{X}_2}$$

For the homophobia study, we have

$$\text{CI}_{.95} = (\overline{X}_1 - \overline{X}_2) \pm t_{.05}s_{\overline{X}_1 - \overline{X}_2} = (24.00 - 16.5) \pm 2.00\sqrt{\frac{144.48}{35} + \frac{144.48}{29}}$$

$$= 7.50 \pm 2.00(3.018) = 7.5 \pm 6.04$$

$$1.46 \leq (\mu_1 - \mu_2) \leq 13.54$$

The probability is .95 that an interval computed as we computed this interval encloses the difference in arousal to homosexual videos between homophobic and nonhomophobic participants. Although the interval is wide, it does not include 0. This is consistent with our rejection of the null hypothesis and allows us to state that homophobic individuals are, in fact, more sexually aroused by homosexual videos than are nonhomophobic individuals. However, I think that we would be remiss if we simply ignored the width of this interval. Although the difference between groups is statistically significant, there is still considerable uncertainty about how large the difference is. In addition, keep in mind that the dependent variable is the "degree of sexual arousal" on an arbitrary scale. Even if your confidence interval were quite narrow, it is difficult to know what to make of the result in absolute terms. To say that the groups differed by 7.5 units in arousal is not particularly informative. Is that a big difference or a little difference? We have no real way to know because the units (mm of penile circumference) are not something that most of us have an intuitive feel for. But when we standardize the measure, as we will in the next section, it is often more informative, as I think it is here.

Effect Size

The confidence interval that we just calculated has shown us that we still have considerable uncertainty about the difference in sexual arousal between groups, even though our statistically significant difference tells us that the homophobic group actually shows more arousal than the nonhomophobic group does. Again, we come to the issue of finding ways to present information to our readers that conveys the magnitude of the difference between our groups. We will use an effect size measure based on Cohen's d. It is very similar to the one that we used in the case of two dependent samples, where we divided the difference between the means by a standard deviation. We will again call this statistic (d). In this case, however, our standard deviation will be the estimated standard deviation of either population. More specifically, we will pool the two variances and take the square root of the result, and that will give us our best estimate of the standard deviation of the populations from which the numbers were drawn.[11] (If we had noticeably different variances, we would most likely use the standard deviation of one sample and note to the reader that this is what we had done.)

For our data on homophobia, we have

$$\hat{d} = \frac{\overline{X}_1 - \overline{X}_2}{s_p} = \frac{24.00 - 16.50}{12.02} = 0.62$$

This result expresses the difference between the two groups in standard deviation units and tells us that the mean arousal for homophobic participants was nearly 2/3 of a standard deviation higher than the arousal of nonhomophobic participants. That strikes me as a big difference. (Using the software by Cumming and Finch [2001], we find that the confidence intervals on d are 0.1155 and 1.125, which is also rather wide. At the same time, even the lower limit on the confidence interval is meaningfully large.)

A word of caution: In the example of homophobia, the units of measurement were largely arbitrary, and a 7.5 difference had no intrinsic meaning to us. Thus, it made more sense to express it in terms of standard deviations because we have at least some understanding of what that means. However, there are many cases wherein the original units are meaningful, and in those cases it may not make much sense to standardize the measure (i.e., report it in standard deviation units). We might prefer to specify the difference between means, or the ratio of means, or some similar statistic. The earlier example of the moon illusion is a case in point. There, it is far more meaningful to speak of the horizon moon appearing approximately half-again as large as the zenith moon, and I see no advantage, and some obfuscation, in converting to standardized units. The important goal is to give the reader an appreciation of the size of a difference, and you should choose that measure that

---

[11] Hedges (1982) was the one who first recommended stating this formula in terms of statistics with the pooled estimate of the standard deviation substituted for the population value. It is sometimes referred to as Hedges' **g**.

best expresses this difference. In one case, a standardized measure such as d is best, and in other cases, other measures, such as the distance between the means, is better.

As you will see in the next chapter, Cohen laid out some very general guidelines for what he considered small, medium, and large effect sizes. He characterized $d = .20$ as an effect that is small, but probably meaningful, an effect size of $d = .50$ as a medium effect that most people would be able to notice (such as a half of a standard deviation difference in IQ), and an effect size of $d = .80$ as large. We should not make too much of Cohen's levels, but they are helpful as a rough guide.

Reporting Results

Reporting results for a t test on two independent samples is basically similar to reporting results for the case of dependent samples. In Adams's et al. study of homophobia, two groups of participants were involved—one group scoring high on a scale of homophobia and the other scoring low. When presented with sexually explicit homosexual videos, the homophobic group actually showed a higher level of sexual arousal (the mean difference = 7.50 units). A t test of the difference between means produced a statistically significant result ($p < .05$), and Cohen's $d = .62$ showed that the two groups differed by nearly 2/3 of a standard deviation. However, the confidence limits on the population mean difference were rather wide ($1.46 \le \mu_1 - \mu_2 \le 13.54$), suggesting that we do not have a tight handle on the size of our difference.

SPSS Analysis

The SPSS analysis of the Adams et al. (1996) data is given in Exhibit 7.2. Notice that SPSS first provides what it calls Levene's test for equality of variances. We will discuss this test shortly, but it is simply a test on our assumption of homogeneity of variance. We do not come close to rejecting the null hypothesis that the variances are homogeneous ($p = .534$), so we don't have to worry about that here. From now on, we will assume equal variances and will focus on the next-to-bottom row of the table.

Group Statistics

GROUP		N	Mean	Std. Deviation	Std. Error Mean
Arousal	Homophobic	35	24.0000	12.2013	2.0624
	Nonhomophobic	29	16.5034	11.7966	2.1906

Independent Samples Test

	Levene's Test for Equality of Variances		t-test for Equality of Means						95% Confidence Interval of the Difference	
	F	Sig.	t	df	Sig. (2-tailed)	Mean Difference	Std. Error Difference	Lower	Upper
Equal variances assumed	.391	.534	2.484	62	.016	7.4966	3.0183	1.4630	13.5301
Equal variances not assumed			2.492	60.495	.015	7.4966	3.0087	1.4794	13.5138

Exhibit 7.2 SPSS analyses of Adams et al. (1996) data

Next note that the t supplied by SPSS is the same as we calculated, and that the probability associated with this value of t (.016) is less than $\alpha = .05$, leading to rejection of the null hypothesis. Note also that SPSS prints the difference between the means and the standard error of that difference, both of which we have seen in our own calculations. Finally, SPSS prints the 95% confidence interval on the difference between means, and it agrees with ours.

7.6 A Final Worked Example

Joshua Aronson has done extensive work on what he refers to as "stereotype threat," which refers to the fact that "members of stereotyped groups often feel extra pressure in situations where their behavior can confirm the negative reputation that their group lacks a valued ability." (Aronson, Lustina, Good, Keough, Steele, & Brown, 1998) This feeling of stereotype threat is then hypothesized to affect their performance, generally by lowering it from what it would have been had they not felt threatened. Considerable work has been done with ethnic groups who are stereotypically reputed to do poorly in some area, but Aronson et al. went a step further to ask if stereotype threat could actually lower the performance of white males—a group not normally associated with stereotype threat.

Aronson et al. (1998) used two independent groups of college students who were known to excel in mathematics, and for whom doing well in math was considered important. The researchers assigned 11 students to a control group that was simply asked to complete a difficult mathematics exam. They assigned 12 students to a threat condition, in which they were told that Asian students typically did better than other students in math tests, and that the purpose of the exam was to help the experimenter to understand why this difference exists. Aronson reasoned that simply telling white students that Asians did better on math tests would arouse feelings of stereotype threat and diminish the students' performance.

The data in Table 7.6 have been constructed to have nearly the same means and standard deviations as Aronson's data. The dependent variable is the number of items correctly solved.

First, we need to specify the null hypothesis, the significance level, and whether we will use a one- or a two-tailed test. We want to test the null hypothesis that the two conditions perform equally well on the test, so we have $H_0: \mu_1 = \mu_2$. We will set alpha at $\alpha = .05$, in line with what we have been using. Finally, we will choose to use a two-tailed test because it is reasonably possible for either group to show superior math performance.

Next, we need to calculate the pooled variance estimate:

$$s_p^2 = \frac{(n_1 - 1)s_1^2 + (n_2 - 1)s_2^2}{n_1 + n_2 - 2} = \frac{10(3.17^2) + 11(3.03^2)}{11 + 12 - 2}$$

$$= \frac{10(10.0489) + 11(9.1809)}{21} = \frac{201.4789}{21} = 9.5942$$

Table 7.6 Data from Aronson et al. (1998)

Control Subjects				Threat Subjects			
4	9	12	8	7	8	7	2
9	13	12	13	6	9	7	10
13	7	6		5	0	10	8
Mean = 9.64				Mean = 6.58			
st. dev. = 3.17				st. dev. = 3.03			
$n_1 = 11$				$n_2 = 12$			

Finally, we can calculate t using the pooled variance estimate:

$$t = \frac{(\overline{X}_1 - \overline{X}_2)}{\sqrt{\dfrac{s_p^2}{n_1} + \dfrac{s_p^2}{n_2}}} = \frac{(9.64 - 6.58)}{\sqrt{\dfrac{9.5942}{11} + \dfrac{9.5942}{12}}} = \frac{3.06}{\sqrt{1.6717}} = \frac{3.06}{1.2929} = 2.37$$

For this example, we have $n_1 + n_2 - 2 = 21$ degrees of freedom. From Appendix t, we find $t_{.025} = 2.080$. Because $2.37 > 2.080$, we will reject H_0 and conclude that the two population means are not equal.

If you were writing up the results of this experiment, you might write something like the following:

> This experiment tested the hypothesis that stereotype threat will disrupt the performance even of a group that is not usually thought of as having a negative stereotype with respect to performance on math tests. Aronson et al. (1998) asked two groups of participants to take a difficult math exam. These were white male college students who reported that they typically performed well in math and that good math performance was important to them. One group of students ($n = 11$) was simply given the math test and asked to do as well as they could. A second, randomly assigned group ($n = 12$) was informed that Asian males often outperformed white males and that the test was intended to help to explain the difference in performance. The test itself was the same for all participants. The results showed that the Control subjects answered a mean of 9.64 problems correctly, whereas the subjects in the Threat group completed only a mean of 6.58 problems. The standard deviations were 3.17 and 3.03, respectively. This represents an effect size (d) of .99, meaning that the two groups differed in the number of items correctly completed by nearly one standard deviation.
>
> Student's t test was used to compare the groups. The resulting $t(21)$ was 2.37, and was significant at $p < .05$, showing that stereotype threat significantly reduced the performance of those subjects to whom it was applied. The 95% confidence interval on the difference in means is $0.3712 \leq \mu_1 - \mu_2 \leq 5.7488$. This is quite a wide interval, but keep in mind that the two sample sizes were 11 and 12. An alternative way of comparing groups is to note that the Threat group answered 32% fewer items correctly than did the Control group.

7.7   Heterogeneity of Variance: The Behrens–Fisher Problem

homogeneity of variance

We have already seen that one of the assumptions underlying the t test for two independent samples is the assumption of **homogeneity of variance** ($\sigma_1^2 = \sigma_2^2 = \sigma^2$). To be more specific, we can say that *when H_0 is true* and *when we have homogeneity of variance*, then, pooling the variances, the ratio

$$t = \frac{(\overline{X}_1 - \overline{X}_2)}{\sqrt{\dfrac{s_p^2}{n_1} + \dfrac{s_p^2}{n_2}}}$$

is distributed as t on $n_1 + n_2 - 2\,df$. If we can assume homogeneity of variance, there is no difficulty, and the techniques discussed in this section are not needed. When we do not have homogeneity of variance, however, this ratio is not, strictly speaking, distributed as t. This leaves us with a problem, but fortunately a solution (or a number of competing solutions) exists.

**heterogeneous
variances**

First, unless $\sigma_1^2 = \sigma_2^2 = \sigma^2$, it makes no sense to pool (average) variances because the reason we were pooling variances in the first place was that we assumed them to be estimating the same quantity. For the case of **heterogeneous variances,** we will first dispense with pooling procedures and define

$$t' = \frac{(\overline{X}_1 - \overline{X}_2)}{\sqrt{\dfrac{s_1^2}{n_1} + \dfrac{s_2^2}{n_2}}}$$

where s_1^2 and s_2^2 are taken to be heterogeneous variances. As noted earlier the expression that I have just denoted as t' is *not* necessarily distributed as t on $n_1 + n_2 - 2\,df$. If we knew what the sampling distribution of t' actually looked like, there would be no problem. We would just evaluate t' against that sampling distribution. Fortunately, although there is no universal agreement, we know at least the approximate distribution of t'.

The Sampling Distribution of t'

**Behrens–Fisher
problem**

One of the first attempts to find the exact sampling distribution of t' was begun by Behrens and extended by Fisher, and the general problem of heterogeneity of variance has come to be known as the **Behrens–Fisher problem.** Based on this work, the Behrens–Fisher distribution of t' was derived and is presented in a table in Fisher and Yates (1953). However, because this table covers only a few degrees of freedom, it is not particularly useful for most purposes.

**Welch–
Satterthwaite
solution**

An alternative solution was developed apparently independently by Welch (1938) and by Satterthwaite (1946). The **Welch–Satterthwaite solution** is particularly important because we will refer back to it when we discuss the analysis of variance. Using this method, t' is viewed as a legitimate member of the t distribution, but for an unknown number of degrees of freedom. The problem then becomes one of solving for the appropriate df, denoted df':

$$df' = \frac{\left(\dfrac{s_1^2}{n_1} + \dfrac{s_2^2}{n_2}\right)^2}{\dfrac{\left(\dfrac{s_1^2}{n_1}\right)^2}{n_1 - 1} + \dfrac{\left(\dfrac{s_2^2}{n_2}\right)^2}{n_2 - 1}}$$

The degrees of freedom (df') are then taken to the nearest integer.[12] The advantage of this approach is that df' is bounded by the smaller of $n_1 - 1$ and $n_2 - 1$ at one extreme and $n_1 + n_2 - 2\,df$ at the other. More specifically, $Min(n_1 - 1, n_2 - 1) \le df' \le n_1 + n_2 - 2$. Because the critical value of t decreases as df increases, we can first evaluate t' as if df' were at its minimum. If the difference is significant, it will certainly be significant for the true df'.

---

[12] Welch (1947) later suggested that letting

$$df' = \left[\frac{\left(\dfrac{s_1^2}{n_1} + \dfrac{s_2^2}{n_2}\right)^2}{\dfrac{\left(\dfrac{s_1^2}{n_1}\right)^2}{n_1 + 1} + \dfrac{\left(\dfrac{s_2^2}{n_2}\right)^2}{n_2 + 1}}\right] - 2$$

might be a more accurate solution, although the difference is negligible.

If the difference is not significant, we can then evaluate t' at its maximum $n_1 + n_2 - 2$. If it is not significant at this point, no reduction in the degrees of freedom (by more accurate calculation of df') would cause it to be significant. Thus, the only time we actually need to calculate df' is when the value of t' would not be significant for $Min(n_1 - 1, n_2 - 1)$ df but would be significant for $n_1 + n_2 - 2$ df.

In this book, we will rely primarily on the Welch–Satterthwaite approximation. It has the distinct advantage of applying easily to problems that arise in the analysis of variance, and it is not noticeably more awkward than the other solutions.

Testing for Heterogeneity of Variance

How do we know whether we even have heterogeneity of variance to begin with? We do not know σ_1^2 and σ_2^2 (if we did we would not be solving for t), so we must in some way test their difference by using our two sample variances (s_1^2 and s_2^2).

A number of solutions have been put forth for testing for heterogeneity of variance. One of the simpler ones was advocated by Levene (1960), who suggested replacing each value of X either by its absolute deviation from the group mean—$d_{ij} = |X_{ij} - \overline{X}_j|$—or by its squared deviation—$d_{ij} = (X_{ij} - \overline{X}_j)^2$—where i and j represent the ith subject in the jth group. He then proposed running a standard two-sample t test on the d_{ij}s. This test makes intuitive sense, because if there is greater variability in one group, the absolute, or squared, values of the deviations will be greater. If t is significant, we would then declare the two groups to differ in their variances. Alternative approaches have been proposed—see, for example, O'Brien (1981)—but they are rarely implemented in standard software, and I will not elaborate on them here.

The procedures just described are suggested as replacements for the more traditional F test, which is a ratio of the larger sample variance to the smaller. This F has been shown by many people to be severely affected by nonnormality of the data and should not be used. The F test is still computed and printed by many of the large computer packages, but I do not recommend using it.

The Robustness of *t* with Heterogeneous Variances

robust

I mentioned that the t test is what is described as **robust,** meaning that it is more or less unaffected by moderate departures from the underlying assumptions. For the t test for two independent samples, we have two major assumptions and one side condition that must be considered. The two assumptions are those of normality of the sampling distribution of differences between means and homogeneity of variance. The side condition is the condition of equal sample sizes versus unequal sample sizes. Although we have just seen how the problem of heterogeneity of variance can be handled by special procedures, it is still relevant to ask what happens if we use the standard approach even with heterogeneous variances.

Box (1953), Norton (1953), Boneau (1960), and many others have investigated the effects of violating, both independently and jointly, the underlying assumptions of t. The general conclusion to be drawn from these studies is that for equal sample sizes, violating the assumption of homogeneity of variance produces very small effects—the nominal value of $\alpha = .05$ is most likely within ±0.02 of the true value of α. By this we mean that if you set up a situation with unequal variances *but with H_0 true* and proceed to draw (and compute t on) a large number of pairs of samples, you will find that somewhere between 3% and 7% of the sample t values actually exceed $\pm t_{.025}$. This level of inaccuracy is not intolerable. The same kind of statement applies to violations of the assumption of normality, provided that the true populations are roughly the same shape or else both are symmetric. If the distributions are markedly skewed (especially in opposite directions), serious problems arise unless their variances are fairly equal.

With unequal sample sizes, however, the results are more difficult to interpret. In Boneau's study, for example, sample variances were pooled in all cases because this is probably the most common procedure in practice (although it is incorrect for heterogeneous variances). Boneau found that when there was heterogeneity of variance *and* unequal sample sizes, the actual and normative probability values differed considerably. Keep in mind, however, that Boneau was pooling variances and evaluating t on $n_1 + n_2 - 2\ df$. We do not know what would have happened had he solved for t' and then evaluated t' on df' degrees of freedom. We do know, however, that had he done so, it would be reasonable to expect that the test would have proven to be robust because the Welch–Satterthwaite solution does not require the homogeneity assumption.

The investigator who has collected data that she thinks may violate one or more of the underlying assumptions should refer to the article by Boneau (1960). This article may be old, but it is quite readable and contains an excellent list of references to other work in the area. A good summary of alternative procedures can be found in Games, Keselman, and Rogan (1981).

Wilcox (1992) has argued persuasively for the use of trimmed samples for comparing group means with heavy-tailed distributions. (Interestingly, statisticians seem to have a fondness for trimmed samples, whereas psychologists and other social science practitioners seem not to have heard of trimming.) He provides results showing dramatic increases in power when compared with more standard approaches. Alternative nonparametric approaches, including "resampling statistics," are discussed in Chapter 18 of this book. These can be very powerful techniques that do not require unreasonable assumptions about the populations from which you have sampled. I suspect that resampling statistics and related procedures will be in the mainstream of statistical analysis in the not-too-distant future.

7.8 Hypothesis Testing Revisited

In Chapter 4, we spent quite a bit of time on examining the process of hypothesis testing. I pointed out that the traditional approach involves setting up a null hypothesis, and then generating a statistic that tells us how likely we are to find the obtained results if, in fact, the null hypothesis is true. In other words, we calculate the probability of the data given the null, and if that probability is very low, we reject the null.

In that chapter, we also looked briefly at a proposal by Jones and Tukey (2000) in which they approached the problem slightly differently. Now that we have several examples, this is a good point to go back and look at their proposal. In discussing the Adams et al. study of homophobia, I suggested that you think about how Jones and Tukey would have approached the issue. I am not going to repeat the traditional approach because that is laid out in each of the examples of how to write up our results.

The study by Adams et al. (1996) makes a good example. I imagine that all of us would be willing to agree that the null hypothesis of equal population means in the two conditions is highly unlikely to be true. Even laying aside the argument about differences in the 10th decimal place, it just seems unlikely that people who differ appreciably in amount of homophobia would show exactly the same mean level of arousal to erotic videos. We don't know which group will show the greater arousal, but one population mean is certain to be larger than the other. So we can rule out the null hypothesis ($H_0: \mu_h - \mu_n = 0$) as a viable possibility. That leaves us with three possible conclusions we could draw as a result of our test. The first is that $\mu_h < \mu_n$, the second is that $\mu_h > \mu_n$, and the third is that we do not have sufficient evidence to draw a conclusion.

Now let's look at the possibilities of error. It could actually be that $\mu_h > \mu_n$, but that we draw the opposite conclusion by deciding that the nonhomophobic participants are more aroused. This is what Jones and Tukey call a "reversal," and the probability of making this error if we use a *one-tailed* test at $\alpha = .05$ is .05. Alternatively, it could be that $\mu_h < \mu_n$ but that we make the error of concluding that the nonhomophobic participants are less aroused. Again with a one-tailed test, the probability of making this error is .05. It is not possible for us to make both of these errors because one of the hypotheses is true, so using a *one-tailed* test (in both directions) at $\alpha = .05$ gives us a 5% error rate. In our particular example, the critical value for a one-tailed test on 62 *df* is approximately 1.68. Because our obtained value of *t* was 2.48, we will conclude that homophobic participants are more aroused, on average, than nonhomophobic participants were. Notice that in writing this paragraph I have not used the phrase "Type I error" because that refers to rejecting a true null, and I have already said that the null can't possibly be true. Notice that my conclusion did not contain the phrase "rejecting the hypothesis." Instead, I referred to "drawing a conclusion." These are subtle differences, but I hope this example clarifies the position taken by Jones and Tukey.

Key Terms

Sampling distribution of the mean (7.1)

Central limit theorem (7.1)

Uniform distribution (7.1)

Standard error (7.2)

Student's *t* distribution (7.3)

Point estimate (7.3)

Confidence limits (7.3)

Confidence interval (7.3)

p level (7.3)

Matched samples (7.4)

Repeated measures (7.4)

Related samples (7.4)

Matched-sample *t* test (7.4)

Difference scores (7.4)

Gain scores (7.4)

Cohen's *d* (7.4)

Sampling distribution of differences between means (7.5)

Variance sum law (7.5)

Standard error of differences between means (7.5)

Weighted average (7.5)

Pooled variance estimate (7.5)

Homogeneity of variance (7.7)

Heterogeneous variances (7.7)

Behrens–Fisher problem (7.7)

Welch–Satterthwaite solution (7.7)

Robust (7.7)

Exercises

7.1   The following numbers represent 100 random numbers drawn from a rectangular population with a mean of 4.5 and a standard deviation of 2.7. Plot the distribution of these digits.

6	4	8	7	8	7	0	8	2	8	5	7
4	8	2	6	9	0	2	6	4	9	0	4
9	3	4	2	8	2	0	4	1	4	7	4
1	7	4	2	4	1	4	2	8	7	9	7
3	7	4	7	3	1	6	7	1	8	7	2
7	6	2	1	8	6	2	3	3	6	5	4
1	7	2	1	0	2	6	0	8	3	2	4
3	8	4	5	7	0	8	4	2	8	6	3
7	3	5	1								

7.2   I drew 50 samples of 5 scores each from the same population that the data in Exercise 7.1 came from, and calculated the mean of each sample. The means are shown here. Plot the distribution of these means.

2.8	6.2	4.4	5.0	1.0	4.6	3.8	2.6	4.0	4.8
6.6	4.6	6.2	4.6	5.6	6.4	3.4	5.4	5.2	7.2
5.4	2.6	4.4	4.2	4.4	5.2	4.0	2.6	5.2	4.0
3.6	4.6	4.4	5.0	5.6	3.4	3.2	4.4	4.8	3.8
4.4	2.8	3.8	4.6	5.4	4.6	2.4	5.8	4.6	4.8

7.3   Compare the means and the standard deviations for the distribution of digits in Exercise 7.1 and the sampling distribution of the mean in Exercise 7.2.

   a.   What would the central limit theorem lead you to expect in this situation?

   b.   Do the data correspond to what you would predict?

7.4   How would the result in Exercise 7.2 differ if you had drawn more samples of size 5?

7.5   How would the result in Exercise 7.2 differ if you had drawn 50 samples of size 15?

7.6   In 1979, the 238 students from North Dakota who took the verbal portion of the SAT exam had a mean score of 525. The standard deviation was not reported.

   a.   Is this result consistent with the idea that the SAT has a mean of 500 and a standard deviation of 100?

   b.   Would you have rejected H_0 had you been looking for evidence that SAT scores in general have been declining over the years from the mean of 500?

   c.   If you rejected H_0 in part (a), you might draw some conclusions about North Dakota's students or our assumption about the general population of students. What are those possible conclusions?

7.7   Why do the data in Exercise 7.6 not really speak to the issue of whether American education in general is in a terrible state?

7.8   In 1979, the 2,345 students from Arizona who took the math portion of the SAT had a mean score of 524. Is this consistent with the notion of a population mean of 500 if we assume that $\sigma = 100$?

7.9   Why does the answer to Exercise 7.8 differ substantially from the answer to Exercise 7.6 even though the means are virtually the same?

7.10   Compute 95% confidence limits on μ for the data in Exercise 7.6.

7.11   Everitt, in Hand et al. (1994), reported on several different therapies as treatments for anorexia. There were 29 girls in a cognitive-behavior therapy condition, and they were weighed before and after treatment. The weight gains of the girls, in pounds, are given here. The scores was obtained by subtracting the Before score from the After score, so that a negative difference represents weight *loss,* and a positive difference represents a *gain.*

1.7	0.7	−0.1	−0.7	−3.5	14.9	3.5	17.1	−7.6	1.6	11.7
6.1	1.1	−4.0	20.9	−9.1	2.1	−1.4	1.4	−0.3	−3.7	−0.8
2.4	12.6	1.9	3.9	0.1	15.4	−0.7				

   a.   What does the distribution of these values look like?

   b.   Did the girls in this group gain a statistically significant amount of weight?

7.12   Compute 95% confidence limits on the weight gain in Exercise 7.11.

7.13   Katz, Lautenschlager, Blackburn, and Harris (1990) examined the performance of 28 students who answered multiple-choice items on the SAT without having read the passages to which the items referred. The mean score (out of 100) was 46.6, with a standard deviation of 6.8. Random guessing would have been expected to result in 20 correct answers.

   a.   Were these students responding at better-than-chance levels?

   b.   If performance is statistically significantly better than chance, does it mean that the SAT test is not a valid predictor of future college performance?

7.14   Compas and others (1994) were surprised to find that young children under stress actually report fewer symptoms of anxiety and depression than we would expect. But they also noticed that their scores on a Lie Scale (a measure of the tendency to give socially desirable answers) were higher than expected. The population mean for the Lie scale on the Children's Manifest Anxiety Scale (Reynolds & Richmond, 1978) is known to be 3.87. For a sample of 36 children under stress, Compas et al. found a sample mean of 4.39, with a standard deviation of 2.61.

    a. How would we test whether this group shows an increased tendency to give socially acceptable answers?

    b. What would the null hypothesis and research hypothesis be?

    c. What can you conclude from the data?

7.15 Calculate the 95% confidence limits for μ for the data in Exercise 7.14. Are these limits consistent with your conclusion in Exercise 7.14?

7.16 Hoaglin, Mosteller, and Tukey (1983) present data on blood levels of beta-endorphin as a function of stress. They took beta-endorphin levels for 19 patients 12 hours before surgery, and again 10 minutes before surgery. The data are presented here, in fmol/ml:

ID	1	2	3	4	5	6	7	8	9	10
12 hours	10.0	6.5	8.0	12.0	5.0	11.5	5.0	3.5	7.5	5.8
10 minutes	6.5	14.0	13.5	18.0	14.5	9.0	18.0	42.0	7.5	6.0

ID	11	12	13	14	15	16	17	18	19
12 hours	4.7	8.0	7.0	17.0	8.8	17.0	15.0	4.4	2.0
10 minutes	25.0	12.0	52.0	20.0	16.0	15.0	11.5	2.5	2.0

Based on these data, what effect does increased stress have on beta-endorphin levels?

7.17 Why would you use a matched-sample t test in Exercise 7.16?

7.18 Construct 95% confidence limits on the true mean difference between beta-endorphin levels at the two times described in Exercise 7.16.

7.19 Hout, Duncan, and Sobel (1987) reported on the relative sexual satisfaction of married couples. They asked each member of 91 married couples to rate the degree to which they agreed with "Sex is fun for me and my partner" on a four-point scale ranging from "never or occasionally" to "almost always." The data appear below (I know it's a lot of data, but it's an interesting question):

Husband	1	1	1	1	1	1	1	1	1	1	1	1	1	1	1	
Wife	1	1	1	1	1	1	1	2	2	2	2	2	2	2	3	
Husband	1	1	1	1	2	2	2	2	2	2	2	2	2	2	2	
Wife	3	4	4	4	1	1	2	2	2	2	2	2	2	2	3	
Husband	2	2	2	2	2	2	2	2	2	3	3	3	3	3	3	
Wife	3	3	4	4	4	4	4	4	4	1	2	2	2	2	2	
Husband	3	3	3	3	3	3	3	3	3	3	3	3	3	4	4	
Wife	3	3	3	3	4	4	4	4	4	4	4	4	4	1	1	
Husband	4	4	4	4	4	4	4	4	4	4	4	4	4	4	4	
Wife	2	2	2	2	2	2	2	2	3	3	3	3	3	3	3	
Husband	4	4	4	4	4	4	4	4	4	4	4	4	4	4	4	4
Wife	3	3	4	4	4	4	4	4	4	4	4	4	4	4	4	4

Start by running a matched-sample t test on these data. Why is a matched-sample test appropriate?

7.20 In the study referred to in Exercise 7.19, what, if anything does your answer to that question tell us about whether couples are sexually compatible? What do we know from this analysis, and what don't we know?

7.21 For the data in Exercise 7.19, create a scatterplot and calculate the correlation between husband's and wife's sexual satisfaction. How does this amplify what we have learned from the analysis in Exercise 7.19. (I do not discuss scatterplots and correlation until Chapter 9, but a quick glance at Chapter 9 should suffice if you have difficulty. SPSS will easily do the calculation.)

7.22   Construct 95% confidence limits on the true mean difference between the Sexual Satisfaction scores in Exercise 7.19, and interpret them with respect to the data.

7.23   Some would object that the data in Exercise 7.19 are clearly discrete, if not ordinal, and that it is inappropriate to run a *t* test on them. Can you think what might be a counter argument? (This is not an easy question, and I really asked it mostly to make the point that there could be controversy here.)

7.24   Give an example of an experiment in which using related samples would be ill advised because taking one measurement might influence another measurement.

7.25   Everitt, in Hand et al. (1994), (see Exercise 7.11) reported on family therapy as a treatment for anorexia. There were 17 girls in this experiment, and they were weighed before and after treatment. The weights of the girls, in pounds, are given here. The row of difference scores was obtained by subtracting the Before score from the After score, so that a negative difference represents weight *loss*, and a positive difference represents a *gain*.

ID	1	2	3	4	5	6	7	8	9	10
Before	83.8	83.3	86.0	82.5	86.7	79.6	76.9	94.2	73.4	80.5
After	95.2	94.3	91.5	91.9	100.3	76.7	76.8	101.6	94.9	75.2
Diff	11.4	11.0	5.5	9.4	13.6	−2.9	−.1	7.4	21.5	−5.3

ID	11	12	13	14	15	16	17	Mean	St. Dev
Before	81.6	82.1	77.6	83.5	89.9	86.0	87.3	83.23	5.02
After	77.8	95.5	90.7	92.5	93.8	91.7	98.0	90.49	8.48
Diff	−3.8	13.4	13.1	9.0	3.9	5.7	10.7	7.26	7.16

a.   What null hypothesis would these data lead you to want to test?

b.   Run the appropriate *t* test and draw the appropriate conclusion.

7.26   What would happen in Exercise 7.25 if I subtracted the After score from the Before score, instead of the other way around?

7.27   Calculate a confidence interval on the weight gain for the girls in Everitt's study in Exercise 7.25.

7.28   Graph the relationship between the Before and After scores to evaluate the degree to which the two sets of scores are related. (I do not discuss scatterplots until Chapter 9, but you should be able to work this out for yourself.)

7.29   In the study referred to in Exercise 7.13, Katz et al. (1990) compared the performance on SAT items of a group of 17 students who were answering questions about a passage after having read the passage with the performance of a group of 28 students who had not seen the passage. The mean and standard deviation for the first group were 69.6 and 10.6, whereas for the second group they were 46.6 and 6.8.

a.   What is the null hypothesis?

b.   What is the alternative hypothesis?

c.   Run the appropriate *t* test.

d.   Interpret the results.

7.30   Many mothers experience a sense of depression shortly after the birth of a child. Design a study to examine postpartum depression and, from material in this chapter, tell how you would estimate the mean increase in depression.

7.31   In Exercise 7.25, we saw data from Everitt that showed that girls receiving family therapy gained weight over the course of that therapy. However, it is possible that they just gained weight because they got older. One way to control for this is to look at the amount of weight gained by the Family Therapy group ($n = 17$) in contrast with the amount gained by girls in a Control group ($n = 26$), who received no therapy. The data on weight gain for the two groups is shown below.

Control		Family Therapy	
−0.5	3.3	11.4	9.0
−9.3	11.3	11.0	3.9
−5.4	0.0	5.5	5.7
12.3	−1.0	9.4	10.7
−2.0	−10.6	13.6	
−10.2	−4.6	−2.9	
−12.2	−6.7	−0.1	
11.6	2.8	7.4	
−7.1	0.3	21.5	
6.2	1.8	−5.3	
−0.2	3.7	−3.8	
−9.2	15.9	13.4	
8.3	−10.2	13.1	

Mean	−0.45		7.26
St Dev.	7.99		7.16
Variance	63.82		51.23

Run the appropriate test to compare the group means. What would you conclude?

7.32 Calculate the confidence interval on $\mu_1 - \mu_2$ for the data in Exercise 7.31.

7.33 In Exercise 7.19, we saw pairs of observations on sexual satisfaction for husbands and wives. Suppose that those data had actually come from unrelated males and females, such that the data are no longer paired. What effect you expect this to have on the analysis?

7.34 Run the appropriate t test on the data in Exercise 7.19 assuming that the observations are independent. What would you conclude?

7.35 Why isn't the difference between the results in Exercises 7.34 and 7.19 greater than it is?

7.36 What is the role of random assignment in the Everitt's anorexia study referred to in Exercise 7.31, and under what conditions might we find it difficult to carry out random assignment?

7.37 The Thematic Apperception Test (TAT) presents subjects with ambiguous pictures and asks them to tell a story about them. These stories can be scored in any number of ways. Werner, Stabenau, and Pollin (1970) asked mothers of 20 Normal and 20 Schizophrenic children to complete the TAT and scored for the number of stories (out of 10) that exhibited a positive parent-child relationship. The data follow:

Normal	8	4	6	3	1	4	4	6	4	2
Schizophrenic	2	1	1	3	2	7	2	1	3	1
Normal	2	1	1	4	3	3	2	6	3	4
Schizophrenic	0	2	4	2	3	3	0	1	2	2

a. What would you assume to be the experimental hypothesis behind this study?

b. What would you conclude with respect to that hypothesis?

7.38 In Exercise 7.37, why might it be smart to look at the variances of the two groups?

7.39 In Exercise 7.37, a significant difference might lead someone to suggest that poor parent-child relationships are the cause of schizophrenia. Why might this be a troublesome conclusion?

7.40 Much has been made of the concept of experimenter bias, which refers to the fact that even the most conscientious experimenters tend to collect data that come out in the desired direction (they see what they want to see). Suppose we use students as experimenters. All the experimenters are told that subjects will be given caffeine before the experiment, but one-half of the experimenters are told that we expect caffeine to lead to good performance and one-half are told that we expect it to lead to poor performance. The dependent variable is the number of simple arithmetic problems the subjects can solve in 2 minutes. The data obtained are as follows:

Expectation good:	19	15	22	13	18	15	20	25	22
Expectation poor:	14	18	17	12	21	21	24	14	

What can you conclude?

7.41   Calculate 95% confidence limits on $\mu_1 - \mu_2$ for the data in Exercise 7.40.

7.42   An experimenter examining decision making asked 10 children to solve as many problems as they could in 10 minutes. One group (5 subjects) was told that this was a test of their innate problem-solving ability; a second group (5 subjects) was told that this was just a time-filling task. The data follow:

Innate ability:	4	5	8	3	7
Time-filling task:	11	6	9	7	9

Does the mean number of problems solved vary with the experimental condition?

7.43   A second investigator repeated the experiment described in Exercise 7.42 and obtained the same results. However, she thought that it would be more appropriate to record the data in terms of minutes per problem (e.g., 4 problems in 10 minutes $= 10/4 = 2.5$ minutes/problem). Thus, her data were as follows:

Innate ability:	2.50	2.00	1.25	3.33	1.43
Time-filling task:	0.91	1.67	1.11	1.43	1.11

Analyze and interpret these data with the appropriate t test.

7.44   What does a comparison of Exercises 7.42 and 7.43 show you?

7.45   I stated earlier that Levene's test consists of calculating the absolute (or squared) differences between individual observations and their group's mean, and then running a t test on those differences. Using any computer software, it is simple to calculate those absolute and squared differences and then to run a t test on them. Calculate both and determine which approach SPSS is using in the example. (*Hint*, $F = t^2$ here, and the F value that SPSS actually calculated was 0.391148, to 6 decimal places.)

7.46   Research on clinical samples (i.e., people referred for diagnosis or treatment) has suggested that children who experience the death of a parent may be at risk for developing depression or anxiety in adulthood. Mireault (1990) collected data on 140 college students who had experienced the death of a parent, 182 students from two-parent families, and 59 students from divorced families. The data are found in the file Mireault.dat and are described in Appendix: Computer Exercises.

   a.   Use any statistical program to run t tests to compare the first two groups on the Depression, Anxiety, and Global Symptom Index t scores from the Brief Symptom Inventory (Derogatis, 1983).

   b.   Are these three t tests independent of one another? (*Hint:* To do this problem you will have to ignore or delete those cases in Group 3 [the Divorced group]. Your instructor or the appropriate manual will explain how to do this for the particular software that you are using.)

7.47   It is commonly reported that women show more symptoms of anxiety and depression than men. Would the data from Mireault's study support this hypothesis?

7.48   Now run separate t tests to compare Mireault's Group 1 versus Group 2, Group 1 versus Group 3, and Group 2 versus Group 3 on the Global Symptom Index. (This is not a good way to compare the three group means, but it is being done here because it leads to more appropriate analyses in Chapter 12.)

7.49   Present meaningful effect sizes estimate(s) for the matched pairs data in Exercise 7.25.

7.50   Present meaningful effect sizes estimate(s) for the two independent group data in Exercise 7.31.

Discussion Questions

7.51 In Chapter 6 (Exercise 6.38), we examined data presented by Hout et al. on the sexual satis-faction of married couples. We did that by setting up a contingency table and computing χ^2 on that table. We looked at those data again in a different way in Exercise 7.19, where we ran a t test comparing the means. Instead of asking subjects to rate their statement "Sex is fun for me and my partner" as "Never, Fairly Often, Very Often, or Almost Always," we converted their categorical responses to a four-point scale from $1 =$ "Never" to $4 =$ "Almost Always."

a. How does the "scale of measurement" issue relate to this analysis?

b. Even setting aside the fact that this exercise and Exercise 6.37 use different statistical tests, the two exercises are asking quite different questions of the data. What are those different questions?

c. What might you do if 15 wives refused to answer the question, although their husbands did, and 8 husbands refused to answer the question when their wives did?

d. How comfortable are you with the t test analysis, and what might you do instead?

7.52 Write a short paragraph containing the information necessary to describe the results of the experiment discussed in Exercise 7.31. This should be an abbreviated version of what you would write in a research article.

CHAPTER 8

Power

Objectives

To introduce the concept of the power of a statistical test and to show how we can calculate the power of a variety of statistical procedures.

Contents

UNTIL RECENTLY, MOST APPLIED STATISTICAL WORK as it is actually carried out in analyzing experimental results was primarily concerned with minimizing (or at least controlling) the probability of a Type I error (α). When designing experiments, people tend to ignore the very important fact that there is a probability (β) of another kind of error, Type II errors. Whereas Type I errors deal with the problem of finding a difference that is not there, Type II errors concern the equally serious problem of not finding a difference that is there. When we consider the substantial cost in time and money that goes into a typical experiment, we could argue that it is remarkably short-sighted of experimenters not to recognize that they may, from the start, have only a small chance of finding the effect they are looking for, even if such an effect does exist in the population.

There are very good historical reasons why investigators have tended to ignore Type II errors. Cohen places the initial blame on the emphasis Fisher gave to the idea that the null hypothesis was either true or false, with little attention to H_1. Although the Neyman–Pearson approach does emphasize the importance of H_1, Fisher's views have been very influential. In addition, until recently, many textbooks avoided the problem altogether, and those books that did discuss power did so in ways that were not easily understood by the average reader (noncentrality parameters and power curves with strange things on the abscissa can be off-putting). Cohen, however, discussed the problem clearly and lucidly in several publications.[1] Cohen (1988) presents a thorough and rigorous treatment of the material. In Welkowitz, Ewen, and Cohen (2000), the material is treated in a slightly simpler way through the use of an approximation technique. That approach is the one adopted in this chapter. Two extremely useful papers that are very accessible and that provide useful methods are by Cohen (1992a, 1992b). You should have no difficulty with either of these sources, or, for that matter, with any of the many excellent papers Cohen published on a wide variety of topics not necessarily directly related to this particular one.

Speaking in terms of Type II errors is a rather negative way of approaching the problem because it keeps reminding us that we might make a mistake. The more positive approach would be to speak in terms of **power,** which is defined as the probability of correctly rejecting a false H_0 when a particular alternative hypothesis is true. Thus, power $= 1 - \beta$. A more powerful experiment is one that has a better chance of rejecting a false H_0 than does a less powerful experiment.

In this chapter, we will take the approach of Welkowitz et al. (2000) and work with an approximation to the true power of a test. This approximation is an excellent one, especially given that we do not really care whether the power is .85 or .83 but, rather, whether it is near .80 or nearer to .30.[2] In recent years, many simple software programs have been developed to automate power calculations. Goldstein (1989) reviewed thirteen of these programs. DataSim (Bradley, 1988) does an excellent job of producing accurate power estimates, is inexpensive, and easily produces data with predefined characteristics. An excellent program named G*Power by Faul and Erdfelder is available on the Internet at http://www.psychologie .uni-duesseldorf.de/aap/projects/gpower/, and there are both Macintosh and DOS programs at that site. The best part is that they are free. In what follows, I am assuming that you will be doing hand calculation by way of tables because that is the most effective way to lay out the relevant concepts. However, you could compare my results with the slightly more accurate results produced by G*Power.

power

---

[1] A somewhat different approach is taken by Murphy and Myors (1998), who base all their power calculations on the F distribution. The F distribution appears throughout this book, and virtually all the statistics covered in this book can be transformed to a F. The Murphy and Myors approach is worth examining and will give results very close to the results we find in this chapter.

[2] Cohen (1988) takes a more detailed approach; rather than working with an approximation, he works with more exact probabilities. That approach requires much more extensive tables but produces answers very close to the ones we get with an approximation.

For expository purposes, we will assume for the moment that we are interested in testing one sample mean against a specified population mean, although the approach will immediately generalize to testing other hypotheses.

8.1   Factors Affecting the Power of a Test

As might be expected, power is a function of several variables. It is a function of (1) α, the probability of a Type I error, (2) the true alternative hypothesis (H_1), (3) the sample size, and (4) the particular test to be employed. With the exception of the relative power of independent versus matched samples, we will avoid this last relationship on the grounds that when the test assumptions are met, most procedures discussed in this book can be shown to be the uniformly most powerful tests of those available to answer the question at hand. It is important to keep in mind, however, that when the underlying assumptions of a test are violated, the nonparametric tests discussed in Chapter 18, and especially the resampling tests, are sometimes more powerful.

The Basic Concept

First, we need a quick review of the material covered in Chapter 4. Consider the two distributions in Figure 8.1. The distribution to the left (labeled H_0) represents the sampling distribution of the mean when the null hypothesis is true and $\mu = \mu_0$. The distribution on the right represents the sampling distribution of the mean that we would have if H_0 were false and the true population mean were equal to μ_1. The placement of this distribution depends entirely on what the value of μ_1 happens to be.

The heavily shaded right tail of the H_0 distribution represents α, the probability of a Type I error, assuming that we are using a one-tailed test (otherwise it represents $\alpha/2$). This area contains the sample means that would result in significant values of t. The second distribution (H_1) represents the sampling distribution of the statistic when H_0 is false and the true mean is μ_1. It is readily apparent that even when H_0 is false, many of the sample means (and therefore the corresponding values of t) will nonetheless fall to the left of the critical value, causing us to fail to reject a false H_0, thus committing a Type II error. The probability of this error is indicated by the lightly shaded area in Figure 8.1 and is labeled β. When H_0 is false and the test statistic falls to the right of the critical value, we will correctly reject a false H_0. The probability of doing this is what we mean by *power* and is shown in the unshaded area of the H_1 distribution.

Power as a Function of α

With the aid of Figure 8.1, it is easy to see why we say that power is a function of α. If we are willing to increase α, our cutoff point moves to the left, thus simultaneously decreasing β and increasing power, although with a corresponding rise in the probability of a Type I error.

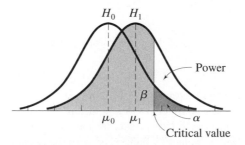

Figure 8.1   Sampling distribution of $\overline{X}$ under H_0 and H_1

Power as a Function of H_1

The fact that power is a function of the true alternative hypothesis (more precisely $[\mu_0 - \mu_1]$, the difference between μ_0 [the mean under H_0] and μ_1 [the mean under H_1]) is illustrated by comparing Figures 8.1 and 8.2. In Figure 8.2, the distance between μ_0 and μ_1 has been increased, and this has resulted in a substantial increase in power, though a still sizeable probability of a Type II error. This is not particularly surprising because all that we are saying is that the chances of finding a difference depend on how large the difference actually is.

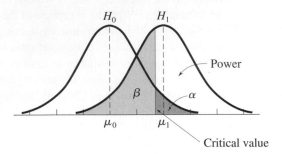

Figure 8.2   Effect on β of increasing $\mu_0 - \mu_1$

Power as a Function of n and σ^2

The relationship between power and sample size (and between power and σ^2) is only a little subtler. Because we are interested in means or differences between means, we are interested in the sampling distribution of the mean. We know that the variance of the sampling distribution of the mean decreases as either n increases or σ^2 decreases because $\sigma_{\bar{X}}^2 = \sigma^2/n$. Figure 8.3 illustrates what happens to the two sampling distributions (H_0 and H_1) as we increase n or decrease σ^2, relative to Figure 8.2. Figure 8.3 also shows that, as $\sigma_{\bar{X}}^2$ decreases, the overlap between the two distributions is reduced with a resulting increase in power. Notice that the two means (μ_0 and μ_1) remain unchanged from Figure 8.2.

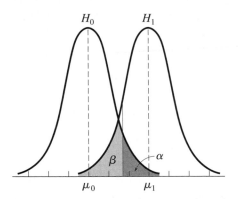

Figure 8.3   Effect on β of decrease in standard error of the mean

If an experimenter concerns himself with the power of a test, then he is most likely interested in those variables governing power that are easy to manipulate. Because n is more easily manipulated than is either σ^2 or the difference ($\mu_0 - \mu_1$), and because tampering with α produces undesirable side effects by increasing the probability of a Type I error, discussions of power are generally concerned with the effects of varying sample size.

8.2   Effect Size

As we saw in Figures 8.1 through 8.3, power depends on the degree of overlap between the sampling distributions under H_0 and H_1. Furthermore, this overlap is a function of both the distance between μ_0 and μ_1 and the standard error. One measure, then, of the degree to which H_0 is false would be the distance from μ_1 to μ_0 expressed in terms of the number of standard errors. The problem with this measure, however, is that it includes the sample size (in the computation of the standard error), when we will usually want to solve for the power associated with a given n or else for that value of n required for a given level of power. For this reason, we will take as our distance measure, or **effect size** (d)

effect size (d)

$$d = \frac{\mu_1 - \mu_0}{\sigma}$$

ignoring the sign of d, and incorporating n later. Thus, d is a measure of the degree to which μ_1 and μ_0 differ in terms of the standard deviation of the parent population. We see that d is estimated independently of n, simply by estimating μ_1, μ_0, and σ. In Chapter 7 we discussed effect size as the standardized difference between two means. This is the same measure here, though one of those means is the mean under the null hypothesis. I will point this out again when we compare the means of two populations.

Estimating the Effect Size

The first task is to estimate d because it will form the basis for future calculations. This can be done in three ways:

1. *Prior research.* On the basis of past research, we can often get at least a rough approximation of d. Thus, we could look at sample means and variances from other studies and make an informed guess at the values we might expect for $\mu_1 - \mu_0$ and for σ. In practice, this task is not as difficult as it might seem, especially when you realize that a rough approximation is far better than no approximation at all.

2. *Personal assessment of which difference is important.* In many cases, an investigator is able to say, I am interested in detecting a difference of at least 10 points between μ_1 and μ_0. The investigator is essentially saying that differences less than this have no important or useful meaning, whereas greater differences do. (This is particularly common in biomedical research, where we are interesting in decreasing cholesterol, for example, by a certain amount, and have no interest in smaller changes.) Here, we are given the value of $\mu_1 - \mu_0$ directly, without needing to know the particular values of μ_1 and μ_0. All that remains is to estimate σ from other data. As an example, the investigator might say that she is interested in finding a procedure that will raise scores on the Graduate Record Exam (GRE) by 40 points above normal. We already know that the standard deviation for this test is 100. Thus, $d = 40/100 = .40$. If our hypothetical experimenter says instead that she wants to raise scores by four-tenths of a standard deviation, she would be giving us d directly.

3. *Use of special conventions.* When we encounter a situation in which there is no way we can estimate the required parameters, we can fall back on a set of conventions proposed by Cohen (1988). Cohen more or less arbitrarily defined three levels of d:

Effect Size	d	Percentage of Overlap
Small	.20	85
Medium	.50	67
Large	.80	53

Thus, in a pinch, the experimenter can simply decide whether she is after a small, medium, or large effect and set *d* accordingly. However, this solution should be chosen *only* when the other alternatives are not feasible. The right column of the table is labeled Percentage of Overlap, and it records the degree to which the two distributions shown in Figure 8.1 overlap. Thus, for example, when $d = 0.50$, two-thirds of the two distributions overlap (Cohen, 1988). This is yet another way of thinking about how big a difference a treatment produces.

Cohen chose a medium effect to be one that would be apparent to an intelligent viewer, a small effect as one that is real but difficult to detect visually, and a large effect as one that is the same distance above a medium effect as "small" is below it. Cohen (1969) originally developed these guidelines only for those who had no other way of estimating the effect size. However, as time went on and he became discouraged by the failure of many researchers to conduct power analyses, presumably because they think them to be too difficult, he made greater use of these conventions (see Cohen, 1992a). In addition, when we think about *d*, as we did in Chapter 7 as a measure of the size of the effect that we have found in our experiment (as opposed to the size we hope to find), Cohen's rules of thumb are being taken as a measure of just how large our obtained difference is. However, Bruce Thompson, of Texas A&M, made an excellent point in this regard. He was speaking of expressing obtained differences in terms of *d*, in place of focusing on the probability value of a resulting test statistic. He wrote, "Finally, it must be emphasized that if we mindlessly invoke Cohen's rules of thumb, contrary to his strong admonitions, in place of the equally mindless consultation of *p* value cutoffs such as .05 and .01, we are merely electing to be thoughtless in a new metric" (Thompson, 2000, personal communication). The point applies to any use of arbitrary conventions for *d*, regardless of whether it is for purposes of calculating power or for purposes of impressing your readers with how large your difference is. Lenth (2001) has argued convincingly that the use of conventions such as Cohen's are dangerous. We need to concentrate on both the value of the numerator and the value of the denominator in *d*, and not just on their ratio. Lenth's argument is really an attempt at making the investigator more responsible for his or her decisions, and I doubt that Cohen would have any disagreement with that.

It may strike you as peculiar that the investigator is being asked to define the difference she is looking for before the experiment is conducted. Most people would respond by saying, "I don't know how the experiment will come out. I just wonder whether there will be a difference." Although many experimenters speak in this way (the author is no virtuous exception), you should question the validity of this statement. Do we really not know, at least vaguely, what will happen in our experiments; if not, why are we running them? Although there is occasionally a legitimate I-wonder-what-would-happen-if experiment, in general, "I do not know" translates to "I have not thought that far ahead."

Recombining the Effect Size and *n*

δ (delta)

We earlier decided to split the sample size from the effect size to make it easier to deal with *n* separately. We now need a method for combining the effect size with the sample size. We use the statistic δ (**delta**) $= d[f(n)]$ to represent this combination where the particular function of *n* (i. e., $f(n)$) will be defined differently for each individual test. The convenient thing about this system is that it will allow us to use the same table of δ for power calculations for all the statistical procedures to be considered.

8.3 Power Calculations for the One-Sample *t*

We will first examine power calculations for the one-sample *t* test. In the preceding section, we saw that δ is based on *d* and some function of *n*. For the one-sample *t*, that function will be $\sqrt{n}$, and δ will then be defined as $\delta = d\sqrt{n}$. Given δ as defined here, we can immediately determine the power of our test from the table of power in Appendix Power.

Assume that a clinical psychologist wants to test the hypothesis that people who seek treatment for psychological problems have higher IQs than the general population. She wants to use the IQs of 25 randomly selected clients and is interested in finding the power of detecting a difference of 5 points between the mean of the general population and the mean of the population from which her clients are drawn. Thus, $\mu_1 = 105$, $\mu_0 = 100$, and $\sigma = 15$.

$$d = \frac{105 - 100}{15} = 0.33$$

then

$$\delta = d\sqrt{n} = 0.33\sqrt{25} = 0.33(5)$$
$$= 1.65$$

Although the clinician expects the sample means to be above average, she plans to use a two-tailed test at $\alpha = .05$ to protect against unexpected events. From Appendix Power, for $\delta = 1.65$ with $\alpha = .05$ (two-tailed), power is between .36 and .40. By crude linear interpolation, we will say that power = .38. This means that, if H_0 is false and μ_1 is really 105, only 38% of the time can our clinician expect to find a "statistically significant" difference between her *sample* mean and that specified by H_0. This is a rather discouraging result because it means that if the true mean really is 105, 62% of the time our clinician will make a Type II error.

Our experimenter was intelligent enough to examine the question of power before she began her experiment, so all is not lost. She still has the chance to make changes that will lead to an increase in power. She could, for example, set α at .10, thus increasing power to approximately .50, but this is probably unsatisfactory. (Journal editors, for example, generally hate to see α set at any value greater than .05.)

Estimating Required Sample Size

Alternatively, the investigator could increase her sample size, thereby increasing power. How large an n does she need? The answer depends on what level of power she desires. Suppose she wants to set power at .80. From Appendix Power, for power = .80, and $\alpha = 0.05$, δ must equal 2.80. Thus, we have δ and can simply solve for n:

$$\delta = d\sqrt{n}$$
$$n = \left(\frac{\delta}{d}\right)^2 = \left(\frac{2.80}{0.33}\right)^2 = 8.48^2$$
$$= 71.91$$

Clients generally come in whole lots, so we will round off to 72. Thus, if the experimenter wants to have an 80% chance of rejecting H_0 when $d = 0.33$ (i.e., when $\mu_1 = 105$), she will have to use the IQs for 72 randomly selected clients. Although this may be more clients than she can test easily, the only alternative is to settle for a lower level of power.

You might wonder why we selected power = .80; with this degree of power, we still run a 20% chance of making a Type II error. The answer lies in the notion of practicality. Suppose, for example, that we had wanted power = .95. A few simple calculations will show that this would require a sample of $n = 119$. For power = .99, you would need approximately 162 subjects. These may well be unreasonable sample sizes for this particular experimental situation, or for the resources of the experimenter. Remember that increases in power are generally bought by increases in n and, at high levels of power, the cost can be very high. If you are taking data from data tapes supplied by the Bureau of the Census, that is quite different from studying teenage college graduates. A value of power = .80 makes a Type II error four times as likely as a Type I error, which some would take as a reasonable reflection of their relative importance.

Noncentrality Parameters

noncentrality
parameter

δ is what most textbooks refer to as a **noncentrality parameter.** The concept is relatively simple, and well worth considering. First, we know that

$$t = \frac{\overline{X} - \mu}{s/\sqrt{n}}$$

is distributed around zero regardless of the truth or falsity of any null hypothesis, *as long as μ is the true mean* of the distribution from which the Xs were sampled. If H_0 states that $\mu = \mu_0$ (some specific value of μ) *and if H_0 is true,* then

$$t = \frac{\overline{X} - \mu_0}{s/\sqrt{n}}$$

will also be distributed around zero. If H_0 is false and $\mu \neq \mu_0$, however, then

$$t = \frac{\overline{X} - \mu_0}{s/\sqrt{n}}$$

will not be distributed around zero because in subtracting μ_0, we have been subtracting the wrong population mean. In fact, the distribution will be centered at the point

$$\delta = \frac{\mu_1 - \mu_0}{\sigma/\sqrt{n}}$$

This shift in the mean of the distribution from zero to δ is referred to as the *degree of noncentrality,* and δ is the noncentrality parameter. (What is δ when $\mu_1 = \mu_0$?) The noncentrality parameter is just one way of expressing just how wrong is the null hypothesis.

The question of power becomes the question of how likely we are to find a value of the noncentral (shifted) distribution that is greater than the critical value that t would have under H_0. In other words, even though larger-than-normal values of t are to be expected because H_0 is false, we will occasionally obtain small values by chance. The percentage of these values that happen to lie within $\pm t_{.025}$ is β, the probability of a Type II error. As we know, we can convert from β to power; power $= 1 - \beta$.

Cohen's contribution can be seen as splitting the noncentrality parameter (δ) into two parts—sample size and effect size. One part (d) depends solely on parameters of the populations, whereas the other depends on sample size. Thus, Cohen has separated parametric considerations (μ_0, μ_1, and σ), about which we can do relatively little, from sample characteristics (n), over which we have more control. Although this produces no basic change in the underlying theory, it makes the concept easier to understand and use.

8.4 Power Calculations for Differences between Two Independent Means

When we want to test the difference between two independent means, the treatment of power is very similar to our treatment of the case that we used for only one mean. In Section 8.3, we obtained d by taking the difference between μ under H_1 and μ under H_0 and dividing by σ. In testing the difference between two independent means, we will do basically the same thing, although this time we will work with mean differences. Thus, we want the difference between the two population means ($\mu_1 - \mu_2$) under H_1 minus the difference ($\mu_1 - \mu_2$) under H_0, divided by σ. (Recall that we assume $\sigma_1^2 = \sigma_2^2 = \sigma^2$.) In all usual applications, however, ($\mu_1 - \mu_2$) under H_0 is zero, so we can drop that term from our formula. Thus,

$$d = \frac{(\mu_1 - \mu_2) - (0)}{\sigma} = \frac{\mu_1 - \mu_2}{\sigma}$$

where the numerator refers to the difference to be expected under H_1 and the denominator represents the standard deviation of the populations. You should recognize that this is the same d that we saw in Chapter 7 where it was also labeled Cohen's d, or sometimes Hedges g. The only difference is that here it is expressed in terms of population means rather than sample means.

With two samples, we must distinguish between experiments involving equal ns and those involving unequal ns. We will treat these two cases separately.

Equal Sample Sizes

Assume we want to test the difference between two treatments and either expect that the difference in population means will be approximately 5 points or else are interested only in finding a difference of at least 5 points. Further assume that from past data we think that σ is approximately 10. Then

$$d = \frac{\mu_1 - \mu_2}{\sigma} = \frac{5}{10} = 0.50$$

Thus, we are expecting a difference of one-half of a standard deviation between the two means, what Cohen (1988) would call a moderate effect.

First, we will investigate the power of an experiment with 25 observations in each of two groups. We will define δ in the two-sample case as

$$\delta = d\sqrt{\frac{n}{2}}$$

where $n =$ the number of cases *in any one sample* (there are $2n$ cases in all). Thus,

$$\delta = (0.50)\sqrt{\frac{25}{2}} = 0.50\sqrt{12.5} = 0.50(3.54)$$
$$= 1.77$$

From Appendix Power, by interpolation for $\delta = 1.77$ with a two-tailed test at $\alpha = .05$, power $= .43$. Thus, if our investigator actually runs this experiment with 25 subjects, and if her estimate of δ is correct, then she has a 43% chance of actually rejecting H_0 if it is false to the extent she expects (and a 57% chance of making a Type II error).

We next turn the question around and ask how many subjects would be needed for power $= .80$. From Appendix Power, this would require $\delta = 2.80$.

$$\delta = d\sqrt{\frac{n}{2}}$$
$$\frac{\delta}{d} = \sqrt{\frac{n}{2}}$$
$$\left(\frac{\delta}{d}\right)^2 = \frac{n}{2}$$
$$n = 2\left(\frac{\delta}{d}\right)^2$$
$$= 2\left(\frac{2.80}{0.50}\right)^2 = 2(5.6)^2$$
$$= 62.72$$

n refers to the number of subjects per sample, so for power $= .80$, we need 63 subjects per sample for a total of 126 subjects.

Unequal Sample Sizes

We just dealt with the case in which $n_1 = n_2 = n$. However, experiments often have two samples of different sizes. This presents difficulties when we try to solve for δ because we need one value for n. What value can we use?

With reasonably large and nearly equal samples, a conservative approximation can be obtained by letting n equal the smaller of n_1 and n_2. This is not satisfactory, however, if the sample sizes are small or if the two ns are quite different. For those cases, we need a more exact solution.

One seemingly reasonable (but incorrect) procedure would be to set n equal to the arithmetic mean of n_1 and n_2. This method would weight the two samples equally, however, when in fact we know that the variance of means is proportional not to n, but to $1/n$. The measure that takes this relationship into account is not the arithmetic mean but the harmonic

harmonic mean   mean. The **harmonic mean** $(\overline{X}_h)$ of k numbers $(X_1, X_2, \ldots, X_k)$ is defined as

$$\overline{X}_h = \frac{k}{\sum \dfrac{1}{X_i}}$$

Thus, for two samples sizes (n_1 and n_2),

$$\overline{n}_h = \frac{2}{\dfrac{1}{n_1} + \dfrac{1}{n_2}} = \frac{2n_1n_2}{n_1 + n_2}$$

We can then use $\overline{n}_h$ in our calculation of δ.

In Chapter 7, we saw an example from Aronson et al. (1998) in which they showed that they could produce a substantial decrement in the math scores of white males just by reminding them that Asian students tend to do better on math exams. This is an interesting difference, and I might have been tempted to use it in a research methods course that I taught, dividing the students in the course into two groups and repeating Aronson's study. Of course, I would not be very happy if I tried out a demonstration experiment on my students and found that it fell flat. I want to be sure that I have sufficient power to have a decent probability of obtaining a statistically significant result in lab.

What Aronson actually found, which is trivially different from the sample data I generated in Chapter 7, were means of 9.58 and 6.55 for the Control and Threatened groups, respectively. Their pooled standard deviation was approximately 3.10. We will assume that Aronson's estimates of the population means and standard deviation are essentially correct. (They almost certainly suffer from some random error, but they are the best guesses that we have of those parameters.) This produces

$$d = \frac{\mu_1 - \mu_2}{\sigma} = \frac{9.58 - 6.55}{3.10} = \frac{3.03}{3.10} = 0.98$$

My class has a lot of students, but only about 30 of them are males, and they are not evenly distributed across the lab sections. Because of the way that I have chosen to run the experiment, assume that I can expect that 18 males will be in the Control group and 12 in the Threat group. Then we will calculate the effective sample size (the sample size to be used in calculating δ) as

$$\overline{n}_h = \frac{2(18)(12)}{18 + 12} = \frac{432}{30} = 14.40$$

effective sample size   We see that the **effective sample size** is less than the arithmetic mean of the two individual sample sizes. In other words, this study has the same power as it would have had we run it with 14.4 subjects per group for a total of 28.8 subjects. Or, to state it differently, with unequal sample sizes it takes 30 subjects to have the same power 28.8 subjects would have in an experiment with equal sample sizes.

To continue,

$$\delta = d\sqrt{\frac{\overline{n}_h}{2}} = 0.98\sqrt{\frac{14.4}{2}} = 0.98\sqrt{7.2}$$

$$= 2.63$$

For $\delta = 2.63$, power $= .75$ at $\alpha = .05$ (two-tailed).

In this case, the power is a bit too low to inspire confidence that the study will work out as a lab exercise is supposed to. I could take a chance and run the study, but the lab might fail and then I'd have to stammer out some excuse in class and hope that people believed that it "really should have worked." I'm not comfortable with that.

An alternative would be to recruit some more students. I will use the 30 males in my course, but I can also find another 20 in another course who are willing to participate. At the risk of teaching bad experimental design to my students by combining two different classes (at least it gives us an excuse to discuss this kind of problem), I will add in those students and expect to get sample sizes of 28 and 22.

These sample sizes would yield $\overline{n}_h = 24.64$. Then

$$\delta = d\sqrt{\frac{\overline{n}_h}{2}} = 0.98\sqrt{\frac{24.64}{2}} = 0.98\sqrt{12.32}$$

$$= 3.44$$

From Appendix Power, we find that power now equals approximately .93, which is certainly sufficient for our purposes.

My sample sizes were unequal, but not seriously so. When we have quite unequal sample sizes, and they are unavoidable, the smaller group should be as large as possible relative to the larger group. You should never throw away subjects to make sample sizes equal. This is just throwing away power.[3]

8.5   Power Calculations for Matched-Sample t

When we want to test the difference between two matched samples, the problem becomes a bit more difficult, and an additional parameter must be considered. For this reason, the analysis of power for this case is frequently impractical. However, the general solution to the problem illustrates an important principle of experimental design, and thus justifies close examination.

With a matched-sample t test, we define d as

$$d = \frac{\mu_1 - \mu_2}{\sigma_{X_1 - X_2}}$$

where $\mu_1 - \mu_2$ represents the expected difference in the means of the two populations of observations (the expected mean of the difference scores). The problem arises because $\sigma_{X_1 - X_2}$ is the standard deviation not of the populations of X_1 and X_2, but of difference scores drawn from these populations. Although we might be able to make an intelligent guess at σ_{X_1} or σ_{X_2}, we probably have no idea about $\sigma_{X_1 - X_2}$.

---

[3] McClelland (1997) has provided a strong argument that when we have more than two groups and the independent variable is ordinal, power may be maximized by assigning disproportionately large numbers of subjects to the extreme levels of the independent variable.

All is not lost, however; it is possible to calculate $\sigma_{X_1 - X_2}$ on the basis of a few assumptions. The variance sum law (discussed in Chapter 7, p. 193) gives the variance for a sum or difference of two variables. Specifically,

$$\sigma^2_{X_1 \pm X_2} = \sigma^2_{X_1} + \sigma^2_{X_2} \pm 2\rho\,\sigma_{X_1}\sigma_{X_2}$$

If we make the general assumption of homogeneity of variance $\sigma^2_{X_1} = \sigma^2_{X_2} = \sigma^2$, for the difference of two variables we have

$$\sigma^2_{X_1 - X_2} = 2\sigma^2 - 2\rho\sigma^2 = 2\sigma^2(1 - \rho)$$
$$\sigma_{X_1 - X_2} = \sigma\sqrt{2(1 - \rho)}$$

where ρ (rho) is the correlation in the population between X_1 and X_2 and can take on values between 1 and -1. It is positive for almost all situations in which we are likely to want a matched-sample t.

Assuming for the moment that we can estimate ρ, the rest of the procedure is the same as that for the one-sample t. We define

$$d = \frac{\mu_1 - \mu_2}{\sigma_{X_1 - X_2}}$$

and

$$\delta = d\sqrt{n}$$

We then estimate $\sigma_{X_1 - X_2}$ as $\sigma\sqrt{2(1 - \rho)}$, and refer the value of δ to the tables.

As an example, assume that I want to use the Aronson study of stereotype threat in class, but this time I want to run it as a matched-sample design. I have 30 male subjects available, and I can first administer the test without saying anything about Asian students typically performing better, and then I can re-administer it in the next week's lab with the threatening instructions. (You might do well to consider how this study could be improved to minimize carryover effects and other contaminants.) Let's assume that we expect the scores to go down in the threatening condition, but that because of the fact that the test was previously given to these same people in the first week, the drop will be from 9.58 down to only 7.55. Assume that the standard deviation will stay the same at 3.10. To solve for the standard error of the difference between means, we need the correlation between the two sets of exam scores, but here we are in luck. Aronson's math questions were taken from a practice exam for the Graduate Record Exam (GRE), and the correlation we seek is estimated simply by the test-retest reliability of that exam. We have a pretty good idea that the reliability of that exam will be somewhere around .92. Then

$$\sigma_{X_1 - X_2} = \sigma\sqrt{2(1 - \rho)} = 3.10\sqrt{2(1 - .92)} = 3.1\sqrt{2(.08)}$$
$$= 1.24$$
$$d = \frac{\mu_1 - \mu_2}{\sigma_{X_1 - X_2}} = \frac{9.58 - 7.55}{1.24} = 1.64$$
$$\delta = d\sqrt{n} = 1.64\sqrt{30} = 8.97$$

Power $= .99$

Notice that I have a smaller effect size than in my first lab exercise because I tried to be honest and estimate that the difference in means would be reduced because of the experimental procedures. However, my power is far greater than it was in my original example because of the added power of matched-sample designs.

Suppose, on the other hand, that we had used a less reliable test, for which $\rho = .40$. We will assume that σ remains unchanged and that we are expecting a 2.03-unit difference between the means. Then

$$\sigma_{X_1-X_2} = 3.10\sqrt{2(1 - .40)} = 3.10\sqrt{2(.60)} = 3.10\sqrt{1.2} = 3.40$$

$$d = \frac{\mu_1 - \mu_2}{\sigma_{X_1-X_2}} = \frac{2.03}{3.40} = 0.60$$

$$\delta = 0.60\sqrt{30} = 3.29$$

$$\text{Power} = .91$$

We see that as ρ drops, so does power. (It is still substantial in this example, but much less than it was.) When $\rho = 0$, our two variables are not correlated, and thus, the matched-sample case has been reduced to very nearly the independent-sample case. The important point here is that for practical purposes, the minimum power for the matched-sample case occurs when $\rho = 0$ and we have independent samples. Thus, for all situations in which we are even remotely likely to use matched samples (when we expect a positive correlation between X_1 and X_2), the matched-sample design is more powerful than the corresponding independent-groups design. This illustrates one of the main advantages of designs using matched samples and was my primary reason for taking you through these calculations.

Remember that we are using an approximation procedure to calculate power. Essentially, we are assuming the sample sizes are sufficiently large that the t distribution is closely approximated by z. If this is not the case, then we have to account for the fact that a matched-sample t has only one-half as many df as the corresponding independent-sample t, and the power of the two designs will not be quite equal when $\rho = 0$. This is not usually a serious problem.

8.6 Power Considerations in Sample Size

Our discussion of power illustrates that reasonably large sample sizes are almost a necessity if one is to run experiments with any decent chance of rejecting H_0 when it is in fact false and the effect is small. (The study by Aronson et al. was an exception here simply because they had such a large effect size. Rarely do our experiments produce such substantial effects. On the other hand, this result testifies to the potential negative effects of cultural stereotypes.) As an illustration of the frequent need for large samples, a few calculations show that if we want to have power $= .80$ and if we accept Cohen's admittedly arbitrary definitions for small, medium, and large effects, our samples may not be small. Table 8.1 presents the total ns required (at power $= .80$, $\alpha = .05$) for small, medium, and large effects for the tests we have been discussing. These figures indicate that power (at least a substantial amount of it) is an expensive commodity, especially for small effects. I could argue that this is a good thing because otherwise the literature would contain many more small (trivial?) results than it already does; however, this assertion is unlikely to comfort most experimenters. The general rule is

Table 8.1 Total sample sizes required for power $= .80$, $\alpha = .05$, two-tailed

Effect Size	d	One-Sample t	Two-Sample t
Small	.20	196	784
Medium	.50	32	126
Large	.80	13	49

either to look for big effects or to use large samples.[4] Interesting articles on the power of published experiments are found in Cohen (1962) and a follow-up study by Sedlmeier and Gigerenzer (1989). An important and very interesting paper arguing in favor of studies of small effect size is Prentice and Miller (1992).

8.7 Retrospective Power

a priori power

In general, the previous discussion has focused on **a priori power,** which is the power that we would calculate before the experiment is conducted. It is based on reasonable estimates of means, variances, correlations, proportions, and so forth that we believe represent the parameters for our population or populations. This is what we generally think of when we consider statistical power.

retrospective (or post hoc) power

In recent years, there has been an increased interest in what is often called **retrospective (or post hoc) power.** For our purposes, retrospective power will be defined as power that is calculated after an experiment has been completed, based on the results of that experiment. For example, retrospective power asks the question, "If the values of the population means and variances were equal to the values found in this experiment, what would be the resulting power?"

One reason why we might calculate retrospective power is to help in the design of future research. Suppose that we have just completed an experiment and want to replicate it, perhaps with a different sample size and a demographically different pool of participants. We can take the results that we just obtained, treat them as an accurate reflection of the population means and standard deviations, and use those values to calculate the estimated effect size. We can then use that effect size to make power estimates. This use of retrospective power, which is, in effect, the a priori power of our next experiment, is relatively noncontroversial. Many statistical packages, including SAS and SPSS, will make these calculations for you.

What is more controversial, however, is to use retrospective power calculations as an explanation of the obtained results. A common suggestion in the literature claims that if the study was not significant, but had high retrospective power, that result speaks to the acceptance of the null hypothesis. This view hinges on the argument that if you had high power, you would have been very likely to reject a false null, and thus nonsignificance indicates that the null is either true or nearly so. That sounds pretty convincing, but as Hoenig and Heisey [2001] point out, there is a false premise here. It is not possible to fail to reject the null and yet have high retrospective power. In fact, a result with p exactly equal to .05 will have a retrospective power of essentially .50, and that retrospective power will decrease for $p > .05$. It is impossible even to create an example of a study that just barely failed to reject the null hypothesis at $\alpha = .05$ that has power of .80. It can't happen.

The argument is sometimes made that retrospective power tells you more than you can learn from the obtained p value. This argument is a derivative of the one in the previous paragraph. However, it is easy to show that for a given effect size and sample size, there is a 1:1 relationship between p and retrospective power. One can be derived from the other. Thus, retrospective power offers no additional information for explaining nonsignificant results.

---

[4] Linda Sorenson (1995, personal communication) objected to the implications of this table on the grounds that in her field (animal behavior) such sample sizes are totally unrealistic. Her point is well taken, but I have two related responses. In the first place, research in her field has been extremely productive with many significant effects. This suggests that the variables they study, given the small sample sizes, are quite large and detectable. The second point is that when sample sizes are, by necessity, small and power is an issue, we need to look elsewhere at ways to maximize power. McClelland (1997) and McClelland and Judd (1993) have explored this issue and discuss ways of increasing power by the choice of the levels of the independent variable that we employ.

As Hoenig and Heisey (2001) argue, rather than focus our energies on calculating retrospective power to try to learn more about what our results have to reveal, we are better off putting that effort into calculating confidence limits on the parameter(s) or the effect size. If, for example, we had a t test on two independent groups with $t(48) = 1.90$, $p = .063$, we would fail to reject the null hypothesis. When we calculate retrospective power, we find it to be .46. When we calculate the 95% confidence interval on $\mu_1 - \mu_2$, we find $-1.10 \leq \mu_1 - \mu_2 \leq 39.1$. The confidence interval tells us more about what we are studying than does the fact that power is only .46. (Even had the difference been slightly greater, and thus significant, the confidence interval shows that we still do not have a very good idea of the magnitude of the difference between the population means.)

Retrospective power can be a useful tool when evaluating studies in the literature, as in a meta-analysis, or planning future work. But retrospective power is not a useful tool for explaining away our own nonsignificant results.

Some current statistical software will produce post-hoc power estimates. A good example is SPSS, which produces power estimates for analyses of variance when you choose the General Linear Model procedure.

8.8   Writing Up the Results of a Power Analysis

We usually don't say very much in a published study about the power of the experiment we just ran. Perhaps that is a holdover from the fact that we didn't even calculate power many years ago. It is helpful, however, to add a few sentences to your Methods section that describes the power of your experiment. For example, after describing the procedures you followed, you could say something like this:

> Based on the work of Jones and others (list references), we estimated that our mean difference would be approximately 8 points, with a standard deviation within each of the groups of approximately 5. This would give us an estimated effect size of $8/11 = .73$. We were aiming for a power estimate of .80, and to reach that level of power with our estimated effect size, we used 30 participants in each of the two groups.

Key Terms

Power (Introduction)

Effect size (d) (8.2)

δ (delta) (8.2)

Noncentrality parameter (8.3)

Harmonic mean ($\overline{X}_h$) (8.4)

Effective sample size (8.4)

A priori power (8.7)

Retrospective (or post hoc) power (8.7)

Exercises

8.1   A large body of literature on the effect of peer pressure has shown that the mean influence score for a scale of peer pressure is 520 with a standard deviation of 80. An investigator would like to show that a minor change in conditions will produce scores with a mean of only 500, and he plans to run a t test to compare his sample mean with a population mean of 520.

　　a.   What is the effect size in question?

　　b.   What is the value of δ if the size of his sample is 100?

　　c.   What is the power of the test?

8.2   Diagram the situation described in Exercise 8.1 along the lines of Figure 8.1.

8.3   In Exercise 8.1, what sample sizes would be needed to raise power to .70, .80, and .90?

8.4   A second investigator thinks that she can show that a quite different manipulation can raise the mean influence score from 520 to 550.

    a.   What is the effect size in question?

    b.   What is the value of δ if the size of her sample is 100?

    c.   What is the power of the test?

8.5   Diagram the situation described in Exercise 8.4 along the lines of Figure 8.1.

8.6   Assume that a third investigator ran both conditions described in Exercises 8.1 and 8.4 and wanted to know the power of the combined experiment to find a difference between the two experimental manipulations.

    a.   What is the effect size in question?

    b.   What is the value of δ if the size of his sample is 50 for both groups?

    c.   What is the power of the test?

8.7   A physiological psychology laboratory has been studying avoidance behavior in rabbits for several years and has published numerous papers on the topic. It is clear from this research that the mean response latency for a particular task is 5.8 seconds with a standard deviation of 2 seconds (based on many hundreds of rabbits). Now the investigators want to induce lesions in certain areas in the rabbits' amygdalae and then demonstrate poorer avoidance conditioning in these animals (i.e., show that the rabbits will repeat a punished response sooner). Investigators expect latencies to decrease by about 1 second, and they plan to run a one-sample t test (of $\mu_0 = 5.8$).

    a.   How many subjects do they need to have at least a 50:50 chance of success?

    b.   How many subjects do they need to have at least an 80:20 chance of success?

8.8   Suppose that the laboratory referred to in Exercise 8.7 decided not to run one group and compare it against $\mu_0 = 5.8$, but instead to run two groups (one with and one without lesions). They still expect the same degree of difference.

    a.   How many subjects do they need (overall) if they are to have power $= .60$?

    b.   How many subjects do they need (overall) if they are to have power $= .90$?

8.9   A research assistant ran the experiment described in Exercise 8.8 without first carrying out any power calculations. He tried to run 20 subjects in each group, but he accidentally tipped over a rack of cages and had to void 5 subjects in the experimental group. What is the power of this experiment?

8.10   We have just conducted a study comparing cognitive development of low- and normal-birthweight babies who have reached 1 year of age. Using a scale we devised, we found that the sample means of the two groups were 25 and 30, respectively, with a pooled standard deviation of 8. Assume that we want to replicate this experiment with 20 subjects in each group. If we assume that the true means and standard deviations have been estimated exactly, what is the a priori probability that we will find a significant difference in our replication?

8.11   Run the t test on the original data in Exercise 8.10. What, if anything, does your answer to this question indicate about your answer to Exercise 8.10?

8.12   Two graduate students recently completed their dissertations. Each used a t test for two independent groups. One found a significant t using 10 subjects per group. The other found a significant t of the same magnitude using 45 subjects per group. Which result impresses you more?

8.13   Draw a diagram (analogous to Figure 8.1) to defend your answer to Exercise 8.12.

8.14   Make up a simple two-group example to demonstrate that for a total of 30 subjects, power increases as the sample sizes become more nearly equal.

8.15   A beleaguered Ph.D. candidate has the impression that he must find significant results if he wants to defend his dissertation successfully. He wants to show a difference in social awareness, as measured by his own scale, between a normal group and a group of ex-delinquents.

He has a problem, however. He has data to suggest that the normal group has a true mean of 38, and he has 50 of those subjects. He has access to 100 high-school graduates who have been classed as delinquent in the past. Or, he has access to 25 high-school dropouts who have a history of delinquency. He suspects that the high-school graduates come from a population with a mean of approximately 35, whereas the dropout group comes from a population with a mean of approximately 30. He can use only one of these groups. Which should he use?

8.16 Generate a table analogous to Table 8.1 for power $= .80$, $\alpha = .01$, two-tailed.

8.17 Generate a table analogous to Table 8.1 for power $= .60$, $\alpha = .05$, two-tailed.

8.18 Assume that we want to test a null hypothesis about a single mean at $\alpha = .05$, one-tailed. Further assume that all necessary assumptions are met. Could there be a case in which we would be more likely to reject a true H_0 than to reject a false one? (In other words, can power ever be less than α?)

8.19 If $\alpha = 15$, $n = 25$, and we are testing $H_0: \mu_0 = 100$ versus $H_1: \mu_0 > 100$, what value of the mean under H_1 would result in power being equal to the probability of a Type II error? (*Hint:* Try sketching the two distributions; which areas are you trying to equate?)

Discussion Questions

8.20 Prentice and Miller (1992) presented an interesting argument that suggested that although most studies do their best to increase the effect size of whatever they are studying (e.g., by maximizing the differences between groups), some research focuses on minimizing the effect and still finding a difference. (For example, although it is well known that people favor members of their own group, it has been shown that even if you create groups on the basis of random assignment, the effect is still there.) Prentice and Miller then state, "In the studies we have described, investigators have minimized the power of an operationalization and, in so doing, have succeeded in demonstrating the power of the underlying process."

   a. Does this seem to you to be a fair statement of the situation? In other words, do you agree that experimenters have run experiments with minimal power?

   b. Does this approach seem reasonable for most studies in psychology?

   c. Is it always important to find large effects? When would it be important to find even quite small effects?

8.21 In the hypothetical study based on Aronson's work on stereotype threat with two independent groups, I could have all male students in a given lab section take the test under the same condition. Then male students in another lab could take the test under the other condition.

   a. What is wrong with this approach?

   b. What alternatives could you suggest?

   c. I have ignored the many women in those labs. What do you think might happen if I used them as well?

8.22 In the modification of Aronson's study to use a matched-sample t test, I always gave the Control condition first, followed by the Threat condition in the next week.

   a. Why would this be a better approach than randomizing the order of conditions?

   b. If I give exactly the same test each week, there should be some memory carrying over from the first presentation. How might I get around this problem?

8.23 Create an example in which a difference is just barely statistically significant at $\alpha = .05$. (*Hint:* Find the critical value for t, invent values for μ_1 and μ_2 and n_1 and n_2, and then solve for the required value of s.) Now calculate the retrospective power of this experiment.

CHAPTER 9

Correlation and Regression

Objectives

To introduce the concepts of correlation and regression and to begin looking at how relationships between variables can be represented.

Contents

relationships

differences

correlation

regression

random variable

fixed variable

linear regression models

bivariate normal models

prediction

IN CHAPTER 7, WE DEALT WITH TESTING HYPOTHESES concerning differences between sample means. In this chapter, we will begin examining questions concerning relationships between variables. Although you should not make too much of the distinction between **relationships** and **differences** (if treatments have *different* means, then means are *related* to treatments), the distinction is useful for the interests of the experimenter and the structure of the experiment. When we are concerned with differences between means, the experiment usually consists of a few quantitative or qualitative levels of the independent variable (e.g., Treatment A and Treatment B) and the experimenter is interested in showing that the dependent variable differs from one treatment to another. When we are concerned with relationships, however, the independent variable (X) usually has many quantitative levels and the experimenter is interested in showing that the dependent variable is some *function* of the independent variable.

This chapter will deal with two interwoven topics: **correlation** and **regression.** Statisticians commonly make a distinction between these two techniques. Although the distinction is frequently not followed in practice, it is important enough to consider briefly. In problems of simple correlation and regression, the data consist of two observations from each of N subjects, one observation on each of the two variables under consideration. If we were interested in the correlation between running speed in a maze (Y) and number of trials to reach some criterion (X) (both common measures of learning), we would obtain a running-speed score and a trials-to-criterion score from each subject. Similarly, if we were interested in the regression of running speed (Y) on the number of food pellets per reinforcement (X), each subject would have scores corresponding to his speed and the number of pellets he received. The difference between these two situations illustrates the statistical distinction between correlation and regression. In both cases, Y (running speed) is a **random variable,** beyond the experimenter's control. We don't know what the rat's running speed will be until we carry out a trial and measure the speed. In the former case, X is also a random variable because the number of trials to criterion depends on how fast the animal learns, and this, too, is beyond the control of the experimenter. Put another way, a replication of the experiment would leave us with different values of both Y and X. In the food pellet example, however, X is a **fixed variable.** The number of pellets is determined by the experimenter (for example, 0, 1, 2, or 3 pellets) and would remain constant across replications.

To most statisticians, the word *regression* is reserved for those situations in which the value of X is *fixed* or specified by the experimenter before the data are collected. In these situations, no sampling error is involved in X, and repeated replications of the experiment will involve the same set of X values. The word *correlation* is used to describe the situation in which both X and Y are random variables. In this case, the Xs, as well as the Ys, vary from one replication to another, and thus, sampling error is involved in both variables. This distinction is basically the distinction between what are called **linear regression models** and **bivariate normal models.** We will consider the distinction between these two models in more detail in Section 9.7.

As mentioned earlier, the distinction between the two models, although appropriate on statistical grounds, tends to break down in practice. A more pragmatic distinction relies on the interest of the experimenter. If the purpose of the research is to allow **prediction** of Y based on knowledge about X, we will speak of regression. If, on the other hand, the purpose is merely to obtain a statistic expressing the degree of relationship between the two variables, we will speak of correlation. Although it is possible to raise legitimate objections to this distinction, it has the advantage of describing the different ways in which these two procedures are used in practice. We will see instances of situations in which regression (rather than correlation) is the goal even when both variables are random.

Having differentiated between correlation and regression, we will now proceed to treat the two techniques together because they are so closely related. The general problem then becomes one of developing an equation to predict one variable from knowledge of the other (regression) and of obtaining a measure of the degree of this relationship (correlation). The only restriction we will impose for the moment is that the relationship between X and Y be linear.

9.1 Scatterplot

scatterplot

scatter diagram

scattergram

predictor

criterion

When we collect measures on two variables for the purpose of examining the relationship between these variables, one of the most useful techniques for gaining insight into this relationship is a **scatterplot** (also called a **scatter diagram** or **scattergram**). In a scatterplot, each experimental subject in the study is represented by a point in two-dimensional space. The coordinates of this point (X_i, Y_i) are the individual's (or object's) scores on variables X and Y, respectively. Examples of three such plots appear in Figure 9.1.

In a scatterplot, the **predictor** variable is traditionally represented on the abscissa, or x-axis, and the **criterion** variable on the ordinate, or y-axis. If the eventual purpose of the study is to predict one variable from knowledge of the other, the distinction is obvious; the criterion variable is the one to be predicted, whereas the predictor variable is the one from which the prediction is made. If the problem is simply one of obtaining a correlation

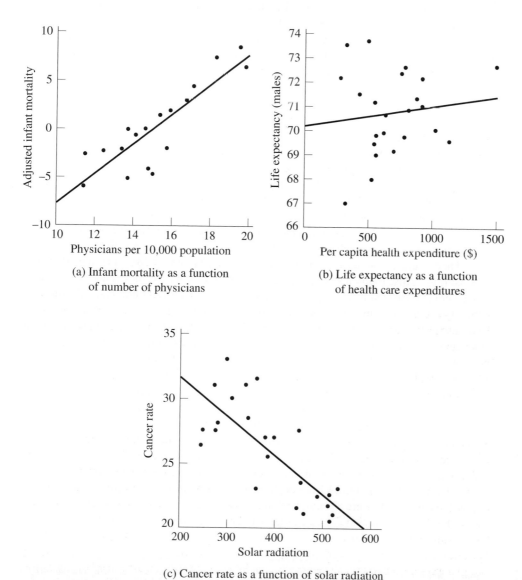

(a) Infant mortality as a function
of number of physicians

(b) Life expectancy as a function
of health care expenditures

(c) Cancer rate as a function of solar radiation

Figure 9.1   Three scatter diagrams

coefficient, the distinction may be obvious (incidence of cancer would be dependent on amount smoked rather than the reverse, and thus, incidence would appear on the ordinate), or it may not (neither running speed nor number of trials to criterion is obviously in a dependent position relative to the other). Where the distinction is not obvious, it is irrelevant which variable is labeled X and which Y.

Consider the three scatter diagrams in Figure 9.1. Figure 9.1a is plotted from data reported by St. Leger, Cochrane, and Moore (1978) on the relationship between infant mortality, adjusted for gross national product, and the number of physicians per 10,000 population.[1] Notice the fascinating result that infant mortality *increases* with the number of physicians. That is certainly an unexpected result, but it is almost certainly not due to chance. (As you look at these data and read the rest of the chapter, you might think about possible explanations for this surprising result.)

The lines superimposed on Figures 9.1a–9.1c represent those straight lines that "best fit the data." How we determine that line will be the subject of much of this chapter. I have included the lines in each of these figures because they help to clarify the relationships. These lines are what we will call the **regression lines** of Y predicted on X (abbreviated "Y on X"), and they represent our best prediction of Y_i for a given value of X_i, for the ith subject or observation. Given any specified value of X, the corresponding height of the regression line represents our best prediction of Y (designated $\hat{Y}$, and read "Y hat"). In other words, we can draw a vertical line from X_i to the regression line and then move horizontally to the y-axis and read $\hat{Y}_i$.

regression lines

correlation (r)

The degree to which the points cluster around the regression line (in other words, the degree to which the actual values of Y agree with the predicted values) is related to the **correlation** (r) between X and Y. Correlation coefficients range between 1 and -1. For Figure 9.1a, the points cluster very closely about the line, indicating that there is a strong linear relationship between the two variables. If the points fell exactly on the line, the correlation would be $+1.00$. As it is, the correlation is actually .81, which represents a high degree of relationship for real variables in the behavioral sciences.

In Figure 9.1b, I have plotted data on the relationship between life expectancy (for males) and per capita expenditure on health care for 23 developed (mostly European) countries. These data are found in Cochrane, St. Leger, and Moore (1978). At a time when there is considerable discussion nationally about the cost of health care, these data give us pause. If we were to measure the health of a nation by life expectancy (admittedly not the only, and certainly not the best, measure), it would appear that the total amount of money we spend on health care bears no relationship to the resultant quality of health (assuming that different countries apportion their expenditures in similar ways). (Several hundred thousand dollars spent on transplanting an organ from a baboon into a 57-year-old male may increase *his* life expectancy by a few years, but it is not going to make a dent in the *nation's* life expectancy. A similar amount of money spent on prevention efforts with young children, however, may eventually have a very substantial effect—hence the inclusion of this example in a text primarily aimed at psychologists.) The two countries with the longest life expectancy (Iceland and Japan) spend nearly the same amount of money on health care as the country with the shortest life expectancy (Portugal). The United States has the second highest rate of expenditure but ranks near the bottom in life expectancy. Figure 9.1b represents a situation in which there is no apparent relationship between the two variables under consideration. If there were absolutely no relationship between the variables, the correlation would be 0.0. As it is, the correlation is only .14, and even that can be shown not to be reliably different from 0.0.

---

[1] Some people have asked how mortality can be negative. The answer is that this is the mortality rate *adjusted for gross national product*. After adjustment, the rate can be negative.

Finally, Figure 9.1c presents data from an article in *Newsweek* (1991) on the relationship between breast cancer and sunshine. For those of us who love the sun, it is encouraging to find that there may be at least some benefit from additional sunlight. Notice that as the amount of solar radiation increases, the incidence of deaths from breast cancer *decreases*. (It has been suggested that perhaps the higher rate of breast cancer with decreased sunlight is attributable to a Vitamin D deficiency.) This is a good illustration of a negative relationship, and the correlation here is $-.76$.

It is important to note that the sign of the correlation coefficient has no meaning other than to denote the direction of the relationship. Correlations of .75 and $-.75$ signify exactly the same *degree* of relationship. Only the direction of that relationship is different. Figures 9.1a and 9.1c illustrate this because the two correlations are nearly the same except for their signs (.81 versus $-.76$).

9.2   The Relationship Between Stress and Health

Psychologists have long been interested in the relationship between stress and health and have accumulated evidence to show that there are very real negative effects of stress on both the psychological and physical health of people. Wagner, Compas, and Howell (1988) investigated the relationship between stress and mental health in first-year college students. Using a scale Wagner et al. developed to measure the frequency, perceived importance, and desirability of recent life events, they created a measure of negative events weighted by the reported frequency and the respondent's subjective estimate of the impact of each event. This served as their measure of the subject's perceived social and environmental stress. They also asked students to complete the Hopkins Symptom Checklist, assessing the presence or absence of 57 psychological symptoms. The stem-and-leaf displays and boxplots for the stress and symptom measures are shown in Table 9.1.

Before we consider the relationship between these variables, we need to study the variables individually. The stem-and-leaf displays for both variables show that the distributions are unimodal but are slightly positively skewed. Except for a few extreme values, there is nothing about either variable that should disturb us, such as extreme skewness or bimodality.[2] Note that there is a fair amount of variability in each variable. This variability is important because if we want to show that different stress scores are associated with differences in symptoms, it is important to have these differences in the first place.

The boxplots in Table 9.1 reveal the presence of outliers on both variables. (The "2" is used to indicate the presence of two overlapping data points.) The existence of outliers should alert us to potential problems that these scores might cause. The first thing we might do is to check the data to see whether these few subjects were responding in unreasonable ways—for example, do they report the occurrence of all sorts of unlikely life events or symptoms, making us question the legitimacy of their responses? (Some subjects have been known to treat psychological experiments with something less than the respect and reverence they deserve! I'm sure that you find that hard to believe, but, sadly, it's true.) The second thing to check is whether the same participant produced outlying data points on both variables. This would suggest that this participant's data, although legitimate, might have a disproportionate influence on the resulting correlation. The third thing to do is to make a scatterplot of the data, again looking for the undue influence of particular extreme data points. (Such a scatterplot will appear later in Figure 9.2, p. 240) Finally, we can run our analyses including and excluding extreme points to see what differences appear in the

---

[2] Bimodality does not make the use of correlation and regression inappropriate, but it is a signal that we should examine the data carefully and think about the interpretation.

Table 9.1   Description of data on the relationship between stress and mental health

Stem-and-Leaf for Stress		Stem-and-Leaf for Symptoms	
0*	01222234	5.	8
0.	556677888899	6*	112234
1*	01112222222333444444	6.	55668
1.	5556666777888999	7*	00012334444
2*	0001111223333334	7.	57788899
2.	556778999	8*	00011122233344
3*	012233444	8.	5666677888899
3.	56677778	9*	0111223344
4*	23444	9.	556679999
4.	55	10*	0001112224
		10.	567799
HI	57, 74	11*	112
		11.	78
		12*	11
Code:	2.|5 = 25	12.	57
		13*	1
		HI	135, 135, 147, 186
		Code:	5.|8 = 58

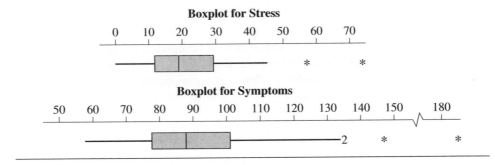

Boxplot for Stress

Boxplot for Symptoms

results. If you carry out each of these steps on the data, you will find nothing to suggest that the outliers we have identified influenced the resulting correlation or regression equation in any important way. However, these steps are important precursors to any good analysis—if only because they give us greater faith in our final result. A more extensive discussion of techniques for examining data to be used in regression analyses will be found in Chapter 15 when we discuss multiple regression.

9.3 The Covariance

covariance

The correlation coefficient we seek to compute on the data[3] in Table 9.2 is itself based on a statistic called the **covariance** (cov_{XY} or s_{XY}). The covariance is basically a number that reflects the degree to which two variables vary together.

---

[3] A copy of the complete data set is available at the website (www.uvm.edu/~dhowell/methods/) in the file named Wagner.dat.

Table 9.2 Data on stress and symptoms for 10 representative participants

Participant	Stress (X)	Symptoms (Y)
1	30	99
2	27	94
3	9	80
4	20	70
5	3	100
6	15	109
7	5	62
8	10	81
9	23	74
10	34	121
⋮	⋮	⋮

$$\sum X = 2297 \qquad \sum Y = 9705$$

$$\sum X^2 = 67{,}489 \qquad \sum Y^2 = 923{,}787$$

$$\overline{X} = 21.467 \qquad \overline{Y} = 90.701$$

$$s_X = 13.096 \qquad s_Y = 20.266$$

$$\sum XY = 222{,}576$$

$$N = 107$$

To define the covariance mathematically, we can write

$$\text{cov}_{XY} = \frac{\sum(X - \overline{X})(Y - \overline{Y})}{N - 1}$$

From this equation, it is apparent that the covariance is similar in form to the variance. If we changed all the Ys in the equation to Xs, we would have s_X^2; if we changed the Xs to Ys, we would have s_Y^2.

For the data on Stress and Symptoms, we would expect that high stress scores will be paired with high symptom scores. Thus, for a stressed participant with many problems, both $(X - \overline{X})$ and $(Y - \overline{Y})$ will be positive and their product will be positive. For a participant experiencing little stress and few problems, both $(X - \overline{X})$ and $(Y - \overline{Y})$ will be negative, but their product will again be positive. Thus, the sum of $(X - \overline{X})(Y - \overline{Y})$ will be large and positive, giving us a large positive covariance.

The reverse would be expected in the case of a strong negative relationship. Here, large positive values of $(X - \overline{X})$ most likely will be paired with large negative values of $(Y - \overline{Y})$, and vice versa. Thus, the sum of products of the deviations will be large and negative, indicating a strong negative relationship.

Finally, consider a situation in which there is no relationship between X and Y. In this case, a positive value of $(X - \overline{X})$ will sometimes be paired with a positive value and sometimes with a negative value of $(Y - \overline{Y})$. The result is that the products of the deviations will be positive about half of the time and negative about half of the time, producing a near-zero sum and indicating no relationship between the variables.

For a given set of data, it is possible to show that cov_{XY} will be at its positive maximum whenever X and Y are perfectly positively correlated ($r = 1.00$), and at its negative maximum whenever they are perfectly negatively correlated ($r = -1.00$). When the two variables are perfectly uncorrelated ($r = 0.00$), cov_{XY} will be zero.

For computational purposes, a simple expression for the covariance is given by

$$\text{cov}_{XY} = \frac{\sum XY - \frac{\sum X \sum Y}{N}}{N - 1}$$

For the full data set represented in Table 9.2, the covariance is

$$\text{cov}_{XY} = \frac{222{,}576 - \frac{(2297)(9705)}{107}}{106} = \frac{222{,}576 - 208{,}340.05}{106} = 134.301$$

9.4 The Pearson Product-Moment Correlation Coefficient (r)

What we said about the covariance might suggest that we could use it as a measure of the degree of relationship between two variables. An immediate difficulty arises, however, because the absolute value of cov_{XY} is also a function of the standard deviations of X and Y. Thus, a value of $\text{cov}_{XY} = 134$, for example, might reflect a high degree of correlation when the standard deviations are small, but a low degree of correlation when the standard deviations are high. To resolve this difficulty, we divide the covariance by the size of the standard deviations and make this our estimate of correlation. Thus, we define

$$r = \frac{\text{cov}_{XY}}{s_X s_Y}$$

The maximum value of cov_{XY} can be shown to be $\pm s_X s_Y$, so it follows that the limits on r are ± 1.00. One interpretation of r, then, is that it is a measure of the degree to which the covariance approaches its maximum.

From Table 9.2 and subsequent calculations, we know that $s_X = 13.096$, $s_Y = 20.266$, and $\text{cov}_{XY} = 134.301$. Then the correlation between X and Y is given by

$$r = \frac{\text{cov}_{XY}}{s_X s_Y}$$

$$r = \frac{134.301}{(13.096)(20.266)} = .506$$

This coefficient must be interpreted cautiously; do not attribute meaning to it that it does not possess. Specifically, $r = .506$ should *not* be interpreted to mean that there is 50.6% of a relationship (whatever that might mean) between stress and symptoms. The correlation coefficient is simply a point on the scale between -1 and 1, and the closer it is to either of those limits, the stronger is the relationship between the two variables. For a more specific interpretation, we can speak in terms of r^2, which will be discussed shortly. It is important to emphasize again that the sign of the correlation merely reflects the direction of the relationship and, possibly, the arbitrary nature of the scale. Changing a variable from "number of items correct" to "number of items incorrect" would reverse the sign of the correlation, but it would have no effect on its absolute value.

Adjusted r

correlation coefficient in the population (ρ) rho

Although the correlation we have just computed is the one we normally report, it is not an unbiased estimate of the **correlation coefficient in the population,** denoted (ρ) **rho.** To see why this would be the case, imagine two randomly selected pairs of points—for example,

(23, 18) and (40, 66). (I pulled those numbers out of the air.) If you plot these points and fit a line to them, the line will fit perfectly because, as you most likely learned in elementary school, two points determine a straight line. Because the line fits perfectly, the correlation will be 1.00, even though the points were chosen at random. Clearly, that correlation of 1.00 does not mean that the correlation in the population from which those points were drawn is 1.00 or anywhere near it. When the number of observations is small, the sample correlation will be a biased estimate of the population correlation coefficient. To correct for this, we can compute what is known as the **adjusted correlation coefficient** (r_{adj}):

adjusted correlation coefficient (r_{adj})

$$r_{adj} = \sqrt{1 - \frac{(1 - r^2)(N - 1)}{N - 2}}$$

This is a relatively unbiased estimate of the population correlation coefficient.

In the example we have been using, the sample size is reasonably large ($N = 107$). Therefore, we would not expect a great difference between r and r_{adj}.

$$r_{adj} = \sqrt{1 - \frac{(1 - .506^2)(106)}{105}} = .499$$

which is very close to $r = .506$. This agreement will not be the case, however, for very small samples.

When we discuss multiple regression, which involves multiple predictors of Y, in Chapter 15, we will see that this equation for the adjusted correlation will continue to hold. The only difference will be that the denominator will be $N - p - 1$, where p stands for the number of predictors. (That is where the $N - 2$ came from in this equation.)

We could draw a parallel between the adjusted r and the way we calculate a sample variance. As I explained earlier, in calculating the variance we divide the sum of squared deviations by $N - 1$ to create an unbiased estimate of the population variance. That is comparable to what we do when we compute an adjusted r. The odd thing is that no one would seriously consider reporting anything but the unbiased estimate of the population variance, whereas we think nothing of reporting a biased estimate of the population correlation coefficient. I don't know why we behave inconsistently like that—we just do. The only reason I even discuss the adjusted value is that most computer software presents both statistics, and students are likely to wonder about the difference and which one they should care about.

9.5 The Regression Line

We have just seen that there is a reasonable degree of positive relationship between stress and psychological symptoms ($r = .506$). We can obtain a better idea of what this relationship is by looking at a scatterplot of the two variables and the regression line for predicting symptoms (Y) on the basis of stress (X). The scatterplot is shown in Figure 9.2, where the best-fitting line for predicting Y on the basis of X has been superimposed. We will see shortly where this line came from, but notice first the way in which the symptom scores increase linearly with increases in stress scores. Our correlation coefficient told us that such a relationship existed, but it is easier to appreciate just what it means when you see it presented graphically. Notice also that the degree of scatter of points about the regression line remains about the same as you move from low values of stress to high values, although, with a correlation of approximately .50, the scatter is fairly wide. We will discuss scatter in more detail when we consider the assumptions on which our procedures are based.

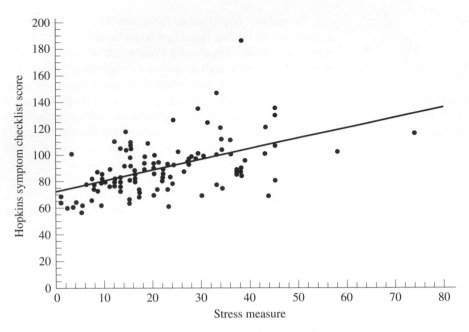

Figure 9.2 Scatterplot of symptoms as a function of stress

As you may remember from high school, the equation of a straight line is an equation of the form $Y = bX + a$. For our purposes, we will write the equation as

$$\hat{Y} = bX + a$$

where

$\hat{Y}$ = the predicted value of Y
b = the **slope** of the regression line (the amount of difference in $\hat{Y}$ associated with a one-unit difference in X)
a = the **intercept** (the value of $\hat{Y}$ when $X = 0$)
X = the value of the predictor variable.

slope

intercept

Our task will be to solve for those values of a and b that will produce the best-fitting linear function. In other words, we want to use our existing data to solve for the values of a and b such that the regression line (the values of $\hat{Y}$ for different values of X) will come as close as possible to the actual obtained values of Y. But how are we to define the phrase "best-fitting"? A logical way would be in terms of **errors of prediction**—that is, in terms of the $(Y - \hat{Y})$ deviations. Because $\hat{Y}$ is the value of the symptom variable that our equation would *predict* for a given level of stress, and Y is a value that we actually *obtained*, $(Y - \hat{Y})$ is the error of prediction, usually called the **residual**. We want to find the line (the set of $\hat{Y}$s) that minimizes such errors. We cannot just minimize the *sum* of the errors, however, because for an infinite variety of lines—any line that goes through the point $(\overline{X}, \overline{Y})$—that sum will always be zero. (We will overshoot some and undershoot others.) Instead, we will look for that line that minimizes the sum of the *squared* errors—that minimizes $\sum(Y - \hat{Y})^2$. (Note that I said much the same thing in Chapter 2 when I was discussing the variance. There I was discussing deviations from the mean, and here I am discussing deviations from the regression line—sort of a floating or changing mean. These two concepts—errors of prediction and variance—have much in common, as we shall see.)[4]

errors of prediction

residual

---

[4] For those who are interested, Rousseeuw and Leroy (1987) present a good discussion of alternative criteria that could be minimized, often to good advantage.

The optimal values of a and b can be obtained by solving for those values of a and b that minimize $\sum(Y - \hat{Y})^2$. The solution is not difficult, and those who want it can find it in earlier editions of this book or in Draper and Smith (1981, p. 13). The solution to the problem yields what are often called the **normal equations**:

normal equations

$$a = \overline{Y} - b\overline{X}$$

$$b = \frac{\text{cov}_{XY}}{s_X^2}$$

We now have solutions for a and b[5] that will minimize $\sum(Y - \hat{Y})^2$. To indicate that our solution was designed to minimize errors in predicting Y from X (rather than the other way around), the constants are sometimes denoted $a_{Y \cdot X}$ and $b_{Y \cdot X}$. When no confusion would arise, the subscripts are usually omitted. (When your purpose is to predict X on the basis of Y (i.e., X on Y), then you can simply reverse X and Y in the previous equations.)

As an example of the calculation of regression coefficients, consider the data in Table 9.2. From that table we know that $\overline{X} = 21.467$, $\overline{Y} = 90.701$, and $s_X = 13.096$. We also know that $\text{cov}_{XY} = 134.301$. Thus,

$$b = \frac{\text{cov}_{XY}}{s_X^2} = \frac{134.301}{13.096^2} = 0.7831$$

$$a = \overline{Y} - b\overline{X} = 90.701 - (0.7831)(21.467) = 73.891$$

$$\hat{Y} = bX + a = (0.7831)(X) + 73.891$$

We have already seen the scatter diagram with the regression line for Y on X superimposed in Figure 9.2. This is the equation of that line.[6]

A word is in order about actually plotting the regression line. To plot the line, you can simply take any two values of X (preferably at opposite ends of the scale), calculate $\hat{Y}$ for each, mark these coordinates on the figure, and connect them with a straight line. For our data, we have

$$\hat{Y}_i = (0.7831)(X_i) + 73.891$$

When $X_i = 0$,

$$\hat{Y}_i = (0.7831)(0) + 73.891 = 73.891$$

and when $X_i = 50$,

$$\hat{Y}_i = (0.7831)(50) + 73.891 = 113.046$$

The line then passes through the points ($X = 0$, $Y = 73.891$) and ($X = 50$, $Y = 113.046$), as shown in Figure 9.2. The regression line will also pass through the points (0, a) and ($\overline{X}$, $\overline{Y}$), which provides a quick check on accuracy.

If you calculate both regression lines (Y on X and X on Y), it will be apparent that the two are not coincident. They do intersect at the point ($\overline{X}$, $\overline{Y}$), but they have different slopes. The fact that they are different lines reflects the fact that they were designed for different purposes—one minimizes $\sum(Y - \hat{Y})^2$ and the other minimizes $\sum(X - \hat{X})^2$. They both go through the point ($\overline{X}$, $\overline{Y}$) because a person who is *average* on one variable would be expected to be *average* on the other, but only when the correlation between the two variables is ± 1.00 will the lines be coincident.

---

[5] An interesting alternative formula for b can be written as $b = r(s_y/s_x)$. This shows explicitly the relationship between the correlation coefficient and the slope of the regression line. Note that when $s_y = s_x$, b will equal r. Can you think of a case where this would happen? (*Answer:* When both variables have a standard deviation of 1, which happens when the variables are standardized.)

[6] An excellent java applet that allows you to enter individual data points and see their effect on the regression line is available at http://www.math.csusb.edu/faculty/stanton/m262/regress/regress.html.

Interpretations of Regression

In certain situations, the regression line is useful in its own right. For example, a college admissions officer might be interested in an equation for predicting college performance on the basis of high-school grade point average (although she would most likely want to include multiple predictors in ways to be discussed in Chapter 15). Similarly, a neuropsychologist might be interested in predicting a patient's response rate based on one or more indicator variables. If the actual rate is well below expectation, we might start to worry about the patient's health (see Crawford, Garthwaite, Howell, & Venneri, 2003). But these examples are somewhat unusual. In most applications of regression in psychology, we are not particularly interested in making an actual prediction. Although we might be interested in knowing the relationship between family income and educational achievement, it is unlikely that we would take any particular child's family-income measure and use that to predict his educational achievement. We are usually much more interested in general principles than in individual predictions. A regression equation, however, can in fact tell us something meaningful about these general principles, even though we may never actually use it to form a prediction for a specific case.

Intercept

We have defined the intercept as that value of $\hat{Y}$ when X equals zero. As such, it has meaning in some situations and not in others, primarily depending on whether or not $X = 0$ has meaning and is near or within the range of values of X used to derive the estimate of the intercept. If, for example, we took a group of overweight people and looked at the relationship between self-esteem (Y) and weight loss (X) (assuming that it is linear), the intercept would tell us what level of self-esteem to expect for an individual who lost 0 pounds. Often, however, there is no meaningful interpretation of the intercept other than a mathematical one. If we are looking at the relationship between self-esteem (Y) and actual weight (X) for adults, it is obviously foolish to ask what someone's self-esteem would be if he weighed 0 pounds. The intercept would appear to tell us this, but it represents such an extreme extrapolation from available data as to be meaningless. (In this case, a nonzero intercept would suggest a lack of linearity over the wider range of weight from 0 to 300 pounds, but we probably are not interested in nonlinearity in the extremes anyway.)

Slope

We have defined the slope as the change in $\hat{Y}$ for a one-unit change in X. As such, it is a measure of the predicted *rate of change* in Y. By definition, then, the slope is often a meaningful measure. If we are looking at the regression of income on years of schooling, the slope will tell us how much of a difference in income would be associated with each additional year of school. Similarly, if an engineer knows that the slope relating fuel economy in miles per gallon (mpg) to weight of the automobile is 0.01, and if she can assume a causal relationship between mpg and weight, then she knows that for every pound that she can reduce the weight of the car, she will increase its fuel economy by 0.01 mpg. Thus, if the manufacturer replaces a 30-pound spare tire with one of those annoying 20-pound temporary ones, the car will gain 0.1 mpg.

Standardized Regression Coefficients

Although we rarely work with standardized data (data that have been transformed so as to have a mean of zero and a standard deviation of one on each variable), it is worth considering what b would represent if the data for each variable were standardized separately. In that case, a difference of one unit in X or Y would represent a difference of one standard deviation. Thus, if the slope were 0.75, for standardized data, we would be able to say that a one standard deviation increase in X will be reflected in three-quarters of a standard deviation increase in $\hat{Y}$. When speaking of the slope coefficient for standardized data, we often refer to the **standardized regression coefficient** as β (beta) to differentiate it from the coefficient for nonstandardized data (b). We will return to the idea of standardized variables when we discuss multiple regression in Chapter 15. (What would the intercept be if the variables were standardized? *Hint:* The line goes through the means.)

standardized regression coefficient β (beta)

Correlation and Beta

What we have just seen with respect to the slope for standardized variables is directly applicable to the correlation coefficient. Recall that r is defined as $\text{cov}_{XY}/s_X s_Y$, whereas b is defined as cov_{XY}/s_X^2. If the data are standardized, $s_X = s_Y = s_X^2 = 1$ and the slope and the correlation coefficient will be equal. Thus, one interpretation of the correlation coefficient is that it is equal to what the slope would be if the variables were standardized. That suggests that a derivative interpretation of $r = .80$, for example, is that one standard deviation difference in X is associated *on the average* with an eight-tenths of a standard deviation difference in Y. In some situations, such an interpretation can be meaningfully applied.

A Note of Caution

What has just been said about the interpretation of b and r must be tempered with a bit of caution. To say that a one-unit difference in family income is associated with 0.75 units difference in academic achievement is not to be interpreted to mean that raising family income for Mary Smith will automatically raise her academic achievement. In other words, we are not speaking about cause and effect. We can say that people who score higher on the income variable also score higher on the achievement variable without in any way implying causation or suggesting what would happen to a given individual if her family income were to increase. Family income is associated (in a correlational sense) with a host of other variables (e.g., attitudes toward education, number of books in the home, access to a variety of environments), and there is no reason to expect all of these to change merely because income changes. Those who argue that eradicating poverty will lead to a wide variety of changes in people's lives often fall into such a cause-and-effect trap. Eradicating poverty is certainly a worthwhile and important goal, but the correlations between income and educational achievement *may* be totally irrelevant to the issue.

9.6   The Accuracy of Prediction

The fact that we can fit a regression line to a set of data does not mean that our problems are solved. On the contrary, they have only begun. The important point is not whether a straight line can be drawn through the data (you can always do that) but whether that line represents a reasonable fit to the data—in other words, whether our effort was worthwhile.

In beginning a discussion of errors of prediction, it is instructive to consider the situation in which we want to predict Y without any knowledge of the value of X.

The Standard Deviation as a Measure of Error

As mentioned earlier, the data plotted in Figure 9.4 represent the number of symptoms shown by students (Y) as a function of the number of stressful life events (X). Assume that you are now given the task of predicting the number of symptoms that will be shown by a particular individual, but that you have no knowledge of the number of stressful life events he or she has experienced. Your best prediction in this case would be the mean number of symptoms ($\overline{Y}$) (averaged across all subjects), and the error associated with your prediction would be the standard deviation of Y (i.e., s_Y) because your prediction is the mean and s_Y deals with deviations around the mean. We know that s_Y is defined as

$$s_Y = \sqrt{\frac{\sum(Y - \overline{Y})^2}{N - 1}}$$

or, in terms of the variance,

$$s_Y^2 = \frac{\sum(Y - \overline{Y})^2}{N - 1}$$

sum of squares of Y (SS_Y)

The numerator is the sum of squared deviations from $\overline{Y}$ (the point you would have predicted in this example) and is what we will refer to as the sum of squares of Y (SS_Y). The denominator is simply the degrees of freedom. Thus, we can write

$$s_Y^2 = \frac{SS_Y}{df}$$

The Standard Error of Estimate

Now suppose we want to make a prediction about symptoms for a student who has a specified number of stressful life events. If we had an infinitely large sample of data, our prediction for symptoms would be the mean of those values of symptoms (Y) that were obtained by all students who had that particular value of stress. In other words, it would be a conditional mean—conditioned on that value of X. We do not have an infinite sample, however, so we will use the regression line. (If all the assumptions that we will discuss shortly are met, the expected value of the Y scores associated with each specific value of X would lie on the regression line.) In our case, we know the relevant value of X and the regression equation, and our best prediction would be $\hat{Y}$. In line with our previous measure of error (the standard deviation), the error associated with the present prediction will again be a function of the deviations of Y about the predicted point, but in this case the predicted point is $\hat{Y}$ rather than $\overline{Y}$. Specifically, a measure of error can now be defined as

$$s_{Y \cdot X} = \sqrt{\frac{\sum(Y - \hat{Y})^2}{N - 2}} = \sqrt{\frac{SS_{\text{residual}}}{df}}$$

standard error of estimate

residual variance

error variance

and again the sum of squared deviations is taken about the prediction ($\hat{Y}$). The sum of squared deviations about $\hat{Y}$ is often denoted SS_{residual} because it represents variability that remains *after* we use X to predict Y.[7] The statistic $s_{Y \cdot X}$ is called the standard error of estimate. It is denoted as $s_{Y \cdot X}$ to indicate that it is the standard deviation of Y predicted from X. It is the most common (although not always the best) measure of the error of prediction. Its square, $s_{Y \cdot X}^2$, is called the residual variance or error variance, and it can be shown to be an unbiased estimate of the corresponding parameter ($\sigma_{Y \cdot X}^2$) in the population. We have $N - 2$ df because we lost two degrees of freedom in estimating our regression line. (Both a and b were estimated from sample data.)

I have suggested that if we had an infinite number of observations, our prediction for a given value of X would be the mean of the Ys associated with that value of X. This idea helps

---

[7] It is also frequently denoted SS_{error} because it is a sum of squared errors of prediction.

us appreciate what $s_{Y \cdot X}$ is. If we had the infinite sample and calculated the variances for the Ys at each value of X, the average of those variances would be the residual variance, and its square root would be $s_{Y \cdot X}$. The set of Ys corresponding to a specific X is called a **conditional distribution** of Y because it is the distribution of Y scores for those cases that meet a certain condition with respect to X. We say that these standard deviations are conditional on X because we calculate them from Y values corresponding to specific values of X. On the other hand, our usual standard deviation of Y (s_Y) is not conditional on X because we calculate it using all values of Y, regardless of their corresponding X values.

conditional distribution

One way to obtain the standard error of estimate would be to calculate $\hat{Y}$ for each observation and then to find $s_{Y \cdot X}$ directly, as has been done in Table 9.3. Finding the standard error using this technique is hardly the most enjoyable way to spend a winter evening. Fortunately, a much simpler procedure exists that provides a way of obtaining the standard error of estimate and leads directly into even more important matters.

Table 9.3   Direct calculation of the standard error of estimate

Subject	Stress (X)	Symptoms (Y)	$\hat{Y}$	$Y - \hat{Y}$
1	30	99	97.383	1.617
2	27	94	95.034	−1.034
3	9	80	80.938	−0.938
4	20	70	89.552	−19.552
5	3	100	76.239	23.761
6	15	109	85.636	23.364
7	5	62	77.806	−15.806
8	10	81	81.721	−0.721
9	23	74	91.902	−17.902
10	34	121	100.515	20.485
⋮	⋮	⋮	⋮	⋮

$$\sum (Y - \hat{Y}) = 0$$
$$\sum (Y - \hat{Y})^2 = 32388.049$$

$$s^2_{Y \cdot X} = \frac{\sum (Y - \hat{Y})^2}{N - 2} = \frac{32388.049}{105} = 308.458 \qquad s_{Y \cdot X} = \sqrt{308.458} = 17.563$$

r^2 and the Standard Error of Estimate

In much of what follows, we will abandon the term *variance* in favor of sums of squares (SS). As you should recall, a variance is a sum of squared deviations from the mean (generally known as a sum of squares) divided by the degrees of freedom. The problem with variances is that they are not additive unless they are based on the same *df*. Sums of squares are additive regardless of the degrees of freedom and, thus, are much easier measures to use.[8]

We earlier defined the residual or error variance as

$$s^2_{Y \cdot X} = \frac{\sum (Y - \hat{Y})^2}{N - 2} = \frac{SS_{\text{residual}}}{N - 2}$$

---

[8] Later in the book when I want to speak about a variance-type measure but do not want to specify whether it is a variance, a sum of squares, or something similar, I will use the vague, wishy-washy term *variation*.

With considerable algebraic manipulation, it is possible to show

$$s_{Y \cdot X} = s_Y \sqrt{(1 - r^2) \frac{N - 1}{N - 2}}$$

For large samples, the fraction $(N - 1)/(N - 2)$ is essentially 1, and we can thus write the equation as it is often found in statistics texts:

$$s_{Y \cdot X}^2 = s_Y^2 (1 - r^2)$$

or

$$s_{Y \cdot X} = s_Y \sqrt{(1 - r^2)}$$

Keep in mind, however, that for small samples these equations are only an approximation and $s_{Y \cdot X}^2$ will underestimate the error variance by the fraction $(N - 1)/(N - 2)$. For samples of any size, however, $SS_{\text{residual}} = SS_Y(1 - r^2)$. This particular formula is going to play a role throughout the rest of the book, especially in Chapters 15 and 16.

Errors of Prediction as a Function of *r*

Now that we have obtained an expression for the standard error of estimate in terms of r, it is instructive to consider how this error decreases as r increases. In Table 9.4, we see the magnitude of the standard error relative to the standard deviation of Y (the error to be expected when X is unknown) for selected values of r.

The values in Table 9.4 are somewhat sobering in their implications. With a correlation of .20, the standard error of our estimate is fully 98% of what it would be if X were unknown. This means that if the correlation is .20, using $\hat{Y}$ as our prediction rather than $\overline{Y}$ (i.e., taking X into account) reduces the standard error by only 2%. Even more discouraging is that if r is .50, as it is in our example, the standard error of estimate is still 87% of the standard deviation. To reduce our error to one-half of what it would be without knowledge of X requires a correlation of .866, and even a correlation of .95 reduces the error by only about two-thirds. All of this is not to say that there is nothing to be gained by using a regression equation as the basis of prediction, only that the predictions should be interpreted with a certain degree of caution. All is not lost, however, because it is often the kinds of relationships we see, rather than their absolute magnitudes, that are of interest to us.

Table 9.4 The standard error of estimate as a function of r

r	$s_{Y \cdot X}$	r	$s_{Y \cdot X}$
.00	s_Y	.60	$0.800 s_Y$
.10	$0.995 s_Y$	.70	$0.714 s_Y$
.20	$0.980 s_Y$	.80	$0.600 s_Y$
.30	$0.954 s_Y$	.866	$0.500 s_Y$
.40	$0.917 s_Y$	.90	$0.436 s_Y$
.50	$0.866 s_Y$	.95	$0.312 s_Y$

r^2 as a Measure of Predictable Variability

From the preceding equation expressing residual error in terms of r^2, it is possible to derive an extremely important interpretation of the correlation coefficient. We have already seen that

$$SS_{\text{residual}} = SS_Y(1 - r^2)$$

Expanding and rearranging, we have

$$SS_{\text{residual}} = SS_Y - SS_Y(r^2)$$

$$r^2 = \frac{SS_Y - SS_{\text{residual}}}{SS_Y}$$

In this equation, SS_Y, which you know to be equal to $\sum(Y - \overline{Y})^2$, is the sum of squares of Y and represents the totals of

1. The part of the sum of squares of Y that is related to X (i.e., $SS_Y(r^2)$)
2. The part of the sum of squares of Y that is independent of X (i.e., SS_{residual})

In the context of our example, we are talking about that part of the number of symptoms people exhibited that is related to how many stressful life events they had experienced, and that part that is related to other things. The quantity SS_{residual} is the sum of squares of Y that is independent of X and is a measure of the amount of error remaining even after we use X to predict Y. These concepts can be made clearer with a second example.

Suppose we were interested in studying the relationship between amount of cigarette smoking (X) and age at death (Y). As we watch people die over time, we notice several things. First, we see that not all die at precisely the same age. There is variability in age at death regardless of smoking behavior, and this variability is measured by $SS_Y = \sum(Y - \overline{Y})^2$. We also notice that some people smoke more than others. This variability in smoking regardless of age at death is measured by $SS_X = \sum(X - \overline{X})^2$. We further find that cigarette smokers tend to die earlier than nonsmokers, and heavy smokers earlier than light smokers. Thus, we write a regression equation to predict Y from X. People differ in their smoking behavior, so they will also differ in their *predicted* life expectancy ($\hat{Y}$), and we will label this variability $SS_{\hat{Y}} = \sum(\hat{Y} - \overline{Y})^2$. This last measure is variability in Y that is directly attributable to variability in X because different values of $\hat{Y}$ arise from different values of X and the same values of $\hat{Y}$ arise from the same value of X—that is, $\hat{Y}$ does not vary unless X varies.

We have one last source of variability: the variability in the life expectancy of those people who smoke exactly the same amount. This is measured by SS_{residual} and is the variability in Y that cannot be explained by the variability in X (because these people do not differ in the amount they smoke). These several sources of variability (sums of squares) are summarized in Table 9.5.

Table 9.5   Sources of variability in regression for the study of smoking and life expectancy

---

SS_X = variability in amount smoked = $\sum(X - \overline{X})^2$

SS_Y = variability in life expectancy = $\sum(Y - \overline{Y})^2$

$SS_{\hat{Y}}$ = variability in life expectancy directly attributable to variability in smoking behavior = $\sum(\hat{Y} - \overline{Y})^2$

SS_{residual} = variability in life expectancy that cannot be attributed to variability in smoking behavior = $\sum(Y - \hat{Y})^2 = SS_Y - SS_{\hat{Y}}$

If we considered the absurd extreme in which all the nonsmokers die at exactly age 72 and all the smokers smoke precisely the same amount and die at exactly age 68, then all of the variability in life expectancy is directly predictable from variability in smoking behavior. If you smoke you will die at 68, and if you don't you will die at 72. Here $SS_{\hat{Y}} = SS_Y$, and $SS_{\text{residual}} = 0$.

As a more realistic example, assume smokers tend to die earlier than nonsmokers, but within each group there is a certain amount of variability in life expectancy. This is a situation in which some of SS_Y is attributable to smoking ($SS_{\hat{Y}}$) and some is not (SS_{residual}). What we want to be able to do is to specify what *percentage* of the overall variability in life expectancy is attributable to variability in smoking behavior. In other words, we want a measure that represents

$$\frac{SS_{\hat{Y}}}{SS_Y} = \frac{SS_Y - SS_{\text{residual}}}{SS_Y}$$

As we have seen, that measure is r^2. In other words,

$$r^2 = \frac{SS_{\hat{Y}}}{SS_Y}$$

This interpretation of r^2 is extremely useful. If, for example, the correlation between amount smoked and life expectancy were an unrealistically high .80, we could say that $.80^2 = 64\%$ of the variability in life expectancy is directly predictable from the variability in smoking behavior. (Obviously, this is an outrageous exaggeration of the real world.) If the correlation were a more likely $r = .10$, we would say that $10^2 = 1\%$ of the variability in life expectancy is related to smoking behavior, whereas the other 99% is related to other factors.

Phrases such as "accounted for by," "attributable to," "predictable from," and "associated with" are *not* to be interpreted as statements of cause and effect. Thus, you could say, "I can predict 10% of the variability of the weather by paying attention to twinges in the ankle that I broke last year—when it aches we are likely to have rain, and when it feels fine the weather is likely to be clear." This does not imply that sore ankles cause rain, or even that rain itself causes sore ankles. For example, it might be that your ankle hurts when it rains because low barometric pressure, which is often associated with rain, somehow affects ankles.

From this discussion, it should be apparent that r^2 is easier to interpret as a measure of correlation than is r because it represents the degree to which the variability in one measure is attributable to variability in the other measure. I recommend that you always square correlation coefficients to get some idea of whether you are talking about anything important. In our symptoms-and-stress example, $r^2 = .506^2 = .256$. Thus, about one-quarter of the variability in symptoms can be predicted from variability in stress. That strikes me as an impressive level of prediction, given all the other factors that influence psychological symptoms.

There is not universal agreement that r^2 is our best measure of the contribution of one variable to the prediction of another, although that is certainly the most popular measure. Judd and McClelland (1989) strongly endorse r^2 because, when we index error in terms of the sum of squared errors, it is the **proportional reduction in error (PRE)**. In other words, when we do not use X to predict Y, our error is SS_Y. When we use X as the predictor, the error is SS_{residual}. Because

proportional reduction in error (PRE)

$$r^2 = \frac{SS_Y - SS_{\text{residual}}}{SS_Y}$$

the value of r^2 can be seen to be the percentage by which error is reduced when X is used as the predictor.[9]

proportional improvement in prediction (PIP)

Others, however, have suggested the **proportional improvement in prediction (PIP)** as a better measure.

$$\text{PIP} = 1 - \sqrt{(1 - r^2)}$$

---

[9] It is interesting to note that r^2_{adj} (defined on p. 239) is nearly equivalent to the ratio of the *variance* terms corresponding to the sums of squares in the equation. (Well, it is interesting to *some* people.)

For large sample sizes, this statistic is the *reduction* in the size of the standard error of estimate (see Table 9.4). Similarly, as we shall see shortly, it is a measure of the reduction in the width of the confidence interval on our prediction.

The choice between r^2 and PIP really depends on how you want to measure error. When we focus on r^2, we are focusing on measuring error in terms of sums of squares. When we focus on PIP, we are measuring error in standard deviation units.

Darlington (1990) has argued for the use of r instead of r^2 as representing the magnitude of an effect. A strong argument in this direction was also made by Ozer (1985), whose paper is well worth reading. In addition, Rosenthal and Rubin (1982) have shown that even small values of r^2 (or almost any other measure of the magnitude of an effect) can be associated with powerful effects, regardless of how you measure that effect (see Chapter 10).

I have discussed r^2 as an index of percentage of variation for a particular reason. There is a very strong movement, at least in psychology, toward more frequent reporting of the magnitude of an effect, rather than just a test statistic and a p value. As I mentioned in Chapter 7, there are two major types of magnitude measures. One type is called effect size, often referred to as the *d*-family of measures, and is represented by Cohen's *d*, which is most appropriate when we have means of two or more groups. The second type of measure, often called the *r*-family, is the "percentage of variation," of which r^2 is the most common representative. We first saw this measure in this chapter, where we found that 25.6% of the variation in psychological symptoms is associated with variation in stress. We will see it again in Chapter 10 when we cover the point-biserial correlation. It will come back again in the analysis of variance chapters (especially Chapters 11 and 13), where it will be disguised as eta-squared and related measures. Finally, it will appear in important ways when we talk about multiple regression. The common thread through all of this is that we want some measure of how much of the variation in a dependent variable is attributable to variation in an independent variable, whether that independent variable be categorical or continuous. I am not as fond of percentage of variation measures as are some people because I don't think that most of us can take much meaning from such measures. However, they are commonly used, and you need to be familiar with them.

9.7 Assumptions Underlying Regression and Correlation

We have derived the standard error of estimate and other statistics without making any assumptions concerning the population(s) from which the data were drawn. Nor do we need such assumptions to use $s_{Y.X}$ as an unbiased estimator of $\sigma_{Y.X}$. If we are to use $s_{Y.X}$ in any meaningful way, however, we will have to introduce certain parametric assumptions. To understand why, consider the data plotted in Figure 9.3. Notice the four statistics labeled $s_{Y.1}^2, s_{Y.2}^2, s_{Y.3}^2$, and $s_{Y.4}^2$. Each represents the variance of the points around the regression line in an array of X (the residual variance of Y conditional on a specific X). As mentioned earlier, the average of these variances, weighted by the degrees of freedom for each array, would be $s_{Y.X}^2$, the residual or error variance. If $s_{Y.X}^2$ is to have any practical meaning, it must be representative of the various terms of which it is an average. This leads us to the assumption of homogeneity of variance in arrays, which is nothing but the assumption that the variance of Y for each value of X is constant (in the population). This assumption will become important when we apply tests of significance using $s_{Y.X}^2$.

One further assumption that will be necessary when we come to testing hypotheses is that of normality in arrays. We will assume that in the population, the values of Y corresponding to any specified value of X—that is, the conditional array of Y for X_i—are normally distributed around $\hat{Y}$. This assumption is directly analogous to the normality

array

homogeneity of variance in arrays

normality in arrays

conditional array

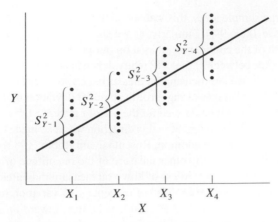

Figure 9.3   Scatter diagram illustrating regression assumptions

assumption we made with the *t* test—that each treatment population was normally distributed around its own mean—and we make it for similar reasons.

To anticipate what we will discuss in Chapter 11, note that our assumptions of homogeneity of variance and normality in arrays are equivalent to the assumptions of homogeneity of variance and normality of populations that we will make in discussing the analysis of
variance. In Chapter 11, we will assume that the treatment populations from which data
were drawn are normally distributed and all have the same variance. If you think of the
levels of *X* in Figure 9.3 as representing different experimental conditions, you can see
the relationship between the regression and analysis of variance assumptions.

The assumptions of normality and homogeneity of variance in arrays are associated
with the regression model, where we are dealing with fixed values of *X*. On the other hand,
when our interest is centered on the correlation between *X* and *Y*, we are dealing with the
bivariate model, in which *X* and *Y* are both random variables. In this case, we are primarily
concerned with using the sample correlation (*r*) as an estimate of the correlation coefficient
in the population (ρ). Here, we will replace the regression model assumptions with the
assumption that we are sampling from a bivariate normal distribution.

The bivariate normal distribution looks roughly like the pictures you see each fall of sur-
plus wheat piled in the main street of some Midwestern town. The way the grain pile falls
off on all sides resembles a normal distribution. (If there were no correlation between *X* and
Y, the pile would look as though all the grain were dropped in the center of the pile and
spread out symmetrically in all directions. When *X* and *Y* are correlated the pile is elongated,
as when grain is dumped along a street and spreads out to the sides and down the ends.)
Imagine that the pile had actually been dumped on top of a huge scattergram, with the main
axis of the pile oriented along the regression line. If you sliced the pile on a line corresponding to any given value of *X*, you would see that the cut end is a normal distribution.
You would also have a normal distribution if you sliced the pile along a line corresponding
to any given value of *Y*. These are called **conditional distributions** because the first repre-
conditional   sents the distribution of *Y* given (conditional on) a specific value of *X*, whereas the second
distributions   represents the distribution of *X* conditional on a specific value of *Y*. If, instead, we looked at
all the values of *Y* regardless of *X* (or all values of *X* regardless of *Y*), we would have what
marginal   is called the **marginal distribution** of *Y* (or *X*). For a bivariate normal distribution, both the
distribution   conditional and the marginal distributions will be normally distributed. (Recall that for the
regression model, we assumed only normality of *Y* in the arrays of *X*—what we now know
as conditional normality of *Y*. For the regression model, there is no assumption of normality of the conditional distribution of *X* or of the marginal distributions.)

9.8   Confidence Limits on Y

Although the standard error of estimate is useful as an overall measure of error, it is not a good estimate of the error associated with any single prediction. When we want to predict a value of Y for a given subject, the error in our estimate will be smaller when X is near $\overline{X}$ than when X is far from $\overline{X}$. (For an intuitive understanding of this, consider what would happen to the predictions for different values of X if we rotated the regression line slightly around the point $\overline{X}, \overline{Y}$. There would be negligible changes near the means, but there would be substantial changes in the extremes.) If we want to predict Y on the basis of X for a new member of the population (someone who was not included in the original sample), the standard error of our prediction is given by

$$s'_{Y \cdot X} = s_{Y \cdot X} \sqrt{1 + \frac{1}{N} + \frac{(X_i - \overline{X})^2}{(N-1)s_X^2}}$$

where $X_i - \overline{X}$ is the deviation of the individual's X score from the mean of X. This leads to the following confidence limits on Y:

$$\text{CI}(Y) = \hat{Y} \pm (t_{\alpha/2})(s'_{Y \cdot X})$$

This equation will lead to elliptical confidence limits around the regression line, which are narrowest for $X = \overline{X}$ and become wider as $|X - \overline{X}|$ increases.

To take a specific example, assume that we wanted to set confidence limits on the number of symptoms (Y) experienced by a student with a stress score of 10—a fairly low level of stress. We know that

$$s_{Y \cdot X} = 17.563$$
$$s_X^2 = 171.505$$
$$\overline{X} = 21.467$$
$$\hat{Y} = 0.7831(10) + 73.891 = 81.722$$
$$t_{.025} = 1.984$$
$$N = 107$$

Then

$$s'_{Y \cdot X} = s_{Y \cdot X} \sqrt{1 + 1/N + \frac{(X_i - \overline{X})^2}{(N-1)s_X^2}}$$

$$s'_{Y \cdot X} = 17.563 \sqrt{1 + 1/107 + \frac{(10 - 21.467)^2}{(106)171.505}}$$

$$= 17.563 \sqrt{1.0166} = 17.708$$

Then

$$\text{CI}(Y) = \hat{Y} \pm (t_{\alpha/2})(s'_{Y \cdot X})$$
$$= 81.722 \pm 1.984(17.708)$$
$$= 81.722 \pm 35.133$$
$$46.589 \leq Y \leq 116.855$$

The confidence interval is 46.589 to 116.855, and the probability is .95 that an interval computed in this way will include the level of symptoms reported by an individual whose stress score is 10. That interval is wide, but it is not as large as the 95% confidence interval of $50.5 \leq Y \leq 130.9$ that we would have had if we had not used X—that is, if we had just based our confidence interval on the obtained values of Y (and s_Y) rather than making it conditional on X.

9.9 A Computer Example Showing the Role of Test-Taking Skills

Most of us can do reasonably well if we study a body of material and then take an exam on that material. But how would we do if we just took the exam without even looking at the material? Katz, Lautenschlager, Blackburn, and Harris (1990) examined that question by asking some students to read a passage and then answer a series of multiple-choice questions, and asking others to answer the questions without having seen the passage. We will concentrate on the second group. The task described here is very much like the task that North American students face when they take the SAT exams for admission to a university. This led the researchers to suspect that students who did well on the SAT would also do well on this task because they both involve test-taking skills such as eliminating unlikely alternatives.

Data with the same sample characteristics as the data obtained by Katz et al. are given in Table 9.6. The variable Score represents the percentage of items answered correctly when the student has not seen the passage, and the variable SATV is the student's verbal SAT score from his or her college application.

Table 9.6   Data based on Katz et al. (1990) for the group that did not read the passage

Score	SATV	Score	SATV
58	590	48	590
48	580	41	490
34	550	43	580
38	550	53	700
41	560	60	690
55	800	44	600
43	650	49	580
47	660	33	590
47	600	40	540
46	610	53	580
40	620	45	600
39	560	47	560
50	570	53	630
46	510	53	620

Exhibit 9.1 illustrates the analysis using SPSS regression. There are a number of things here to point out. First, we must decide which is the dependent variable and which is the independent variable. This would make no difference if we just wanted to compute the correlation between the variables, but it is important in regression. In this case, I have made a relatively arbitrary decision that my interest lies primarily in seeing whether people who do well at making intelligent guesses also do well on the SAT. Therefore, I am using SATV as the dependent variable, even though it was actually taken before the experiment. The first two panels of Exhibit 9.1 illustrate the menu selections required for SPSS. The means and standard deviations are found in the middle of the output, and you can see that we are

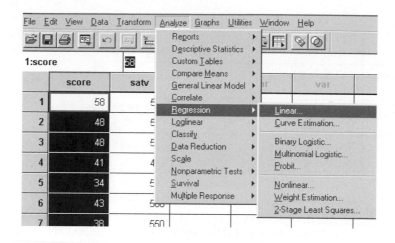

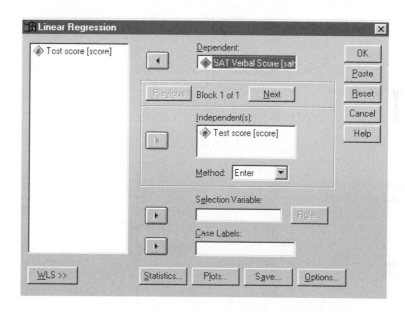

Descriptive Statistics

	Mean	Std. Deviation	N
SAT Verbal Score	598.57	61.57	28
Test Score	46.21	6.73	28

(continued)

Exhibit 9.1   SPSS output on Katz et al. (1990) study of test-taking behavior

Correlations

		SAT Verbal Score	Test Score
Pearson Correlation	SAT Verbal Score	1.000	.532
	Test Score	.532	1.000
Sig. (1-tailed)	SAT Verbal Score	.	.002
	Test Score	.002	.
N	SAT Verbal Score	28	28
	Test Score	28	28

Model Summary

Model	R	R Square	Adjusted R Square	Std. Error of the Estimate
1	.532	.283	.255	53.13

ANOVA[b]

Model		Sum of Squares	df	Mean Square	F	Sig.
1	Regression	28940.123	1	28940.123	10.251	.004[a]
	Residual	73402.734	26	2823.182		
	Total	102342.9	27			

[a] Predictors: (Constant), Test score

[b] Dependent Variable: SAT Verbal Score

Coefficients[a]

Model		Unstandardized Coefficients		Standardized Coefficients	t	Sig.
		B	Std. Error	Beta		
1	(Constant)	373.736	70.938		5.269	.000
	Test score	4.865	1.520	.532	3.202	.004

[a] Dependent Variable: SAT Verbal Score

Exhibit 9.1 (*continued*)

dealing with a group that has high achievement scores (the mean is almost 600, with a standard deviation of about 60. This puts them about 100 points above the average for the SAT. They also do quite well on Katz' test, getting nearly 50% of the items correct. Below these statistics you see the correlation between Score and SATV, which is .532. We will test this correlation for significance in a moment.

In the section labeled Model Summary, you see both R and R^2. The "R" here is capitalized because if there were multiple predictors, it would be a multiple correlation, and we always capitalize that symbol. One thing to note is that R here is calculated as the square root of R^2, and as such it will always be positive, even if the relationship is negative. This is a result of the fact that the procedure is applicable for multiple predictors.

The ANOVA table is a test of the null hypothesis that the correlation is .00 in the population. We will discuss hypothesis testing next, but what is most important here is that the test statistic is F, and that the significance level associated with that F is $p = .004$. Because p is less than .05, we will reject the null hypothesis and conclude that the variables are not linearly independent. In other words, there is a linear relationship between how well students score on a test that reflects test-taking skills and how well they perform on the SAT. The exact nature of this relationship is shown in the next part of the printout. Here we have a table labeled "Coefficients," and this table gives us the intercept and the slope. The intercept is labeled here as "Constant" because it is the constant that you add to every prediction. In this case, it is 373.736. Technically, it means that if a student answered 0 questions correctly on Katz' test, we would expect that student to have an SAT of approximately 370. Because a score of 0 would be so far from the scores these students actually obtained (and it is hard to imagine anyone earning a 0 even by guessing), I would not pay very much attention to that value.

In this table, the slope is labeled by the name of the predictor variable. (All software solutions do this because if there were multiple predictors we would have to know which variable goes with which slope. The easiest way to do this is to use the variable name as the label.) In this case, the slope is 4.865, which means that two students who differ by 1 point on Katz' test would be predicted to differ by 4.865 on the SAT. Our regression equation would now be written as $\hat{Y} = 4.865 \times \text{Score} + 373.736$.

The standardized regression coefficient is shown as .532. This means that a one standard deviation difference in test scores is associated with approximately a one-half standard deviation difference in SAT scores. Note that, because we have only one predictor, this standardized coefficient is equal to the correlation coefficient.

To the right of the standardized regression coefficient you will see t and p values for tests on the significance of the slope and intercept. We will discuss the test on the slope shortly. The test on the intercept is rarely of interest, but its interpretation should be evident from what I say about testing the slope.

9.10   Hypothesis Testing

In this chapter, we have seen how to calculate r as an estimate of the relationship between two variables and how to calculate the slope (b) as a measure of the rate of change of Y as a function of X. In addition to estimating r and b, we often want to perform a significance test on the null hypothesis that the corresponding population parameters equal zero. The fact that a value of r or b calculated from a sample is not zero is not in itself evidence that the corresponding parameters in the population are also nonzero.

Testing the Significance of r

The most common hypothesis that we test for a sample correlation is that the correlation between X and Y in the population, denoted ρ (rho), is zero. This is a meaningful test because the null hypothesis being tested is really the hypothesis that X and Y are linearly independent. Rejection of this hypothesis leads to the conclusion that they are not independent and that there is some linear relationship between them.

It can be shown that when $\rho = 0$, for large N, r will be approximately normally distributed around zero.

A legitimate t test can be formed from the ratio

$$t = \frac{r\sqrt{N-2}}{\sqrt{1-r^2}}$$

which is distributed as t on $N - 2$ df.[10] Returning to the example in Exhibit 9.1, $r = .532$ and $N = 28$. Thus,

$$t = \frac{.532\sqrt{26}}{\sqrt{1 - .532^2}} = \frac{.532\sqrt{26}}{\sqrt{.717}} = 3.202$$

This value of t is significant at $\alpha = .05$ (two-tailed), and we can thus conclude that there is a significant relationship between SAT scores and scores on Katz' test. In other words, we can conclude that differences in SAT are associated with differences in test scores, although this does not necessarily imply a causal association.

In Chapter 7, we saw a brief mention of the F statistic, about which we will have much more to say in Chapters 11 through 16. You should know that any t statistic on d degrees of freedom can be squared to produce an F statistic on 1 and d degrees of freedom. Many statistical packages use the F statistic instead of t to test hypotheses. In this case, you simply take the square root of that F to obtain the t statistics we are discussing here. (From Exhibit 9.1, we find an F of 10.251. The square root of this is 3.202, which agrees with the t we have just computed for this test.)

As a second example, if we go back to our data on stress and psychological symptoms in Table 9.1, and the accompanying text, we find $r = .506$ and $N = 107$. Thus,

$$t = \frac{.506\sqrt{105}}{\sqrt{1 - .506^2}} = \frac{.506\sqrt{105}}{\sqrt{.744}} = 6.011$$

Here again we will reject $H_0: \rho = 0$. We will conclude that there is a significant relationship between stress and symptoms. Differences in stress are associated with differences in reported psychological symptoms.

The fact that we have a hypothesis test for the correlation coefficient does not mean that the test is always wise. There are many situations where statistical significance, although perhaps comforting, is not particularly meaningful. If I have established a scale that purports to predict academic success, but it correlates only $r = .25$ with success, that test is not going to be very useful to me. It matters not whether $r = .25$ is statistically significantly different from .00, it explains so little of the variation that it is unlikely to be of any use. And anyone who is excited because a test-retest reliability coefficient is statistically significant hasn't really thought about what they are doing.

Testing the Significance of *b*

If you think about the problem for a moment, you will realize that a test on b is equivalent to a test on r in the one-predictor case we are discussing in this chapter. If it is true that X and Y are related, then it must also be true that Y varies with X—that is, that the slope is nonzero. This suggests that a test on b will produce the same answer as a test on r, and we could dispense with a test for b altogether. However, because regression coefficients play an important role in multiple regression, and because in multiple regression a significant correlation does not necessarily imply a significant slope for each predictor variable, the exact form of the test will be given here.

We will represent the parametric equivalent of b (the slope we would compute if we had X and Y measures on the whole population) as b^*.[11]

---

[10] This is the same Student's t that we saw in Chapter 7.

[11] Many textbooks use β instead of b^*, but that would lead to confusion with the standardized regression coefficient.

It can be shown that b is normally distributed about b^* with a standard error approximated by[12]

$$s_b = \frac{s_{Y \cdot X}}{s_X \sqrt{N-1}}$$

Thus, if we want to test the hypothesis that the true slope of the regression line in the population is zero ($H_0: b^* = 0$), we can simply form the ratio

$$t = \frac{b - b^*}{s_b} = \frac{b}{\dfrac{s_{Y \cdot X}}{s_X \sqrt{N-1}}} = \frac{(b)(s_X)(\sqrt{N-1})}{s_{Y \cdot X}}$$

which is distributed as t on $N - 2$ df.

For our sample data on SAT performance and test-taking ability, $b = 4.865$, $s_X = 6.73$, and $s_{Y \cdot X} = 53.127$. Thus

$$t = \frac{(4.865)(6.73)(\sqrt{27})}{53.127} = 3.202$$

which is the same answer we obtained when we tested r. Because $t_{obt} = 3.202$ and $t_{.025}(26) = 2.056$, we will reject H_0 and conclude that our regression line has a nonzero slope. In other words, higher levels of test-taking skills are associated with higher predicted SAT scores.

From what we know about the sampling distribution of b, it is possible to set up confidence limits on b^*,

$$CI(b^*) = b \pm (t_{\alpha/2}) \left[\frac{(s_{Y \cdot X})}{s_X \sqrt{N-1}} \right]$$

where $t_{\alpha/2}$ is the two-tailed critical value of t on $N - 2$ df.

For our data, the 95% confidence limits are

$$CI(b^*) = 4.865 \pm 2.056 \left[\frac{53.127}{6.73 \sqrt{27}} \right]$$
$$= 4.865 \pm 3.123 = 1.742 \leq b^* \leq 7.988$$

Thus, the chances are 95 out of 100 that the limits constructed in this way will encompass the true value of b^*. Note that the confidence limits do not include zero. This is in line with the results of our t test, which rejected $H_0: b^* = 0$.

Testing the Difference Between Two Independent *bs*

This test is less common than the test on a single slope, but the question that it is designed to ask is often a very meaningful question. Suppose we have two sets of data on the relationship between the amount that a person smokes and life expectancy. One set is made up of females and the other of males. We have two separate data sets rather than one large one because we do not want our results to be contaminated by normal differences in life expectancy between

---

[12] There is surprising disagreement concerning the best approximation for the standard error of b. Its denominator is variously given as $s_x \sqrt{N}$, $s_x \sqrt{N-1}$, $s_x \sqrt{N-2}$.

males and females. Suppose further that we obtained the following data:

	Males	Females
b	-0.40	-0.20
$s_{Y \cdot X}$	2.10	2.30
s_X^2	2.50	2.80
N	101	101

It is apparent that for our data the regression line for males is steeper than the regression line for females. If this difference is significant, it means that males decrease their life expectancy more than do females for any given increment in the amount they smoke. If this were true, it would be an important finding, and we are therefore interested in testing the difference between b_1 and b_2.

The t test for differences between two independent regression coefficients is directly analogous to the test of the difference between two independent means. If H_0 is true ($H_0 : b_1^* = b_2^*$), the sampling distribution of $b_1 - b_2$ is normal with a mean of zero and a standard error of

$$s_{b_1 - b_2} = \sqrt{s_{b_1}^2 + s_{b_2}^2}$$

This means that the ratio

$$t = \frac{b_1 - b_2}{\sqrt{s_{b_1}^2 + s_{b_2}^2}}$$

is distributed as t on $N_1 + N_2 - 4$ df. We already know that the standard error of b can be estimated by

$$s_b = \frac{s_{Y \cdot X}}{s_X \sqrt{N - 1}}$$

and therefore can write

$$s_{b_1 - b_2} = \sqrt{\frac{s_{Y \cdot X_1}^2}{s_{X_1}^2 (N_1 - 1)} + \frac{s_{Y \cdot X_2}^2}{s_{X_2}^2 (N_2 - 1)}}$$

where $s_{Y \cdot X_1}^2$ and $s_{Y \cdot X_2}^2$ are the error variances for the two samples. As was the case with means, if we assume homogeneity of error variances, we can pool these two estimates, weighting each by its degrees of freedom:

$$s_{Y \cdot X}^2 = \frac{(N_1 - 2)s_{Y \cdot X_1}^2 + (N_2 - 2)s_{Y \cdot X_2}^2}{N_1 + N_2 - 4}$$

For our data,

$$s_{Y \cdot X}^2 = \frac{99(2.10^2) + 99(2.30^2)}{101 + 101 - 4} = 4.85$$

Substituting this pooled estimate into the equation, we obtain

$$s_{b_1 - b_2} = \sqrt{\frac{s_{Y \cdot X_1}^2}{s_{X_1}^2 (N_1 - 1)} + \frac{s_{Y \cdot X_2}^2}{s_{X_2}^2 (N_2 - 1)}}$$

$$= \sqrt{\frac{4.85}{(2.5)(100)} + \frac{4.85}{(2.8)(100)}} = 0.192$$

Given $s_{b_1-b_2}$, we can now solve for t:

$$t = \frac{b_1 - b_2}{s_{b_1-b_2}} = \frac{(-0.40) - (-0.20)}{0.192} = -1.04$$

on 198 *df*. Because $t_{.025}(198) = \pm 1.97$, we would fail to reject H_0 and would therefore conclude that we have no reason to doubt that life expectancy decreases as a function of smoking at the same rate for males as for females.

It is worth noting that although $H_0: b^* = 0$ is equivalent to $H_0: \rho = 0$, it does not follow that $H_0: b_1^* - b_2^* = 0$ is equivalent to $H_0: \rho_1 - \rho_2 = 0$. If you think about it for a moment, it should be apparent that two scatter diagrams could have the same regression line ($b_1^* = b_2^*$) but different degrees of scatter around that line, hence, $\rho_1 \neq \rho_2$. The reverse also holds—two different regression lines could fit their respective sets of data equally well.

Testing the Difference Between Two Independent *r*s

When we test the difference between two independent *r*s, a minor difficulty arises. When $\rho \neq 0$, the sampling distribution of *r* is not approximately normal (it becomes more and more skewed as $\rho \Rightarrow \pm 1.00$), and its standard error is not easily estimated. The same holds for the difference $r_1 - r_2$. This raises an obvious problem because, as you can imagine, we will need to know the standard error of a difference between correlations if we are to create a *t* test on that difference. Fortunately, R. A. Fisher provided the solution.

Fisher (1921) showed that if we transform *r* to

$$r' = (0.5) \log_e \left| \frac{1 + r}{1 - r} \right|$$

then r' is approximately normally distributed around ρ' (the transformed value of ρ) with standard error

$$s_{r'} = \frac{1}{\sqrt{N - 3}}$$

(Fisher labeled his statistic "*z*," but "*r'*" is often used to avoid confusion with the standard normal deviate.) Because we know the standard error, we can now test the null hypothesis that $\rho_1 - \rho_2 = 0$ by converting each *r* to r' and solving for

$$z = \frac{r_1' - r_2'}{\sqrt{\dfrac{1}{N_1 - 3} + \dfrac{1}{N_2 - 3}}}$$

Note that our test statistic is *z* rather than *t* because our standard error does not rely on statistics computed from the sample (other than *N*) and is therefore a parameter.

Appendix r' tabulates the values of r' for different values of *r*, which eliminates the need to solve the equation for r'.

To take a simple example, assume that for a sample of 53 males, the correlation between number of packs of cigarettes smoked per day and life expectancy was .50. For a sample of 43 females, the correlation was .40. (These are unrealistically high values for *r*, but they better illustrate the effects of the transformation.) The question of interest is, Are these two coefficients significantly different, or are the differences in line with what we would expect when sampling from the same bivariate population of *X*, *Y* pairs?

	Males	Females
r	.50	.40
r'	.549	.424
N	53	53

$$z = \frac{.549 - .424}{\sqrt{\dfrac{1}{53 - 3} + \dfrac{1}{53 - 3}}} = \frac{.125}{\sqrt{\dfrac{2}{50}}} = \frac{.125}{\dfrac{1}{5}} = 0.625$$

Because $z_{obt} = 0.625$ is less than $z_{.025} = \pm 1.96$, we fail to reject H_0 and conclude that with a two-tailed test at $\alpha = .05$, we have no reason to doubt that the correlation between smoking and life expectancy is the same for males as it is for females.

Testing the Hypothesis That ρ Equals Any Specified Value

Now that we have discussed the concept of r', we are in a position to test the null hypothesis that ρ is equal to any value, not just to zero. You probably can't think of many situations in which you would like to do that, and neither can I. But the ability to do so allows us to establish confidence limits on ρ, a more interesting procedure.

As we have seen, for any value of ρ, the sampling distribution of r' is approximately normally distributed around ρ' (the transformed value of ρ) with a standard error of $\frac{1}{\sqrt{N-3}}$. From this it follows that

$$z = \frac{r' - \rho'}{\sqrt{\dfrac{1}{N-3}}}$$

is a standard normal deviate. Thus, if we want to test the null hypothesis that a sample r of .30 (with $N = 103$) came from a population where $\rho = .50$, we proceed as follows:

$$r = .30 \qquad r' = .310$$
$$\rho = .50 \qquad \rho' = .549$$
$$N = 103 \qquad s_{r'} = 1/\sqrt{N-3} = 0.10$$

$$z = \frac{.310 - .549}{0.10} = -0.239/0.10 = -2.39$$

Because $z_{obt} = -2.39$ is more extreme than $z_{.025} = \pm 1.96$, we reject H_0 at $\alpha = .05$ (two-tailed) and conclude that our sample did not come from a population where $\rho = .50$.

Confidence Limits on ρ

We can easily establish confidence limits on ρ by solving the previous equation for ρ instead of z. To do this, we first solve for confidence limits on ρ', and then convert ρ' to ρ.

$$z = \frac{r' - \rho'}{\sqrt{\dfrac{1}{N-3}}}$$

therefore

$$\sqrt{\dfrac{1}{N-3}}(\pm z) = r' - \rho'$$

and thus

$$CI(\rho') = r' \pm z_{\alpha/2}\sqrt{\frac{1}{N-3}}$$

For our stress example, $r = .506$ ($r' = .557$) and $N = 107$, so the 95% confidence limits are

$$CI(\rho') = .557 \pm 1.96\sqrt{\frac{1}{104}}$$
$$= .557 \pm 1.96(0.098) = .557 \pm 0.192$$
$$= .365 \le \rho' \le .749$$

Converting from ρ' back to ρ and rounding,

$$.350 \le \rho \le .635$$

Thus, the limits are $\rho = .350$ and $\rho = .635$. The probability is .95 that limits obtained in this way encompass the true value of ρ. Note that $\rho = 0$ is not included within our limits, thus offering a simultaneous test of $H_0: \rho = 0$, should we be interested in that information.

Confidence Limits versus Tests of Significance

At least in the behavioral sciences, most textbooks, courses, and published research have focused on tests of significance, and paid scant attention to confidence limits. In some cases that really is probably appropriate, but in other cases, it leaves the reader short.

In this chapter, we have repeatedly referred to an example on stress and psychological symptoms. For the first few people who investigated this issue, it really was an important question whether there was a significant relationship between these two variables. But now that everyone believes it, a more appropriate question becomes how large the relationship is. And for that question, a suitable answer is provided by a statement such as the correlation between the two variables was .506, with a 95% confidence interval of $.350 \le \rho \le .635$. (A comparable statement from the public opinion polling field would be something like $r = .506$ with a 95% margin of error of $\pm.15$(approx.).[13]

Testing the Difference Between Two Nonindependent *r*s

Occasionally, we come across a situation in which we want to test the difference between two correlations that are not independent. (I am probably asked this question a couple of times per year.) One case arises when two correlations share one variable in common. We will see such an example later. Another case arises when we correlate two variables at Time 1 and then again at some later point (Time 2), and we want to ask whether there has been a significant change in the correlation over time. I will not cover that case, but a very good discussion of that particular issue can be found at core.ecu.edu/psyc/wuenschk/StatHelp/ZPF.doc and in a paper by Raghunathan, Rosenthal, and Rubin (1996).

As an example of correlations that share a common variable, Reilly, Drudge, Rosen, Loew, and Fischer (1985) administered two intelligence tests (the WISC-R and the McCarthy) to first-grade children, and then administered the Wide Range Achievement Test (WRAT) to

---

[13] I had to insert the label "approx." here because the limits, as we saw earlier, are not exactly symmetrical around r.

those same children 2 years later. Reilly et al. obtained, among other findings, the following correlations:

	WRAT	WISC-R	McCarthy
WRAT	1.00	.80	.72
WISC-R		1.00	.89
McCarthy			1.00

Note that the WISC-R and the McCarthy are highly correlated but that the WISC-R correlates somewhat more highly with the WRAT (reading) than does the McCarthy. It is of interest to ask whether this difference between the WISC-R–WRAT correlation (.80) and the McCarthy–WRAT correlation (.72) is significant, but to answer that question requires a test on nonindependent correlations because they both have the WRAT in common and they are based on the same sample.

When we have two correlations that are not independent—as these are not, because the tests were based on the same 26 children—we must take into account this lack of independence. Specifically, we must incorporate a term representing the degree to which the two tests are themselves correlated. Hotelling (1931) proposed the traditional solution, but a better test was devised by Williams (1959) and endorsed by Steiger (1980). This latter test takes the form

$$t = (r_{12} - r_{13}) \sqrt{\dfrac{(N-1)(1+r_{23})}{2\left(\dfrac{N-1}{N-3}\right)|R| + \dfrac{(r_{12}+r_{13})^2}{4}(1-r_{23})^3}}$$

where

$$|R| = \left(1 - r_{12}^2 - r_{13}^2 - r_{23}^2\right) + (2r_{12}r_{13}r_{23})$$

This ratio is distributed as t on $N - 3$ df. In this equation, r_{12} and r_{13} refer to the correlation coefficients whose difference is to be tested, and r_{23} refers to the correlation between the two predictors. $|R|$ is the determinant of the 3×3 matrix of intercorrelations, but you can calculate it as shown without knowing anything about determinants.

For our example, let

r_{12} = correlation between the WISC-R and the WRAT = .80

r_{13} = correlation between the McCarthy and the WRAT = .72

r_{23} = correlation between the WISC-R and the McCarthy = .89

$N = 26$

then

$$|R| = (1 - .80^2 - .72^2 - .89^2) + (2)(.80)(.72)(.89) = .075$$

$$t = (.80 - .72) \sqrt{\dfrac{(25)(1 + .89)}{2\left(\dfrac{25}{23}\right)(.075) + \dfrac{(.80 + .72)^2}{4}(1 - .89)^3}}$$

$$= 1.36$$

A value of $t_{obt} = 1.36$ on 23 df is not significant. Although this does not prove the argument that the tests are equally effective in predicting third-grade children's performance on the reading scale of the WRAT, because you cannot prove the null hypothesis, it is consistent with that argument and thus supports it.

9.11   The Role of Assumptions in Correlation and Regression

There is considerable confusion in the literature concerning the assumptions underlying the use of correlation and regression techniques. Much of the confusion stems from the fact that the correlation and regression models, although they lead to many of the same results, are based on different assumptions. Confusion also arises because statisticians tend to make all their assumptions at the beginning and fail to point out that some of these assumptions are not required for certain purposes.

linearity of regression

curvilinear

The major assumption that underlies both the linear-regression and bivariate-normal models and all our interpretations is that of **linearity of regression.** We assume that whatever the relationship between X and Y, it is a linear one—meaning that the line that best fits the data is a straight one. We will later refer to measures of **curvilinear** (nonlinear) relationships, but standard discussions of correlation and regression assume linearity unless otherwise stated. (We do occasionally fit straight lines to curvilinear data, but we do so on the assumption that the line will be sufficiently accurate for our purpose—although the standard error of prediction might be poorly estimated. There are other forms of regression besides linear regression, but we will not discuss them here.)

As mentioned earlier, whether or not we make various assumptions depends on what we want to do. If our purpose is simply to describe data, no assumptions are necessary. The regression line and r best describe the data at hand, without the necessity of any assumptions about the population from which the data were sampled.

If our purpose is to assess the degree to which variance in Y is linearly attributable to variance in X, we again need make no assumptions. This is true because s_Y^2 and $s_{Y \cdot X}^2$ are both unbiased estimators of their corresponding parameters, independent of any underlying assumptions, and

$$\frac{SS_Y - SS_{\text{residual}}}{SS_Y}$$

is algebraically equivalent to r^2.

If we want to set confidence limits on b or Y, or if we want to test hypotheses about b^*, we will need to make the conditional assumptions of homogeneity of variance and normality in arrays of Y. The assumption of homogeneity of variance is necessary to ensure that $s_{Y \cdot X}^2$ is representative of the variance of each array, and the assumption of normality is necessary because we use the standard normal distribution.

If we want to use r to test the hypothesis that $\rho = 0$, or if we want to establish confidence limits on ρ, we will have to assume that the (X, Y) pairs are a random sample from a bivariate-normal distribution.

9.12   Factors That Affect the Correlation

The correlation coefficient can be substantially affected by characteristics of the sample. Two such characteristics are the restriction of the range (or variance) of X or Y and the use of heterogeneous subsamples.

The Effect of Range Restrictions

range restrictions

A common problem concerns restrictions on the range over which X and Y vary. The effect of such **range restrictions** is to alter the correlation between X and Y from what it would have been if the range had not been so restricted. Depending on the nature of the data, the correlation may either rise or fall as a result of such restriction, although most commonly r is reduced.

With the exception of very unusual circumstances, restricting the range of X will increase r only when the restriction results in eliminating some curvilinear relationship. For example, if we correlated reading ability with age, where age ran from 0 to 70 years, the data would be decidedly curvilinear (flat to about age 4, rising to about 17 years of age and then leveling off) and the correlation, which measures *linear* relationships, would be relatively low. If, however, we restricted the range of ages to 5 to 17 years, the correlation would be quite high because we would have eliminated those values of Y that were not varying linearly as a function of X.

The more usual effect of restricting the range of X or Y is to reduce the correlation. This problem is especially pertinent in the area of test construction because here criterion measures (Y) may be available for only the higher values of X. Consider the hypothetical data in Figure 9.4. This figure represents the relation between college GPAs and scores on some standard achievement test (such as the SAT) for a hypothetical sample of students. In the ideal world of the test constructor, all people who took the exam would then be sent on to college and earn a GPA, and the correlation between achievement test scores and GPAs would be computed. As can be seen from Figure 9.4, this correlation would be reasonably high. In the real world, however, not everyone is admitted to college. Colleges take only the more able students, whether this classification is based on achievement test scores, high-school performance, or whatever. This means that GPAs are available mainly for students who had relatively high scores on the standardized test. Suppose that this has the effect of allowing us to evaluate the relationship between X and Y for only those values of X that are greater than 400. For the data in Figure 9.4, the correlation will be relatively low, not because the test is worthless, but because the range has been restricted. In other words, when we use the entire sample of points in Figure 9.4, the correlation is .65. However, when we restrict the sample to those students having test scores of at least 400, the correlation drops to only .43. (This is easier to see if you cover all data points for $X < 400$.)

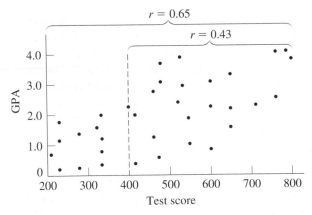

Figure 9.4 Hypothetical data illustrating the effect of restricted range

We must consider the effect of range restrictions whenever we see a correlation coefficient based on a restricted sample. The coefficient might be inappropriate for the question at hand. Essentially, what we have done is to ask how well a standardized test predicts a person's suitability for college, but we have answered that question by referring only to those people who were actually admitted to college.

The Effect of Heterogeneous Subsamples

heterogeneous subsamples

Another important consideration in evaluating the results of correlational analyses deals with heterogeneous subsamples. This point can be illustrated with a simple example involving the relationship between height and weight in male and female subjects. These variables may appear to have little to do with psychology, but considering the important role both variables play in the development of people's images of themselves, the example is not as far afield as you might expect. The data plotted in Figure 9.5, using Minitab, come from sample data from the Minitab manual (Ryan, Joiner, & Ryan, 1985). These are actual data from 92 college students who were asked to report height, weight, gender, and several other variables. (Keep in mind that these are self-report data, and there may be systematic reporting biases.)

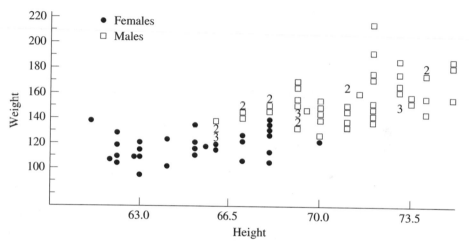

Figure 9.5   Relationship between height and weight for males and females combined

When we combine the data from both males and females, the relationship is strikingly good, with a correlation of .78. When you look at the data from the two genders separately, however, the correlations fall to .60 for males and .49 for females. (Males and females have been plotted using different symbols, with data from females primarily in the lower left.) The important point is that the high correlation we found when we combined genders is not due purely to the relation between height and weight. It is also due largely to the fact that men are, on average, taller and heavier than women. In fact, a little doodling on a sheet of paper will show that you could create artificial, and improbable, data where within each gender, weight is negatively related to height, whereas the relationship is positive when you collapse across gender. The point I am making here is that experimenters must be careful when they combine data from several sources. The relationship between two variables may be obscured or enhanced by the presence of a third variable. Such a finding is important in its own right.

A second example of heterogeneous subsamples that makes a similar point is the relationship between cholesterol consumption and cardiovascular disease in men and women. If you collapse across both genders, the relationship is not impressive. But when you separate the data by male and female, there is a distinct trend for cardiovascular disease to increase with increased cholesterol level. This relationship is obscured in the combined data because men, regardless of cholesterol level, have an elevated level of cardiovascular disease compared with women.

9.13 Power Calculation for Pearson's *r*

Consider the problem of the individual who wants to demonstrate a relationship between television violence and aggressive behavior. Assume that he has surmounted all the very real problems associated with designing this study and has devised a way to obtain a correlation between the two variables. He believes that the correlation coefficient in the population (ρ) is approximately .30. (This correlation may seem small, but it is impressive when you consider all the variables involved in aggressive behavior. This value is in line with the correlation obtained in a study by Eron, Huesmann, Lefkowitz, and Walden, 1972.) Our experimenter wants to conduct a study to find such a correlation but wants to know something about the power of his study before proceeding. Power calculations are easy to make in this situation.

As you should recall, when we calculate power we first define an effect size (d). We then introduce the sample size and compute δ, and finally we use δ to compute the power of our design from Appendix Power.

We begin by defining

$$d = \rho_1' - \rho_0' = \rho_1' - 0 = \rho_1'$$

where ρ_1' is Fisher's transformation of the correlation in the population defined by H_1—in this case, .31. We next define

$$\delta = d\sqrt{N-1} = \rho_1'\sqrt{N-1}$$

For a sample of size 50,

$$\delta = .31\sqrt{50-1} = 2.2$$

From Appendix Power, for $\delta = 2.2$ and $\alpha = .05$ (two-tailed), power $= .60$.

A power coefficient of .60 does not please the experimenter, so he casts around for a way to increase power. He wants power $= .80$. From Appendix Power, we see that this will require $\delta = 2.8$. Therefore,

$$\delta = \rho_1'\sqrt{N-1}$$
$$2.8 = .31\sqrt{N-1}$$

Squaring both sides,

$$2.8^2 = .31^2(N-1)$$
$$\left(\frac{2.8}{.31}\right)^2 + 1 = N = 82$$

Thus, to obtain power $= .80$, the experimenter will have to collect data on over 80 participants.

Key Terms

Relationships (Introduction)

Differences (Introduction)

Correlation (Introduction)

Regression (Introduction)

Random variable (Introduction)

Fixed variable (Introduction)

Linear-regression models (Introduction)

Bivariate-normal models (Introduction)

Prediction (Introduction)

Scatterplot (9.1)

Scatter diagram (9.1)

Scattergram (9.1)

Predictor (9.1)

Criterion (9.1)

Regression lines (9.1)

Correlation (*r*) (9.1)

Covariance (cov_{XY} or s_{XY}) (9.3)

Correlation coefficient in the population ρ (rho) (9.4)

Adjusted correlation coefficient (r_{adj}) (9.4)

Slope (9.5)

Intercept (9.5)

Errors of prediction (9.5)

Residual (9.5)

Normal equations (9.5)

Standardized regression coefficient β (beta) (9.5)

Sum of squares of Y (SS_Y) (9.6)

Standard error of estimate (9.6)

Residual variance (9.6)

Error variance (9.6)

Conditional distribution (9.6)

Proportional reduction in error (PRE) (9.6)

Proportional improvement in prediction (PIP) (9.6)

Array (9.7)

Homogeneity of variance in arrays (9.7)

Normality in arrays (9.7)

Conditional array (9.7)

Conditional distributions (9.7)

Marginal distribution (9.7)

Linearity of regression (9.11)

Curvilinear (9.11)

Range restrictions (9.12)

Heterogeneous subsamples (9.12)

Exercises

9.1 The State of Vermont is divided into 10 Health Planning Districts, which correspond roughly to counties. The following data for 1980 represent the percentage of births of babies under 2,500 grams (Y), the fertility rate for females younger than 18 or older than 34 years of age (X_1), and the percentage of births to unmarried mothers (X_2) for each district.[14]

District	Y	X_1	X_2
1	6.1	43.0	9.2
2	7.1	55.3	12.0
3	7.4	48.5	10.4
4	6.3	38.8	9.8
5	6.5	46.2	9.8
6	5.7	39.9	7.7
7	6.6	43.1	10.9
8	8.1	48.5	9.5
9	6.3	40.0	11.6
10	6.9	56.7	11.6

a. Make a scatter diagram of Y and X_1.

b. Draw on your scatter diagram (by eye) the line that appears to best fit the data.

9.2 Calculate the correlation between Y and X_1 in Exercise 9.1.

9.3 Calculate the correlation between Y and X_2 in Exercise 9.1.

9.4 Use a t test to test H_0: $\rho = 0$ for the answers to Exercises 9.2 and 9.3.

9.5 Draw scatter diagrams for the following sets of data. Note that the same values of X and Y are involved in each set.

1		2		3	
X	Y	X	Y	X	Y
2	2	2	4	2	8
3	4	3	2	3	6
5	6	5	8	5	4
6	8	6	6	6	2

---

[14] Both X_1 and X_2 are known to be risk factors for low birthweight.

9.6 Calculate the covariance for each set in Exercise 9.5.

9.7 Calculate the correlation for each data set in Exercise 9.5. How can the values of Y in Exercise 9.5 be rearranged to produce the smallest possible positive correlation?

9.8 Assume that a set of data contains a slightly curvilinear relationship between X and Y (the best-fitting line is slightly curved). Would it ever be appropriate to calculate r on these data?

9.9 An important developmental question concerns the relationship between severity of cerebral hemorrhage in low-birthweight infants and cognitive deficit in the same children at age 5 years.

    a. Suppose we expect a correlation of .20 and are planning to use 25 infants. How much power does this study have?

    b. How many infants would be required for power to be .80?

9.10 From the data in Exercise 9.1, compute the regression equation for predicting the percentage of births of infants under 2,500 grams (Y) on the basis of fertility rate for females younger than 18 or older than 34 years of age (X_1). (X_1 is known as the "high-risk fertility rate.")

9.11 Calculate the standard error of estimate for the regression equation from Exercise 9.10.

9.12 Calculate confidence limits on b^* for Exercise 9.10.

9.13 If as a result of ongoing changes in the role of women in society, the age at which women tend to bear children rose such that the high-risk fertility rate defined in Exercise 9.10 jumped to 70, what would you predict for incidence of babies with birthweights less than 2,500 grams?

9.14 Should you feel uncomfortable making a prediction if the rate in Exercise 9.13 were 70? Why or why not?

9.15 Using the information in Table 9.2 and the computed coefficients, predict the number of symptoms for a stress score of 8.

9.16 The mean stress score for the data in Table 9.2 was 21.467. What would your prediction for symptoms be for someone who had that stress score? How does this compare with $\overline{Y}$?

9.17 Calculate an equation for the 95% confidence interval in $\hat{Y}$ for predicting psychological symptoms—you can overlay the confidence limits on Figure 9.2.

9.18 Within a group of 200 faculty members who have been at a well-known university for less than 15 years (i.e., since before the salary curve levels off) the equation relating salary (in thousands of dollars) to years of service is $\hat{Y} = 0.9X + 15$. For 100 administrative staff at the same university, the equation is $\hat{Y} = 1.5X + 10$. Assuming that all differences are significant, interpret these equations. How many years must pass before an administrator and a faculty member earn roughly the same salary?

9.19 In 1886, Sir Francis Galton, an English scientist, spoke about "regression toward mediocrity," which we more charitably refer to today as regression toward the mean. The basic principle is that those people at the ends of any continuum (e.g., height, IQ, or musical ability) tend to have children who are closer to the mean than they are. Use the concept of r as the regression coefficient (slope) with standardized data to explain Galton's idea.

9.20 You want to demonstrate a relationship between the amount of money school districts spend on education and the performance of students on a standardized test such as the SAT. You are interested in finding such a correlation only if the true correlation is at least .40. What are your chances of finding a significant sample correlation if you have 30 school districts?

9.21 In Exercise 9.20, how many districts would you need for power = .80?

9.22 Guber (1999) actually assembled the data to address the basic question referred to in Exercises 9.20 and 9.21. She obtained the data for all 50 states on several variables associated with school performance, including expenditures for education, SAT performance, percentage of students taking the SAT, and other variables. We will look more extensively at these data later, but the following table contains the Minitab and SPSS computer printouts for Guber's data.

SPSS

Model Summary[b]

Model	R	R Square	Adjusted R Square	Std. Error of the Estimate
1	.453[a]	.205	.188	65.49

[a] Predictors: (Constant), Current expenditure per pupil—1994–95
[b] Dependent Variable: Average combined SAT 1994–95

ANOVA[b]

Model		Sum of Squares	df	Mean Square	F	Sig.
1	Regression	50920.767	1	50920.767	11.872	.001[a]
	Residual	197303.0	46	4289.197		
	Total	248223.8	47			

[a] Predictors: (Constant), Current expenditure per pupil—1994–95
[b] Dependent Variable: Average combined SAT 1994–95

Coefficients[a]

Model		Unstandardized Coefficients		Standardized Coefficients	t	Sig.
		B	Std. Error	Beta		
1	(Constant)	1112.769	42.341		26.281	.000
	Current expenditure per pupil—1994–95	−23.918	6.942	−.453	−3.446	.001

[a] Dependent Variable: Average combined SAT 1994–1995

Minitab

Regression Analysis

The regression equation is
Combined = 1113 − 23.9 Expend

48 cases used 2 cases contain missing values

Predictor	Coef	StDev	T	P
Constant	1112.77	42.34	26.28	0.000
Expend	−23.918	6.942	−3.45	0.001

S = 65.49   R − Sq = 20.5%   R − Sq(adj) = 18.8%

Analysis of Variance

Source	DF	SS	MS	F	P
Regression	1	50921	50921	11.87	0.001
Residual Error	46	197303	4289		
Total	47	248224			

Unusual Observations

Obs	Expend	Combined	Fit	StDev Fit	Residual	St Resid
33	5.19	854.00	988.56	10.80	−134.56	−2.08 R
40	9.77	898.00	879.00	28.21	19.00	0.32 X
41	9.62	892.00	882.61	27.22	9.39	0.16 X

R denotes an observation with a large standardized residual
X denotes an observation whose X value gives it large influence.

a. What information do you find on the Minitab output that is not on the SPSS output?

b. What information do you find on the SPSS output that is not on the Minitab output? (*Note:* You can ignore the reference to "multiple" regression; with one predictor, multiple regression and simple regression are the same thing.)

c. These data do not really reveal the pattern that we would expect. What do they show? (In Chapter 15, we will see that the expected pattern actually is there if we control for other variables.)

9.23 In the study by Katz et al. (1990) used in this chapter and in Exercises 7.13 and 7.29, we saw that students who were answering reading comprehension questions on the SAT without first reading the passages performed at better-than-chance levels. This does not necessarily mean that the SAT is not a useful test. Katz et al. went on to calculate the correlation between the actual SAT Verbal scores on their participants' admissions applications and performance on the 100-item test. For those participants who had read the passage, the correlation was .68 ($N = 17$). For those who had not read the passage, the correlation was .53 ($N = 28$), as we have seen.

a. Were these correlations significantly different?

b. What would you conclude from these data?

9.24 Katz et al. replicated their experiment using subjects whose SAT Verbal scores showed considerably more within-group variance than had those in the first study. In this case, the correlation for the group that read the passage was .88 ($N = 52$), whereas for the nonreading group it was .72 ($N = 74$). Were these correlations significantly different?

9.25 What conclusions can you draw from the difference between the correlations in Exercises 9.23 and 9.24?

9.26 Make up your own example along the lines of the "smoking versus life expectancy" example given on pages 247–248 to illustrate the relationship between r^2 and accountable variation.

9.27 Moore and McCabe (1989) found some interesting data on the consumption of alcohol and tobacco that illustrate an important statistical concept. Their data, taken from the Family Expenditure Survey of the British Department of Employment, follow. The dependent variables are the average weekly household expenditures on the product.

Region	Alcohol	Tobacco
North	6.47	4.03
Yorkshire	6.13	3.76
Northeast	6.19	3.77
East Midlands	4.89	3.34
West Midlands	5.63	3.47
East Anglia	4.52	2.92
Southeast	5.89	3.20
Southwest	4.79	2.71
Wales	5.27	3.53
Scotland	6.08	4.51
Northern Ireland	4.02	4.56

a. What is the relationship between these two variables?

b. Popular stereotypes have the Irish as heavy drinkers. Do the data support that belief?

c. What effect does the inclusion of Northern Ireland have on our results? (A scatterplot would be helpful.)

9.28 Using the data from Mireault (1990) in the file Mireault.dat (www.uvm.edu/~dhowell/methods/), is there a relationship between how well a student performs in college (as assessed by GPA) and that student's psychological symptoms (as assessed by GSIT)?

9.29 Using the data referred to in Exercise 9.28,

a. Calculate the correlations among all of the Brief Symptom Inventory subscales. (*Hint:* Virtually all statistical programs are able to calculate these correlations in one statement. You don't have to calculate each one individually.)

b. What does the answer to (a) tell us about the relationships among the separate scales?

9.30 One of the assumptions lying behind our use of regression is the assumption of homogeneity of variance in arrays. One way to examine the data for violations of this assumption is to calculate predicted values of Y and the corresponding residuals $(Y - \hat{Y})$. If you plot the residuals against the predicted values, you should see a more or less random collection of points. The vertical dispersion should not increase or decrease systematically as you move from right to left, nor should there be any other apparent pattern. Create the scatterplot for the data from Cancer.dat. Most computer packages let you request this plot. If not, you can easily generate the appropriate variables by first determining the regression equation and then feeding that equation back into the program in a "compute statement" (e.g., "set Pred = 0.256 * GSIT + 4.65," and "set Resid = TotBPT − Pred").

9.31 The following data represent the actual heights and weights referred to earlier for male college students.

a. Make a scatterplot of the data.

b. Calculate the regression equation of weight predicted from height for these data. Interpret the slope and the intercept.

Height	Weight	Height	Weight
70	150	73	170
67	140	74	180
72	180	66	135
75	190	71	170
68	145	70	157
69	150	70	130
71.5	164	75	185
71	140	74	190
72	142	71	155
69	136	69	170
67	123	70	155
68	155	72	215
66	140	67	150
72	145	69	145
73.5	160	73	155
73	190	73	155
69	155	71	150
73	165	68	155
72	150	69.5	150
74	190	73	180
72	195	75	160
71	138	66	135
74	160	69	160
72	155	66	130
70	153	73	155
67	145	68	150
71	170	74	148
72	175	73.5	155
69	175		

c. What is the correlation coefficient for these data?

d. Are the correlation coefficient and the slope significantly different from zero?

9.32 The following data are the actual heights and weights, referred to in this chapter, of female college students.

    a. Make a scatterplot of the data.

    b. Calculate the regression coefficients for these data. Interpret the slope and the intercept.

    c. What is the correlation coefficient for these data? Is the slope significantly different from zero?

Height	Weight	Height	Weight
61	140	65	135
66	120	66	125
68	130	65	118
68	138	65	122
63	121	65	115
70	125	64	102
68	116	67	115
69	145	69	150
69	150	68	110
67	150	63	116
68	125	62	108
66	130	63	95
65.5	120	64	125
66	130	68	133
62	131	62	110
62	120	61.75	108
63	118	62.75	112
67	125		

9.33 Using your own height and the appropriate regression equation from Exercise 9.31 or 9.32, predict your own weight. (If you are uncomfortable reporting your own weight, predict mine—I am 5'8" and weigh 146 pounds.)

    a. How much is your actual weight greater than or less than your predicted weight? (You have just calculated a residual.)

    b. What effect will biased reporting on the part of the students who produced the data play in your prediction of your own weight?

9.34 Use your scatterplot of the data for students of your own gender and observe the size of the residuals. (*Hint:* You can see the residuals in the vertical distance of points from the line.) What is the largest residual for your scatterplot?

9.35 Given a male and a female student who are both 5'6", how much would they be expected to differ in weight? (*Hint:* Calculate a predicted weight for each of them using the regression equation specific to their gender.)

9.36 The slope (*b*) used to predict the weights of males from their heights is greater than the slope for females. Is this significant, and what would it mean if it were?

9.37 In Chapter 2, I presented data on the speed of deciding whether a briefly presented digit was part of a comparison set and gave data from trials on which the comparison set had contained one, three, or five digits. Eventually, I would like to compare the three conditions (using only the data from trials on which the stimulus digit had in fact been a part of that set), but I worry that the trials are not independent. If the subject (myself) was improving as the task went along, he would do better on later trials, and how he did would in some way be related to the number of the trial. If so, we would not be able to say that the responses were independent. Using only the data from the trials labeled *Y* in the condition in which there were five digits in the comparison set, obtain the regression of response on trial number. Was performance improving significantly over trials? Can we assume that there is no systematic linear trend over time?

Discussion Question

9.38   In a recent email query, someone asked about how they should compare two air pollution monitors that sit side by side and collect data all day. They had the average reading per monitor for each of 50 days and wanted to compare the two monitors; their first thought was to run a *t* test between the means of the readings of the two monitors. This question would apply equally well to psychologists and other behavioral scientists if we simply substitute two measures of Extraversion for two measures of air pollution and collect data using both measures on the same 50 subjects. How would you go about comparing the monitors (or measures)? What kind of results would lead you to conclude that they are measuring equivalently or differently? This is a much more involved question than it might first appear, so don't just say you would run a *t* test or obtain a correlation coefficient. Sample data that might have come from such a study are to be found on the website (www.uvm.edu/~dhowell/methods/) in a file named AirQual.dat in case you want to play with data.

CHAPTER 10

Alternative Correlational Techniques

Objectives

To discuss correlation and regression with regard to dichotomous variables and ranked data and to present measures of association between categorical variables.

Contents

THE PEARSON PRODUCT-MOMENT CORRELATION COEFFICIENT (r) is only one of many available correlation coefficients. It generally applies to those situations in which the relationship between two variables is basically linear, where both variables are measured on a more or less continuous scale, and where some sort of normality and homogeneity of variance assumptions can be made. As this chapter will point out, r can be meaningfully interpreted in other situations as well, although for those cases it is given a different name and many people fail to recognize it for what it actually is.

In this chapter, we will discuss a variety of coefficients that apply to different kinds of data. For example, the data might represent rankings, one or both of the variables might be dichotomous, or the data might be categorical. Depending on the assumptions we are willing to make about the underlying nature of our data, different coefficients will be appropriate in different situations. Some of these coefficients will turn out to be calculated as if they were Pearson rs, and some will not. The important point is that they all represent attempts to obtain some measure of the relationship between two variables and fall under the general heading of *correlation* rather than *regression*.

When we speak of relationships between two variables without any restriction on the nature of these variables, we have to distinguish between **correlational measures** and **measures of association.** When at least some sort of order can be assigned to the levels of each variable, such that higher scores represent more (or less) of some quantity, then it makes sense to speak of correlation. We can speak meaningfully of increases in one variable being associated with increases in another variable. In many situations, however, different levels of a variable do not represent an orderly increase or decrease in some quantity. For example, we could sort people on the basis of their membership in different campus organizations, and then on the basis of their views on some issue. We might then find that there is in fact an association between people's views and their membership in organizations, and yet neither of these variables represents an ordered continuum. In cases such as this, the coefficient we will compute is not a correlation coefficient. We will instead speak of it as a measure of association.

We might be interested in calculating any type of coefficient of correlation for three basic reasons. The most obvious, but not necessarily the most important, reason is to obtain an estimate of ρ, the correlation in the population. Thus, someone interested in the **validity** of a test actually cares about the true correlation between his test and some criterion and approaches the calculation of a coefficient with this purpose in mind. This use is the one for which the alternative techniques are least satisfactory, although they can serve this purpose.

A second use of correlation coefficients occurs with such techniques as multiple regression and factor analysis. In this situation, the coefficient is not in itself an end product; rather, it enters into the calculation of further statistics. For these purposes, many of the coefficients to be discussed are satisfactory.

The final reason for calculating a correlation coefficient is to use its square as a measure of the variation in one variable accountable for by variation in the other variable. This is a measure of effect size (from the r-family of measures) and is often useful as a way of conveying the magnitude of the effect that we found. Here again, the coefficients to be discussed are in many cases satisfactory for this purpose. I will specifically discuss the creation of r-family effect size measures in what follows.

correlational measures

measures of association

validity

10.1 Point-Biserial Correlation and Phi: Pearson Correlations by Another Name

In the previous chapter, I discussed the standard Pearson product-moment correlation coefficient (r) in terms of variables that are relatively continuous on both measures. However, that same formula also applies to a pair of variables that are dichotomous (having two levels) on one or both measures. We may need to be somewhat cautious in our interpretation, and there are some interesting relationships between those correlations and other statistics we have discussed, but the same basic procedure is used for these special cases as we used for the more general case.

Point-Biserial Correlation (r_{pb})

dichotomy

Frequently, variables are measured in the form of a **dichotomy,** such as male-female, pass-fail, experimental group–control group, and so on. Ignoring for the moment that these variables are seldom measured numerically (a minor problem), it is also quite apparent that they are not measured continuously. There is no way we can assume that a continuous distribution, such as the normal distribution, for example, will represent the obtained scores on the dichotomous variable male-female. If we want to use r as a measure of relationship between variables, we have a problem—for r to have certain desirable properties as an estimate of ρ, we need to assume at least an approximation of normality in the joint (bivariate) population of X and Y.

The difficulty regarding the numerical measurement of X turns out to be trivial for dichotomous variables. If X represents married versus unmarried, for example, then we can legitimately score married as 0 and unmarried as 1, or vice versa. (Actually, *any* two values will do. Thus, all married subjects could be given a score of 7 on X, and all unmarried subjects could receive a score of 18, without affecting the correlation in the least. We use 0 and 1, or sometimes 1 and 2, for the simple reason that this makes the arithmetic easier.) Given such a system of quantification, it should be apparent that the sign of the correlation will depend solely on the arbitrary way in which we choose to assign 0 and 1 and is therefore meaningless for most purposes.

point-biserial coefficient (r_{pb})

If we set aside until the end of the chapter the problem of r as an estimate of ρ, things begin to look brighter. For any other purpose, we can proceed as usual to calculate the standard Pearson correlation coefficient (r), although we will label it the **point-biserial coefficient (r_{pb}).** Thus, algebraically, $r_{pb} = r$, where one variable is dichotomous and the other is roughly continuous and more or less normally distributed in arrays.[1]

Calculating r_{pb}

One of the more common questions among statistical discussion groups on the Internet is "Does anyone know of a program that will calculate a point-biserial correlation?" The answer is very simple—any statistical package I know of will calculate the point-biserial correlation because it is simply Pearson's r applied to a special kind of data.

As an example of the calculation of the point-biserial correlation, we will use the data in Table 10.1. These are the first 12 cases of male (Sex $= 0$) weights and the first 15 cases of female (Sex $= 1$) weights from Exercises 9.31 and 9.32 in Chapter 9. I have chosen unequal numbers of males and females just to show that it is possible to do so. Keep in mind that these are actual self-report data from real subjects.

The scatterplot for these data is given in Figure 10.1, with the regression line superimposed. There are fewer than 27 data points here simply because some points overlap. Notice that the regression line passes through the mean of each array. Thus, when $X = 0$, $\hat{Y}$ is the intercept and equals the mean weight for males, and when $X = 1$, $\hat{Y}$ is the mean weight for females. These values are shown in Table 10.1, along with the correlation coefficient. The slope of the line is negative because we have set "female" $= 1$ and therefore plotted females to the right of males. If we had reversed the scoring, the slope would have been

---

[1] When there is a clear criterion variable and when that variable is the one that is dichotomous, you might want to consider logistic regression (see Chapter 15).

Table 10.1 Calculation of point-biserial correlation for weights of males and females

Sex	Weight	Sex	Weight
0	150	1	130
0	140	1	138
0	180	1	121
0	190	1	125
0	145	1	116
0	150	1	145
0	164	1	150
0	140	1	150
0	142	1	125
0	136	1	130
0	123	1	120
0	155	1	130
1	140	1	131
1	120		

$\text{Mean}_{\text{male}} = 151.25$ $\text{Mean}_{\text{female}} = 131.4$

$s_{\text{male}} = 18.869$ $s_{\text{female}} = 10.979$

$\text{Mean}_{\text{weight}} = 140.222$ $\text{Mean}_{\text{sex}} = 0.556$

$s_{\text{weight}} = 17.792$ $s_{\text{sex}} = 0.506$

$\text{cov}_{XY} = -5.090$

$$r = \frac{\text{cov}_{XY}}{s_X s_Y} = \frac{-5.090}{(0.506)(17.792)} = -.565$$

$$b = \frac{\text{cov}_{XY}}{s_x^2} = \frac{-5.090}{(0.506)^2} = -19.85$$

$$a = \overline{Y} - b\overline{X} = 151.25$$

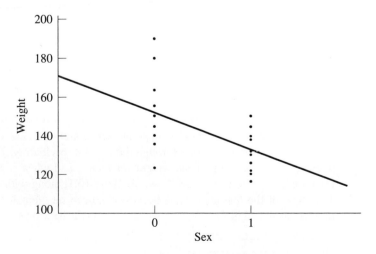

Figure 10.1   Weight as a function of Sex

positive. The fact that the regression line passes through the two Y means will assume particular relevance when we later consider eta squared (η^2), where the regression line is deliberately drawn to pass through several array means.

From Table 10.1, you can see that the correlation between weight and sex is $-.565$. As noted, we can ignore the sign of this correlation because the decision about coding sex is arbitrary. A negative coefficient indicates that the mean of the group coded 1 is less than the mean of the group coded 0, whereas a positive correlation indicates the reverse. We can still interpret r^2 as usual, however, and say that $-.565^2 = 32\%$ of the variability in weight can be accounted for by sex. We are not speaking here of cause and effect. One of the more immediate causes of weight is the additional height of males, which is certainly related to sex, but a lot of other sex-linked characteristics also enter the picture.

Another interesting fact illustrated in Figure 10.1 concerns the equation for the regression line. Recall that the intercept is the value of $\hat{Y}$ when $X = 0$. In this case, $X = 0$ for males and $\hat{Y} = 151.25$. In other words, the mean weight of the group coded 0 is the intercept. Moreover, the slope of the regression line is defined as the change in $\hat{Y}$ for a one-unit change in X. Because a one-unit change in X corresponds to a change from male to female, and the predicted value ($\hat{Y}$) changes from the mean weight of males to the mean weight of females, the slope (-19.85) will represent the difference in the two means. We will return to this idea in Chapter 16, but it is important to notice it here in a simple context.

The Relationship Between r_{pb} and t

The relationship between r_{pb} and t is very important. It can be shown, although the proof will not be given here, that

$$r_{pb}^2 = \frac{t^2}{t^2 + df}$$

where t is obtained from the t test of the difference of means (for example, between the mean weights of males and females) and $df =$ the degrees of freedom for t, namely $N_1 + N_2 - 2$. For example, if we were to run a t test on the difference in mean weight between male and female subjects, using a t for two independent groups with unequal sample sizes,

$$s_p^2 = \frac{(N_1 - 1)s_1^2 + (N_2 - 1)s_2^2}{N_1 + N_2 - 2}$$

$$= \frac{11(18.869^2) + 14(10.979^2)}{12 + 15 - 2} = 224.159$$

$$t = \frac{\overline{X}_1 - \overline{X}_2}{\sqrt{\dfrac{s_p^2}{N_1} + \dfrac{s_p^2}{N_2}}}$$

$$= \frac{151.25 - 131.4}{\sqrt{\dfrac{224.159}{12} + \dfrac{224.159}{15}}}$$

$$= \frac{19.85}{5.799} = 3.42$$

With 25 df, the difference between the two groups is significant. We now calculate

$$r_{pb}^2 = \frac{t^2}{t^2 + df} = \frac{3.42^2}{3.42^2 + 25} = .319$$

$$r_{pb} = \sqrt{.319} = .565$$

which, with the exception of the arbitrary sign of the coefficient, agrees with the more direct calculation.

What is important about the equation linking r_{pb}^2 and t is that it demonstrates that the distinction between relationships and differences is not as clear-cut as you might at first think. More important, we can use r_{pb}^2 and t together to obtain a rough estimate of the practical, as well as the statistical, significance of a difference. Thus, a $t = 3.42$ is evidence in favor of the experimental hypothesis that the two sexes differ in weight. At the same time, r_{pb}^2 (which is a function of t) tells us that sex accounts for 32% of the variation in weight. Finally, the equation shows us how to calculate r from the research literature when only t is given, and vice versa.

Testing the Significance of r_{pb}^2

A test of r_{pb} against the null hypothesis $H_0: \rho = 0$ is simple to construct. Because r_{pb} is a Pearson product-moment coefficient, it can be tested in the same way as r. Namely,

$$t = \frac{r_{pb}\sqrt{N-2}}{\sqrt{1 - r_{pb}^2}}$$

on $N - 2$ df. Furthermore, this equation can be derived directly from the definition of r_{pb}^2, so the $t = 3.42$ obtained here is the same (except possibly for the sign) as a t test between the two levels of the dichotomous variable. This makes sense when you realize that a statement that males and females differ in weight is the same as the statement that weight varies with sex.

r_{pb}^2 and Effect Size

We can take one more important step. Elsewhere, we have considered a measure of effect size put forth by Cohen (1988), who defined

$$d = \frac{\mu_1 - \mu_2}{\sigma}$$

as a measure of the effect of one treatment compared with another. We have to be a bit careful here because Cohen originally expressed effect size in terms of parameters (i.e., in terms of population means and standard deviations). Others (Glass (1976) and Hedges (1981) expressed their statistics (g' and g, respectively) in terms of sample statistics, where Hedges used the pooled estimate of the population variance as the denominator (see Chapter 7 for the pooled estimate). The nice thing about any of these effect size measures is that they express the difference between means in terms of the size of a standard deviation. Although it is nice to be correct, it is also nice, and sometimes clearer, to be consistent. As I have done elsewhere, I am going to continue to refer to our effect size measure as d, with apologies to Hedges and Glass.

There is a direct relationship between the squared point-biserial correlation coefficient and d.

$$d = \frac{\overline{X}_1 - \overline{X}_2}{s_{pooled}} = \sqrt{\frac{df(n_1 + n_2)r_{pb}^2}{n_1 n_2(1 - r_{pb}^2)}}$$

For our data on weights of males and females, we have

$$d = \frac{\overline{X}_1 - \overline{X}_2}{s_{pooled}} = \sqrt{\frac{df(n_1 + n_2)r_{pb}^2}{n_1 n_2 \left(1 - r_{pb}^2\right)}}$$

$$= \frac{151.25 - 131.4}{14.972} = 1.33 = \sqrt{\frac{25(12 + 15)(-.565)^2}{12 * 15(1 - .565^2)}} = \sqrt{1.758} = 1.33$$

We can now conclude that the difference between the average weights of males and females is about $1^1/_3$ standard deviations. To me, that is more meaningful than saying that sex accounts for about 32% of the variation in weight.[2]

An important point here is to see that these statistics are related in meaningful ways. We can go from r_{pb}^2 to d, and vice versa, depending on which seems to be a more meaningful statistic. With the increased emphasis on the reporting of effect sizes and similar measures, it is important to recognize these relationships.

The Phi Coefficient (φ)

The point-biserial correlation coefficient deals with the situation in which one of the variables is a dichotomy. When both variables are dichotomies, we will want a different statistic. For example, we might be interested in the relationship between gender and employment, where individuals are scored as either male or female and as employed or unemployed. Similarly, we might be interested in the relationship between employment status (employed-unemployed) and whether an individual has been arrested for drunken driving. As a final example, we might want to know the correlation between smoking (smokers versus non-smokers) and death by cancer (versus death by other causes). Unless we are willing to make special assumptions concerning the underlying continuity of our variables, the most appropriate correlation coefficient is the **φ (phi) coefficient.** This is the same φ that we considered briefly in Chapter 6.

φ (phi) coefficient

Calculating φ

Table 10.2 contains a small portion of the data from Gibson and Leitenberg (2000) (referred to in Exercise 6.33) on the relationship between sexual abuse training in school, (which some of you may remember as "stranger danger" or "good touch–bad touch") and subsequent sexual abuse. Both variables have been scored as 0, 1 variables—an individual received instruction, or she did not, and she was either abused, or she was not.

Table 10.2 Calculation of φ for Gibson's data

X:	0 = Instruction
	1 = No Instruction
Y:	0 = Sexual Abuse
	1 = No Sexual Abuse

Partial data:

X: 0 0 0 1 0 1 0 0 0 1 0 0 1 0
Y: 0 0 1 0 1 0 0 1 1 0 0 1 0 0

Calculations (based on full data set):

$\overline{X} = 0.3888$ $s_X = 0.4878$ $\text{cov}_{XY} = -0.0169$

$\overline{Y} = 0.8863$ $s_Y = 0.3176$ $N = 818$

$$\phi = r = \frac{\text{cov}_{XY}}{s_X s_y} = \frac{-0.0169}{(.4878)(.3176)} = -.1094$$

$\phi^2 = .012$

---

[2] If you then want to calculate confidence limits on d, consult Kline (2004).

The appropriate correlation coefficient is the ϕ coefficient, which is equivalent to Pearson's r calculated on these data. Again, special formulae exist for those people who can be bothered to remember them, but they will not be considered here.

From Table 10.2, we can see that the correlation between whether students receive instruction on how to avoid sexual abuse in school, and whether they are subsequently abused, is $-.1094$, with a $\phi^2 = .012$. The correlation is in the right direction, but it does not look terribly impressive. But that may be misleading. (I chose to use these data precisely because what looks like a very small effect from one angle looks like a much larger effect from another angle.) We will come back to this issue shortly.

Significance of ϕ

Having calculated ϕ, we are likely to want to test it for statistical significance. The appropriate test of ϕ against $H_0: \rho = 0$ is a chi-square test because $N\phi^2$ is distributed as χ^2 on 1 df. For our data,

$$\chi^2 = N\phi^2 = 818(-.1094^2) = 9.79$$

which, on one df, is clearly significant. We would therefore conclude that we have convincing evidence of a relationship between sexual abuse training and subsequent abuse.

The Relationship Between ϕ and χ^2

The data that form the basis of Table 10.2 could be recast in another form, as shown in Table 10.3. The two tables (10.2 and 10.3) contain the same information; they merely display it differently. You will immediately recognize Table 10.3 as a contingency table. From it, you could compute a value of χ^2 to test the null hypothesis that the variables are independent. In doing so, you would obtain a χ^2 of 9.79—which, on 1 df, is significant. It is also the same value for χ^2 that we computed in the previous subsection.

It should be apparent that in calculating ϕ and χ^2, we have been asking the same question in two different ways: Not surprisingly, we have come to the same conclusion. When we calculated ϕ and tested it for significance, we were asking whether there was any correlation (relationship) between X and Y. When we ran a chi-square test on Table 10.3, we were also asking whether the variables are related (correlated). Because these questions are the same, we would hope that we would come to the same answer, which we did. On the one hand, χ^2 relates to the statistical significance of a relationship. On the other, ϕ measures the degree or magnitude of that relationship.

Table 10.3 Calculation of χ^2 for Gibson's data on sexual abuse (χ^2 is shown as "approximate" simply because of the effect of rounding error in the table)

	Training	No Training	
Abused	43 (56.85)	50 (36.15)	93
Not Abused	457 (443.15)	268 (281.85)	725
	500	318	818

$$\chi^2 = \frac{(43 - 56.85)^2}{56.85} + \frac{(50 - 36.15)^2}{36.15} + \frac{(457 - 443.15)^2}{443.15} + \frac{(268 - 281.85)^2}{281.85}$$

$$= 9.79 \,(\text{approx.})$$

It will come as no great surprise that there is a linear relationship between ϕ^2 and χ^2. From the fact that $\chi^2 = N\phi^2$, we can deduce that

$$\phi = \sqrt{\frac{\chi^2}{N}}$$

For our example,

$$\phi = \sqrt{\frac{9.79}{818}} = \sqrt{0.0120} = .1095$$

(again, with a bit of correction for rounding), which agrees with our previous calculation.

ϕ^2 as a Measure of the Practical Significance of χ^2

The fact that we can go from χ^2 to ϕ means that we have one way of evaluating the practical significance (importance) of the relationship between two dichotomous variables. We have already seen that for Gibson's data, the conversion from χ^2 to ϕ^2 showed that our χ^2 of 9.79 accounted for about 1.2% of the variation. As I said, that does not look very impressive, even if it is significant.

Rosenthal and Rubin (1982) have argued that psychologists and others in the "softer sciences" are too ready to look at a small value of r^2 or ϕ^2 and label an effect as unimportant. They maintain that very small values of r^2 can in fact be associated with important effects. It is easiest to state their case with respect to ϕ, which is why their work is discussed here.

Rosenthal and Rubin pointed to a large-scale evaluation (called a meta-analysis) of more than 400 studies of the efficacy of psychotherapy. The authors, Smith and Glass (1977), reported an effect equivalent to a correlation of .32 between presence or absence of psychotherapy and presence or absence of improvement, by whatever measure. A reviewer subsequently squared this correlation ($r^2 = .1024$) and deplored the fact that psychotherapy accounted for only 10% of the variability in outcome. Rosenthal and Rubin were not impressed by the reviewer's perspicacity. They pointed out that if we took 100 people in a control group and 100 people in a treatment group and dichotomized them as improved or not improved, a correlation of $\phi = .32$ would correspond to $\chi^2 = 20.48$. This can be seen by computing

$$\phi = \sqrt{\chi^2/N}$$
$$\phi^2 = \chi^2/N$$
$$.1024 = \chi^2/200$$
$$\chi^2 = 20.48$$

The interesting fact is that such a χ^2 would result from a contingency table in which 66 of the 100 subjects in the treatment group improved whereas only 34 of the 100 subjects in the control group improved. (You can easily demonstrate this for yourself by computing χ^2 on such a table.) That is a dramatic difference in improvement rates.

But I have two more examples, one of which is the Gibson study. Rosenthal (1990) pointed to a well-known study of (male) physicians who took a daily dose of either aspirin or a placebo to reduce the incidence of heart attacks. (We considered this study briefly in earlier chapters, but for a different purpose.) This study was terminated early because the review panel considered the results so clearly in favor of the aspirin group that it would have been unethical to continue to give the control group a placebo. But, said Rosenthal, what was the correlation between aspirin and heart attacks that was so dramatic as to cut short such a study? Would you believe $\phi = .034$ ($\phi^2 = .001$)?

I include Rosenthal's work to make the point that one does not require large values of r^2 (or ϕ^2) to have an important effect. Small values in certain cases can be quite impressive. For further examples, see Rosenthal (1990).

To return to what appears to be a small effect in Gibson's sexual abuse data, we will take an approach adopted in Chapter 6 with odds ratios. In Gibson's data, 50 of 318 children who received no instruction were subsequently abused, which makes the odds of abuse for this group to be $50/268 = 0.187$. On the other hand, 43 of 500 children who received training were subsequently abused, for odds of $43/457 = 0.094$. This gives us an odds ratio (the ratio of the two calculated odds) of $0.187/0.094 = 1.98$. A child who does not receive sexual abuse training in school is nearly twice as likely to be subsequently abused as one who does. That looks quite a bit different from a squared correlation of only .012, which illustrates why we must be careful in the statistic we select.

At this point, perhaps you are thoroughly confused. I began by showing that you can calculate a correlation between two dichotomous variables. I then showed that this correlation could either be calculated as a Pearson correlation coefficient, or it could be derived directly from a chi-square test on the corresponding contingency table because there is a nice relationship between ϕ and χ^2. I argued that ϕ or ϕ^2 can be used to provide an r-family effect size measure (a measure of variation accounted for) of the effectiveness of the independent variable. But then I went a step further and said that when you calculate ϕ^2, you may be surprised by how small it is. In that context, I pointed to the work of Rosenthal and Rubin and to Gibson's data, showing in two different ways that accounting for only small amounts of the variation can still be impressive and important. I am mixing different kinds of measures of "importance" (statistical significance, percentage of accountable variation, effect sizes [d], and odds ratios), and, even though that may be confusing, it is the nature of the problem. Statistical significance is a good thing, but it certainly isn't everything. Percentage of variation is an important kind of measure, but it is not very intuitive and may be small in important situations. The d-family measures of effect sizes have the advantage of presenting a difference in concrete terms (distance between means in terms of standard deviations). Odds ratios are very useful when you have a 2×2 table, but less so with more complex or with simpler situations.

10.2 Biserial and Tetrachoric Correlation: Non-Pearson Correlation Coefficients

In considering the point-biserial and phi coefficients, we were looking at data where one or both variables were measured as a dichotomy. We might even call this a "true dichotomy" because we often think of those variables as "either-or" variables. A person is a male or a female, not halfway in between. Those are the coefficients we will almost always calculate with dichotomous data, and nearly all computer software will calculate those coefficients by default.

biserial correlation (r_b)

tetrachoric correlation (r_t)

Two other coefficients, to which you are likely to see reference, but are most unlikely to use, are the **biserial correlation** (r_b) and the **tetrachoric correlation** (r_t). In earlier editions of this book, I showed how to calculate those coefficients, but there does not seem to be much point in doing so anymore. I will simply explain how they differ from the coefficients I have discussed.

As I have said, we usually treat people as male or female, as if they pass or they fail a test, or as if they are abused or not abused. But we know that those dichotomies, especially the last two, are somewhat arbitrary. People fail miserably, or barely fail, or barely pass, and so on. People suffer varying degrees of sexual abuse, and although all abuse is bad, some is worse than others. If we are willing to take this underlying continuity into account, we can make an estimate of what the correlation would have been if the variable (or variables) had been normally distributed instead of dichotomously distributed. We have special formulae to make these estimates, but I will not give them here.

The biserial correlation is the direct analog of the point-biserial correlation, except that the biserial assumes underlying normality in the dichotomous variable. The tetrachoric correlation is the direct analog of ϕ, where we assume underlying normality on both variables. That is all you really need to know about these two coefficients.

10.3   Correlation Coefficients for Ranked Data

In some experiments, the data naturally occur in the form of ranks. For example, we might ask judges to rank objects in order of preference under two different conditions and want to know the correlation between the two sets of rankings. Cities are frequently ranked in terms of livability, and we might want to correlate those rankings with rankings given 10 years later. Usually, we are most interested in these correlations when we want to assess the reliability of some ranking procedure, though in the case of the city ranking example, we are interested in the stability of rankings.

A related procedure, which has frequently been recommended in the past, is to rank sets of measurement data when we have serious reservations about the nature of the underlying scale of measurement. In this case, we are substituting ranks for raw scores. Although we could seriously question the necessity of ranking measurement data (for reasons mentioned in the discussion of measurement scales in Section 1.3 of Chapter 1), this is nonetheless a fairly common procedure.

Ranking Data

ranking

Students occasionally experience difficulty in **ranking** a set of measurement data, and this section is intended to present the method briefly. Assume we have the following set of data, which have been arranged in increasing order:

5, 8, 9, 12, 12, 15, 16, 16, 16, 17

The lowest value (5) is given the rank of 1. The next two values (8 and 9) are then assigned ranks 2 and 3. We then have two tied values (12) that must be ranked. If they were untied, they would be given ranks 4 and 5, so we split the difference and rank them both 4.5. The sixth number (15) is now given rank 6. Three values (16) are tied for ranks 7, 8, and 9; the mean of these ranks is 8. Thus, all are given ranks of 8. The last value is 17, which has rank 10. The data and their corresponding ranks are given here:

X:	5	8	9	12	12	15	16	16	16	17
Ranks:	1	2	3	4.5	4.5	6	8	8	8	10

Spearman's Correlation Coefficient for Ranked Data (r_s)

Spearman's correlation coefficient for ranked data (r_s)

Spearman's rho

Whether data naturally occur in the form of ranks (as, for example, when we looking at the rankings of 20 cities on two different occasions) or whether ranks have been substituted for raw scores, an appropriate correlation is **Spearman's correlation coefficient for ranked data (r_s)**. (This statistic is sometimes referred to as **Spearman's rho**.)

Calculating r_s

The easiest way to calculate r_s is to apply Pearson's original formula to the ranked data. Alternative formulae do exist, but they have been designed to give exactly the same answer as Pearson's formula as long as there are no ties in the data. When there are ties, the alternative formula leads to a wrong answer unless a correction factor is applied. That correction factor brings you back to where you would have been had you used Pearson's formula to begin with, so why bother with alternative formulae?

The Significance of r_s

Recall that in Chapter 9, we imposed normality and homogeneity assumptions to provide a test on the significance of r (or to set confidence limits). With ranks, the data clearly cannot be normally distributed. There is no generally accepted method for calculating the standard error of r_S for small samples. As a result, computing confidence limits on r_S is not practical. Numerous textbooks contain tables of critical values of r_S, but for $N \geq 28$, these tables are themselves based on approximations. Keep in mind in this connection that a typical judge has difficulty ranking a large number of items, and therefore, in practice N is usually small when we are using r_S.

Kendall's Tau Coefficient (τ)

Kendall's τ

A serious competitor to Spearman's r_S is **Kendall's τ.** Whereas Spearman treated the ranks as scores and calculated the correlation between the two sets of ranks, Kendall based his statistic on the number of *inversions* in the rankings.

We will take as our example a dataset from the Data and Story Library (DASL) website, found at http://lib.stat.cmu.edu/DASL/Stories/AlcoholandTobacco.html. These data describe the average weekly spending on alcohol and tobacco in 11 regions of Great Britain. The data follow, and I have organized the rows to correspond to increasing expenditures on Alcohol. Though it is not apparent from looking at either the Alcohol or Tobacco variable alone, in a bivariate plot it is clear that Northern Ireland is a major outlier. Similarly, the distribution of Alcohol expenditures is decidedly nonnormal, whereas the ranked data on alcohol, like all ranks, are rectangularly distributed.

Region	Alcohol	Tobacco	RankA	RankT	Inversions
Northern Ireland	4.02	4.56	1	11	10
East Anglia	4.52	2.92	2	2	1
Southwest	4.79	2.71	3	1	0
East Midlands	4.89	3.34	4	4	1
Wales	5.27	3.53	5	6	2
West Midlands	5.63	3.47	6	5	1
Southeast	5.89	3.20	7	3	0
Scotland	6.08	4.51	8	10	3
Yorkshire	6.13	3.76	9	7	0
Northeast	6.19	3.77	10	8	0
North	6.47	4.03	11	9	0

Notice that when the entries are listed in the order of rankings given by Alcohol, there are reversals (or inversions) of the ranks given by Tobacco (rank 11 of tobacco comes before all lower ranks, and rank 10 of tobacco comes before 3 lower ranks). I can count the number of inversions just by going down the Tobacco column and counting the number of times a ranking further down the table is lower than one further up the table. For instance, looking at tobacco expenditures, row 1 has 10 inversions because all 10 values below it are lower. Row 2 has only one inversion because only the rank of "1" is lower than a rank of 2. And so on.

If there were a perfect ordinal relationship between these two sets of ranks, we would not expect to find any inversions. The region that spent the most money on alcohol would spend the most on tobacco, the region with the next highest expenditures on alcohol would be second highest on tobacco, and so on. Inversions of this form are the basis for Kendall's statistic.

Calculating τ

We will use the number of inversions (I) in the ranks to create a measure of correlation. That correlation is essentially 1.0 minus the percentage of possible inversions. There are $n(n-1)/2 = 11(10)/2 = 55$ pairs of rankings. Eighteen of those rankings are inversions (the sum of the right-most column).

Kendall defined

$$\tau = 1 - \frac{2(\text{Number of inversions})}{\text{Number of pairs of objects}}$$

It is well known that the number of pairs of N objects is given by $N(N-1)/2$. For our data

$$\tau = 1 - \frac{2I}{\dfrac{N(N-1)}{2}}$$

$$= 1 - \frac{2(18)}{\dfrac{10(11)}{2}} = 1 - \frac{36}{55} = .345$$

Thus, as a measure of the agreement between rankings on Alcohol and Tobacco, Kendall's $\tau = .345$.

Significance of τ

Unlike Spearman's r_S, there is an accepted method for estimation of the standard error of Kendall's τ.

$$s_\tau = \sqrt{\frac{2(2N+5)}{9N(N-1)}}$$

Moreover, τ is approximately normally distributed for $N \geq 10$. This allows us to approximate the sampling distribution of Kendall's τ using the normal approximation.

$$z = \frac{\tau}{S_\tau} = \frac{\tau}{\sqrt{\dfrac{2(2N+5)}{9N(N-1)}}} = \frac{.345}{\sqrt{\dfrac{2(27)}{9(11)(10)}}} = \frac{.345}{.2335} = 1.48$$

For a two-tailed test, $p = .139$, which is not statistically significant.

With a standard error of 0.2335, the confidence limits on Kendall's τ, assuming normality of τ, would be

$$CI = \tau \pm 1.96 s_\tau = \tau \pm 1.96\left(\sqrt{\frac{2(2N+5)}{9N(N-1)}}\right) = \tau \pm 1.96(.2335)$$

For our example, this would produce confidence limits of $-.11 \leq t \leq .80$.

Kendall's τ has generally been given preference over Spearman's r_S because it is a better estimate of the corresponding population parameter, and its standard error is known.

Although there is evidence that Kendall's τ holds up better than Pearson's r to nonnormality in the data, that seems to be true only at quite extreme levels. In general, Pearson's r on the raw data has been, and remains, the coefficient of choice.

10.4   Analysis of Contingency Tables with Ordered Variables

In Chapter 6 on chi-square, I referred to the problem that arises when the independent variables are ordinal variables. The traditional chi-square analysis does not take this ordering into account, but it is important for a proper analysis. As I said in Chapter 6, this section was motivated by a question sent to me by Jennifer Mahon at the University of Leicester, in England, who has graciously allowed me to use her data for this example. Ms. Mahon was interested in the question of whether the likelihood of dropping out of a study on eating disorders was related to the number of traumatic events the participants had experienced in childhood.

The data from this study follow. I have taken the liberty of altering them very slightly so that I don't have to deal with the problem of small expected frequencies at the same time that I am trying to show how to make use of the ordinal nature of the data. The altered data are still a faithful representation of the effects that she found.

	\multicolumn{5}{c}{Number of Traumatic Events}	
	0	1	2	3	4+	Total
Dropout	25	13	9	10	6	63
Remain	31	21	6	2	3	63
Total	56	34	15	12	9	126

At first glance, we might be tempted to apply a standard chi-square test to these data, testing the null hypothesis that dropping out of treatment is independent of the number of traumatic events the person experienced during childhood. If we do that, we find a chi-square of 9.459 on 4 *df*, which has an associated probability of .051. Strictly speaking, this result does not allow us to reject the null hypothesis, and we might conclude that traumatic events are not associated with dropping out of treatment. However, that answer is a bit too simplistic.

Notice that Trauma represents an ordered variable. Four traumatic events are more than 3, 3 traumatic events are more than 2, and so on. If we look at the percentage of participants who dropped out of treatment, as a function of the number of traumatic events they had experienced as children, we see that there is a general, though not a monotonic, increase in dropouts as we increase the number of traumatic events. However, this trend was not allowed to play any role in our calculated chi-square. What we want is a statistic that does take order into account.

A Correlational Approach

There are several ways we can accomplish what we want, but they all come down to assigning some kind of ordered metric to our independent variables. Dropout is not a problem because it is a dichotomy. We could code dropout as 1 and remain as 2, or dropout as 1 and remain as 0, or any other two values we like. The result will not be affected by our choice of values. When it comes to the number of traumatic events, we could simply use the numbers 0, 1, 2, 3, and 4. Alternatively, if we thought that 3 or 4 traumatic events would be much more important than 1 or 2, we might use 0, 1, 2, 4, 6. In practice, as long as we chose numbers that are monotonically increasing and are not very extreme, the result will not change much as a function of our choice. I will choose to use 0, 1, 2, 3, and 4.

Now that we have established a metric for each independent variable, there are several different ways that we could go. We'll start with one that has good intuitive appeal. We will simply correlate our two variables.[3] Each participant will have a score of 0 or 1 on Dropout, and a score between 0 and 4 on Trauma. The standard Pearson correlation between those two measures is .215, which has an associated probability under the null of .016. This correlation is significant, and we can reject the null hypothesis of independence.

Some people may be concerned about the use of Pearson's r in this situation because "number of traumatic events" is such a discrete variable. Actually, that is not a problem for Pearson's r, and no less an authority than Agresti (2002) recommends that approach. Perhaps you are unhappy with the idea of specifying a particular metric for Trauma, although you do agree that it is an ordered variable. If so, you could calculate Kendall's τ instead of Pearson's r. τ would be the same for any set of values you assign to the levels of Trauma, assuming that they increased across the levels of that variable. For our data, τ would be .169, with a probability of .04. So the relationship would still be significant even if we are only confident about the order of the independent variable(s). (The appeal to Kendall's τ as a possible replacement for Pearson's r is the reason why I included this material here rather than in Chapter 9. Agresti, however, has pointed out that if the cell frequencies are very different, there are negative consequences to using either Kendall's τ or Spearman's r_S. I recommend strongly that you simply use r.)

Agresti (2002, p. 87) presents the approach that we have just adopted and shows that we can compute a chi-square statistic from the correlation. He gives

$$M^2 = (N - 1)r^2$$

where M^2 is a chi-square statistic on 1 degree of freedom, r is the Pearson correlation between Dropout and Trauma, and N is the sample size. For our example, this becomes

$$M^2 = \chi^2(1) = (N - 1)r^2$$

$$\chi^2(1) = 125(0.215^2) = 5.757$$

which has an associated probability under the null hypothesis of .016.

---

[3] Many articles in the literature refer to Maxwell (1961) as a source for dealing with ordinal data. With one minor exception, Maxwell's approach is the one advocated here, though it is difficult to tell that from his description because his formulae were selected for computational ease.

The probability value was already given by the test on the correlation, so that is nothing new. But we can go one step further. We know that the overall Pearson chi-square on 4 *df* is 9.459. We also know that we have just calculated a chi-square is 5.757 on 1 *df* associated with the *linear* relationship between the two variables. That linear relationship is part of the total chi-square, and if we subtract the linear component from the overall chi-square we obtain

	df	Chi-square
Pearson	4	9.459
Linear	1	5.757
Deviation from linear	3	3.702

The departure from linearity is itself a chi-square equal to 3.702 on 3 *df*, which has a probability under the null of .295. Thus, we do not have any evidence that there is anything other than a linear trend underlying these data. (In other words, the relationship between Trauma and Dropout is not curvilinear.)

Agresti (1996, 2002) has an excellent discussion of the approach taken here, and he makes the interesting point that for small to medium sample sizes, the standard Pearson chi-square is more sensitive to the negative effects of small sample size than is the ordinal chi-square that we calculated. In other words, although some of the cells in the contingency table are small, I am more confident of the ordinal (linear) chi-square value of 5.757 than I can be of the Pearson chi-square of 9.459.

You can calculate the chi-square for linearity using SPSS. If you request the chi-square statistic from the statistics dialog box, your output will include the Pearson chi-square, the likelihood ratio chi square, and linear-by-linear association. The SPSS printout of the results for Mahon's data follows. You will see that the Linear-by-Linear Association measure of 5.757 is the same as the χ^2 that we calculated using $(N - 1)r^2$.

Chi-Square Tests

	Value	df	Asymp. Sig. (2-sided)
Pearson Chi-Square	9.459[a]	4	.051
Likelihood Ratio	9.990	4	.041
Linear-by-Linear Association	5.757	1	.016
N of Valid Cases	126		

[a] 2 cells (20.0%) have expected count less than 5. The minimum expected count is 4.50.

There are a number of other ways to approach the problem of ordinal variables in a contingency table. In some cases, only one of the variables is ordinal and the other is nominal. (Remember that dichotomous variables can always be treated as ordinal without affecting the analysis.) In other cases, one of the variables is clearly an independent variable whereas the other is a dependent variable. An excellent discussion of some of these methods can be found in Agresti (1996, 2002).

10.5   Kendall's Coefficient of Concordance (*W*)

All the statistics we have been concerned with in this chapter have dealt with the relationship between two sets of scores (*X* and *Y*). But suppose that instead of having two judges rank a set of objects, we had six judges doing the ranking. What we need is some measure

Kendall's coefficient of concordance (W)

of the degree to which the six judges agree. Such a measure is afforded by **Kendall's coefficient of concordance (W).**

Suppose, as an example, that we asked six judges to rank order the pleasantness of eight colored patches and obtained the data in Table 10.4. If all the judges had agreed that Patch B was the most pleasant, they would all have assigned it a rank of 1, and the column total for that patch across six judges would have been 6. Similarly, if A had been ranked second by everyone, its total would have been 12. Finally, if every judge assigned the highest rank to Patch H, its total would have been 48. In other words, the column totals would have shown considerable variability.

Table 10.4   Judge's rankings of pleasantness of colored patches

Judges	Colored Patches							
	A	B	C	D	E	F	G	H
1	1	2	3	4	5	6	7	8
2	2	1	5	4	3	8	7	6
3	1	3	2	7	5	6	8	4
4	2	1	3	5	4	7	8	6
5	3	1	2	4	6	5	7	8
6	2	1	3	6	5	4	8	7
$\sum$	11	9	18	30	28	36	45	39

On the other hand, if the judges showed no agreement, each column would have had some high ranks and some low ranks assigned to it, and the column totals would have been roughly equal. Thus, the variability of the column totals, given disagreement (or random behavior) among judges, would be low.

Kendall used the variability of the column totals in deriving his statistic. He defined W as the ratio of the variability among columns to the maximum possible variability.

$$W = \frac{\text{Variance of column totals}}{\text{Maximum possible variance of column totals}}$$

We are dealing with ranks, so we know what the maximum variance of the totals will be. With a bit of algebra, we can define

$$W = \frac{12 \sum T_j^2}{k^2 N(N^2 - 1)} - \frac{3(N + 1)}{N - 1}$$

where T_j represents the column totals, N = the number of items to be ranked, and k = the number of judges doing the ranking. For the data in Table 10.4,

$$\sum T_j^2 = 11^2 + 9^2 + 18^2 + 30^2 + 28^2 + 36^2 + 45^2 + 39^2 = 7052$$

$$W = \frac{12 \sum T_j^2}{k^2 N(N^2 - 1)} - \frac{3(N + 1)}{N - 1}$$

$$= \frac{12(7052)}{6^2(8)(63)} - \frac{3(9)}{7} = \frac{84624}{18144} - \frac{27}{7}$$

$$= .807$$

As you can see from the definition of W, it is not a standard correlation coefficient. It does have an interpretation in terms of a familiar statistic, however: It can be viewed as a function of the average Spearman correlation computed on the rankings of all possible pairs of judges. Specifically,

$$\bar{r}_s = \frac{kW - 1}{k - 1}$$

For our data,

$$\bar{r}_s = \frac{kW - 1}{k - 1} = \frac{6(.807) - 1}{5} = .768$$

Thus, if we took all possible pairs of rankings and computed r_S for each, the average r_S would be .768.

Hays (1981) recommends reporting W but converting to $\bar{r}_S$ for interpretation. Indeed, it is hard to disagree with that recommendation because no intuitive meaning attaches to W itself. W does have the advantage of being bounded by zero and one, whereas $\bar{r}_S$ does not, but it is difficult to attach much practical meaning to the statement that the variance of column totals is 80.7% of the maximum possible variance. Whatever its faults, $\bar{r}_S$ seems preferable.

A test on the null hypothesis that there is no agreement among judges is possible under certain conditions. If $k \geq 7$, the quantity

$$\chi^2_{(N-1)} = k(N - 1)W$$

is approximately distributed as χ^2 on $N - 1$ degrees of freedom. Such a test is seldom used, however, because W is usually calculated in those situations in which we seek a level of agreement substantially above the minimum level required for significance, and we rarely have seven or more judges.

Key Terms

Correlational measures (Introduction)

Measures of association (Introduction)

Validity (Introduction)

Dichotomy (10.1)

Point-biserial coefficient (r_{pb}) (10.1)

ϕ (phi) coefficient (10.1)

Biserial correlation coefficient (r_b) (10.2)

Tetrachoric correlation (r_t) (10.2)

Ranking (10.3)

Spearman's correlation coefficient for Ranked data (r_S) (10.3)

Spearman's rho (10.3)

Kendall's τ (10.3)

Kendall's coefficient of concordance (W) (10.5)

Exercises

10.1   Some people think that they do their best work in the morning, whereas others claim that they do their best work at night. We have dichotomized 20 office workers into morning or evening people (0 = morning, 1 = evening) and have obtained independent estimates of the quality of work they produced on some specified morning. The ratings were based on a 100-point scale and appear here.

Peak time of day:	0	0	0	0	0	0	0	0	0	0
Performance rating:	65	80	55	60	55	70	60	70	55	70

Peak time of day:	0	0	0	1	1	1	1	1	1	1
Performance rating:	40	70	50	40	60	50	40	50	40	60

a. Plot these data and fit a regression line.

b. Calculate r_{pb} and test it for significance.

c. Interpret the results.

10.2   Because of a fortunate change in work schedules, we were able to reevaluate the subjects referred to in Exercise 10.1 for performance on the same tasks in the evening. The data follow:

Peak time of day:	0	0	0	0	0	0	0	0	0	0
Performance rating:	40	60	40	50	30	40	50	50	20	30
Peak time of day:	0	0	0	1	1	1	1	1	1	1
Performance rating:	40	50	30	30	50	50	40	50	40	60

a. Plot these data and fit a regression line.

b. Calculate r_{pb} and test it for significance.

c. Interpret the results.

10.3   Compare the results you obtained in Exercises 10.1 and 10.2. What can you conclude?

10.4   Why would it not make sense to calculate a biserial correlation on the data in Exercises 10.1 and 10.2?

10.5   Perform a t test on the data in Exercise 10.1 and show the relationship between this value of t and r_{pb}.

10.6   A graduate-school admissions committee is concerned about the relationship between an applicant's GPA in college and whether the individual eventually completes the requirements for a doctoral degree. The committee first looked at the data on 25 randomly selected students who entered the program 7 years ago, assigning a score of 1 to those who completed the Ph.D. program, and of 0 to those who did not. The data follow:

GPA:	2.0	3.5	2.75	3.0	3.5	2.75	2.0	2.5	3.0	2.5
Ph.D.:	0	0	0	0	0	0	0	0	1	1
GPA:	3.5	3.25	3.0	3.0	2.75	3.25	3.0	3.33	2.5	2.75
Ph.D.:	1	1	1	1	1	1	1	1	1	1
GPA:	2.0	4.0	3.0	3.25	2.5					
Ph.D.:	1	1	1	1	1					

a. Plot these data.

b. Calculate r_{pb}.

c. Calculate r_b.

d. Is it reasonable to look at r_b in this situation? Why or why not?

10.7   Compute the regression equation for the data in Exercise 10.6. Show that the line defined by this equation passes through the means of the two groups.

10.8   What do the slope and the intercept obtained in Exercise 10.7 represent?

10.9   Assume that the committee in Exercise 10.6 decided that a GPA-score cutoff of 3.00 would be appropriate. In other words, they classed everyone with a GPA of 3.00 or higher as acceptable and those with a GPA below 3.00 as unacceptable. They then correlated this with completion of the Ph.D. program.

a. Rescore the data in Exercise 10.6 as indicated.

b. Run the correlation.

c. Test this correlation for significance.

10.10   Visualize the data in Exercise 10.9 as fitting into a contingency table.

a. Compute the chi-square on this table.

b. Show the relationship between chi-square and ϕ.

10.11 An investigator is interested in the relationship between alcoholism and a childhood history of attention deficit disorder (ADD). He has collected the following data, where a 1 represents the presence of the relevant problem.

ADD:	0	1	0	0	1	1	0	0	0	1	0	0	1	0	0	1
Alcoholism:	0	1	0	0	0	1	0	0	0	1	1	0	0	0	0	1

ADD:	1	1	0	0	0	0	0	0	0	1	0	0	1	0	0	0
Alcoholism:	0	1	0	0	0	0	0	0	0	1	0	0	1	0	1	0

a. What is the correlation between these two variables?

b. Is the relationship significant?

10.12 An investigator wants to arrange the 15 items on her scale of language impairment on the basis of the order in which language skills appear in development. Not being entirely confident that she has selected the correct ordering of skills, she asks another professional to rank the items from 1 to 15 in the order in which he thinks they should appear. The data follow:

Investigator:	1	2	3	4	5	6	7	8	9	10	11	12	13	14	15
Consultant:	1	3	2	4	7	5	6	8	10	9	11	12	15	13	14

a. Use Pearson's formula (r) to calculate Spearman's r_S.

b. Discuss what the results tell you about the ordering process.

10.13 For the data in Exercise 10.12,

a. Compute Kendall's τ.

b. Test τ for significance.

10.14 In a study of diagnostic processes, entering clinical graduate students are shown a 20-minute videotape of children's behavior and asked to rank order 10 behavioral events on the tape in the order of the importance each has for a behavioral assessment (1 = most important). The data are then averaged to produce an average rank ordering for the entire class. The same thing was then done using experienced clinicians. The data follow:

Events:	1	2	3	4	5	6	7	8	9	10
Experienced clinicians:	1	3	2	7	5	4	8	6	9	10
New students:	2	4	1	6	5	3	10	8	7	9

Use Spearman's r_S to measure the agreement between experienced and novice clinicians.

10.15 Rerun the analysis on Exercise 10.14 using Kendall's τ.

10.16 Assume in Exercise 10.14 that there were five entering clinical students. They produced the following data:

Student 1:	1	4	2	6	5	3	9	10	7	8
Student 2:	4	3	2	5	7	1	10	8	6	9
Student 3:	1	5	2	6	4	3	8	10	7	9
Student 4:	2	5	1	7	4	3	10	8	6	9
Student 5:	2	5	1	4	6	3	9	7	8	10

Calculate Kendall's W and $\bar{r}_S$ for these data as a measure of agreement. Interpret your results.

10.17 On page 283, I noted that Rosenthal and Rubin showed that an r^2 of .1024 actually represented a pretty impressive effect. They demonstrated that this would correspond to a χ^2 of 20.48, and with 100 subjects in each of two groups, the 2×2 contingency table would have a 34:66 split for one row and a 66:34 split for the other row.

a. Verify this calculation with your own 2×2 table.

b. What would that 2×2 table look like if there were 100 subjects in each group, but if the r^2 were .0512? (This may require some trial and error in generating 2×2 tables and computing χ^2 on each.)

10.18 Using Mireault's data (Mireault.dat), calculate the point-biserial correlation between Gender and the Depression T score. Compare the relevant aspects of this question to the results you obtained in Exercise 7.46. (See "The Relationship Between r_{pb} and t" within Section 10.1.)

10.19 In Exercise 7.48 using Mireault.dat, we compared the responses of students who had lost a parent and students who had not lost a parent in terms of their responses on the Global Symptom Index T score (GSIT), among other variables. An alternative analysis would be to use a clinically meaningful cutoff on the GSIT, classifying anyone over that score as a clinical case (showing a clinically significant level of symptoms) and everyone below that score as a noncase. Derogatis (1983) has suggested a score of 63 as the cutoff (e.g., if GSIT > 63 then ClinCase = 1; else ClinCase = 0).

    a. Use any statistical package to create the variable of ClinCase, as defined by Derogatis. Then cross-tabulate ClinCase against Group. Compute chi-square and Cramer's ϕ_C.

    b. How does the answer to part (a) compare to the answers obtained in Chapter 7?

    c. Why might we prefer this approach (looking at case versus noncase) to the procedure adopted in Chapter 7?

    (*Hint:* Minitab will use the "Tables" command; SAS will require Proc Freq; and SPSS will use CrossTabs. The appropriate manuals will help you set up the commands.)

10.20 Repeat the analysis shown in Exercise 10.19, but this time cross-tabulate ClinCase against Gender.

    a. Compare this answer with the results of Exercise 10.18.

    b. How does this analysis differ from the one in Exercise 10.18 on roughly the same question?

Discussion Question

10.21 Rosenthal and others (cited earlier) have argued that small effects, as indexed by a small r^2, for example, can be important in certain situations. We would probably all agree that small effects could be trivial in other situations.

    a. Can an effect that is not statistically significant ever be important if it has a large enough r^2?

    b. How will the sample size contribute to the question of the importance of an effect?

CHAPTER 11

Simple Analysis of Variance

Objectives

To introduce the analysis of variance as a procedure for testing differences among two or more means.

Contents

analysis of variance (ANOVA)

THE **ANALYSIS OF VARIANCE** (**ANOVA**) has long enjoyed the status of being the most used (some would say abused) statistical technique in psychological research. The popularity and usefulness of this technique can be attributed to two sources. First, the analysis of variance, like t, deals with differences between or among sample means; unlike t, it imposes no restriction on the number of means. Instead of asking whether two means differ, we can ask whether three, four, five, or k means differ. The analysis of variance also allows us to deal with two or more independent variables simultaneously, asking about the individual effects of each variable separately and about the interacting effects of two or more variables.

This chapter will be concerned with the underlying logic of the analysis of variance (which is really quite simple) and the analysis of results of experiments employing only one independent variable. We will also examine a number of related topics that are most easily

one-way analysis

understood in the context of a **one-way** (one-variable) **analysis.** Subsequent chapters will deal with comparisons among individual sample means, with the analysis of experiments involving two or more independent variables, and with designs in which repeated measurements are made on each subject.

11.1 An Example

Many features of the analysis of variance can be best illustrated by a simple example, so we will begin with a study by M. W. Eysenck (1974) on recall of verbal material as a function of the level of processing. The data we will use have the same group means and standard deviations as those reported by Eysenck, but the individual observations are fictional. The study may be an old one, but it still has important things to tell us.

Craik and Lockhart (1972) proposed as a model of memory that the degree to which verbal material is remembered by the subject is a function of the degree to which it was processed when it was initially presented. Thus, for example, if you were trying to memorize a list of words, repeating a word to yourself (a low level of processing) would not lead to as good recall as thinking about the word and trying to form associations between that word and some other word. Eysenck (1974) was interested in testing this model and, more important, in looking to see whether it could help to explain reported differences between young and old subjects in their ability to recall verbal material. An examination of Eysenck's data on age differences will be postponed until Chapter 13; we will concentrate here on differences caused by the level of processing.

Eysenck randomly assigned 50 subjects between the ages of 55 and 65 years to one of five groups—four incidental-learning groups and one intentional-learning group. (Incidental learning is learning in the absence of the expectation that the material will later need to be recalled.) The Counting group was asked to read through a list of words and simply count the number of letters in each word. This involved the lowest level of processing because subjects did not need to deal with each word as anything more than a collection of letters. The Rhyming group was asked to read each word and think of a word that rhymed with it. This task involved considering the sound of each word, but not its meaning. The Adjective group had to process the words to the extent of giving an adjective that could reasonably be used to modify each word on the list. The Imagery group was instructed to try to form vivid images of each word. This was assumed to require the deepest level of processing of the four incidental conditions. None of these four groups were told that they would later be asked for recall of the items. Finally, the Intentional group was told to read through the list and to memorize the words for later recall. After subjects had gone through the list of 27 items three times, they were given a sheet of paper and asked to write down all of the words they could remember. If learning involves nothing more than being exposed to the material (the way most of us read a newspaper or, heaven forbid, a class assignment), then the five groups

Table 11.1   Number of words recalled as a function of level of processing

	Counting	Rhyming	Adjective	Imagery	Intentional	Total
	9	7	11	12	10	
	8	9	13	11	19	
	6	6	8	16	14	
	8	6	6	11	5	
	10	6	14	9	10	
	4	11	11	23	11	
	6	6	13	12	14	
	5	3	13	10	15	
	7	8	10	19	11	
	7	7	11	11	11	
Mean	7.00	6.90	11.00	13.40	12.00	10.06
St. Dev.	1.83	2.13	2.49	4.50	3.74	4.01
Variance	3.33	4.54	6.22	20.27	14.00	16.058

should have shown equal recall—after all, they all saw all of the words. If the level of processing of the material is important, then there should have been noticeable differences among the group means. The data are presented in Table 11.1.

11.2   The Underlying Model

The analysis of variance, as all statistical procedures, is built on an underlying model. I am not going to beat the model to death and discuss all of its ramifications, but a general understanding of that model is important for understanding what the analysis of variance is all about.

To start with an example that has a clear physical referent, suppose that the average height of all American adults is 5'7" and that adult males tend to be about 2 inches taller than adults in general. Suppose further that you are a male older than 18. I could break your height into three components, one of which is the mean height of all American adults, one of which is your sex, and one of which is your own unique contribution. Thus, I could specify that your height is 5'7" plus 2 inches extra for being a male, plus or minus a couple of inches to account for the variability in height for males. (We could make this model even more complicated by allowing for height differences among different nationalities, but we won't do that here.) We can write this model as

Height = 5'7" + 2" + uniqueness

where "uniqueness" represents your deviation from the average for males. Another way to write it would be

Height = grand mean + gender component + uniqueness

If we want to represent the previous statement in more general terms, we can let μ stand for the mean height of the population of all American adults, τ_{male} stand for the extra component of being a male ($\tau_{male} = \mu_{male} - \mu$), and ε_{you} be your unique contribution to the model. Then our model becomes

$$X_{ij} = \mu + \tau_{male} + \varepsilon_{you}$$

Now let's move from our physical model of height to one that more directly underlies the analysis of variance. We will look at this model in terms of Eysenck's experiment on the recall of verbal material. Here X_{ij} represents the score of Person$_i$ in Condition$_j$ (e.g., X_{32} represents the third person in the Rhyming condition). We let μ represent the mean of all subjects who could theoretically be run in Eysenck's experiment, regardless of condition. The symbol μ_j represents the population mean of Condition$_j$ (e.g., μ_2 is the mean of the Rhyming condition), and τ_j is the degree to which the mean of Condition$_j$ deviates from the grand mean ($\tau_j = \mu_j - \mu$). Finally, ε_{ij} is the amount by which Person$_i$ in Condition$_j$ deviates from the mean of his or her group ($\varepsilon_{ij} = X_{ij} - \mu_j$). Imagine that you were a subject in the memory study by Eysenck that was just described. We can specify your score on that retention test as a function of these components.

$$X_{ij} = \mu + (\mu_j - \mu) + \varepsilon_{ij}$$
$$= \mu + \tau_j + \varepsilon_{ij}$$

structural model

This is the **structural model** that underlies the analysis of variance. In future chapters, we will extend the model to more complex situations, but the basic idea will remain the same. Of course, we do not know the values of the various parameters in this structural model, but that doesn't stop us from positing such a model.

Assumptions

As we know, Eysenck was interested in studying the level of recall under the five conditions. We can represent these conditions in Figure 11.1, where μ_j and σ_j^2 represent the mean and variance of whole populations of scores that would be obtained under each of these conditions. The analysis of variance is based on certain assumptions about these populations and their parameters.

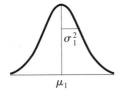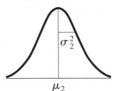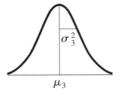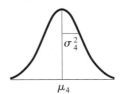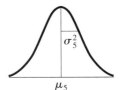

$$\sigma_1^2 \qquad \sigma_2^2 \qquad \sigma_3^2 \qquad \sigma_4^2 \qquad \sigma_5^2$$

$$\mu_1 \qquad \mu_2 \qquad \mu_3 \qquad \mu_4 \qquad \mu_5$$

Figure 11.1   Graphical representation of populations of recall scores

Homogeneity of Variance

A basic assumption underlying the analysis of variance is that each of our populations has the same variance. In other words,

$$\sigma_1^2 = \sigma_2^2 = \sigma_3^2 = \sigma_4^2 = \sigma_5^2 = \sigma_e^2$$

homogeneity of variance

homoscedasticity

error variance

where the notation σ_e^2 is used to indicate the common value held by the five population variances. This assumption is called the assumption of **homogeneity of variance,** or, if you like long words, **homoscedasticity.**

The subscript "e" stands for error, and this variance is the **error variance**—the variance unrelated to any treatment differences, which is variability of scores within the same condition. Homogeneity of variance would be expected to occur if the effect of a treatment is to add a constant to everyone's score—if, for example, everyone who thought of adjectives in Eysenck's study recalled five more words than they would otherwise have recalled.

heterogeneity
of variance

heteroscedasticity

As we will see later, under certain conditions the assumption of homogeneity of variance can be relaxed without substantially damaging the test. However, there are cases where **heterogeneity of variance,** or **heteroscedasticity** (populations having *different* variances) is a problem.

Normality

A second assumption of the analysis of variance is that the recall scores for each condition are normally distributed around their mean. In other words, each of the distributions in Figure 11.1 is normal. Because e_{ij} represents the variability of each person's score around the mean of that condition, our assumption really boils down to saying that error is normally distributed within conditions. Thus, you will often see the assumption stated in terms of "the normal distribution of error."

To anticipate, I should point out that moderate departures from normality are not usually fatal. We said much the same thing when looking at the *t* test for two independent samples, which is really just a special case of the analysis of variance.

Independence of Observations

Our third important assumption is that the observations are independent of one another. (Technically, this assumption really states that the error components [e_{ij}] are independent, but that amounts to the same thing.) Thus, for any two observations within an experimental treatment, we assume that knowing how one of these observations stands relative to the treatment (or population) mean tells us nothing about the other observation. This is one of the important reasons why subjects are randomly assigned to groups. Violation of the independence assumption can have serious consequences for an analysis (see Kenny & Judd, 1986).

The Null Hypothesis

As we know, Eysenck was interested in testing the *research* hypothesis that the level of recall varies with the level of processing. Support for such a hypothesis would come from rejection of the standard *null* hypothesis

$$H_0 : \mu_1 = \mu_2 = \mu_3 = \mu_4 = \mu_5$$

The null hypothesis could be false in a number of ways (e.g., all means could be different from each other, the first two could be equal to each other but different from the last three, and so on), but for now we are going to be concerned only with whether the null hypothesis is completely true or is false. In Chapter 12, we will deal with the problem of whether subsets of means are equal or unequal.

11.3 The Logic of the Analysis of Variance

The logic underlying the analysis of variance is really very simple, and once you understand it, the rest of the discussion will make considerably more sense. Consider for a moment the effect of our three major assumptions—normality, homogeneity of variance, and the independence of observations. By making the first two of these assumptions, we have said that the five distributions represented in Figure 11.1 have the same shape and dispersion. As a result, the only way left for them to differ is in their means. (Recall that the normal distribution depends only on two parameters, μ and σ.)

We will begin by making no assumption concerning H_0—it may be true or false. For any one treatment, the variance of the 10 scores in that group would be an estimate of the variance of the population from which the scores were drawn. Because we have assumed that all populations have the same variance, it is also one estimate of the common population variance σ_e^2. If you prefer, you can think of

$$\sigma_1^2 \doteq s_1^2, \qquad \sigma_2^2 \doteq s_2^2, \qquad \ldots, \qquad \sigma_e^2 \doteq s_e^2$$

where $\doteq$ is read as "is estimated by." Because of our homogeneity assumption, all these are estimates of σ_e^2. For the sake of increased reliability, we can pool the five estimates by taking their mean, if $n_1 = n_2 = \cdots = n_5$, and thus

$$\sigma_e^2 \doteq s_e^2 \doteq \bar{s}_j^2 \doteq \sum s_j^2 / k$$

MS_error

MS_within

where $k =$ the number of treatments (in this case, five).[1] This gives us one estimate of the population variance that we will later refer to as MS_{error} (read "mean square error"), or, sometimes, MS_{within}. It is important to note that this estimate does not depend on the truth or falsity of H_0 because s_j^2 is calculated on each sample separately. For the data from Eysenck's study, our pooled estimate of σ_e^2 will be

$$\sigma_e^2 \doteq (3.33 + 4.54 + 6.22 + 20.27 + 14.00)/5 = 9.67$$

Now let us assume that H_0 is true. If this is the case, then our five samples of 10 cases can be thought of as five independent samples from the same population (or, equivalently, from five identical populations), and we can produce another possible estimate of σ_e^2. Recall from Chapter 7 that the central limit theorem states that the variance of means drawn from the same population equals the variance of the population divided by the sample size. If H_0 is true, the sample means have been drawn from the same population (or identical ones, which amounts to the same thing), and therefore the variance of our five sample means estimates σ_e^2/n.

$$\frac{\sigma_e^2}{n} \doteq s_{\bar{X}}^2$$

where n is the size of each sample. Thus, we can reverse the usual order of things and calculate the variance of our sample means $(s_{\bar{X}}^2)$ to obtain the second estimate of σ_e^2:

$$\sigma_e^2 \doteq n s_{\bar{X}}^2$$

MS_treatment

This term is referred to as $MS_{\text{treatment}}$ often abbreviated as MS_{treat}; we will return to it shortly.

We now have two estimates of the population variance (σ_e^2). One of these estimates (MS_{error}) is independent of the truth or falsity of H_0. The other ($MS_{\text{treatment}}$) is an estimate of σ_e^2 only as long as H_0 is true (only as long as the conditions of the central limit theorem are met; namely, that the means are drawn from one population or several identical populations). Thus, if the two estimates agree, we will have support for the truth of H_0, and if they disagree, we will have support for the falsity of H_0.[2]

From the preceding discussion, we can concisely state the logic of the analysis of variance. To test H_0, we calculate two estimates of the population variance—one that is

---

[1] If the sample sizes were not equal, we would still average the five estimates, but in this case we would weight each estimate by the number of degrees of freedom for each sample—just as we did in Chapter 7.

[2] Students often have trouble with the statement that "means are drawn from the same population" when we know in fact that they are often drawn from logically distinct populations. It seems silly to speak of means of males and females as coming from one population when we know that these are really two different populations of people. However, if the population of scores for females is exactly the same as the population of scores for males, then we can legitimately speak of these as being the identical (or same) population of *scores,* and we can behave accordingly.

independent of the truth or falsity of H_0, and another that is dependent on H_0. If the two estimates agree, we have no reason to reject H_0. If they disagree sufficiently, we conclude that underlying treatment differences must have contributed to our second estimate, inflating it and causing it to differ from the first. Therefore, we reject H_0.

Variance Estimation

treatment effect

It might be helpful at this point to state without proof the two values that we are really estimating. We will first define the **treatment effect,** denoted τ_j, as $(\mu_j - \mu)$, the difference between the mean of treatment$_j$ (μ_j) and the grand mean (μ), and we will define σ_τ^2 as the variance of the true populations' means $(\mu_1, \mu_2, \ldots, \mu_5)$.[3]

$$\sigma_\tau^2 = \frac{\sum (\mu_j - \mu)^2}{k - 1} = \frac{\sum \tau_j^2}{k - 1}$$

expected value

In addition, recall that we defined the **expected value** of a statistic [written $E(\ )$] as its long-range average—the average value that statistic would assume over repeated sampling, and thus, our best guess as to its value on any particular trial. With these two concepts, we can state

$$E(MS_{error}) = \sigma_e^2$$

$$E(MS_{treat}) = \sigma_e^2 + \frac{n \sum \tau_j^2}{k - 1}$$
$$= \sigma_e^2 + n\sigma_\tau^2$$

where σ_e^2 is the variance within each population and σ_τ^2 is the variance of the population means (μ_j).

Now, if H_0 is true and $\mu_1 = \mu_2 = \cdots = \mu_5 = \mu$, then the population means don't vary and $\sigma_\tau^2 = 0$,

$$E(MS_{error}) = \sigma_e^2$$

and

$$E(MS_{treat}) = \sigma_e^2 + n(0) = \sigma_e^2$$

and thus

$$E(MS_{error}) = E(MS_{treat})$$

Keep in mind that these are expected values; rarely in practice will the two sample-based mean squares be numerically equal.

If H_0 is false, however, the σ_τ^2 will not be zero, but some positive number. In this case,

$$E(MS_{error}) < E(MS_{treat})$$

because MS_{treat} will contain a nonzero term representing the true differences among the μ_j.

11.4   Calculations in the Analysis of Variance

At this point, we will use the example from Eysenck to illustrate the calculations used in the analysis of variance. Even though you may think that you will always use computer software to run analyses of variance, it is very important to understand how you would carry out

---

[3] Technically, σ_τ^2 is not actually a variance because, having the actual parameter (μ), we should be dividing by k instead of $k - 1$. Nonetheless, we lose very little by thinking of it as a variance, as long as we keep in mind precisely what we have done.

the calculations using a calculator. First, it helps you to understand the basic procedure. In addition, it makes it much easier to understand some of the controversies and alternative analyses that are proposed. Finally, no computer program will do everything you want it to do, and you must occasionally resort to direct calculations. So bear with me on the calculations, even if you think that I am wasting my time.

Sum of Squares

sums of squares

In the analysis of variance, much of our computation deals with **sums of squares.** As we saw in Chapter 9, a sum of squares is merely the sum of the squared deviations about the mean $\left(\sum (X - \overline{X})^2 \right)$ or, more often, some multiple of that. When we first defined the sample variance, we saw that

$$s_X^2 = \frac{\sum (X - \overline{X})^2}{n - 1} = \frac{\sum X^2 - \left(\sum X \right)^2 / n}{n - 1}$$

Here, the numerator is the *sum of squares* of X and the denominator is the degrees of freedom. Sums of squares have the advantage of being additive, whereas mean squares and variances are additive only if they happen to be based on the same number of degrees of freedom.

The Data

The data are reproduced in Table 11.2, along with the calculations and the results, which we will discuss in detail. Because these actual data points are fictitious (although the means and variances are not), there is little to be gained by examining the distribution of observations within individual groups—the data were actually drawn from a normally distributed population. With real data, however, it is important to examine these distributions first to make sure that they are not seriously skewed or bimodal and, even more important, that they are not skewed in different directions. Even for this example, it is useful to examine the individual group variances as a check on the assumption of homogeneity of variance. Although the variances are not as similar as we might like (the variance for Imagery is noticeably larger than the others), they do not appear to be so drastically different as to cause concern. As we will see later, the analysis of variance is robust against violations of assumptions, especially when we have the same number of observations in each group.

Table 11.2b shows the calculations required to perform a one-way analysis of variance. These calculations require some elaboration.

SS_{total}

SS_{total}

The SS_{total} (read "sum of squares total") represents the sum of squares of all the observations, regardless of which treatment produced them. Letting $\overline{X}_{..}$ represent the grand mean, the definitional formula is

$$SS_{total} = \sum (X_{ij} - \overline{X}_{..})^2$$

This is a term we saw much earlier when we were calculating the variance of a set of numbers and is the numerator for the variance. (The denominator was the degrees of freedom). This formula, like the ones that follow, is probably not the formula we would use if we were to do the hand calculations for this problem. The formulae are very susceptible to the effects of rounding error. However, they are perfectly correct formulae, and represent the way that we normally think about the analysis. For those who prefer more traditional hand-calculation formulae, they can be found in earlier editions of this book.

Table 11.2 Data and computations for example from Eysenck (1974)

(a) Data

	Counting	Rhyming	Adjective	Imagery	Intentional	Total
	9	7	11	12	10	
	8	9	13	11	19	
	6	6	8	16	14	
	8	6	6	11	5	
	10	6	14	9	10	
	4	11	11	23	11	
	6	6	13	12	14	
	5	3	13	10	15	
	7	8	10	19	11	
	7	7	11	11	11	
Mean	7.00	6.90	11.00	13.40	12.00	10.06
St. Dev.	1.83	2.13	2.49	4.50	3.74	4.01
Variance	3.33	4.54	6.22	20.27	14.00	16.058

(b) Computations

$$SS_{total} = \sum (X_{ij} - \overline{X}_{..})^2 = (9 - 10.06)^2 + (8 - 10.06)^2 + \cdots + (11 - 10.06)^2$$
$$= 786.82$$

$$SS_{treat} = n \sum (\overline{X}_j - \overline{X}_{..})^2 = 10\big((7 - 10.06)^2 + (6.90 - 10.06)^2 + \cdots + (12 - 10.06)^2\big)$$
$$= 10(35.152) = 351.52$$

$$SS_{error} = SS_{total} - SS_{treat} = 786.82 - 351.52 = 435.30$$

(c) Summary Table

Source	df	SS	MS	F
Treatments	4	351.52	87.88	9.08
Error	45	435.30	9.67	
Total	49	786.82		

SS_{treat}

SS_{treat}

The definitional formula for SS_{treat} is framed in the context of deviations of group means from the grand mean. Here we have

$$SS_{treat} = n \sum (\overline{X}_j - \overline{X}_{..})^2$$

You can see that SS_{treat} is just the sum of squared deviations of the treatment means around the grand mean, multiplied by n to give us an estimate of the population variance.

SS_{error}

SS_{error}

In practice, SS_{error} is obtained by subtraction. Because it can be easily shown that

$$SS_{total} = SS_{treat} + SS_{error}$$

then it must also be true that

$$SS_{error} = SS_{total} - SS_{treat}$$

This is the procedure presented in Table 11.2, and it makes our calculations easier. To present SS_{error} in terms of deviations from means, we can write

$$SS_{error} = \sum(X_{ij} - \overline{X}_j)^2$$

Here you can see that SS_{error} is simply the sum over groups of the sums of squared deviation of scores around their group's mean. This approach is illustrated in the following, where I have calculated the sum of squares within each of the groups. Notice that for each group there is absolutely no influence of data from other groups, and therefore, the truth or falsity of the null hypothesis is irrelevant to the calculations.

$$SS_{within\,Counting} = \sum\left((9 - 7.00)^2 + (8 - 7.00)^2 + \cdots + (7 - 7.00)^2\right) = \qquad 30.00$$

$$SS_{within\,Rhyming} = \sum\left((7 - 6.90)^2 + (9 - 6.90)^2 + \cdots + (7 - 6.90)^2\right) = \qquad 40.90$$

$$SS_{within\,Adjective} = \sum\left((11 - 11.00)^2 + (13 - 11.00)^2 + \cdots + (11 - 11.00)^2\right) = \quad 56.00$$

$$SS_{within\,Imagery} = \sum\left((12 - 13.4)^2 + (11 - 13.4)^2 + \cdots + (11 - 13.4)^2\right) = \qquad 182.40$$

$$SS_{within\,Intentional} = \sum\left((10 - 12.00)^2 + (19 - 12.00)^2 + \cdots + (11 - 12.00)^2\right) = \underline{126.00}$$

$$SS_{error} = \qquad\qquad\qquad\qquad\qquad\qquad\qquad\qquad\qquad\qquad 435.30$$

When we sum these individual terms, we obtain 435.30, which agrees with the answer we obtained in Table 11.2.

The Summary Table

summary table

Table 11.2c shows the **summary table** for the analysis of variance. It is called a summary table for the rather obvious reason that it summarizes a series of calculations, making it possible to tell at a glance what the data have to offer. In older journal articles, you will often find the complete summary table displayed. More recently, primarily to save space, usually just the resulting Fs (to be defined) and the degrees of freedom are presented.

Sources of Variation

The first column of the summary table contains the sources of variation—the word "variation" being synonymous with the phrase "sum of squares." As can be seen from the table, there are three sources of variation: the variation caused by treatments (variation among treatment means), the variation caused by error (variation within the treatments), and the total variation. These sources reflect the fact that we have partitioned the total sum of squares into two portions, one representing variability within the individual groups and the other representing variability among the several group means.

Degrees of Freedom

df_{total}

df_{treat}

df_{error}

The degrees of freedom column in Table 11.2c represents the allocation of the total number of degrees of freedom between the two sources of variation. With 49 df overall (i.e., $N - 1$), four of these are associated with differences among treatment means and the remaining 45 are associated with variability within the treatment groups. The calculation of df is probably the easiest part of our task. The total number of degrees of freedom (df_{total}) is always $N - 1$, where N is the total number of observations. The number of degrees of freedom between treatments (df_{treat}) is always $k - 1$, where k is the number of treatments. The number of degrees of freedom for error (df_{error}) is most easily thought of as what is left over and is obtained by subtracting df_{treat} from df_{total}. However, df_{error} can be calculated more directly as the sum of the degrees of freedom within each treatment.

To put this in a slightly different form, the total variability is based on N scores and therefore has $N - 1$ df. The variability of treatment means is based on k means and therefore has $k - 1$ df. The variability within any one treatment is based on n scores, and thus has $n - 1$ df, but because we sum k of these within-treatment terms, we will have k times $n - 1$ or $k(n - 1)$ df.

Mean Squares

We will now go to the MS column in Table 11.2c. (There is little to be said about the column labeled SS; it simply contains the sums of squares obtained in Table 11.2b.)

The column of mean squares contains our two estimates of σ_e^2. These values are obtained by dividing the sums of squares by their corresponding df. Thus, $351.52/4 = 87.88$ and $435.30/45 = 9.67$. We typically do not calculate MS_{total} because we have no use for it. If we were to do so, this term would equal $786.82/49 = 16.058$, which, as you can see from Table 11.2a, is the variance of all N observations, regardless of treatment. Although it is true that mean squares are variance estimates, it is important to keep in mind what variances these terms are estimating. Thus, MS_{error} is an estimate of the population variance (σ_e^2), regardless of the truth or falsity of H_0, and is actually the average of the variances within each group when the sample sizes are equal:

$$MS_{\text{error}} = (3.33 + 4.54 + 6.22 + 20.27 + 14.00)/5 = 9.67$$

However, MS_{treat} is not the variance of treatment means but, rather, is the variance of those means corrected by n to produce a second estimate of the population variance (σ_e^2).

The *F* Statistic

The last column in Table 11.2c, labeled F, is the most important one for testing the null hypothesis. F is obtained by dividing MS_{treat} by MS_{error}. There is a precise way and a sloppy way to explain why this ratio makes sense, and we will start with the latter. As mentioned earlier, MS_{error} is an estimate of the population variance (σ_e^2). I also said that MS_{treat} is an estimate of the population variance (σ_e^2) *if* H_0 is true, but not if it is false. If H_0 is true, then MS_{error} and MS_{treat} are both estimating the same thing, and as such they should be approximately equal. If this is the case, the ratio of one to the other will be approximately 1, give or take a certain amount for sampling error. Thus, all we have to do is to compute the ratio and determine whether it is close enough to 1 to indicate support for the null hypothesis.

So much for the informal way of looking at F. A more precise approach starts with the *expected mean squares* for error and treatments. From earlier in the chapter, we know

$$E(MS_{\text{error}}) = \sigma_e^2$$
$$E(MS_{\text{treat}}) = \sigma_e^2 + n\sigma_\tau^2$$

If we now form the ratio

$$\frac{E(MS_{\text{treat}})}{E(MS_{\text{error}})} = \frac{\sigma_e^2 + n\sigma_\tau^2}{\sigma_e^2}$$

The only time this ratio would have an expectation of 1 is when $\sigma_\tau^2 = 0$—that is, when H_0 is true and $\mu_1 = \cdots = \mu_5$.[4] When $\sigma_\tau^2 > 0$, the expectation will be greater than 1.

---

[4] As an aside, note that the expected value of F is not precisely 1 under H_0, although

$\frac{E(MS_{\text{treat}})}{E(MS_{\text{error}})} = 1$ if $\sigma_\tau^2 = 0$. To be exact, under H_0, $E(F) = \frac{df_{\text{error}}}{df_{\text{error}} - 2}$

For all practical purposes, nothing is sacrificed by thinking of F as having an expectation of 1 under H_0 and greater than 1 under H_1 (the alternative hypothesis).

The question that remains, however, is, How large a ratio will we accept without rejecting H_0 when we use not *expected* values but *obtained* mean squares, which are computed from data and are therefore subject to sampling error? The answer to this question lies in the fact that we can show that the ratio

$$F = MS_{\text{treat}}/MS_{\text{error}}$$

is distributed as F on $k - 1$ and $k(n - 1)$ df. This is the same F distribution discussed earlier in conjunction with testing the ratio of two variances (which in fact is what we are doing here). Note that the degrees of freedom represent the df associated with the numerator and denominator, respectively.

For our example, $F = 9.08$. We have 4 df for the numerator and 45 df for the denominator, and we can enter the F table (Appendix F) with these values. Appendix F, a portion of which is shown in Table 11.3, gives the critical values for $\alpha = .05$ and $\alpha = .01$. For our particular case, we have 4 and 45 df and, with linear interpolation, $F_{.05}(4,45) = 2.58$. Thus, if we have chosen to work at $\alpha = .05$, we would reject H_0 and conclude that there are significant differences among the treatment means.

Conclusions

On the basis of a significant value of F, we have rejected the null hypothesis that the treatment means in the population are equal. Strictly speaking, this conclusion indicates that at least one of the population means is different from at least one other mean, but we don't know exactly which means are different from which other means. We will pursue that topic in Chapter 12. By examining the data in Table 11.2, however, we can see that increased processing of the material is associated with increased levels of recall. For example, a strategy that involves associating images with items to be recalled leads to nearly twice the level of recall as does merely counting the letters in the items.

Results such as these give us important hints about how to go about learning any material, and highlight the poor recall to be expected from passive studying. Good recall, whether it be lists of words or of complex statistical concepts, requires active and "deep" processing of the material, which is in turn facilitated by noting associations between the to-be-learned material and other material that you already know. You have probably noticed that sitting in class and dutifully recording everything that the instructor says doesn't usually lead to the grades that you think such effort deserves. Now you know a bit about why.

11.5   Writing Up the Results

Reporting results for an analysis of variance is somewhat more complicated than reporting the results of a t test. This is because we not only want to indicate whether the overall F is significant, but we probably also want to make statements about the differences between individual means. We won't discuss tests on individual means until the next chapter, so this example will be incomplete. We will come back to it in Chapter 12. An abbreviated version of a statement about the results follows.

> In a test of the hypothesis that memory depends upon the level of processing of the material to be recalled, participants were divided into five groups of 10 participants each. The groups differed in the amount of processing of verbal material required by the instructions, varying from simply counting the letters in the words to be recalled to forming mental images evoked by each word. After going through the list of 27 words three times, participants were asked to recall as many items on the list as possible. A one-way analysis of variance revealed that there were significant differences among the means of the

Table 11.3   Abbreviated version of Appendix *F*, Critical Values of the *F* Distribution Where $\alpha = .05$

	Degrees of Freedom for Numerator									
df denom.	1	2	3	4	5	6	7	8	9	10
1	161.4	199.5	215.8	224.8	230.0	233.8	236.5	238.6	240.1	242.1
2	18.51	19.00	19.16	19.25	19.30	19.33	19.35	19.37	19.38	19.40
3	10.13	9.55	9.28	9.12	9.01	8.94	8.89	8.85	8.81	8.79
4	7.71	6.94	6.59	6.39	6.26	6.16	6.09	6.04	6.00	5.96
5	6.61	5.79	5.41	5.19	5.05	4.95	4.88	4.82	4.77	4.74
6	5.99	5.14	4.76	4.53	4.39	4.28	4.21	4.15	4.10	4.06
7	5.59	4.74	4.35	4.12	3.97	3.87	3.79	3.73	3.68	3.64
8	5.32	4.46	4.07	3.84	3.69	3.58	3.50	3.44	3.39	3.35
9	5.12	4.26	3.86	3.63	3.48	3.37	3.29	3.23	3.18	3.14
10	4.96	4.10	3.71	3.48	3.33	3.22	3.14	3.07	3.02	2.98
11	4.84	3.98	3.59	3.36	3.20	3.09	3.01	2.95	2.90	2.85
12	4.75	3.89	3.49	3.26	3.11	3.00	2.91	2.85	2.80	2.75
13	4.67	3.81	3.41	3.18	3.03	2.92	2.83	2.77	2.71	2.67
14	4.60	3.74	3.34	3.11	2.96	2.85	2.76	2.70	2.65	2.60
15	4.54	3.68	3.29	3.06	2.90	2.79	2.71	2.64	2.59	2.54
16	4.49	3.63	3.24	3.01	2.85	2.74	2.66	2.59	2.54	2.49
17	4.45	3.59	3.20	2.96	2.81	2.70	2.61	2.55	2.49	2.45
18	4.41	3.55	3.16	2.93	2.77	2.66	2.58	2.51	2.46	2.41
19	4.38	3.52	3.13	2.90	2.74	2.63	2.54	2.48	2.42	2.38
20	4.35	3.49	3.10	2.87	2.71	2.60	2.51	2.45	2.39	2.35
22	4.30	3.44	3.05	2.82	2.66	2.55	2.46	2.40	2.34	2.30
24	4.26	3.40	3.01	2.78	2.62	2.51	2.42	2.36	2.30	2.25
26	4.23	3.37	2.98	2.74	2.59	2.47	2.39	2.32	2.27	2.22
28	4.20	3.34	2.95	2.71	2.56	2.45	2.36	2.29	2.24	2.19
30	4.17	3.32	2.92	2.69	2.53	2.42	2.33	2.27	2.21	2.16
40	4.08	3.23	2.84	2.61	2.45	2.34	2.25	2.18	2.12	2.08
50	4.03	3.18	2.79	2.56	2.40	2.29	2.20	2.13	2.07	2.03
60	4.00	3.15	2.76	2.53	2.37	2.25	2.17	2.10	2.04	1.99
120	3.92	3.07	2.68	2.45	2.29	2.18	2.09	2.02	1.96	1.91
200	3.89	3.04	2.65	2.42	2.26	2.14	2.06	1.98	1.93	1.88
500	3.86	3.01	2.62	2.39	2.23	2.12	2.03	1.96	1.90	1.85
1000	3.85	3.01	2.61	2.38	2.22	2.11	2.02	1.95	1.89	1.84

five groups ($F(4, 45) = 9.08$, $p < .05$). Visual inspection of the group means revealed that the level of recall generally increased with the level of processing required, as predicted by the theory.

(*Note:* Further discussion of these differences will have to wait until Chapter 12.)

11.6   Computer Solutions

Most analyses of variance are now done using standard computer software, and Exhibit 11.1(a and b) contains examples of output from two major programs. These programs were chosen because they illustrate quite different ways of printing out the same information.

The first of these (Minitab) gives us the kind of printout we would most expect, with an appropriate array of descriptive information.

Notice the menu selections used to request the necessary analysis from Minitab. You could almost certainly have figured these out for yourself just by experimenting with the choices. Notice also that the output looks very similar to the summary table we calculated. In addition to the summary table, Minitab prints the means and confidence limits on those

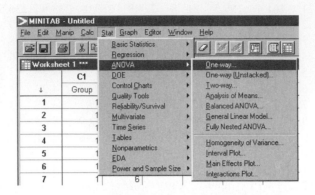

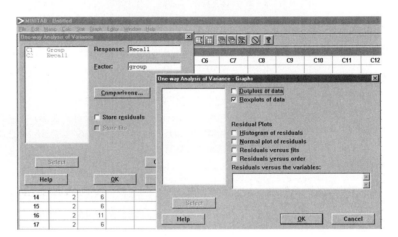

Analysis of Variance for Recall

Source	DF	SS	MS	F	P
Group	4	351.52	87.88	9.08	0.000
Error	45	435.30	9.67		
Total	49	786.82			

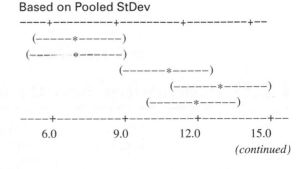

Individual 95% CIs for Mean
Based on Pooled StDev

Level	N	Mean	StDev				
				----+---------+---------+---------+--			
1	10	7.000	1.826	(------*------)			
2	10	6.900	2.132	(----- *------)			
3	10	11.000	2.494			(------*-----)	
4	10	13.400	4.502				(------*-----)
5	10	12.000	3.742			(------*-----)	
				----+---------+---------+---------+--			

Pooled StDev = 3.110   6.0   9.0   12.0   15.0

(continued)

Exhibit 11.1(a)   Minitab printout

Boxplots of Recall by Group
(means are indicated by solid circles)

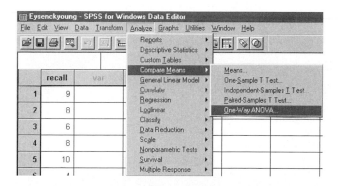

Exhibit 11.1(a)   *(continued)*

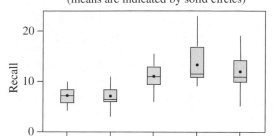

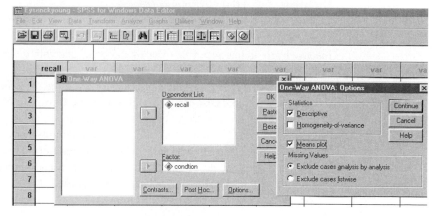

Descriptives

RECALL

	N	Mean	Std. Deviation	Std. Error	95% Confidence Interval for Mean		Minimum	Maximum
					Lower Bound	Upper Bound		
Counting	10	7.00	1.83	.58	5.69	8.31	4	10
Rhyming	10	6.90	2.13	.67	5.38	8.42	3	11
Adjective	10	11.00	2.49	.79	9.22	12.78	6	14
Imagery	10	13.40	4.50	1.42	10.18	16.62	9	23
Intentional	10	12.00	3.74	1.18	9.32	14.68	5	19
Total	50	10.06	4.01	.57	8.92	11.20	3	23

Exhibit 11.1(b)   SPSS One-Way printout

(continued)

ANOVA

RECALL

	Sum of Squares	df	Mean Square	F	Sig.
Between Groups	351.520	4	87.880	9.085	.000
Within Groups	435.300	45	9.673		
Total	786.820	49			

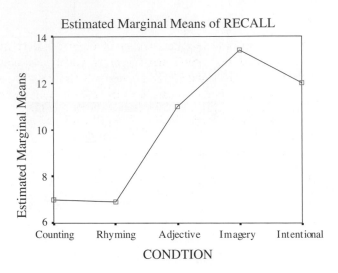

Exhibit 11.1(b)   *(continued)*

means. I also requested a boxplot of the data, and this gives a clear picture of what is happening.

Exhibit 11.1b gives the SPSS printout for the same data, along with the required menu selections. I used the **One-Way** selection from the **Compare Means** menu.

The output here also looks like what we computed, although you can see that it is presented somewhat differently from the way Minitab presents its output. You would get the same general results if you had selected **Analyze/General Linear Model/Univariate** from the menus, although the summary table would contain additional lines of information that I won't discuss until the end of this chapter and Chapter 13.

11.7   Derivation of the Analysis of Variance

Although I will not develop the statistical theory of the analysis of variance in detail, some insight into the operation of the underlying structural model on which the analysis is based is useful. For the student who wants a more complete understanding of the theory, Winer, Brown, and Michels (1991) cover this material in more depth but at a reasonable level. The student who wants a *less* complete understanding of the theory can skip directly to the next section, provided that his or her instructor agrees.

Given the standard assumptions of normality, homogeneity of variance, and independence of error, the model we are discussing leads directly to the statement made earlier that $SS_{total} = SS_{treat} + SS_{error}$, and this in turn produces the analysis of variance. To show this, we will start with the basic model

$$X_{ij} = \mu + \tau_j + e_{ij}$$

If we substitute the relevant statistics in place of the parameters, we obtain

$$X_{ij} = \overline{X}_{..} + (\overline{X}_j - \overline{X}_{..}) + (X_{ij} - \overline{X}_j)$$

where $\overline{X}_{..}$ represents the grand mean. This expression is an identity, as is readily seen by removing the parentheses and collecting terms. Subtracting $\overline{X}_{..}$ from both sides, we have

$$X_{ij} - \overline{X}_{..} = (\overline{X}_j - \overline{X}_{..}) + (X_{ij} - \overline{X}_j)$$

This equation states that the deviation of a subject's score from the grand mean is equal to the deviation of his group's mean from the grand mean plus the deviation of his score from the group mean, and again is an identity. These last two equations are illustrated geometrically in Figure 11.2. If we square both sides of the equation, we obtain

$$(X_{ij} - \overline{X}_{..})^2 = (\overline{X}_j - \overline{X}_{..})^2 + (X_{ij} - \overline{X}_j)^2 + 2(\overline{X}_j - \overline{X}_{..})(X_{ij} - \overline{X}_j)$$

Summing this expression over subjects (i) and treatments (j) produces

$$\sum\sum_{ij} (X_{ij} - \overline{X}_{..})^2 = \sum\sum_{ij} (\overline{X}_j - \overline{X}_{..})^2 + \sum\sum_{ij} (X_{ij} - \overline{X}_j)^2$$
$$+ 2\sum\sum_{ij} (\overline{X}_j - \overline{X}_{..})(X_{ij} - \overline{X}_j)$$

The last term will be zero and will drop out, leaving

$$\sum\sum_{ij} (X_{ij} - \overline{X}_{..})^2 = \sum\sum_{ij} (\overline{X}_j - \overline{X}_{..})^2 + \sum\sum_{ij} (X_{ij} - \overline{X}_j)^2$$

But within any treatment $\overline{X}_j - \overline{X}_{..}$ is a constant, and as a result, we have

$$\sum\sum_{ij} (X_{ij} - \overline{X}_{..})^2 = n\sum_j (\overline{X}_j - \overline{X}_{..})^2 + \sum\sum_{ij} (X_{ij} - \overline{X}_j)^2$$

$$SS_{\text{total}} \quad = \quad SS_{\text{treat}} \quad + \quad SS_{\text{error}}$$

Thus we have shown not only that the total sum of squares is completely partitioned into SS_{treat} and SS_{error}, but also that all of this can be derived on the basis of our underlying structural model using simple algebra. From here, we could proceed to obtain our mean squares and F.

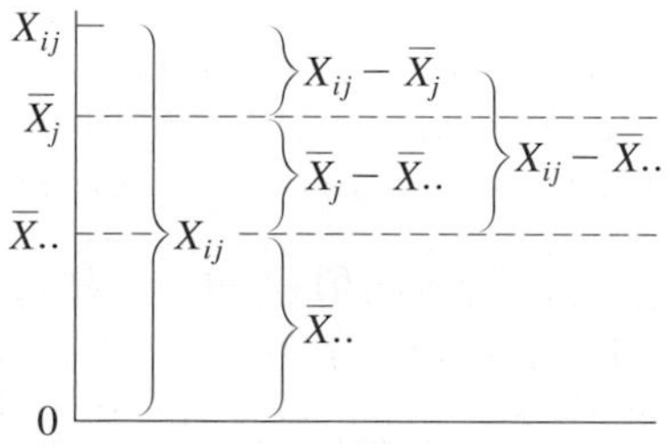

Figure 11.2   Geometric representation of the analysis of variance model

11.8 Unequal Sample Sizes

balanced designs

Most experiments are originally designed with the idea of collecting the same number of observations in each treatment. (Such designs are generally known as **balanced designs.**) Frequently, however, things do not work out that way. Subjects fail to arrive for testing, or are eliminated because they fail to follow instructions. Animals occasionally become ill during an experiment from causes that have nothing to do with the treatment. There is even a case in the literature in which an animal was eliminated from the study for repeatedly biting the experimenter (Sgro & Weinstock, 1963). Moreover, studies conducted on intact groups, such as school classes, have to contend with the fact that such groups nearly always vary in size.

If the sample sizes are not equal, the analysis discussed earlier needs to be modified. For the case of one independent variable, however, this modification is relatively minor.[5]

Earlier we defined

$$SS_{treat} = n \sum (\overline{X}_j - \overline{X}_{..})^2$$

We were able to multiply the deviations by n because n was common to all treatments. If the sample sizes differ, however, and we define n_j as the number of subjects in the jth treatment $\left(\sum n_j = N \right)$, we can rewrite the expression as

$$SS_{treat} = \sum \left[n_j (\overline{X}_j - \overline{X}_{..})^2 \right]$$

which, when all n_j are equal, reduces to the original equation. This expression shows us that with unequal ns, the deviation of each treatment mean from the grand mean is weighted by the sample size. Thus, the larger the size of one sample relative to the others, the more it will contribute to SS_{treat}, all other things being equal.

Effective Therapies for Anorexia

The following example is taken from a study by Everitt that compared the effects of two therapy conditions and a control condition on weight gain in anorexic girls. The data are reported in Hand et al. (1994). Everitt used a control condition that received no intervention, a cognitive-behavioral treatment condition, and a family therapy condition. The dependent variable analyzed here was the gain in weight over a fixed period. The data are given in Table 11.4.

The computation of the analysis of variance follows, and you can see that the change required by the presence of unequal sample sizes is minor. I should hasten to point out that unequal sample sizes will not be so easily dismissed when we come to more complex designs, but there is no particular difficulty with the one-way design.

$$\begin{aligned}
SS_{total} &= \sum (X_{ij} - \overline{X}_{..})^2 \\
&= \left[(-0.5 - 2.76)^2 + (-9.3 - 2.76)^2 + \cdots + (10.7 - 2.76)^2 \right] \\
&= 4525.386
\end{aligned}$$

$$\begin{aligned}
SS_{treat} &= \sum n_j (\overline{X}_j - \overline{X}_{..})^2 \\
&= 26 \times (-0.45 - 2.76)^2 + 29 \times (3.01 - 2.76)^2 + (17 \times (7.26 - 2.76)^2) \\
&= 614.644
\end{aligned}$$

$$\begin{aligned}
SS_{error} &= SS_{total} - SS_{treat} = 4525.386 - 614.644 \\
&= 3910.742
\end{aligned}$$

---

[5] Discussion of alternative approaches to missing data can be found at http://www.uvm.edu/~dhowell/StatPages/ under the heading of "New Material." The approach presented in this chapter is the most common approach to missing data, but the issues become more involved when we move to more complex designs.

Table 11.4   Data from Everitt (1994) on the treatment of anorexia in young girls

	Control	Cognitive-Behav Therapy	Family Therapy	Total
	−.5	1.7	11.4	
	−9.3	.7	11.0	
	−5.4	−.1	5.5	
	12.3	−.7	9.4	
	−2.0	−3.5	13.6	
	−10.2	14.9	−2.9	
	−12.2	3.5	−.1	
	11.6	17.1	7.4	
	−7.1	−7.6	21.5	
	6.2	1.6	−5.3	
	−.2	11.7	−3.8	
	−9.2	6.1	13.4	
	8.3	1.1	13.1	
	3.3	−4.0	9.0	
	11.3	20.9	3.9	
	.0	−9.1	5.7	
	−1.0	2.1	10.7	
	−10.6	−1.4		
	−4.6	1.4		
	−6.7	−.3		
	2.8	−3.7		
	.3	−.8		
	1.8	2.4		
	3.7	12.6		
	15.9	1.9		
	−10.2	3.9		
		.1		
		15.4		
		−.7		
Mean	−0.45	3.01	7.26	2.76
St. Dev.	7.989	7.308	7.157	7.984
Variance	63.819	53.414	51.229	63.738
n	26	29	17	72

The summary table for this analysis follows:

Source	df	SS	MS	F
Treatments	2	614.644	307.322	5.422*
Error	69	3910.742	56.677	
Total	71	4525.386		

* $p < .05$

From the summary table you can see that there is a significant effect due to treatment. The presence of this effect is clear in Figure 11.3, where the control group showed no

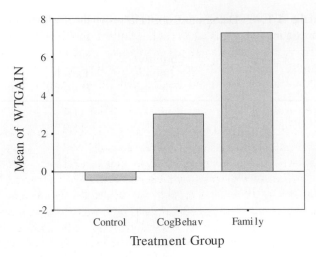

Figure 11.3 Mean weight gain in Everitt's three groups

appreciable weight gain, whereas the other two groups showed substantial gain. We do not yet know whether the Cognitive-behavior group and the Family therapy group were significantly different, or whether they both differed from the Control group, but we will reserve that problem until the next chapter.

11.9 Violations of Assumptions

As we have seen, the analysis of variance is based on the assumptions of normality and homogeneity of variance. In practice, however, the analysis of variance is a robust statistical procedure, and the assumptions frequently can be violated with relatively minor effects. This is especially true for the normality assumption. For studies dealing with this problem, see Boneau (1960), Box (1953, 1954a, 1954b), Bradley (1964), and Grissom (2000). The latter reference is somewhat more pessimistic than the others, but there is still reason to believe that normality is not a crucial assumption and that the homogeneity of variance assumption can be violated without terrible consequences.

In general, if the populations can be assumed to be symmetric, or at least similar in shape (e.g., all negatively skewed), and if the largest variance is no more than four times the smallest, the analysis of variance is most likely to be valid. It is important to note, however, that heterogeneity of variance and unequal sample sizes do not mix. If you have reason to anticipate unequal variances, make every effort to keep your sample sizes as equal as possible. This is a serious issue, and people tend to forget that noticeably unequal sample sizes make the test appreciably less robust to heterogeneity of variance.

In Chapter 7, we considered the Levene (1960) test for heterogeneity of variance, and I mentioned the existence of the test by O'Brien (1981). The Levene test is essentially a *t* test on the deviations (absolute or squared) of observations from their sample mean or median. If one group has a larger variance than another, then the deviations of scores from the mean or median will also, on average, be larger than for a group with a smaller variance. Thus, a significant *t* test on the absolute values of the deviations represents a test on group variances. Each of those tests can be readily extended to the case of more than two groups in obvious ways. The only difference is that with multiple groups the *t* test on the deviations would be replaced by an analysis of variance on those deviations. Evidence suggests that the Levene test is the weaker of the two, but it is the one traditionally reported by most statistical software. Wilcox (1987b) reports that this test appears to be conservative.

If you are not willing to ignore the existence of heterogeneity or nonnormality in your data, there are alternative ways of handling the problems that result. Box (1954a) has shown that with unequal variances the appropriate F distribution against which to compare F_{obt} is a regular F with altered degrees of freedom. If we define the true critical value of F (adjusted for heterogeneity of variance) as F'_α, then Box has proven that

$$F_\alpha(1, n-1) \geq F'_\alpha \geq F_\alpha[k-1, k(n-1)]$$

In other words, the true critical value of F lies somewhere between the critical value of F on 1 and $(n-1)$ df and the critical value of F on $(k-1)$ and $k(n-1)$ df. This latter limit is the critical value we would use if we met the assumptions of normality and homogeneity of variance. Box suggested a conservative test by comparing F_{obt} to $F_\alpha(1, n-1)$. If this leads to a significant result, then the means are significantly different regardless of the equality, or inequality, of variances. (For those of you who raised your eyebrows when I cavalierly declared the variances in Eysenck's study to be "close enough," it is comforting to know that even Box's conservative approach would lead to the conclusion that the groups are significantly different: $F_{.05}(1, 9) = 5.12$, whereas our obtained F was 9.08.)

The only difficulty with Box's approach is that it is extremely conservative. There are, however, alternative methods. Box (1954a) presented formulae for estimating the actual number of degrees of freedom to use in evaluating F. (These formulae may be found in Myers [1979].) A different approach is one proposed by Welch (1951), which we will consider in the next section, and which is implemented by much of the statistical software that we use.

Wilcox (1987b) has argued that, in practice, variances frequently differ by more than a factor of four, which is often considered a reasonable limit on heterogeneity. He has some strong opinions concerning the consequences of heterogeneity of variance. He recommends Welch's procedure with samples having different variances, especially when the sample sizes are unequal. Tomarken and Serlin (1986) have investigated the robustness and power of Welch's procedure and the procedure proposed by Brown and Forsythe (1974). They have shown Welch's test to perform well under several conditions. The Brown and Forsythe test also has advantages in certain situations. The Tomarken and Serlin paper is a good reference for those concerned with heterogeneity of variance.

The Welch Procedure

Kohr and Games (1974) and Keselman, Games, and Rogan (1979) have investigated alternative approaches to the treatment of samples with heterogeneous variances (including the one suggested by Box) and have shown that a procedure proposed by Welch (1951) has considerable advantages both in power and protection against Type I errors, at least when sampling from normal populations. The formulae and calculations are somewhat awkward, but not particularly difficult, and you should use them whenever a test, such as Levene's (discussed in Chapter 7), indicates heterogeneity of variance—especially when you have unequal sample sizes.

Define

$$w_k = \frac{n_k}{s_k^2}$$

$$\overline{X}' = \frac{\sum w_k \overline{X}_k}{\sum w_k}$$

Then

$$F'' = \frac{\dfrac{\sum w_k(\overline{X}_k - \overline{X}')^2}{k-1}}{1 + \dfrac{2(k-2)}{k^2-1} \sum \left(\dfrac{1}{n_k-1}\right)\left(1 - \dfrac{w_k}{\sum w_k}\right)^2}$$

This statistic (F'') is approximately distributed as F on $k - 1$ and df' degrees of freedom, where

$$df' = \frac{k^2 - 1}{3 \sum \left(\frac{1}{n_k - 1}\right)\left(1 - \frac{w_k}{\sum w_k}\right)^2}$$

These formulae are messy, but they are not impossible to use. If you collect all of the terms (such as w_k) first and then work systematically through the problem, you should have no difficulty. (Formulae like this are actually very easy to implement if you have access to any spreadsheet program.) When you have only two groups, it is probably easier to fall back on a t test with heterogeneous variances, using the approach (also attributable to Welch) taken in Chapter 7.

11.10 Transformations

In the preceding section, we considered one approach to the problem of heterogeneity of variance—calculate F'' on the heterogeneous data and evaluate it against the usual F distribution on an adjusted number of degrees of freedom. This procedure has been shown to work well when samples are drawn from normal populations. But little is known about its behavior with nonnormal populations. An alternative approach is to transform the data to a form that yields homogeneous variances and then run a standard analysis of variance on the transformed values.

Most people find it difficult to accept the idea of transforming data. It somehow seems dishonest to decide that you do not like the data you have and therefore to change them into data you like better or, even worse, to throw out some of them and pretend they were never collected. When you think about it, however, there is really nothing unusual about transforming data. We frequently transform data. We sometimes measure the *time* it takes a rat to run through a maze, but then look for group differences in running *speed*, which is the reciprocal of time (a nonlinear transformation). We measure sound in terms of physical energy, but then report it in terms of decibels, which represents a logarithmic transformation. We ask a subject to adjust the size of a test stimulus to match the size of a comparison stimulus, and then take the radius of the test patch setting as our dependent variable—but the *radius* is a function of the square root of the *area* of the patch, and we could just as legitimately use area as our dependent variable. On some tests, we calculate the number of items that a student answered correctly, but then report scores in percentiles—a decidedly nonlinear transformation. Who is to say that speed is a "better" measure than time, that decibels are better than energy levels, that radius is better than area, or that a percentile is better than the number correct? Consider a study by Conti and Musty (1984) on the effects of THC (the most psychoactive ingredient in marijuana) on locomotor activity in rats. Conti and Musty measured activity by reading the motion of the cage from a transducer that represented that motion in voltage terms. In what way could their electrically transduced measure of test-chamber vibration be called the "natural" measure of activity? More important, they took postinjection activity as a percentage of pre-injection activity as their dependent variable, but would you leap out of your chair and cry "Foul!" because they had used a transformation? Of course you wouldn't—but it was a transformation nonetheless.

As pointed out earlier in this book, our dependent variables are only convenient and imperfect indicators of the underlying variables we want to study. No sensible experimenter ever started out with the serious intention of studying, for example, the "number of stressful life events" that a subject reports. The real purpose of such experiments has always been to study *stress*, and the number of reported events is merely a convenient measure of stress. Actually, stress probably does not vary in a linear fashion with number of events. It is quite possible that it varies exponentially—you can take a few stressful events in stride, but once you have a few on your plate, additional ones start having greater and greater effects. If this is true, the number of events raised to some power—for example, $Y = (\text{number of events})^2$—might be a more appropriate variable.

The point of this fairly extended, but necessary, digression is to encourage flexibility. You should not place blind faith in your original numbers; you must be willing to consider possible transformations. Tukey probably had the right idea when he called these calculations "reexpressions" rather than "transformations." You are merely reexpressing what the data have to say in other terms.

Having said that, it is important to recognize that conclusions that you draw on transformed data do not always transfer neatly to the original measurements. Grissom (2000) reports that the means of transformed variables can occasionally reverse the difference of means of the original variables. This is disturbing, and it is important to think about the meaning of what you are doing, but it is not, in itself, a reason to rule out the use of transformations.

If you are willing to accept that it is permissible to transform one set of measures into another—for example, $Y_i = \log(X_i)$ or $Y_i = \sqrt{X_i}$—then many possibilities become available for modifying our data to fit more closely the underlying assumptions of our statistical tests. The nice thing about most of these transformations is that when we transform the data to meet one assumption, we often come closer to meeting other assumptions as well. Thus, a square-root transformation may help us equate group variances, and because it compresses the upper end of a distribution more than it compresses the lower end, it may also have the effect of making positively skewed distributions more nearly normal in shape.

A word is in order about reporting transformed data. Although it is legitimate and proper to run a statistical test, such as the analysis of variance, on the transformed values, we often report means in the units of the untransformed scale. This is especially true when the original units are intrinsically meaningful.

One example is the salaries of baseball players from different teams. People who work with salary figures routinely perform their analyses on log(salary). However, log(salary) is not a meaningful measure to most of us. A better approach would be to convert all data to logs (assuming you have chosen to use a logarithmic transformation), find the mean of those log values, and then take the antilog to convert that mean back to the original units. This converted mean almost certainly will not equal the mean of the original values, but it is this converted mean that should be reported. But I would urge you to look at both the converted and unconverted means and make sure that they are telling the same basic story. Do not convert standard deviations—you will do serious injustice if you try that. And be sure to indicate to your readers what you have done.

In this chapter, we will consider only the most common transformations because they are the ones that will be most useful to you. Excellent discussions of the whole approach to transformations can be found in Tukey (1977), Hoaglin, Mosteller, and Tukey (1985), and Grissom (2000). Although the first two presentations are framed in the language of exploratory data analysis, you should not have much difficulty following them if you invest a modest amount of time in learning the terminology.

Logarithmic Transformation

The logarithmic transformation is useful whenever the standard deviation is proportional to the mean. It is also useful when the data are markedly positively skewed. The easiest way to appreciate why both of these statements are true is to recall what logarithms do. (Remember that a logarithm is a power—$\log_{10}(25)$ is the power to which 10 must be raised to give

25; therefore, $\log_{10}(25) = 1.39794$ because $10^{1.39794} = 25$.) If we take the numbers 10, 100, and 1000, their logs are 1, 2, and 3. Thus, the distance between 10 and 100, in log units, is now equivalent to the distance between 100 and 1,000. In other words, the right side of the distribution (more positive values) will be compressed more than will the left side by taking logarithms. (Why do you suppose that the salaries of baseball players offer a good example?) This means that positively skewed distributions tend toward symmetry under logarithmic transformations, and it means that if a set of relatively large numbers has a large standard deviation whereas a set of small numbers has a small standard deviation, taking logs will reduce the standard deviation of the sample with large numbers more than it will reduce the standard deviation of the sample with small numbers.

Table 11.5 contains an example from a study by Conti and Musty (1984) on activity levels in rats following administration of THC, the active ingredient in marijuana. I have

Table 11.5   Original and transformed data from Conti and Musty (1984)

(a) Original Data

	Control	0.1 µg	0.5 µg	1 µg	2 µg	
	130	93	510	229	144	
	94	444	416	475	111	
	225	403	154	348	217	
	105	192	636	276	200	
	92	67	396	167	84	
	190	170	451	151	99	
	32	77	376	107	44	
	64	353	192	235	84	
	69	365	384		284	
	93	422			293	
Mean	109.40	258.60	390.56	248.50	156.00	
						$r = .88$
S.D.	58.50	153.32	147.68	118.74	87.65	
Variance	3421.82	23,506.04	21,809.78	14,098.86	7682.22	

(b) Log Data

	Control	0.1 µg	0.5 µg	1 µg	2 µg	
	2.11	1.97	2.71	2.36	2.16	
	1.97	2.65	2.62	2.68	2.04	
	2.35	2.60	2.19	2.54	2.34	
	2.02	2.28	2.80	2.44	2.30	
	1.96	1.83	2.60	2.22	1.92	
	2.28	2.23	2.65	2.18	2.00	
	1.50	1.89	2.58	2.03	1.64	
	1.81	2.55	2.28	2.37	1.92	
	1.84	2.56	2.58		2.45	
	1.97	2.62			2.47	
Mean	1.981	2.318	2.557	2.353	2.124	
						$r = -.33$
S.D.	0.241	0.324	0.197	0.208	0.268	
Variance	0.058	0.105	0.039	0.043	0.072	

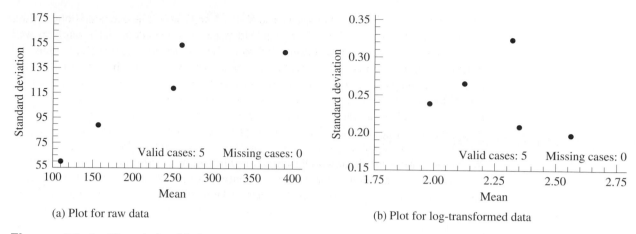

Figure 11.4   The relationship between means and standard deviations for original and transformed values of the data in Table 11.5

reported the activity units (on an arbitrary scale) for each animal over the 10-minute postinjection period, whereas Conti and Musty reported postinjection activity as a percentage of baseline activity. From the data in Table 11.5, you can see that the variances are unequal: The largest variance is nearly seven times the smallest. This is partly a function of the well-established fact that drugs tend to increase variability as well as means. Not only are the variances unequal, but the standard deviations appear to be proportional to the means. This is easily seen in Figure 11.4a, where I have plotted the standard deviations on the ordinate and the means on the abscissa. The linear relationship between these two statistics ($r = .88$) suggests that a logarithmic transformation might be useful. In Table 11.5b, the data have been transformed to logarithms to the base 10. (I could have used any base and still had the same effect. I chose base 10 because of its greater familiarity.) Here the means and the standard deviations are no longer correlated, as can be seen in Figure 11.4b ($r = -.33$: nonsignificant). We have broken up the proportionality between the mean and the standard deviation, and the largest group variance is now less than three times the smallest.

An analysis of variance could now be run on these transformed data. In this case, we would find $F(4,42) = 7.2$, which is clearly significant. Conti and Musty chose to run their analysis of variance on the proportion measures, as I said earlier, both for theoretical reasons and because that is standard practice in this area of research. A case might be made, however, that a logarithmic transformation of the original units might be a more appropriate one for future analyses, especially if problems occur with respect to either the shapes of the distributions or heterogeneity of variance.

As noted earlier, it makes no difference what base you use for a logarithmic transformation, and most statisticians tend to use $\log_e$. Regardless of the base, however, there are problems when the original values (X_i) are negative or near zero because logs are only defined for positive numbers. In this case, you should add a constant to make all X values positive before taking the log. In general, when you have near-zero values, you should use $\log(X_i + 1)$ instead of $\log(X_i)$. If the numbers themselves are less than -1, add whatever constant is necessary to make them all greater than zero.

Square-Root Transformation

When the data are in the form of counts (e.g., number of bar presses), the mean is often proportional to the *variance* rather than to the standard deviation. In this case, $Y = \sqrt{X}$ is sometimes useful for stabilizing variances and decreasing skewness. If the values of X are fairly small (i.e., less than 10), then $Y = \sqrt{X + 0.5}$ or $Y = \sqrt{X} + \sqrt{X + 1}$ is often better for stabilizing variances. For the Conti and Musty data, the mean correlates nearly as well with the variance as it does with the standard deviation (because standard deviations and variances are themselves highly correlated if the range of values is not large [in this case $r_{s \cdot s^2} = .99$]). Therefore, you might want to investigate how a square-root transformation affects the data.

Reciprocal Transformation

When you have a distribution with very large values in the positive tail, a reciprocal transformation may dramatically reduce the influence of those extreme values. For example, animals in a maze or straight alley often seem to forget their jobs and stop to sniff at all the photocells and such that they find along the way. Once an animal has been in the apparatus for 30 seconds, it does not matter to us if he takes another 300 seconds to complete the run. One approach was referred to in Chapter 2—if there are several trials per day, you might take the daily median time as your measure. An alternative approach is to use all the data but to take the reciprocal of time (i.e., speed) because it has the effect of nearly equating long times. Suppose that we collected the following times:

[10, 11, 13, 14, 15, 45, 450]

The reciprocals of these times are

[0.100, 0.091, 0.077, 0.071, 0.067, 0.022, 0.002]

Notice that the differences among the longer times are much reduced from what they were in the original units. Moreover, the outliers will have considerably less effect on the size of the standard deviation than they had before the transformation. Similar kinds of effects are found when we apply reciprocal transformations to reaction times, where long reaction times probably indicate less about information-processing speeds than they do about the fact that the subject was momentarily not paying attention or missed the response key that she was supposed to hit.

The Arcsine Transformation

In Chapter 5, we saw that for the binomial distribution, $\mu = Np$ and $\sigma^2 = Npq$. In this case, then, because both the mean and the variance are dependent on p, the variance will be a direct function of the mean. Suppose that for some experiment our dependent variable was the proportion of items recalled correctly. Then each item can be thought of as a Bernoulli trial with probability p of being correct (and probability $1 - p$ of being incorrect), and the whole set of items can be thought of as a series of Bernoulli trials. In other words, the results would have a binomial distribution where the variance is dependent on the mean. If this is so, groups with different means would necessarily have different variances, and we would have a problem. For this situation, the arcsine transformation is often helpful. The usual form of this transformation is $Y = 2 \arcsin \sqrt{p}$. In this case, p is the proportion correct and Y will be twice the angle whose sine equals the square root of p.[6] The arcsine transformation can be obtained with most calculators and is presented in any handbook of statistical tables.

---

[6] The arcsine transformation is often referred to as an "angular" transformation because of this property. When p is close to 0 or 1, we often take $2 \arcsin \sqrt{p \pm 1/2n}$, where the $+$ is used when p is close to 0, and the minus when p is close to 1.

Both the square-root and arcsine transformations are suitable when the variance is proportional to the mean. There is, however, a difference between them. The square-root transformation compresses the upper tail of the distribution, whereas the arcsine transformation stretches out both tails relative to the middle. Normally the arcsine is more helpful when you are dealing with proportions.

Trimmed Samples

heavy-tailed distributions

Rather than transforming each of your raw scores to achieve homogeneity of variance or normality, an alternative approach with **heavy-tailed distributions** (relatively flat distributions that have an unusual number of observations in the tails) is to use trimmed samples. In Chapter 2, a *trimmed sample* was defined as a sample from which a fixed percentage of the extreme values in each tail has been removed. Thus, with 40 cases, a 5% trimmed sample will be the sample with two of the observations in each tail eliminated. When comparing several groups, as in the analysis of variance, you would trim each sample by the same amount. Although trimmed samples have been around in statistics for a very long time, they have recently received a lot of attention because of their usefulness in dealing with distributions with occasional outliers. You will probably see more of them in the future.

Winsorized samples

Closely related to trimmed samples are **Winsorized samples,** in which the trimmed values are replaced by the most extreme value remaining in each tail. Thus, a 10% Winsorization of

3	7	12	15	17	17	18	19	19	19
20	22	24	26	30	32	32	33	36	50

would replace the two lowest values (3 and 7) by 12s and the two highest values (36 and 50) by 33s, leaving

12	12	12	15	17	17	18	19	19	19
20	22	24	26	30	32	32	33	33	33

The variance is based on $N - 2$ *df*, but subsequent standard errors must be adjusted to account for the fact that pseudovalues have replaced actual data and artificially narrowed the distribution of the data. We will see this adjustment in Chapter 12 (page 363). Experiments with samples containing an unusual number of outliers may profit from trimming or "Winsorizing." When you run an analysis of variance on trimmed data, however, you should base the MS_{error} on the variance of the corresponding Winsorized sample and not on the variance of the trimmed sample. Yuen and Dixon (1973) conducted a fairly readable study of the effect of applying *t* tests (and, by extension, the analysis of variance) to trimmed samples; you should read it before running such analyses. You should also look at papers by Wilcox (1993, 1995). A useful reference when we come to multiple comparisons in Chapter 12 is Keselman, Holland, and Cribbie (2005, pp. 1918–1919).

When to Transform and How to Choose a Transformation

You should not get the impression that transformations should be applied routinely to all of your data. As a rule of thumb, "If it's not broken, don't fix it." If your data are reasonably distributed (i.e., are more or less symmetrical and have few, if any, outliers) and if your variances are reasonably homogeneous, there is probably nothing to be gained by applying a transformation. If you have markedly skewed data or heterogeneous variances, however, some form of transformation may be useful. Furthermore, it is perfectly legitimate to shop around for a transformation that makes the necessary changes to the variance or shape. If a logarithmic transformation does not do what you want (stabilize the variances or improve shape), then consider the square-root (or cubed-root) transformation. If you have near-zero values and $Y = \sqrt{X + 0.5}$ does not work, try $Y = \sqrt{X} + \sqrt{X + 1}$. The only thing that you should *not* do is to try out every transformation, looking for one that gives you a significant result. (You are trying to optimize the *data,* not the resulting F.) Finally, if you are considering using transformations, it would be a good idea to look at Tukey (1977) or Hoaglin et al. (1983).

Resampling

An old approach to statistical hypothesis testing that is beginning to win a lot of adherents is known as "resampling statistics." I say a great deal about this approach in Chapter 18, but before leaving methods for dealing with violations of assumptions, I should at least mention that resampling methods offer the opportunity to avoid some of the assumptions required in the analysis of variance. These methods essentially create a population that exactly resembles the distribution of obtained data. Then the computer creates samples by drawing randomly, without replacement, from this population as if the null hypothesis were true, and calculates a test statistic, such as F, for that sample. This process is then repeated a very large number of times, producing a whole distribution of F values that would be expected with a true null hypothesis. It is then simple to calculate how many of these Fs were more extreme than the one from your data, and reject, or fail to reject, depending on the result. Students interested in this approach can jump to Chapter 18, which should not be too difficult to understand.

11.11 Fixed versus Random Models

We have not said anything about how we choose the levels of our independent variable; we have simply spoken of "treatments." Actually, if you think about it, we could obtain the levels of the treatment variable in at least two different ways: We could, and usually do, deliberately select them or we could sample them at random. The way in which the levels are derived has implications for the generalizations we might draw from our study.

Assume that we were hired as consultants by the Food and Drug Administration (FDA) and asked to run a study to compare the four most popular pain relievers. We will have four treatment levels (corresponding to the four pain relievers) that were selected by the FDA. If we chose to **replicate** the study (run it over again to verify our results), we would use exactly the same levels (drugs). In a sense, the treatment levels actually used have exhausted the levels of interest. The important point here is that the levels are *fixed* in the sense that they do not change randomly from one replication of the study to another. The analysis of such an experiment is referred to as a **fixed-model analysis of variance.**

Now assume that we are hired by the FDA again, but this time they merely tell us to compare a number of pain relievers to see whether "one brand is as good as the next." In this case, it would make sense to select *randomly* the pain relievers to be compared from the population of all available pain relievers. Here, the treatment levels are the result of a random process, and the population of interest with respect to pain relievers is quite large (probably over 50). Moreover, if we replicated this study we would again choose the brands randomly and would most likely have a whole new set of brands to compare. Because of the process by which treatment levels are obtained, we speak of treatments as a random variable and of the analysis as a **random-model analysis of variance.**

We will have more to say about fixed and random models as we go on. The important point at this time is that in a fixed model, the treatment levels are deliberately selected and

replicate

fixed-model analysis of variance

random-model analysis of variance

will remain constant from one replication to another. In our example of a fixed model, we actually set out to compare, for example, Bayer Aspirin with Anacin. In a random model, treatment levels are obtained by a random process and would be expected to vary across replications. In our example of a random model, we were studying *pain relievers,* and the ones that we happened to use were just random samples of pain relievers in general. For a one-way analysis of variance, the distinction is not particularly critical, but it can become quite important when we use more complex designs.

11.12 Magnitude of Experimental Effect

That an analysis of variance has produced a significant F simply tells us that there are differences among the means of treatments that cannot be attributed to error. It says nothing about whether these differences are of any practical importance. For this reason, we must look beyond the value of F to define an additional measure reflecting the "importance" of the difference. In previous chapters, I have made a distinction between the *d*-family of measures, which relate directly to differences among means, and the *r*-family of measures, which are based on correlations between the independent and dependent variables. When we are considering the omnibus F, which looks for any differences among the full set of means, *d*-family measures are not generally appropriate. They will become very appropriate, however, when we discuss multiple comparisons in Chapter 12. For the omnibus test of all means, the *r*-family of measures is more appropriate, and that is what I will focus on here. I must admit, however, that I don't find *r*-family measures particularly appealing because it is difficult to know what is a large, or a small, value for that measure. In some situations, explaining 5% of the variation may be very important, but in others 5% might be trivial.

The set of measures discussed here are often classed as "magnitude of effect" measures and are related to r^2. They represent how much of the overall variability in the dependent variable can be attributed to the treatment effect. In the next section, when we discuss power, we will look at what we have earlier called "effect size" measures, such as Cohen's *d*. In some ways, *d* is a more satisfactory measure of how important the differences in treatment means are than is an r^2 type measure, though *d* is particularly useful when we are examining contrasts among individual means or pairs of means.

magnitude of the experimental effect

At last count, there were at least six measures of the **magnitude of the experimental effect**—all different and most claiming to be less biased than some other measure. In this section, we will focus on only the two most common measures (η^2 and ω^2) because they have the strongest claim to our attention.

Eta-Squared (η^2)

eta-squared (η^2)

correlation ratio

Eta-squared is probably the oldest measure of the strength of an experimental effect. Although it is certainly not the best, it has several points to recommend it. As you will see, **eta-squared (η^2),** sometimes called the **correlation ratio,** has a certain intuitive appeal. Moreover, it forms a strong link between the traditional analysis of variance and multiple regression, as we will see in Chapter 16.

curvilinear regression

In some textbooks, eta (η) is defined as the correlation coefficient associated with **curvilinear regression**—that is, regression where the best-fitting line is not a straight line. Suppose that I proposed to calculate the correlation between the recall scores and the treatment levels (counting, rhyming, adjective, imagery, and intentional) for Eysenck's data from Table 11.2. The first criticism that would be raised is that the names Counting, . . . , intentional are merely labels for treatments and bear no relationship to anything. This would be true even if we called them treatment 1, 2, . . . , 5. True enough, but that will not stop us. The next objection raised might be that the treatments are not ordered on any particular underlying scale, and therefore we would not know in what order to place them if we were to plot the data. Again, true enough, and again that will not stop us. The next objection could be that the regression might not be linear. True again, but we can get around this problem by calling the coefficient η instead of r. Having cavalierly brushed aside all the objections, we set about plotting the data anyway, as shown in Figure 11.5. (The numbers 2, 3, and 4 in Figure 11.5 indicate the number of overlapping data points.) As you may recall from high school (or may not), a kth-order polynomial will exactly fit $k + 1$ points, which means that if we did try to fit a fourth-order polynomial to the five points represented by the treatment *means*, it would fit perfectly. (This is just an extension of the phrase "two points determine a straight line.") We do not particularly care what the equation would look like, but we can represent the line (as in Figure 11.5) simply by connecting the array means.

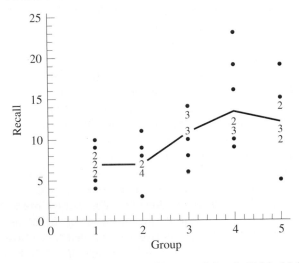

Figure 11.5   Scatter diagram of data in Table 11.2

You should recall that in Chapter 9 we saw that

$$r^2 = \frac{SS_{\text{total}} - SS_{\text{residual}}}{SS_{\text{total}}} = \frac{\sum (Y_{ij} - \overline{Y})^2 - \sum (Y_{ij} - \hat{Y}_{ij})^2}{\sum (Y_{ij} - \overline{Y})^2}$$

(Don't be confused by the fact that we routinely represent the dependent variable in regression discussions as Y, and the dependent variable in analysis of variance discussions as X. It really makes no difference what we call them.) We can apply this formula to the case of multiple groups by realizing that for each group, the predicted score for subjects in that group is the group mean. Thus, we can replace $\hat{Y}_{ij}$ with $\overline{Y}_j$. Doing this allows us to rewrite the equation as follows, substituting η^2 for r^2.

$$\eta^2 = \frac{SS_{\text{total}} - SS_{\text{error}}}{SS_{\text{total}}} = \frac{\sum (Y_{ij} - \overline{Y})^2 - \sum (Y_{ij} - \overline{Y}_j)^2}{\sum (Y_{ij} - \overline{Y})^2}$$

Note that I have relabeled SS_{residual} as SS_{error} in line with the terminology we use in talking about the analysis of variance and substituted $\overline{Y}_j$ for $\hat{Y}_{ij}$.[7]

---

[7] You will often see eta-squared given in computer printouts, such as SPSS General Linear Model, though it is usually labeled R^2.

Because $SS_{\text{total}} - SS_{\text{error}}$ is really $SS_{\text{treatment}}$, we can rewrite the last expression as

$$\eta^2 = \frac{SS_{\text{treatment}}}{SS_{\text{total}}}$$

We have now defined η^2 in terms of the sums of squares in the summary table of our analysis of variance. Applying η^2 to Eysenck's data in Table 11.2, we have

$$\eta^2 = \frac{SS_{\text{treatment}}}{SS_{\text{total}}} = \frac{351.52}{786.82} = .447$$

The equation for η^2 provides a simple way to estimate the maximum squared correlation between the independent variable and the dependent variable. Its derivation also points out that it can be treated as any other squared correlation coefficient, indicating the proportion of the variation accounted for by the independent variable. For Eysenck's data, 44.7% of the variation in recall scores can be attributed to differences in the instructions given to the groups, and therefore, presumably, to the depth to which the items were processed. This is an unusually large amount of explained variation, reflecting the extreme nature of group differences.

Proportional Reduction in Error (PRE)

There is another way to look at η^2 that derives directly from the last formula and that has been viewed recently as a desirable feature of any measure of the magnitude of effect. In the terminology popularized by Judd and McClelland (1989), η^2 is the **Proportional Reduction in Error (PRE).** If we did not take group membership into account, the error associated with our recall scores would be SS_{total}. But when we know which group a subject is in, the error associated with our predictions is reduced to SS_{error}, the average variation within groups. But the difference between SS_{total} and SS_{error} is $SS_{\text{treatment}}$. Thus, $SS_{\text{treatment}}$ divided by SS_{total} is the percentage by which the error of our prediction has been *reduced* by considering group membership. For our example, without attending to group membership, we had 786.82 units of error. After attending to group membership, we only have 435.30 units of error. Thus, we have reduced our error by $786.82 - 435.30 = 351.52$ points, or by $351.52/786.82 = 44.7\%$.

It is important to realize that η^2 assumes that the true regression line passes through the individual treatment means. When the data are treated as a population, the assumption is correct. When the data are treated as a sample from some larger population, however, bias is introduced. These means are really sample means, so they are subject to sampling error, and η^2 will be biased upward—whatever the *true* regression line through the population means, it will probably not pass exactly through each sample mean. Although all measures we discuss will be biased, η^2 is the most biased. Thus, although it has the advantage of simplicity and is intuitively appealing, we will generally prefer to use a less biased estimate when our interest is in making general statements about our variables. If we are interested in making statements only about our particular set of data, or if we want a rough idea of the magnitude of the effect, then η^2 is a perfectly good measure. Moreover, η^2 and other, less biased, measures converge as sample sizes increase.

Omega-Squared (ω^2)

omega-squared (ω^2)

An alternative, and for many purposes better, method of assessing the magnitude of the experimental effect with balanced (equal ns) or nearly balanced designs is **omega-squared** (ω^2). This statistic has been discussed by Hays (1994) and developed extensively by Fleiss (1969), Vaughan and Corballis (1969), and Dodd and Schultz (1973). The derivation of ω^2 is based on the underlying structural model that we discussed earlier, and there are two different formulae for ω^2, depending on whether the independent variable is fixed or random. A random independent variable is so rare in one-way designs that we will ignore that version here, though it will become meaningful in the more complex designs in Chapters 13 and 14. I will also omit any derivation here, but the interested reader can find a derivation in the earlier editions of this book (Howell, 1997) and in Winer (1971).

For the fixed-model analysis of variance, a relatively unbiased estimate of the magnitude of experimental effect is given by

$$\omega^2 = \frac{SS_{treat} - (k-1)MS_{error}}{SS_{total} + MS_{error}}$$

Applying this to our data from Table 11.2, we have

$$\omega^2 = \frac{SS_{treat} - (k-1)MS_{error}}{SS_{total} + MS_{error}} = \frac{351.52 - 4(9.67)}{786.82 + 9.67} = \frac{312.84}{796.49} = .393$$

The estimate of ω^2 in this case (.393) is noticeably less than the estimate of $\eta^2 = .447$, reflecting the fact that the latter is more biased.

We have discussed two measures of the degree of association between the dependent and independent variables. These are only two of the many approaches that have been suggested. In general, ω^2 is probably the best. Fowler (1985) presents evidence on the bias of six different estimates and shows that ω^2 performs well. A point worth noting is that one measure that partially corrects η^2 for bias is closely related to what is called the **squared intraclass correlation,** which has been put forward as yet another measure of the magnitude of effect. Actually, one version of the intraclass correlation coefficient is nothing but ω^2 for the random model.

squared intraclass correlation

Aside from their concern about whether one statistic is more or less biased than another, researchers have raised questions regarding the interpretation of magnitude of effect measures in general. Rosenthal and Rubin (1982) present an interesting argument that quite small values of r^2 (the squared correlation coefficient) can represent important and dramatic effects. O'Grady (1982) presents several arguments why magnitude-of-effect measures may not be good measures of whatever it is we mean by "importance." Even an important variable may, for several reasons, account for small percentages of variation, and, more commonly, a large value of η^2 may simply mean that we have studied a trivial variable (such as the difference in height between elementary-school children and college students). (Even if not for what O'Grady says about the magnitude of effect, his excellent paper is worth reading for what it has to say about the psychometric and methodological considerations behind all the studies psychologists run.) Lane and Dunlap (1978) raise some important reservations about the routine reporting of magnitude measures and their interpretation given that journals mainly publish studies with significant results. Finally, Cohen (1973) outlines some important considerations in the calculation and interpretation of magnitude measures. Although Cohen is primarily concerned with factorial designs (to be discussed in Chapter 13), the philosophy behind his comments is relevant even here. All the papers cited are clear and readable, and I recommend them.

11.13   Power

Estimating power in the analysis of variance is a straightforward extension of the power analysis for t, although the notation is different, as are the tables. Just as with t, we will define a statistic, phi prime (ϕ'), based on the expected differences among the μ_j then derive a second statistic, phi (ϕ), which is a function of n and ϕ', and then calculate power from tables of the **noncentral F distribution.** A more complete treatment of power can be found in Cohen (1988) and Koele (1982).

noncentral F distribution

We already know (see p. 303) that

$$\frac{E(MS_{\text{treat}})}{E(MS_{\text{error}})} = \frac{\sigma_e^2 + n \sum \tau_j^2/(k-1)}{\sigma_e^2}$$

If H_0 is true, $\sum \tau_j^2 = 0$ and the ratio $F = MS_{\text{treat}}/MS_{\text{error}}$ will be distributed as the usual (central) F distribution. The mean of this distribution is $df_{\text{error}}/(df_{\text{error}} - 2)$, which is very close to 1.00 for reasonable sample sizes. (See footnote, p. 307). If H_0 is false, this ratio will depart from the central F distribution by a factor of

$$\frac{n \sum \tau_j^2}{\sigma_e^2(k-1)}$$

noncentrality parameter

which is called a **noncentrality parameter (ncP)**.[8]

The noncentrality parameter tells us that when H_0 is false, $E(F)$ will be approximately equal to

$$\left(1 + \frac{n \sum \tau_j^2}{\sigma_e^2(k-1)} \right) \left(\frac{df_{\text{error}}}{df_{\text{error}} - 2} \right)$$

You can see that the noncentrality parameter simply displaces the F distribution in a positive direction away from one, with the amount of displacement depending on the true differences among the population means.

These formulae may not convey a lot of meaning, but recall that $\tau = \overline{X}_j - \overline{X}_{..}$, which is the deviation of a group mean from the grand mean. As such, it is a measure of how much the means differ. Similarly, σ_τ is the standard deviation of group means, and, as such, is an excellent measure of differences between groups. One way of calculating power is to define a standardized measure[9] of effect size

$$\phi' = \frac{\sigma_\tau}{\sigma_e} = \sqrt{\frac{\sum (\mu_j - \mu)^2/k}{\sigma_e^2}}$$

This statistic (ϕ') is the same as Cohen's (1988) measure of effect size, which he labels f. (If we had two groups, it would be numerically equal to what we have previously called $d/2$.) You should recall that when we were calculating power for a t test on two independent groups, we took an effect size measure (d) and incorporated the sample size. That is just what we will do here. We define

$$\phi = \phi' \sqrt{n}$$

This way we can estimate ϕ' without regard to n, and then include the sample size when we come to estimating ϕ. This just makes our life a bit easier. We can then look ϕ up in the tables of the noncentral F distribution, given the level of α and the degrees of freedom for the numerator and denominator in F. (It is useful to note that ϕ is $\sqrt{ncP \cdot (k-1)/k}$, which is simply another way to see ϕ as a function of the noncentrality parameter.)

---

[8] There are a number of different quantities labeled "noncentrality parameter," but this is one of the more common.

[9] To say a measure is a "standardized measure" is just to say that we have divided a quantity by a standard deviation, thus scaling the result in standard deviation units. (This is analogous to dividing 87 *inches* by 12, getting 7.25, and declaring the result to be 7.25 *feet*.)

An Example

Before we proceed, let's work with an example that will illustrate several of the points made here and lead to some further elaboration. Suppose that we take the original data from the Conti and Musty (1984) experiment referred to earlier. They analyzed postinjection activity as a percentage of pre-injection activity, rather than the raw activity measures themselves. We can treat Conti and Musty's sample means and the average sample variance (MS_{error}) as if they were the actual population values.[10] We can then draw a large number of sets of samples from populations with those characteristics and observe what kind of sample means and variances we actually obtain. This will give you a handle on the kinds of variability you can expect from experiment to experiment, even with the kind of robust effect that Conti and Musty found. We can go a step further and compute an F on each set of samples and observe the variability in F values that we obtain. Finally, we could ask how many of those F values exceed the critical value of F, thus leading to a significant result. In other words, what we're saying here is, "Assume that the populations really have means like the ones Conti and Musty obtained. How often would we obtain sample data from such populations that would lead us to reject H_0?" This is what power is all about.

Using an excellent, and inexpensive, data generating and analysis program called DataSim (Bradley, 1988), I drew five samples at a time from Conti and Musty's five populations and ran an analysis of variance on the result. I simplified the problem slightly by assuming that we ran 10 subjects in each group, rather than the unequal numbers of subjects in their groups. I then repeated this procedure 10,000 times. The first 10 sets of means, along with their associated F and p values, are shown in Table 11.6 so that you can get some idea of the natural variability of these statistics even in a case where we know that the null hypothesis is false.

Notice that the eighth set of means has an F of 2.54, which just barely misses being significant. This illustrates the point that even having quite different population means does not guarantee that each replication of the experiment will reject the null hypothesis.

Another way to look at these results is to collect all 10,000 F values that were produced and plot them. If the null hypothesis had been true, the Fs would be distributed around a mean of approximately $df_{error}/(df_{error} - 2) = 45/43 = 1.046$. Instead, this empirical F

Table 11.6   Means of 10 computer replications of Conti and Musty (1984)

Control	0.1μg	0.5μg	1μg	2μg	F	p
34.30	57.60	65.00	47.62	37.10	4.10	.0064
25.30	54.80	61.22	49.37	39.70	8.60	<.0001
26.90	44.80	56.44	53.43	31.30	7.45	.0001
27.40	49.50	59.89	46.12	37.70	7.34	.0001
31.20	50.80	61.22	47.37	35.10	9.73	<.0001
32.70	47.60	62.33	56.00	43.10	3.73	.0105
30.10	47.70	62.44	59.87	26.90	8.87	<.0001
39.60	57.00	60.44	52.12	53.40	2.54	.0527
26.70	52.30	60.33	47.25	32.90	4.77	.0027
36.70	42.70	60.00	58.62	46.20	5.00	.0020

---

[10] For this dependent variable, their sample means were 34.00, 50.80, 60.33, 48.50, and 38.10, with a grand mean of 46.346. The average sample variance was 240.35.

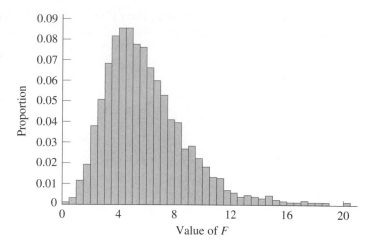

Figure 11.6   Empirical sampling distribution of F when noncentrality parameter equals 5.839

distribution, shown in Figure 11.6, has a mean considerably above 1.046, reflecting the fact that the noncentrality parameter is not 0.00. The mean of this distribution is 5.833. I said earlier that when H_0 is not true, the expected value of F is

$$E(F) = \left(1 + \frac{n \sum \tau_j^2}{\sigma_e^2(k-1)}\right)\left(\frac{df_{\text{error}}}{df_{\text{error}} - 2}\right)$$

where the first part of this equation contains the noncentrality parameter. Using the means and variance given in the footnote on page 330, and treating Conti and Musty's means as population means, and their MS_{error} as our estimate of σ_e^2, would give an expected value of $(1 + 4.58)(1.046) = 5.838$, which is very close to the actual mean of this distribution (5.833).

Finally, the critical value is $F_{4,45} = 2.58$. You can see that most of this distribution is above that point. In fact, 92.16% of the values exceed 2.58, meaning that given these population parameters, the probability of rejecting H_0 (i.e., the power of the test) is .9216. This value agrees closely with the values we could calculate exactly using DataSim (.9207), or approximately using the tables of the noncentral F distribution.

This example illustrates the use of repeated sampling to directly investigate the power of a test, the variability of sample means over replication, and the meaning of the noncentrality parameter. It also shows you graphically the idea of power looked at from the point of view of the sampling distribution of a statistic (in this case, F). You even have the opportunity to see a Type II error in action because the eighth case in Table 11.6, whose probability was greater than .05 under H_0, is a Type II error.

Power Calculations

I can illustrate the calculations of ϕ and ϕ' by assuming that the population values correspond exactly to those that Conti and Musty found in their experiment. I will again simplify the problem slightly by assuming that we plan to run 10 subjects in each group, rather than the unequal numbers of subjects they had.

I defined

$$\phi' = \frac{\sigma_\tau}{\sigma_e} = \sqrt{\frac{\sum(\mu_j - \mu)^2/k}{\sigma_e^2}}$$

Our presumed population means are [34.00, 50.80, 60.33, 48.50, 38.10] and, with equal ns, the grand mean would be the mean of the means $= 46.346$. Then

$$\phi' = \sqrt{\frac{\sum (\mu_j - \mu)^2 / k}{\sigma_e^2}}$$

$$= \sqrt{\frac{(34.00 - 46.346)^2 + \cdots + (38.10 - 46.346)^2 / 5}{240.35}}$$

$$= \sqrt{\frac{88.0901}{240.35}} = \sqrt{0.3665} = 0.6054$$

Each of our samples will contain 10 subjects, so $n = 10$. Then

$$\phi = \phi' \sqrt{n} = 0.6054 \sqrt{10} = 1.91$$

To use the table of the noncentral F distribution (Appendix ncF) we must enter it with ϕ, df_t and df_e, where df_t is the df for treatments and df_e is the df for error. For our example, $df_t = 4$, $df_e = 45$, and $\phi = 1.91$. Because this table does not contain all possible values of ϕ, df_t, and df_e, we either have to interpolate or else round off to the nearest value. For purposes of illustration, we will round off every value in the conservative direction. Thus, we will take $\phi = 1.8$, $df_t = 4$, and $df_e = 30$. The entry in the table for $F(df_t, df_e; \phi) = F(4, 30; 1.8)$ is .14 at $\alpha = .05$. This value is β, the probability of a Type II error. Power $= 1 - \beta = 1 - .14 = .86$, which is a conservative estimate given the way we have rounded off.

Perhaps we are willing to sacrifice some power to save on the number of subjects we use. To calculate the required sample sizes for a different degree of power, we simply need to work the problem backward. Suppose that we would be satisfied with power $= .80$. Then $\beta = .20$, and we simply need to find that value of ϕ for which $\beta = .20$. A minor complication arises because we cannot enter Appendix ncF without f_e and we cannot calculate f_e without knowing n. This is not a serious problem, however, because whether df_e is 30, 50, 180, or whatever will not make any really important difference in the tables. We will therefore make the arbitrary decision that $df_e = 30$ because we already know that it will have to be less than 45, and 30 is the closest value. With $df_t = 4$, $df_e = 30$, and $\beta = .20$, we find from the table that ϕ will have to be 1.68 (by interpolation).

Given

$$\phi = \phi' \sqrt{n}$$

then

$$n = \phi^2 / \phi'^2$$

$$= 1.68^2 / .6054^2$$

$$= 7.70 \approx 8 \text{ subjects per group}$$

Thus, we would need 8 subjects per group to have an 80% chance of rejecting H_0 if it is false to the extent that we believe it to be.

For those readers who were disturbed by my cavalierly setting $df_e = 30$, it might be instructive to calculate power for $n = 8$:

$$\phi = \phi' \sqrt{n} = 0.6054 \sqrt{8} = 1.71$$

$$df_t = 4$$

$$df_e = 5(7) = 35$$

From Appendix ncF for $F(4, 30; 1.71)$, we see that $\beta = .19$ (by interpolation) and our power $= .81$. This is quite close to the power $= .80$ that we sought.

In an effort to give some guidance in situations where little is known about the likely values of parameters, Cohen (1988) has defined a small effect as $\phi' = 0.10$, a medium effect as $\phi' = 0.25$, and a large effect as $\phi' = 0.40$. Cohen meant these to be used as a last resort when no other estimates were possible, but it has been my experience over the past few years that those who calculate power most often fall back on these conventions. They have tended to become starting points for a power analysis rather than a route of last resort. I have found myself using them because I was either too lazy or too ignorant to estimate noncentrality parameters directly, and I know of many others who fall in the same camp. They have also become rules of thumb for deciding whether effect sizes based on sample means should be classed as small, medium, or large. If we go back to Eysenck's data on recall as a function of depth of processing, we would calculate $\phi' = .95$. By Cohen's rule of thumb, this is a very large effect. When we looked at $\omega^2 = .393$, we were saying that depth of processing accounted for 39% of the variability in recall. Both of these statistics are giving us useful information on the meaning of the differences.

Cohen's 1988 book on power has become the standard by which psychologists and others calculate power, and I recommend it highly. The terminology takes a bit of getting used to, and Cohen uses his own tables rather than those of the noncentral F distribution, but there are many examples and the book is well written. Bradley, Russell, and Reeve (1996) have shown that Cohen's power estimates tend to be conservative for more complex designs, but they are certainly good enough for a rough estimate.

A number of software programs are available to calculate power (e.g., Borenstein and Cohen [1988]), and many statistical analysis packages (e.g., JMP and DataSim) contain the necessary routines. I particularly recommend DataSim (Bradley, 1988) because it will calculate power and quickly run simulations that allow you to see the sampling variability and power directly. An excellent program called G*Power is available free at http://www.psychologie.uni-duesseldorf.de/app/projects/gpower/. I recommend it highly. It is easy to use, gives quick results, and lets you experiment with alternative assumptions and sample sizes.

G*Power

Having recommended G*Power as an excellent program for calculating power, I have used it to produce the printout in Exhibit 11.2. The first screen shows the results of calculating the effect size. I have specified that I want power for an analysis of variance and have entered the means and sample sizes for the five groups. The program automatically computes the effect size. In this case, it is 0.605407, which is the same answer that we calculated earlier. I then clicked on the Calc & Copy button and the program moves to the next screen to calculate power. I told it to calculate post hoc power because I am using the actual sample means and error term from the Conti and Musty data. I also had to enter the total sample size, which is 50. When I again click on the Calculate button I obtain the next screen shown.

You will notice that the calculated power is .9207, which is higher than our calculation. But remember that we had to use 30 df for the error term in our calculation because the tables of the noncentral F distribution did not allow us to use the true value of 45 df.

Koele (1982) presents methods for calculating the power of random models. Random models, although not particularly common in a one-way layout, are more common in higher order designs and present particular problems because they generally have a low level of power. For these models, two random processes are involved—random sampling of participants and random sampling of treatment levels. As Koele phrased it, "Not only should there be many observations per level, but also many levels per treatment (independent variable). Experiments that have random factors with only two or three levels must be considered as absurd as t tests on samples with two or three observations" (p. 516). This is important

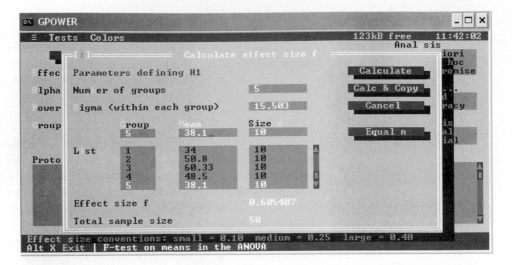

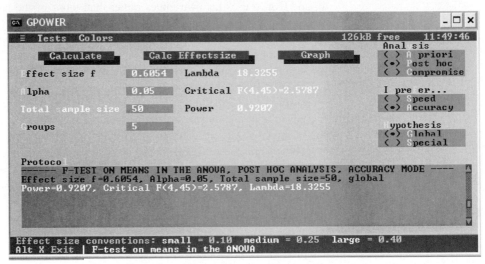

Exhibit 11.2   Results of G*Power power estimation

advice to keep in mind when you are considering random models. We will say more about this in Chapter 13.

One final point should be made about power and design considerations. McClelland (1997) has argued persuasively that with fixed variables we often use far more levels of our independent variable than we need. For example, if he were running the Eysenck (1974) experiment on recall as a function of levels of processing, I suspect that he would run only the two extreme groups (Counting and Imagery), or perhaps three groups, adding the Adjective condition. He would argue that to use five groups dilutes the effect across four degrees of freedom. Similarly, he would probably use only the 0, 0.5 μg, and 2 μg groups in the Conti and Musty (1984) study, putting the same number of subjects in the 0.5 μg group as in the other two conditions combined. I recommend this paper to those who are concerned about maximizing power and good experimental design. It is important and very readable.

An excellent piece of software written by Tor Tosteson at Dartmouth as a Java applet is available on the Web at http://www.dartmouth.edu/~matc/X10/java/anova/Anova.html. You can enter a set of sample means and MS_{error}, and have the program generate data a large number of times, counting the number of data sets with $p < .05$. This is an excellent way to see what is happening when you work with power. Power calculators are available on the Web that will calculate power for analysis of variance designs. Other power-computing software is available at http://members.aol.com/johnp71/javastat.html, which also contains many other programs that you should examine.

11.14   Computer Analyses

Exhibit 11.3 contains printout for the SPSS analysis of Everitt's data on the treatment of anorexic girls. Instead of choosing the one-way procedures from **Analyze/Compare means/One-Way Anova,** I have used the **Analyze/General linear model/Univariate procedure.** (Menu selections are not shown, but they should be evident.) We will use this procedure in Chapters 13 and 14 and produce the same answers as the one-way procedure. This procedure also produces some output that will not be familiar to you, which is explained later.

Notice in the summary table that the first line is labeled "Corrected Model." If there were two or more independent variables (e.g., Group and Sex), then this line would represent the combined effects of those variables. Because there is only one independent variable, the Model and the Group effects will be exactly the same.

The line labeled "Intercept" refers to a test on the null hypothesis that the grand mean is equal to 0 in the population. We very rarely care about this test, although in this case it is a test of a meaningful question about whether the girls in this study, averaged across groups, gained weight.

The lines labeled "Group," "Error," and "Corrected total" are the same the results we saw in Exhibit 11.1.

Descriptive Statistics

Dependent Variable: WTGAIN

Treatment Group	Mean	Std. Deviation	N
Control	−.450000	7.988705	26
CogBehav	3.006897	7.308504	29
Family	7.264706	7.157421	17
Total	2.763889	7.983598	72

Tests of Between-Subjects Effects

Dependent Variable: WTGAIN

Source	Type III Sum of Squares	df	Mean Square	F	Sig.	Eta Squared	Noncent. Parameter	Observed Power[a]
Corrected Model	614.644[b]	2	307.322	5.422	.006	.136	10.845	.830
Intercept	732.075	1	732.075	12.917	.001	.158	12.917	.943
GROUP	614.644	2	307.322	5.422	.006	.136	10.845	.830
Error	3910.742	69	56.677					
Total	5075.400	72						
Corrected Total	4525.386	71						

[a] Computed using alpha = .05

[b] R Squared = .136 (Adjusted R Squared = .111)

Exhibit 11.3   SPSS general linear model analysis of Everitt's data on treatment of anorexia

Notice that the entry of eta-squared for the Group effect is the same as the "R-squared" given at the bottom of the table. This tells us that 14% of the variation in weight gain could be attributable to differences between treatments.

SPSS then calculates observed power, treating the obtained means as parameters, and the obtained MS_{error} as an accurate estimate of the population variance. Because there are unequal sample sizes in this example, you will have difficulty reproducing these values exactly.

Key Terms

Analysis of variance (ANOVA) (Introduction)

One-way analysis (Introduction)

Structural model (11.2)

homogeneity of variance (11.2)

homoscedasticity (11.2)

Error variance (11.2)

Heterogeneity of variance (11.2)

heteroscedasticity (11.2)

MS_{error} (11.3)

MS_{within} (11.3)

$MS_{treatment}$ (11.3)

Treatment effect (11.3)

Expected value (11.3)

Sums of squares (11.4)

SS_{total} (11.4)

SS_{treat} (11.4)

SS_{error} (11.4)

Summary table (11.4)

df_{total} (11.4)

df_{treat} (11.4)

df_{error} (11.4)

Balanced designs (11.8)

Heavy-tailed distributions (11.10)

Winsorized samples (11.10)

Replicate (11.11)

Fixed-model analysis of variance (11.11)

Random-model analysis of variance (11.11)

Magnitude of the experimental effect (11.12)

Eta-squared (η^2) (11.12)

Correlation ratio (11.12)

Curvilinear regression (11.12)

Proportional reduction in error (PRE) (11.12)

Omega-squared (ω^2) (11.12)

Squared intraclass correlation (11.12)

Noncentral F distribution (11.13)

Noncentrality parameter (11.13)

Exercises

11.1 To investigate the maternal behavior of laboratory rats, we move the rat pup a fixed distance from the mother and record the time (in seconds) required for the mother to retrieve the pup to the nest. We run the study with 5-, 20-, and 35- day old pups. The data are given below for six pups per group.

5 days:	15	10	25	15	20	18
20 days:	30	15	20	25	23	20
35 days:	40	35	50	43	45	40

Run a one-way analysis of variance on the data.

11.2 Another aspect of the study by Eysenck (1974), referred to earlier, compared Younger and Older subjects on their ability to recall material in the face of instructions telling them that they would be asked to memorize the material for later recall. (Presumably this task required a high level of processing.) The data follow, where the dependent variable is the number of items recalled.

Younger:	21	19	17	15	22	16	22	22	18	21
Older:	10	19	14	5	10	11	14	15	11	11

a. Run the analysis of variance comparing the means of these two groups.

b. Run an independent groups t test on the data and compare the results with those you obtained in part (a).

11.3   Another way of looking at the data from Eysenck's (1974) study is to compare four groups of subjects. One group consisted of Younger subjects who were presented the words to be recalled in a condition that elicited a Low level of processing. A second group involved Younger subjects who were given a task requiring the Highest level of processing (as in Exercise 11.2). The two other groups were Older subjects who were given tasks requiring either Low or High levels of processing. The data follow.

Younger/Low:	8	6	4	6	7	6	5	7	9	7
Younger/High:	21	19	17	15	22	16	22	22	18	21
Older/Low:	9	8	6	8	10	4	6	5	7	7
Older/High:	10	19	14	5	10	11	14	15	11	11

   a.   Run a one-way analysis of variance on these data.

   b.   Now run a one-way analysis of variance on treatments 1 and 3 combined ($n = 20$) versus treatments 2 and 4 combined. What question are you answering?

   c.   Why might your answer to part (b) be difficult to interpret?

11.4   Refer to Exercise 11.1. Assume that, for reasons beyond our control, neither the data for the last pup in the 5-day group nor the data for the last two pups in the 35-day group could be used. Rerun the analysis of variance with the remaining data.

11.5   Refer to Exercise 11.2. Suppose that we collected additional data and had two more subjects in the Younger group, with scores of 13 and 15.

   a.   Rerun the analysis of variance.

   b.   Run an independent groups t test without pooling the variances.

   c.   Run an independent groups t test after pooling the variances.

   d.   For (b) and (c), which of these values of t corresponds (after squaring) to the F in (a)?

11.6   Calculate η^2 and ω^2 for the data in Exercise 11.2. Would you assume a fixed or a random model?

11.7   Calculate η^2 and ω^2 for the data in Exercise 11.3.

11.8   Foa, Rothbaum, Riggs, and Murdock (1991) conducted a study evaluating four different types of therapy for rape victims. The Stress inoculation therapy (SIT) group received instructions on coping with stress. The Prolonged exposure (PE) group went over the events in their minds repeatedly. The Supportive counseling (SC) group was taught a general problem-solving technique. Finally, the Waiting list (WL) control group received no therapy. The data follow, where the dependent variable was the severity rating of a series of symptoms.

Group	n	Mean	S.D.
SIT	14	11.07	3.95
PE	10	15.40	11.12
SC	11	18.09	7.13
WL	10	19.50	7.11

   a.   Run the analysis of variance, ignoring any problems with heterogeneity of variance, and draw whatever conclusions are warranted.

   b.   Apply Welch's procedure for heterogeneous variances. Does this affect your conclusions?

   c.   Draw a graph showing the means of the four groups.

   d.   What does rejection of H_0 mean in this case?

11.9   Calculate η^2 and ω^2 for the data in Exercise 11.8 and interpret the results.

11.10   What would happen if the sample sizes in Exercise 11.8a were twice as large as they actually were, but all other statistics remained the same?

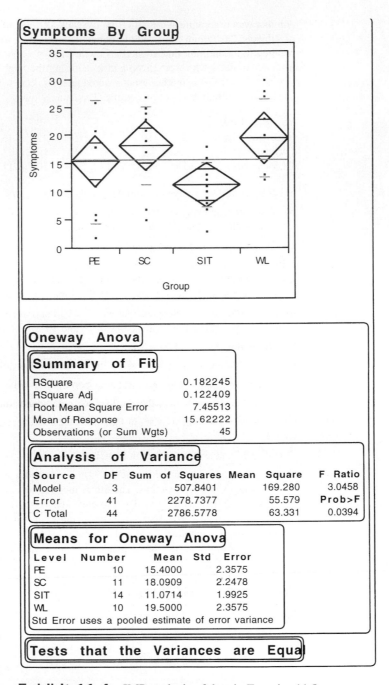

Exhibit 11.4 JMP analysis of data in Exercise 11.8

11.11 The computer printout in Exhibit 11.4 is from a JMP analysis of the data in Exercise 11.8.

    a. Compare the results with those you obtained in Exercise 11.8a.

    b. What can you tell from this printout that you cannot tell from a standard summary table?

11.12 The following results from JMP include tests for heterogeneity of variance and Welch's modification to allow for heterogeneity of variance. How does this result compare with your answer to Exercise 11.8b? From what you know from this chapter and from Chapter 7, what are the F ratios in the middle table?

Tests that the Variances are Equal

Level	Count	Std Dev	Mean. Abs Dif to Mean	Mean. Abs Dif to Median
PE	10	11.11755	9.600000	9.600000
SC	11	7.13379	5.537190	5.636364
SIT	14	3.95094	3.071429	3.071429
WL	10	7.10634	6.100000	6.100000

Test	F Ratio	DF Num	DF Den	Prob > F
O'Brien [.5]	5.1694	3	41	0.0040
Brown-Forsythe	6.3543	3	41	0.0012
Levene	6.6330	3	41	0.0009
Bartlett	3.5390	3	*	0.0140

Welch Anova testing Means Equal, allowing Std's Not Equal

F Ratio	DF Num	DF Den	Prob > F
5.3749	3	19.128	0.0075

11.13   Write an appropriate statistical model for Exercise 11.1.

11.14   Write an appropriate statistical model for Exercise 11.2.

11.15   Write an appropriate statistical model for Exercise 11.3. Save it for later use in Chapter 13.

11.16   When F is less than 1, we usually write "<1" rather than the actual value. What meaning can be attached to an F appreciably less than 1? Can we speak intelligently about an F "significantly" less than 1? Include $E(MS)$ in your answer.

11.17   Howell and Huessy (1981) classified children as exhibiting (or not exhibiting) attention deficit disorder (ADD)–related behaviors in second, fourth, and fifth grades. The subjects were then sorted on the basis of the year(s) in which the individual was classed as exhibiting such behavior. Howell and Huessy then looked at GPA for these children when the latter were in high school. The data are given in terms of mean GPA per group.

	Never ADD	Second Only	Fourth Only	Second and Fourth
Mean	2.6774	1.6123	1.9975	2.0287
S.D.	0.9721	1.0097	0.7642	0.5461
n	201	13	12	8

	Fifth Only	Second and Fifth	Fourth and Fifth	Second, Fourth, and Fifth
Mean	1.7000	1.9000	1.8986	1.4225
S.D.	0.8788	1.0318	0.3045	0.5884
n	14	9	7	8

Run the analysis of variance and draw the appropriate conclusion.

11.18   Rerun the analysis of Exercise 11.17, leaving out the Never ADD group. In what way does this analysis clarify the interpretation of the data?

11.19   Apply a square-root transformation to the data in Table 11.5.

11.20   Run the analysis of variance for the transformed data you obtained in Exercise 11.19.

11.21 Calculate η^2 and ω^2 for the data in Exercise 11.17.

11.22 Darley and Latané (1968) recorded the speed with which subjects summoned help for a person in trouble. Subjects thought either that they were the only one listening to the person (Group 1, $n = 13$), that one other person was listening (Group 2, $n = 26$), or that four other people were listening (Group 3, $n = 13$). The dependent variable was the speed with which the person summoned help ($= 1/\text{time} \times 100$). The mean speed scores for the three groups were 0.87, 0.72, and 0.51, respectively. The MS_{error} was 0.053. Reconstruct the analysis of variance summary table. What can you conclude?

11.23 In Exercise 11.22, the data were transformed from their original units, which were in seconds. What effect would this have on the shape of the distributions?

11.24 Would a transformation of Eysenck's data in Table 11.2 be useful for equalizing the variances? What transformation would you suggest applying, if any?

11.25 Suppose that we wanted to run a study comparing recall of nouns and verbs. We present each subject with 25 nouns or 25 verbs and later ask for recall of the list. We look at both differences between parts of speech and between different words within the category of "noun." What variable is a fixed variable and what is a random variable?

11.26 Give an example of a study in which the main independent variable would be a random variable.

11.27 Davey, Startup, Zara, MacDonald, and Field (2003) were interested in the role of mood on the degree of compulsive checking in which a person engaged. (Compulsive checking is involved in a number of psychopathologies.) Three groups of 10 participants each listened to music designed to induce a positive, negative, or neutral mood. They were then asked to "list as many things around your home that you should check for safety or security reasons before you go away" for three weeks. The dependent variable was the number of things listed. The actual data follow:

Induced Mood		
Negative	Positive	None
7	9	8
5	12	5
16	7	11
13	3	9
13	10	11
24	4	10
20	5	11
10	4	10
11	7	7
7	9	5

a. Run the appropriate analysis of variance and draw you own conclusion.

b. Which column means would you be interested in comparing for theoretical reasons when we discuss multiple comparisons in the next chapter?

Computer Exercises

11.28 In Exercise 7.46, you had data on students who had lost a parent through death, who came from a divorced household, or who grew up with two parents. You then ran three separate t tests comparing those groups.

a. Now reanalyze those data using an analysis of variance with GSIT as the dependent variable.

b. How does your answer to this question differ from your answer in Chapter 7?

Use the following material to answer Exercises 11.29–11.31.

Introini-Collison and McGaugh (1986) examined the hypothesis that hormones normally produced in the body can play a role in memory. Specifically, they looked at the effect of posttraining injections of epinephrine on retention of a previously learned discrimination. To oversimplify the experiment, they first trained mice to escape mild shock by choosing the left arm of a Y maze. Immediately after training, the researchers injected the mice with either 0.0, 0.3, or 1.0 mg/kg of epinephrine. (The first group was actually injected with saline.) Introini-Collison and McGaugh predicted that low doses of epinephrine would facilitate retention, whereas high doses would inhibit it.

Either 1 day, 1 week, or 1 month after original training, each mouse was again placed in the Y maze, but this time was required to run to the right arm of the maze to escape shock. Presumably, the stronger the memory of the original training was, the more it would interfere with the learning of this new task and the more errors the subject would make.

There are two data sets for this experiment, and they are described in Appendix Computer Data Sets and are available at the website (www.uvm.edu/~dhowell/methods/). The original study used 18 animals in the three dosage groups tested after 1 day, and 12 animals in each group tested after intervals of 1 week and 1 month. Hypothetical data that closely reproduce the original results are contained in Epinuneq.dat, although for our purposes there are data for only 7 subjects in the 1.0 mg/kg dose at the 1-month test. A second data set was created with 12 observations in each of the 9 cells, and is called Epineq.dat. In both cases, the need to create data that were integers led to results that are slightly conservative relative to the actual data, but the conclusions with respect to H_0 are the same.

11.29   On the reasonable assumption that there are no important differences from one interval to the next, combine the data by ignoring the Interval variable and run the analysis of variance on Dosage. Use the data in Epinuneq.dat. (You will have 42 observations for the 0.0 and 0.3 mg/kg doses and 37 subjects for the 1.0 mg/kg dose.)

11.30   Use the data in Epinuneq.dat to run three separate one-way analyses of variance, one at each retention interval. In each case, test the null hypothesis that the three dosage means are equal. Have your statistical package print out the means and standard deviations of the three dosage groups for each analysis. Now run a separate analysis testing the hypotheses that the three Interval means are equal. In this case, you will simply ignore Dosage.

11.31   Rerun Exercise 11.30, this time using Epineq.dat. (The results will differ somewhat because the data are different.) Calculate the average of the three error terms (MS_{error}) and show that this is equal to the average of the variances within each of the nine groups in the experiment. Save this value to use in Chapter 13.

Discussion Questions

11.32   Some experimenters have a guilty conscience whenever they transform data. Construct a reasoned argument why transformations are generally perfectly acceptable.

11.33   In the study by Conti and Musty (1984) on the effects of THC on activity, the means clearly do not increase linearly with dosage. What effect, if any, should this have on the magnitude-of-effect measure?

11.34   With four groups, you could have the means equally spaced along some continuum, or you could have three means approximately equal to each other and a fourth one different, or you could have two means approximately equal but different from two other approximately equal means, or some other pattern. Using very simple data that you create yourself (holding within-groups variance constant), how does the F statistic vary as a function of the pattern of means?

CHAPTER 12

Multiple Comparisons Among Treatment Means

Objectives

To extend the analysis of variance by examining ways of making comparisons within a set of means.

Contents

A SIGNIFICANT F IN AN ANALYSIS of variance is simply an indication that not all the population means are equal. It does not tell us which means are different from which other means. As a result, the overall analysis of variance often raises more questions than it answers. We now face the problem of examining differences among individual means, or sets of means, for the purpose of isolating significant differences or testing specific hypotheses. We want to be able to make statements of the form $\mu_1 = \mu_2 = \mu_3$, and $\mu_4 = \mu_5$, but the first three means are different from the last two, and all of them are different from μ_6.

Many different techniques for making comparisons among means are available, and the list grows monthly. Here, we will consider the most common and useful ones. A thorough discussion of this topic can be found in Miller (1981), Hochberg and Tamhane (1987), and Toothaker (1991). Keselman, Holland, and Cribbie (2005) offer a review of some of the newer methods. The papers by Games (1978a, 1978b) are also helpful, as is the paper by Games and Howell (1976) on the treatment of unequal sample sizes.

12.1 Error Rates

The major issue in any discussion of multiple-comparison procedures is the question of the probability of Type I errors. Most differences among alternative techniques result from different approaches to the question of how to control these errors.[1] The problem is in part technical, but it is really much more a subjective question of how you want to define the error rate and how large you are willing to let the maximum possible error rate be.

We will distinguish two basic ways of specifying error rates, or the probability of Type I errors.[2] In doing so, we shall use the terminology that has become more or less standard since an extremely important unpublished paper by Tukey in 1953. (See also O'Neil & Wetherill, 1971 and Ryan, 1959.)

Error Rate per Comparison (*PC*)

error rate per comparison (*PC*)

We have used the **error rate per comparison (*PC*)** in the past and it requires little elaboration. It is the probability of making a Type I error on any given comparison. If, for example, we make a comparison by running a t test between two groups and we reject the null hypothesis because our t exceeds $t_{.05}$, then we are working at a per comparison error rate of .05.

Familywise Error Rate (*FW*)

When we have completed running a set of comparisons among our group means, we will arrive at a set (often called a *family*) of conclusions. For example, the family might consist of the statements

$$\mu_1 < \mu_2$$
$$\mu_3 < \mu_4$$
$$\mu_1 < (\mu_3 + \mu_4)/2$$

---

[1] Some authors choose among tests on the basis of power and are concerned with the probability of finding any or all significant differences among pairs of means (any-pairs power and all-pairs power). In this chapter, however, we will focus on the probability of Type I errors and how different test procedures deal with these error rates.

[2] There is a third error rate called the error rate per experiment (*PE*), which is the expected *number* of Type I errors in a set of comparisons. The error rate per experiment is not a probability, and we typically do not attempt to control it directly. We can easily calculate it, however, as $PE = c\alpha$, where c is the number of comparisons and α is the per comparison error rate.

familywise error
rate (*FW*)

The probability that this family of conclusions will contain *at least* one Type I error is called the **familywise error rate (*FW*).**[3] Many of the procedures we will examine are specifically directed at controlling the *FW* error rate, and even those procedures that are not intended to control *FW* are still evaluated relative to what the level of *FW* is likely to be.

In an experiment in which only one comparison is made, both error rates will be the same. As the number of comparisons increases, however, the two rates diverge. If we let α' represent the error rate for any one comparison and c represent the number of comparisons, then

Error rate per comparison (*PC*): $\alpha = \alpha'$

Familywise error rate (*FW*): $\alpha = 1 - (1 - \alpha')^c$

(if comparisons are independent)

If the comparisons are not independent, the per comparison error rate remains unchanged, but the familywise rate is affected. In most situations, however, $1 - (1 - \alpha')^c$ still represents a reasonable approximation to *FW*. It is worth noting that the limits on *FW* are $PC \leq FW \leq c\alpha$, and in most reasonable cases, *FW* is in the general vicinity of $c\alpha$. This becomes important when we consider the Bonferroni tests.

The Null Hypothesis and Error Rates

omnibus null
hypothesis

We have been speaking as if the null hypothesis in question were what is usually called the *complete, or omnibus, null hypothesis* ($\mu_1 = \mu_2 = \mu_3 = \cdots = \mu_k$). Actually, this is the null hypothesis tested by the overall analysis of variance. In many experiments, however, nobody is seriously interested in the complete null hypothesis; rather, people are concerned about a few more restricted null hypotheses, such as ($\mu_1 = \mu_2 = \mu_3$, $\mu_4 = \mu_5$, $\mu_6 = \mu_7$), with differences among the various subsets. If this is the case, the problem becomes more complex, and it is not always possible to specify *FW* without knowing the pattern of population means. We will need to consider this in designating the error rates for the different tests we shall discuss.

A Priori versus Post Hoc Comparisons

a priori
comparisons

post hoc
comparisons

In the first five editions of this book, I have carefully distinguished between **a priori comparisons,** which are chosen before the data are collected, and **post hoc comparisons,** which are planned after the experimenter has collected the data, looked at the means, and noted which of the latter are far apart and which are close together. This is a traditional distinction, but one that seems to be less and less important to people who run such comparisons. In practice, the real distinction seems to come down to the difference between deliberately making a few comparisons that are chosen because of their theoretical or practical nature and making comparisons among all possible pairs of means. I am going to continue to make the a priori/post hoc distinction because it organizes the material nicely, but keep in mind that the distinction is a rather fuzzy one. To take a simple example, consider a situation in which you have five means. In this case, there are 10 possible comparisons involving pairs of means (e.g., $\overline{X}_1$ versus $\overline{X}_2$, $\overline{X}_1$ versus $\overline{X}_3$, and so on). Assume that the complete null hypothesis is true but that by chance two of the means are far enough apart to lead us erroneously to reject H_0: $\mu_i = \mu_j$. In other words, the data contain one Type I error. If you have

---

[3] This error rate is frequently referred to, especially in older sources, as the "experimentwise" error rate. However, Tukey's term "familywise" has become more common. In more complex analyses of variance, the experiment often may be thought of as comprising several different families of comparisons.

to plan your single comparison in advance, you have a probability of .10 of hitting the 1 comparison out of 10 that will involve a Type I error. If you look at the data first, however, you are certain to make a Type I error, assuming that you are not so dim that you test anything other than the largest difference. In this case, you are implicitly making all 10 comparisons in your head, even though you perform the arithmetic for only the largest one. For some post hoc tests, we will actually adjust the error rate as if you literally made all 10 comparisons.

This simple example demonstrates that if comparisons are planned in advance (*and are a subset of all possible comparisons*), the probability of a Type I error is smaller than if the comparisons are arrived at on a post hoc basis. It should not surprise you, then, that we will treat a priori and post hoc comparisons separately. It is important to realize that when we speak of a priori tests, we commonly mean a relatively small set of comparisons. If you are making *all* possible pairwise comparisons among several means, for example, it won't make any difference whether that was planned in advance or not. (I would wonder, however, if you really wanted to make all possible comparisons.)

Significance of the Overall *F*

Some controversy surrounds the question of whether one should insist that the overall *F* on treatments be significant before conducting multiple comparisons between individual group means. In the past, the general advice was that without a significant group effect, individual comparisons were inappropriate. The rationale underlying the error rates for Fisher's least significant different test, to be discussed in Section 12.6, required overall significance.

The logic behind most of our multiple comparison procedures, however, does not require overall significance before making specific comparisons. First, the hypotheses tested by the overall test and a multiple-comparison test are quite different, with quite different levels of power. For example, the overall *F* actually distributes differences among groups across the number of degrees of freedom for groups. This has the effect of diluting the overall *F* in the situation where several group means are equal to each other but different from some other mean. Second, requiring overall significance will actually change the *FW*, making the multiple comparison tests conservative. The tests were designed, and their significance levels established, without regard to the overall *F*.

Wilcox (1987a) has considered this issue and suggested, "There seems to be little reason for applying the (overall) *F* test at all" (p. 36). Wilcox would jump straight to multiple-comparisons without even computing the *F*. Others have said much the same thing. That position may have seemed a bit extreme in the past, but it does emphasize the point. However, it does not seem as extreme today as it did 20 years ago. If you recognize that typical multiple-comparison procedures do not require a significant overall *F*, you will examine group differences regardless of the value of that *F*. Why, then, do we even need that *F* except to provide a sense of closure? The only reason I can think of is "tradition," and that is a powerful force. You would need to go as far as calculating MS_{error} anyway, so you might as well take the extra step and calculate the omnibus *F*.

12.2 Multiple Comparisons in a Simple Experiment on Morphine Tolerance

In discussing the various procedures, it will be helpful to have a data set to which each of the approaches can be applied. We will take as an example a study similar to an important experiment on morphine tolerance by Siegel (1975). Although the data are fictitious and a good deal of liberty has been taken in describing the conditions, the means (and the significance of the differences among the means) are the same as those in Siegel's paper. It will be necessary to describe this study in some detail, but the example is worth the space required. It will be to your advantage to take the time to understand the hypotheses and the treatment labels.

Morphine is a drug that is frequently used to alleviate pain. Repeated administrations of morphine, however, lead to morphine tolerance, in which morphine has a decreasing effect (pain reduction) over time. (You may have experienced the same thing if you eat spicy food very often. You will find that the more you eat it, the hotter you have to make it so it tastes the way it did when you started.) A common experimental task that demonstrates morphine tolerance involves placing a rat on an uncomfortably warm surface. When the heat becomes too uncomfortable, the rat will lick its paws, and the latency of the paw-lick is used as a measure of the rat's sensitivity to pain. A rat that has received a single morphine injection typically shows a longer paw-lick latency, indicating a reduced pain sensitivity. The development of morphine tolerance is indicated by a progressive shortening of paw-lick latencies (indicating increased sensitivity) with repeated morphine injections.

Siegel noted that there are a number of situations involving drugs other than morphine in which *conditioned* (learned) drug responses are opposite in direction to the unconditioned (natural) effects of the drug. For example, an animal injected with atropine will usually show a marked decrease in salivation. However, if physiological saline (which should have no effect whatsoever) is suddenly injected (*in the same physical setting*) after repeated injections of atropine, the animal will show an *increase* in salivation. It is as if the animal were compensating for the anticipated effect of atropine. In such studies, it appears that a learned compensatory mechanism develops over trials and counterbalances the effect of the drug. (You experience the same thing if you leave the seasoning out of food that you normally add seasoning to. It will taste unusually bland, though the Grape Nuts cereal you eat for breakfast does not taste bland—and I hope that you don't put salt on Grape Nuts.)

Siegel theorized that such a process might help to explain morphine tolerance. He reasoned that if you administered a series of pretrials in which the animal was injected with morphine and placed on a warm surface, morphine tolerance would develop. Thus, if you again injected the subject with morphine on a subsequent test trial, the animal would only be as sensitive to pain as would be a naive animal (one who had never received morphine) because of the tolerance that has developed. Siegel further reasoned that if on the test trial you instead injected the animal with physiological saline *in the same test setting* as the normal morphine injections, the conditioned hypersensitivity that results from the repeated administration of morphine would not be counterbalanced by the presence of morphine, and the animal would show very short paw-lick latencies. Siegel also reasoned that if you gave the animal repeated morphine injections in one setting but then tested it with morphine in a *new* setting, the new setting would not elicit the conditioned compensatory hypersensitivity to counterbalance the morphine. As a result, the animal would respond as would an animal that was being injected for the first time. Heroin is a morphine derivative. Imagine a heroin addict who is taking large doses of heroin because he has built up tolerance to it. If his response to this now large dose were suddenly that of a first-time (instead of a tolerant) user because of a change of setting, the result could be, and often is, lethal. We're talking about a serious issue here.

Our version of Siegel's experiment is based on the predictions just outlined. The experiment involved five groups of rats. Each group received four trials, but the data for the analysis come from only the critical fourth (test) trial. The groups are designated by indicating the treatment on the first three trials and then the treatment on the fourth trial. Group M-M received morphine on the first three trials in the test setting and then again on the fourth trial in the same test setting. This is the standard morphine-tolerant group, and, because morphine tolerance develops very quickly, we would expect to see normal, or at least near-normal, levels of pain sensitivity on that fourth trial. Group M-S received morphine on the first three trials but then received saline on the fourth trial (in the same test setting). These animals would be expected to be hypersensitive to the pain stimulus because the conditioned hypersensitivity would not be balanced by any compensating effects of morphine. Group M(cage)-M (abbreviated Mc-M) received morphine on the first three trials in their home cage but then received morphine on the fourth trial in the standard test setting, which was new to them. For this group, cues originally associated with morphine injection were not present on the test trial, and therefore, according to Siegel's model, the animals should not exhibit morphine tolerance on that trial. The fourth group (group S-M) received saline on the first three trials (in the test setting) and morphine on the fourth trial. These animals would be expected to show the least sensitivity to pain because there has been no opportunity for morphine tolerance to develop. Finally, group S-S received saline on all four trials.

If Siegel's model is correct, group S-M should show the longest latencies (indicating least sensitivity), whereas group M-S should show the shortest latency (most sensitivity). Group Mc-M should resemble group S-M because cues associated with group Mc-M's first three trials would not be present on the test trial. Groups M-M and S-S should be intermediate. Whether group M-M will be equal to group S-S will depend on the rate at which morphine tolerance develops. The pattern of anticipated results is

$$S\text{-}M = Mc\text{-}M > M\text{-}M \ ? \ S\text{-}S > M\text{-}S$$

The "?" indicates no prediction. The dependent variable is the latency (in seconds) of paw-licking.

The results of this experiment are presented in Table 12.1a, and the overall analysis of variance is presented in Table 12.1b. Notice that the within-group variances are more or less equal (a test for heterogeneity of variance was not significant), and there are no obvious outliers. The overall analysis of variance is clearly significant, indicating differences among the five treatment groups.

Table 12.1   Data and analysis on morphine tolerance

(a) Data

	M-S	M-M	S-S	S-M	Mc-M
	3	2	14	29	24
	5	12	6	20	26
	1	13	12	36	40
	8	6	4	21	32
	1	10	19	25	20
	1	7	3	18	33
	4	11	9	26	27
	9	19	21	17	30
Mean	4.00	10.00	11.00	24.00	29.00
St. Dev	3.16	5.13	6.72	6.37	6.16

(b) Summary Table

Source	df	SS	MS	F
Treatment	4	3497.60	874.40	27.33*
Error	35	1120.00	32.00	
Total	39	4617.60		

*$p < .05$

Magnitude of Effect

We can calculate η^2 for these data as $SS_{treat}/SS_{total} = 3497.60/4617.60 = .76$, indicating that treatment differences account for 76% of the variation in the study. A nearly unbiased estimate would be ω^2, which would be

$$\omega^2 = \frac{SS_{treat} - (k-1)MS_{error}}{SS_{total} + MS_{error}} = \frac{3497.60 - 4(32)}{4617.60 + 32} = \frac{3369.6}{4649.6} = 0.72$$

Both estimates indicate that group differences account for a very substantial proportion of the variation in this study.

12.3 A Priori Comparisons

There are two reasons for starting our discussion with t tests. First, standard t tests between pairs of means can, in a limited number of situations, be a perfectly legitimate method of comparison. Second, the basic formula for t, and minor modifications on it, apply to a large number of procedures, and a review at this time is useful.

contrasts
As we have seen, a priori comparisons (also called **contrasts**) are planned before the data have been collected. There are several different kinds of a priori comparison procedures, and we will discuss them in turn.

Multiple *t* Tests

One of the simplest methods of running preplanned comparisons is to use individual t tests between pairs of groups. In running individual t tests, if the assumption of homogeneity of variance is tenable, we usually replace the individual variances, or the pooled variance estimate, with MS_{error} from the overall analysis of variance and evaluate the t on df_{error} degrees of freedom. When the variances are heterogeneous but the sample sizes are equal, we do not use MS_{error} but, instead, use the individual sample variances and evaluate t on $2(n-1)$ degrees of freedom. Finally, when we have heterogeneity of variance and unequal sample sizes, we use the individual variances and correct the degrees of freedom using the Welch–Satterthwaite approach (see Chapter 7). (For an evaluation of this approach, albeit for a slightly different test statistic, see Games and Howell, 1976.)

The indiscriminate use of multiple t tests is typically brought up as an example of a terrible approach to multiple comparisons. In some ways, this is an unfair criticism. It *is* a terrible thing to jump into a set of data and lay waste all around you with t tests on each and every pair of means that looks as if it might be interesting. The familywise error rate will be outrageously high. However, if you have only one or two comparisons to make and if those comparisons were truly planned in advance (you cannot cheat and say, "Oh well, I would have planned to make them if I had thought about it"), the t test approach has much to recommend it. With only two comparisons, for example, the maximum FW would be approximately 0.10 if each comparison were run at $\alpha = .05$ and would be approximately 0.02 if each comparison were run at $\alpha = .01$.

In the study on morphine tolerance described previously, we would probably not use multiple t tests simply because too many important comparisons should be considered. (Actually, we would probably use one of the post hoc procedures for making all pairwise comparisons unless we can restrict ourselves to relatively few comparisons.) For the sake of an example, however, consider two fundamental comparisons that were clearly predicted by the theory and that can be tested easily with a t test. The theory predicted that a rat that had received three previous morphine trials and was then tested in the same environment using a saline injection would show greater pain sensitivity than would an animal that had always been tested using saline. This involves a comparison of group M-S with group S-S. Furthermore, the theory predicted that group Mc-M would show less sensitivity to pain than would group M-M because the former would be tested in an environment different from the one in which it had previously received morphine. Because the sample variances are similar and the sample sizes are equal, we will use MS_{error} as the pooled variance estimate and will evaluate the result on df_{error} degrees of freedom.

Our general formula for t, replacing individual variances with MS_{error}, will then be

$$t = \frac{\overline{X}_i - \overline{X}_j}{\sqrt{\dfrac{MS_{error}}{n} + \dfrac{MS_{error}}{n}}} = \frac{\overline{X}_i - \overline{X}_j}{\sqrt{\dfrac{2MS_{error}}{n}}}$$

Substituting the data from our example, the contrast of group M-S with group S-S yields

$$\overline{X}_{M\text{-}S} = 4.00 \qquad \overline{X}_{S\text{-}S} = 11.00 \qquad MS_{error} = 32.00$$

$$t = \frac{\overline{X}_{M\text{-}S} - \overline{X}_{S\text{-}S}}{\sqrt{\dfrac{2MS_{error}}{n}}} = \frac{4.00 - 11.00}{\sqrt{\dfrac{2(32.00)}{8}}} = \frac{-7}{\sqrt{8}} = -2.47$$

And group Mc-M versus group M-M yields

$$\overline{X}_{Mc\text{-}M} = 29.00 \qquad \overline{X}_{M\text{-}M} = 10.00 \qquad MS_{error} = 32.00$$

$$t = \frac{\overline{X}_{Mc\text{-}M} - \overline{X}_{M\text{-}M}}{\sqrt{\dfrac{2MS_{error}}{n}}} = \frac{29.00 - 10.00}{\sqrt{\dfrac{2(32.00)}{8}}} = \frac{19}{\sqrt{8}} = 6.72$$

Both of these obtained values of t would be evaluated against $t_{.025}(35) = \pm 2.03$, and both would lead to rejection of the corresponding null hypothesis. We can conclude that with two groups of animals tested with saline, the group that had previously received morphine in the same situation will show a heightened sensitivity to pain. We can also conclude that changing the setting in which morphine is given significantly reduces, if it does not eliminate, the conditioned morphine-tolerance effect. Because we have tested two null hypotheses, each with $\alpha = .05$ per comparison, the FW will approach .10 if both null hypotheses are true, which seems quite unlikely.

The basic t test that we have just used is the basis for almost everything to follow. I may tweak the formula here or there, and I will certainly use a number of different tables and decision rules, but it remains your basic t test—even when I change the formula and call it q.

Linear Contrasts

The use of individual t tests is a special case of a much more general technique involving what are known as linear contrasts.[4] In particular, t tests allow us to compare one group with another group, whereas linear contrasts allow us to compare one group *or set of groups* with another group or set of groups. Although we can use the calculational procedures of linear contrasts with post hoc tests as well as with a priori tests, they are discussed here under a priori tests because that is where they are most commonly used.

---

[4] The words "contrast" and "comparison" are used pretty much interchangeably in this context.

linear combination

To define linear contrasts, we must first define a **linear combination**. A linear combination of means takes the form

$$L = a_1\overline{X}_1 + a_2\overline{X}_2 + \cdots + a_k\overline{X}_k = \sum a_j\overline{X}_j$$

This equation simply states that a linear combination is a weighted sum of treatment means. If, for example, the a_j were all equal to 1, L would just be the sum of the means. If, on the other hand, the a_j were all equal to $1/k$, then L would be the mean of the means.

linear contrast

When we impose the restriction that $\sum a_j = 0$, a linear combination becomes what is called a **linear contrast**. By convention, we designate the fact that it is a linear contrast by replacing "L" with the Greek psi (ψ). With the proper selection of the a_j, a linear contrast is very useful. It can be used, for example, to compare one mean with another mean, or the mean of one condition with the combined mean of several conditions. As an example, consider three means ($\overline{X}_1$, $\overline{X}_2$, and $\overline{X}_3$). Letting $a_1 = 1$, $a_2 = -1$, and $a_3 = 0$, $\sum a_j = 0$,

$$\psi = (1)\overline{X}_1 + (-1)\overline{X}_2 + (0)\overline{X}_3 = \overline{X}_1 - \overline{X}_2$$

In this case, ψ is simply the difference between the means of group 1 and group 2, with the third group left out. If, on the other hand, we let $a_1 = 1/2$, $a_2 = 1/2$, and $a_3 = -1$, then

$$\psi = (1/2)\overline{X}_1 + (1/2)\overline{X}_2 + (-1)\overline{X}_3 = \frac{\overline{X}_1 + \overline{X}_2}{2} - \overline{X}_3$$

in which case ψ represents the difference between the mean of the third treatment and the average of the means of the first two treatments.

Sum of Squares for Contrasts

One of the advantages of linear contrasts is that they can be converted to sums of squares very easily and can represent the sum of squared differences between the means of sets of treatments. If we write

$$\psi = a_1\overline{X}_1 + a_2\overline{X}_2 + \cdots + a_k\overline{X}_k = \sum a_j\overline{X}_j$$

it can be shown that

$$SS_{\text{contrast}} = \frac{n\psi^2}{\sum a_j^2} = \frac{n\left(\sum a_j\overline{X}_j\right)^2}{\sum a_j^2}$$

is a component of the overall SS_{treat} on 1 *df*, where n represents the number of scores per treatment.[5]

Suppose we have three treatments such that

$$n = 10 \qquad \overline{X}_1 = 1.5 \qquad \overline{X}_2 = 2.0 \qquad \overline{X}_3 = 3.0$$

For the overall analysis of variance,

$$SS_{\text{treat}} = n\sum(\overline{X}_j - \overline{X}_{..})^2 = 10[(1.5 - 2.167)^2 + (2 - 2.167)^2 + (3 - 2.167)^2]$$
$$= 10[0.4449 + 0.0278 + 0.6939] = 11.667$$

Suppose we wanted to compare the average of treatments 1 and 2 with treatment 3. Let $a_1 = 1/2$, $a_2 = 1/2$, $a_3 = -1$. Then

$$\psi = \sum a_j\overline{X}_j = \left(\tfrac{1}{2}\right)(1.5) + \left(\tfrac{1}{2}\right)(2.0) + (-1)(3.0) = -2.5$$

$$SS_{\text{contrast}} = \frac{n\psi^2}{\sum a_j^2} = \frac{10(-2.5)^2}{6} = \frac{62.5}{6} = 10.417$$

---

[5] For unequal sample sizes, $SS_{\text{contrast}} = \dfrac{\psi^2}{\sum\left(a_j^2/n_j\right)}$

This sum of squares is a component of the overall SS_{treat} on 1 df. We have 1 df because we are really comparing two quantities (the mean of the first two treatments with the mean of the third treatment).

Now suppose we obtain an additional linear contrast comparing treatment 1 with treatment 2. Let $a_1 = 1$, $a_2 = -1$, and $a_3 = 0$. Then

$$\psi = \sum a_j \overline{X}_j = (1)(1.5) + (-1)(2.0) + (0)(3.0) = -0.5$$

$$SS_{\text{contrast}} = \frac{n\psi^2}{\sum a_j^2} = \frac{10(-0.5)^2}{2} = \frac{2.5}{2} = 1.25$$

This SS_{contrast} is also a component of SS_{treat} on 1 df. In addition, because of the particular contrasts that we chose to run,

$$SS_{\text{treat}} = SS_{\text{contrast}_1} + SS_{\text{contrast}_2}$$
$$11.667 = 10.417 + 1.25$$

and thus, the two contrasts account for all of the SS_{treat} and all the df attributable to treatments. We say that we have *completely* **partitioned** SS_{treat}.

partitioned

The Choice of Coefficients

In the previous example, it should be reasonably clear why we chose the coefficients we did. They weight the treatment means in what seems to be a logical way to perform the contrast in question. Suppose, however, that we have five groups of equal size and want to compare the first three with the last two. We need a set of coefficients (a_j) that will accomplish this task and for which $\sum a_j = 0$. The simplest rule is to form the two sets of treatments and to assign as weights to one set the reciprocal of the number of treatment groups in that set, and vice versa. One arbitrary set of coefficients is then given a minus sign. For example, take the means

$$\overline{X}_1 \quad \overline{X}_2 \quad \overline{X}_3 \quad \overline{X}_4 \quad \overline{X}_5$$

We want to compare $\overline{X}_1$, $\overline{X}_2$, and $\overline{X}_3$ combined with $\overline{X}_4$ and $\overline{X}_5$ combined. The first set contains three means, so for $\overline{X}_1$, $\overline{X}_2$, and $\overline{X}_3$ the $a_j = \frac{1}{3}$. The second set contains two means, so for $\overline{X}_4$ and $\overline{X}_5$ the $a_j = \frac{1}{2}$. We will let the $\frac{1}{2}$s be negative. Then we have

Means:	$\overline{X}_1$	$\overline{X}_2$	$\overline{X}_3$	$\overline{X}_4$	$\overline{X}_5$	
a_j:	$\frac{1}{3}$	$\frac{1}{3}$	$\frac{1}{3}$	$-\frac{1}{2}$	$-\frac{1}{2}$	$\sum a_j = 0$

Then $\sum a_j \overline{X}_j$ reduces to $\frac{1}{3}(\overline{X}_1 + \overline{X}_2 + \overline{X}_3) - \frac{1}{2}(\overline{X}_4 + \overline{X}_5)$.

(If you go back to Siegel's experiment on morphine, lump the first three groups together and the last two groups together, and look at the means of the combined treatments, you will get an idea of why this system makes sense.)[6]

---

[6] If we have different numbers of subjects in the several groups, we *may* need to obtain our coefficients somewhat differently. If the sample sizes differ in non-essential ways, such as when a few subjects are missing at random, this approach will be the appropriate one. It will not weight one group mean more than another just because the group happens to have a few more subjects. However, if the sample sizes are systematically different, not just different at random, and *if* we want to give more weight to the means from the larger groups, then we need to do something different. Because there really are very few cases where I can imagine wanting the different sample sizes to play an important role, I have dropped that approach from this edition of the book. However, you can find it in earlier editions and on the Web pages referred to earlier. (You may even send me a note if you need the information, and I will send it to you.)

There are other ways of setting up the coefficients using whole numbers, and for many purposes you will arrive at the same result. I used to like alternative approaches because I find fractions messy, but using fractional values as I did here, where the sum of the absolute values of all coefficients is equal to 2, has some important implications when it comes to estimating effect sizes.

The Test of Significance

We have seen that linear contrasts can be easily converted to sums of squares on 1 degree of freedom. These sums of squares can be treated exactly like any other sums of squares. They happen also to be mean squares because they always have 1 degree of freedom and can thus be divided by MS_{error} to produce an F. Because *all* contrasts have 1 degree of freedom

$$F = \frac{MS_{contrast}}{MS_{error}} = \frac{n\psi^2 / \sum a_j^2}{MS_{error}} = \frac{n\psi^2}{\sum a_j^2 MS_{error}}$$

This F will have one and df_{error} degrees of freedom. And if you feel more comfortable with t, you can take the square root of F and have a t on df_{error} degrees of freedom.

For our example, suppose we had planned (a priori) to compare the mean of the two groups for whom the morphine should be maximally effective, either because they had never had morphine (Condition S-M) or because they had received morphine in a different context (Mc-M) with the mean of the other three groups (M-M, S-S, and M-S). We also planned to compare group Mc-M with group M-M, and group M-S with group S-S, for the same reasons given in the discussion of individual t tests. Finally, we planned to compare group M-M with group S-S to see whether morphine tolerance developed to such an extent that animals that always received morphine were no different after only four trials from animals that always received saline. (As we will see shortly, these four contrasts are not independent, but they answer substantive questions.) The analysis is shown in Table 12.2.

Each of these F values can be evaluated against $F_{.05}(1, 35) = 4.12$. As expected, the first three contrasts are significant. The fourth contrast, comparing M-M with S-S, is not significant, indicating that complete morphine tolerance seems to develop in as few as four trials. (Be careful here, as I am acting as if I can prove the null hypothesis, when we know that such is not possible.) Note that contrasts 2 and 3 test the same hypotheses that we tested using individual t tests. If you take the square root of the Fs for these two contrasts, they will equal 6.72 and 2.47, which are precisely the values we obtained for t earlier. This simply illustrates the fact that t tests are a special case of linear contrasts.

With four contrasts, we have an FW approaching .20. This error rate is uncomfortably high, although some experimenters would accept it, especially for a priori contrasts. One way of reducing the error rate would be to run each comparison at a more stringent level of α; for example, $\alpha = .01$. Another alternative would be to use a different a priori procedure, the Bonferroni procedure, which amounts to almost the same thing as the first alternative but is conducted in a more precise manner. We will consider this procedure after we briefly discuss a special type of linear contrast, called orthogonal contrasts. Yet a third way to control FW is to run fewer contrasts. For example, the comparison of M-M with S-S is probably not very important. Whether complete tolerance develops on the fourth trial or on the sixth or seventh trial is of no great theoretical interest. By eliminating that contrast, we could reduce the maximum FW to .15. You should never choose to run contrasts the way you eat peanuts or climb mountains—just because they are there. In general, if a contrast is not important, do not run it.

Table 12.2   A set of a priori comparisons on morphine data

Groups:	M-S	M-M	S-S	S-M	Mc-M
Means:	4.00	10.00	11.00	24.00	29.00

	Coefficient					$\sum a_j^2$	$\psi = \sum a_j \overline{X}_j$
a_j	$-1/3$	$-1/3$	$-1/3$	$1/2$	$1/2$	0.833	18.167
b_j	0	-1	0	0	1	2	19
c_j	-1	0	1	0	0	2	7
d_j	0	1	-1	0	0	2	-1

$$SS_{\text{contrast}_1} = \frac{n\left(\sum a_j \overline{X}_j\right)^2}{\sum a_j^2} = \frac{8(18.17)^2}{0.8333} = \frac{2641.19}{0.8333} = 3169.42$$

$$F = \frac{MS_{\text{contrast}}}{MS_{\text{error}}} = \frac{3169.42}{32.00} = 99.04$$

$$SS_{\text{contrast}_2} = \frac{n\left(\sum b_j \overline{X}_j\right)^2}{\sum b_j^2} = \frac{8(19)^2}{2} = \frac{2888}{2} = 1444.00$$

$$F = \frac{MS_{\text{contrast}}}{MS_{\text{error}}} = \frac{1444.00}{32.00} = 45.125$$

$$SS_{\text{contrast}_3} = \frac{n\left(\sum c_j \overline{X}_j\right)^2}{\sum c_j^2} = \frac{8(7)^2}{2} = \frac{392}{2} = 196.00$$

$$F = \frac{MS_{\text{contrast}}}{MS_{\text{error}}} = \frac{196.00}{32.00} = 6.125$$

$$SS_{\text{contrast}_4} = \frac{n\left(\sum d_j \overline{X}_j\right)^2}{\sum d_j^2} = \frac{8(-1)^2}{2} = \frac{8}{2} = 4.00$$

$$F = \frac{MS_{\text{contrast}}}{MS_{\text{error}}} = \frac{4.00}{32.00} = 0.125$$

Orthogonal Contrasts

Linear contrasts as they have been defined allow us to test a series of hypotheses about treatment differences. Sometimes contrasts are independent of one another, and sometimes they are not. For example, knowing that $\overline{X}_1$ is greater than the average of $\overline{X}_2$ and $\overline{X}_3$ tells you nothing about whether $\overline{X}_4$ is likely to be greater than $\overline{X}_5$. These two contrasts are independent. However, knowing that $\overline{X}_1$ is greater than the average of $\overline{X}_2$ and $\overline{X}_3$ suggests that there is a better than 50:50 chance that $\overline{X}_1$ is greater than $\overline{X}_2$. These two contrasts are not independent. When members of a set of contrasts are independent of one another, they are called **orthogonal contrasts,** and the sums of squares of a complete set of orthogonal contrasts sum to SS_{treat}. (If the contrasts are not orthogonal, they contain overlapping amounts of information and do not have this additivity property.) From a calculational point of view, what sets orthogonal contrasts apart from other types of contrasts we might choose is the relationship between the coefficients for one contrast and the coefficients for other contrasts in the set.

orthogonal contrasts

Orthogonal Coefficients

Given that sample sizes are equal, for contrasts to be orthogonal the coefficients must meet the following criteria:

1. $\sum a_j = 0$
2. $\sum a_j b_j = 0$

where a_j and b_j are the sets of coefficients for different contrasts. Furthermore, for the SS_{contrast} to sum to SS_{treat}, we need to add a third criterion:

3. Number of comparisons = number of df for treatments

The first restriction has been discussed already; it results in the contrast's being a sum of squares. The second restriction ensures that the contrasts are independent of (or orthogonal to) one another and, thus, that we are summing nonoverlapping components. The third restriction says nothing more than that if you want the parts to sum to the whole, you need to have all the parts.

At first glance, it would appear that finding sets of coefficients satisfying the requirement $\sum a_j b_j = 0$ would require that we either undertake a frustrating process of trial and error or else solve a set of simultaneous equations. Actually, a simple rule exists for finding orthogonal sets of coefficients; although the rule will not find all possible sets, it will lead to most of them. The rule for forming the coefficients visualizes the process of breaking down SS_{treat} in terms of a tree diagram. The overall F for five treatments deals with all five treatment means simultaneously. That is the trunk of the tree. If we then compare the combination of treatments 1 and 2 with the combination of treatments 3, 4, and 5, we have formed two branches of our tree, one representing treatments 1 and 2 and the other representing treatments 3, 4, and 5. As discussed earlier, the value of a_j for the treatment means on the left will be equal to the reciprocal of the number of treatments in that set, and vice versa, with one of the sets being negative. Thus, the coefficients are ($\frac{1}{2}$, $\frac{1}{2}$, $-\frac{1}{3}$, $-\frac{1}{3}$, $-\frac{1}{3}$) for the five treatments, respectively.

Now that we have formed two limbs or branches of our tree, we can never compare treatments on one limb with treatments on another limb, although we can compare treatments on the same limb. Thus, comparing treatment 3 with the combination of treatments 4 and 5 is an example of a legitimate comparison. The coefficients in this case would be (0, 0, 1, $-\frac{1}{2}$, $-\frac{1}{2}$). Treatments 1 and 2 have coefficients of 0 because they are not part of this comparison. Treatment 3 has a coefficient of 1 because it contains one treatment. Treatments 4 and 5 received coefficients of $-\frac{1}{2}$ because there are two treatments in that set. The negative signs can be arbitrarily assigned to either side of the comparison.

The previous procedure could be carried on until we have exhausted all possible sets of comparisons. This will occur when we have made as many comparisons as there are df for treatments. As a result of this procedure, we might arrive at the comparisons and coefficients shown in Figure 12.1. To show that these coefficients are orthogonal, we need to show only that all *pairwise* products of the coefficients sum to zero. For example,

$$\sum a_j b_j = \left(\tfrac{1}{2}\right)(1) + \left(\tfrac{1}{2}\right)(-1) + \left(-\tfrac{1}{3}\right)(0) + \left(-\tfrac{1}{3}\right)(0) + \left(-\tfrac{1}{3}\right)(0) = 0$$

and

$$\sum a_j c_j = \left(\tfrac{1}{2}\right)(0) + \left(\tfrac{1}{2}\right)(0) + \left(-\tfrac{1}{3}\right)(2) + \left(-\tfrac{1}{3}\right)(-1) + \left(-\tfrac{1}{3}\right)(-1) = 0$$

Thus, we see that the first and second and the first and third contrasts are both independent. Similar calculations will show that all the other contrasts are also independent of one another.

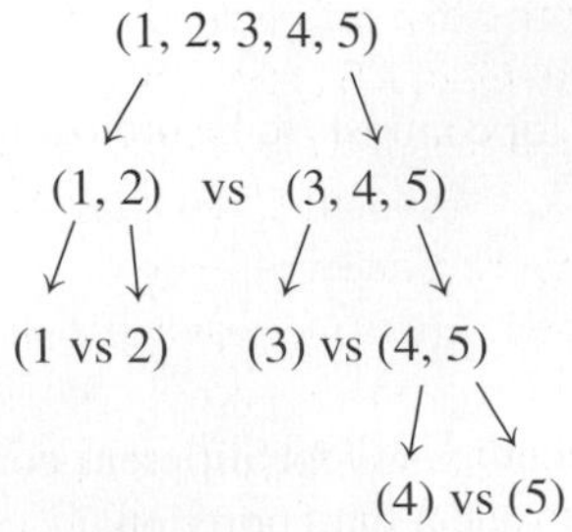

Figure 12.1   Tree diagram illustrating orthogonal partition of SS_{treat}

These coefficients will lead to only one of many possible sets of orthogonal contrasts. If we had begun by comparing treatment 1 with the combination of treatments 2, 3, 4, and 5, the resulting set of contrasts would have been entirely different. The experimenter must decide which contrasts she considers important, and to plan accordingly.

The actual computation of F with orthogonal contrasts is the same as when we are using nonorthogonal contrasts. Because of this, there is little to be gained by working through an example here. It would be good practice, however, for you to create a complete set of orthogonal contrasts and to carry out the arithmetic. You can check your answers by showing that the sum of the sums of squares equals SS_{treat}.

When I first started teaching and writing about statistics, orthogonal contrasts were a big deal, just as was the distinction between a priori and post hoc tests. Authors went out of their way to impress on you the importance of orthogonality and the need to feel somewhat guilty if you ran comparisons that were not orthogonal. That attitude has changed over the years. Although it is nice to have a set of orthogonal comparisons, in part because they sum to SS_{treat}, people are far more willing to run nonorthogonal contrasts. I would certainly not suggest that you pass up an important contrast just because it is not orthogonal to others that you ran. In fact, the contrasts that I ran earlier are not orthogonal to each other, and that does not worry me over much. They address important questions; (well, possibly not S-S versus M-M, as I said). Nor should you use a contrast in which you have no interest, just because it is part of an orthogonal set. But keep in mind that being nonorthogonal means that these contrasts are not independent of each other.

Bonferroni *t* (Dunn's Test)

I suggested earlier that one way to control the familywise error rate when using linear contrasts is to use a more conservative level of α for each comparison. The proposal that you might want to use $\alpha = .01$ instead of $\alpha = .05$ was based on the fact that our statistical tables are set up that way. (We do not usually have critical values of t for α between .05 and .01.) A formal way of controlling *FW* more precisely by manipulating the per comparison error rate can be found in a test proposed by Dunn (1961), which is particularly appropriate when you want to make only a few of all possible comparisons. Although this test had been known for a long time, Dunn was the first person to formalize it and to present the necessary tables, and it is sometimes referred to as **Dunn's test.** It now more commonly goes under the name **Bonferroni *t*.** The Bonferroni *t* test is based on what is known as the **Bonferroni inequality,** which states that the probability of occurrence of one *or more* events can never exceed the sum of their individual probabilities. This means that when we make three comparisons, each with a probability of $\alpha = .05$ of a Type I error, the probability of *at least* one Type I error can never exceed $3 \times .05 = .15$. In more formal terms, if c represents the number of comparisons and α' represents the probability of a Type I error for each comparison, then

Dunn's test

Bonferroni *t*

Bonferroni inequality

FW is less than or equal to *ca'*. From this, it follows that if we set $\alpha' = \alpha/c$ for each comparison, where α = the desired maximum *FW*, then $FW \leq c\alpha' = c(\alpha/c) = \alpha$. Dunn (1961) used this inequality to design a test in which each comparison is run at $\alpha' = \alpha/c$, leaving the $FW \leq \alpha$ for the set of comparisons. This can be accomplished by using the standard *t* test procedure but referring the result to modified *t* tables.

The problem that you immediately encounter when you attempt to run each comparison at $\alpha' = \alpha/c$ is that standard tables of Student's *t* do not provide critical values for the necessary levels of α. If you want to run each of three comparisons at $\alpha' = \alpha/c = .05/3 = .0167$, you would need tables of critical values of *t* at $\alpha = .0167$. Dunn's major contribution was to provide such tables. (Although such tables are less crucial now that virtually all computer programs report exact probability values for each *F* or *t*, they still have a role to play, and her table can be found in the Appendix *t'*.)

For the Bonferroni test on pairwise comparisons (i.e., comparing one mean with one other mean), define

$$t' = \frac{\overline{X}_i - \overline{X}_j}{\sqrt{\dfrac{MS_{error}}{n} + \dfrac{MS_{error}}{n}}} = \frac{\overline{X}_i - \overline{X}_j}{\sqrt{\dfrac{2MS_{error}}{n}}}$$

and evaluate *t'* against the critical value of *t'* taken from Dunn's tables in Appendix *t'*. Notice that we still use the standard formula for *t*. The only difference between *t'* and a standard *t* is the tables used in their evaluation. With unequal sample sizes but homogeneous variances, replace the *n*s in the leftmost equation with n_i and n_j. With heterogeneity of variance, see the solution by Games and Howell later in this chapter.

To write a general expression that allows us to test any comparison of means, pairwise or not, we can express *t'* in terms of linear contrasts.

$$\psi = \sum a_j \overline{X}_j \qquad \text{and} \qquad t' = \frac{\psi}{\sqrt{\dfrac{\sum a_j^2 MS_{error}}{n}}}$$

This represents the most general form for the Bonferroni *t*, and it can be shown that if ψ is *any* linear combination (not necessarily even a linear contrast, requiring $\sum a_j = 0$), the *FW* with *c* comparisons is at most α (Dunn, 1961).[7] To put it most simply, the Bonferroni *t* runs a regular *t* test but evaluates the result against a modified critical value of *t* that has been chosen to limit *FW*.

Dunn-Šidák test

A variation on the Bonferroni procedure was proposed by Šidák (1967). His test is based on the multiplicative inequality $p(FW) \leq 1 - (1 - \alpha)^c$ and evaluates *t'* at $\alpha' = 1 - (1 - \alpha)^{1/c}$. (This is often called the **Dunn-Šidák test.**) A comparison of the power of the two tests shows only very minor differences in favor of the Šidák approach, and we will stick with the Bonferroni test because of its much wider use. Many computer software programs, however, provide this test. (For four comparisons, the Šidák approach would test each comparison at $\alpha' = 1 - (1 - \alpha)^{1/4} = 1 - .95^{.25} = 0.0127$ level, whereas the Bonferroni approach would test at $\alpha/c = .05/4 = .0125$. You can see that there is not a lot of difference in power.)

When we considered linear contrasts earlier in this section, we ran four comparisons, which had an *FW* of nearly .20. (Our test of each of those contrasts involved an *F* statistic, but, because each contrast involves 1 *df*, we can go from *t* to *F* and vice versa by means of the relationship $t = \sqrt{F}$.) If we want to run those same comparisons but keep *FW* at a maximum of .05 instead of $4 \times (.05) = .20$, we can use the Bonferroni *t* test. In each case,

---

[7] Note the similarity between the right side of the equation and our earlier formula for *F* with linear contrasts. The resemblance is not accidental; one is just the square of the other.

we will solve for t' and refer that to Dunn's tables. Taking the pairwise tests first, the calculations follow.

Mc-M versus M-M:

$$t' = \frac{\overline{X}_i - \overline{X}_j}{\sqrt{\dfrac{2MS_{\text{error}}}{n}}} = \frac{29.00 - 10.00}{\sqrt{\dfrac{(2)(32.00)}{8}}} = \frac{19}{\sqrt{8}} = 6.72$$

S-S versus M-S:

$$t' = \frac{\overline{X}_i - \overline{X}_j}{\sqrt{\dfrac{2MS_{\text{error}}}{n}}} = \frac{11.00 - 4.00}{\sqrt{\dfrac{(2)(32.00)}{8}}} = \frac{7}{\sqrt{8}} = 2.47$$

M-M versus S-S:

$$t' = \frac{\overline{X}_i - \overline{X}_j}{\sqrt{\dfrac{2MS_{\text{error}}}{n}}} = \frac{10.00 - 11.00}{\sqrt{\dfrac{(2)(32.00)}{8}}} = \frac{-1}{\sqrt{8}} = -0.35$$

The calculations for the more complex contrast, letting the $a_j = 1/3, 1/3, 1/3, -1/2, -1/2$ as before, follow.

S-M and Mc-M versus M-M, S-S and M-S:

$$t' = \frac{\sum a_j \overline{X}_j}{\sqrt{\dfrac{\sum a_j^2 MS_{\text{error}}}{n}}} = \frac{\left(\frac{1}{2}\right)(24) + \cdots + \left(-\frac{1}{3}\right)(4)}{\sqrt{\dfrac{(0.833)(32.00)}{8}}} = \frac{18.167}{\sqrt{3.3333}} = 9.95$$

From Appendix t', with $c = 4$ and $df_{\text{error}} = 35$, we find by interpolation $t'_{.05}(35) = 2.64$. In this case, the first and last contrasts are significant, but the other two are not.[8] Whereas we earlier rejected the hypothesis that groups S-S and M-S were sampled from populations with the same mean, using the more conservative Bonferroni t test, we are no longer able to reject that hypothesis. Here we cannot conclude that prior morphine injections lead to hypersensitivity to pain. The difference in conclusions between the two procedures is a direct result of our use of the more conservative familywise error rate. If we want to concentrate on per comparison error rates, ignoring FW, then we evaluate each t (or F) against the critical value at $\alpha = .05$. On the other hand, if we are primarily concerned with controlling FW, then we evaluate each t, or F, at a more stringent level. The difference is not in the arithmetic of the test; it is in the critical value we choose to use. The choice is up to the experimenter.

Multistage Bonferroni Procedures

The Bonferroni multiple-comparison procedure has a number of variations. Although these are covered here in the context of the analysis of variance, they can be applied equally well whenever we have multiple hypothesis tests for which we want to control the familywise error rate. These procedures have the advantage of setting a limit on the FW error rate at α against any set of possible null hypotheses, as does the Tukey Honestly Significant Difference (HSD) (to be discussed shortly), while at the same time being less conservative than

---

[8] The actual probabilities would be .000, .073, 1.00, and .000.

Tukey's test *when our interest is in a specific subset of contrasts*. In general, however, multistage procedures would not be used as a substitute when making all pairwise comparisons among a set of means.

As you saw, the Bonferroni test is based on the principle of dividing *FW* for a family of contrasts among each of the individual contrasts. Thus, if we want *FW* to be .05 and we want to test four contrasts, we test each one at $\alpha = .05/4 = .0125$. The multistage tests follow a similar principle, the major difference being in the way they choose to partition α.

Holm and Larzelere and Mulaik Tests

Both Holm (1979) and Larzelere and Mulaik (1977) have proposed a multistage test that adjusts the denominator (*c*) in $\alpha' = \alpha/c$ depending on the number of null hypotheses remaining to be tested. Holm's test is generally referred to when speaking about the analysis of variance, whereas the Larzelere and Mulaik test is best known as a test of significance for a large set of correlation coefficients. The logic of the two tests is the same, though the method of calculation is different.

In the Holm procedure, we calculate values of t' just as we did with the Bonferroni *t* test. For the equal *n* case, we compute

$$t' = \frac{\overline{X}_i - \overline{X}_j}{\sqrt{\dfrac{2MS_{\text{error}}}{n}}}$$

For the unequal *n* case, or when we are concerned about heterogeneity of variance, we compute

$$t' = \frac{\overline{X}_i - \overline{X}_j}{\sqrt{\dfrac{s_i^2}{n_i} + \dfrac{s_j^2}{n_j}}}$$

We calculate t' for all contrasts of interest and then arrange the t' values in increasing order *without regard to sign*. This ordering can be represented as $|t'_1| \leq |t'_2| \leq |t'_3| \leq \cdots \leq |t'_c|$, where *c* is the total number of contrasts to be tested.

The first significance test is carried out by evaluating t'_c, the largest of the t', against the critical value in Dunn's table corresponding to *c* contrasts. In other words, t'_c is evaluated at $\alpha' = \alpha/c$. If this largest t' is significant, then we test the next largest t' (*i.e.* t'_{c-1}) against the critical value in Dunn's table corresponding to $c - 1$ contrasts. Thus, t'_{c-1} is evaluated at $\alpha' = \alpha/(c - 1)$. The same procedure continues for $t'_{c-2}, t'_{c-3}, t'_{c-4}, \ldots$ until the test returns a nonsignificant result. At that point, we stop testing. Holm has shown that such a procedure continues to keep $FW \leq \alpha$, while offering a more powerful test.

The rationale behind the test is that when we reject the null for t'_c, we are declaring that null hypothesis to be false. If it is false, that only leaves $c - 1$ possibly true null hypotheses, and so we only need to protect against $c - 1$ contrasts. A similar logic applies as we carry out additional tests. This logic makes particular sense when you know, even before the experiment is conducted, that several of the null hypotheses are almost certain to be false. If they are false, there is no point in protecting yourself from erroneously rejecting them.

To illustrate the use of Holm's test, consider our example on morphine tolerance. With the standard Bonferroni *t* test, we evaluated four contrasts with the following results, arranged by increasing magnitude of t', as in Table 12.3.

If we were using the Bonferroni test, each of these t's would be evaluated against $t'_{.05} = 2.64$, which is actually Student's *t* at $\alpha = 0.0125$. For Holm's test, we vary the critical value in stages, depending on the number of contrasts that have not been tested. This

Table 12.3   Application of Holm's test to the morphine data

Contrast	Order (i)	t'	t'_{crit}
M-M vs. S-S	1	$t' = -0.35$	2.03
S-S vs. M-S	2	$t' = -2.47$	2.35*
Mc-M vs. M-M	3	$t' = -6.72$	2.52*
S-M, Mc-M vs. M-S, M-M, S-S	4	$t' = -9.95$	2.64*

$*p < .05$

number is indexed by "Order (i)" in the Table 12.3. These critical values are presented in the right column of the table. They were taken, with interpolation, from Dunn's tables for $c = i$ and 35 degrees of freedom. For example, the critical value of 2.35 corresponds to the entry in Dunn's tables for $c = 2$ and $df = 35$. For the smallest t', the critical value came from the standard Student t distribution (Appendix t).

From this table, you can see that the test on the complex contrast S-M, Mc-M vs. M-S, M-M, S-S required a t' of 2.64 or above to reject H_0. Because the absolute value of t' was 9.95, the difference was significant. The next largest t' was 6.72 for Mc-M vs. M-M, and that was also significant, exceeding the critical value of 2.52. The contrast S-S vs. M-S is tested as if there were only two contrasts in the set, and thus, t' must exceed 2.35 for significance. Again, this test is significant. If it had not been, we would have stopped at this point. But because it is, we continue and test M-M vs. S-S, which is not significant. Because of the increased power of Holm's test over the Bonferroni t test, we have rejected one null hypothesis (S-S vs. M-S) that was not rejected by the Bonferroni.

Larzelere and Mulaik Test

Larzelere and Mulaik (1977) proposed a test equivalent to Holm's test, but their primary interest was in using that test to control *FW* when examining a large set of correlation coefficients. As you might suspect, something that controls error rates in one situation will tend to control them in another. (When you are testing all possible correlation coefficients in a correlation matrix, it is conceptually the same as testing all possible pairwise differences in a set of means. This would mean that perhaps we really should consider the Larzelere and Mulaik test as being a post hoc test and discuss it in that section of the chapter. But because it essentially is the same as the Holm procedure, I am discussing it here.)

I will consider the Larzelere and Mulaik test with respect to correlation coefficients rather than the analysis of variance because such an example will prove useful to those who conduct research that yields large numbers of such coefficients. However, as you will see when you look at the calculations, the test would be applied in the same way whenever you have a number of test statistics with their associated probability values. (Your test statistic could be t, F, χ^2, or any other test statistic, as long as you can calculate its probability under the null.) If you had never heard of Larzelere and Mulaik, you could still accomplish the same thing with Holm's test. However, the different calculational approach is instructive. It is worth noting that when these tests are applied in an analysis of variance setting, we usually have a small number of comparisons. However, when they are used in a regression/correlation setting, we commonly test all pairwise correlations.

Compas, Howell, Phares, Williams, and Giunta (1989) investigated the relationship between daily stressors, parental levels of psychological symptoms, and adolescent behavior problems (as measured by Achenbach's Youth Self-Report Form [YSR] and by the Child Behavior Checklist [CBCL]). The study represented an effort to understand risk factors for emotional and behavioral problems in adolescents. Among the analyses of the study was the

Table 12.4   Correlations among behavioral and stress measures

	(1)	(2)	(3)	(4)	(5)	(6)	(7)
Mother							
(1) Stress	1.00	.69	.48	.37	−.02	.30	.03
(2) Symptoms		1.00	.38	.42	.12	.39	.19
Father							
(3) Stress			1.00	.62	.07	.22	.07
(4) Symptoms				1.00	.00	.24	.20
Adolescent							
(5) Stress					1.00	.11	.44
(6) CBCL						1.00	.23
(7) YSR							1.00

set of intercorrelations between these variables at Time 1. These correlations are presented in Table 12.4.

Most standard correlation programs print a t statistic for each of these correlations. However, we know that with 21 hypothesis tests, the probability of a Type I error based on that standard t test, if all null hypotheses were true, would be high. It would still be high if only a reduced set of them were true. For this reason, we will apply the modified Bonferroni test proposed by Larzelere and Mulaik. There are two ways to apply this test to this set of correlations. For the first method, we could calculate a t value for each coefficient, based on

$$t = \frac{r\sqrt{(N-2)}}{\sqrt{(1-r^2)}}$$

(or take the t from a standard computer printout) and then proceed exactly as we did for the Holm procedure. Alternatively, we could operate directly on the two-tailed p values associated with the t test on each correlation. These p values can be taken from standard computer printouts, or they can be calculated using commonly available programs. For purposes of an example, I will use the p-value approach.

Table 12.5 shows the correlations to be tested from Table 12.4 as well as the associated p values. The p values have been arranged in increasing numerical order. (Note that the sign of the correlation is irrelevant—only the absolute value matters.)

The right column gives the value of α' required for significance. For example, if we consider 21 contrasts to be of interest, $\alpha' = \alpha/(k - i + 1) = .05/21 = .00238$. By the time we have rejected the first four correlations and want to test the fifth largest, we are going to behave as if we want a Bonferroni t adjusted for just the $k - i + 1 = 21 - 5 + 1 = 21 - 4 = 17$ remaining correlations. This correlation will be tested at $\alpha' = \alpha/(k - i + 1) = .05/17 = .00294$.

Each correlation coefficient is tested for significance by comparing the p value associated with that coefficient with the entry in the final column. For example, for the largest correlation coefficient out of a set of 21 coefficients to be significant, it must have a probability (under H_0: $\rho = 0$) less than .00238. Because the probability for $r = .69$ is given as .0000 (there are no nonzero digits until the sixth decimal place), we can reject H_0 and declare that correlation to be significant.

Having rejected H_0 for the largest coefficient, we then move down to the second row, comparing the obtained p value against $p = .00250$. Again, we reject H_0 and move on to the third row. We continue this procedure until we find a row at which the obtained p value in column 4 exceeds the critical p value in column 5. At that point, we declare that correlation

Table 12.5   Significance tests for correlations in Table 12.4

Pair	i	Correlation	p value	$\alpha(k - i + 1)$
1 vs. 2	1	.69	.0000	.00238*
3 vs. 4	2	.62	.0000	.00250*
1 vs. 3	3	.48	.0000	.00263*
5 vs. 7	4	.44	.0000	.00278*
2 vs. 4	5	.42	.0000	.00294*
2 vs. 6	6	.39	.0001	.00313*
2 vs. 3	7	.38	.0001	.00333*
1 vs. 4	8	.37	.0002	.00357*
1 vs. 6	9	.30	.0028	.00385*
4 vs. 6	10	.24	.0179	.00417
6 vs. 7	11	.23	.0236	.00455
3 vs. 6	12	.22	.0302	.00500
4 vs. 7	13	.20	.0495	.00556
2 vs. 7	14	.19	.0618	.00625
2 vs. 5	15	.12	.2409	.00714
5 vs. 6	16	.11	.2829	.00833
3 vs. 5	17	.07	.4989	.01000
3 vs. 7	18	.07	.4989	.01250
1 vs. 7	19	.03	.7724	.01667
1 vs. 5	20	−.02	.8497	.02500
4 vs. 5	21	.00	1.0000	.05000

*$p < .5$

to be nonsignificant and stop testing. All correlations below that point are likewise classed as nonsignificant. For our data, those correlations equal to or greater than .30 are declared significant, and those below .30 are nonsignificant. The significant correlations are indicated with an asterisk in the table.

Had we used a standard Bonferroni test, we would have set $\alpha' = .05/21 = .00238$, and a correlation less than .37 would not have been significant. In this particular case, the multistage test made only a small difference. But often the difference is substantial for the number of coefficients that are declared significant.

Trimmed Means

I want to include one more approach that is very general and can be shown to be more powerful than standard procedures when the data come from long-tailed distributions. This is the use of trimmed means. The nice thing about this approach is that it can be adapted to carry out any of the procedures in this chapter, simply be substituting the appropriate trimmed means and squared standard errors.

I will assume that you have reasonably large sample sizes because we will trim those samples from each end. Wilcox recommends 20% trimming, which results in a sizable drop in the effective sample size, but with a corresponding gain in power. For convenience assume that we have 40 observations in each of several groups and that we will go along with Wilcox's suggestion of 20% trimming. That means that we will omit the lowest $(.20)(40) = 8$ observations and the highest 8 observations, leaving us with a sample of 24 observations for each condition. The trimmed means will be the means of those 24 observations in each group. To calculate the variance, we will use Winsorized samples, in which the lowest

8 scores are replace with the ninth lowest score and the highest 8 scores are replaced with the ninth highest score. This leaves us with samples of $n_i = 40$ scores, but only $h_i = 24$ of those are independent observations from the ith sample. If we let $s_{W_i}^2$ represent the variance of the Winsorized sample of 40 scores, then the squared standard error of the mean for that sample would be

$$s_{W_{\overline{X}_i}}^2 = \frac{(n-1)s_{W_i}^2}{h(h-1)}$$

and the robust pairwise t test on the difference between two means can be written as

$$t_W = \frac{\overline{Y}_{ti} - \overline{Y}_{tj}}{\sqrt{s_{W_{\overline{i}}}^2 + s_{W_{\overline{Xj}}}^2}}$$

Notice that we are not doing anything very surprising here. We are replacing means with trimmed means and variances with variances that are based on Winsorized samples, but with an adjustment to n_i to account for the trimming. Other than that, we have a standard t test, and it can be used as a replacement for the t in any of the procedures we have discussed, or will discuss, in this chapter. There is one complication, however, and that refers to the estimated degrees of freedom. The degrees of freedom are estimated as

$$df_W = \frac{\left(s_{W_{\overline{X}_i}}^2 + s_{W_{\overline{X}_j}}^2\right)^2}{s_{W_{\overline{X}_i}}^2(h_i - 1) + s_{W_{\overline{X}_j}}^2(h_j - 1)}$$

That is a messy formula, but not very difficult to work out. As Keselman et al. (2005) noted, "When researchers feel they are dealing with nonnormal data, they can replace the usual least squares estimators of central tendency and variability with robust estimators and apply these estimators in any of the previously recommended" multiple comparison procedures.

One More Comment

I want to emphasize one more time that the Bonferroni test and its variants are completely general. They are not the property of the analysis of variance or of any other statistical procedure. If you have several tests that were carried by any statistical procedure (and perhaps by different procedures), you can use the Bonferroni approach to control FW. For example, I recently received an e-mail message in which someone asked how they might go about applying the Bonferroni to logistic regression. You would do it the same way you would do it for the analysis of variance. Take the set of statistical tests that came from your logistic regression, divide α by the number of tests you ran, and declare a test to be significant only if its resulting probability was less than α/c. You don't even need to know anything about logistic regression to do that.

12.4   Confidence Intervals and Effect Sizes for Contrasts

Having run a statistical significance test on the data from an experiment and looked at individual comparisons, often called "individual contrasts," we will generally want to look at some measure of the amount of difference between group means. In Chapter 11, I said that when we have the omnibus F, which compares all means together, the best measure is a member of the r-family measures, such as η^2 or ω^2. However, when we are looking at comparisons of individual means, or sets of means, it generally makes more sense to calculate confidence limits on our differences or to use a *d*-family measure of the effect size.

We could approach *d*-family measures in several ways. One very simple way is to go back to Chapter 7, which discusses *t* tests, and apply the measures that were discussed there. We will come out at the same place, however, if we approach the problem through linear contrasts. Remember that when you are looking at two groups, it makes no difference whether you run a *t* test between those groups, or compute a linear contrast and then an *F*, and take the square root of that *F*. The advantage of going with linear contrasts is that they are more general, allowing us to compare means of sets of groups rather than just two individual groups.

We will take an example from our morphine study by Siegel. One contrast that really interests me is the contrast between Group M-M and Group Mc-M. If their means are statistically significantly different, then that tells us that there is something important about changing the physical context in which the morphine is given. The two means are given here:

Condition	M-M	Mc-M
Mean	10.00	29.00
St. Dev	5.13	6.06
Variance	26.32	37.95
MS_{error}	32.00	

The coefficients for the linear contrast of these two groups would be "−1" for M-M, "+1" for Mc-M, and "0" for the other three conditions.

Confidence Interval

Let us first compute a confidence interval on the difference between conditions. The general formula for a confidence interval on a contrast of two means is

$$CI_{.95} = (\overline{X}_i - \overline{X}_j) \pm t_{.05}s_{\overline{X}_i - \overline{X}_j}$$

or, if we let "ψ_j" represent the value of the contrast, where $\psi_j = \sum a_i \overline{X}_i$, then

$$CI_{.95} = (\psi_j) \pm t_{.05}s_{error}$$

s_{error} is the standard error of the contrast, which is $\sqrt{MS_{error}(\frac{2}{n})}$. For our confidence interval on the difference between the two conditions of interest, I have

$$CI_{.95} = (-1(10) + 1(29)) \pm 2.03\sqrt{8}$$
$$= 19 \pm 2.03(2.828) = 19 \pm 5.74$$
$$13.26 \le \mu_{M-M} - \mu_{Mc-M} \le 24.74$$

The probability is .95 that an interval formed as I have formed this one will include the true difference between the population means.

When it comes time to form our effect size measure, we have a choice of what we will use as the error term—the standard deviation in the equation. I could choose to use the square root of MS_{error} from the overall analysis because that represents the square root of the average variance within each group. Kline (2004) recommends this approach. I have two other perfectly reasonable alternatives, however. First, I could take the square root of the average sample variance of the two groups in question (perhaps weighted if the sample sizes were unequal). In this case, it would be $(26.32 + 37.95)/2 = 32.135$; $\sqrt{32.135} = 5.669$. Alternatively, I could consider one of the groups to be a control group and use its standard deviation as my error term. Here, I might argue that M-M is sort of a control group because the conditions don't change on trial 4. In this case, I would let $s_{\text{error}} = 5.13$. I think that my general preference would be to base my estimate on the average of the variances of the groups in question. If there is heterogeneity of variance across the groups, then this would be a more representative estimate. If the variances are homogeneous across all five groups, then the average of the groups in question won't deviate much from the average of the variances of all five groups, so I haven't lost much. Others might take a different view.

Effect Size

We have just seen that the confidence interval on the difference between Mc-M and M-M is $13.26 \leq (\mu_{\text{Mc-M}} - \mu_{\text{m-m}}) \leq 24.74$. Both limits are on the same side of 0, reflecting the fact that the difference was statistically significant. However, the dependent variable here is the length of time before the animal starts to lick its paws, and I don't suppose that any of us have a strong intuitive understanding of what a long or short interval is for this case. A difference of at least thirteen seconds seems pretty long, but I would like some better understanding of what is happening. One way to compute that would be to calculate an effect size on the difference between these means.

Our effect size measure will be essentially the same as it was in the case for t tests for independent samples. However, I will write it slightly differently because doing so will generalize to more complex comparisons. We have just seen that ψ represents a contrast between two means or sets of means, so it is really just a difference in means. We will take this difference and standardize it, which simply says that we want to represent the difference in group means in standard deviation units. (That is what we did in Chapter 7 as well.)

In Chapter 7, we defined

$$\hat{d} = \frac{\overline{X}_i - \overline{X}_j}{s_p}$$

where s_p is the square root of our pooled variance estimate and is a measure of the average standard deviation within the groups. We are going to calculate essentially the same thing here, but I will write its expression as

$$\hat{d} = \frac{\psi}{s_e} = \frac{\sum(a_i \overline{X}_i)}{s_e}$$

The numerator is a simple linear contrast, and the denominator is some estimate of the within groups standard deviation.

The preceding formula raises two points. In the first place, the coefficients must form what is sometimes called a "standard set." This simply means that the absolute values of the coefficients must sum to 2. For example, if we want to compare the mean of two groups with the mean of a third, we could use coefficients of $(\frac{1}{2} \ \frac{1}{2} \ -1)$ to form our contrast. Alternatively, we would get to the same place as far as our test of significance is concerned by using $(1 \ 1 \ -2)$ or $(3 \ 3 \ -6)$. The resulting F would be the same. But only the first would give us a numerical answer for the contrast that is the difference between the mean of the first two groups and the mean of the third. This is easily seen when you write

$$\psi = \left(\tfrac{1}{2}\right)(\overline{X}_1) + \left(\tfrac{1}{2}\right)\overline{X}_2 + (-1)\overline{X}_3$$
$$= \frac{\overline{X}_1 + \overline{X}_2}{2} - \overline{X}_3$$

You can see that we are taking the difference between the mean of the first two groups and the third group.

The second question raised by our equation for $\hat{d}$ is the choice of the denominator. There are at least three possible estimates. We could use the square root of MS_{error}, or the square root of the average of the variances in the groups being contrasted, or we could conceive of one of the groups as a control group, and use its standard deviation as our estimate. The most common approach seems to be to use the square root of MS_{error}, though, for reasons given earlier, I would prefer to base my estimate on only the groups in question.

Earlier, we looked at four contrasts that seemed to be of interest for theoretical reasons. Holm's procedure showed that three of the contrasts were statistically significant, but the fourth was not. Computation of the effect sizes for these contrasts is shown in Table 12.6. In these calculations, I have used the square root of MS_{error} as my denominator.

Because the Holm test showed that the last contrast was not nearly statistically significant, our best approach would probably be to treat that effect size as 0.00. There are no differences between groups. An interesting question arises about what we would do if the test statistic had been nearly large enough to be significant. In that case, I would present my effect size measure but caution that the corresponding hypothesis test was not significant.

You can see that the other effect sizes are substantial, all showing a difference of at least one standard deviation. I will speak about these effects in the following section.

12.5 Reporting Results

We have run several different tests on these data, and the following is a report based on Holm's procedure:

> This experiment examined the phenomenon of morphine tolerance in rats placed on a warm surface. The underlying hypothesis was that with repeated injections of morphine, animals develop a hypersensitivity to pain, which reduces the effect of the drug. When animals are then tested without the drug, or with the drug in a different context, this hypersensitivity will be expressed in a shorter paw-lick latency.
>
> The omnibus F from the overall analysis was statistically significant ($F(4, 35) = 27.33$, $p < .05$). Subsequent contrasts using Holm's adaptation of the Bonferroni test revealed that morphine's effects were as predicted. The groups receiving morphine on the test trial after having received either saline or morphine in a different context on trials 1 to 3 showed longer reaction times than the average of groups who (1) never received morphine on any trials, (2) received morphine on all trials and had the opportunity to develop tolerance, and (3) switched from morphine to saline on the test trial and were predicted to show hypersensitivity. ($t(35) = 9.95$, $t_{.0125} = 2.64$). The standardized effect size was 3.21, indicating a difference of nearly $3\frac{1}{4}$ standard deviations between the means of the two sets of groups.
>
> The effect of context is seen in a statistically longer mean paw-lick latency in the Mc-M ($\overline{X} = 29$) condition than in the M-M condition ($\overline{X} = 10$) ($t(35) = 6.72$, $t_{.0167} = 2.52$). The standardized effect size here was 3.36.
>
> The hypersensitivity effect of morphine can be seen in the contrast of group M-S with group S-S, where group M-S had statistically significantly shorter reaction times than S-S ($t(35) = 2.475$, $t_{.025} = 2.35$). Here we have a standardized effect size estimate of 1.24, indicating that animals that were switched from morphine to saline were nearly one and a quarter standard deviations faster in paw-lick latency than animals who had

Table 12.6 Means of conditions in our morphine example

Groups:	M-S	M-M	S-S	S-M	Mc-M		
Means:	4.00	10.00	11.00	24.00	29.00		
		Coefficient				$\sum a_j^2$	$\psi = \sum a_j \overline{X}_j$
a_j	$-\frac{1}{2}$	$\frac{1}{3}$	$-\frac{1}{2}$	$\frac{1}{3}$	$\frac{1}{3}$	0.833	18.167
b_j	0	-1	0	0	1	2	19
c_j	-1	0	1	0	0	2	7
d_j	0	1	-1	0	0	2	-1

M-S, S-S versus M-M, S-M, Mc-M

$$\hat{d}_1 = \frac{\sum a_i \overline{X}_i}{s_{\text{error}}} = \frac{\left(-\frac{1}{3}\right)\overline{X}_{\text{M-S}} + \left(-\frac{1}{3}\right)\overline{X}_{\text{M-M}} + \left(-\frac{1}{3}\right)\overline{X}_{\text{S-S}} + \left(\frac{1}{2}\right)\overline{X}_{\text{S-M}} + \left(\frac{1}{2}\right)\overline{X}_{\text{Mc-M}}}{\sqrt{MS_{\text{error}}}}$$

$$= \frac{\left(-\frac{1}{3}\right)4.00 + \left(-\frac{1}{3}\right)10.00 + \left(-\frac{1}{3}\right)11.00 + \left(\frac{1}{2}\right)24.00 + \left(\frac{1}{2}\right)29.00}{\sqrt{32}}$$

$$= \frac{\dfrac{4.00 + 10.00 + 11.00}{3} + \dfrac{24.00 + 29.00}{2}}{\sqrt{32}} = \frac{-8.333 + 26.5}{5.657} = \frac{18.167}{5.657} = 3.21$$

M-M versus Mc-M

$$\hat{d}_2 = \frac{\sum b_i \overline{X}_i}{s_{\text{error}}} = \frac{-1\overline{X}_{\text{M-M}} + (1)\overline{X}_{\text{Mc-M}}}{\sqrt{MS_{\text{error}}}}$$

$$= \frac{(-1)10.00 + (1)29.00}{\sqrt{32}}$$

$$= \frac{-10.00 + 29.00}{32} = \frac{19}{5.657} = 3.36$$

M-S versus S-S

$$\hat{d}_3 = \frac{\sum c_i \overline{X}_i}{s_{\text{error}}} = \frac{-1\overline{X}_{\text{M-S}} + (1)\overline{X}_{\text{S-S}}}{\sqrt{MS_{\text{error}}}}$$

$$= \frac{(-1)4.00 + (1)11.00}{\sqrt{32}}$$

$$= \frac{-4 + 11}{32} = \frac{7}{5.657} = 1.24$$

M-M versus S-S

$$\hat{d}_4 = \frac{\sum d_i \overline{X}_i}{s_{\text{error}}} = \frac{1\overline{X}_{\text{M-M}} + (-1)\overline{X}_{\text{S-S}}}{\sqrt{MS_{\text{error}}}}$$

$$= \frac{(-1)10.00 - (1)11.00}{\sqrt{32}}$$

$$= \frac{10 - 11}{32} = \frac{1}{5.657} = 0.177$$

never had morphine. Finally, the complete development of morphine tolerance in four trials is suggested by a nonsignificant difference between the means of conditions M-M and S-S ($t(35) = 0.35$, $t_{.05} = 2.03$).

12.6 Post Hoc Comparisons

There is much to recommend the use of linear contrasts and the Bonferroni t test when a relatively small number of comparisons can be specified a priori. However, many experiments involve many hypotheses[9] or hypotheses that are arrived at only after the data have been examined. In this situation, a number of a posteriori or post hoc techniques are available.

Fisher's Least Significant Difference Procedure

Fisher's least significant difference (LSD)

One of the oldest methods for making post hoc comparisons is known as **Fisher's least significant difference (LSD)** test (also known as Fisher's protected t). The only difference between the post hoc LSD procedure and the a priori multiple t test procedure discussed earlier is that the LSD requires a significant F for the overall analysis of variance. When the complete null hypothesis is true (all population means are equal), the requirement for a significant overall F ensures that the familywise error rate will equal α. Unfortunately, if the complete null hypothesis is *not* true but some other more limited null hypotheses involving subsets of means are true, the overall F no longer affords protection for FW. For this reason, many people recommend that you not use this test, although Carmer and Swanson (1973) have shown it to be the most powerful of the common post hoc multiple-comparison procedures. If your experiment involves three means, the LSD procedure is a good one because FW will stay at α, and you will gain the added power of using standard t tests. (The FW error rate will be α with three means because if the complete null hypothesis is true, you have a probability equal to α of making a Type I error with your overall F, and any subsequent Type I errors you might commit with a t test will not affect FW. If the complete null is not true but a more limited one is, with three means there can be only one null difference among the means and, therefore, only one chance of making a Type I error, again with a probability equal to α.) You should generally be reluctant to use the LSD for more than three means unless you have good reason to believe that there is at most one true null hypothesis hidden in the means.

The Studentized Range Statistic (q)

Studentized range statistic (q)

Because many of the post hoc tests we are about to discuss are based on the Studentized range statistic or special variants of it, we will consider this statistic before proceeding. The **Studentized range statistic (q)** is defined as

$$q_r = \frac{\overline{X}_l - \overline{X}_s}{\sqrt{\dfrac{MS_{\text{error}}}{n}}}$$

where $\overline{X}_l$ and $\overline{X}_s$ represent the largest and smallest of a set of treatment means and r is the number of treatments in the set. You probably have noticed that the formula for q is very

---

[9] If there are many hypotheses to be tested, regardless of whether they were planned in advance, the procedures discussed here are usually more powerful than is the Bonferroni t test.

similar to the formula for t. In fact,

$$q_r = \frac{\overline{X}_l - \overline{X}_s}{\sqrt{\dfrac{MS_{\text{error}}}{n}}}$$

$$t = \frac{\overline{X}_i - \overline{X}_j}{\sqrt{\dfrac{2(MS_{\text{error}})}{n}}}$$

and the only difference is that the formula for t has a "$\sqrt{2}$" in the denominator. Thus, q is a linear function of t and we can always go from t to q by the relation $q = t\sqrt{2}$. The real difference between q and t tests comes from the fact that the tables of q (Appendix q) are set up to allow us to adjust the critical value of q for the number of means involved, as will become apparent shortly. When there are only two treatments, whether we solve for t or q is irrelevant as long as we use the corresponding table.

When we have only two means or when we want to compare two means chosen *at random* from the set of available means, t is an appropriate test.[10] Suppose, however, that we looked at a set of means and deliberately selected the largest and smallest means for testing. It is apparent that we have drastically altered the probability of a Type I error. Given that H_0 is true, the largest and smallest means certainly have a greater chance of being called "significantly different" than do means that are adjacent in an ordered series of means. This is the point at which the Studentized range statistic becomes useful. It was designed for just this purpose.

To use q, we first rank the means from smallest to largest. We then consider the number of steps between the means to be compared. For adjacent means, no change is made and $q_{.05} = t_{.05}\sqrt{2}$. For means that are not adjacent, however, the critical value of q increases, growing in magnitude as the number of intervening steps between means increases.

As an example of the use of q, consider the data on morphine tolerance. The means are

$\overline{X}_1$	$\overline{X}_2$	$\overline{X}_3$	$\overline{X}_4$	$\overline{X}_5$
4	10	11	24	29

with $n = 8$, $df_{\text{error}} = 35$, and $MS_{\text{error}} = 32.00$. The largest mean is 29 and the smallest is 4, and there are a total (r) of 5 means in the set (in the terminology of most tables, we say that these means are $r = 5$ steps apart).

$$q_5 = \frac{\overline{X}_l - \overline{X}_s}{\sqrt{\dfrac{MS_{\text{error}}}{n}}} = \frac{29 - 4}{\sqrt{\dfrac{32.00}{8}}} = \frac{25}{\sqrt{4}} = 12.5$$

Notice that r is not involved in the calculation. It is involved, however, when we go to the tables. From Appendix q, for $r = 5$ and $df_{\text{error}} = 35$, $q_{.05}(5, 35) = 4.07$. Because $12.5 > 4.07$, we will reject H_0 and conclude that there is a significant difference between the largest and smallest means.

An alternative to solving for q_{obt} and referring q_{obt} to the sampling distribution of q would be to solve for the smallest difference that would be significant and then to compare our actual difference with the minimum significant difference. This approach is frequently

---

[10] With only two means, we obtain all of the information we need from the F in the analysis of variance table and have no need to run any contrast.

taken by post hoc procedures, so I cover it here, but I really don't find that it saves any time. Because

$$q_r = \frac{\overline{X}_1 - \overline{X}_s}{\sqrt{\dfrac{MS_{\text{error}}}{n}}}$$

then

$$\overline{X}_1 - \overline{X}_s = q_{.05}(r, df_{\text{error}})\sqrt{\frac{MS_{\text{error}}}{n}}$$

where $\overline{X}_1 - \overline{X}_s$ is the minimum difference between two means that will be found to be significant.

We know that with five means the critical value of $q_{.05}(5,35) = 4.07$. Then, for our data,

$$\overline{X}_1 - \overline{X}_s = 4.07\sqrt{\frac{32}{8}} = 8.14$$

Thus, a difference in means equal to or greater than 8.14 would be judged significant, whereas a smaller difference would not. Because the difference between the largest and smallest means in the example is 25, we would reject H_0.

Although q could be used in place of an overall F (i.e., instead of running the traditional analysis of variance, we would test the difference between the two extreme means), there is rarely an occasion to do so. In most cases, F is more powerful than q. However, where you expect several control group means to be equal to each other but different from an experimental treatment mean (i.e., $\mu_1 = \mu_2 = \mu_3 = \mu_4 \neq \mu_5$), q might well be the more powerful statistic.

Although q is seldom a good substitute for the overall F, it is a very important statistic when it comes to making multiple comparisons among individual treatment means. It forms the basis for the next several tests.

12.7 Tukey's Test

Tukey test

Tukey's HSD (honestly significant difference) test

Much of the work on multiple comparisons has been based on the original work of Tukey, and an important test bears his name.[11] The **Tukey test,** also called the **Tukey's HSD (honestly significant difference) test,** uses the Studentized q statistic for its comparisons, except that q_{HSD} is always taken as the maximum value of q_r. In other words, if there are five means, *all* differences are tested as if they were five steps apart. The effect is to fix the familywise error rate at α against all possible null hypotheses, not just the complete null hypothesis, although with a loss of power. The Tukey HSD is the favorite pairwise test for many people because of the control it exercises over α.

If we apply the Tukey HSD to the data on morphine tolerance, we first arrange the means in the order of increasing magnitude, as follows.

M-S	M-M	S-S	S-M	Mc-M
4	10	11	24	29

From Appendix q, we find that with 35 df for MS_{error} and r set at 5, the critical value of q equals 4.07. If we use $r = 5$ for all comparisons, we can calculate the minimal difference

---

[11] A second test (the WSD), which is a modification on the HSD test, was proposed by Tukey as less conservative. However, I have never seen it used and have omitted it from discussion.

we will need between means for the difference to be declared significant.

$$\overline{X}_i - \overline{X}_j = q_{.05}(r, df)\sqrt{\frac{MS_{\text{error}}}{n}}$$

$$= 4.07\sqrt{\frac{32}{8}} = 8.14$$

Thus, we declare all mean differences $(\overline{X}_i - \overline{X}_j)$ to be significant if they exceed 8.14 and to be not significant if they are less than 8.14. For our data, the difference between $\overline{X}_{\text{M-M}}$ and $\overline{X}_{\text{S-S}} = 10 - 4 = 6$, and the difference between $\overline{X}_{\text{S-S}}$ and $\overline{X}_{\text{M-S}}$ is $11 - 4 = 7$. The Tukey HSD test would declare them not significant because 6 and 7 are less than 8.14. The difference between M-S and S-M is 20 and between M-S and Mc-M is 25, both of which exceed 8.14. Thus, M-S is not significantly different from M-M and from S-S. Neither, of course, is the difference between M-M and S-S, which is a difference of 1. Therefore, the first three means (M-S, M-M, and S-S) form a homogeneous set, which is different from S-M and Mc-M. Furthermore, S-M differs from Mc-M by 5 points, which again is not significant, yielding another homogeneous set. We can write these as

$$(\text{M-S} = \text{M-M} = \text{S-S}) \neq (\text{S-M} = \text{Mc-M})$$

The equal signs indicate simply that we could not reject the null hypothesis of equality, not that we have proven the means to be equal.

Unequal Sample Sizes and Heterogeneity of Variance

The Tukey procedure was designed primarily for the case of equal sample sizes $(n_1 = n_2 = \cdots = n_k = n)$. Frequently, however, experiments do not work out as planned, and we find ourselves with unequal numbers of observations and still want to carry out a comparison of means. A good bit of work has been done on this problem with respect to the Tukey HSD test (see particularly Games & Howell, 1976; Keselman & Rogan, 1977; Games, Keselman, & Rogan, 1981).

One solution, known as the Tukey–Kramer approach, is to replace $\sqrt{MS_{\text{error}}/n}$ with

$$\sqrt{\frac{\dfrac{MS_{\text{error}}}{n_i} + \dfrac{MS_{\text{error}}}{n_j}}{2}}$$

and otherwise conduct the test the same way you would if the sample sizes were equal.

An alternative, and generally preferable, test was proposed by Games and Howell (1976). The Games and Howell procedure uses what was referred to as the Behrens–Fisher approach to t tests in Chapter 7. The authors suggest that a critical difference between means (i.e., W_r) be calculated separately for every pair of means using

$$W_r = \overline{X}_i - \overline{X}_j = q_{.05}(r, df')\sqrt{\frac{s_i^2/n_i + s_j^2/n_j}{2}}$$

where $q_{.05}(r, df')$ is taken from the tables of the Studentized range statistic on

$$df' = \frac{\left(\dfrac{s_i^2}{n_i} + \dfrac{s_j^2}{n_j}\right)^2}{\dfrac{\left(\dfrac{s_i^2}{n_i}\right)^2}{n_i - 1} + \dfrac{\left(\dfrac{s_j^2}{n_j}\right)^2}{n_j - 1}}$$

degrees of freedom. This is basically the solution referred to earlier in the discussion of multiple t tests, although here we are using the Studentized range statistic instead of t. This solution is laborious, but the effort involved is still small compared with that of designing the study and collecting the data. The need for special procedures arises from the fact that the analysis of variance and its attendant contrasts are especially vulnerable to violations of the assumption of homogeneity of variance when the sample sizes are unequal. Moreover, regardless of the sample sizes, if the sample variances are nearly equal you may replace s_i^2 and s_j^2 in the formula for W_r with MS_{error} from the overall analysis of variance. And regardless of the sample size, if the variances are heterogeneous, you should probably use the Games and Howell procedure.

The Newman–Keuls Test

Newman–Keuls test

The **Newman–Keuls test** is a controversial test. I covered this procedure in the first five editions, but have finally given in to those who argue with its underlying logic. All I will say here is that the Newman-Keuls, often called the Student-Newman-Keuls, does not test all comparisons as if $r = 5$, but, instead, continually readjusts r depending on the means being compared. This allows for means that are closer in an ordered series to be tested with a smaller critical value than can means that are further apart. (As a result, the Newman–Keuls concludes that group M-S is different from all other groups, with M-M and S-S forming a homogeneous subset). Unfortunately, this adjustment to r and the critical value allows FW to exceed .05, which many people find a critical flaw. In general, the *maximum FW* for the Newman–Keuls is approximately α times the maximum number of null hypotheses that could be true, which is equal to the number of pairs involving different means. Therefore,

$$FW_{\text{max}} \cong \begin{cases} \dfrac{\alpha k}{2} & \text{if } k \text{ is even} \\[2ex] \dfrac{\alpha(k-1)}{2} & \text{if } k \text{ is odd} \end{cases}$$

This means that with three means, $FW = .05$ because there is at most one true null hypothesis to be falsely declared "significant," whereas with four or five means, $FW \cong .10$ (there *at most* two true null hypotheses). I don't feel that allowing FW to rise to approximately .10 when you have five means is going to mean the end of civilization as we know it, but I bow to the general consensus. We will have little to say about the Newman–Keuls test after this, although it is produced by most statistical software.

12.8 The Ryan Procedure (REGWQ)

As we have seen, the Tukey procedure controls the familywise error rate at α regardless of the number of true null hypotheses (not just for the overall null hypothesis), whereas the Newman–Keuls allows the familywise error rate to rise as the number of true null hypotheses increases. The Tukey test, then, provides a firm control over Type I errors, but at some loss in power. The Newman–Keuls tries to maximize power, but with some loss in control over the familywise error rate. A compromise, which holds the familywise error rate at α but which also allows the critical difference between means to shrink as r (the number of means in a set) decreases, was proposed by Ryan (1960) and subsequently modified by others.

The effect of the Newman–Keuls approach was to allow the critical values to grow as r increases, but they actually grow too slowly to keep the familywise error rate at α when multiple null hypotheses are true. Ryan (1960) also proposed modifying the value of α for each step size, but in such a way that the overall familywise error rate would remain unchanged at α. For k means and a step size of r, Ryan proposed using critical values of q_r at the

$$\alpha_r = \frac{\alpha}{k/r} = \frac{r\alpha}{k}$$

level of significance, rather than always using q_r at the α level of significance. Einot and Gabriel (1975) then modified this suggestion to set

$$\alpha_r = 1 - (1 - \alpha)^{1/(k/r)} = 1 - (1 - \alpha)^{r/k}$$

and then Welsch (1977) modified it to keep the Einot and Gabriel suggestion but to allow α_r to remain at α for $r = k$, and $r = k - 1$. These changes hold the overall familywise error rate at α while giving greater power than does Tukey to some comparisons. (Notice the similarity in the first two of these suggestions to the way α is adjusted by the Bonferroni and the Dunn-Šidák procedures.)

What these proposals really do is to allow you to continue to use the tables of the Studentized Range Distribution, but instead of always looking for q_r at $\alpha = .05$, for example, you look for q_r at $\alpha = \alpha_r$, which is likely to be some unusual fractional value. The problem is that you don't have tables that give q_r at any values other than $\alpha = .05$ or $\alpha = .01$. Computer software can compute the necessary values easily, and, because almost all computations are done using software, there is no particular problem.

Ryan procedure

REGWQ

One way that you can run the **Ryan procedure** (or the Ryan/Einot/Gabriel/Welsch procedure) is to use SPSS or SAS and request multiple comparisons using the **REGWQ** method. (The initials refer to the authors and to the fact that it uses the Studentized Range Distribution [q].) For those who have access to SPSS or other software that will implement this procedure, I recommend it over either the Newman–Keuls or the Tukey because it appears to be the most powerful test generally available that still keeps the familywise error rate at α. Those who don't have access to the necessary software will have to fall back on one of the more traditional tests. The SAS output for the REGWQ procedure (along with the Student–Newman–Keuls, the Tukey, and the Scheffé tests) are presented later in the chapter so that you can examine the results. In this situation, the conclusions to be drawn from the REGWQ and Tukey tests are the same, although you can see the difference in their critical ranges.

12.9   The Scheffé Test

Scheffé test

The post hoc tests we have considered all primarily involve pairwise comparisons of means, although they can be extended to more complex contrasts. One of the best-known tests, which is both broader and more conservative, was developed by Scheffé. The **Scheffé test,** which uses the F distribution rather than the Studentized range statistic, sets the familywise error rate at α against all possible linear contrasts, not just pairwise contrasts. If we let

$$\psi = \sum a_j \overline{X}_j \qquad \text{and} \qquad SS_{\text{contrast}} = \frac{n\psi^2}{\sum a_j^2}$$

then

$$F = \frac{n\psi^2}{\sum a_j^2 MS_{\text{error}}}$$

Scheffé has shown that if F_{obt} is evaluated against $(k - 1)F_\alpha(k - 1, df_{\text{error}})$—rather than against $F_\alpha(1, df_{\text{error}})$—the FW is at most α. (Note that all that we have done is to calculate F on a standard linear contrast, but we have evaluated that F against a modified critical value.) Although this test has the advantage of holding constant *FW* for all possible linear contrasts—not just pairwise ones—it pays a price; it has the least power of all the tests we have discussed. Partly to overcome this objection, Scheffé proposed that people may prefer to run his test at $\alpha = .10$. He further showed that the test is much less sensitive than the Tukey HSD for pairwise differences but is more sensitive than the Tukey HSD for complex comparisons (Scheffé, 1953, 1959). In general, the Scheffé test should never be used to make a set of solely pairwise comparisons, nor should it normally be used for comparisons that were thought of a priori. The test was specifically designed as a post hoc test, and its use on a limited set of comparisons that were planned before the data were collected would generally be foolish. Although most discussions of multiple-comparison procedures include the Scheffé, and many people recommend it, perhaps out of habit, it is not often seen as the test of choice in research reports because of its conservative nature. I can't imagine when I would ever use it, but I have to include it here because it is such a standard test.

12.10 Dunnett's Test for Comparing All Treatments with a Control

Dunnett's test

In some experiments, the important comparisons are between one control treatment and each of several experimental treatments. In this case, the most appropriate test is **Dunnett's test.** This is more powerful (in this situation) than are any of the other tests we have discussed that seek to hold the familywise error rate at or below α.

We will let t_d represent the critical value of a modified t statistic. This statistic is found in tables supplied by Dunnett (1955, 1964) and reproduced in Appendix t_d. We can either run a standard t test between the appropriate means (using MS_{error} as the variance estimate and evaluating the t against the tables of t_d) or solve for a critical difference between means. For a difference between means $\overline{X}_c$ and $\overline{X}_j$ (where $\overline{X}_c$ represents the mean of the control group) to be significant, the difference must exceed

$$\text{Critical value}\,(\overline{X}_c - \overline{X}_j) = t_d\sqrt{\frac{2MS_{\text{error}}}{n}}$$

Applying this test to our data, letting group S-S from Table 12.1 be the control group,

$$\text{Critical value}\,(\overline{X}_c - \overline{X}_j) = t_d\sqrt{\frac{2(32.00)}{8}}$$

We enter Appendix t_d with $k = 5$ means and $df_{\text{error}} = 35$. The resulting value of t_d is 2.56.

$$\text{Critical value}\,(\overline{X}_c - \overline{X}_j) = 2.56\sqrt{\frac{2(32.00)}{8}} = 2.56(2.828) = 7.24$$

Thus, whenever the difference between the control group mean (group S-S) and one of the other group means exceeds ±7.24, that difference will be significant. The $k - 1$ statements we will make concerning this difference will have an *FW* of $\alpha = .05$.

S-S versus M-S $= 11 - 4 = 7$

S-S versus M-M $= 11 - 10 = 1$

S-S versus S-M $= 11 - 24 = -13$

S-S versus Mc-M $= 11 - 29 = -18$

Because we have a two-tailed test (t_d was taken from two-tailed tables), the sign of the difference is irrelevant. The last two differences exceed ±7.24 and are therefore declared to be significant.

In the case in which the groups have unequal sample sizes or heterogeneous variances, a test on the difference in treatment means is given by the same general procedure we used with the Tukey.

Comparison of Dunnett's Test and the Bonferroni *t*

Because the Bonferroni *t* test allows the experimenter to make any a priori test, it is reasonable to ask what would happen if we decided a priori to apply that test to the differences between the control mean and the experimental treatment means. If we did this for our data, we would find that the required critical difference would be 7.47 for the Bonferroni instead of the 7.24 required for Dunnett's test. Thus, we would have a less powerful test because a larger difference is needed for rejection of H_0. Both the Bonferroni *t* and Dunnett's test are based on inequalities of the form $FW \leq \alpha$, but Dunnett's test uses a sharper inequality (Miller, 1981). To put this rather crudely, in Dunnett's case there is more of the *equal to* and less of the *less than* involved in the relationship between FW and α. For this reason, it is a more powerful test whenever you want simply to compare one treatment (it does not really have to be called a "control" treatment) with each of the others.

12.11 Comparison of the Alternative Procedures

Because the multiple-comparison techniques we have been discussing were designed for different purposes, there is no truly fair basis on which they can be compared. There is something to be gained, however, from summarizing their particular features and comparing the critical differences they require for the same set of data. Table 12.7 lists the tests, the error rate most commonly associated with them, the kinds of comparisons they are primarily designed to test, the type of test (range test, *F* test, or *t*—modified or not in each case), and whether the test is generally thought of as an a priori or post hoc test.

If we compare the tests in terms of the critical values they require, we are being somewhat unfair to the a priori tests. To say that the Bonferroni *t* test, for example, requires a large critical value when making all possible pairwise comparisons is not really doing the test justice because it was designed to make relatively few individual comparisons and not

Table 12.7 Comparison of alternative multiple-comparison procedures

Test	Error Rate	Comparison	Type	A Priori/ Post Hoc
Individual *t* tests	PC	Pairwise	*t*	A priori
Linear contrasts	PC	Any contrasts	*F*	A priori
Bonferroni *t*	FW	Any contrasts	$t^{\ddagger}$	A priori
Holm: Larzelere & Mulaik	FW	Any contrasts	$t^{\ddagger}$	Either
Fisher's LSD	$FW^{\dagger}$	Pairwise	*t*	Post hoc
Newman–Keuls	$FW^{\dagger}$	Pairwise	Range	Post hoc
Ryan (REGWQ)	FW	Pairwise	Range	Post hoc
Tukey HSD	FW	Pairwise[§]	Range[‡]	Post hoc
Scheffé test	FW	Any contrasts	$F^{\ddagger}$	Post hoc
Dunnett's test	FW	With control	$F^{\ddagger}$	Post hoc

[†] Against complete null hypothesis

[‡] Modified

[§] Tukey HSD can be used for all contrasts, but is poor for this purpose

Table 12.8 Comparison of critical differences for alternative procedures

	W_2	W_3	W_4	W_5
Individual t tests	5.74	5.74	5.74	5.74
Bonferroni t tests*	7.47	7.47	7.47	7.47
Holm**	5.74	6.64	7.13	7.47
Newman–Keuls	5.74	6.93	7.63	8.13
Ryan (REGWQ)	6.88	7.54	7.63	8.13
Tukey HSD	8.13	8.13	8.13	8.13
Scheffé test	9.19	9.19	9.19	9.19

* Assuming only four pairwise comparisons are desired.
** Assuming significance at each preceding level.

to be limited to pairwise contrasts. With this word of caution, Table 12.8 compares the critical differences (W_r) for each test. Linear contrasts have been omitted because they are not appropriate to the structure of the table, and the critical values for pairwise comparisons would be the same as for the individual t tests. Dunnett's test has also been omitted because it does not fit with the structure of the table.

12.12 Which Test?

Choosing the most appropriate multiple-comparison procedure for your specific situation is not easy. Many tests are available, and they differ in a number of ways. The choice is a bit easier if we consider the two extremes first.

If you have planned your test in advance and you want to run only one comparison, I would suggest that you run a standard t test (correcting for heterogeneity of variance if necessary), or, if you have a complex comparison, a linear contrast. If you have several a priori contrasts to run, not necessarily pairwise, the multistage Bonferroni t proposed by Holm does a good job of controlling *FW* while maximizing power.

If you have a large number of groups and want to make many comparisons, whether or not you are interested in all of the possible pairwise comparisons, you would probably be better off using the Ryan REGWQ if you have it available or, if not, the Tukey. In the past I recommended the Newman–Keuls, because it does a fairly good job when you have five or fewer groups, but I have found myself in a distinct minority and have decided to bail out. With three groups, the Newman–Keuls and the REGWQ test will be the same anyway, given Welsch's modification to that test, which earned him a place in its initials. I can't think of a situation where I would personally recommend the Scheffé, but I presented it here because it is a common test and real hard-liners like it.

People often fail to realize that in selecting a test, it is perfectly acceptable to compare each of several tests on your own data in terms of the size of the critical values and to select a test on that basis. For example, if you are interested in only a few pairwise comparisons, the critical values for the Holm-modified Bonferroni test may be smaller than the critical values for the REGWQ. In that case, go with modified Bonferroni. On the other hand, you may discover that even though you do not want to make all possible pairwise comparisons, the REGWQ (or the Tukey) gives smaller critical values than the modified Bonferroni, in which case you would waste power to go with the Bonferroni. The important point is that these decisions have to be based on a consideration of the critical values, and not the final results. You can't just try out every test you have and choose the one that gives you the answers you like.

12.13   Computer Solutions

Most software packages will perform multiple comparison procedures, but not all packages have all procedures available. Exhibit 12.1 contains the results of an analysis of the morphine data using SAS. I chose SAS because it has a broad choice of procedures and is one of the major packages. It also has more information in its printout than do SPSS and Minitab, and is thus somewhat more useful for our purpose.

Exhibit 12.1 begins with the program commands and the overall analysis of variance. ("Condition" is spelled as "Condtion" because SAS allows only 8 characters in a name.) This analysis agrees with the summary table shown in Table 12.1. The $R^2 = .757$ is simply η^2. You can see that our experimental manipulation accounts for a substantial portion of the variance. The remainder of the exhibit includes the results of the Newman–Keuls, Ryan, Tukey, and Scheffé tests.

The Newman–Keuls, as the least conservative test, reports the most differences between conditions. If you look first at the means and "SNK Grouping" at the end of that portion of the printout, you will see a column consisting of the letters A, B, and C. Conditions that share the same letter are judged to not differ from one another. Thus, the means of Conditions Mc-M and S-M are not significantly different from one another, but, because they don't have a letter in common with other conditions, they are different from the means of S-S, M-M, and M-S. Similarly, Conditions S-S and M-M share the letter B and their means are thus not significantly different from each other, but are different from the means of the other three conditions. Finally, the mean of Condition M-S is different from the means of all other conditions.

If you look a bit higher in the table, you will see a statement about how this test deals with the familywise (here called "experimentwise") error rate. As was said earlier, the Newman–Keuls holds the familywise error rate at α against the complete null hypothesis, but allows it to rise in the case where a subset of null hypotheses are true. You next see a statement saying that the test is being run at $\alpha = .05$, that we have 35 df for the error term, and that $MS_{\text{error}} = 32.00$. Following this information, you see the critical ranges. These are the minimum differences between means that would be significant for different values of r. The critical ranges are equal to

$$W_r = q_{.05}(r, df_e)\sqrt{\frac{MS_{\text{error}}}{n}}$$

For example, when $r = 3$ (a difference between the largest and smallest of three means)

$$W_3 = q_{.05}(3, df_e)\sqrt{\frac{MS_{\text{error}}}{n}} = 3.46\sqrt{\frac{32}{8}} = 3.46(2) = 6.92$$

Because all three step differences (e.g., $29 - 11 = 18$; $24 - 10 = 14$; $11 - 4 = 7$) are greater than 6.92, they will all be declared significant.

The next section of Exhibit 12.1 shows the results of the Ryan (REGWQ) test. Notice that the critical ranges for $r = 2$ and $r = 3$ are larger than they were for the Newman–Keuls (though smaller than they will be for the Tukey). As a result, for $r = 3$ we need to exceed a difference of 7.54, whereas the difference between 11 and 4 is only 7. Thus, this test will not find Group 1 (M-S) to be different from Group 3 (S-S), whereas it was different for the more liberal Newman–Keuls. However, the familywise error rate for this set of comparisons is $\alpha = .05$, whereas it would be nearly $\alpha = .10$ for the Newman–Keuls.

The Tukey test is presented slightly differently, but you can see that Tukey requires all differences between means to exceed a critical range of 8.1319 to be declared significant, regardless of where they lie in an ordered series. For this specific set of data, our conclusions are the same as they were for the Ryan test, although that will certainly not always be the case.

Although the Scheffé test is run quite differently from the others, it is possible to compute a critical range for all pairwise comparisons. From Exhibit 12.1, we can see that this

```
Options LineSize = 78;

Data Siegel;
      Infile 'Alexander:SAS610:Data Files:Siegel.dat';
      Input Condtion Latency;
run;

Proc GLM Data = Siegel;
      Class Condtion;
      Model Latency = Condtion/SS3;
      Means Condtion /SNK Tukey REGWQ Scheffe;
Run;
```

The SAS System   1

September 22, 2005

General Linear Models Procedure

Dependent Variable: LATENCY

Source	DF	Sum of Squares	Mean Square	F Value	Pr > F
Model	4	3497.600000	874.400000	27.33	0.0001
Error	35	1120.000000	32.000000		
Corrected Total	39	4617.600000			

$\eta^2 \longrightarrow$

R-Square	C.V.	Root MSE	LATENCY Mean
0.757450	36.26189	5.656854	15.60000

Dependent Variable: LATENCY

Source	DF	Type III SS	Mean Square	F Value	Pr > F
CONDTION	4	3497.600000	874.400000	27.33	0.0001

F for Condition

Student-Newman-Keuls test for variable: LATENCY

NOTE: This test controls the type I experimentwise error rate under the complete null hypothesis but not under partial null hypotheses.

Alpha= 0.05   df= 35   MSE= 32

Number of Means	2	3	4	5
Critical Range	5.7420598	6.921941	7.6279952	8.1319061

$w_5 = q_5 \sqrt{\dfrac{MS_{error}}{n}}$

Means with the same letter are not significantly different.

SNK Grouping	Mean	N	CONDTION
A	29.000	8	Mc-M
A	24.000	8	S-M
B	11.000	8	S-S
B	10.000	8	M-M
C	4.000	8	M-S

Ryan-Einot-Gabriel-Welsch Multiple Range Test for variable: LATENCY

NOTE: This test controls the type I experimentwise error rate.

Alpha= 0.05   df= 35   MSE= 32

Number of Means	2	3	4	5
Critical Range	6.8765473	7.5391917	7.6279952	8.1319061

Larger than for SNK ⟶   ⟵ Same as SNK

(continued)

Exhibit 12.1

Means with the same letter are not significantly different.

REGWQ Grouping		Mean	N	CONDTION
	A	29.000	8	Mc-M
	A	24.000	8	S-M
B		11.000	8	S-S
B		10.000	8	M-M
B		4.000	8	M-S

Tukey's Studentized Range (HSD) for variable: LATENCY

NOTE: This test controls the type I experimentwise error rate, but generally has a higher type II error rate than REGWQ.

Alpha= 0.05   df= 35   MSE= 32
Critical Value of Studentized Range= 4.066
Minimum Significant Difference = (8.1319) ←——— Critical range for all differences

Means with the same letter are not significantly different.

Tukey Grouping		Mean	N	CONDTION
	A	29.000	8	Mc-M
	A	24.000	8	S-M
B		11.000	8	S-S
B		10.000	8	M-M
B		4.000	8	M-S

Scheffe's test for variable: LATENCY

NOTE: This test controls the type I experimentwise error rate but generally has a higher type II error rate than REGWF for all pairwise comparisons

Alpha= 0.05   df= 35   MSE= 32
Critical Value of F= 2.64147
Minimum Significant Difference = (9.1939) ←——— Critical range for all differences

Means with the same letter are not significantly different.

Scheffe Grouping		Mean	N	CONDTION
	A	29.000	8	Mc-M
	A	24.000	8	S-M
B		11.000	8	S-S
B		10.000	8	M-M
B		4.000	8	M-S

Exhibit 12.1 *(continued)*

range is 9.1939, almost a full point larger than the critical range for Tukey. This reflects the extreme conservatism of the Scheffé procedure, especially with just pairwise contrasts, and illustrates my major objection to the use of this test.

SAS will also produce a number of other multiple comparison tests, including the Bonferroni and the Dunn-Šidák. I do not show those here because it is generally foolish to use either of those tests when you want to make *all possible* pairwise comparisons among means. The Ryan or Tukey test is almost always more powerful and still controls the familywise error rate. I suppose that if I had a limited number of pairwise contrasts that I was interested in, I could use the Bonferroni procedure in SAS (BON) and promise not to look at the contrasts that were not of interest. But I'd first have to "cross my heart and hope to die," and even then I'm not sure if I'd trust myself.

12.14 Trend Analysis

trend

The analyses we have been discussing are concerned with identifying differences among group means, whether these comparisons represent complex contrasts among groups or simple pairwise comparisons. Suppose, however, that the groups defined by the independent variable are ordered along some continuum. An example might be a study of the beneficial effects of aspirin in preventing heart disease. We could ask subjects to take daily doses of 1, 2, 3, 4, or 5 grains of aspirin, where 1 grain is equivalent to what used to be called "baby aspirin" and 5 grains is the standard tablet. In this study, we would not be concerned so much with whether a 4-grain dose was better than a 2-grain dose, for example, as with whether the beneficial effects of aspirin increase with increasing the dosage of the drug. In other words, we are concerned with the **trend** in effectiveness rather than multiple comparisons among specific means.

To continue with the aspirin example, consider two possible outcomes. In one outcome, we might find that the effectiveness increases linearly with dosage. In this case, the more aspirin you take, the greater the effect, at least within the range of dosages tested. A second, alternative, finding might be that effectiveness increases with dosage up to some point, but then the curve relating effectiveness to dosage levels off and perhaps even decreases. This would be either a "quadratic" relationship or a relationship with both linear and quadratic components. It would be important to discover such relationships because they would suggest that there is some optimal dose, with low doses being less effective and high doses adding little, if anything, to the effect.

quadratic functions

Typical linear and **quadratic functions** are illustrated in Figure 12.2. (They were produced using JMP on a Macintosh.) It is difficult to characterize quadratic functions neatly because the shape of the function depends both on the sign of the coefficient of X^2 and on the sign of X (the curve changes direction when X passes from negative to positive, and for positive values of X, the curve rises if the coefficient is positive and falls if it is negative). Also included in Figure 12.2 is a function with both linear and quadratic components. Here you can see that the curvature imposed by a quadratic function is superimposed on a rising linear trend.

Tests of trend differ in an important way from the comparison procedures we have been discussing. In all the previous examples, the independent variable was generally qualitative. Thus, for example, we could have written down the groups in the morphine-tolerance example in any order we chose. Moreover, the F or t values for the contrasts depended only on the numerical value of the means, not on which particular groups went with which particular means. In the analysis we are now considering, F or t values will depend on both the group means and the particular ordering of those means. To put this slightly differently using the aspirin example, a REGWQ test between the largest and the smallest means will not be affected by which group happens to have each mean. However, in trend analysis the results would be quite different if the 1-grain and 5-grain groups had the smallest and largest means than if the 4- and 2-grain groups had the smallest and largest means, respectively. (A similar point was made in Section 6.7 in discussing the nondirectionality of the chi-square test.)

Boring Is Attractive

A useful example of trend analysis comes from a study by Langlois and Roggman (1990), which examined the question of what makes a human face attractive. They approached the problem from both an evolutionary and a cognitive perspective. Modern evolutionary theory would suggest that average values of some trait would be preferred to extreme ones,

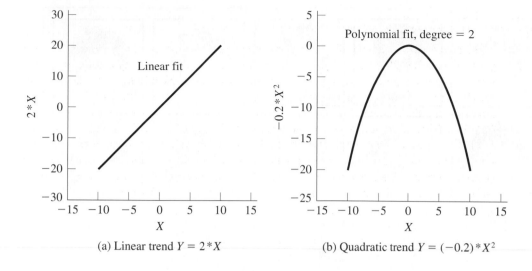

(a) Linear trend $Y = 2*X$

(b) Quadratic trend $Y = (-0.2)*X^2$

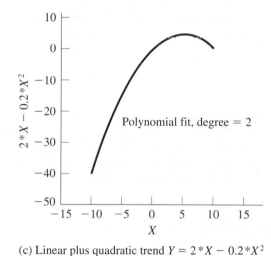

(c) Linear plus quadratic trend $Y = 2*X - 0.2*X^2$

Figure 12.2   Typical linear and quadratic functions

and cognitive theory suggests that both adults and children respond to prototypes of objects more positively than to objects near the extremes on any dimension. A prototype, by definition, possesses average values of the object along important dimensions. (A prototype of a cat is one that is not too tall or too short, not too fat or too thin, and doesn't purr too loudly or too quietly.)

Langlois and Roggman took facial photographs of 336 males and 214 females. They then created five groups of composite photographs by computer-averaging the individual faces. Thus, for one group the computer averaged 32 randomly selected same-gender faces, producing a quite recognizable face with average width, height, eyes, nose length, and so on. For the other groups, the composite faces were averaged over either 2, 4, 8, or 16 individual faces. An example of composite faces can be seen in Figure 12.3. The label Composite will be used to represent the five different groups. That is not an ideal name for the independent variable, but neither I nor the study's authors have a better suggestion. Within each group of composite photographs were three male and three female faces, but we will

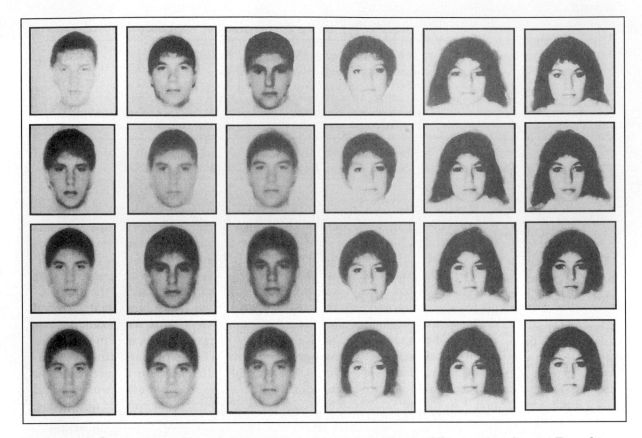

Figure 12.3   Composite faces. Faces from left to right represent the six different composite sets. Faces from top to bottom represent composite levels of 4 faces, 8 faces, 16 faces, and 32 faces

From "Attractive faces are only average" by J. H. Langlois and L. A. Roggman in *Psychological Science,* 1, p. 115–121. © 1990 Blackwell Publishers.

Table 12.9   Data on rated attractiveness (from left to right the groups represent averaging across 2, 4, 8, 16, or 32 faces)

	Group 1	Group 2	Group 3	Group 4	Group 5
	2.201	1.893	2.906	3.233	3.200
	2.411	3.102	2.118	3.505	3.253
	2.407	2.355	3.226	3.192	3.357
	2.403	3.644	2.811	3.209	3.169
	2.826	2.767	2.857	2.860	3.291
	3.380	2.109	3.422	3.111	3.290
Mean	2.6047	2.6450	2.8900	3.1850	3.2600

ignore gender for this example. (There were no significant gender differences, and the overall test on group differences is not materially affected by ignoring that variable.)

Langlois and Roggman presented different groups of subjects with composite faces and asked them to rate the attractiveness of the faces on a 1–5 scale, where 5 represents "very attractive." The individual data points in their analysis were actually the means averaged across raters for the six different composites in each condition. The data are given in Table 12.9. These data are fictional, but they have been constructed to have the same mean and variance as those reported by Langlois and Roggman, so the overall *F* and the tests on trend will be the same as those they reported.

A standard one-way analysis of variance on these data would produce the following summary table:

Source	df	SS	MS	F
Composite	4	2.1704	0.5426	3.13*
Error	25	4.3281	0.1731	
Total	29	6.4985		

*$p < .05$

From the summary table, we can see that there are significant differences among the five groups, but it is not clear how these differences are manifested. One way to examine these differences would be to plot the group means as a function of the number of individual pictures that were averaged to create the composite. An important problem that arises if we try to do this concerns the units on the abscissa. We could label the groups as "2, 4, 8, 16, and 32," on the grounds that these values correspond to the number of elements over which the average was taken. However, it seems unlikely that rated attractiveness would increase directly with those values. We might expect that a picture averaged over 32 items would be more attractive than one averaged over 2 items, but I doubt that it would be 16 times more attractive. But notice that each value of the independent variable is a power of 2. In other words, the values of 2, 4, 8, 16, and 32 correspond to 2^1, 2^2, 2^3, 2^4, and 2^5. (Put another way, taking the $\log_2$ of 2, 4, 8, 16, and 32 would give us 1, 2, 3, 4, and 5.) For purposes of analyzing these data, I am going to represent the groups with the numbers 1 to 5 and refer to these as measuring the degree of the composite. (If you don't like my approach, and there is certainly room to disagree, be patient, and we will soon see a solution using unequally spaced values of the independent variable. The example will be simpler statistically if the units on the abscissa are evenly spaced.) The group means using my composite measure on the abscissa are plotted in Figure 12.4, where you can see that the rated attractiveness does increase with increasing levels of Composite.

Our first question asks whether a nonhorizontal straight line provides a good fit to the data. A glance at Figure 12.4 would suggest that this is the case. We will then follow that question by asking whether systematic residual (non-error) variance remains in the data after fitting a linear function, and, if so, whether this residual variance can be explained by a quadratic function.

To run a trend analysis, we will return to the material we discussed under the headings of linear and orthogonal contrasts. (Don't be confused by the use of the word *linear* in the last sentence. We will use the same approach when it comes to fitting a quadratic function. Linear in this sense simply means that we will form a linear combination of coefficients and means, where nothing is raised to a power.)

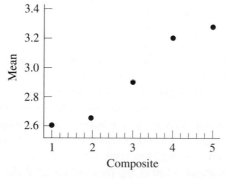

Figure 12.4   Scatterplot of mean versus composite group

In Section 12.3, we defined a linear contrast as

$$\psi = a_1 \overline{X}_1 + a_2 \overline{X}_2 + a_3 \overline{X}_3 + \cdots + a_k \overline{X}_k = \sum a_j \overline{X}_j$$

The only difference between what we are doing here and what we did earlier will be in the coefficients we use. In the case in which there are equal numbers of subjects in the groups and the values on the abscissa are equally spaced, the coefficients for linear, quadratic, and higher-order functions **(polynomial trend coefficients)** are easily tabled and are found in Appendix Polynomial. From Appendix Polynomial, we find that for five groups, the linear and quadratic coefficients are

polynomial trend coefficients

Linear:	−2	−1	0	1	2
Quadratic:	2	−1	−2	−1	2

We will not be using the cubic and quartic coefficients shown in the appendix, but their use will be evident from what follows. Notice that like any set of orthogonal linear coefficients, the requirements that $\sum a_j = 0$ and $\sum a_i b_j = 0$ are met. The coefficients do not form a "standard set" because the sum of the absolute values of the coefficients does not equal 2. That is not a problem here.

As you should recall from Section 12.3, we calculate a sum of squares for the contrast as

$$SS_{\text{contrast}} = \frac{n\psi^2}{\sum a_j^2}$$

In our case,

$$\psi_{\text{linear}} = (-2)2.6047 + (-1)2.6450 + (0)2.8900 + (1)3.1850 + (2)3.2600$$
$$= 1.8506$$

$$SS_{\text{linear}} = \frac{n\psi^2}{\sum a_j^2} = \frac{6(1.8506^2)}{10}$$
$$= 2.0548$$

Like all contrasts, this contrast has a single degree of freedom, and therefore $SS_{\text{linear}} = MS_{\text{linear}}$. As you probably suspect from what you already know, we can convert this mean square for the contrast to an F by dividing by MS_{error}:

$$F = \frac{MS_{\text{linear}}}{MS_{\text{error}}}$$
$$= \frac{2.0548}{0.1731}$$
$$= 11.8706$$

This is an F on 1 and 25 degrees of freedom, and from Appendix F we find that $F_{.05}(1,25) = 4.245$. Because the F for the linear component (11.87) exceeds 4.245, we will reject H_0 and conclude that there is a significant linear trend in our means. In other words, we will conclude that attractiveness varies linearly with increasing levels of Composite. Notice here that a significant F means that the trend component we are testing is significantly different from 0.

It is conceivable that we could have a significant linear trend in our data and still have residual variance that can be explained by a higher-order term. For example, we might have both linear and quadratic, or linear and cubic, components. In fact, it would be reasonable to expect a quadratic component in addition to a linear one because it seems unlikely that judged attractiveness will keep increasing indefinitely as we increase the number of individual photographs we average to get the composite. There will presumably be some diminishing returns, and the curve should level off.

The next step is to ask whether the residual variance remaining after we fit the linear component is significantly greater than the error variance that we already know is present.

If SS_{linear} accounted for virtually all of $SS_{Composite}$, there would be little or nothing left for higher-order terms to explain. On the other hand, if SS_{linear} were a relatively small part of $SS_{Composite}$, then it would make sense to look for higher-order components. From our previous calculations, we obtain

$$SS_{residual} = SS_{Composite} - SS_{linear}$$
$$= 2.1704 - 2.0548$$
$$= 0.1156$$

$$df_{residual} = df_{Composite} - df_{linear}$$
$$= 4 - 1$$
$$= 3$$

$$MS_{residual} = \frac{SS_{residual}}{df_{residual}}$$
$$= \frac{0.1156}{3}$$
$$= 0.0385$$

$$F_{residual} = \frac{MS_{residual}}{MS_{error}}$$
$$= \frac{0.0385}{0.1731}$$
$$< 1$$

Because F for the residual is less than 1, we know automatically that it is not significant. This tells us that there is no significant variability left to be explained beyond that accounted for by the linear component. We would, therefore, normally stop here. However, for purposes of an example I will go ahead and calculate the quadratic component. The calculations will be shown without discussion because the discussion would essentially be the same as earlier with the word *quadratic* substituted for *linear*.

$$\psi_{quadratic} = (2)2.6047 + (-1)2.6450 + (-2)2.8900 + (-1)3.1850 + (2)3.2600$$
$$= 0.1194$$

$$SS_{quadratic} = \frac{n\psi^2}{\sum b_j^2}$$
$$= \frac{6(0.1194^2)}{14}$$
$$= 0.0061$$

$$F = \frac{MS_{quadratic}}{MS_{error}}$$
$$= \frac{0.0061}{0.1731}$$
$$< 1$$

As our test on the residual suggested, there is no significant quadratic component on our plot of the group means. Thus, there is no indication, over the range of values used in this study, that the means are beginning to level off. Therefore, we would conclude from these data that attractiveness increases linearly with Composite, at least given the definition of Composite used here.

A word of caution is in order at this point. You might be tempted to go ahead and apply the cubic and quartic coefficients that you find in Appendix Polynomial. You might also observe that having done this, the four sums of squares ($SS_{linear}, \ldots, SS_{quartic}$) will sum to $SS_{Composite}$, and be very impressed that you have accounted for all the sums of squares between groups. Before you get too impressed, think about how proud you would be if you showed that you could draw a straight line that exactly fit two points. The same idea applies here. Regardless of the data, you know before you begin that a polynomial of order $k - 1$ will exactly fit k points. That is one reason why I was not eager to go much beyond fitting the linear components to the data at hand. A quadratic was stretching things a bit. Moreover, if you were to fit a fourth-order polynomial and found that the quartic component was significant, what would you have to say about the results? A linear or quadratic component would make some sense, but a quartic component could not be explained by any theory I know.

Unequal Intervals

In the preceding section, we assumed that the levels of the independent variable are equally spaced along some continuum. I actually transformed the independent variable into a scale called Composite to fulfill that requirement. It is possible to run a trend analysis when we do not have equal intervals, and the arithmetic is the same. The only problem comes when we try to obtain the trend coefficients because we cannot take our coefficients from Appendix Polynomial unless the intervals are equal.

Calculating quadratic coefficients is not too difficult, and a good explanation can be found in Keppel (1973). For higher-order polynomials, the calculations are more laborious, but a description of the process can be found in Robson (1959). Most people will carry out their analyses with standard statistical software, and that software will often handle the problem of unequal spacing. Without diving deeply into the manuals, it is often difficult to determine how your software handles the spacing problem. The simplest thing to do, using the attractiveness data as an example, would be to code the independent variable as 1, 2, 3, 4, and 5, and then recode it as 2, 4, 8, 16, 32. If the software is making appropriate use of the levels of the independent variable, you should get different answers. Then the problem is left up to you to decide which answer you want, when both methods of coding make sense. For example, if you use the SPSS ONE-WAY procedure and ask for polynomial contrasts, *where the independent variable is coded 1, 2, 3, 4, 5*, you will obtain the same results as previously. If you code the variable 2, 4, 8, 16, 32, you will obtain slightly different results. However, if you use the SPSS General Linear Model/Univariate procedure, the way in which you code the independent variable will not make any difference—both will produce results as if the coding were 1, 2, 3, 4, 5. It always pays to check.

An example containing both a quadratic and a cubic component can be found in Exercise 12.25. Working through that exercise can teach you a lot about trend analysis.

Key Terms

Error rate per comparison (PC) (12.1)

Familywise error rate (FW) (12.1)

Omnibus null hypothesis (12.1)

A priori comparisons (12.1)

Post hoc comparisons (12.1)

Contrasts (12.3)

Linear combination (12.3)

Linear contrast (12.3)

Partition (12.3)

Orthogonal contrasts (12.3)

Dunn's test (12.3)

Bonferroni t (12.3)

Bonferroni inequality (12.3)

Dunn–Šidák test (12.3)

Fisher's least significant difference (LSD) (12.6)

Studentized range statistic (q) (12.6)

Tukey's test (12.7)

Tukey's HSD (honestly significant difference) test (12.7)

Newman–Keuls test (12.7)

Ryan procedure (REGWQ) (12.8)

Scheffé test (12.9)

Dunnett's test (12.10)

Trend (12.14)

Quadratic function (12.14)

Polynomial trend coefficients (12.14)

Exercises

12.1 Assume that the data that follow represent the effects of food or water deprivation on be-havior in a learning task. Treatments 1 and 2 represent control conditions in which the animal received ad lib food and water (1) or else food and water twice per day (2). In treatment 3, animals were food deprived; in treatment 4, they were water deprived; and in treatment 5, they were deprived of both food and water. The dependent variable is the num-ber of trials to reach a predetermined criterion. Assume that before running our experiment we decided that we wanted to compare the combined control groups (treatments 1 and 2) with the combined experimental groups, the control groups with each other, the singly de-prived treatments with the doubly deprived treatment, and the singly deprived treatments with each other.

Ad Lib Control	Two per Day Control	Food Deprived	Water Deprived	Food and Water Deprived
18	20	6	15	12
20	25	9	10	11
21	23	8	9	8
16	27	6	12	13
15	25	11	14	11
90	120	40	60	55

a. Analyze the data using linear contrasts (*Note:* I am not asking for linear polynomials (trend) here, just standard contrasts).

b. Show that the contrasts are orthogonal.

c. Show that the sums of squares for the contrasts sum to SS_{treat}.

12.2 Using the data from Exercise 11.1, compute the linear contrasts for 5 versus (20 and 35) days and 20 versus 35 days, using $\alpha = .05$ for each contrast. *Note*: This and subsequent exercises refer to exercises in Chapter 11, not this chapter.

12.3 What would be the per comparison and familywise error rates in Exercise 12.2? (*Hint*: Are the contrasts orthogonal?)

12.4 Compute F for the linear contrast on the two groups in Exercise 11.2. Is this a waste of time? Why or why not?

12.5 Compute the Studentized range statistic for the two groups in Exercise 11.2, and show that it is equal to $t\sqrt{2}$ (where t is taken from Exercise 11.2b).

12.6 Compute the Fs for the following linear contrasts in Exercise 11.3. Save the results for use in Chapter 13.

a. 1 and 2 versus 3 and 4

b. 1 and 3 versus 2 and 4

c. 1 and 4 versus 2 and 3

d. What questions do the contrasts in (a), (b), and (c) address?

12.7 Run the Bonferroni t test on the data for Exercise 11.1, using the contrasts supplied in Exercise 12.2. Set the maximum FW at .05.

12.8 Repeat Exercise 12.7, using Holm's multistage test. What differences do you find between these answers and the answers to Exercise 12.7?

12.9 Apply Holm's multistage test to Exercise 12.1.

12.10 Run a REGWQ test on the example given in Table 11.2 (page 305) and interpret the results.

12.11 Calculate Tukey's test on the data in the example in Table 11.2, and compare your results with those you obtained for Exercise 12.10.

12.12 Why might you be more interested in running specific contasts on the data referred to in Exercises 12.10 and 12.11?

12.13 Run the Games and Howell (1976) approach to the Tukey's HSD procedure for unequal sample sizes on the data in Exercise 12.12.

Group	1	2	3	4	5
$\overline{X}_j$	10	18	19	21	29
n_j	8	5	8	7	9
s_j^2	7.4	8.9	8.6	7.2	9.3

12.14 Use the Scheffé test on the data in Exercise 12.13 to compare groups 1, 2, and 3 (combined) with groups 4 and 5 (combined). Then compare group 1 with groups 2, 3, and 4 (combined). (*Hint*: You will need to go back to the section in which unequal sample sizes are discussed.) See www.uvm.edu/~dhowell/methods/Errata/Unequal_n's_contrasts.html

12.15 Apply Tukey's procedure to the log transformed THC data from Table 11.5 (page 320). What is the maximum *FW* for this procedure?

12.16 Apply Dunnett's test to the log transformed data in Table 11.5.

12.17 How could a statistical package that did not have a Bonferroni command be used to run the Bonferroni *t* test on the data in Exercise 12.7?

12.18 The Holm test is referred to as a modified sequentially rejective procedure. Why?

12.19 Fit linear and quadratic trend components to the Conti and Musty (1984) log-transformed data in Table 11.5. The control condition received 0 µg of THC. For purposes of this example, assume that there were 10 subjects in all groups. (You could add a 2.56 to the 0.5 µg group and a 2.35 and 2.36 to the 1 µg group without altering the results significantly.) The linear coefficients (calculated with unequal spacing on the independent variable) are [−0.72, −0.62, −0.22, 0.28, 1.28]. The quadratic coefficients are [0.389, 0.199, −0.362, −0.612, 0.387].

Verify your answers using SPSS One-Way if you have it available.

12.20 Use any statistical package to compute Fisher's LSD procedure on all three pairs of means (even though the overall *F* was not significant) for GSIT from Mireault's data (Mireault.dat). (This is based on the analysis of variance in Exercise 11.27.) Compare these results with the individual *t* tests that you ran for Exercise 7.46. Interpret the results.

12.21 Use any statistical package to apply the REGWQ (if available), and Scheffé procedures to the data from Introini-Collison and McGaugh (1986), described in the exercises for Chapter 11 (page 341). Do these analyses for both Epineq.dat and Epinuneq.dat, which are on the website (www.uvm.edu/~dhowell/methods/). Do not combine across the levels of the interval variable.

12.22 In Exercise 12.21, it would not have made much of a difference whether we combined the data across the three intervals or not. Under what conditions would you expect that it would make a big difference?

12.23 Using the data in Epineq.dat, compute both the linear and quadratic trend tests on the three drug dosages. Do this separately for each of the three intervals. (*Hint*: The linear coefficients are [−0.597110, −0.183726, 0.780836], and the quadratic coefficients are [0.556890, −0.795557, 0.238667].)

12.24 Interpret the results in Exercise 12.23.

12.25 Stone, Rudd, Ragozzino, and Gold (1992) investigated the role that glucose plays in memory. Mice were raised with a 12-hour light-on/light-off cycle, starting at 6:00 a.m. During training, mice were placed in the lighted half of an experimental box and given foot shocks when they moved into the dark half. The mice quickly learned to stay in the lighted half. The day/night cycle was then advanced by 4 hours for all mice, which is known to interfere with memory of the original training. Three days later, mice were retested 30 minutes after being injected with 0, 1, 10, 100, 250, or 500 mg/kg of sucrose. The purpose was to see whether sucrose would reduce the disruptive effects of changing the diurnal cycle, and whether different doses would have different effects. Data that have been generated to loosely mimic the results of Stone et al. follow, where the dependent variable is the latency to enter the dark chamber.

Glucose Level in mg/kg

0	1	10	100	250	500
295	129	393	653	379	521
287	248	484	732	530	241
91	350	308	570	364	162
260	278	112	434	385	197
193	150	132	690	355	156
52	195	414	679	558	384

a. Plot these data using both the actual dosage, and the values 1, 2, 3, 4, 5, 6 as the values of X.

b. Run a trend analysis using SPSS One-Way, if available, with the actual dosage as the independent variable.

c. Repeat part (b) using the 1, 2, 3, 4, 5, 6 coding as the independent variable.

d. Interpret your results. How might these results have something to say to students who stay up all night studying for an exam?

e. Why might you, or Stone et al., prefer one coding system to another?

12.26 Using the data from Exercise 12.1, compute confidence interval for the first comparison (contrast) described in that question. Interpret your answer. (If you use SPSS, use the **Compare Means/One-Way ANOVA** procedure, which allows you to specify coefficients.)

12.27 Using the data from Exercise 12.1, compute effect sizes on all of the contrasts that you ran with that question. How would you interpret these effect sizes? Why are these called standardized effect sizes, and what would an unstandardized effect size be?

12.28 Write up a brief report of the results computed for Exercises 12.1, 12.26, and 12.27.

12.29 Using the data from Exercise 11.27, perform the appropriate test(s) to draw meaningful conclusions from the study by Davey et al. (2003).

Discussion Questions

12.30 Students often have difficulty seeing why a priori and post hoc tests have different family-wise error rates. Make up an example (not necessarily from statistics) that would help to explain the difference to others.

12.31 Find an example in the research literature of a study that used at least five different conditions, and create a data set that might have come from this experiment. Apply several of the techniques we have discussed, justifying their use, and interpret the results. (You would never apply several different techniques to a set of data except for an example such as this.) (*Hint*: You can generate data with a given mean and variance by taking any set of numbers [make them at least unimodal and symmetrical], standardizing them, multiplying the standard scores by the desired standard deviation, and then adding the desired mean to the result. Do this *for each group separately* and you will have your data.)

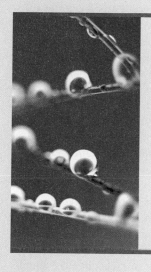

Factorial Analysis of Variance

IN THE PREVIOUS TWO CHAPTERS, we dealt with a one-way analysis of variance in which we had only one independent variable. In this chapter, we will extend the analysis of variance to the treatment of experimental designs involving two or more independent variables. For purposes of simplicity, we will be concerned primarily with experiments involving two or three variables, although the techniques discussed can be extended to more complex designs.

In Chapter 11, we considered a study by Eysenck (1974) in which he asked participants to recall lists of words to which they had been exposed under one of several different conditions. In that example, we were interested in determining whether recall was related to the level at which material was processed initially. Eysenck's study was actually more complex. He was interested in whether level-of-processing notions could explain differences in recall between older and younger participants. If older participants do not process information as deeply, they might be expected to recall fewer items than would younger participants, especially in conditions that entail greater processing. This study now has two independent variables, which we shall refer to as **factors:** Age and Recall condition (hereafter referred to simply as Condition). Thus, the experiment is an instance of what is called a **two-way factorial design.**

factors

two-way factorial design

factorial design

An experimental design in which every level of every factor is paired with every level of every other factor is called a **factorial design.** In other words, a factorial design is one in which we include all *combinations* of the levels of the independent variables. In the factorial designs discussed in this chapter, we will consider only the case in which different participants serve under each of the treatment combinations. For instance, in our example, one group of younger participants will serve in the Counting condition, a different group of younger participants will serve in the Rhyming condition, and so on. We have 10 combinations of our two factors (5 recall Conditions × 2 Ages), so we will have 10 different groups of participants. When the research plan calls for the *same* participant to be included under more than one treatment combination, we will speak of **repeated-measures designs.** Repeated-measures designs will be discussed in Chapter 14.

repeated-measures designs

Factorial designs have several important advantages over one-way designs. First, factorial designs allow greater generalizability of the results. Consider Eysenck's study for a moment. If we were to run a one-way analysis using the five Conditions with only the older participants, as in Chapter 11, then our results would apply only to older participants. When we use a factorial design with both older and younger participants, we are able to determine whether differences between Conditions apply to younger participants as well as older ones. We are also able to determine whether age differences in recall apply to all tasks, or whether younger (or older) participants excel on only certain kinds of tasks. Thus, factorial designs allow for a much broader interpretation of the results, while giving us the ability to say something meaningful about the results for each of the independent variables separately. An interesting discussion of this issue, though from the perspective of engineering, can be found in Czitrom (1999).

interaction

The second important feature of factorial designs is that they allow us to look at the **interaction** of variables. We can ask whether the effect of Condition is independent of Age or whether there is some interaction between Condition and Age. For example, we would have an interaction if younger participants showed much greater (or smaller) differences between the five recall conditions than did older participants. Interaction effects are often among the most interesting results we obtain.

A third advantage of a factorial design is its economy. We are going to average the effects of one variable across the levels of the other variable, so a two-variable factorial will require fewer participants than would two one-ways for the same degree of power.

Essentially, we are getting something for nothing. Suppose we had no reason to expect an interaction of Age and Condition. Then, with 10 old participants and 10 young participants in each Condition, we would have 20 scores for each of the five conditions. If we instead ran a one-way with young participants and then another one-way with old participants, we would need twice as many participants overall for each of our experiments to have the same power to detect Condition differences—that is, each experiment would have to have 20 participants per condition, and we would have two experiments.

Factorial designs are labeled by the number of factors involved. A factorial design with two independent variables, or factors, is called a two-way factorial, and one with three factors is called a three-way factorial. An alternative method of labeling designs is by the number of levels of each factor. Eysenck's study had two levels of Age and five levels of Condition. As such, it is a **2 × 5 factorial.** A study with three factors, two of them having three levels and one having four levels, would be called a 3 × 3 × 4 factorial. The use of such terms as "two-way" and "2 × 5" are both common ways of designating designs, and both will be used throughout this book.

In much of what follows, we will concern ourselves primarily with the two-way analysis. Higher-order analyses follow almost automatically once you understand the two-way, and many of the related problems we will discuss are most simply explained in terms of two factors. For the moment, we will also limit our discussion to fixed—as opposed to random—models, as these were defined in Chapter 11.

2 × 5 factorial

Notation

Consider a hypothetical experiment with two variables, A and B. A design of this type is illustrated in Table 13.1. The number of levels of A is designated by a, and the number of levels of B is designated by b. Any combination of one level of A and one level of B is called a **cell,** and the number of observations per cell is denoted n, or, more precisely, n_{ij}. The total number of observations is $N = \sum n_{ij} = abn$. When any confusion might arise, an individual observation (X) can be designated by three subscripts, X_{ijk}, where the subscript i refers to the number of the row (level of A), the subscript j refers to the number of the column

cell

Table 13.1 Representation of factorial design

	B_1	B_2	...	B_b	
	X_{111}	X_{121}	...	X_{1b1}	
	X_{112}	X_{122}		X_{1b2}	
A_1	...	...		...	$\overline{X}_{1.}$
	X_{11n}	X_{12n}		X_{1bn}	
	$\overline{X}_{11}$	$\overline{X}_{12}$		$\overline{X}_{1b}$	
	X_{211}	X_{221}	...	X_{2b1}	
	X_{212}	X_{222}		X_{2b2}	
A_2	...	...		...	$\overline{X}_{2.}$
	X_{21n}	X_{22n}		X_{2bn}	
	$\overline{X}_{21}$	$\overline{X}_{22}$		$\overline{X}_{2b}$	
	...	...	...	...	...
	X_{a11}	X_{a21}		X_{ab1}	
	Xa_{12}	Xa_{22}		X_{ab2}	
A_a	...	...		...	$\overline{X}a.$
	X_{a1n}	X_{a2n}		X_{abn}	
	$\overline{X}_{a1}$	$\overline{X}_{a2}$		$\overline{X}_{ab}$	
	$\overline{X}_{.1}$	$\overline{X}_{.2}$	...	$\overline{X}_{.b}$	$\overline{X}_{..}$

(level of *B*), and the subscript *k* refers to the *k*th observation in the *ij*th cell. Thus, X_{234} is the fourth participant in the cell corresponding to the second row and the third column. Means for the individual levels of *A* are denoted as $\overline{X}_A$ or $\overline{X}_{i.}$, and for the levels of *B* are denoted $\overline{X}_B$ or $\overline{X}_{.j}$. The cell means are designated $\overline{X}_{ij}$, and the grand mean is symbolized by $\overline{X}_{..}$. Needless subscripts are often a source of confusion, and whenever possible they will be omitted.

The notation outlined here will be used throughout the discussion of the analysis of variance, and it is important that you understand it thoroughly before proceeding. The advantage of the present system is that it is easily generalized to more complex designs. Thus, if participants recalled at three different times of day, it should be self-evident to what $\overline{X}_{\text{Time 1}}$ refers.

13.1 An Extension of the Eysenck Study

As mentioned earlier, Eysenck actually conducted a study varying Age as well as Recall Condition. The study included 50 participants in the 18-to-30-year age range, as well as 50 participants in the 55-to-65-year age range. The data in Table 13.2 have been created to have the same means and standard deviations as those reported by Eysenck. The table contains

Table 13.2 Data and computations for example from Eysenck (1974)

(a) Data:

	Recall Conditions					$\text{Mean}_{i.}$
	Counting	Rhyming	Adjective	Imagery	Intention	
Old	9	7	11	12	10	
	8	9	13	11	19	
	6	6	8	16	14	
	8	6	6	11	5	
	10	6	14	9	10	
	4	11	11	23	11	
	6	6	13	12	14	
	5	3	13	10	15	
	7	8	10	19	11	
	7	7	11	11	11	
Mean_{1j}	7.0	6.9	11.0	13.4	12.0	10.06
Young	8	10	14	20	21	
	6	7	11	16	19	
	4	8	18	16	17	
	6	10	14	15	15	
	7	4	13	18	22	
	6	7	22	16	16	
	5	10	17	20	22	
	7	6	16	22	22	
	9	7	12	14	18	
	7	7	11	19	21	
Mean_{2j}	6.5	7.6	14.8	17.6	19.3	13.16
$\textbf{Mean}_{.j}$	6.75	7.25	12.9	15.5	15.65	11.61

(continued)

Table 13.2   *(continued)*

(b) Calculations:

$$SS_{total} = \sum (X - \overline{X}_{..})^2$$
$$= (9 - 11.61)^2 + (8 - 11.61)^2 + \cdots + (21 - 11.61)^2$$
$$= 2667.79$$

$$SS_A = nc \sum (\overline{X}_{i.} - \overline{X}_{..})^2$$
$$= 10 \times 5[(10.06 - 11.61)^2 + (13.16 - 11.61)^2]$$
$$= 240.25$$

$$SS_C = na \sum (\overline{X}_{.j} - \overline{X}_{..})^2$$
$$= 10 \times 2[(6.75 - 11.61)^2 + (7.25 - 11.61)^2 + \cdots + (15.65 - 11.61)^2]$$
$$= 1514.94$$

$$SS_{cells} = n \sum (\overline{X}_{ij} - \overline{X}_{..})^2$$
$$= 10[(7.0 - 11.61)^2 + (6.9 - 11.61)^2 + \cdots + (19.3 \quad 11.61)^2]$$
$$= 1945.49$$

$$SS_{AC} = SS_{cells} - SS_A - SS_C = 1945.49 - 240.25 - 1514.94 = 190.30$$
$$SS_{error} = SS_{total} - SS_{cells} = 2667.79 - 1945.49 = 722.30$$

(c) Summary table

Source	df	SS	MS	F
A (Age)	1	240.25	240.250	29.94*
C (Condition)	4	1514.94	378.735	47.19*
AC	4	190.30	47.575	5.93*
Error	90	722.30	8.026	
Total	99	2667.79		

* $p < .05$

all the calculations for a standard analysis of variance, and we will discuss each of these in turn. Before beginning the analysis, it is important to note that the data themselves are approximately normally distributed with acceptably equal variances. The boxplots are not given in the table because the individual data points are artificial, but for real data, it is well worth your effort to compute them. You can tell from the cell and marginal means that recall appears to increase with greater processing, and younger participants seem to recall more items than do older participants. Notice also that the difference between younger and older participants seems to depend on the task, with greater differences for those tasks that involve deeper processing. We will have more to say about these results after we consider the analysis itself.

It will avoid confusion later if I take the time here to define two important terms. As I have said, we have two factors in this experiment—Age and Condition. If we look at the differences between means of older and younger participants, *ignoring the particular conditions,* we are dealing with what is called the **main effect** of Age. Similarly, if we look

main effect

at differences among the means of the five conditions, ignoring the Age of the participants, we are dealing with the main effect of Condition.

An alternative method of looking at the data would be to compare means of older and younger participants for only the data from the Counting task, for example. Or we might compare the means of older and younger participants on the Intentional task. Finally, we might compare the means on the five conditions for only the older participants. In each of these three examples we are looking at the effect of one factor for the data at only *one* level of the other factor. When we do this, we are dealing with a **simple effect**—the effect of one factor at one level of the other factor. A main effect, on the other hand, is that of a factor *ignoring* the other factor. If we say that tasks that involve more processing lead to better recall, we are speaking of a main effect. If we say that for younger participants, tasks that involve more processing lead to better recall, we are speaking about a simple effect. Simple effects are conditional on the level of the other variable. We will have considerably more to say about simple effects and their calculation shortly. For now, it is important only that you understand the terminology.

simple effect

Calculations

The calculations for the sums of squares appear in Table 13.2b. Many of these calculations should be familiar because they resemble the procedures used with a one-way analysis. For example, SS_{total} is computed the same way it was in Chapter 11, which is the same way it is always computed. We sum all the squared deviations of the observations from the grand mean.

The sum of squares for the Age factor (SS_A) is nothing but the SS_{treat} that we would obtain if this were a one-way analysis of variance without the Condition factor. In other words, we simply sum the squared deviations of the Age means from the grand mean and multiply by nc. We use nc as the multiplier here because each age has n participants at each of c levels. (Don't try to remember that multiplier as a formula. Just keep in mind that it is the number of scores upon which the relevant means are based.) The same procedures are followed in the calculation of SS_C, except that here we ignore the presence of the Age variable.

Having obtained SS_{total}, SS_A, and SS_C, we come to an unfamiliar term, SS_{cells}. This term represents the variability of the individual cell means and is only a dummy term; it will not appear in the summary table. It is calculated just like any other sum of squares. We take the deviations of the cell means from the grand mean, square and sum them, and multiply by n, the number of observations per mean. Although it might not be readily apparent why we want this term, its usefulness will become clear when we calculate a sum of squares for the interaction of Age and Condition. (It may be easier to understand the calculation of SS_{cells} if you think of it as what you would have if you viewed this as a study with 10 "groups" and calculated SS_{treat}.)

SS_{cells}

The SS_{cells} is a measure of how much the cell means differ. Two cell means may differ for any of three reasons, other than sampling error: (1) because they come from different levels of A (Age), (2) because they come from different levels of C (Condition), or (3) because of an interaction between A and C. We already have a measure of how much the cells differ because we know SS_{cells}. SS_A tells us how much of this difference can be attributed to differences in Age, and SS_C tells us how much can be attributed to differences in Condition. Whatever cannot be attributed to Age or Condition must be attributable to the interaction between Age and Condition (SS_{AC}). Thus, SS_{cells} has been partitioned into its three constituent parts—SS_A, SS_C, and SS_{AC}. To obtain SS_{AC}, we simply subtract SS_A and SS_C from SS_{cells}. Whatever is left over is SS_{AC}. In our example,

$$SS_{AC} = SS_{cells} - SS_A - SS_C$$
$$= 1945.49 - 240.25 - 1514.94 = 190.30$$

All that we have left to calculate is the sum of squares due to error. Just as in the one-way analysis, we will obtain this by subtraction. The total variation is represented by SS_{total}. Of this total, we know how much can be attributed to A, C, and AC. What is left over represents unaccountable variation or error. Thus,

$$SS_{\text{error}} = SS_{\text{total}} - (SS_A + SS_C + SS_{AC})$$

However, because $SS_A + SS_C + SS_{AC} = SS_{\text{cells}}$, it is simpler to write

$$SS_{\text{error}} = SS_{\text{total}} - SS_{\text{cells}}$$

This provides us with our sum of squares for error, and we now have all the necessary sums of squares for our analysis.

A more direct, but tiresome, way to calculate SS_{error} exists, and it makes explicit just what the error sum of squares is measuring. SS_{error} represents the variation within each cell, and as such can be calculated by obtaining the sum of squares for each cell separately. For example,

$$SS_{\text{cell}_{11}} = (9 - 7)^2 + (8 - 7)^2 + \cdots + (7 - 7)^2 = 30$$

We could perform a similar operation on each of the remaining cells, obtaining

$$SS_{\text{cell}_{11}} = 30.0$$
$$SS_{\text{cell}_{12}} = 40.9$$
$$\cdots \quad \cdots$$
$$\frac{SS_{\text{cell}_{25}}}{SS_{\text{error}}} = \frac{64.1}{722.30}$$

The sum of squares within each cell is then summed over the 10 cells to produce SS_{error}. Although this is the hard way of computing an error term, it demonstrates that SS_{error} is the sum of within-cell variation.

Table 13.2c shows the summary table for the analysis of variance. The source column and the sum of squares column are fairly obvious from what has already been said. Note, however, that we could organize the summary table somewhat differently, although we would seldom do so in practice. Thus, we could have

Source	df	SS
Between cells	9	1945.49
A	1	240.25
C	4	1514.94
AC	4	190.30
Within cells (Error)	90	722.30
Total	99	2667.79

This alternative summary table makes it clear that we have partitioned the total variation into variation among the cell means and variation within the cells. The former is then further partitioned into A, C, and AC.

Returning to Table 13.2c, look at the degrees of freedom. The calculation of df is straightforward. The total number of degrees of freedom (df_{total}) is always equal to $N - 1$. The degrees of freedom for Age and Condition are the number of levels of the variable minus 1. Thus, $df_A = a - 1 = 1$ and $df_C = c - 1 = 4$. The number of degrees of freedom for any interaction is simply the product of the degrees of freedom for the components of that interaction. Thus, $df_{AC} = df_A \times df_C = (a-1)(c-1) = 1 \times 4 = 4$. These three rules apply to *any* analysis of variance, no matter how complex. The degrees of freedom for error can be obtained either by subtraction $(df_{\text{error}} = df_{\text{total}} - df_A - df_C - df_{AC})$, or by realizing that the error term represents variability within each cell. Because each cell has $n - 1$ df, and because there are ac cells, $df_{\text{error}} = ac(n-1) = 2 \times 5 \times 9 = 90$.

Just as with the one-way analysis of variance, the mean squares are again obtained by dividing the sums of squares by the corresponding degrees of freedom. This same procedure is used in any analysis of variance.

Finally, to calculate F, we divide each MS by MS_{error}. Thus for Age, $F_A = MS_A/MS_{\text{error}}$; for Condition, $F_C = MS_C/MS_{\text{error}}$; and for AC, $F_{AC} = MS_{AC}/MS_{\text{error}}$. To appreciate why MS_{error} is the appropriate divisor in each case, we will digress briefly in a moment and consider the underlying structural model and the expected mean squares. First, however, we need to consider what the results of this analysis tell us.

Interpretation

From the summary table in Table 13.2c, you can see that there were significant effects for Age, Condition, and their interaction. In conjunction with the means, we can see that younger participants recall more items overall than do older participants. We can also see that those tasks involving greater depth of processing lead to better recall overall than do tasks involving less processing. This is in line with the differences we found in Chapter 11. The significant interaction tells us that the effect of one variable depends on the level of the other variable. For example, differences between older and younger participants on the easier tasks such as counting and rhyming are less than age differences on those tasks, such as imagery and intentional, that involve greater depths of processing. Another view is that differences among the five conditions are less extreme for the older participants than they are for the younger ones.

These results support Eysenck's hypothesis that older participants do not perform as well as younger participants on tasks that involve a greater depth of processing of information, but perform about equally with younger participants when the task does not involve much processing. These results do not mean that older participants are not *capable* of processing information as deeply. Older participants simply may not make the effort that younger participants do. Whatever the reason, however, they do not perform as well on those tasks.

13.2 Structural Models and Expected Mean Squares

Recall that in discussing a one-way analysis of variance, we employed the structural model

$$X_{ij} = \mu + \tau_j + e_{ij}$$

where $\tau_j = \mu_j - \mu$ represented the effect of the jth treatment. In a two-way design, we have two "treatment" variables (call them A and B) and their interaction. These can be represented in the model by α, β, and $\alpha\beta$, producing a slightly more complex model. This model can be written as

$$X_{ijk} = \mu + \alpha_i + \beta_j + \alpha\beta_{ij} + e_{ijk}$$

where

X_{ijk} = any observation

μ = the grand mean

$$\alpha_i = \text{the effect of Factor } A_i = \mu_{A_i} - \mu$$

$$\beta_j = \text{the effect of Factor } B_j = \mu_{B_j} - \mu$$

$$\alpha\beta_{ij} = \text{the interaction effect of Factor } A_i \text{ and Factor } B_j$$

$$= \mu - \mu_{A_i} - \mu_{B_j} + \mu_{ij}; \sum_i \alpha\beta_{ij} = \sum_j \alpha\beta_{ij} = 0$$

$$e_{ijk} = \text{the unit of error associated with observation } X_{ijk}$$

$$= N(0, \sigma_e^2)$$

From this model, it can be shown that with fixed variables, the expected mean squares are those given in Table 13.3. It is apparent that the error term is the proper denominator for each F ratio because the $E(MS)$ for any effect contains only one term other than σ_e^2.

Table 13.3   Expected mean squares for two-way analysis of variance (fixed)

Source	E(MS)
A	$\sigma_e^2 + nb\sigma_\alpha^2$
B	$\sigma_e^2 + na\sigma_\beta^2$
AB	$\sigma_e^2 + n\sigma_{\alpha\beta}^2$
Error	σ_e^2

Consider for a moment the test of the effect of Factor A:

$$\frac{E(MS_A)}{E(MS_{\text{error}})} = \frac{\sigma_e^2 + nb\sigma_\alpha^2}{\sigma_e^2}$$

If H_0 is true, then $\mu_{A_1} = \mu_{A_2} = \mu$ and σ_α^2, and thus $nb\sigma_\alpha^2$, will be 0. In this case, F will have an expectation of approximately 1 and will be distributed as the standard (central) F distribution. If H_0 is false, however, σ_α^2 will not be 0 and F will have an expectation greater than 1 and will not follow the central F distribution. The same logic applies to tests on the effects of B and AB.

13.3   Interactions

One of the major benefits of factorial designs is that they allow us to examine the interaction of variables. Indeed, in many cases, the interaction term may well be of greater interest than are the main effects (the effects of factors taken individually). Consider, for example, the study by Eysenck. The means are plotted in Figure 13.1 for each age group separately. Here you can see clearly what I referred to in the interpretation of the results when I said that the differences due to Conditions were greater for younger participants than for older ones. The fact that the two lines are not parallel is what we mean when we speak of an interaction. If Condition differences were the same for the two Age groups, then the lines would be parallel—whatever differences between Conditions existed for younger participants would be equally present for older participants. This would be true regardless of whether younger participants were generally superior to older participants or whether the two groups were comparable. Raising or lowering the entire line for younger participants would change the main effect of Age, but it would have no effect on the interaction because it would not affect the degree of parallelism between the lines.

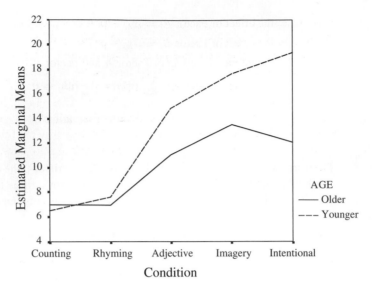

Figure 13.1   Cell means for data in Table 13.2

It may make the situation clearer if you consider several plots of cell means that represent the presence or absence of an interaction. In Figure 13.2, the first three plots represent the case in which there is no interaction. In all three cases, the lines are parallel, even when they are not straight. Another way of looking at this is to say that the simple effect of Factor B at A_1 is the same as it is at A_2 and at A_3. In the second set of three plots, the lines clearly are not parallel. In the first, one line is flat and the other rises. In the second, the lines actually cross. In the third, the lines do not cross, but they move in opposite directions. In every case, the simple effect of B is *not* the same at the different levels of A. Whenever the lines are (significantly) nonparallel, we say that we have an interaction.

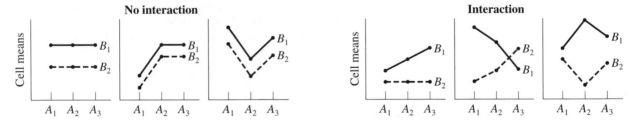

Figure 13.2   Illustration of possible noninteractions and interactions

Many people will argue that if you find a significant interaction, the main effects should be ignored. It is not reasonable, however, automatically to exclude interpretation of main effects in the presence of *any* significant interaction. In the Eysenck study, we had a significant interaction, but for both younger and older participants the tasks that involved greater processing led to greater recall. That this effect was more pronounced in the younger group does not negate the fact that it was also clearly present in the older participants. Here, it is perfectly legitimate to speak about the main effect of Condition, even in the presence of an interaction. However, had the younger group shown better recall with more demanding tasks whereas the older group had shown poorer recall, then it might actually not be of interest whether the main effect of Condition was significant or not, and we would instead concentrate on discussing only the simple effects of difference among Conditions for the younger and older participants separately. (Interactions in which group differences reverse their sign

**"disordinal"
interactions**

**"ordinal"
interaction**

at some level of the other variable are sometimes referred to as **"disordinal" interactions.** When one group is consistently above the other group we have an **"ordinal" interaction.**) In general, the interpretation depends on common sense. If the main effects are clearly meaningful, then it makes sense to interpret them, whether or not an interaction is present. However, if the main effect does not really have any meaning, then it should be ignored.

This discussion of the interaction effects has focused on examining cell means. I have taken that approach because it is the easiest to see and has the most to say about the results of the experiment. Rosnow and Rosenthal (1989) have pointed out that a more accurate way to look at an interaction is to first remove any row and column effects from the data. They raise an interesting point, but most interactions are probably better understood in terms of the previous explanation.

13.4 Simple Effects

I earlier defined a simple effect as the effect of one factor (independent variable) at one level of the other factor—for example, the differences among Conditions for the younger participants. The analysis of simple effects can be an important technique for analyzing data that contain significant interactions. In a very real sense, it allows us to "tease apart" interactions.

I will use the Eysenck data to illustrate how to calculate and interpret simple effects. Table 13.4 shows the cell means and the summary table reproduced from Table 13.2. The table also contains the calculations involved in obtaining all the simple effects.

The first summary table in Table 13.4c reveals significant effects caused by Age, Condition, and their interaction. We already discussed these results earlier in conjunction

Table 13.4 Illustration of calculation of simple effects (data taken from Table 13.2)

(a) Cell means ($n = 10$)

	Counting	Rhyming	Adjective	Imagery	Intention	Mean
Older	7.0	6.9	11.0	13.4	12.0	10.06
Younger	6.5	7.6	14.8	17.6	19.3	13.16
Mean	6.75	7.25	12.90	15.50	15.65	11.61

(b) Calculations:

Conditions at Each Age

$$SS_{C \text{ at Old}} = 10 \times [(7.0 - 10.06)^2 + (6.9 - 10.06)^2 + \cdots + (12 - 10.06)^2]$$
$$= 351.52$$
$$SS_{C \text{ at Young}} = 10 \times [(6.5 - 13.16)^2 + (7.6 - 13.16)^2 + \cdots + (19.3 - 13.16)^2]$$
$$= 1353.72$$

Age at Each Condition

$$SS_{A \text{ at Counting}} = 10 \times [(7.0 - 6.75)^2 + (6.5 - 6.75)^2] = 1.25$$
$$SS_{A \text{ at Rhyming}} = 10 \times [(6.9 - 7.25)^2 + (7.6 - 7.25)^2] = 2.45$$
$$SS_{A \text{ at Adjective}} = 10 \times [(11.0 - 12.9)^2 + (14.8 - 12.9)^2] = 72.2$$
$$SS_{A \text{ at Imagery}} = 10 \times [(13.4 - 15.5)^2 + (17.6 - 15.5)^2] = 88.20$$
$$SS_{A \text{ at Intentional}} = 10 \times [(12.0 - 15.65)^2 + (19.3 - 15.65)^2] = 266.45$$

(continued)

Table 13.4 *(continued)*

(c) Summary Tables

Overall Analysis

Source	df	SS	MS	F
A (Age)	1	240.25	240.25	29.94*
C (Condition)	4	1514.94	378.735	47.19*
AC	4	190.30	47.575	5.93*
Error	90	722.30	8.026	
Total	99	2667.79		

*$p < .05$

Simple Effects

Source	df	SS	MS	F
Conditions				
C at Old	4	351.52	87.88	10.95*
C at Young	4	1353.72	338.43	42.15*
Age				
A at Counting	1	1.25	1.25	<1
A at Rhyming	1	2.45	2.45	<1
A at Adjective	1	72.20	72.20	9.00*
A at Imagery	1	88.20	88.20	10.99*
A at Intentional	1	266.45	266.45	33.20*
Error	90	722.30	8.03	

*$p < .05$

with the original analysis. As I said there, the presence of an interaction means that there are different Condition effects for the two Ages, and there are different Age effects for the five Conditions. It thus becomes important to ask whether our general Condition effect really applies for older as well as younger participants, and whether there really are Age differences under all Conditions. The analysis of these simple effects is found in Table 13.4b and the second half of Table 13.4c. I have shown all possible simple effects for the sake of completeness of the example, but in general you should calculate only those effects in which you are interested.

Calculation

In Table 13.4b, you can see that $SS_{C \text{ at Old}}$ is calculated in the same way as any sum of squares. We simply calculate SS_C *using only the data for the older participants*. If we consider only those data, the five Condition means are 7.0, 6.9, 11.0, 13.4, and 12.0. Thus, the sum of squares will be

$$SS_{C \text{ at Old}} = n \sum (\overline{X}_{1j} - \overline{X}_{1.})^2$$
$$= 10 \times [(7 - 10.06)^2 + (6.9 - 10.06)^2 + \cdots + (12 - 10.06)^2] = 351.52$$

The other simple effects are calculated in the same way, by ignoring all data in which you are not at the moment interested. Notice that the sum of squares for the simple effect of Condition for older participants (351.52) is the same value as that we obtained in Chapter 11 when we ran a one-way analysis of variance on only the data from older participants.

Table 13.5   Expected mean squares for simple effects

Source	$E(MS)$
Simple Effects of A	
A at B_1	$\sigma_e^2 + n\sigma_{\alpha \text{ at } \beta_1}^2$
A at B_2	$\sigma_e^2 + n\sigma_{\alpha \text{ at } \beta_2}^2$
A at B_3	$\sigma_e^2 + n\sigma_{\alpha \text{ at } \beta_3}^2$
Simple Effect of B	
B at A_1	$\sigma_e^2 + n\sigma_{\beta \text{ at } \alpha_1}$
B at A_2	$\sigma_e^2 + n\sigma_{\beta \text{ at } \alpha_2}$
Error	σ_e^2

The degrees of freedom for the simple effects are calculated in the same way as for the corresponding main effects. This makes sense because the number of means we are comparing remains the same. Whether we use all of the participants or only some of them, we are still comparing five conditions and have $5 - 1 = 4$ df for Conditions.

To test the simple effects, we generally use the error term from the overall analysis (MS_{error}). The expected mean squares are presented in Table 13.5, and they make it clear why this is the appropriate error term. The expected mean square for each simple effect contains only one effect other than error (e.g., $n\sigma_{\alpha \text{ at } \beta_j}^2$), whereas MS_{error} is an estimate of error variance (σ_e^2). Actually, the only difference between what I have done in Table 13.4 and what I would do if I ran a standard one-way analysis of variance on the Old participants' data (which is the way I usually calculate sums of squares for simple effects when I use computer software) is the error term. MS_{error} continues to be based on all the data because it is a better estimate with more degrees of freedom.

Interpretation

From the column labeled F in the bottom table in Table 13.4c, we can see that differences due to Conditions occur for both ages although the sum of squares for the older participants is only about one-quarter of what it is for the younger ones. With regard to the Age effects, however, no differences occur on the lower-level tasks of counting and rhyming, but differences do occur on the higher-level tasks. In other words, differences between age groups show up on only those tasks involving higher levels of processing. This is basically what Eysenck set out to demonstrate.

In general, we seldom look at simple effects unless a significant interaction is present. However, this practice must be governed by common sense. It is not difficult to imagine data for which an analysis of simple effects would be warranted even in the face of a nonsignificant interaction, or to imagine studies in which the simple effects are the prime reason for conducting the experiment.

Additivity of Simple Effects

All sums of squares in the analysis of variance (other than SS_{total}) represent a partitioning of some larger sum of squares, and the simple effects are no exception. The simple effect of Condition at each level of Age represents a partitioning of SS_C and $SS_{A \times C}$, whereas the effects of Age at each level of Condition represent a partitioning of SS_A and $SS_{A \times C}$. Thus,

$$\sum SS_{C \text{ at } A} = 351.52 + 1353.72 = 1705.24$$
$$SS_C + SS_{AC} = 1514.94 + 190.30 = 1705.24$$

and

$$\sum SS_{A \text{ at } C} = 1.25 + 2.45 + 72.20 + 88.20 + 266.45 = 430.55$$
$$SS_A + SS_{AC} = 240.25 + 190.30 = 430.55$$

A similar additive relationship holds for the degrees of freedom. The fact that the sums of squares for simple effects sum to the combined sums of squares for the corresponding main effect and interaction affords us a quick and simple check on our calculations.

Error Rates

In testing simple effects, keep in mind that we are generally working with an error rate per comparison of α. In the present example, we have calculated seven Fs, each at $\alpha = .05$. If the Fs were independent, FW would actually be .30, and even though they are not independent, the FW error rate approaches .30. This is an uncomfortably high error rate and is not one to be recommended without careful consideration. As a general rule, it is important to balance the gain to be expected from testing all simple effects against the danger to be incurred from an unpleasantly high error rate. In practice, you will usually find that only a few of the possible simple effects are of a priori interest, and only these few should be tested.

13.5 Analysis of Variance Applied to the Effects of Smoking

This next example is based on a study by Spilich, June, and Renner (1992), who investigated the effects of smoking on performance. They used three tasks that differed in the level of cognitive processing that was required to perform them, with different participants serving in each task. The first task was a Pattern recognition task in which the participants had to locate a target on a screen. The second was a Cognitive task in which the participants were required to read a passage and then recall it at a later time. The third task was a Driving simulation video game. In each case, the dependent variable was the number of errors that the participant committed. (This isn't really true for all tasks, but it allows me to treat Task as an independent variable. I am not seriously distorting the results that Spilich et al. obtained.)

Participants were further divided into three Smoking groups. Group AS was composed of people who actively smoked during or just before carrying out the task. Group DS participants were regular smokers who had not smoked for 3 hours before the task (D stands for delay). Group NS was nonsmokers.

The data follow, but before you look at those data you should make some predictions about the kinds of effects that you might find for Task, Smoking, and about their interaction.

						Pattern Recognition									
NS:	9	8	12	10	7	10	9	11	8	10	8	10	8	11	10
DS:	12	7	14	4	8	11	16	17	5	6	9	6	6	7	16
AS:	8	8	9	1	9	7	16	19	1	1	22	12	18	8	10

						Cognitive Task									
NS:	27	34	19	20	56	35	23	37	4	30	4	42	34	19	49
DS:	48	29	34	6	18	63	9	54	28	71	60	54	51	25	49
AS:	34	65	55	33	42	54	21	44	61	38	75	61	51	32	47

Driving Simulation

NS:	15	2	2	14	5	0	16	14	9	17	15	9	3	15	13
DS:	7	0	6	0	12	17	1	11	4	4	3	5	16	5	11
AS:	3	2	0	0	6	2	0	6	4	1	0	0	6	2	3

I will omit hand calculations here on the assumption that you can carry them out yourself, and it would be good practice to do so. In Exhibit 13.1, you will find the analysis of these data using Minitab.

(a) Summary table and confidence limits on means

Two-way Analysis of Variance

Analysis of Variance for Errors

Source	DF	SS	MS	F	P
Task	2	28662	14331	132.90	0.000
SmokeGrp	2	355	177	1.64	0.197
Interaction	4	2729	682	6.33	0.000
Error	126	13587	108		
Total	134	45332			

```
                          Individual 95% CI
Task        Mean    ---------+---------+---------+---------+----
Cognitiv    38.8                                      (--*--)
Driving      6.4    (--*--)
PattRec      9.6       (--*--)
                    ---------+---------+---------+---------+----
                        10.0      20.0      30.0      40.0
```

```
                        Individual 95% CI
SmokeGrp    Mean    ---------+---------+---------+---------+----
AS          19.9                  (-------------*-------------)
DS          18.8              (-------------*------------)
NS          16.1    (-------------*------------)
                    ---------+---------+---------+---------+----
                        15.0      17.5      20.0      22.5
```

(b) Main effects

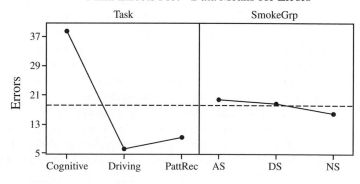

Main Effects Plot - Data Means for Errors

Exhibit 13.1   Analysis of Spilich et al. data

(continued)

(c) Interaction

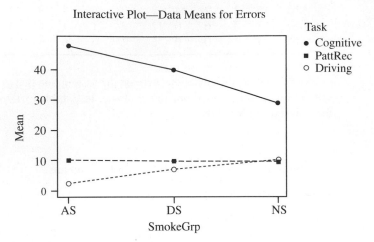

Interactive Plot—Data Means for Errors

Exhibit 13.1 *(continued)*

The summary table reveals that there are significant effects due to Task and to the interaction of Task and SmokeGrp, but there is no significant effect due to the SmokeGrp variable. The Task effect is of no interest because it simply says that people make more errors on some kinds of tasks than others. This is like saying that your basketball team scored more points in yesterday's game than did your soccer team.

Next, you see the means and confidence limits on those means. This kind of printout is very common in Minitab. Finally, I have plotted the two main effects and the interaction. It is pretty clear that smoking makes a difference for the cognitive task but appears to make no difference for the pattern recognition task. It is unclear whether there would be a significant smoking effect for the driving task. I will leave these simple effect tests up to you.

13.6 Multiple Comparisons

All the multiple-comparison procedures discussed in Chapter 12 are applicable to the analysis of factorial designs. Thus, we can test the differences among the five Condition means in the Eysenck example, or the three SmokeGrp means in the Spilich example using the Bonferroni t test, the Tukey test, Ryan's REGWQ, or any other procedure. Keep in mind, however, that we must interpret the "n" that appears in the formulae in Chapter 12 to be the number of observations on which each treatment mean was based. The Condition means are based on $(a \times n)$ observations, so that is the value that you would enter into the formula, not n.

On rare occasions, it is relevant to compare the individual cell means rather than the means attributable to the main effects. (Be sure that this is really what you want to do. People often ask me how to do it, but when I ask them why they want to, they quickly discover that they really do not want to do that after all.) If you do want to compare cell means, just treat them as if they came from a large one-way design with $c \times a$ groups and operate accordingly. For example, for the Eysenck study, you could compare the mean of younger participants in the Counting condition with the mean of the older participants in the Adjective condition (though I can't imagine why you would want to) by acting as if you had a one-way design with 10 groups. You could then use the standard multiple-comparison procedures to make your comparison.[1] The problem with making such comparisons is that if you did find a difference between the two

---

[1] For a much more complete discussion of the use of multiple-comparison techniques with factorial designs, see Keppel (1973).

specified cells, you could not tell whether the difference was due to the different conditions, the different ages, or a combination of the two factors. Here Age and Condition are confounded. That is why I recommend such a procedure only in unusual circumstances.

In the Spilich smoking example, there is no significant effect due to SmokeGrp, so you would probably not want to run contrasts among the three levels of that variable. Because the dependent variable (errors) is not directly comparable across groups, it makes no sense to look for specific group differences there. We could do so, but no one would be likely to care. (Remember the basketball and soccer teams referred to earlier.) However, the interaction suggests that you might want to run multiple comparisons on simple effects. In particular, you would probably want to know whether there are significant differences among the three SmokeGrp conditions when looking only at the Cognitive task. You could run these tests by restricting yourself just to the data from the Cognitive task. However, I would suggest making these contrasts using MS_{error} from the overall analysis, assuming that you have no reason to think that you have heterogeneity of variance. If you run your analysis using standard computer software, you will have to recalculate your effects by substituting MS_{error} from the main summary table. Unless you were smart enough *in advance* to plan to run these analyses only on the Cognitive task (in which case one of the Bonferroni tests would be appropriate), I would recommend the Ryan (REGWQ) procedure if you have access to it, or the Tukey procedure if you don't.

The analysis of SmokeGrp differences on the Cognitive task gives a frequent, but unwelcome, result. Ryan's procedure (and Tukey's) shows that the Nonsmoking group performs significantly better than the Active group, but not significantly better than the Delayed group. The Delayed group is also not significantly different from the Active group. Representing this graphically, we have

Nonsmoking   Delayed   Active
_____
                    _____

where the lines underline groups that do not differ significantly.

If you just came from your class in Logic 132, you know that it does not make sense to say A = B, B = C, but A ≠ C. But, don't confuse Logic, which is in some sense exact, with Statistics, which is probabilistic. Don't forget that a failure to reject H_0 does not mean that the means are equal. It just means that they are not sufficiently different for us to know that they're different (and which one is larger). Here, we don't have enough evidence to conclude that Delayed is different from Nonsmoking, but we *do* have enough evidence (i.e., power) to conclude that there is a significant difference between Active and Nonsmoking. This kind of result occurs frequently with multiple-comparison procedures, and we just have to learn to live with a bit of uncertainty.

13.7 Power Analysis for Factorial Experiments

Calculating power for fixed-variable factorial designs is basically the same as it was for one-way designs. In the one-way design, we defined

$$\phi' = \sqrt{\frac{\sum \tau_j^2}{k\sigma_e^2}}$$

and

$$\phi = \phi' \sqrt{n}$$

where $\sum \tau_j^2 = \sum (\mu_j - \mu)^2$, k = the number of treatments, and n = the number of observations in each treatment. In the two-way and higher-order designs, we have more than one "treatment," but this does not alter the procedure in any important way. If we let $\alpha_i = \mu_{i.} - \mu$, and $\beta_j = \mu_{.j} - \mu$, where $\mu_{i.}$ represents the parametric mean of Treatment A_i (across all levels of B) and $\mu_{.j}$ represents the parametric mean of Treatment B_j (across all levels of A), then we can define the following terms:

$$\phi'_\alpha = \sqrt{\frac{\sum \alpha_j^2}{a\sigma_e^2}}$$

$$\phi_\alpha = \phi'_\alpha \sqrt{nb}$$

and

$$\phi'_\beta = \sqrt{\frac{\sum \beta_j^2}{b\sigma_e^2}}$$

$$\phi_\beta = \phi'_\beta \sqrt{na}$$

Examination of these formulae reveals that to calculate the power against a null hypothesis concerning A, we act as if variable B did not exist. To calculate the power of the test against a null hypothesis concerning B, we similarly act as if variable A did not exist.

Calculating the power against the null hypothesis concerning the interaction follows the same logic. We define

$$\phi'_{\alpha\beta} = \sqrt{\frac{\sum \alpha\beta_{ij}^2}{ab\sigma_e^2}}$$

$$\phi_{\alpha\beta} = \phi'_{\alpha\beta} \sqrt{n}$$

where $\alpha\beta_{ij}$ is defined as for the underlying structural model ($\alpha\beta_{ij} = \mu - \mu_{i.} - \mu_{.j} + \mu_{ij}$). Given $\phi_{\alpha\beta}$, we can simply obtain the power of the test just as we did for the one-way design.

Calculating power for the random model is more complicated, and for the mixed model requires a set of rather unrealistic assumptions. To learn how to obtain estimates of power with these models, see Winer (1971, p. 334).

In certain situations, a two-way factorial is more powerful than are two separate one-way designs, in addition to the other advantages that accrue to factorial designs. Consider two hypothetical studies, where the number of participants per treatment is held constant across both designs.

In Experiment 1, an investigator wants to examine the efficacy of four different treatments for posttraumatic stress disorder (PTSD) in rape victims. She has chosen to use both male and female therapists. Our experimenter is faced with two choices. She can run a one-way analysis on the four treatments, ignoring the sex of therapist (SexTher) variable entirely, or she can run a 4×2 factorial analysis on the four treatments and two sexes. In this case, the two-way has more power than the one-way. In the one-way design, we would ignore any differences due to SexTher and the interaction of Treatment with SexTher, and these would go toward increasing the error term. In the two-way, we would consider differences that can be attributed to SexTher and to the interaction between Treatment and SexTher, thus removing them from the error term. The error term for the two-way would thus be smaller than for the one-way, giving us greater power.

For Experiment 2, consider the experimenter who had originally planned to use only female therapists in her experiment. Her error term would not be inflated by differences among SexTher and by the interaction because neither of those exist. If she now *expanded* her study to include male therapists, SS_{total} would increase to account for additional effects due to the new independent variable, but the error term would remain constant because the extra variation would be accounted for by the extra terms. Because the error term would remain constant, she would have no increase in power in this situation over the power she would have had in her original study, except for an increase in n.

As a general rule, a factorial design is more powerful than a one-way design only when the extra factors can be thought of as refining or purifying the error term. In other words, when extra factors or variables account for variance that would normally be incorporated into the error term, the factorial design is more powerful. Otherwise, all other things being equal, it is not, although it still possesses the advantage of allowing you to examine the interactions and simple effects.

You need to be careful about one thing, however. When you add a factor that is a random variable (e.g., Classroom) you may actually decrease the power of your test. As you will see in a moment, the fixed factor, which may well be the one in which you are most interested, will probably have to be tested using $MS_{\text{interaction}}$ as the error term instead of MS_{error}. This is likely to cost you a considerable amount of power. And you can't just pretend that the Classroom factor didn't exist because then you will run into problems with the independence of errors. For a discussion of this issue, see Judd, McClelland, and Culhane (1995).

We need to discuss one additional consideration in terms of power. McClelland and Judd (1993) have shown that power can be increased substantially using what they call "optimal" designs. These are designs in which sample sizes are apportioned to the cells unequally to maximize power. McClelland has argued that we often use more levels of the independent variables than we need, and we frequently assign equal numbers of participants to each cell when we would be better off with fewer (or no) participants in some cells (especially the central levels of ordinal independent variables). For example, imagine two independent variables that can take on as many as five levels, denoted as $A_1, A_2, A_3, A_4,$ and A_5 for Factor A, and $B_1, B_2, B_3, B_4,$ and B_5 for Factor B. McClelland and Judd (1993) show that a 5×5 design using all five levels of each variable is only 25% as efficient as a design using only A_1 and A_5, and B_1 and B_5. A 3×3 design using $A_1, A_3,$ and A_5, and $B_1, B_3,$ and B_5 is 44% as efficient. I recommend a close reading of their paper.

13.8 Expected Mean Squares

Although fixed and random models led to the same F test when we were discussing the one-way analysis of variance, this is not the case in more complex designs. The denominator in an F ratio is a function of the type of model we are considering, so when we come to the factorial designs, we find that MS_{error} is not always the appropriate denominator for F. Although you might think that most experiments you will run will involve fixed factors, there are enough examples where at least one independent variable is a random factor to require that we cover random and mixed models. Along the way, we will discover some messy formulae, but they are not particularly overwhelming if you take them as they come.

Recall that a variable is defined as a *fixed* variable if we *select* the levels of that variable, and as a *random* variable if we obtain the levels by random sampling. Designs that consist of one or more fixed variables and one or more random variables are referred to as **mixed-model designs.** The difference between a fixed and a random term in any design becomes clearer if we first define a sampling fraction. The **sampling fraction** for a variable is the ratio of the number of levels of a given variable that *actually* are used to the potential number of levels that *could have been* used. We will use lowercase letters to represent the number of levels used and uppercase letters to represent the potential number of levels available.

mixed-model designs

sampling fraction

As pointed out in Chapter 11, for a *fixed* variable, the population of levels is limited to the levels actually used. This means that if variable A is fixed, $a = A$ and $a/A = 1$. For a fixed variable, the sampling fraction is always 1. For a random variable, however, this is not the case because the number of potential levels is generally very large, and a would be very much smaller than A, meaning that a/A would approach the limit of 0 as A approaches

Table 13.6 Sampling fractions

Variable	Sampling Fraction
A (fixed)	$a/A = 1$
A (random)	$a/A = 0$
B (fixed)	$b/B = 1$
B (random)	$b/B = 0$
Subjects (random)	$n/N = 0$

infinity. From this discussion, we can write the sampling fractions shown in Table 13.6. We will always treat *participants* (subjects[2]) as if they were sampled at random from a large population.

Although it is possible to have sampling fractions between 0 and 1—as, for example, when the variable is a high-school class, which has only four possible levels—these rarely occur in practice. When the number of potential levels is small, that variable is almost always treated as a fixed variable anyway. When peculiar sampling fractions do arise, the investigator will have to work out the $E(MS)$ expressions for herself.

Given the concept of a sampling fraction, it is possible to define the expected mean squares for all models. The expected means squares are given in Table 13.7 for a two-way factorial.

Table 13.7 Expected mean squares for all models of a two-way factorial

Source	$E(MS)$
A	$(1 - n/N)\sigma_e^2 + n(1 - b/B)\sigma_{\alpha\beta}^2 + nb\sigma_\alpha^2$
B	$(1 - n/N)\sigma_e^2 + n(1 - a/A)\sigma_{\alpha\beta}^2 + na\sigma_\beta^2$
AB	$(1 - n/N)\sigma_e^2 + n\sigma_{\alpha\beta}^2$
Error	$(1 - n/N)\sigma_e^2$

Each $E(MS)$ in Table 13.7 contains the term

$$\left(1 - \frac{n}{N}\right)\sigma_e^2$$

Because participants are almost always assumed to be chosen at random (at least we look around innocently and pretend that they are), and because the population of potential participants is huge, n/N vanishes and we are left with σ_e^2 for the first term under each $E(MS)$. Similar reasoning applies to the other terms in Table 13.7, where, for example, $1 - a/A$ is 0 for fixed effects and 1 for random effects. If we substitute values of 0 and 1 for fixed and random effects, we arrive at the results presented in Table 13.8.

Table 13.8 shows that the expected mean squares are heavily dependent on the underlying structural model. This in turn means that the denominators for our F ratios will also depend on the model we adopt.

---

[2] The American Psychological Association Publication Manual, which is commonly used as a general style manual, advocates the use of words like "participants" in describing those people who participated in the study, but retains the word "subjects" in relationship to statistical analyses. So it is still proper to speak of "subject differences," or the sum of squares due to subjects.

Table 13.8 Expected mean squares for fixed, random, and mixed models

Source	Fixed A fixed B fixed	Random A random B random	Mixed A random B fixed	Mixed A fixed B random
A	$\sigma_e^2 + nb\sigma_\alpha^2$	$\sigma_e^2 + n\sigma_{\alpha\beta}^2 + nb\sigma_\alpha^2$	$\sigma_e^2 + nb\sigma_\alpha^2$	$\sigma_e^2 + n\sigma_{\alpha\beta}^2 + nb\sigma_\alpha^2$
B	$\sigma_e^2 + na\sigma_\beta^2$	$\sigma_e^2 + n\sigma_{\alpha\beta}^2 + na\sigma_\beta^2$	$\sigma_e^2 + n\sigma_{\alpha\beta}^2 + na\sigma_\beta^2$	$\sigma_e^2 + na\sigma_\beta^2$
AB	$\sigma_e^2 + n\sigma_{\alpha\beta}^2$	$\sigma_e^2 + n\sigma_{\alpha\beta}^2$	$\sigma_e^2 + n\sigma_{\alpha\beta}^2$	$\sigma_e^2 + n\sigma_{\alpha\beta}^2$
Error	σ_e^2	σ_e^2	σ_e^2	σ_e^2

Consider first the usual fixed model. From Table 13.8, we can see that MS_{error} will always form a suitable test term because the other $E(MS)$s differ from $E(MS_{\text{error}})$ only by the parameter in question. Thus, for example,

$$\frac{E(MS_A)}{E(MS_{\text{error}})} = \frac{\sigma_e^2 + nb\sigma_\alpha^2}{\sigma_e^2}$$

will have an expectation appreciably greater than 1 only if $\sigma_\alpha^2 \neq 0$. Fortunately, this is by far the most common case.

For the random model, it is apparent from the expected mean squares that MS_{error} is not appropriate for testing the main effects. For example, consider

$$\frac{E(MS_A)}{E(MS_{\text{error}})} = \frac{\sigma_e^2 + n\sigma_{\alpha\beta}^2 + nb\sigma_\alpha^2}{\sigma_e^2}$$

This ratio would have an expectation appreciably greater than 1 if *either* σ_α^2 or $\sigma_{\alpha\beta}^2$ were greater than 0, and a significant F would not indicate which H_0 should be rejected. However, the interaction term does provide us with a proper test because

$$\frac{E(MS_A)}{E(MS_{AB})} = \frac{\sigma_e^2 + n\sigma_{\alpha\beta}^2 + nb\sigma_\alpha^2}{\sigma_e^2 + n\sigma_{\alpha\beta}^2}$$

would have an expectation of approximately 1 unless σ_α^2 were greater than 0. In this case, a significant F would have an unequivocal interpretation. Thus, to test the two main effects we would use MS_{AB} as our denominator. Unfortunately, the interaction terms usually have few degrees of freedom.

Even with the random model, MS_{error} does serve as the test term against the null hypothesis concerning the interaction, as is obvious from the ratio

$$\frac{E(MS_{AB})}{E(MS_{\text{error}})} = \frac{\sigma_e^2 + n\sigma_{\alpha\beta}^2}{\sigma_e^2}$$

where a significant F would occur only if $\sigma_{\alpha\beta}^2$ is greater than 0 (except for Type I errors).

For the mixed models, the situation is more complex because one main effect will be tested against MS_{error} and the other against MS_{AB}. The interaction will again be tested against MS_{error}. These tests can be illustrated for the case in which A is random and B is fixed.

$$\frac{E(MS_A)}{E(MS_{\text{error}})} = \frac{\sigma_e^2 + nb\sigma_\alpha^2}{\sigma_e^2}$$

$$\frac{E(MS_B)}{E(MS_{AB})} = \frac{\sigma_e^2 + n\sigma_{\alpha\beta}^2 + na\sigma_\beta^2}{\sigma_e^2 + n\sigma_{\alpha\beta}^2}$$

$$\frac{E(MS_{AB})}{E(MS_{\text{error}})} = \frac{\sigma_e^2 + n\sigma_{\alpha\beta}^2}{\sigma_e^2}$$

No, I did not make an error! You can see that if you work out the algebra. In a mixed model, the *fixed* term is tested against MS interaction and the *random* term is tested against MS_{error}. Although this looks backward, it follows from what we have said about the role of the sampling fraction.

Pooling Error Terms

In both the random and mixed models, we often find that an important variable is tested against $MS_{interaction}$. Interactions usually have relatively few degrees of freedom compared with MS_{error}, so this may result in a substantial loss in power, turning what might otherwise be a significant result into a nonsignificant one.

One way out of this difficulty lies in first showing that there is no evidence to cause us to doubt that $\sigma^2_{\alpha\beta} = 0$, and then dropping the interaction term from the model. If this is possible, we may now pool MS_{AB} and MS_{error}, forming a new error term, and use this to test the main effects.

We started out with the model $X_{ijk} = \mu + \alpha_i + \beta_j + \alpha\beta_{ij} + e_{ijk}$. If there is no interaction between A and B, the model is unnecessarily complicated by the inclusion of $\alpha\beta_{ij}$. We might therefore begin by testing the null hypothesis $H_0: \sigma^2_{\alpha\beta} = 0$. To reduce the risk of a Type II error, this test should be run at a relatively high level of α—for example, $\alpha = .25$. If we cannot reject H_0 at this level, we can then be reasonably confident about deleting $\alpha\beta_{ij}$ from our model, leaving $X_{ijk} = \mu + \alpha_i + \beta_j + e_{ijk}$. The effect of deleting $\alpha\beta_{ij}$ from the model is to delete all terms of the form $n\sigma^2_{\alpha\beta}$ from the table of expected mean squares (Table 13.8), with the result that both MS_{AB} and MS_{error} will now be estimates of σ^2_e. We can then form a new test by combining the error and interaction mean squares as $MS_{residual}$.

$$MS_{residual} = \frac{SS_{AB} + SS_{error}}{df_{AB} + df_{error}}$$

This new term, on $df_{AB} + df_{error}$ degrees of freedom, is now used to test the main effects.

Although we might run a preliminary test on the interaction at $\alpha = .25$, we would not declare the interaction to be significant (in terms of our final conclusions about the data) unless we could also reject H_0 at $\alpha = .05$. In other words, our two tests represent two different strategies (accepting versus rejecting H_0), and our levels of α must reflect these differing strategies.

The entire procedure of pooling mean squares is usually relevant only for the random and mixed models. For fixed models, the MS_{error} is always an appropriate error term.

13.9 Measures of Association and Effect Size

We can look at the magnitude of an effect in two different ways, just as we did with the one-way analysis. We can either calculate a measure of association, such as η^2, or we can calculate an effect size. Normally, when we are examining an omnibus F, we use a measure of association. However, when we are looking at a contrast between means, it is usually more meaningful to calculate an effect size estimate (d).

r-Family Measures

As with the one-way design, it is both possible and desirable to calculate the magnitude of effect associated with each independent variable. The easiest, but also the most biased, way to do this is to calculate η^2. Here we would simply take the relevant sum of squares and divide by SS_{total}. Thus, the magnitude of effect for variable A is $\eta^2_\alpha = SS_A/SS_{total}$ and for variable B is $\eta^2_\beta = SS_B/SS_{total}$, whereas the magnitude of effect for the interaction is $\eta^2_{\alpha\beta} = SS_{AB}/SS_{total}$.

There are two difficulties with the measure that we have just computed. In the first place, η^2 is a biased estimate of the true magnitude of effect in the population. To put this somewhat differently, η^2 is a very good descriptive statistic, but a poor inferential statistic. Second, η^2, as we calculated it here, may not measure what we want to measure. We will speak about that shortly when we discuss partial η^2.

Although ω^2 is also biased, the bias is much less than for η^2. In addition, the statistical theory underlying ω^2 allows us to differentiate between fixed, random, and mixed models and to act accordingly.

To develop ω^2 for two-way and higher-order designs, we begin with the set of expected mean squares, derive estimates of σ_α^2, σ_β^2, $\sigma_{\alpha\beta}^2$, and σ_e^2, and then form ratios of each of these components relative to the total variance. Rather than derive the formulae for calculating ω^2 for the three different models, as I have done in previous editions of this book, I will present the results in a simple table. I strongly suspect that no student remembered the derivation five minutes after he or she read it, and that many students were so numb by the end of the derivation that they missed the final result.

For a factorial analysis of variance, the basic formula to estimate ω^2 remains the same whether we are looking at fixed or random variables. The only difference is in how we calculate the components of that formula. We will start by letting $\hat{\sigma}_{\text{effect}}^2$ refer to the estimate of the variance of the independent variable we care about at the moment, such as A, B, or AB, and $\hat{\sigma}_{\text{total}}^2$ refer to the sum of all sources of variance. Then if we know the value of these terms we can estimate ω_{effect}^2 as

$$\hat{\omega}_{\text{effect}}^2 = \frac{\hat{\sigma}_{\text{effect}}^2}{\hat{\sigma}_{\text{total}}^2}$$

For the main effect of A, for example, this becomes

$$\omega_\alpha^2 = \frac{\hat{\sigma}_\alpha^2}{\hat{\sigma}_{\text{total}}^2} = \frac{\hat{\sigma}_\alpha^2}{\hat{\sigma}_\alpha^2 + \hat{\sigma}_\beta^2 + \hat{\sigma}_{\alpha\beta}^2 + \hat{\sigma}_e^2}$$

All we have to do is to know how to calculate the variance components ($\hat{\sigma}_{\text{effect}}^2$).

Table 13.9 contains the variance components for fixed and random variables for the one-way, two-way, and three-way factorial, where the subscripts in the leftmost column stand for fixed (f) or random (r) variables. (This table is based on the work of Vaughan & Corballis [1969]. Anyone who tries to memorize this table is out of his or her mind—it is here for reference purposes, and I trust that you will have occasion to refer to it later.) You simply calculate each of these terms as given, and then form the appropriate ratio. This procedure is illustrated using the summary table from the study of cognitive processing by Eysenck (Table 13.2). Although in Eysenck's study, both of the variables were fixed (for good reason), I will base my calculations on the assumption that Condition was fixed but that Age was random.[3] I do this just to make a more useful example.

The summary table for Eysenck's study is reproduced here for convenience:

Source	df	SS	MS	F
A (Age)	1	240.25	240.250	29.94*
C (Condition)	4	1514.94	378.735	47.19*
AC	4	190.30	47.575	5.93*
Error	90	722.30	8.026	
Total	99	2667.79		

*$p < .05$

---

[3] Some authors do as I do and use ω^2 for effects of both random and fixed factors. Others use ω^2 to refer to effects of fixed factors and ρ^2 (the squared intraclass correlation coefficient) to refer to effects of random factors.

Table 13.9 Estimates of variance components in one-way, two-way, and three-way designs

Model	Variance Component
A_f	$\hat{\sigma}_\alpha^2 = (a-1)(MS_A - MS_e)/na$
	$\hat{\sigma}_e^2 = MS_e$
A_r	$\hat{\sigma}_\alpha^2 = (MS_A - MS_e)/n$
	$\hat{\sigma}_e^2 = MS_e$
A_fB_f	$\hat{\sigma}_\alpha^2 = (a-1)(MS_A - MS_e)/nab$
	$\hat{\sigma}_\beta^2 = (b-1)(MS_B - MS_e)/nab$
	$\hat{\sigma}_{\alpha\beta}^2 = (a-1)(b-1)(MS_{AB} - MS_e)/nab$
	$\hat{\sigma}_e^2 = MS_e$
A_rB_f	$\hat{\sigma}_\alpha^2 = (MS_A - MS_e)/nb$
	$\hat{\sigma}_\beta^2 = (b-1)(MS_B - MS_{AB})/nab$
	$\hat{\sigma}_{\alpha\beta}^2 = (b-1)(MS_{AB} - MS_e)/nb$
	$\hat{\sigma}_e^2 = MS_e$
A_rB_r	$\hat{\sigma}_\alpha^2 = (MS_A - MS_{AB})/nb$
	$\hat{\sigma}_\beta^2 = (MS_B - MS_{AB})/na$
	$\hat{\sigma}_{\alpha\beta}^2 = (MS_{AB} - MS_e)/n$
	$\hat{\sigma}_e^2 = MS_e$
$A_fB_fC_f$	$\hat{\sigma}_\alpha^2 = (a-1)(MS_A - MS_e)/nabc$
	$\hat{\sigma}_\beta^2 = (b-1)(MS_B - MS_e)/nabc$
	$\hat{\sigma}_\gamma^2 = (c-1)(MS_C - MS_e)/nabc$
	$\hat{\sigma}_{\alpha\beta}^2 = (a-1)(b-1)(MS_{AB} - MS_e)/nabc$
	$\hat{\sigma}_{\alpha\gamma}^2 = (a-1)(c-1)(MS_{AC} - MS_e)/nabc$
	$\hat{\sigma}_{\beta\gamma}^2 = (b-1)(c-1)(MS_{BC} - MS_e)/nabc$
	$\hat{\sigma}_{\alpha\beta\gamma}^2 = (a-1)(b-1)(c-1)(MS_{ABC} - MS_e)/nabc$
	$\hat{\sigma}_e^2 = MS_e$
$A_rB_fC_f$	$\hat{\sigma}_\alpha^2 = (MS_A - MS_e)/nbc$
	$\hat{\sigma}_\beta^2 = (b-1)(MS_B - MS_{AB})/nabc$
	$\hat{\sigma}_\gamma^2 = (c-1)(MS_C - MS_{AC})/nabc$
	$\hat{\sigma}_{\alpha\beta}^2 = (b-1)(MS_{AB} - MS_e)/nbc$
	$\hat{\sigma}_{\alpha\gamma}^2 = (c-1)(MS_{AC} - MS_e)/nbc$
	$\hat{\sigma}_{\beta\gamma}^2 = (b-1)(c-1)(MS_{BC} - MS_{ABC})/nabc$
	$\hat{\sigma}_{\alpha\beta\gamma}^2 = (MS_{ABC} - MS_e)/n$
	$\hat{\sigma}_e^2 = MS_e$

(continued)

Table 13.9 *(continued)*

Model	Variance Component
$A_r B_r C_f$	$\hat{\sigma}_{\alpha}^2 = (MS_A - MS_{AB})/nbc$
	$\hat{\sigma}_{\beta}^2 = (MS_B - MS_{AB})/nac$
	$\hat{\sigma}_{\gamma}^2 = (c-1)(MS_C - MS_{AC} - MS_{BC} + MS_{ABC})/nabc$
	$\hat{\sigma}_{\alpha\beta}^2 = (MS_{AB} - MS_e)/nc$
	$\hat{\sigma}_{\alpha\gamma}^2 = (MS_{AC} - MS_{ABC})/nb$
	$\hat{\sigma}_{\beta\gamma}^2 = (MS_{BC} - MS_{ABC})/na$
	$\hat{\sigma}_{\alpha\beta\gamma}^2 = (MS_{ABC} - MS_e)/n$
	$\hat{\sigma}_{e}^2 = MS_e$
$A_r B_r C_r$	$\hat{\sigma}_{\alpha}^2 = (MS_A - MS_{AB} - MS_{AC} + MS_{ABC})/nbc$
	$\hat{\sigma}_{\beta}^2 = (MS_B - MS_{AB} - MS_{BC} + MS_{ABC})/nac$
	$\hat{\sigma}_{\gamma}^2 = (MS_C - MS_{AC} - MS_{BC} + MS_{ABC})/nab$
	$\hat{\sigma}_{\alpha\beta}^2 = (MS_{AB} - MS_{ABC})/nc$
	$\hat{\sigma}_{\alpha\gamma}^2 = (MS_{AC} - MS_{ABC})/nb$
	$\hat{\sigma}_{\beta\gamma}^2 = (MS_{BC} - MS_{ABC})/na$
	$\hat{\sigma}_{\alpha\beta\gamma}^2 = (MS_{ABC} - MS_e)/n$
	$\hat{\sigma}_{e}^2 = MS_e$

If we let α represent the random effect of Age and β represent the fixed effect of Condition, then we have (using the formulae in Table 13.9)

$$\hat{\sigma}_{\alpha}^2 = (MS_A - MS_{\text{error}})/nb$$
$$= (240.250 - 8.026)/(10)(5) = 4.6445$$
$$\hat{\sigma}_{\beta}^2 = (b-1)(MS_B - MS_{AB})/nab$$
$$= (5-1)(378.735 - 47.575)/(10)(2)(5) = 13.2464$$
$$\hat{\sigma}_{\alpha\beta}^2 = (b-1)(MS_{AB} - MS_{\text{error}})/nb$$
$$= 4(47.575 - 8.026)/(10)(5) = 3.1639$$
$$\hat{\sigma}_{e}^2 = MS_{\text{error}} = 8.026$$

Thus

$$\hat{\sigma}_{\text{total}}^2 = \hat{\sigma}_{\alpha}^2 + \hat{\sigma}_{\beta}^2 + \hat{\sigma}_{\alpha\beta}^2 + \hat{\sigma}_{e}^2$$
$$= 4.6445 + 13.2464 + 3.1639 + 8.026 = 29.0808$$

We can now estimate ω^2 for each effect:

$$\hat{\omega}_{\text{Age}}^2 = \frac{\hat{\sigma}_{\alpha}^2}{\hat{\sigma}_{\text{total}}^2} = \frac{4.6445}{29.0808} = 0.16$$

$$\hat{\omega}_{\text{Condition}}^2 = \frac{\hat{\sigma}_{\beta}^2}{\hat{\sigma}_{\text{total}}^2} = \frac{13.2464}{29.0808} = 0.46$$

$$\hat{\omega}_{\text{Age} \times \text{Condition}}^2 = \frac{\hat{\sigma}_{\alpha\beta}^2}{\hat{\sigma}_{\text{total}}^2} = \frac{3.1639}{29.0808} = 0.11$$

As I said, I assumed that Age was a random variable in the previous calculations just to show how they are carried out. Actually Age was fixed at Older and Younger. If we treat both Age and Condition as fixed variables, the resulting measures of association are

$$\hat{\omega}_A^2 = 0.087$$

$$\hat{\omega}_C^2 = 0.554$$

$$\hat{\omega}_{AC}^2 = 0.059$$

Notice that these values differ considerably from what we obtained with a mixed model.

Partial Effects

Both η^2 and ω^2 represent the size of an effect (SS_{effect}) relative to the total variability in the experiment (SS_{total}). Often it makes more sense just to consider one factor separately from the others. For example, in the Spilich et al (1992) study of the effects of smoking under different kinds of tasks, the task differences were huge and of relatively limited interest in themselves. If we want a measure of the effect of smoking, we probably don't want to dilute that measure with irrelevant variance. Thus, we might want to estimate the effect of smoking relative to a total variability based only on smoking and error. This can be written

$$\text{partial } \omega^2 = \frac{\hat{\sigma}_{\text{effect}}^2}{\hat{\sigma}_{\text{effect}}^2 + \hat{\sigma}_e^2}$$

partial effect

We then simply calculate the necessary terms and divide. For example, in the case of the **partial effect** of the smoking by task interaction, treating both variables as fixed, we would have

$$\hat{\sigma}_{SxT}^2 = (s - 1)(t - 1)(MS_{ST} - MS_e)/nst$$

$$= (3 - 1)(3 - 1)(682 - 108)/(15)(3)(3) = \frac{5166}{135} = 38.26$$

$$\hat{\sigma}_e = MS_{\text{error}} = 108$$

$$\omega_{ST(\text{partial})}^2 = \frac{\hat{\sigma}_{ST}}{\hat{\sigma}_{ST} + \hat{\sigma}_{\text{error}}} = \frac{38.26}{38.26 + 108} = 0.26$$

This is a reasonable sized effect.

d-Family Measures

The *r*-family measures (η^2 and ω^2) make some sense when we are speaking about an omnibus *F* test involving several levels of one of the independent variables, but when we are looking closely at differences among individual groups or sets of groups, the *d*-family of measures often is more useful and interpretable. Effect sizes (*d*) are a bit more complicated when it comes to factorial experiments, primarily because you have to decide what to consider "error." They also becomes more complicated when we have unequal sample sizes **unbalanced** (called an "**unbalanced design**"). In this chapter, we will deal only with estimation with **design** balanced, or nearly balanced, designs. The reader is referred to Kline (2004) for a more thorough discussion of these issues.

As was the case with *t* tests and the one-way analysis of variance, we will define our effect size as

$$\hat{d} = \frac{\hat{\Psi}}{\hat{s}}$$

where the "hats" indicate that we are using estimates based on sample data. There is no real difficulty in estimating ψ because it is just a linear contrast. You will see an example in a minute in case you have forgotten what that is, but it is really just a difference between means of two groups or sets of groups. On the other hand, our estimate of the appropriate standard deviation will depend on our variables. Some variables normally vary in the population (e.g., amount of caffeine a person drinks in a day) and are, at least potentially, what Glass, McGraw, and Smith (1981) call a "variable of theoretical interest." Gender, extraversion, metabolic rate, and hours of sleep are other examples. On the other hand, many experimental variables, such as the number of presentations of a stimulus, area of cranial stimulation, size of a test stimulus, and presence or absence of a cue during recall do not normally vary in the population, and are of less theoretical interest. I am very aware that the distinction is a slippery one, and if a manipulated variable is not of theoretical interest, why are we manipulating it?

It might make more sense if we look at the problem slightly differently. Suppose that I ran a study to investigate differences among three kinds of psychotherapy. If I just ran that as a one-way design, my error term would include variability due to all sorts of things, one of which would be variability between men and women in how they respond to different kinds of therapy. Now suppose that I ran the same study but included gender as an independent variable. In effect, I am controlling for gender, and MS_{error} would not include gender differences because I have "pulled them out" in my analysis. So MS_{error} would be smaller here than in the one-way. That's a good thing in terms of power, but it may not be a good thing if I use the square root of MS_{error} in calculating the effect size. If I did, I would have a different sized effect due to psychotherapy in the one-way experiment than I have in the factorial experiment. That doesn't seem right. The effect of therapy ought to be pretty much the same in the two cases. So what I will do instead is to put that gender variability, and the interaction of gender with therapy, back into error when it comes to computing an effect size.

But suppose that I ran a slightly different study where I examined the same three different therapies, but also included, as a second independent variable, whether or not the patient sat in a cold tub of water during therapy. Now patients don't normally sit in a cold tub of water, but it would certainly be likely to add variability to the results. That variability would not be there in the one-way design because we can't imagine some patients bringing in a tub of water and sitting in it. And it is variability that I wouldn't want to add back into the error term because it is in some way artificial. The point is that I would like the effect size for types of therapy to be the same whether I used a one-way or a factorial design. To accomplish that, I would add effects due to Gender and the Gender X Therapy interaction back into the error term in the first study, and withhold the effects of Water and its interaction with Therapy in the second example. What follows is an attempt to do that. The interested reader is referred to Glass et al. (1981) for further discussion.

We will return to working with the example from Eysenck's (1974) study. The means and the analysis of variance summary table are presented here for easy reference.

	Counting	Rhyming	Adjective	Imagery	Intention	Mean
Older	7.0	6.9	11.0	13.4	12.0	10.06
Younger	6.5	7.6	14.8	17.6	19.3	13.16
Mean	6.75	7.25	12.90	15.50	15.65	11.61

Source	df	SS	MS	F
A (Age)	1	240.25	240.25	29.94*
C(Condition	4	1514.94	378.735	47.19*
AC	4	190.30	47.575	5.93*
Error	90	722.30	8.026	
Total	99	2667.79		

*$p < .05$

One of the questions that would interest me is the contrast between the two lower levels of processing (Counting and Rhyming) and the two high levels (Adjective and Imagery). I don't have any particular thoughts about the Intentional group, so we will ignore that. My coefficients, then, are

Counting	Rhyming	Adjective	Imagery	Intention
$-\frac{1}{2}$	$-\frac{1}{2}$	$\frac{1}{2}$	$\frac{1}{2}$	0

$$\hat{\psi} = \left(-\frac{1}{2}\right)(6.75) + \left(-\frac{1}{2}\right)(7.25) + \left(\frac{1}{2}\right)(12.90) + \left(\frac{1}{2}\right)(15.50) + (0)(11.61) = 7.20$$

The test on this contrast is

$$t = \frac{\hat{\psi}}{\sqrt{\dfrac{\left(\sum a_i^2\right) MS_{\text{error}}}{n}}} = \frac{7.20}{\sqrt{\dfrac{(1)(8.026)}{10}}} = \frac{7.20}{0.896} = 8.04$$

This t is clearly significant, showing that higher levels of processing lead to greater levels of recall. But I want an effect size for this difference.

I am looking for an effect size on a difference between two sets of conditions, but I need to consider the error term. Age is a normal variable in our world, and it leads to variability in people's responses. (If I had just designed this experiment as a one-way on Conditions, and ignored the age of my participants, that age variability would have been a normal part of MS_{error}). I need to have any Age effects contributing to error when it comes to calculating an effect size. So I will add SS_{age} and $SS_{A \times C}$ back into the error.

$$s_{\text{error}} = \sqrt{\frac{SS_{\text{error}} + SS_{\text{Age}} + SS_{A \times C}}{df_{\text{error}} + df_{\text{Age}} + df_{A \times C}}} = \sqrt{\frac{722.30 + 240.25 + 190.30}{90 + 1 + 4}}$$

$$= \sqrt{\frac{1152.85}{95}} = \sqrt{12.135} = 3.48$$

Having computed our error term for this effect, we find

$$\hat{d} = \frac{\hat{\psi}}{\hat{s}} = \frac{7.20}{3.48} = 2.07$$

The difference between recall with high levels of processing and recall with low levels of processing is about two standard deviations, which is a considerable difference. Thinking about material certainly helps you to recall it.

Now suppose that you wanted to look at the effects of Age. Because we can guess that people vary in the levels of processing that they normally bring to a memory task, then we should add the main effect of Condition and its interaction with Age to the error term in calculating the effect size. Thus,

$$s_{\text{error}} = \sqrt{\frac{SS_{\text{error}} + SS_{\text{Condition}} + SS_{A \times C}}{df_{\text{error}} + df_{\text{Condition}} + df_{A \times C}}} = \sqrt{\frac{722.30 + 1514.94 + 190.30}{90 + 4 + 4}}$$

$$= \sqrt{\frac{2427.54}{98}} = \sqrt{24.77} = 4.98$$

Because we only have two ages, the contrast (Ψ) is just the difference between the two means, which is $(13.16 - 10.06) = 3.10$.

$$\hat{d} = \frac{\hat{\psi}}{\hat{s}} = \frac{3.10}{4.98} = 0.62$$

In this case, younger subjects differ from older participants by nearly two-thirds of a standard deviation.

Simple Effects

The effect sizes for simple effects are calculated in ways directly derived from the way we calculate main effects. The error term in these calculations is the same error term as that used for the corresponding main effect. Thus, for the simple effect of Age at highest level of processing (Imagery) is

$$\hat{d} = \frac{\hat{\Psi}}{\hat{s}} = \frac{(17.6 - 13.4)}{4.98} = \frac{4.20}{4.98} = 0.84$$

Similarly, for the contrast of low levels of processing versus high levels among young participants, we would have

$$\psi = \left(-\frac{1}{2}\right)(6.5) + \left(-\frac{1}{2}\right)(7.6) + \left(\frac{1}{2}\right)(14.8) + \left(\frac{1}{2}\right)(17.6) + (0)(19.3) = 9.15$$

and the effect size is

$$\hat{d} = \frac{\hat{\psi}}{\hat{\sigma}} = \frac{9.15}{3.48} = 2.63$$

which means that for younger participants, there is nearly a 2⅔ standard deviation difference in recall between the high and low levels of processing.

13.10   Reporting the Results

We have carried out a number of calculations to make various points, and I would certainly not report all them when writing up the results. What follows is the basic information that I think needs to be presented:

In an investigation of the effects of different levels of information processing on the retention of verbal material, participants were instructed to process verbal material in one of four ways, ranging from the simple counting of letters in words to forming a visual image of each word. Participants in a fifth condition were not given any instructions about what to do with the items. A second dimension of the experiment compared Younger and Older participants on recall, thus forming a 2×5 factorial design.

The dependent variable was the number of items recalled after three presentations. There was a significant Age effect ($F(1, 90) = 29.94, p < .05, \omega^2 = .087$), with younger participants recalling more items than did older participants. There was also a significant effect due to Condition ($F(4, 90) = 47.19, p < .05, \omega^2 = .554$), and visual inspection of the means shows that there was greater recall for conditions in which there was a greater degree of processing. Finally, the Age by Condition interaction was significant ($F(4, 90) = 5.93, p < .05, \omega^2 = .059$), with a stronger effect of Condition for the younger participants.

A contrast of lower levels of processing (Counting and Rhyming) with higher levels of processing (Adjective and Imagery) produced a clearly statistically significant effect in favor of higher levels of processing ($t(90) = 8.04, p < .05$). This corresponds to an effect size of $\hat{d} = 2.07$, indicating that participants with higher levels of processing outperform those with lower levels of processing by more than two standard deviations. This effect is even greater if we look only at the younger participants, where $\hat{d} = 2.63$.

13.11   Unequal Sample Sizes

Although many (but certainly not all) experiments are designed with the intention of having equal numbers of observations in each cell, the cruel hand of fate frequently intervenes to upset even the most carefully laid plans. Participants fail to arrive for testing, animals die, data are lost, apparatus fails, patients drop out of treatment, and so on. When such problems arise, we are faced with several alternative solutions, with the choice depending on the nature of the data and the reasons why data are missing.

When we have a plain one-way analysis of variance, the solution is simple and we have already seen how to carry that out. When we have more complex designs, the solution is not simple. With unequal sample sizes in factorial designs, the row, column, and interaction effects are no longer independent. This lack of independence produces difficulties in interpretation, and deciding on the best approach depends both on why the data are missing and how we conceive of our model.

A great deal has been written about the treatment of unequal sample sizes, and we won't see any true resolution of this issue for a long time. (That is partly because there is no single answer to the complex questions that arise.) However, some approaches seem more reasonable than others for the general case. Unfortunately, the most reasonable and the most common approach is available only using standard computer packages, and a discussion of what is going on will have to wait until Chapter 15. I will, however, discuss a pencil-and-paper solution that is a very good approximation to the solution that I prefer, and that approach will give you a good sense of what the preferred method is all about and what hypotheses it is testing.

The Problem

I think you can easily see what our problem is if we take a very simple 2×2 factorial where we know what is happening. Suppose that we propose testing vigilance on a simple driving task when participants are either sober or are under the influence of alcohol. The task involves using a driving simulator and having to respond when cars suddenly come out of driveways and when pedestrians suddenly step into the street. We would probably expect that sober drivers would make many fewer errors on this task than would drivers who had been plied with alcohol. We will have two investigators working together on this problem, one from Michigan and one from Arizona, and each of them will run half of the participants in their own facilities. We have absolutely no reason to believe that participants in Michigan are any different from participants in Arizona, nor do we have any reason to believe that there would be an interaction between State and Alcohol condition.

Suppose that we obtained the quite extreme data shown in Table 13.10 with unequal numbers of participants in the four cells. The dependent variable is the number of errors

Table 13.10   Illustration of the contaminating effects of unequal sample sizes

	Non-Drinking	Drinking	Row Means
Michigan	13 15 14 16 12	18 20 22 19 21 23 17 18 22 20	$\overline{X}_{1.} = 18.0$
	$\overline{X}_{11} = 14$	$\overline{X}_{12} = 20$	
Arizona	13 15 18 14 10 12 16 17 15 10 14	24 25 17 16 18	$\overline{X}_{2.} = 15.9$
	$\overline{X}_{21} = 14$	$\overline{X}_{22} = 20$	
Col Means	$\overline{X}_{.1} = 14$	$\overline{X}_{.2} = 20$	

each driver made in one half-hour session. From the cell means in this table, you can see that the data came out as expected. The Drinking participants made, on average, 6 more errors than the participants in the Non-Drinking condition, and they did so whether they came from Michigan or Arizona. Similarly, you can see that there are no differences between Michigan and Arizona participants, whether you look at the Drinking or the Non-Drinking column. So what's wrong with this picture?

Well, if you look at the column means you see what you expect, but if you look at the row means you find that the mean for Michigan is 18, whereas the mean for Arizona is only 15.9. It looks as if we have a difference between States, even after we went to such pains to make sure there wasn't one here. What you are seeing is really a Drinking effect disguised as a State effect. And that is allowed to happen only because you have unequal numbers of participants in the cells. Michigan's mean is relatively high because they have more Drinking participants, and Arizona's mean is relatively low because they have more Non-Drinking participants. Now I suppose that if we had used actual people off the street, and Michigan had more drunks, perhaps a higher mean for Michigan would make some sort of sense. But that isn't what we did, and we don't usually want State effects contaminated by Condition effects. So what do we do?

The most obvious thing to do would be to calculate row and column means *ignoring* the differing cell sizes. We could simply average cell means, paying no attention to how many participants are in each cell. If we did this, the means for both Michigan and Arizona would be $(14 + 20)/2 = 17$, and there would be no difference due to States. In the analysis that follows, you will see that I will compute row and column means exactly this way, and I will calculate an average cell size (as the harmonic mean $\overline{X}_h$ of the cell sizes) and use that average n as if every cell had that many participants. When we get to Chapter 15, we will see a better way of accomplishing this, though the answers will be slightly different. We will also see other ways of running the analysis if we want to make different assumptions about the role that cell sizes should play. In general, we do not want sample sizes to influence how we interpret differences between means, and we use techniques that minimize or eliminate their effect. This does not mean that we *always* want to rule out their influence, but only that we usually do. It is possible to think of experiments where the mean of all the observations in a row is more important than the average of the cell means, but you have to work pretty hard to come up with a meaningful example.

Proportional Cell Frequencies

Before I go further, I should point out that the approach that we took with the one-way design will work in one special case, but it will not work in general. As you should recall, in the one-way design with unequal sample sizes, we took each squared deviation of the group mean from the grand mean and multiplied it by its own sample size as we went along, rather than multiplying by a common value of n at the end. Suppose that we modified the previous example so that we had 36 students from Michigan. Suppose further that we put 24 in the Non-Drinking condition and 12 in the Drinking condition. Also suppose that we had 24 cases from Arizona, and put 16 in the Non-Drinking condition and 8 in the Drinking condition. Then for each state, the *proportion* of cases in the two conditions is always 2:1. A more general rule that says the same thing is that for each cell, $n_{ij} = n_i n_j / N$. When this

proportionality   situation holds we have **proportionality,** and you can multiply by the appropriate cell sizes as you go along without doing injustice to your analysis. But you can do that only in the proportional case. And you will still get the same answer you would have had if you used the procedure that I'm about to describe.

Unweighted-Means Solution

In the previous table, when I calculated the State and Condition means I simply added up the scores in each row or column and divided by the number of scores in that row or column.

But notice what this really entails (letting M, D, and ND stand for Michigan, Drinking, and Non-Drinking):

$$\overline{X}_{\text{Michigan}} = \frac{\sum X_{M,ND} + \sum X_{M,D}}{N_M}$$

$$= \frac{n_{M,ND}\overline{X}_{M,ND} + n_{M,D}\overline{X}_{M,D}}{N_M}$$

$$= \frac{5 \cdot 14 + 10 \cdot 20}{15} = 18.0$$

weighted means

unweighted means

Here we see that the individual contribution of the ND and D means is a function of the sample sizes, and thus, our value of 18 is a **weighted** combination of **means.** But when I simply averaged the means in each row or column, I did not weight each mean by n_{ij} but, rather, averaged them without weighting. This gave me a set of **unweighted means,** and those are what I usually care about.

When cell frequencies are not proportional, one appropriate method of analysis is the unweighted-means solution. As indicated, the term *unweighted means* is actually a misnomer because what we really have are *equally weighted means*. (Unfortunately, the term "equally weighted means" was appropriated years ago for a slightly different procedure, so we can't use that phrase here.) As we saw in the previous section, multiplying each squared deviation by its own sample size as we go along—that is, $\sum (n_{i.}(\overline{X}_{i.} - \overline{X}_{..})^2)$ —amounts to weighting each mean in proportion to its sample size. In this section, we will weight all means equally by using a form of average sample size (the harmonic mean of the n_{ij}).

We will take as our example data from a study by Klemchuk, Bond, and Howell (1990) on role-taking in children. In this study, children between the ages of 2 and 5 years were administered a battery of role-taking tasks. (For example, a stimulus card with a different picture on each side was placed upright between the experimenter and the participant, and the participant was asked to identify what he or she thought that the *experimenter* saw.) Participants were classified into a group who had had no previous daycare experience and a group who had had extensive daycare experience (children who had had intermediate levels of daycare were not involved in this analysis). Children were also sorted into two age groups (2 to 3 years and 4 to 5 years). The investigator's hypothesis was that children with daycare experience would perform better on role-taking tasks than would children without daycare experience because of the former group's greater opportunity for social development. It was expected that older children would outperform younger children and that there would be no Age by Daycare interaction. The dependent variable was a score on a role-taking factor, and it has no intuitive meaning other than that a higher score represents better performance. As you will see, the sample sizes are distinctly unequal. There is no logical reason why fewer older children had participated in some sort of daycare experiences—this is just the sort of result that we could happily have done without. With such unequal sample sizes, it is important to interpret the results with caution.

The data are presented in Table 13.11. Table 13.11a lists the raw data, the individual cell frequencies, the individual cell means, and the harmonic mean of the n_{ij} ($\overline{n}_h$). The harmonic mean of k observations ($X_1, \ldots, X_k$) is defined as

$$\overline{X}_h = \frac{k}{\dfrac{1}{X_1} + \dfrac{1}{X_2} + \dfrac{1}{X_3} + \cdots + \dfrac{1}{X_k}}$$

The row, column, and grand means are obtained simply by averaging the relevant cell means (e.g. $(-1.2089 + 0.750)/2 = -0.5669$.)

Table 13.11 Data from Klemchuk, Bond, and Howell (1990)

(a) Data:

	Younger	Older
No Daycare	−0.139	−0.167
	−2.002	−0.285
	−1.631	0.851
	−2.173	−0.397
	0.179	0.351
	−0.829	−0.240
	−1.503	0.160
	0.009	−0.535
	−1.934	−0.102
	−1.448	0.273
	−1.470	0.277
	−1.545	0.714
	−0.137	
	−2.302	
Daycare	−1.412	0.859
	−0.681	0.782
	0.638	0.851
	−0.222	−0.158
	0.668	
	−0.896	
	−0.464	
	−1.659	
	−2.096	
	0.493	

Cell, Row, and Column, and Sample Size means

	Young	Older	Mean
No Daycare	1.2089	0.0750	−0.5669
Daycare	−0.5631	0.5835	0.0102
Mean	−0.8860	0.3292	−0.2784

$$\overline{n}_h = \frac{4}{\dfrac{1}{14} + \dfrac{1}{12} + \dfrac{1}{10} + \dfrac{1}{4}} = 7.925$$

This means that there will be 2×7.925 participants in each row and column.

(b) Computations

$$SS_{\text{Daycare}} = an_h \sum (\overline{X}_{i.} - \overline{X}_{..})^2$$
$$= 2 \times 7.925[(-0.5669 - (-.2784))^2 + (0.0102 - (-.2784))^2]$$
$$= 2 \times 7.925 \times .0.1665 = 2.639$$

$$SS_{\text{Age}} = dn_h \sum (\overline{X}_{.j} - \overline{X}_{..})^2$$
$$= 2 \times 7.925[(-0.8860 - (-.2784))^2 + (0.3292 - (-.2784))^2]$$
$$= 2 \times 7.925 \times 0.7384 = 11.704$$

(continued)

Table 13.11 *(continued)*

$$SS_{cells} = n_h \sum (\overline{X}_{ij} - \overline{X}_{..})^2$$

$$= 7.925 \left[\begin{array}{l} (-1.2089 - (-.2784))^2 + (0.0750 - (-.2784))^2 \\ + (-0.5631 - (-.2784))^2 + (0.5835 - (-.2784))^2 \end{array} \right]$$

$$= 7.925 \times 1.8146 \times 14.3811$$

$$SS_{AD} = SS_{cells} - SS_D - SS_A = 14.381 - 2.639 - 11.704 = 0.038$$

$$SS_{error} = \sum SS_{ij} = 9.630 + 2.087 + 8.596 + 0.737 = 21.050$$

(c) Summary table

Source	df	SS	MS	F
A (Age)	1	11.704	11.704	20.02*
D (Daycare)	1	2.639	2.639	4.51*
AD	1	0.038	0.038	<1
Error	36	21.050	0.585	
Total	39			

*$p < .05$

Once we have calculated the harmonic mean of the sample sizes, we will now act as though every cell contained $\overline{n}_h$ observations. Thus, the number of participants per row will be $\overline{n}_h$ times the number of cells in that row, the number of participants per column will be $\overline{n}_h$ times the number of cells in that column, and the overall sample size will be $\overline{n}_h$ times the number of cells overall. The means and the unweighted row and column means are shown, along with the harmonic mean of the sample sizes.

From here on, calculating main effects and the interaction proceeds just as in the equal-*n* case, substituting the unweighted means as the row and column means and the harmonic mean of the sample sizes for *n*. The calculations are shown in Table 13.11b. Note that the error term (SS_{error}) is not obtained by subtraction; instead, we calculate $SS_{within cell}$ for each cell of the design (not shown) and then sum these terms to obtain the sum of squares due to error. An alternative approach is to calculate MS_{error} directly by taking a weighted average of the individual cell variances—that is, we weight each cell variance by its *df*, just as we did when we calculated a pooled variance estimate in Chapter 7.

The summary table appears in Table 13.11c, where it is apparent that the two main effects are significant. Thus, role-taking ability increases with age (hardly a surprising finding) and with daycare experience (an important finding). There is no interaction, indicating that the difference between Daycare and No-Daycare children is the same for each age group.

Note that the summary table does not contain SS_{total}. The separate sums of squares do not usually sum to SS_{total} in the case of unequal sample sizes. I will expand on this point considerably in Chapter 16, but for now I will simply state that when the sample sizes are unequal, the main effects and interaction will account for overlapping portions of the overall variation. If we were to add them together, we would be summing some parts of the variation two or more times.

The point of this example is to show that with the unweighted-means solution, each cell mean contributes equally to the calculation of each of the sums of squares. We do not give greater weight to cells (or rows or columns) with larger sample sizes. The actual calculations are given to allow you to see how the harmonic mean is used, and how the cells receive equal weight. This should help to make clear what we will do in Chapter 16.

13.12   Analysis for Unequal Sample Sizes Using SAS

We have seen how to run this analysis by hand, and now we can look at what you would find if you used a standard computer package to run the analysis. I have chosen to demonstrate SAS, rather than SPSS or Minitab, because it readily makes clear that there is more than one way to do the analysis, and the answers usually will differ if we have unequal ns. In Exhibit 13.2, you will see a SAS program and some of the results it produces. In the program itself, the notation Daycare | Age is a shorthand way to tell the program to fit a model that includes Daycare, Age, and their interaction. The LSMeans statement produces the unweighted means that we have discussed, although they are not shown here.

In the printout, you will note that the main effects and interaction are tested two different ways. One of these tests is labeled Type I SS, and the other is labeled Type III SS. I always tell my students that unless they have a specific reason to the contrary, they should always examine and interpret the Type III SS. (Actually, I usually end my Model statement with "/SS3" just to get rid of the Type I SS.) You will see that the Type III SS results are virtually identical to the ones you obtained by hand. Notice, however, that the Type I SS, which we are going to ignore, produced a completely different conclusion with respect to

```
Data Helen;
      Infile "Calvin:Statistics:DataSets-Other:Klemchuk.dat";
      Input Daycare Age RoleTake;
Run;
Proc GLM Data = Helen;
      Class Daycare Age;
      Model RoleTake = Daycare | Age;
      LSMeans Daycare | Age;
Run;
```

Dependent Variable: ROLETAKE

Source	DF	Sum of Squares	Mean Square	F Value	Pr > F
Model	3	15.72782907	5.24260969	8.97	0.0001
Error	36	21.04971683	0.58471436		
Corrected Total	39	36.77754590			

R-Square	C.V.	Root MSE	ROLETAKE Mean
0.427648	−158.2996	0.764666	−0.483050

General Linear Models Procedure

Dependent Variable: ROLETAKE

Source	DF	Type I SS	Mean Square	F Value	Pr > F
DAYCARE	1	1.31989852	1.31989852	2.26	0.1417
AGE	1	14.37056812	14.37056812	24.58	0.0001
DAYCARE*AGE	1	0.03736244	0.03736244	0.06	0.8019

Source	DF	Type III SS	Mean Square	F Value	Pr > F
DAYCARE	1	2.63980787	2.63980787	4.51	0.0405
AGE	1	11.70347659	11.70347659	20.02	0.0001
DAYCARE*AGE	1	0.03736244	0.03736244	0.06	0.8019

Exhibit 13.2   Analysis of Daycare data

the main effect of Daycare. (And they would produce another different result if we reversed the order of our main effects in the Model statement.)

13.13   Higher-Order Factorial Designs

All of the principles concerning a two-way factorial design apply equally well to a three-way or higher-order design. With one additional piece of information, you should have no difficulty running an analysis of variance on any factorial design imaginable, although the arithmetic becomes increasingly more tedious as variables are added. We will take a simple three-way factorial as an example because it is the easiest to use.

The only major way in which the three-way differs from the two-way is in the presence of more than one interaction term. To see this, we must first look at the underlying structural model for a factorial design with three variables:

$$X_{ijkl} = \mu + \alpha_i + \beta_j + \gamma_k + \alpha\beta_{ij} + \alpha\gamma_{ik} + \beta\gamma_{jk} + \alpha\beta\gamma_{ijk} + e_{ijkl}$$

first-order interactions

second-order interaction

In this model, we have main effects, symbolized by α_i, β_j, and γ_k, and two kinds of interaction terms. The two-variable or **first-order interactions** are $\alpha\beta_{ij}$, $\alpha\gamma_{ik}$, and $\beta\gamma_{jk}$, which refer to the interaction of variables A and B, A and C, and B and C, respectively. We also have a **second-order interaction** term, $\alpha\beta\gamma_{ijk}$, which refers to the joint effect of all three variables. We have already examined the first-order interactions in discussing the two-way. The second-order interaction can be viewed in several ways. Probably the easiest way to view the ABC interaction is to think of the AB interaction itself interacting with variable C. Suppose that we had two levels of each variable and plotted the AB interaction separately for each level of C. We might have the result shown in Figure 13.3. Notice that for C_1 we have one AB interaction, whereas for C_2 we have a different one. Thus, AB depends on C, producing an ABC interaction. This same kind of reasoning could be invoked using the AC interaction at different levels of B, or the BC interaction at different levels of A. The result would be the same.

As I have said, the three-way factorial is merely an extension of the two-way, with a slight twist. The twist comes about in obtaining the interaction sums of squares. In the two-way, we took an $A \times B$ table of cell means, calculated SS_{cells}, subtracted the main effects, and were left with SS_{AB}. In the three-way, we have several interactions, but we will calculate them using techniques analogous to those employed earlier. Thus, to obtain SS_{BC} we will take a $B \times C$ table of cell means (summing over A), obtain $SS_{\text{cells}\,BC}$, subtract the main effects of B and C, and end up with SS_{BC}. The same applies to SS_{AB} and SS_{AC}. We also follow the same procedure to obtain SS_{ABC}, but here we need to begin with an $A \times B \times C$ table of cell means, obtain $SS_{\text{cells}\,ABC}$, and then subtract the main effects *and* the lower-order interactions to arrive at SS_{ABC}. In other words, for each interaction, we start with a different table of cell means, summing over the variable(s) in which we are not at the moment interested. We then obtain an SS_{cells} for that table and subtract from it any main effects and lower-order interactions that involve terms included in that interaction.

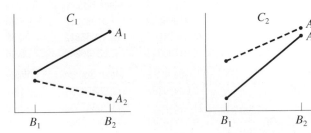

Figure 13.3   Plot of second-order interaction

Variables Affecting Driving Performance

For an example, consider a hypothetical experiment concerning the driving ability of two different types of drivers—inexperienced (A_1) and experienced (A_2). These people will drive on one of three types of roads—first class (B_1), second class (B_2), or dirt (B_3), under one of two different driving conditions—day (C_1) and night (C_2). Thus, we have a $2 \times 3 \times 2$ factorial. The experiment will include four participants per condition (for a total of 48 participants), and the dependent variable will be the number of steering corrections in a one-mile section of roadway. The raw data are presented in Table 13.12a.

Table 13.12 Illustration of calculations for $2 \times 3 \times 2$ factorial design

(a) Data

	C_1			C_2		
	B_1	B_2	B_3	B_1	B_2	B_3
A_1	4	23	16	21	25	32
	18	15	27	14	33	42
	8	21	23	19	30	46
	10	13	14	26	20	40
A_2	6	2	20	11	23	17
	4	6	15	7	14	16
	13	8	8	6	13	25
	7	12	17	16	12	12

Cell Means

	C_1			C_2			Means
	B_1	B_2	B_3	B_1	B_2	B_3	
A_1	10.000	18.000	20.000	20.000	27.000	40.000	22.500
A_2	7.500	7.000	15.000	10.000	15.500	17.500	12.083
Means	8.750	12.500	17.500	15.000	21.250	28.750	17.292

More Cell Means

	AB Cells					AC Cells		
	B_1	B_2	B_3	Means		C_1	C_2	Means
A_1	15.000	22.500	30.000	22.500	A_1	16.000	29.000	22.500
A_2	8.750	11.250	16.250	12.083	A_2	9.833	14.333	12.083
Means	11.875	16.875	23.125	17.292	Means	12.917	21.667	17.292

	BC Cells			
	B_1	B_2	B_3	Means
C_1	8.750	12.500	17.500	12.917
C_2	15.000	21.250	28.750	21.667
Means	11.875	16.875	23.125	17.292

(continued)

Table 13.12 *(continued)*

(b) Calculations

$$SS_{total} = \sum(X - \overline{X}_{...})^2 = (4 - 17.292)^2 + \cdots + (12 - 17.292)^2 = 4727.92$$

$$SS_A = nbc\sum(\overline{X}_{i..} - \overline{X}_{...})^2 = 4 \times 3 \times 2[(22.50 - 17.292)^2 + (12.083 - 17.292)^2]$$
$$= 1302.08$$

$$SS_B = nac\sum(\overline{X}_{.j.} - \overline{X}_{...})^2 = 4 \times 2 \times 2[(11.875 - 17.292)^2 + \cdots$$
$$+ (23.125 - 17.292)^2] = 1016.67$$

$$SS_C = nab\sum(\overline{X}_{..k} - \overline{X}_{...})^2 = 4 \times 2 \times 3[(12.917 - 17.292)^2 + (21.667 - 17.292)^2]$$
$$= 918.75$$

$$SS_{Cell\,AB} = nc\sum(\overline{X}_{ij.} - \overline{X}_{...})^2 = 4 \times 2[(15.00 - 17.292)^2 + \cdots + (16.25 - 17.292)^2]$$
$$= 2435.42$$

$$SS_{AB} = SS_{Cell\,AB} - SS_A - SS_B = 2435.42 - 1302.08 - 1016.67 = 116.67$$

$$SS_{Cell\,AC} = nb\sum(\overline{X}_{i.k} - \overline{X}_{...})^2 = 4 \times 3[(16.00 - 17.292)^2 + \cdots + (14.333 - 17.292)^2]$$
$$= 2437.58$$

$$SS_{AC} = SS_{Cell\,AC} - SS_A - SS_C = 2437.58 - 1302.08 - 918.75$$
$$= 216.75$$

$$SS_{Cell\,BC} = na\sum(\overline{X}_{.jk} - \overline{X}_{...})^2 = 4 \times 2\left[(8.75 - 17.292)^2 + ... + (28.75 - 17.292)^2\right]$$
$$= 1985.42$$

$$SS_{BC} = SS_{Cell\,BC} - SS_B - SS_C = 1985.42 - 1016.67 - 918.75$$
$$= 50.00$$

$$SS_{Cell\,ABC} = n\sum(\overline{X}_{ijk} - \overline{X}_{...})^2 = 4[(10.00 - 17.292)^2 + \cdots + (17.50 - 17.292)^2]$$
$$= 3766.92$$

$$SS_{ABC} = SS_{Cell\,ABC} - SS_A - SS_B - SS_C - SS_{AB} - SS_{AC} - SS_{BC}$$
$$= 3766.92 - 1302.08 - 1016.67 - 918.75 - 116.67 - 216.75 - 50.00$$
$$= 146.00$$

$$SS_{error} = SS_{total} - SS_{Cell\,ABC} = 4727.92 - 3766.92 = 961.00$$

(c) Summary table

Source	df	SS	MS	F
A (Experience)	1	1302.08	1302.08	48.78*
B (Road)	2	1016.67	508.33	19.04*
C (Conditions)	1	918.75	918.75	34.42*
AB	2	116.67	58.33	2.19
AC	1	216.75	216.75	8.12*
BC	2	50.00	25.00	<1
ABC	2	146.00	73.00	2.73
Error	36	961.00	26.69	
Total	47	4727.92		

*$p < .05$

The lower part of Table 13.12a contains all the necessary matrices of cell means for the subsequent calculation of the interaction sums of squares. These matrices are obtained simply by averaging across the levels of the irrelevant variable. Thus, the upper left cell of the AB summary table contains the sum of all scores obtained under the treatment combination AB_{11}, regardless of the level of C (i.e., $ABC_{111} + ABC_{112}$). (*Note:* You should be aware that I have rounded everything to two decimals for the tables, but the computations were based on more decimals. Beware of rounding error.[4])

Table 13.12b shows the calculations of the sums of squares. For the main effects, the sums of squares are obtained exactly as they would be for a one-way. For the first-order interactions, the calculations are just as they would be for a two-way, taking two variables at a time. The only new calculation is for the second-order interaction, and the difference is only a matter of degree. Here we first obtain the SS_{cells} for the three-dimensional matrix. This sum of squares represents all the variation among the cell means in the full-factorial design. From this, we must subtract all the variation that can be accounted for by the main effects *and* by the first-order interactions. What remains is the variation that can be accounted for by only the joint effect of all three variables, namely SS_{ABC}.

The final sum of squares is SS_{error}. This is most easily obtained by subtracting $SS_{cells\,ABC}$ from SS_{total}. Because $SS_{cells\,ABC}$ represents all the variation that can be attributable to differences among cells ($SS_{cells\,ABC} = SS_A + SS_B + SS_C + SS_{AB} + SS_{AC} + SS_{BC} + SS_{ABC}$), subtracting it from SS_{total} will leave us with only that variation within the cells themselves.

The summary table for the analysis of variance is presented in Table 13.12c. From this, we can see that the three main effects and the $A \times C$ interaction are significant. None of the other interactions is significant.[5]

Simple Effects

We have a significant interaction, so the main effects of A and C should be interpreted with caution. To this end, the AC interaction has been plotted in Figure 13.4. When plotted, the data show that for the inexperienced driver, night conditions produce considerably more steering corrections than do day conditions, whereas for the experienced driver, the difference in the number of corrections made under the two conditions is relatively slight. Although the data do give us confidence in reporting a significant effect for A (the difference between experienced and inexperienced drivers), they should leave us a bit suspicious about differences due to variable C. At a quick glance, it would appear that there is a significant C effect for the inexperienced drivers, but possibly not for the experienced drivers. To examine this question more closely, we must consider the simple effects of C under A_1 and A_2 separately. This analysis is presented in Table 13.13, from which we can see that there is a significant effect between day and night conditions, not only for the

---

[4] The fact that substantial rounding error accumulates when you work with means is one major reason why formulae intended for use with calculators worked with totals. I am using the definitional formulae in these chapters because they are clearer, but that means that we need to put up with occasional rounding errors. Good computing software uses very sophisticated formulae optimized to minimize rounding error.

[5] You will notice that this analysis of variance included seven F values and, thus, seven hypothesis tests. With so many hypothesis tests, the familywise error rate would be quite high. Most people ignore the problem and simply test each F at a per-comparison error rate of $\alpha = .05$. However, if you are concerned about error rates, it would be appropriate to employ the equivalent of either the Bonferroni or multistage Bonferroni t procedure. This is generally practical only when you have the probability associated with each F and can compare this probability against the probability required by the Bonferroni (or multistage Bonferroni) procedure. An interesting example of this kind of approach is found in Rosenthal and Rubin (1984). I suspect that most people will continue to evaluate each F on its own, and not worry about familywise error rates.

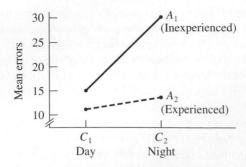

Figure 13.4   *AC* interaction for data in Table 13.12

Table 13.13   Simple effects for data in Table 13.12

(a) Data

	C_1	C_2	Mean
A_1	16.000	29.000	22.500
A_2	9.833	14.333	12.083

(b) Computations

$$SS_{C \text{ at } A_1} = nb\sum(\overline{X}_{1.k} - \overline{X}_{1..})^2$$
$$= 4 \times 3[(16.000 - 22.500)^2 + (29.000 - 22.500)^2] = 1014.00$$

$$SS_{C \text{ at } A_2} = nb\sum(\overline{X}_{2.k} - \overline{X}_{2..})^2$$
$$= 4 \times 3[(9.833 - 12.083)^2 + (14.333 - 12.083)^2] = 121.50$$

(c) Summary table

Source	df	SS	MS	F
C at A_1	1	1014.00	1014.00	37.99*
C at A_2	1	121.50	121.50	4.55*
Error	36	961.00	26.69	

*$p < .05$

(d) Decomposition of sums of squares

$$SS_{C \text{ at } A_1} + SS_{C \text{ at } A_2} = SS_C + SS_{AC}$$
$$1014.00 + 121.50 = 918.75 + 216.75$$
$$1135.50 = 1135.50$$

inexperienced drivers, but also for the experienced drivers. (Note that we can again check the accuracy of our calculations; the simple effects should sum to $SS_C + SS_{AC}$.)

From this hypothetical experiment, we would conclude that there are significant differences between the three types of roadway and between experienced and inexperienced drivers. We would also conclude that there is a significant difference between day and night conditions, for both experienced and inexperienced drivers.

Simple Interaction Effects

simple main effects

simple interaction effect

With the higher-order factorials, we can look at the effects of one variable at individual levels of some other variable (what we have called simple effects but what should more accurately be called **simple main effects**), and we can look at the interaction of two variables at individual levels of some third variable. We will refer to this as a **simple interaction effect.**

Although our second-order interaction (ABC) was not significant, you might have a theoretical reason to expect an interaction between Experience (A) and Road (B) under night conditions because driving at night is more difficult, but would expect no AB interaction during the day. As an example, I will break down the ABC interaction to get at those two simple interaction effects. (I should stress, however, that it is not good practice to test everything in sight just because it is possible to do so.)

In Figure 13.5, the AB interaction has been plotted separately for each level of C. It appears that there is no AB interaction under C_1, but there may be an interaction under C_2. We can test this hypothesis by calculating the AB interaction at each level of C, in a manner logically equivalent to the test we used for simple main effects. Essentially, all we need to do is treat the C_1 (day) and C_2 (night) data separately, calculating SS_{AB} for C_1 data and then for C_2 data. These simple interaction effects are then tested using MS_{error} from the overall analysis. This has been done in Table 13.14.

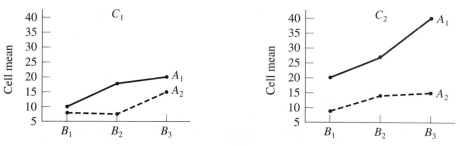

Figure 13.5 ABC interaction for data in Table 13.12

From the analysis of the simple interaction effects, it is apparent that the AB interaction is not significant for the day data, but it is for the night data. When night conditions (C_2) and dirt roads (B_3) occur together, differences between experienced (A_2) and inexperienced (A_1) drivers are magnified.

Although there is nothing to prevent someone from examining simple interaction effects in the absence of a significant higher-order interaction, cases for which this would make any logical sense are rare. If, however, the experimenter has a particular reason for looking at, for example, the AB interaction at each level of C, he is perfectly free to do so. On the other hand, if a higher-order interaction is significant, the experimenter should cast a wary eye on all lower-order effects and consider testing the important simple effects. However, to steal a line from Winer (1971, p. 442), "Statistical elegance does not necessarily imply scientifically meaningful inferences." Common sense is at least as important as statistical manipulations.

13.14 A Computer Example

The following example illustrates the analysis of a three-way factorial design with unequal numbers of participants in the different cells. It is roughly based on a study by Seligman, Nolen-Hoeksema, Thornton, and Thornton (1990), although the data are contrived and one of the independent variables (Event) is fictitious. The main conclusions of the example are in line with the results reported.

Table 13.14 Simple interaction effects for data in Table 13.12

(a) Data and Computations for $SS_{AB \text{ at } C_1}$

	C_1 Means			
	B_1	B_2	B_3	Mean
A_1	10.00	18.00	20.00	16.000
A_2	7.50	7.00	15.00	9.833
	8.75	12.50	17.50	12.917

$$SS_{A \text{ at } C_1} = nb\sum(\overline{X}_{i.1} - \overline{X}_{..1})^2$$
$$= 4 \times 3[(16.000 - 12.917)^2 + (9.833 - 12.917)^2] = 228.17$$

$$SS_{B \text{ at } C_1} = na\sum(\overline{X}_{.j1} - \overline{X}_{..1})^2$$
$$= 4 \times 2[(8.750 - 12.917)^2 + \cdots + (17.500 - 12.917)^2] = 308.33$$

$$SS_{\text{Cells } AB \text{ at } C_1} = n\sum(\overline{X}_{ij1} - \overline{X}_{..1})^2$$
$$= 4 \times [(10.000 - 12.917)^2 + \cdots + (15.000 - 12.917)^2] = 612.83$$

$$SS_{AB \text{ at } C_1} = SS_{\text{Cells } AB \text{ at } C_1} - SS_{A \text{ at } C_1} - SS_{B \text{ at } C_1}$$
$$= 612.83 - 228.17 - 308.33 = 76.33$$

(b) Data and Computations for $SS_{AB \text{ at } C_2}$

	C_2 Means			
	B_1	B_2	B_3	Mean
A_1	20.00	27.00	40.00	29.000
A_2	10.00	15.50	17.50	14.333
Mean	15.00	21.25	28.75	21.667

$$SS_{A \text{ at } C_2} = nb\sum(\overline{X}_{i.2} - \overline{X}_{..2})^2$$
$$= 4 \times 3[(29.00 - 21.667)^2 + (14.33 - 21.667)^2] = 1290.67$$

$$SS_{B \text{ at } C_2} = na\sum(\overline{X}_{.j2} - \overline{X}_{..2})^2$$
$$= 4 \times 2[(15.00 - 21.667)^2 + \cdots + (28.75 - 21.667)^2] = 758.33$$

$$SS_{\text{Cells } AB \text{ at } C_2} = n\sum(\overline{X}_{ij2} - \overline{X}_{..3})^2$$
$$= 4 \times [(20.00 - 21.667)^2 + \cdots + (15.00 - 21.667)^2] = 2235.33$$

$$SS_{AB \text{ at } C_2} = SS_{\text{Cells } BC \text{ at } C_2} - SS_{A \text{ at } C_2} - SS_{B \text{ at } C_2}$$
$$= 2235.33 - 1290.67 - 758.33 = 186.33$$

The study involved collegiate swimming teams. At a team practice, all participants were asked to swim their best event as fast as possible, but in each case the time that was reported was falsified to indicate poorer than expected performance. Thus, each swimmer was disappointed at receiving a poor result. Half an hour later, each swimmer was asked to perform the same event, and their times were again recorded. The authors predicted that on the second trial more pessimistic swimmers would do worse than on their first trial, whereas optimists would do better.

Participants were classified by their explanatory Style (optimism versus pessimism), Sex, and the preferred Event. The dependent variable was the ratio of $Time_2/Time_1$, so a value greater than 1.00 means that the swimmer did better on the second trial. The data and results are given in Table 13.15. The results were obtained using SPSS. (A short primer on SPSS can be found on the website [www.uvm.edu/~dhowell/methods/].)

A few lines of the preceding SPSS printout might cause confusion. The first line in the summary table is labeled Corrected Model, and it tests the null hypothesis that performance is a function of the full set of effects (the main effects and the interactions). If this example had equal sample sizes, SS_{model} would be exactly equal to the sum of the main and interaction

Table 13.15 Analysis of variance on responses to failure by optimists and pessimists

(a) Data

| | Optimists | | | | | | Pessimists | | | | | |
| | Male | | | Female | | | Male | | | Female | | |
	Free	Breast	Back	Free	Breast	Back	Free	Breast	Back	Free	Breast	Back
	0.986	1.026	1.009	1.108	1.048	1.004	0.983	0.962	0.936	0.997	1.045	1.045
	1.108	1.045	1.065	0.985	1.027	0.936	0.947	0.944	0.995	0.983	1.095	0.864
	1.080	0.996	1.053	1.001		1.040	0.932	0.941	0.872	1.105	0.944	0.982
	0.952	0.923		0.924			1.078	0.831		1.116	1.039	0.915
	0.998	1.000		0.968			0.914			0.997	0.927	1.047
	1.017	1.003					0.955			0.960	0.988	
	1.080	0.934									1.015	
$\overline{X}$	1.032	0.990	1.042	0.997	1.038	0.993	0.968	0.920	0.934	1.026	1.008	0.971

(b) Summary Table from SPSS

Tests of Between-Subjects Effects

Dependent Variable: PERFORM

Source	Type III Sum of Squares	df	Mean Square	F	Sig.
Corrected Model	6.804E-02[a]	11	6.186E-03	1.742	.094
Intercept	48.779	1	48.779	13738.573	.000
OPTIM	2.412E-02	1	2.412E-02	6.793	.012
SEX	7.427E-03	1	7.427E-03	2.092	.155
STROKE	4.697E-03	2	2.348E-03	.661	.521
OPTIM * SEX	1.631E-02	1	1.631E-02	4.594	.037
OPTIM * STROKE	5.612E-03	2	2.806E-03	.790	.460
SEX * STROKE	1.142E-02	2	5.708E-03	1.608	.211
OPTIM * SEX * STROKE	1.716E-03	2	8.578E-04	.242	.786
Error	.163	46	3.550E-03		
Total	57.573	58			
Corrected Total	.231	57			

[a.] R Squared D .294 (Adjusted R Squared D .125)

(continued)

Table 13.15   *(continued)*

(c) Plot by Sex × Optim interaction

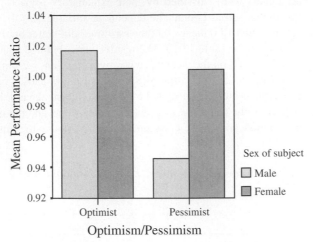

effects. Because the sample sizes are unequal, the effects are not independent and will not sum to SS_{model}. The effect labeled "intercept" is a test on the null hypothesis that the grand mean is equal to 0. We rarely care about such a test, and most software doesn't even perform that test. The line labeled "total" is $\sum X^2$. This is also not a line that we care very much about.

From the SPSS computer output, you can see that there is a significant effect due to the attributional style, with Optimists showing slightly improved performance after a perceived failure, and pessimists doing worse. The difference in means may appear to be small, but when you consider how close a race of this type usually is, even a tiny difference is important. You can also see that there is an Optim × Sex interaction. Looking at the means, we see that there is almost no difference between Optimistic males and females, but this is not true of pessimists. Pessimistic males appear in these data to be much more affected by a perceived loss than are females. This Optim × Sex interaction is plotted following the summary table. This plot has collapsed across Event, because that variable had no effect.[6]

---

[6] To be fair to Seligman et al. (1990), I should say that this is not a result they appeared to have analyzed for, and therefore not one they found. I built it in to illustrate a point.

Key Terms

Exercises

The following problems can all be solved by hand, but any of the standard computer software packages will produce the same results.

13.1 In a study of mother-infant interaction, mothers are rated by trained observers on the quality of their interactions with their infants. Mothers are classified on the basis of whether or not this was their first child (primiparous versus multiparous) and on the basis of whether this was a low-birthweight (LBW) infant or normal-birthweight (NBW) infant. Mothers of LBW infants were further classified on the basis of whether or not they were under 18 years old. The data represent a score on a 12-point scale; a higher score represents better mother-infant interaction. Run and interpret the appropriate analysis of variance.

Primiparous			Multiparous			Primiparous			Multiparous		
LBW <18	LBW >18	NBW	LBW <18	LBW >18	NBW	LBW <18	LBW >18	NBW	LBW <18	LBW >18	NBW
4	6	8	3	7	9	7	6	2	7	2	10
6	5	7	4	8	8	4	2	5	1	1	9
5	5	7	3	8	9	5	6	8	4	9	8
3	4	6	3	9	9	4	5	7	4	9	7
3	9	7	6	8	3	4	5	7	4	8	10

13.2 In Exercise 13.1, the design may have a major weakness from a practical point of view. Notice the group of multiparous mothers under 18 years of age. Without regard to the data, would you expect this group to lie on the same continuum as the others?

13.3 Refer to Exercise 13.1. It seems obvious that the sample sizes do not reflect the relative frequency of age and parity characteristics in the population. Under what conditions would this be a relevant consideration, and under what conditions would it not be?

13.4 Use simple effects to compare the three groups of multiparous mothers in Exercise 13.1.

13.5 In a study of memory processes, animals were tested in a one-trial avoidance-learning task. The animals were presented with a fear-producing stimulus on the learning trial as soon as they stepped across a line in the test chamber. The dependent variable was the time it took them to step across the line on the test trial. Three groups of animals differed in the areas in which they had electrodes implanted in their cortex (Neutral site, Area A, or Area B). Each group was further divided and given electrical stimulation either 50, 100, or 150 milliseconds after crossing the line and being presented with the fear-inducing stimulus. If the brain area that was stimulated is involved in memory, stimulation would be expected to interfere with memory consolidation and retard learning of the avoidance response, and the animal should not show any hesitancy in recrossing the line. The data on latency to recross the line are as follows:

	Stimulation Area							
Neutral Site			Area A			Area B		
50	100	150	50	100	150	50	100	150
25	30	28	11	31	23	23	18	28
30	25	31	18	20	28	30	24	21
28	27	26	26	22	35	18	9	30
40	35	20	15	23	27	28	16	30
20	23	35	14	19	21	23	13	23

Run the analysis of variance.

13.6   Plot the cell means in Exercise 13.5.

13.7   For the study in Exercise 13.5, to what would α_1 refer (if A were used to represent Area)?

13.8   Use simple effects to clarify the results for the Area factor in Exercise 13.5. Show that these simple effects sum to the correct figure.

13.9   Use the Bonferroni test to compare the neutral site with each of the other areas in Exercise 13.5, ignoring the length of stimulation. (*Hint:* Follow the procedures outlined in Chapters 11 and 12, but be sure that you take n as the number of scores on which the mean is based.)

13.10 Use simple effects to examine the effect of delay of stimulation in area A for the data in Exercise 13.5.

13.11 Refer to Exercise 11.3a. You will see that it forms a 2×2 factorial. Run the factorial analysis and interpret the results.

13.12 In Exercise 11.3, you ran a test between Groups 1 and 3 combined versus Groups 2 and 4 combined. How does that compare with testing the main effect of level of processing in Exercise 13.11? Is there any difference?

13.13 Make up a set of data for a 2×2 design that has two main effects but no interaction.

13.14 Make up a set of data for a 2×2 design that has no main effects but does have an interaction.

13.15 Describe a reasonable experiment for which the primary interest would be in the interaction effect.

13.16 Assume that in Exercise 13.1 the last three participants in cell$_{12}$ (Primiparous, LBW > 18) and the last two participants in cell$_{23}$ (Multiparous, NBW) refused to give consent for their data to be used. Rerun the analysis.

13.17 An experimenter was interested in hospital patients' responses to two different forms of physical therapy. For her own reasons, she thought that it was important to have greater precision in estimates of means for Treatment B (we need not be concerned with why), and therefore she allocated two-thirds of the participants to Treatment B and one-third to Treatment A. She ran the study at two different hospitals, one of which had more patients available. The data are given in terms of ratings of recovery of function:

	Treatment	
	A	B
Hospital 1	5	10
	8	12
	6	14
		12
		10
		8
Hospital 2	10	15
	12	28
		32
		34

Run the analysis of variance using an unweighted means approach.

13.18 Use any standard computer software to analyze the data in Exercise 13.17. Compare your results with those you obtained previously.

13.19 Calculate η^2 and $\hat{\omega}^2$ for Exercise 13.1

13.20 Calculate $\hat{d}$ for the data in Exercise 13.1

13.21 Calculate η^2 and $\hat{\omega}^2$ for Exercise 13.5.

13.22 Calculate $\hat{d}$ for the data in Exercise 13.5.

13.23 To study the effects of early experience on conditioning, an experimenter raised four groups of rats in the presence of (1) no special stimuli, (2) a tone stimulus, (3) a vibratory stimulus, and (4) both a tone and a vibratory stimulus. The rats were later classically conditioned using either a tone or a vibratory stimulus as the conditioned stimulus and one of three levels of foot shock as the unconditioned stimulus. This is a $4 \times 2 \times 3$ factorial design. The cell means, rather than the raw data, follow. The $SS_{total} = 41,151.00$ and $n_{ijk} = 5$. The dependent variable was the number of trials to a predetermined criterion.

| | Conditioned Stimulus | | | | | |
| | Tone | | | Vibration | | |
	High	Med	Low	High	Med	Low
Control	11	16	21	19	24	29
Tone	25	28	34	21	26	31
Vibration	6	13	20	40	41	52
Tone and Vibration	22	30	30	35	38	48

Analyze the data and interpret the results.

13.24 In Chapter 2, I wrote at length about Sternberg's experiment on the time it takes to report whether a test stimulus was part of a prior stimulus display. The independent variables were the number of stimuli in the display (1, 3, or 5) and whether the test stimulus had been included in the display (Yes or No). The data are found in RxTime.dat on the website (www .uvm.edu/~dhowell/methods/). This is a two-way analysis of variance. Run the analysis and interpret the results, including mention and interpretation of effect sizes.

13.25 Use any statistical package to run the two-way analysis of variance on Interval and Dosage for the data in Epineq.dat. Compare the results you obtain here with the results you obtained in Chapter 11, Exercises 11.28–11.30.

13.26 In Exercise 11.30, you calculated the average of the nine cell variances. How does that answer compare to the MS_{error} from Exercise 13.25?

13.27 Obtain the Tukey test for Dosage from the analysis of variance in Exercise 13.25. Interpret the results.

13.28 The data for the three-way analysis of variance given in Table 13.12 are found on the website. They are named Tab13–12.dat. The first three entries in each record represent the coding for A (Experience), B (Road), and C (Conditions). The fourth entry is the dependent variable. Use any analysis of variance package to reproduce the summary table found in Table 13.12c.

13.29 Using the data from Exercise 13.28, reproduce the simple effects shown in Table 13.13.

Discussion Questions

13.30 In the analysis of Seligman et al. (1990) data on explanatory style (Table 13.15) you will note that there are somewhat more males than females in the Optimist group and more females than males in the Pessimist group. Under what conditions might this affect the way you would want to deal with unequal sample sizes, and when might you want to ignore it?

13.31 Think of a nonstatistical example that could be used to explain to a student in an introductory statistics course why it is possible with any of the pairwise comparison tests to show that Group 1 is not different from Group 2, Group 2 is not different from Group 3, but Group 1 is different from Group 3.

13.32 Find an example of a three-way factorial in the research literature in which at least one of the interactions is significant and meaningful. Then create a data set that mirrors those results.

13.33 Write up the results of the analyses in Exercises 13.28 and 13.29 as you would in a research report.

CHAPTER 14

Repeated-Measures Designs

Objectives

To discuss the analysis of variance by considering experimental designs
in which the same subject is measured under all levels of one or more
independent variables.

Contents

IN OUR DISCUSSION OF THE ANALYSIS OF VARIANCE, we have concerned ourselves with experimental designs that have different subjects in the different cells. More precisely, we have been concerned with designs in which the cells are independent, or uncorrelated. (Under the assumptions of the analysis of variance, *independent* and *uncorrelated* are synonymous in this context.) In this chapter, we are going to be concerned with the problem of analyzing data where some or all of the cells are not independent. Such designs are somewhat more complicated to analyze, and the formulae become more complex. Most, or perhaps even all, readers will approach the problem using computer software such as SPSS, Minitab, or SAS. However, to understand what you are seeing, you need to know something about how you would approach the problem by hand, and that leads to lots and lots of formulae. I urge you to treat the formulae lightly, and not feel that you have to memorize any of them. This chapter needs to be complete, and that means we have to go into the analysis at some depth, but remember that you can always come back to the formulae when you need them, and don't worry about the calculations too much until you do need them.

If you think of a typical one-way analysis of variance with different subjects serving under the different treatments, you would probably be willing to concede that the correlations between treatments 1 and 2, 1 and 3, and 2 and 3 have an expectation of zero.

Treatment 1	Treatment 2	Treatment 3
X_{11}	X_{21}	X_{31}
X_{12}	X_{22}	X_{32}
$\cdots$	$\cdots$	$\cdots$
X_{1n}	X_{2n}	X_{3n}

However, suppose that in the design diagrammed here the same subjects were used in all three treatments. Thus, instead of $3n$ subjects measured once, we have n subjects measured three times. In this case, we would be hard put to believe that the intercorrelations of the three treatments would have expectancies of zero. On the contrary, the better subjects under treatment 1 would probably also perform well under treatments 2 and 3, and the poorer subjects under treatment 1 would probably perform poorly under the other conditions, leading to significant correlations among treatments.

partition

partialling out

repeated-measures designs

This lack of independence among the treatments would cause a serious problem except that we can separate out, or **partition,** and remove the dependence imposed by repeated measurements on the same subjects. (To use a term that will become much more familiar in Chapter 15, we can say that we are **partialling out** effects that cause the dependence.) Actually, one of the main advantages of **repeated-measures designs** is that they allow us to reduce overall variability by using a common subject pool for all treatments, while allowing us to remove subject differences from our error term, leaving the error components independent from treatment to treatment or cell to cell.

As an illustration, consider the highly exaggerated set of data on four subjects over three treatments presented in Table 14.1. Here, the dependent variable is the number of trials to criterion on some task. If you look first at the treatment means, you will see some slight differences, but nothing to get too excited about. There is so much variability within each treatment that it would at first appear that the means differ only by chance. But look at the subject means. It is apparent that subject 1 learns quickly under all conditions, and that subjects 3 and 4 learn remarkably slowly. These differences among the subjects are producing most of the differences *within* treatments, and yet they have nothing to do with the treatment effect. If we could remove these subject differences, we would have a better (and smaller) estimate of error. At the same time, the subject differences are creating the high positive intercorrelations among the treatments, and these too we will partial out by forming a separate term for subjects.

Table 14.1   Hypothetical data for simple repeated-measures designs

Subject	Treatment			
	1	2	3	Mean
1	2	4	7	4.33
2	10	12	13	11.67
3	22	29	30	27.00
4	30	31	34	31.67
Mean	16	19	21	18.67

One laborious way to do this would be to put all the subjects' contributions on a common footing by equating subject means without altering the relationships among the scores obtained by that particular subject. Thus, we could set $X'_{ij} = X_{ij} - \overline{X}_i$, where $\overline{X}_i$ is the mean of the ith subject. Now subjects would all have the same means ($\overline{X}'_{i.} = 0$), and any remaining differences among the scores could be attributable only to error or to treatments. Although this approach would work, it is not practical. An alternative, and easier, approach is to calculate a sum of squares between subjects (denoted as either $SS_{\text{between subj}}$ or SS_S) and remove this from SS_{total} before we begin. This can be shown to be algebraically equivalent to the first procedure and is essentially the approach we will adopt.

$SS_{\text{between subj}}$
SS_S

The problem is represented diagrammatically in Figure 14.1. Here, we partition the overall variation into variation between subjects and variation within subjects. We do the same with the degrees of freedom. Some of the variation within a subject is attributable to the fact that the scores come from different treatments, and some is attributable to error; this further partitioning of variation is shown in the third line of the figure. We will always think of a repeated-measures analysis as *first* partitioning the SS_{total} into $SS_{\text{between subj}}$ and $SS_{\text{within subj}}$. Depending on the complexity of the design, one or both of these partitions may then be further partitioned.

$SS_{\text{within subj}}$

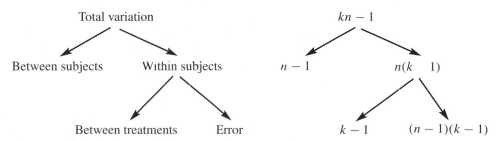

| **Partition of Sums of Squares** | **Partition of Degrees of Freedom** |

Total variation → Between subjects, Within subjects → Between treatments, Error

$kn - 1$ → $n - 1$, $n(k\ 1)$ → $k - 1$, $(n - 1)(k - 1)$

Figure 14.1   Partition of sums of squares and degrees of freedom

The following discussion of repeated-measures designs can only begin to explore the area. For historical reasons, the statistical literature has underemphasized the importance of these designs. As a result, they have been developed mostly by social scientists, particularly psychologists. By far the most complete coverage of these designs is found in Winer (1971). His treatment of repeated-measures designs is excellent and extensive, and much of this chapter reflects the influence of his work.

14.1   The Structural Model

First, some theory to keep me happy. Two structural models could underlie the analysis of data like those shown in Table 14.1. The simplest model is

$$X_{ij} = \mu + \pi_i + \tau_j + e_{ij}$$

where

μ = the grand mean

π_i = a constant associated with the ith person or subject, representing how much that person differs from the average person

τ_j = a constant associated with the jth treatment, representing how much that treatment mean differs from the average treatment mean

e_{ij} = the experimental error associated with the ith subject under the jth treatment

The variables π_i and e_{ij} are assumed to be independently and normally distributed around zero within each treatment. In addition, σ_π^2 and σ_e^2 are assumed to be homogeneous across treatments. With these assumptions, it is possible to derive the expected mean squares shown in Model I of Table 14.2.

Table 14.2   Expected mean squares for simple repeated-measures designs

Model I		Model II	
$X_{ij} = \mu + \pi_i + \tau_j + e_{ij}$		$X_{ij} = \mu + \pi_i + \tau_j + \pi\tau_{ij} + e_{ij}$	
Source	E(MS)	Source	E(MS)
Subjects	$\sigma_e^2 + k\sigma_\pi^2$	Subjects	$\sigma_e^2 + k\sigma_\pi^2$
Treatments	$\sigma_e^2 + n\sigma_\tau^2$	Treatments	$\sigma_e^2 + \sigma_{\pi\tau}^2 + n\sigma_\tau^2$
Error	σ_e^2	Error	$\sigma_e^2 + \sigma_{\pi\tau}^2$

An alternative and probably more realistic model is given by

$$X_{ij} = \mu + \pi_i + \tau_j + \pi\tau_{ij} + e_{ij}$$

Here, we have added a Subject × Treatment interaction term to the model. The assumptions of the first model will continue to hold, and we will also assume the $\pi\tau_{ij}$ to be distributed around zero independently of the other elements of the model. This second model gives rise to the expected mean squares shown in Model II of Table 14.2.

The discussion of these two models and their expected mean squares may look as if it is designed to bury the solution to a practical problem (comparing a set of means) under a mountain of statistical theory. However, it is important to an explanation of how we will run our analyses and where our tests come from. You'll need to bear with me only a little longer.

14.2   F Ratios

The expected mean squares in Table 14.2 indicate that the model we adopt influences the F ratios we employ. If we are willing to assume that there is no Subject × Treatment interaction, we can form the following ratios:

$$\frac{E(MS_{\text{between subj}})}{E(MS_{\text{error}})} = \frac{\sigma_e^2 + k\sigma_\pi^2}{\sigma_e^2}$$

and

$$\frac{E(MS_{\text{treat}})}{E(MS_{\text{error}})} = \frac{\sigma_e^2 + n\sigma_\tau^2}{\sigma_e^2}$$

Given an additional assumption about sphericity, which we will discuss in the next section, both of these lead to respectable F ratios that can be used to test the relevant null hypotheses.

Usually, however, we are cautious about assuming that there is no Subject $\times$ Treatment interaction. In much of our research, it seems more reasonable to assume that different subjects will respond differently to different treatments, especially when those "treatments" correspond to phases of an ongoing experiment. As a result, we usually prefer to work with the more complete model. (Tukey [1949] developed a test for additivity of effects that is useful in choosing between the two models. This test is discussed in Kirk [1968] and Winer [1962, 1971], but Hsu [personal communication]) has recently pointed out problems with it.

The full model (which includes the interaction term) leads to the following ratios:

$$\frac{E(MS_{\text{between subj}})}{E(MS_{\text{error}})} = \frac{\sigma_e^2 + k\sigma_\pi^2}{\sigma_e^2 + \sigma_{\pi\tau}^2}$$

and

$$\frac{E(MS_{\text{treat}})}{E(MS_{\text{error}})} = \frac{\sigma_e^2 + \sigma_{\pi\tau}^2 + n\sigma_\tau^2}{\sigma_e^2 + \sigma_{\pi\tau}^2}$$

Although the resulting F for treatments is appropriate, the F for subjects is biased. If we did form this latter ratio and obtained a significant F, we would be fairly confident that subject differences really did exist. However, if the F were not significant, the interpretation would be ambiguous. A nonsignificant F could mean either that $k\sigma_\pi^2 = 0$ or that $k\sigma_\pi^2 > 0$ but $\leq \sigma_{\pi\tau}^2$. Because we usually prefer this second model, and hate ambiguity, we seldom test the effect due to Subjects. This represents no great loss, however, because we have little to gain by testing the Subject effect. The main reason for obtaining $SS_{\text{between subj}}$ in the first place is to absorb the correlations between treatments and thereby remove subject differences from the error term. A test on the Subject effect, if it were significant, would merely indicate that people are different—hardly a momentous finding. The important thing is that both underlying models show that we can use MS_{error} as the denominator for F in this situation.

14.3 The Covariance Matrix

A very important assumption that is required for any F ratio in a repeated-measures design to be distributed as the central (tabled) F is that of compound symmetry of the covariance matrix.[1] To understand what this means, consider a matrix $(\hat{\Sigma})$ representing the covariances among the three treatments for the data given in Table 14.1.

$$\hat{\Sigma} = \begin{array}{c|ccc} & A_1 & A_2 & A_3 \\ \hline A_1 & 154.67 & 160.00 & 160.00 \\ A_2 & 160.00 & 176.67 & 170.67 \\ A_3 & 160.00 & 170.67 & 170.00 \end{array}$$

---

[1] This assumption is overly stringent and will shortly be relaxed somewhat. It is nonetheless a sufficient assumption, and it is made often.

main diagonal

off-diagonal elements

compound symmetry

covariance matrix

sphericity

multivariate analysis of variance (MANOVA)

multivariate procedure

On the **main diagonal** of this matrix are the variances within each treatment ($\hat{\sigma}^2_{A_j}$). Notice that they are all or less equal, indicating that we have met the assumption of homogeneity of variance. The **off-diagonal elements** represent the covariances among the treatments (cov_{12}, cov_{13}, and cov_{23}). Notice that these are also more or less equal. (The fact that they are also of the same magnitude as the variances is irrelevant, reflecting merely the very high intercorrelations among treatments.) A pattern of constant variances on the diagonal and constant covariances off the diagonal is referred to as **compound symmetry.** (Again, the relationship between the variances and covariances is irrelevant.) The assumption of compound symmetry of the (*population*) **covariance matrix** ($\sum$), of which $\hat{\sum}$ is an estimate, represents a sufficient condition underlying a repeated-measures analysis of variance. The more general condition is known as **sphericity,** and you will often see references to that broader assumption. If we have compound symmetry we will meet the sphericity assumption, but it is possible, though not likely in practice, to have sphericity without compound symmetry. (Older textbooks, including my own, generally make reference to compound symmetry, even though that is too strict an assumption. In recent years, the trend has been toward reference to "sphericity," and that is how we will generally refer to it here.) Without this sphericity assumption, the F ratios may not have a distribution given by the distribution of F in the tables. Although this assumption applies to any analysis of variance design, when the cells are independent, the covariances are always zero, and there is no problem—we merely need to assume homogeneity of variance. With repeated-measures designs, however, the covariances will not be zero and we need to assume that they are all equal. This has led some people (e.g., Hays, 1981) to omit serious consideration of repeated-measures designs. However, when we do have sphericity, the Fs are valid, and when we do not, we can use either very good approximation procedures (to be discussed later in this chapter) or alternative methods that do not depend on assumptions about $\sum$. One alternative procedure that does not require any assumptions about the covariance matrix is **multivariate analysis of variance (MANOVA).** This is a **multivariate procedure,** which is essentially one that deals with multiple dependent variables simultaneously. A brief introduction to multivariate analysis of variance for repeated-measures designs can be found in Section 14.14.

14.4 Analysis of Variance Applied to Relaxation Therapy

As an example of a simple repeated-measures design, we will consider a study of the effectiveness of relaxation techniques in controlling migraine headaches. The data described here are fictitious, but they are in general agreement with data collected by Blanchard, Theobald, Williamson, Silver, and Brown (1978), who ran a similar, although more complex, study.

In this experiment, we have recruited nine migraine sufferers and have asked them to record the frequency and duration of their migraine headaches. After 4 weeks of baseline recording during which no training was given, we had a 6-week period of relaxation training. (Each experimental subject participated in the program at a different time, so such things as changes in climate and holiday events should not systematically influence the data.) For our example, we will analyze the data for the last 2 weeks of baseline and the last 3 weeks of training. The dependent variable is the duration (hours/week) of headaches in each of those 5 weeks. The data and the calculations are shown in Table 14.3.[2] It is important to note that I have identified the means with a subscript naming the variable. Thus, instead of using the

---

[2] Because I have rounded the means to three decimal places, there is rounding error in the answers. The answers given here have been based on more decimal places.

Table 14.3   Analysis of data on migraine headaches

(a) Data

Subject	Baseline		Training			Subject Means
	Week 1	Week 2	Week 3	Week 4	Week 5	
1	21	22	8	6	6	12.6
2	20	19	10	4	4	11.4
3	17	15	5	4	5	9.2
4	25	30	13	12	17	19.4
5	30	27	13	8	6	16.8
6	19	27	8	7	4	13.0
7	26	16	5	2	5	10.8
8	17	18	8	1	5	9.8
9	26	24	14	8	9	16.2
Week Means	22.333	22.000	9.333	5.778	6.778	13.244

(b) Calculations

$$SS_{total} = \sum(X - \overline{X}_{..})^2 = (21 - 13.244)^2 + \cdots + (9 - 13.244)^2 = 3166.31$$

$$SS_{subjects} = w\sum(\overline{X}_S - \overline{X}_{..})^2 = 5[(12.6 - 13.244)^2 + \cdots + (16.2 - 13.244)^2] = 486.71$$

$$SS_{weeks} = n\sum(\overline{X}_W - \overline{X}_{..})^2 = 9[(22.333 - 13.244)^2 + \cdots + (6.778 - 13.244)^2] = 2449.20$$

$$SS_{error} = SS_{total} - SS_{subjects} - SS_{weeks} = 3166.31 - 486.71 - 2449.20 = 230.40$$

(c) Summary table

Source	df	SS	MS	F
Between subjects	8	486.71		
Weeks	4	2449.20	612.30	85.04*
Error	32	230.40	7.20	
Total	44	3166.31		

*$p < .05$

standard "dot notation" (e.g., $\overline{X}_{i.}$ for the Week means), I have used the letter indicating the variable name as the subscript (e.g., the means for Weeks are denoted $\overline{X}_W$ and the means for Subjects are denoted $\overline{X}_S$). As usual, the grand mean is denoted $\overline{X}_{..}$, and X represents the individual observations.

Look first at the data in Table 14.3a. Notice that there is a great deal of variability, but much of that variability comes from the fact that some people have more or longer-duration headaches than do others, which really has very little to do with the intervention program. As I have said, what we are able to do with a repeated-measures design but were not able to do with between-subjects designs is to remove this variability from SS_{error}, producing a smaller MS_{error} than we would otherwise have.

From Table 14.3b, you can see that SS_{total} is calculated in the usual manner. Similarly, $SS_{subjects}$ and SS_{weeks} are calculated just as main effects always are (take the sum of the squared deviations from the grand mean and multiply by the appropriate constant [i.e., the number of observations contributing to each mean]). Finally, the error term is obtained by subtracting SS_{subjects} and SS_{weeks} from SS_{total}.

The summary table is shown in Table 14.3c. Notice that I have computed an F for Weeks but not for subjects, for the reasons given earlier. The F value for Weeks is based on 4 and 32 degrees of freedom, and $F_{.05}(4, 32) = 2.68$. We can therefore reject $H_0: \mu_1 = \mu_2 = \cdots = \mu_5$ and conclude that the relaxation program led to a reduction in the duration per week of headaches reported by subjects. Examination of the means in Table 14.3 reveals that during the last three weeks of training, the amount of time per week involving headaches was about one-third of what it was during baseline.

You may have noticed that no Subject × Weeks interaction is shown in the summary table. With only one score per cell, the interaction term *is* the error term, and some people prefer to label it $S \times W$ instead of error. To put this differently, in the design discussed here it is impossible to separate error from any possible Subject × Weeks interaction because they are completely confounded. As we saw in the discussion of structural models, both of these effects, if present, are combined in the expected mean square for error.

I spoke earlier of the assumption of sphericity, or compound symmetry. For the data in the example, the variance–covariance matrix follows, represented by the notation $\hat{\sum}$, where the ^ is used to indicate that this is an estimate of the population variance–covariance matrix $\sum$.

$$\hat{\sum} = \begin{matrix} 21.000 & 11.750 & 9.250 & 7.833 & 7.333 \\ 11.750 & 28.500 & 13.750 & 16.375 & 13.375 \\ 9.250 & 13.750 & 11.500 & 8.583 & 8.208 \\ 7.833 & 16.375 & 8.583 & 11.694 & 10.819 \\ 7.333 & 13.375 & 8.208 & 10.819 & 16.945 \end{matrix}$$

Visual inspection of this matrix suggests that the assumption of sphericity is reasonable. The variances on the diagonal range from 11.5 to 28.5, whereas the covariances off the diagonal range from 7.333 to 16.375. Considering that we have only nine subjects, these values represent an acceptable level of constancy. (Keep in mind that the variances do not need to be equal to the covariances; they seldom are.) A statistical test of this assumption of sphericity was developed by Mauchly (1940) and is given in Winer (1971, p. 596). It would show that we have no basis for rejecting the sphericity hypothesis. Box (1954b), however, showed that regardless of the form of $\sum$, a conservative test on null hypotheses in the repeated-measures analysis of variance is given by comparing F_{obt} against $F_{.05}(1, n - 1)$—that is, by acting as though we had only two treatment levels. This test is exceedingly conservative, however, and for most situations, you will be better advised to evaluate F in the usual way. We will return to this problem later when we consider Greenhouse and Geisser's (1959) extension of Box's work.

As already mentioned, one of the major advantages of the repeated-measures design is that it allows us to reduce the error term by using the same subject for all treatments. Suppose for a moment that the data illustrated in Table 14.3 had actually been produced by five independent groups of subjects. For such an analysis, SS_{error} would equal 717.11. In this case, we would not be able to pull out a subject term because $SS_{\text{between subj}}$ would be synonymous with SS_{total}. (A subject total and an individual score are identical.) As a result, differences among subjects would be inseparable from error, and SS_{error} would be the sum of what, for the repeated-measures design, are SS_{error} and $SS_{\text{between subj}}$ ($= 230.4 + 486.71 = 717.11$ on $32 + 8 = 40$ df). This would lead to

$$F = \frac{MS_{\text{weeks}}}{MS_{\text{error}}} = \frac{612.30}{17.93} = 34.15$$

which, although still significant, is less than one-half of what it was in Table 14.3.

To put it succinctly, subjects differ. When subjects are observed only once, these subject differences contribute to the error term. When subjects are observed repeatedly, we can obtain an estimate of the degree of subject differences and partial these differences out of the error term. In general, the greater the differences among subjects, the higher the correlations between pairs of treatments. The higher the correlations among treatments, the greater the relative power of repeated-measures designs.

We have been speaking of the simple case in which we have one independent variable (other than subjects) and test each subject on every level of that variable. In actual practice, there are many different ways in which we could design a study using repeated measures. For example, we could set up an experiment using two independent variables and test each subject under all combinations of both variables. Alternatively, each subject might serve under only one level of one of the variables, but under all levels of the other. If we had three variables, the possibilities are even greater. In this chapter, we will discuss only a few of the possible designs. If you understand the designs discussed here, you should have no difficulty generalizing to even the most complex problems.

14.5 Contrasts and Effect Sizes in Repeated-Measures Designs

As we did with one-way and factorial designs, we need to consider how to run contrasts among means of repeated measures variables. Fortunately, there is not really much that is new here. We will again be comparing the mean of a condition or set of conditions against the mean of another condition or set of conditions, and we will be using the same kinds of coefficients that we have used all along.

In our example, the first two weeks were Baseline measures, and the last three weeks were Training measures. Our omnibus F told us that there were statistically significant differences between the five Weeks, but not where those differences lie. Now I would like to contrast the means of the set of Baseline weeks with the mean of the set of Training weeks. The coefficients that will do this are shown here, along with the means.

	Week 1	Week 2	Week 3	Week 4	Week 5
Coefficient	1/2	1/2	−1/3	−1/3	−1/3
Mean	22.333	22.000	9.333	5.778	6.778

Just as we have been doing, we will define our contrast as

$$\hat{\psi} = \sum a_i \overline{X}_i$$
$$= \left(\tfrac{1}{2}\right)(22.333) + \left(\tfrac{1}{2}\right)(22.000) + \left(-\tfrac{1}{3}\right)(9.333) + \left(-\tfrac{1}{3}\right)(5.778) + \left(-\tfrac{1}{3}\right)(6.778)$$
$$= \frac{22.333 + 22.000}{2} - \frac{9.333 + 5.778 + 6.778}{3} = \frac{44.333}{2} - \frac{21.889}{3} = 22.166 - 7.296$$
$$= 14.870$$

We can test this contrast with either a t or an F, but I will use t here. (F is just the square of t.)

$$t = \frac{\hat{\psi}}{\sqrt{\dfrac{\left(\sum a_i^2\right) MS_{\text{error}}}{n}}} = \frac{14.870}{\sqrt{\dfrac{0.833(7.20)}{9}}} = \frac{14.870}{\sqrt{0.667}} = \frac{14.870}{0.816} = 18.21$$

This is a t on $df_{\text{error}} = 32\ df$, and is clearly statistically significant.

Notice that in calculating my t, I used the MS_{error} from the overall analysis. And this was the same error term that was used to test the Weeks effect. I point that out only because when we come to more complex analyses we will have multiple error terms, and the one to use for a specific contrast is the one that was used to test the main effect of that independent variable.

Effect Sizes

Although there was a direct translation from one-way designs to repeated measures designs in testing contrasts among means, the situation is a bit more complicated when it comes to estimating effect sizes. We will continue to define our effect size as

$$\hat{d} = \frac{\hat{\psi}}{s_{error}}$$

There should be no problem with $\hat{\psi}$ because it is the same contrast that we computed earlier—the difference between the mean of the baseline weeks and the mean of the training weeks. But there are several choices for s_{error}. Kline (2004) gives three possible choices for our denominator, but points out that two of these are unsatisfactory either because they ignore the correlation between weeks or because they standardize $\hat{\psi}$ by a standard deviation that is not particularly meaningful. What we will actually do is create an error term that is unique to the particular contrast. We will form a contrast for each subject. That means that for each subject we will calculate the difference between his mean on the baseline weeks and his mean on the training weeks. These are difference scores, which are analogous to the difference scores we computed for a paired sample t test. The standard deviation of these difference scores is analogous to the denominator we used for computing effect size with paired data when we just had two repeated measures with the t test.

For our migraine example, the first subject would have a difference score of $(21 + 22)/2 - (8 + 6 + 6)/3 = 21.5 - 6.667 = 14.833$. The complete set of difference scores would be

[14.833, 13.500, 11.333, 13.500, 19.500, 16.667, 17.000, 12.833, 14.667]

The standard deviation of these difference scores is 2.49. Then our effect size measure is

$$\hat{d} = \frac{\hat{\psi}}{s_{error}} = \frac{14.87}{2.49} = 5.97$$

This tells us that the severity of headaches during baseline is nearly 6 standard deviations greater than the severity of headaches during training. That is a very large difference, and we can see that just by looking at the data. Remember, in calculating this effect size we have eliminated the variability between participants (subjects) for headache severity. We are in a real sense comparing each individual to his or her self.

14.6 Writing Up the Results

In writing up the results of this experiment, we could simply say

> To investigate the effects of relaxation therapy on the severity of migraine headaches, 9 participants rated the severity of headaches on each of 2 weeks before receiving relaxation therapy and for 3 weeks while receiving therapy. An overall analysis of variance for repeated measures showed a significant difference between weeks ($F(4, 32) = 85.04$, $p < .05$). The mean severity rating during baseline weeks was 22.166, which dropped to a mean of 7.296 during training, for a difference of 14.87. A contrast on this difference was significant ($t(32) = 18.21$, $p < .05$). Using the standard deviation of contrast differences for each participant produced an effect size measure of $d = 5.97$, documenting the importance of relaxation therapy in treating migraine headaches.

14.7   One Between-Subjects Variable and One Within-Subjects Variable

Consider the data presented in Table 14.4. These are actual data from a study by King (1986). This study in some ways resembles the one on morphine tolerance by Siegel (1975) that we examined in Chapter 12. King investigated motor activity in rats following injection of the drug midazolam. The first time that this drug is injected, it typically leads to a distinct decrease in motor activity. Like morphine, however, a tolerance for midazolam develops

Table 14.4   Ambulatory behavior by Group and Trial

(a) Data

		Interval						
		1	2	3	4	5	6	Mean
Control		150	44	71	59	132	74	88.333
		335	270	156	160	118	230	211.500
		149	52	91	115	43	154	100.667
		159	31	127	212	71	224	137.333
		159	0	35	75	71	34	62.333
		292	125	184	246	225	170	207.000
		297	187	66	96	209	74	154.833
		170	37	42	66	114	81	85.000
	Mean	213.875	93.250	96.500	128.625	122.875	130.125	130.875
Same		346	175	177	192	239	140	211.500
		426	329	236	76	102	232	233.500
		359	238	183	123	183	30	186.000
		272	60	82	85	101	98	116.333
		200	271	263	216	241	227	236.333
		366	291	263	144	220	180	244.000
		371	364	270	308	219	267	299.833
		497	402	294	216	284	255	324.667
	Mean	354.625	266.250	221.000	170.000	198.625	178.625	231.521
Different		282	186	225	134	189	169	197.500
		317	31	85	120	131	205	148.167
		362	104	144	114	115	127	161.000
		338	132	91	77	108	169	152.500
		263	94	141	142	120	195	159.167
		138	38	16	95	39	55	63.500
		329	62	62	6	93	67	103.167
		292	139	104	184	193	122	172.333
	Mean	290.125	98.250	108.500	109.000	123.500	138.625	144.667
	Interval mean	286.208	152.583	142.000	135.875	148.333	149.125	169.021

(continued)

Table 14.4 *(continued)*

(b) Calculations

$$SS_{total} = \sum(X - \overline{X}_{...})^2 = (150 - 169.021)^2 + \cdots + (122 - 169.021)^2 = 1{,}432{,}292.9$$

$$SS_{subj} = i\sum(\overline{X}_S - \overline{X}_{...})^2 = 6[(88.333 - 169.021)^2 + \cdots + (172.333 - 169.021)^2] = 670{,}537.1$$

$$SS_{groups} = ni\sum(\overline{X}_G - \overline{X}_{...})^2 = 8 \times 6[(130.875 - 169.021)^2 + \cdots + (144.667 - 169.021)^2] = 285{,}815.0$$

$$SS_{intervals} = ng\sum(\overline{X}_I - \overline{X}_{...})^2 = 8 \times 3[(286.208 - 169.021)^2 + \cdots + (149.125 - 169.021)^2] = 399{,}736.5$$

$$SS_{cells} = n\sum(\overline{X}_{GI} - \overline{X}_{...})^2 = 8[(212.875 - 169.021)^2 + \cdots + (138.625 - 169.021)^2] = 766{,}371.5$$

$$SS_{I \times G} = SS_{cells} - SS_{interval} - SS_{groups} = 766{,}371.5 - 285{,}815.0 - 399.736.5 = 80{,}820.0$$

(c) Summary Table

Source	df	SS	MS	F
Between subjects	23	670,537.1		
Groups	2	285,815.0	142,907.5	7.80*
Ss w/in groups**	21	384,722.0	18,320.1	
Within subjects**	120	761,755.8		
Intervals	5	399,736.5	79,947.3	29.85*
I × G	10	80,820.0	8,082.0	3.02*
I × Ss w/in groups**	105	281,199.3	2,678.1	
Total	143	1,432,292.9		

* $p < .05$

** Calculated by subtraction

rapidly. King wanted to know whether that acquired tolerance could be explained on the basis of a *conditioned* tolerance related to the physical context in which the drug was administered, as in Siegel's work. He used three groups, collecting the crucial data (presented in Table 14.4) on only the last day, which was the test day. During pretesting, two groups of animals were repeatedly injected with midazolam over several days, whereas the Control group was injected with physiological saline. On the test day, one group—the "Same" group—was injected with midazolam in the *same* environment in which it had earlier been injected. The "Different" group was also injected with midazolam, but in a *different* environment. Finally, the Control group was injected with midazolam for the first time. This Control group should thus show the typical initial response to the drug (decreased ambulatory behavior), whereas the Same group should show the normal tolerance effect—that is, they should decrease their activity little or not at all in response to the drug on the last trial. If King is correct, however, the Different group should respond similarly to the Control group because, although they have had several exposures to the drug, they are receiving it in a novel context and any conditioned tolerance that might have developed will not have the necessary cues required for its elicitation. The dependent variable in Table 14.4 is a measure of ambulatory behavior, in arbitrary units. Again, the first letter of the name of a variable is used as a subscript to indicate what set of means we are referring to.

Because the drug is known to be metabolized over a period of approximately 1 hour, King recorded his data in 5-minute blocks, or Intervals. We would expect to see the effect of the drug increase for the first few intervals and then slowly taper off. Our analysis uses the first six blocks of data. The design of this study can then be represented diagrammatically as shown in Figure 14.2.

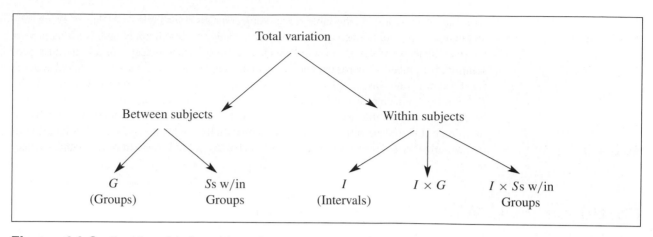

Figure 14.2   Partition of design with one between-subject variable and one within-subject variable

Here we have distinguished those effects that represent differences between subjects from those that represent differences within subjects. When we consider the between-subjects term, we can partition it into differences between groups of subjects (G) and differences between subjects in the same group (Ss w/in groups). The within-subject term can similarly be subdivided into three components—the main effect of Intervals (the repeated measure) and its interactions with the two partitions of the between-subject variation. You will see this partitioning represented in the summary table when we come to it.

Partitioning the Between-Subjects Effects

Let us first consider the partition of the between-subjects term in more detail. From the design of the experiment, we know that this term can be partitioned into two parts. One of these parts is the main effect of Groups (G) because the treatments (Control, Same, and Different) involve different groups of subjects. This is not the only source of differences among subjects, however. We have eight different subjects within the control group, and differences among them are certainly between-subjects differences. The same holds for the subjects within the other groups. Here, we are speaking of differences among subjects in the same group—that is, Ss within groups.

If we temporarily ignore intervals entirely (e.g., we simply collect our data over the entire session rather than breaking it down into 5-minute intervals), we can think of the study as producing the following data:

Control	Same	Different
88.333	211.500	197.500
211.500	233.500	148.167
100.667	186.000	161.000
137.333	116.333	152.500
62.333	236.333	159.167
207.000	244.000	63.500
154.833	299.833	103.167
85.000	324.667	172.333
130.875	231.521	144.667

where the "raw scores" in this table are the subject means from Table 14.4. Because each subject is represented only once in these totals, the analysis we will apply here is the same as a one-way analysis of variance on independent groups. Indeed, except for a constant representing the number of scores per subject (which cancels out in the end), the sums of squares for the simple one-way on these data would be the same as those in the actual analysis. The F that tests the main effect of Groups if this were a simple one-way on subject totals would be equal to the one that we will obtain from the full analysis. Thus, the between-subjects partition of the total variation can be seen as essentially a separate analysis of variance, with its own error term (sometimes referred to as **error**$_{between}$) independent of the within-subject effects.

error$_{between}$

Partitioning the Within-Subjects Effects

Next consider the within-subjects element of the partition of SS_{total}. As we have already seen, this is itself partitioned into three terms. A comparison of the six intervals involves comparisons of scores from the same subject, and thus, Intervals is a within-subjects term—it depends on differences within each subject. Intervals is a within-subjects term, so the interaction of Intervals with Groups is also a within-subjects effect. The third term (Intervals × Ss within groups) is sometimes referred to as **error**$_{within}$ because it is the error term for the within-subjects effects. The $SS_{Intervals \times Ss\,w/in\,groups}$ term is actually the sum of the sums of squares for the $I \times S$ interactions calculated separately for each group. Thus, it can be seen as logically equivalent to the error term used in the previous design.

error$_{within}$

The Analysis

Before considering the analysis in detail, it is instructive to look at the general pattern of results. Although there are not enough observations in each cell to examine the distributions in any serious way, it is apparent that on any given interval there is substantial variability within groups. For example, for the second interval in the control group, scores range from 0 to 270. There do not appear to be any extreme outliers, however, as often happens in this kind of research, and the variances within cells, although large, are approximately equal. You can also see that there are large individual differences, with some of the animals consistently showing relatively little ambulatory behavior and some showing a great deal. These are the kinds of differences that will be partialled out by our analysis. Looking at the Interval means, you will see that, as expected, behavior decreased substantially after the first 5-minute interval and then increased slightly during the rest of the session. Finally, looking at the difference between the means for the Control and Same groups, you will see the anticipated tolerance effect, and looking at the Different group, you see that it is much more like the Control group than it is like the Same group. This is the result that King predicted.

Very little needs to be said about the actual calculations in Table 14.4b because they are really no different from the usual calculations of main and interaction effects. Whether a factor is a between-subjects or within-subjects factor has no bearing on the calculation of its sum of squares, although it does affect its placement in the summary table and the ultimate calculation of the corresponding F.

In the summary table in Table 14.4c, the source column reflects the design of the experiment, with SS_{total} first partitioned into $SS_{between\,subj}$ and $SS_{w/in\,subj}$. Each of these sums of squares is further subdivided. The double asterisks next to the three terms show that we calculate these by subtraction ($SS_{w/in\,subj}$, $SS_{Ss\,w/in\,groups}$, and $SS_{I \times Ss\,w/in\,groups}$), based on the fact that sums of squares are additive and the whole must be equal to the sum of its parts.

This simplifies our work considerably. Thus

$$SS_{\text{w/in subj}} = SS_{\text{total}} - SS_{\text{between subj}}$$

$$SS_{Ss\text{ w/in groups}} = SS_{\text{between subj}} - SS_{\text{groups}}$$

$$SS_{I \times Ss\text{ w/in groups}} = SS_{\text{w/in subj}} - SS_{\text{intervals}} - SS_{IG}$$

These last two terms will become error terms for the analysis.

The degrees of freedom are obtained in a relatively straightforward manner. For each of the main effects, the number of degrees of freedom is equal to the number of levels of the variable minus 1. Thus, for Subjects, there are $24 - 1 = 23$ df; for Groups, there are $3 - 1 = 2$ df; and for Intervals, there are $6 - 1 = 5$ df. As for all interactions, the df for $I \times G$ is equal to the product of the df for the component terms. Thus, $df_{IG} = (6 - 1)(3 - 1) = 10$. The easiest way to obtain the remaining degrees of freedom is by subtraction, just as we did with the corresponding sums of squares.

$$df_{\text{w/in subj}} = df_{\text{total}} - df_{\text{between subj}}$$

$$df_{Ss\text{ w/in groups}} = df_{\text{between subj}} - df_{\text{groups}}$$

$$df_{I \times Ss\text{ w/in groups}} = df_{\text{w/in subj}} - df_{\text{intervals}} - df_{IG}$$

These df can also be obtained directly by considering what these terms represent. Within each subject, we have $6 - 1 = 5$ df. With 24 subjects, this amounts to $(5)(24) = 120$ $df_{\text{w/in subj}}$. Within each level of the Groups factor, we have $8 - 1 = 7$ df between subjects, and with three Groups, we have $(7)(3) = 21$ $df_{\text{w/in groups}}$. $I \times Ss$ w/in groups is really an interaction term, and as such its df is simply the product of df_I and $df_{Ss\text{ w/in groups}} = (5)(21) = 105$.

Skipping over the mean squares, which are merely the sums of squares divided by their degrees of freedom, we come to F. From the column of F, it is apparent that, as we anticipated, Groups and Intervals are significant. The interaction is also significant, reflecting the fact that the Different group was at first intermediate between the Same and the Control group, but that by the second 5-minute interval, it had come down to be equal to the Control group. This finding can be explained by a theory of conditioned tolerance. The really interesting finding is that, at least for the later intervals, simply injecting an animal in an environment different from the one in which it had been receiving the drug was sufficient to overcome the tolerance that had developed. These animals respond almost exactly as do animals that had never experienced midazolam. We will return to the comparison of Groups at individual Intervals later.

Assumptions

For the F ratios actually to follow the F distribution, we must invoke the usual assumptions of normality, homogeneity of variance, and sphericity of $\hat{\sum}$. For the *between-subjects* term(s), this means that we must assume that the variance of subject means within any one level of A is the same as the variance of subject means within every other level of A. If necessary, this assumption can be tested by calculating each of the variances and testing using either F_{max} on $(g, n - 1)$ df or, preferably, the test proposed by Levene (1960) or O'Brien (1981), which were referred to in Chapter 7. In practice, however, the analysis of variance is relatively robust against reasonable violations of this assumption (see Collier, Baker, & Mandeville, 1967; and Collier, Baker, Mandeville, & Hayes, 1967). Because the groups are independent, compound symmetry, and thus sphericity, of the covariance matrix is ensured if we have homogeneity of variance because all off-diagonal entries will be zero.

For the *within-subjects* terms, we must also consider the usual assumptions of homogeneity of variance and normality. The homogeneity of variance assumption in this case is that the $I \times S$ interactions are constant across the Groups, and here again, this can be tested using F_{max} on g and $(n-1)(i-1)$ df. (You would simply calculate an $I \times S$ interaction for each group—equivalent to the error term in Table 14.3—and test the largest against the smallest.) For the within-subjects effects, we must also make assumptions concerning the covariance matrix.

There are two assumptions on the covariance matrix (or matrices). Again, we will let $\hat{\sum}$ represent the matrix of variances and covariances among the levels of I (Intervals). Thus, with six intervals,

	I_1	I_2	I_3	I_4	I_5	I_6
	$\hat{\sigma}_{11}$	$\hat{\sigma}_{12}$	$\hat{\sigma}_{13}$	$\hat{\sigma}_{14}$	$\hat{\sigma}_{15}$	$\hat{\sigma}_{16}$
	$\hat{\sigma}_{21}$	$\hat{\sigma}_{22}$	$\hat{\sigma}_{23}$	$\hat{\sigma}_{24}$	$\hat{\sigma}_{25}$	$\hat{\sigma}_{26}$
$\hat{\sum} =$	$\hat{\sigma}_{31}$	$\hat{\sigma}_{32}$	$\hat{\sigma}_{33}$	$\hat{\sigma}_{34}$	$\hat{\sigma}_{35}$	$\hat{\sigma}_{36}$
	$\hat{\sigma}_{41}$	$\hat{\sigma}_{42}$	$\hat{\sigma}_{43}$	$\hat{\sigma}_{44}$	$\hat{\sigma}_{45}$	$\hat{\sigma}_{46}$
	$\hat{\sigma}_{51}$	$\hat{\sigma}_{52}$	$\hat{\sigma}_{53}$	$\hat{\sigma}_{54}$	$\hat{\sigma}_{55}$	$\hat{\sigma}_{56}$
	$\hat{\sigma}_{61}$	$\hat{\sigma}_{62}$	$\hat{\sigma}_{63}$	$\hat{\sigma}_{64}$	$\hat{\sigma}_{65}$	$\hat{\sigma}_{66}$

For each Group, we would have a separate population variance-covariance matrix $\sum_{G_i}$. ($\sum$ and $\sum_{G_i}$ are estimated by $\hat{\sum}$ and $\hat{\sum}_{G_i}$, respectively.) For $MS_{I \times Ss\,w/in\,groups}$ to be an appropriate error term, we will first assume that the individual variance-covariance matrices ($\sum_{G_i}$) are the same for all levels of G. This can be thought of as an extension (to covariances) of the common assumption of homogeneity of variance.

The second assumption concerning covariances deals with the overall matrix $\sum$, where $\sum$ is the pooled average of the $\sum_{G_i}$. (For equal sample sizes in each group, an entry in $\sum$ will be the average of the corresponding entries in the individual $\sum_{G_i}$ matrices.) A common and sufficient, but not necessary, assumption is that the matrix exhibits compound symmetry—meaning, as I said earlier, that all the variances on the main diagonal are equal, and all the covariances off the main diagonal are equal. Again, the variances do not have to equal the covariances, and usually will not. This assumption is actually more stringent than necessary. All that we really need to assume is that the standard errors of the differences between pairs of Interval means are constant—in other words, that $\sigma^2_{\bar{I}_i - \bar{I}_j}$ is constant for all i and $j\,(j \neq i)$. This sphericity requirement is met automatically if $\sum$ exhibits compound symmetry, but other patterns of $\sum$ will also have this property. For a more extensive discussion of the covariance assumptions, see Huynh and Feldt (1970) and Huynh and Mandeville (1979); a particularly good discussion can be found in Edwards (1985, pp. 327–329, 336–339).

Adjusting the Degrees of Freedom

Box (1954a) and Greenhouse and Geisser (1959) considered the effects of departure from this sphericity assumption on $\sum$. They showed that regardless of the form of $\sum$, the F ratio from the within-subjects portion of the analysis of variance will be approximately distributed as F on

$$(i-1)\varepsilon, \; g(n-1)(i-1)\varepsilon$$

df for the Interval effect and

$$(g-1)(i-1)\varepsilon, \; g(n-1)(i-1)\varepsilon$$

df for the $I \times G$ interaction, where $i =$ the number of intervals and ε is estimated by

$$\hat{\varepsilon} = \frac{i^2(\bar{s}_{jj} - \bar{s})^2}{(i-1)\left(\sum s_{jk}^2 - 2i \sum \bar{s}_j^2 + i^2\bar{s}^2\right)}$$

Here,

$\bar{s}_{jj}$ = the mean of the entries on the main diagonal of $\hat{\sum}$

$\bar{s}$ = the mean of all entries in $\hat{\sum}$

s_{jk} = the jkth entry in $\hat{\sum}$

$\bar{s}_j$ = the mean of all entries in the jth row of $\hat{\sum}$

The effect of using $\hat{\varepsilon}$ is to decrease both df_{effect} and df_{error} from what they would normally be. Thus, $\hat{\varepsilon}$ is simply the proportion by which we reduce them. Greenhouse and Geisser recommended that we adjust our degrees of freedom using $\hat{\varepsilon}$. They further showed that when the sphericity assumptions are met, $\varepsilon = 1$, and as we depart more and more from sphericity, ε approaches $1/(i-1)$ as a minimum. They therefore suggested that a very conservative test can be made by setting $\hat{\varepsilon} = 1/(i-1)$, which reduces to setting the df for the test on I to

$$1, g(n-1)$$

and the df for the $I \times G$ interaction to

$$(g-1), g(n-1)$$

This suggestion is so conservative that I cannot recommend its general adoption. You are always better off if you calculate and use $\hat{\varepsilon}$. It does point out, however, that very large values of F are significant regardless of the form of $\sum$. Most statistical software already produces the Greenhouse and Geisser estimate of $\hat{\varepsilon}$, along with an estimate by Huynh and Feldt to be discussed next, so there is no need to use the very conservative approach.

There is some suggestion that for large values of ε, even using $\hat{\varepsilon}$ to adjust the degrees of freedom can lead to a conservative test. Huynh and Feldt (1976) investigated this correction and recommended a modification of $\hat{\varepsilon}$ when there is reason to believe that the true value of ε lies near or above 0.75. Huynh and Feldt define

$$\tilde{\varepsilon} = \frac{ng(i-1)\hat{\varepsilon} - 2}{(i-1)[g(n-1) - (i-1)\hat{\varepsilon}]}$$

We then use $\hat{\varepsilon}$ or $\tilde{\varepsilon}$, depending on our estimate of the true value of ε. (Under certain circumstances, $\tilde{\varepsilon}$ will exceed 1, at which point it is set to 1.)

A test on the assumption of sphericity has been developed by Mauchly (1940) and evaluated by Huynh and Mandeville (1979) and by Keselman, Rogan, Mendoza, and Breen (1980), who point to its extreme lack of robustness. This test is available on SPSS, SAS, and other software and is routinely printed out. Because tests of sphericity are likely to have serious problems when we need them the most, it has been suggested that we *always* use the correction to our degrees of freedom afforded by $\hat{\varepsilon}$ or $\tilde{\varepsilon}$, whichever is appropriate, or use a multivariate procedure to be discussed later. This is a reasonable suggestion and one worth adopting.

For our data, the F value for Intervals ($F = 29.85$) is such that its interpretation would be the same regardless of the value of ε because the Interval effect will be significant even for the lowest possible df. If the assumption of sphericity is found to be invalid, however, alternative treatments would lead to different conclusions with respect to the $I \times G$ interaction. For King's data, the Mauchly sphericity test, as found from SPSS, indicates that the assumption has been violated, and therefore it is necessary to deal with the problem resulting from this violation.

We can calculate $\hat{\varepsilon}$ and $\tilde{\varepsilon}$ and evaluate F on the appropriate df. The pooled variance-covariance matrix (averaged across the separate matrices) is presented in Table 14.5. (I have not presented the variance-covariance matrices for the several groups because they are roughly equivalent and because each of the elements of the matrix is based on only eight observations.)

Table 14.5   Variance-covariance matrix and calculation of $\hat{\varepsilon}$ and $\tilde{\varepsilon}$

Interval

1	2	3	4	5	6	Mean
6388.173	4696.226	2240.143	681.649	2017.726	1924.066	2991.330
4696.226	7863.644	4181.476	2461.702	2891.524	3531.869	4271.074
2240.143	4181.476	3912.380	2696.690	2161.690	3297.762	3081.690
681.649	2461.702	2696.690	4601.327	2248.600	3084.589	2629.093
2017.726	2891.524	2161.690	2248.600	3717.369	989.310	2337.703
1924.066	3531.869	3297.762	3084.589	989.310	5227.649	3009.208

$$\bar{s}_{jj} = \frac{6388.173 + 7863.644 + \cdots + 5227.649}{6} = 5285.090$$

$$\bar{s} = \frac{6388.173 + 4696.226 + \cdots + 989.310 + 5227.649}{36} = 3053.350$$

$$\sum s_{jk}^2 = 6388.173^2 + 4696.226^2 + \cdots + 5227.649^2 = 416{,}392{,}330$$

$$\sum \bar{s}_j^2 = 2991.330^2 + \cdots + 3009.208^2 = 58{,}119{,}260$$

$$\hat{\varepsilon} = \frac{i^2(\bar{s}_{jj} - \bar{s})^2}{(i-1)\left(\sum s_{jk}^2 - 2i\sum \bar{s}_j^2 + i^2\bar{s}^2\right)}$$

$$= \frac{36(5285.090 - 3053.350)^2}{(6-1)[416{,}392{,}330 - (2)(6)(58{,}119{,}260) + (36)(3053.350^2)]}$$

$$= \frac{179{,}303{,}883}{5[416{,}392{,}330 - 697{,}431{,}120 + 335{,}626{,}064]} = 0.6569$$

$$\tilde{\varepsilon} = \frac{ng(i-1)\hat{\varepsilon} - 2}{(i-1)[g(n-1) - (i-1)\hat{\varepsilon}]}$$

$$= \frac{(8)(3)(5)(0.6569) - 2}{5[3(7) - 5(0.6569)]} = \frac{76.828}{5[21 - 5(0.6569)]} = 0.8674$$

From Table 14.5, we can see that our values of $\hat{\varepsilon}$ and $\tilde{\varepsilon}$ are .6569 and .8674, respectively. These are in the neighborhood of .75, so we will follow Huynh and Feldt's suggestion and use $\tilde{\varepsilon}$. In this case, the degrees of freedom for the interaction are

$$(g-1)(i-1)(.8674) = 8.674$$

and

$$g(n-1)(i-1)(.8674) = 91.077$$

The exact critical value of $F_{.05}(8.674, 91.077)$ is 2.002, which means that we will reject the null hypothesis for the interaction. Thus, regardless of any problems with sphericity, all the effects in this analysis are significant. (They would also be significant if we used $\hat{\varepsilon}$ instead of $\tilde{\varepsilon}$.)

Simple Effects

The Interval × Group interaction is plotted in Figure 14.3; the interpretation of the data is relatively clear. It is apparent that the Same group consistently performs above the level of the other two groups—that is, the conditioned tolerance to midazolam leads to greater activity in that group than in the other groups. It is also clear that activity decreases noticeably after the

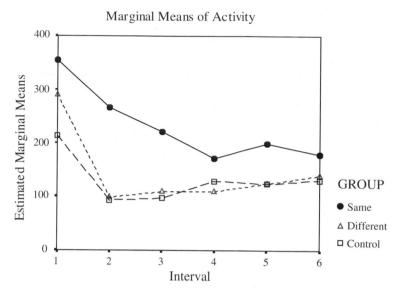

Figure 14.3   Interval × Group interaction for data from Table 14.4

first 5-minute interval (during which the drug is presumably having its effect). The interaction appears to be produced by the fact that the Different group is intermediate between the other two groups during the first interval, but it is virtually indistinguishable from the Control group thereafter. In addition, the Same group continues declining until at least the fourth interval, whereas the other two groups drop precipitously and then level off. Simple effects will prove useful in interpreting these results, especially for examining group differences during the first and the last intervals. Simple effects will also be used to test for differences between intervals within the Control group, but only for purposes of illustration—it should be clear that Interval differences exist within each group.

As I have suggested earlier, the Greenhouse and Geisser and the Huynh and Feldt adjustments to degrees of freedom appear to do an adequate job of correcting for problems with the sphericity assumption when testing for overall main effects or interactions. However, a serious question about the adequacy of the adjustment arises when we consider within-subjects simple effects (Boik, 1981; Harris, 1985). The traditional approach to testing simple effects (see Howell, 1987) involves testing individual within-subjects contrasts against a pooled error term ($MS_{I \times Ss \text{ w/in groups}}$). If there are problems with the underlying assumption, this error term will sometimes underestimate and sometimes overestimate what would be the proper denominator for F, playing havoc with the probability of a Type I error. For that reason, we are going to adopt a different, and in some ways simpler, approach.

The approach we will take follows Boik's advice that a separate error term be derived for each tested effect. Thus, when we look at the simple effect of Intervals for the Control condition, for example, the error term will speak specifically to that effect and will not pool other error terms that apply to other simple effects. In other words, it will be based solely on the Control group. We can test the Interval simple effects quite easily by running separate repeated-measures analyses of variance for each of the groups. For example, we can run a one-way repeated-measures analysis on Intervals for the Control group, as discussed in Section 14.4. We can then turn around and perform similar analyses on Intervals for the Same and Different groups separately. These results are shown in Table 14.6. In each case, the Interval differences are significant, even after we correct the degrees of freedom using $\hat{\varepsilon}$ or $\tilde{\varepsilon}$, whichever is appropriate.

If you look at the within-subject analyses in Table 14.6, you will see that the average MS_{error} is $(2685.669 + 3477.571 + 1871.026)/3 = 2678.089$, which is $MS_{I \times Ss \text{ w/in groups}}$

Table 14.6 Calculation of within-subjects simple effects for data from King (1986)

(a) Interval at Control

Source	df	SS	MS	F
Between subjects	7	134,615.58		
Interval	5	76,447.25	15,289.45	5.69*
Error	35	93,998.42	2685.67	
Total	47	305,061.25		

*$p < .05$; $\hat{\varepsilon} = .404$; $\tilde{\varepsilon} = .570$

(b) Interval at Same

Source	df	SS	MS	F
Between subjects	7	175,600.15		
Interval	5	193,090.85	38,618.17	11.10*
Error	35	121,714.98	3477.57	
Total	47	490,405.98		

*$p < .05$; $\hat{\varepsilon} = .578$; $\tilde{\varepsilon} = 1.00$

(c) Interval at Different

Source	df	SS	MS	F
Between subjects	7	74,506.33		
Interval	5	211,018.42	42,203.68	22.56*
Error	35	65,485.92	1871.03	
Total	47	351,010.67		

*$p < .05$; $\hat{\varepsilon} = .598$; $\tilde{\varepsilon} = 1.00$

from the overall analysis found on page 450. Here, these denominators for the F ratios are noticeably different from what they would have been had we used the pooled term, which is the traditional approach. You can also verify with a little work that the MS_{Interval} terms for each analysis are the same as those that we would compute if we followed the usual procedures for obtaining simple effects mean squares.

For the between-subjects simple effects (e.g., Groups at Interval 1), the procedure is more complicated. Although we could follow the within-subject example and perform separate analyses at each Interval, we would lose considerable degrees of freedom unnecessarily. Here, it is usually legitimate to pool error terms, and it is generally wise to do so.

For this example, we will examine the simple effects of Group at Interval 1 and Group at Interval 6. The original data can be found in Table 14.4 on page 449. The sums of squares for these effects are

$$SS_{G \text{ at Int.1}} = 8[(213.875 - 286.208)^2 + (354.625 - 286.208)^2$$
$$+ (290.125 - 286.208)^2]$$
$$= 79,426.33$$
$$SS_{G \text{ at Int.6}} = 8[(130.125 - 149.125)^2 + (178.625 - 149.125)^2$$
$$+ (138.625 - 149.125)^2]$$
$$= 10,732.00$$

Testing the simple effects of between-subjects terms is a little trickier. Consider for a moment the simple effect of Group at Interval 1. This is essentially a one-way analysis of variance with no repeated measures because the Group means now represent the average of single—rather than repeated—observations on subjects. Thus, subject differences are confounded with experimental error. In this case, the appropriate error sum of squares is $SS_{\text{w/in cell}}$, where, from Table 14.4,

$$SS_{\text{w/in cell}} = SS_{Ss\,\text{w/in group}} + SS_{I \times Ss\,\text{w/in groups}}$$
$$= 384,722.03 + 281,199.34 = 665,921.37$$

and

$$MS_{\text{w/in cell}} = \frac{SS_{\text{w/in cell}}}{df_{Ss\,\text{w/in group}} + df_{I \times Ss\,\text{w/in group}}}$$
$$= \frac{665,921.37}{21 + 105} = 5285.09$$

It may be easier for you to understand why we need this special $MS_{\text{w/in cell}}$ error term if you think about what it really represents. If you were presented with only the data for Interval 1 in Table 14.4 and wanted to test the differences among the three groups, you would run a standard one-way analysis of variance, and the MS_{error} would be the average of the variances within each of the three groups. Similarly, if you had only the data from Interval 2, Interval 3, and so on, you would again average the variances within the three treatment groups. The $MS_{\text{w/in cell}}$ that we have just finished calculating is in reality the average of the error terms for these six different sets (Intervals) of data. As such, it is the average of the variance within each of the 18 cells.

We can now proceed to form our F ratios:

$$F_{G\,\text{at Int.1}} = \frac{MS_{G\,\text{at Int.1}}}{MS_{\text{w/in cell}}} = \frac{79,426.33/2}{5285.09} = 7.51$$

$$F_{G\,\text{at Int.6}} = \frac{MS_{G\,\text{at Int.6}}}{MS_{\text{w/in cell}}} = \frac{10,732/2}{5285.09} = 1.02$$

A further difficulty arises in the evaluation of F. $MS_{\text{w/in cell}}$ also represents the sum of two *heterogeneous* sources of error—as can be seen by examination of the $E(MS)$ for Ss w/in groups and $I \times Ss$ w/in groups—so our F will not be distributed on 2 and 126 df. We will get ourselves out of this difficulty in the same way we did when we faced a similar problem concerning t in Chapter 7. We will simply calculate the relevant df against which to evaluate F—more precisely, we will calculate a statistic denoted as f' and evaluate F_{obt} against $F_{.05}(a - 1, f')$. In this case, Welch (1938) and Satterthwaite (1946) give the value of f' as

$$f' = \frac{(u + v)^2}{\dfrac{u^2}{df_u} + \dfrac{v^2}{df_v}}$$

where

$$u = SS_{Ss\,\text{w/in groups}}$$
$$v = SS_{I \times Ss\,\text{w/in groups}}$$

and df_u and df_v are the corresponding degrees of freedom. For our example,

$$u = 384,722.03 \quad df_u = 21$$
$$v = 281,199.34 \quad df_v = 105$$

$$f' = \frac{(384,722.03 + 281,199.34)^2}{\dfrac{384,722.03^2}{21} + \dfrac{281,199.34^2}{105}} = 56.84$$

Rounding to the nearest integer gives $f' = 57$. Thus, our F is distributed on $(g - 1, f') = (2, 57)$ df under H_0. For 2 and 57 df, $F_{.05} = 3.16$. Only the difference at Interval 1 is significant. By the end of 30 minutes, the three groups were performing at equivalent levels. It is logical to conclude that somewhere between the first and the sixth interval the three groups become nonsignificantly different, and many people test at each interval to find that point. However, I strongly recommend against this practice as a general rule. We have already run a number of significance tests, and running more of them only increases the error rate. Unless there is an important theoretical reason to determine the point at which the group differences become nonsignificant—and I suspect that there are very few such cases—then there is nothing to be gained by testing each interval. Tests should be carried out to answer important questions, not to address idle curiosity or to make the analysis look "complete."

Multiple Comparisons

Several studies have investigated the robustness of multiple-comparison procedures for testing differences among means on the within-subjects variable. Maxwell (1980) studied a simple repeated-measures design with no between-subject component and advised adopting multiple-comparison procedures that do not use a pooled error term. We discussed such a procedure (the Games–Howell procedure) in Chapter 12. (I did use a pooled error term in the analysis of the migraine study, but there it was reasonable to assume homogeneity of variance and I was using all of the weeks. If I had only been running a contrast involving three of the weeks, I would seriously consider calculating an error term based on just the data from those weeks.)

Keselman and Keselman (1988) extended Maxwell's work to designs having one between-subject component and made a similar recommendation. In fact, they showed that when the Groups are of different sizes and sphericity is violated, familywise error rates can become very badly distorted. In the simple effects procedures that we have just considered, I recommended using separate error terms by running one-way repeated-measures analyses for each of the groups. For subsequent multiple-comparison procedures exploring those simple effects, especially with unequal sample sizes, it would probably be wise to employ the Games–Howell procedure using those separate covariance matrices. In other words, to compare Intervals 3 and 4 for the Control group, you would generate your error term using only the Intervals 3 and 4 data from just the Control group.

Myers (1979) has suggested making post hoc tests on a repeated measure using paired t tests and a Bonferroni correction. (This is essentially what I did for the migraine example, though a Bonferroni correction was not necessary because I ran only one contrast.) Maxwell (1980) showed that this approach does a good job of controlling the familywise error rate, and Baker and Lew (1987) showed that it generally compared well against Tukey's test in power. Baker proposed a simple modification of the Bonferroni (roughly in line with that of Holm) that had even greater power.

14.8 Two Between-Subjects Variables and One Within-Subjects Variable

The basic theory of repeated-measures analysis of variance has already been described in the discussion of the previous designs. However, experimenters commonly plan experiments with three or more variables, some or all of which represent repeated measures on the same subjects. We will briefly discuss the analysis of these designs. The calculations are basically very simple because the sums of squares for main effects and interactions are obtained in the usual way and the error terms are obtained by subtraction.

We will not consider the theory behind these designs at any length. Essentially, it amounts to the extrapolation of what has already been said about the two-variable case. For an excellent discussion of the underlying statistical theory, see Winer (1971) or Maxwell and Delaney (1990).

I will take as an example a study by St. Lawrence and colleagues (1995) on an intervention program to reduce the risk of HIV infection among African American adolescents. The study involved comparing two approaches, one of which was a standard 2-hour educational program used as a control condition (EC) and the other was an 8-week behavioral skills training program (BST). Subjects were Male and Female adolescents, and measures were taken at Pretest, Posttest, and 6 and 12 months follow-up (FU6 and FU12). There were multiple dependent variables in the study, but the one that we will consider is $\log(\text{freq} + 1)$, where freq is the frequency of condom-protected intercourse.[3] This is a $2 \times 2 \times 4$ repeated-measures design, with Intervention and Sex as between-subjects factors and Time as the within-subjects factor. This design may be diagrammed as follows, where G_i represents the ith group of subjects.

	Behavioral Skills Training				Educational Control			
	Pretest	Posttest	FU6	FU12	Pretest	Posttest	FU6	FU12
Male	G_1	G_1	G_1	G_1	G_2	G_2	G_2	G_2
Female	G_3	G_3	G_3	G_3	G_4	G_4	G_4	G_4

The raw data and the necessary summary tables of cell totals are presented in Table 14.7a. (These data have been generated to closely mimic the data reported by St. Lawrence et al., though they had many more subjects. Decimal points have been omitted.) Table 14.7b shows the calculations for the main effects and interactions. Here, as elsewhere, the calculations are carried out exactly as they are for any main effects and interactions.

The summary table for the analysis of variance is presented in Table 14.7c. In this table, the ** indicate terms that were obtained by subtraction. Specifically,

$$SS_{\text{w/in subj}} = SS_{\text{total}} - SS_{\text{between subj}}$$

$$SS_{Ss \text{ w/in groups}} = SS_{\text{between subj}} - SS_G - SS_S - SS_{GS}$$

$$SS_{T \times Ss \text{ w/in groups}} = SS_{\text{w/in subj}} - SS_T - SS_{TG} - SS_{TS} - SS_{TGS}$$

These last two terms are the error terms for between-subjects and within-subjects effects, respectively. That these error terms are appropriate is shown by examining the expected mean squares presented in Table 14.8 on page 464. For the expected mean squares of random and mixed models, see Kirk (1968) or Winer (1971).

---

[3] The authors used a logarithmic transformation here because the original data were very positively skewed. They took the log of $(X + 1)$ instead of X because $\log(0)$ is not defined.

From the column of F in the summary table of Table 14.7c, we see that the main effect of Sex is significant, as is the Time $\times$ Group interaction. Both of these results are meaningful. As you will recall, the dependent variable is a measure of the frequency of use of condoms (log(freq + 1)). Examination of the means reveals adolescent girls report a lower frequency of use than do adolescent boys. That could mean either that they have a lower

Table 14.7   Data and analysis of study by St. Lawrence et al. (1995)

(a) Data

	Male				Female			
	Pretest	Posttest	FU6	FU12	Pretest	Posttest	FU6	FU12
Behavioral Skill Training	7	22	13	14	0	6	22	26
	25	10	17	24	0	16	12	15
	50	36	49	23	0	8	0	0
	16	38	34	24	15	14	22	8
	33	25	24	25	27	18	24	37
	10	7	23	26	0	0	0	0
	13	33	27	24	4	27	21	3
	22	20	21	11	26	9	9	12
	4	0	12	0	0	0	14	1
	17	16	20	10	0	0	12	0
Educational Control	0	0	0	0	15	28	26	15
	69	56	14	36	0	0	0	0
	5	0	0	5	6	0	23	0
	4	24	0	0	0	0	0	0
	35	8	0	0	25	28	0	16
	7	0	9	37	36	22	14	48
	51	53	8	26	19	22	29	2
	25	0	0	15	0	0	5	14
	59	45	11	16	0	0	0	0
	40	2	33	16	0	0	0	0

Group $\times$ Sex $\times$ Time means

		Pretest	Posttest	FU6	FU12	Mean
BST	Male	19.7	20.7	24.0	18.1	20.625
BST	Female	7.2	9.8	13.6	10.2	10.200
EC	Male	29.5	18.8	7.5	15.1	17.725
EC	Female	10.1	10.0	9.7	9.5	9.825
Mean		16.625	14.825	13.700	13.225	14.594

Group $\times$ Sex means

	Male	Female	Mean
BST	20.625	10.200	15.412
EC	17.725	9.825	13.775
Mean	19.175	10.012	14.594

(continued)

Table 14.7   *(continued)*

(b) Calculations

$$SS_{total} = \sum(X - \overline{X})^2 = (7 - 14.594)^2 + \cdots + (0 - 14.594)^2 = 35404.594$$

$$SS_{subj} = t\sum(\overline{X}_{Subj} - \overline{X})^2 = 4[(14 - 14.594)^2 + \cdots + (0 - 14.594)^2] = 21490.344$$

$$SS_{group} = nts\sum(\overline{X}_G - \overline{X})^2 = 10 \times 4 \times 2[(15.412 - 14.594)^2 + (13.775 - 14.594)^2] = 107.256$$

$$SS_{Sex} = ntg\sum(\overline{X}_{Sex} - \overline{X})^2 = 10 \times 4 \times 2[(19.175 - 14.594)^2 + (10.012 - 14.594)^2] = 3358.056$$

$$SS_{cells\,GS} = nt\sum(\overline{X}_{cells\,GS} - \overline{X})^2 = 10 \times 4[(20.625 - 14.594)^2 + \cdots + (9.825 - 14.594)^2] = 3529.069$$

$$SS_{GS} = SS_{cells\,GS} - SS_G - SS_S = 3529.069 - 107.256 - 3358.056 = 63.757$$

$$SS_{time} = ngs\sum(\overline{X}_T - \overline{X})^2 = 10 \times 2 \times 2[(16.625 - 14.594)^2 + \cdots + (13.225 - 14.594)^2] = 274.069$$

$$SS_{cells\,TG} = ns\sum(\overline{X}_{cells\,TG} - \overline{X})^2 = 10 \times 2[(13.45 - 14.594)^2 + \cdots + (12.300 - 14.594)^2] = 1759.144$$

$$SS_{TG} = SS_{cells\,TG} - SS_T - SS_G = 1759.144 - 274.069 - 107.256 = 1377.819$$

$$SS_{cells\,TS} = ng\sum(\overline{X}_{cells\,TS} - \overline{X})^2 = 10 \times 2[(24.60 - 14.594)^2 + \cdots + (9.85 - 14.594)^2] = 4412.044$$

$$SS_{TS} = SS_{cells\,TS} - SS_T - SS_S = 4412.044 - 274.069 - 3358.056 = 779.919$$

$$SS_{cells\,GTS} = n\sum(\overline{X}_{cells\,GTS} - \overline{X})^2 = 10[(19.7 - 14.594)^2 + \cdots + (9.50 - 14.594)^2] = 6437.294$$

$$SS_{GTS} = SS_{cells\,GTS} - SS_G - SS_T - SS_S - SS_{GT} - SS_{GS} - SS_{TS}$$
$$= 6437.294 - 107.256 - 274.069 - 3358.056 - 1377.819 - 63.757 - 779.919 = 476.419$$

(c) Summary Table

Source	df	SS	MS	F
Between subjects	39	21,490.344		
Group (Condition)	1	107.256	107.256	0.21
Sex	1	3358.056	3358.056	6.73*
$G \times S$	1	63.757	63.757	0.13
Ss w/in groups**	36	17,961.275	498.924	
Within subjects**	120	13,914.250		
Time	3	274.069	91.356	0.90
$T \times G$	3	1377.819	459.273	4.51*
$T \times S$	3	779.919	259.973	2.55
$T \times G \times S$	3	476.419	158.806	1.56
$T \times$ Ss w/in groups**	108	11,006.025	101.908	
Total	159	35,404.594		

* $p < .05$
** Obtained by subtraction

frequency of intercourse, or that they use condoms a lower percentage of the time. Supplementary data supplied by St. Lawrence et al. show that females do report using condoms a lower percentage of the time than males do, but not enough to account for the difference that we see here. Apparently, what we are seeing is a reflection of the reported frequency of intercourse.

The most important result in this summary table is the Time × Group interaction. This is precisely what we would be looking for. We don't really care about a Group effect because

Table 14.8 Expected mean squares with A, B, and C fixed

Source	df	SS
Between subjects	$abn - 1$	
A	$a - 1$	$\sigma_e^2 + c\sigma_\pi^2 + nbc\sigma_\alpha^2$
B	$b - 1$	$\sigma_e^2 + c\sigma_\pi^2 + nac\sigma_\beta^2$
AB	$(a - 1)(b - 1)$	$\sigma_e^2 + c\sigma_\pi^2 + nc\sigma_{\alpha\beta}^2$
Ss w/in groups	$ab(n - 1)$	$\sigma_e^2 + c\sigma_\pi^2$
Within subjects	$abn(c - 1)$	
C	$c - 1$	$\sigma_e^2 + \sigma_{\gamma\pi}^2 + nab\sigma_\gamma^2$
AC	$(a - 1)(c - 1)$	$\sigma_e^2 + \sigma_{\gamma\pi}^2 + nb\sigma_{\alpha\gamma}^2$
BC	$(b - 1)(c - 1)$	$\sigma_e^2 + \sigma_{\gamma\pi}^2 + na\sigma_{\beta\gamma}^2$
ABC	$(a - 1)(b - 1)(c - 1)$	$\sigma_e^2 + \sigma_{\gamma\pi}^2 + n\sigma_{\alpha\beta\gamma}^2$
$C \times Ss$ w/in groups	$ab(n - 1)(c - 1)$	$\sigma_e^2 + \sigma_{\gamma\pi}^2$
Total	$N - 1$	

we would like the groups to be equal at pretest, and that equality would dilute any overall group difference. Nor do we particularly care about a main effect of Time because we expect the Educational Control group not to show appreciable change over time, and that would dilute any Time effect. What we really want to see is that the BST group increases its use over time, whereas the EC group remains constant. That is an interaction, and that is what we found.

Simple Effects for Complex Repeated-Measures Designs

In the previous example, we saw that tests on within-subjects effects were occasionally disrupted by violations of the sphericity assumption, and we took steps to work around this problem. We will have much the same problem with this example.

The cell means plotted in Figure 14.4 reveal the way in which frequency of condom use changes over time for the two treatment conditions and for males and females separately. We can see from this figure that the data do not tell a simple story.

We are again going to have to distinguish between simple effects on between-subject factors and simple effects on within-subject factors. We will start with between-subject simple effects. We have three different between-subjects simple effects that we could examine—namely, the simple main effects of Condition and Sex at each Time, and the Sex × Condition simple interaction effect at each Time. For example, we might want to check that the two Conditions (BST and EC) do not differ at pretest. Again, we might also want to test that they do differ at FU6 or at FU12. Here, we are really dissecting the Condition × Time interaction effect, which we know from Table 14.7 to be significant.

By far the easiest way to test these between-subjects effects is to run separate two-way (Condition × Sex) analyses at each level of the Time variable. These four analyses will give you all three simple effects at each Time with only minor effort. You can then accept the F values from these analyses, as I have done here for convenience, or you can pool the error terms from the four separate analyses and use that pooled error term in testing the mean square for the relevant effect. If these terms are heterogeneous, you would be wise not to

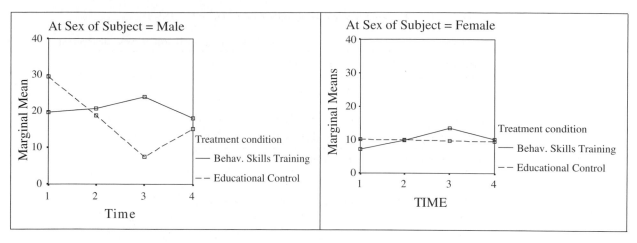

Figure 14.4   Frequency of condom use as a function of Sex and Condition

pool them. On the other hand, if they represent homogeneous sources of variance, they may be pooled, giving you more degrees of freedom for error. For these effects, you don't need to worry about sphericity because each simple effect is calculated on only one level of the repeated-measures variable.

The within-subjects simple effects are handled in much the same way. For example, there is some reason to look at the simple effects of Time for each Condition separately to see whether the EC condition shows changes over time in the absence of a complete intervention. Similarly, we would like to see how the BST condition changes with time. However, we want to include Sex as an effect in both of these analyses so we don't inflate the error term unnecessarily. We also want to use a separate error term for each analysis, rather than pooling these across Conditions.

The relevant analyses are presented in Table 14.9 for simple effects at one level of the other variable. Tests at the other levels would be carried out in the same way. Although this table has more simple effects than we care about, they are presented to illustrate the way in which tests were constructed. You would probably be foolish to consider all the tests that result from this approach because you would seriously inflate the familywise error rate. Decide what you want to look at before you run the analyses, and then stick to that decision. If you really want to look at a large number of simple effects, consider adopting one of the Bonferroni approaches discussed in Chapter 12.

From the between-subjects analysis in Table 14.9a, we see that at Time 1 (Pretest) there was a significant difference between males and females (females show a lower frequency of use). But there were no Condition effects nor was there a Condition × Sex interaction. Males exceed females by just about the same amount in each Condition. The fact that there is no Condition effect is reassuring because it would not be comforting to find that our two conditions differed before we had applied any treatment.

From the results in Table 14.9b, we see that for the BST condition there is again a significant difference due to Sex, but there is no Time effect, nor a Time × Sex interaction. This is discouraging: It tells us that when we average across Sex, there is no change in frequency of condom use as a result of our intervention. This runs counter to the conclusion that we might have drawn from the overall analysis where we saw a significant Condition by Time interaction, and speaks to the value of examining simple effects. That an effect we seek is significant does not necessarily mean that it is significant in the direction we desire.

Table 14.9   Analysis of simple effects

(a) Between-subjects effects (Condition, Sex, and Condition × Sex) at Pretest

Source	df	SS	MS	F
Condition	1	403.225	403.225	1.45
Sex	1	2544.025	2544.025	9.13*
Condition × Sex	1	119.025	119.025	0.43
Error	36	10027.100	278.530	
Total	39	13093.375		

(b) Within-subject effects (Sex, Time, Time × Sex) at BST

Source	df	SS	MS	F
Between subjects	19	7849.13		
Sex	1	2173.61	2173.61	6.89*
Error (between)	18	5675.52	315.30	
Within subjects	60	3646.26		
Time	3	338.94	112.98	1.88
$T \times S$	3	54.54	18.18	0.30
Error (within)	54	3252.78	60.24	
Total	79	11495.39		

* $p < .05$

14.9 Two Within-Subjects Variables and One Between-Subjects Variable

The design we just considered can be seen as a straightforward extension of the case of one between- and one within-subjects variable. All that we needed to add to the summary table was another main effect and the corresponding interactions. However, when we examine a design with two within-subjects main effects, the problem becomes slightly more complicated because of the presence of additional error terms. To use a more generic notation, we will label the independent variables as A, B, and C.

Suppose that as a modification of the previous study we continued to use different subjects for the two levels of variable A, but we ran each subject under all combinations of variables B (Condition) and C (Time). I know that this is not practical in that example, but humor me. For many studies, such a design would be practical and appropriate. This design can be diagrammed as

	A_1			A_2		
	C_1	C_2	C_3	C_1	C_2	C_3
B_1	G_1	G_1	G_1	G_2	G_2	G_2
B_2	G_1	G_1	G_1	G_2	G_2	G_2
B_3	G_1	G_1	G_1	G_2	G_2	G_2

Before we consider an example, we will examine the expected mean squares for this design. These are presented in Table 14.10 for the case of the model in which all factors other

Table 14.10   Expected mean squares

Source	df	E(MS)
Between subjects	$an - 1$	
A (groups)	$a - 1$	$\sigma_e^2 + bc\sigma_\pi^2 + nbc\sigma_\alpha^2$
Ss w/in groups	$a(n - 1)$	$\sigma_e^2 + bc\sigma_\pi^2$
Within subjects	$na(bc - 1)$	
B	$b - 1$	$\sigma_e^2 + c\sigma_{\beta\pi}^2 + nac\sigma_\beta^2$
AB	$(a - 1)(b - 1)$	$\sigma_e^2 + c\sigma_{\beta\pi}^2 + nc\sigma_{\alpha\beta}^2$
B × Ss w/in groups	$a(b - 1)(n - 1)$	$\sigma_e^2 + c\sigma_{\beta\pi}^2$
C	$c - 1$	$\sigma_e^2 + b\sigma_{\gamma\pi}^2 + nab\sigma_\gamma^2$
AC	$(a - 1)(c - 1)$	$\sigma_e^2 + b\sigma_{\gamma\pi}^2 + nb\sigma_{\alpha\gamma}^2$
C × Ss w/in groups	$a(c - 1)(n - 1)$	$\sigma_e^2 + b\sigma_{\gamma\pi}^2$
BC	$(b - 1)(c - 1)$	$\sigma_e^2 + n\sigma_{\beta\gamma\pi}^2 + na\sigma_{\beta\gamma}^2$
ABC	$(a - 1)(b - 1)(c - 1)$	$\sigma_e^2 + n\sigma_{\beta\gamma\pi}^2 + n\sigma_{\alpha\beta\gamma}^2$
BC × Ss w/in groups	$a(b - 1)(c - 1)(n - 1)$	$\sigma_e^2 + n\sigma_{\beta\gamma\pi}^2$
Total	$N - 1$	

than subjects are fixed. From the expected mean squares, it is evident that we will have four error terms for this design. As before, the $MS_{Ss\,w/in\,groups}$ is used to test the between-subjects effect. When it comes to the within-subjects terms, however, B and the interaction of B with A are tested by $B \times Ss$ within groups; C and its interaction with A are tested by $C \times Ss$ within groups; and BC and its interaction with A are tested by $BC \times Ss$ within groups. Why this is necessary is apparent from the expected mean squares. It may be easier to appreciate what is happening, however, if we rewrite the source column of the summary table as in Table 14.11.

Table 14.11   Alternative partition of total variation

Source	df
Between subjects	$an - 1$
A (groups)	$a - 1$
Ss w/in groups	$a(n - 1)$
Within subjects	$na(bc - 1)$
B	$b - 1$
B × S	$(b - 1)(an - 1)$
AB	$(a - 1)(b - 1)$
B × Ss w/in groups	$a(b - 1)(n - 1)$
C	$c - 1$
C × S	$(c - 1)(an - 1)$
AC	$(a - 1)(c - 1)$
C × Ss w/in groups	$a(c - 1)(n - 1)$
BC	$(b - 1)(c - 1)$
BC × S	$(b - 1)(c - 1)(an - 1)$
ABC	$(a - 1)(b - 1)(c - 1)$
BC × Ss w/in groups	$a(b - 1)(c - 1)(n - 1)$
Total	$N - 1$

Here, we can see that each within-subjects effect is considered as interacting with the between-subjects term, and consequently with the partitions of the between-subjects variation. (These new terms are often called *dummy* terms because they play no real role in the analysis.) A similar kind of table could have been drawn up for the previous design, but there was no necessity for it. Here, we must consider the form of Table 14.11 because it gives us a hint as to how we will compute our within-subjects error terms.

Direct computation of error terms in this design would be a most unpleasant undertaking. However, the calculation of the (dummy) terms $SS_{B \times S}$ and $SS_{C \times S}$ is relatively straightforward because they are calculated just as is any interaction term. Because we know from Table 14.11 that $SS_{B \times S}$ is partitioned into SS_{AB} and $SS_{B \times Ss\,w/in\,groups}$, then

$$SS_{B \times Ss\,w/in\,groups} = SS_{B \times S} - SS_{AB}$$

The same reasoning holds for $SS_{C \times Ss\,w/in\,groups}$ and $SS_{BC \times Ss\,w/in\,groups}$. We do not need to calculate $SS_{BC \times S}$ directly, however, because it is readily obtained by subtraction:

$$SS_{BC \times S} = SS_{w/in\,subj} - SS_B - SS_{B \times S} - SS_C - SS_{C \times S} - SS_{BC}$$

An Analysis of Data on Conditioned Suppression

Assume that a tiny "click" on your clock radio always slightly precedes your loud and intrusive alarm going off. Over time that click (psychologists would call it a "CS") could come to elicit the responses normally produced by the alarm (the "US"). Moreover, it is possible that simply presenting the click might lead to the suppression of an ongoing behavior, even if that click is not accompanied by the alarm. (If you were lying there reading, you might pause in your reading.) In a laboratory investigation of how the click affects (suppresses) ongoing behavior, Bouton and Swartzentruber (1985) investigated the degree to which a tone, which had previously been paired with shock, would suppress the rate of an ongoing bar-pressing response in rats. Suppression was measured by taking the ratio of the number of bar presses during a 1-minute test period following the tone to the total number of bar presses during both a baseline period and the test period. For all groups, behavior was assessed in two Phases—a Shock phase (shock accompanied the tone) and a No-shock phase (shock did not accompany the tone) repeated over a series of four Cycles of the experiment.

It may be easier to understand the design of the study if you first glance at the layout of Table 14.12. During Phase I, Group *A-B* was placed in Box *A*. After a 1-minute baseline interval, during which the animal bar-pressed for food, a tone was presented for 1 minute and was followed by a mild shock. The degree of suppression of the bar-pressing response when the tone was present (a normal fear response) was recorded. The animal was then placed in Box *B* for Phase II of the cycle, where, after 1 minute of baseline bar-pressing, only the tone stimulus was presented. Because the tone was previously paired with shock, it should suppress bar-pressing behavior to some extent. Over a series of *A-B* cycles, however, the subject should learn that shock is never administered in Phase II and that Box *B* is therefore a "safe" box. Thus, for later cycles there should be less suppression on the no-shock trials.

Group *L-A-B* was treated in the same way as Group *A-B* except that these animals previously had had experience with a situation in which a light, rather than a tone, had been paired with shock. Because of this previous experience, the authors expected the animals to perform slightly better (less suppression during Phase II) than did the other group, especially on the first cycle or two.

Group *A-A* was also treated in the same way as Group *A-B* except that both Phases were carried out in the same box—Box *A*. Because there were no differences in the test boxes to serve as cues (i.e., animals had no way to distinguish the no-shock from the shock phases), this group would be expected to show the most suppression during the No-shock phases.

Bouton and Swartzentruber predicted that overall there would be a main effect due to Phase (i.e., a difference between shock and no-shock Phases), a main effect due to Groups (*A-B* and *L-A-B* showing less suppression than *A-A*), and a main effect due to Cycles (animals tested in Box *B* would learn over time that it was a safe location). The researchers also predicted that each of the interactions would be significant. (One reason I chose to use this example, even though it is difficult to describe concisely, is that it is one of those rare studies in which all effects are predicted to be significant and meaningful.)

Table 14.12   Analysis of conditioned suppression (Lower scores represent greater suppression.)

(a₁) Data

	Cycle								
	1		2		3		4		
	Phase		Phase		Phase		Phase		Subject
Group	I	II	I	II	I	II	I	II	Mean
A-B	1*	28	22	48	22	50	14	48	29.125
	21	21	16	40	15	39	11	56	27.375
	15	17	13	35	22	45	1	43	23.875
	30	34	55	54	37	57	57	68	49.000
	11	23	12	33	10	50	8	53	25.000
	16	11	18	34	11	40	5	40	21.875
	7	26	29	40	25	50	14	56	30.875
	0	22	23	45	18	38	15	50	26.375
Mean_AB	12.625	22.750	23.500	41.125	20.000	46.125	15.625	51.750	29.188
A-A	1	6	16	8	9	14	11	33	12.250
	37	59	28	36	34	32	26	37	36.125
	18	43	38	50	39	15	29	18	31.250
	1	2	9	8	6	5	5	15	6.375
	44	25	28	42	47	46	33	35	37.500
	15	14	22	32	16	23	32	26	22.500
	0	3	7	17	6	9	10	15	8.375
	26	15	31	32	28	22	16	15	23.125
Mean_AA	17.750	20.875	22.375	28.125	23.125	20.750	20.250	24.250	22.188
L-A-B	33	43	40	52	39	52	38	48	43.125
	4	35	9	42	4	46	23	51	26.750
	32	39	38	47	24	44	16	40	35.000
	17	34	21	41	27	50	13	40	30.375
	44	52	37	48	33	53	33	43	42.875
	12	16	9	39	9	59	13	45	25.250
	18	42	3	62	45	49	60	57	42.000
	13	29	14	44	9	50	15	48	27.750
Mean_LAB	21.625	36.250	21.375	46.875	23.750	50.375	26.375	46.500	34.141
Total	17.333	26.625	22.417	38.708	22.292	39.083	20.750	40.833	28.505

* Decimal points have been omitted in the table, but included in the calculations.

(continued)

Table 14.12 *(continued)*

(a₂) Subtables of means

Phase × Cycle

	Cycle				
	1	2	3	4	Mean
Phase I (Shock)	17.333	22.417	22.292	20.750	20.698
Phase II (No-shock)	26.625	38.708	39.083	40.833	36.312
Mean	21.979	30.562	30.687	30.792	28.505

Group × Cycle

	Cycle				
	1	2	3	4	Mean
A-B	17.688	32.313	33.062	33.688	29.188
A-A	19.313	25.250	21.937	22.250	22.188
L-A-B	28.937	34.125	37.062	36.437	34.141
Mean	21.979	30.562	30.687	30.792	28.505

Phase × Group

	Group			
	A-B	A-A	L-A-B	Mean
Phase I (Shock)	17.938	20.875	23.281	20.698
Phase II (No-shock)	40.438	23.500	45.000	36.312
Mean	29.188	22.188	34.141	28.505

Cycle × Subject

	1	2	3	4	5	6	7	8	9	10	11	12
Cycle 1	14.5	21.0	16.0	32.0	17.0	13.5	16.5	11.0	3.5	48.0	30.5	1.5
Cycle 2	35.0	28.0	24.0	54.5	22.5	26.0	34.5	34.0	12.0	32.0	44.0	8.5
Cycle 3	36.0	27.0	33.5	47.0	30.0	25.5	37.5	26.0	11.5	33.0	27.0	5.5
Cycle 4	31.0	33.5	22.0	62.5	30.5	22.5	35.0	32.5	22.0	31.5	23.5	10.0

	13	14	15	16	17	18	19	20	21	22	23	24
Cycle 1	34.5	14.5	1.5	20.5	38.0	19.5	35.5	25.5	48.0	14.0	30.0	21.0
Cycle 2	35.0	27.0	12.0	31.5	46.0	25.5	42.5	31.0	42.5	24.0	32.5	29.0
Cycle 3	46.5	19.5	7.5	25.0	45.5	25.0	34.0	38.5	43.0	34.0	47.0	29.5
Cycle 4	34.0	29.0	12.5	15.5	43.0	37.0	28.0	26.5	38.0	29.0	58.5	31.5

(continued)

Phase × Subject

	Subject											
	1	2	3	4	5	6	7	8	9	10	11	12
Phase 1	14.75	15.75	12.75	44.75	10.25	12.50	18.75	14.00	9.25	31.25	31.00	5.25
Phase 2	43.50	39.00	35.00	53.25	39.75	31.25	43.00	38.75	15.25	41.00	31.50	7.50
	13	14	15	16	17	18	19	20	21	22	23	24
Phase 1	38.00	21.25	5.75	25.25	37.50	10.00	27.50	19.50	36.75	10.75	31.50	12.75
Phase 2	37.00	23.75	11.00	21.00	48.75	43.50	42.50	41.25	49.00	39.75	52.50	42.75

(b) Calculations

(*Notation:* Subjects; $n = 8$: Groups; $g = 3$: Phase; $p = 2$: Cycles; $c = 4$)

$$SS_{\text{total}} = \sum(X - \overline{X})^2 = 5.2400$$

$$SS_{\text{subjects}} = pc\sum(\overline{X}_{\text{subj}} - \overline{X})^2 = 2 \times 4[(29.125 - 28.505)^2 + \cdots + (27.750 - 28.505)^2] = 2.0340$$

$$SS_{\text{groups}} = ncp\sum(\overline{X}_G - \overline{X})^2 = 8 \times 4 \times 2[(29.188 - 28.505)^2 + \cdots + (34.141 - 28.505)^2] = 0.4617$$

$$SS_{\text{Ss w/in groups}} = SS_{\text{subjects}} - SS_G = 2.0340 - 0.4617 = 1.5723$$

$$SS_{\text{cycle}} = ngp\sum(\overline{X}_C - \overline{X})^2 = 8 \times 3 \times 2[(21.979 - 28.505)^2 + \cdots + (30.792 - 28.505)^2] = 0.2727$$

$$SS_{\text{cells}(CG)} = np\sum(\overline{X}_{\text{cells}(CG)} - \overline{X})^2 = 8 \times 2[(17.688 - 28.505)^2 + \cdots + (36.437 - 28.505)^2] = 0.8391$$

$$SS_{CG} = SS_{\text{cells}(CG)} - SS_C - SS_G = 0.8391 - 0.2727 - 0.4617 = 0.1047$$

$$SS_{\text{cells}(CS)} = p\sum(\overline{X}_{\text{cells}(CS)} - \overline{X})^2 = 2[(14.500 - 28.505)^2 + \cdots + (31.500 - 28.505)^2] = 2.8876$$

$$SS_{CS} = SS_{\text{cells}(CS)} - SS_C - SS_{\text{subj}} = 2.8876 - 0.2727 - 2.0340 = 0.5809$$

$$SS_{C \times \text{Ss w/in groups}} = SS_{CS} - SS_{CG} = 0.5809 - 0.1047 = 0.4762$$

$$SS_{\text{phase}} = ncg\sum(\overline{X}_P - \overline{X})^2 = 8 \times 4 \times 3[(20.698 - 28.505)^2 + (36.312 - 28.505)^2] = 1.1703$$

$$SS_{\text{cells}(PG)} = nc\sum(\overline{X}_{\text{cells}(PG)} - \overline{X})^2 = 8 \times 4[(17.938 - 28.505)^2 + \cdots + (45.000 - 28.505)^2] = 2.0374$$

$$SS_{PG} = SS_{\text{cells}(PG)} - SS_P - SS_G = 2.0374 - 1.1703 - 0.4617 = 0.4054$$

$$SS_{\text{cells}(PS)} = c\sum(\overline{X}_{\text{cells}(PS)} - \overline{X})^2 = 4[(14.750 - 28.505)^2 + \cdots + (42.750 - 28.505)^2] = 3.7990$$

$$SS_{PS} = SS_{\text{cells}(PS)} - SS_P - SS_{\text{subj}} = 3.7990 - 1.1703 - 2.0340 = 0.5947$$

$$SS_{P \times \text{Ss w/in groups}} = SS_{PS} - SS_{PG} = 0.5947 - 0.4054 = 0.1893$$

$$SS_{\text{cells}(CP)} = ng\sum(\overline{X}_{\text{cells}(CP)} - \overline{X})^2 = 8 \times 3[(17.333 - 28.505)^2 + \cdots + (40.833 - 28.505)^2] = 1.5172$$

$$SS_{CP} = SS_{\text{cells}(CP)} - SS_C - SS_P = 1.5172 - 0.2727 - 1.1703 = 0.0742$$

$$SS_{\text{cells}(CPG)} = n\sum(\overline{X}_{\text{cells}(CPG)} - \overline{X})^2 = 8[(12.625 - 28.505)^2 + \cdots + (46.500 - 28.505)^2] = 2.6164$$

$$SS_{CPG} = SS_{\text{cells}(CPG)} - SS_C - SS_P - SS_G - SS_{CP} - SS_{CG} - SS_{PG}$$
$$= 2.6164 - 0.2727 - 1.1703 - 0.4617 - 0.0742 - 0.1047 - 0.4054 = 0.1274$$

$$SS_{CP \times \text{Ss w/in groups}} = SS_{\text{total}} - \sum(\text{all other effects}) = 5.2400 - 0.4617 - 1.5723 - 0.2727$$
$$- 0.1047 - 0.4762 - 1.1703 - 0.4054 - 0.1893 - 0.0742 - 0.1274$$
$$= 0.3858$$

(*continued*)

Table 14.12   *(continued)*

(c) Summary Table

Source	df	SS	MS	F
Between subjects	23	2.0340		
Group (Condition)	2	0.4617	0.2308	3.08
*S*s w/in groups**	21	1.5723	0.0749	
Within subjects**	168	3.2060		
Cycle	3	0.2727	0.0909	11.96*
$C \times G$	6	0.1047	0.0175	2.30*
$C \times S$s w/in groups	63	0.4762	0.0076	
Phase	1	1.1703	1.1703	130.03*
$P \times G$	2	0.4054	0.2027	22.52*
$P \times S$s w/in groups	21	0.1893	0.0090	
$C \times P$	3	0.0742	0.0247	4.05*
$C \times P \times G$	6	0.1274	0.0212	3.48*
$CP \times S$s w/in groups**	63	0.3858	0.0061	
Total	191	5.2400		

* $p < .05$
** Obtained by subtraction

The data and analysis of variance for this study are presented in Table 14.12. The analysis has not been elaborated in detail because it mainly involves steps that you already know how to do. The results are presented graphically in Figure 14.5 for convenience, and for the most part they are clear-cut and in the predicted direction. Keep in mind that for these data a lower score represents more suppression—that is, the animals are responding more slowly. Calculating the sums of squares for the error terms may be somewhat easier to understand if you compare this summary table and Table 14.11. I have given an intermediate

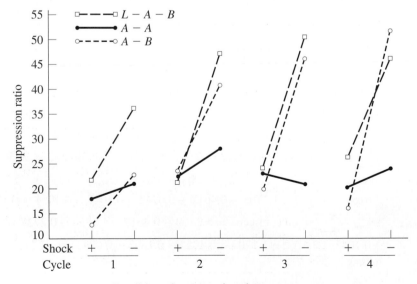

Figure 14.5   Conditioned suppression data

step so that if you have any question about how each term is calculated, you can work it out and compare your answer with mine. If your answer matches mine, you are almost certainly performing the calculations correctly. Beware that the data are given without decimals, to conserve space, but the calculations were carried out with the decimals in place. (If you do the calculations without the decimals, your *SS* and *MS* will be 10,000 times mine, though your *F*s will be the same.)

From the summary table in Table 14.12c, we can see that nearly all the predictions were supported. The only effect that was not significant was the main effect of Groups, but that effect is not crucial because it represents an average across the shock and the no-shock phases, and the experimenters had predicted little or no group differences in the shock phase. In this context, the Phase × Group interaction is of more interest, and it is clearly significant.

The presence of an interpretable three-way interaction offers the opportunity to give another example of the use of simple interaction effects. We would have predicted that all groups would show high levels of suppression of the shock trials on all Cycles because anticipated shock is clearly disruptive. On no-shock trials, however, Groups *A-B* and *L-A-B* should show less suppression (higher scores) than Group *A-A*, and this latter difference should increase with Cycles. In other words, there should be a Groups × Cycles interaction for the no-shock trials, but no such interaction for the shock trials. The simple effects are shown in Table 14.13. The calculation of the appropriate tests was carried out the same way

Table 14.13   Simple interaction effects

(a) Within-subject effects (Group £ Cycle at Phase I)

Source	df	SS	MS	F
Between subjects	23	1.2154		
Group (Condition)	2	0.0458	0.0229	<1
*S*s w/in groups	21	1.1696	0.0557	
Within subjects	72	0.5691		
Cycle	3	0.0404	0.0135	1.74
$C \times G$	6	0.0416	0.0069	<1
$C \times S$s w/in groups	63	0.4871	0.0077	
Total	95	1.7845		

$\hat{\varepsilon} = .7971; \tilde{\varepsilon} = .9922$

(b) Within-subject effects (Group × Cycle at Phase II)

Source	df	SS	MS	F
Between subjects	23	1.4133		
Group (Condition)	2	0.8213	0.4106	14.57*
*S*s w/in groups	21	0.5920	0.0282	
Within subjects	72	0.8719		
Cycle	3	0.3065	0.1022	17.03*
$C \times G$	6	0.1905	0.0318	5.30*
$C \times S$s w/in groups	63	0.3749	0.0060	
Total	95	2.2852		

* $p < .05$; $\hat{\varepsilon} = .7583$; $\tilde{\varepsilon} = .9363$

it was earlier, by running a reduced analysis of variance at each level of the Phase variable. Here again, we are using separate error terms to test the Shock and No-shock effects, thus reducing problems with the sphericity assumption. Because we still have a within-subjects term in each of our analyses, I have included $\hat{\varepsilon}$ and $\tilde{\varepsilon}$, although the conclusions would not be affected by adjusting the degrees of freedom in this case. (Again, just because the analyses also give simple effects due to Groups and Cycles is no reason to feel an obligation to interpret them. If they don't speak to issues raised by the experimental hypotheses, they should neither be reported nor interpreted unless you take steps to minimize the increase in the experimentwise error rate.)

From the simple interaction effects of Group × Cycle at each level of Phase, you can see that Bouton and Swartzentruber's predictions were upheld. There is no Cycle × Group interaction on Shock trials, but there is a clear interaction on No-shock trials.

14.10 Intraclass Correlation

One of the important issues in designing experiments in any field is the question of the reliability of the measurements. Most of you would probably expect that the *last* place to look for anything about reliability is in a discussion of the analysis of variance, but that is exactly where you will find it. (For additional material on the intraclass correlation, go to http://www.uvm.edu/~dhowell/StatPages/More_Stuff/icc/icc.html)

Suppose that we are interested in measuring the reliability with which judges rate the degree of prosocial behavior in young children. We might investigate this reliability by having two or more judges each rate a behavior sample of a number of children, assigning a number from 1 to 10 to reflect the amount of prosocial behavior in each behavior sample. I will demonstrate the procedure with some extreme data that were created to make a point. Look at the data in Table 14.14.

Table 14.14 Data for intraclass correlation examples

	(a) Judge			(b) Judge			(c) Judge		
Child	I	II	III	I	II	III	I	II	III
1	1	1	2	1	0	3	1	3	7
2	3	3	3	3	2	5	3	1	5
3	5	5	5	5	4	7	5	7	4
4	5	6	6	5	4	7	5	5	5
5	7	7	7	7	6	8	7	6	7

In Table 14.14a, the judges are in almost perfect agreement. They all see wide differences between children, they all agree on which children show high levels of prosocial behavior and which show low levels, *and* they are nearly in agreement on how high or low those levels are. In this case, nearly all of the variability in the data involves differences among children—there is almost no variability among judges and almost no random error.

In Table 14.14b, we see much the same pattern, but with a difference. The judges do see overall differences among the children, and they do agree on which children show the highest (and lowest) levels of the behavior. But the judges disagree in terms of the amount of prosocial behavior they see. Judge II sees slightly less behavior than Judge I (his mean is 1 point lower), and Judge III sees relatively more behavior than do the others. In other words, although the judges agree on *ordering* children, they disagree on *level*. Here, the data involve both variability among children and variability among judges. However, the random error component is still very small. This is often the most realistic model of how people rate behavior because each of us has a different understanding of how much behavior is required to earn a rating of "7," for example. Our assessment of the reliability of a rating system must normally consider variability between judges.

Finally, Table 14.14c shows a pattern where the judges disagree in level and disagree in ordering children. A large percentage of the variability in these data is error variance.

So what do we do when we want to talk about reliability? One way to measure reliability when judges use only a few levels or categories is to calculate the percentage of times that two judges agree on their rating, but this measure is biased because of high levels of chance agreement whenever one or two categories predominate. (But see the discussion earlier of Cohen's kappa in Chapter 6.) Another common approach is to correlate the ratings of two judges, and perhaps average pairwise correlations if you have multiple judges. But this approach will not consider differences between judges. (If one judge always rates five points higher than another judge, the correlation will be 1.00, but the judges are saying different things about the subjects.) A third way is to calculate what is called the **intraclass correlation,** accounting for differences between judges. That is what we will do here.

intraclass correlation

You can calculate an intraclass correlation coefficient in a number of different ways, depending on whether you treat judges as a fixed or random variable and whether judges evaluate the same or different subjects. Shrout and Fleiss (1979) discuss several alternative approaches. I am going to discuss only the most common approach here, one in which we consider our judges to be a random sample of all judges we could have used and in which each judge rates the same set of subjects once. (In what follows, I am assuming that judges are rating "subjects," but they could be rating pictures, cars, or the livability of cities. Take the word "subject" as a generic term for whatever is being rated.)

We will start by assuming that the data in Table 14.14 can be represented by the following model:

$$X_{ij} = \mu + \alpha_i + \pi_j + \alpha\pi_{ij} + e_{ij}$$

In this model, α_i stands for the effect of the ith judge, π_j stands for the effect of the jth subject (person), $\alpha\pi_{ij}$ is the interaction between the ith judge and the jth subject (the degree to which the judge changes his or her rating system when confronted with that subject), and e_{ij} stands for the error associated with that specific rating. Because each judge rates each subject only once, it is not possible in this model to estimate $\alpha\pi_{ij}$ and e_{ij} separately, but it is necessary to keep them separate in the model.

If you look back to the previous chapter, you will see that when we calculated a magnitude-of-effect measure (which was essentially an r-family measure), we took the variance estimate for the effect in question (in this case differences among subjects) relative to the sum of the estimates of the several sources of variance. That is precisely what we are going to do here. We will let

Intraclass correlation $= \sigma_\pi^2 / \left(\sigma_\alpha^2 + \sigma_\pi^2 + \sigma_{\alpha\pi}^2 + \sigma_e^2\right)$

If most of the variability in the data is due to differences between subjects, with only a small amount due to differences between judges, the interaction of judges and subjects, and error, then this ratio will be close to 1.00. If judges differ from one another in how high or low they rate people in general, or if there is a judge by subject interaction (different judges rate different people differently), or if there is a lot of error in the ratings, the denominator will be substantially larger than the numerator and the ratio will be much less than 1.00.

To compute the intraclass correlation, we are first going to run a Subjects × Judges analysis of variance with Judges as a repeated measure. Because each judge rates each subject only once, there will not be an independent estimate of error, and we will have to use the Judge × Subject interaction as the error term. From the summary table that results, we will compute our estimate of the intraclass correlation as

$$\text{Intraclass correlation} = \frac{MS_{\text{Subjects}} - MS_{J \times S}}{MS_{\text{Subjects}} + (j-1)MS_{J \times S} + j(MS_{\text{Judge}} - MS_{J \times S})/n}$$

where j represents the number of judges and n represents the number of subjects.

To illustrate this, I have run the analysis of variance on the data in Table 14.14b, which is the data set where I have deliberately built in some differences due to subjects and judges. The summary table for this analysis follows.

Source	df	SS	MS	F
Between subjects	4	57.067	14.267	
Within subjects	10	20.666	2.067	
Judge	2	20.133	10.067	150.25
Judge × Subjects	8	0.533	0.067	
Total	14	77.733		

We can now calculate the intraclass correlation as

$$\text{Intraclass correlation} = \frac{14.267 - 0.067}{14.267 + (3-1)0.067 + 3(10.067 - 0.067)/5}$$
$$= \frac{14.200}{14.267 + 0.134 + 6} = \frac{14.2}{20.401} = .70$$

Thus, our measure of reliability is .70, which is probably not as good as we would like to see it. But we can tell from the calculation that the main thing that contributed to low reliability was not error, but differences among judges. This would suggest that we need to have our judges work together to decide on a consistent scale where a "7" means the same thing to each judge.

14.11 Other Considerations

Sequence Effects

sequence effects

carryover effects

Repeated-measures designs are notoriously susceptible to **sequence effects** and **carryover** (practice) **effects.** Whenever the possibility exists that exposure to one treatment will influence the effect of another treatment, the experimenter should consider very seriously before deciding to use a repeated-measures design. In certain studies, carryover effects are desirable. In learning studies, for example, the basic data represent what is carried over from one trial to another. In most situations, however, carryover effects (and especially differential carryover effects) are considered a nuisance—something to be avoided.

The statistical theory of repeated-measures designs assumes that the order of administration is randomized separately for each subject. In some situations, however, it makes more sense to assign testing sequences by means of a **Latin square** or some other device. Although this violates the assumption of randomization, in some situations, the gains outweigh the losses. What is important, however, is that random assignment, Latin squares, and so on do not in themselves eliminate sequence effects. Ignoring analyses in which the data are *analyzed* by means of a Latin square or a related statistical procedure, any system of

Latin square

assignment simply distributes sequence and carryover effects across the cells of the design, with luck lumping them into the error term(s). The phrase "with luck" implies that if this does not happen, the carryover effects will be confounded with treatment effects and the results will be very difficult, if not impossible, to interpret. For those students particularly interested in examining sequence effects, Winer (1971), Kirk (1968), and Cochran and Cox (1957) present excellent discussions of Latin square and related designs.

Unequal Group Sizes

One of the pleasant features of repeated-measures designs is that when a subject fails to arrive for an experiment, it usually means that that subject is missing from every cell in which he was to serve. This has the effect of keeping the cell sizes proportional, even if unequal. From this, it follows that the solution for proportionally unequal sample sizes is possible, if the experimenter thinks the solution's treatment of sample sizes is appropriate. Otherwise, the unweighted-means solution is available, and most often preferable, for most cases. If you are so unlucky as to have a subject for whom you have partial data, the best procedure would probably be to eliminate that subject from the analysis. If, however, only one or two scores are missing, it is possible to replace them with estimates, and in many cases, this is a satisfactory approach. For a discussion of this topic, see Federer (1955, pp. 125–126, 133ff) and especially Little and Rubin (1987).

Matched Samples and Related Problems

randomized blocks designs

matched samples

In discussing repeated-measures designs, we have spoken in terms of repeated measurements on the same subject. Although this represents the most common instance of the use of these designs, it is not the only one. The specific fact that a subject is tested several times really has nothing to do with the matter. Technically, what distinguishes repeated-measures designs (or, more generally, **randomized blocks designs,** of which repeated-measures designs are a special case) from the common factorial designs with equal ns is that for repeated-measures designs, the off-diagonal elements of $\sum$ do not have an expectation of zero—that is, the treatments are correlated. Repeated use of the same subject leads to such correlations, but so does use of **matched samples** of subjects. Thus, for example, if we formed 10 sets of three subjects each, with the subjects matched on driving experience, and then set up an experiment in which the first subject under each treatment came from the same matched triad, we would have correlations among treatments and would thus have a repeated-measures design. Any other data-collection procedure leading to nonzero correlations (or covariances) could also be treated as a repeated-measures design.

14.12 A Computer Analysis Using a Traditional Approach

SPSS is one of the most widely used statistical programs, and its printout for a repeated-measures analysis of variance is one of the most unusual. For this reason, I have presented the analysis of the data in Table 14.4 in Exhibit 14.1 You need to work through this printout carefully because it is not laid out the way you would expect it to be. Notice that the analysis contains a test of sphericity on the matrix of the orthogonal components of error and computes both the Greenhouse and Geisser $\hat{\varepsilon}$ and the Huynh and Feldt $\tilde{\varepsilon}$. (I have chosen to ignore the matrix of orthogonal components of error because of the complexity of

Menu selections

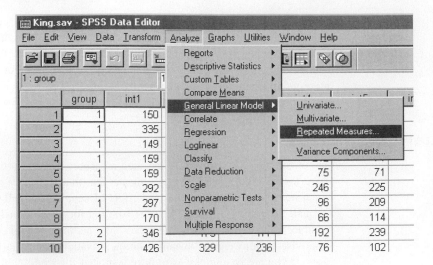

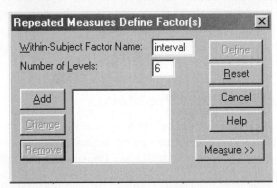

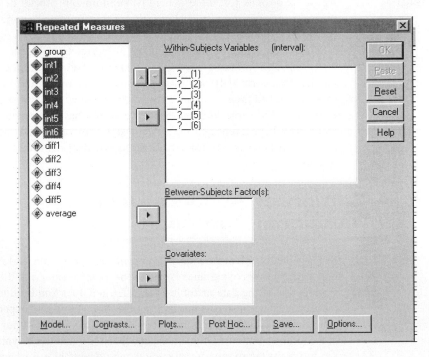

(continued)

Exhibit 14.1 SPSS printout for data from Table 14.4

Selected output

Tests of Within-Subjects Effects

Measure: MEASURE_1

Source		Type III Sum of Squares	df	Mean Square	F	Sig.
INTERVAL	Sphericity Assumed	399736.6	5	79947.313	29.852	.000
	Greenhouse-Geisser	399736.6	3.285	121695.6	29.852	.000
	Huynh-Feldt	399736.6	4.337	92166.251	29.852	.000
	Lower-bound	399736.6	1.000	399736.6	29.852	.000
INTERVAL*GROUP	Sphericity Assumed	80819.958	10	8081.996	3.018	.002
	Greenhouse-Geisser	80819.958	6.569	12302.398	3.018	.009
	Huynh-Feldt	80819.958	8.674	9317.227	3.018	.004
	Lower-bound	80819.958	2.000	40409.979	3.018	.070
Error(INTERVAL)	Sphericity Assumed	281199.3	105	2678.089		
	Greenhouse-Geisser	281199.3	68.979	4076.581		
	Huynh-Feldt	281199.3	91.080	3087.401		
	Lower-bound	281199.3	21.000	13390.443		

Tests of Between-Subjects Effects

Measure: MEASURE_1

Transformed Variable: Average

Source	Type III Sum of Squares	df	Mean Square	F	Sig.
Intercept	4113798	1	4113798	224.551	.000
Group	285815.0	2	142907.5	7.801	.003
Error	384722.1	21	18320.098		

Exhibit 14.1   *(continued)*

the explanation, but the sphericity test on that matrix is a test of the necessary assumption of sphericity.) Alternatively, you can use one of the programs shown in Section 14.13 on MANOVA. Although the printout is somewhat more difficult to read, you do have the advantage of seeing the values for both the MANOVA and standard repeated-measures procedure. As long as you are looking at overall effects (or simple effects with separate error terms), I suggest relying on the standard design with correction for degrees of freedom.

Notice that, unlike the summary tables in this book, SPSS prints out the tests on the within-subject effects first, and then the between-subjects effects. This certainly isn't wrong, but it is different. Mauchly's test of sphericity (not shown) is significant. This gives us even more reason to correct the degrees of freedom using $\hat{\varepsilon}$ or $\tilde{\varepsilon}$.

The within-subjects part of the summary table comes first and includes the corrected p values using $\hat{\varepsilon}$ or $\tilde{\varepsilon}$. Using the Greenhouse and Geisser correction (as the more conservative of the two for this sample size), we can see that there are significant effects for Interval and the Group $\times$ Interval interaction. From the next table in the printout, which reports the analysis for the between-subjects part of the analysis, there is a significant difference due to Group. These analyses agree completely with the analysis given in Table 14.4.

14.13 Multivariate Analysis of Variance for Repeated-Measures Designs

Earlier in the chapter I said that the standard repeated-measures analysis of variance requires an assumption about the variance-covariance matrix known as *sphericity*, a specific form of which is known as *compound symmetry*. When we discussed $\hat{\varepsilon}$ and $\tilde{\varepsilon}$, we were concerned with correction factors that we could apply to the degrees of freedom to circumvent some of the problems associated with a failure of the sphericity assumption.

There is a considerable literature on repeated-measures analyses and their robustness in the face of violations of the underlying assumptions. Although there is not universal agreement that the adjustments proposed by Greenhouse and Geisser and by Huynh and Feldt are successful, the adjustments work reasonably well as long as we stick to overall main or interaction effects. Where we encounter serious trouble is when we try to run individual contrasts or simple effects analyses. Boik (1981) has shown that in these cases the repeated-measures analysis is remarkably sensitive to violations of the sphericity assumption unless we adopt separate error terms for each contrast. In several sections of this chapter, I have indicated how to obtain those separate error terms.

A number of authors have suggested that we would be much further ahead if we used the multivariate analysis of variance (MANOVA) to analyze data having repeated measurements. MANOVA is considerably more complex and in some cases has less power, especially for studies with small numbers of observations, than a repeated-measures analysis, but it does not require the restrictive assumption of sphericity. Although not all authors agree that MANOVA is generally preferable to the standard repeated-measures analysis, psychologists can no longer continue to ignore that more complex approach in the hopes that it will go away. This section is designed to provide only an introduction to the topic, but I have included enough material to allow you to do meaningful analyses.

A multivariate analysis of variance is basically an analysis of variance that deals with more than one dependent variable at the same time. In the most common type of multivariate analysis of variance, we might want to compare three groups differing with respect to the type of therapy they received, on four different measures of depression. In this case, each subject has scores on all four depression measures, and we want to treat those four scores in the *same* analysis rather than running a separate analysis of variance on each measure. A second example of MANOVA, and the one to be elaborated on in this chapter, is the specific case in which the multiple dependent variables are, originally, repeated measurements as we normally think of that term (e.g., trials). In particular, I am assuming that you have measured the same dependent variable at multiple points in time, and that you want to deal primarily with how that variable changes over time.

Essentially, you can think of MANOVA as extracting a linear combination of the dependent variables (which is optimal in some sense) and applying a standard analysis of variance to that linear composite. How those linear composites are obtained and how the significance test is evaluated are beyond the scope of this book, but I will attempt to provide a simple example to illustrate the basic approach. More complete coverage can be found in Harris (1985), Stevens (1992), and Tabachnick and Fidell (1996) (see particularly their chapter on profile analysis). A very good nontechnical discussion can be found in O'Brien and Kaiser (1985). Most textbook discussions are designed to provide a thorough understanding of the logic behind the tests and the possibilities that exist for structuring contrasts to ask meaningful questions. However, many of the standard computer programs have defaults that make setting up the analysis considerably easier than the discussions might lead you to expect. In particular, SAS and SPSS allow you to obtain the MANOVA results in what looks more like a standard repeated-measures summary table, without going through most of the calculations discussed later. However, it is worth taking the time to understand the underlying process because you can run more powerful analyses if you carefully design them to address the questions of interest to you.

We will use the example of drug tolerance and ambulatory behavior found in Table 14.4. If you have statistical software with a "MANOVA" command, it would be tempting to enter the data into a data set with one variable representing group membership, and then to run a straightforward MANOVA on the data, treating the different intervals as different dependent variables. (This could easily be done using **SPSS/General Linear Model/Multivariate.**) But this analysis would not really address the most important questions. It would simply ask whether the three groups differed on the set of dependent variables. It would not get at differences among the different dependent variables (Intervals), nor would it get at the Group × Interval interaction, which is probably the most interesting effect. We are going to need to restructure the data in such a way as to get at all three effects. In what follows, I am using the **SPSS/General Linear Model/Multivariate** procedure to run the analyses, but I am not running that on the original variables, for reasons given earlier.

One of the things we are most interested in with this experiment is change over time. Therefore, we will start by creating new variables that represent the amount of change from interval to interval. For example, define

$$\text{Diff}_1 = \text{Interval}_2 - \text{Interval}_1$$
$$\text{Diff}_2 = \text{Interval}_3 - \text{Interval}_2$$
$$\cdots$$
$$\text{Diff}_5 = \text{Interval}_6 - \text{Interval}_5$$

Each of these new change or difference variables measures how much change has taken place at each interval, and the set can be used to directly assess the Group × Interval interaction. Suppose that there is no interaction. Then each group will change just as much as any other group during a particular time period. The means on the original variables don't have to remain constant from interval to interval, and if there is an Interval effect, they won't. However, if there is no interaction, each group would be expected to show as much, or as little, change as every other group. In other words, in the absence of an interaction, there would be no Group effect for the change scores. Group differences on the *change* scores really represent our test for interaction.

The difference scores also offer a test on the Interval effect. If there is no effect due to Interval, then we would expect the mean change from one interval to another to be 0.0 for each difference score. Thus, in the absence of an Interval effect, the grand mean of the change scores would be 0. All we need to do is test H_0: $\mu = 0$, where μ represents the grand mean. Such a test is provided by an F on the mean. Depending on the program we use, such a test is labeled as a test on the MEAN, CONSTANT, or INTERCEPT.

Using the difference scores, we have found ways to test both of the within-subjects terms. The only term left is the Group difference, which is our only between-subjects effect. As you might suspect, the Group effect is a test on group differences in the average score that each rat received. In other words, we average over the six intervals. The simplest way to do this is to create one more variable called Average:

$$\text{Average} = \frac{\text{Interval}_1 + \text{Interval}_2 + \cdots + \text{Interval}_5 + \text{Interval}_6}{6}$$

We now ask whether there is a difference between the three groups on Average. This test requires a separate analysis, however. Because each subject has only one score (Average), we can use a simple **univariate** one-way analysis of variance. (An analysis of variance with only one dependent variable is often referred to as a univariate analysis, in contrast to the

univariate

multivariate analysis of variance with multiple dependent variables.) We have to do this here because SPSS will not allow us to run MANOVA with only one dependent variable. The results are shown in a very truncated form in Exhibit 14.2.

For each effect, you will see four separate test statistics, and for the interaction effect, there are four different values of F, depending on which statistic you chose. Although Harris

Menu selections

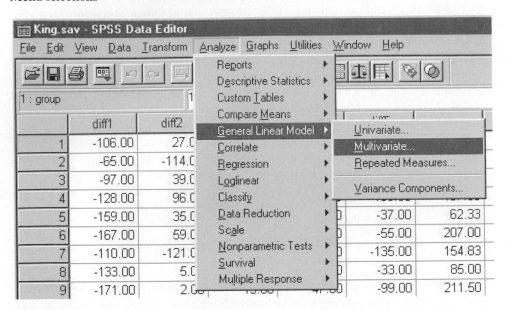

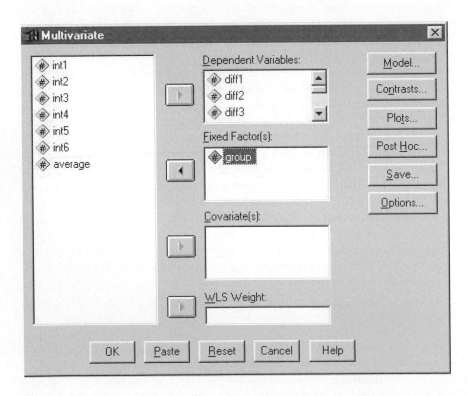

(continued)

Exhibit 14.2   Output of SPSS-MANOVA program for analysis of data in Table 14.4

MANOVA tests on difference scores

Multivariate Tests[c]

Effect		Value	F	Hypothesis df	Error df	Sig.
Intercept	Pillai's Trace	.849	19.110[a]	5.000	17.000	.000
	Wilks' Lambda	.151	19.110[a]	5.000	17.000	.000
	Hotelling's Trace	5.621	19.110[a]	5.000	17.000	.000
	Roy's Largest Root	5.621	19.110[a]	5.000	17.000	.000
GROUP	Pillai's Trace	.753	2.173	10.000	36.000	.043
	Wilks' Lambda	.367	2.216[a]	10.000	34.000	.041
	Hotelling's Trace	1.403	2.244	10.000	32.000	.041
	Roy's Largest Root	1.109	3.992[b]	5.000	18.000	.013

[a] Exact statistic

[b] The statistic is an upper bound on F that yields a lower bound on the significance level.

[c] Design: Intercept + GROUP

Tests of Between-Subjects Effects

Dependent Variable: AVERAGE

Source	Type III Sum of Squares	df	Mean Square	F	Sig.
Corrected Model	47635.840[a]	2	23817.920	7.801	.003
Intercept	685633.0	1	685633.0	224.551	.000
GROUP	47635.840	2	23817.920	7.801	.003
Error	64120.344	21	3053.350		
Total	797389.2	24			
Corrected Total	111756.2	23			

[a] R Squared = .426 (Adjusted R Squared = .372)

Exhibit 14.2   *(continued)*

(1985) has argued strongly for the greatest characteristic root (Roy's) statistic, the literature generally appears to support the use of Pillai's trace. In our case, it doesn't really matter because each of these would lead to the same conclusion. We will reject all three null hypotheses and conclude that there is a significant Group effect, a significant Interval effect, and a significant Group × Interval interaction. These are the same conclusions that we reached in conjunction with the analysis in Table 14.4, but the magnitudes of the *F* values for the within-subjects components are noticeably different. The *F* for Intervals has dropped from 29.85 to 19.11, whereas the *F* for the Group × Interval interaction has dropped from 3.02 to 2.17 (using Pillai's trace statistic). Earlier, I pointed out that if we took the most conservative approach to possible violations of sphericity and evaluated the interaction on $(g - 1)$ and $g(n - 1)$ *df*, we would have declared it nonsignificant, whereas when we calculated $\hat{\varepsilon}$ and $\tilde{\varepsilon}$, the effect was significant. Here, we have a test that does not depend on the validity of the sphericity assumption and it also tells us that the difference is significant. The finding is reassuring. Notice that the *F* on the between-subjects term is unchanged, as it should be.

If you ran a standard *repeated measures* analysis of variance using SPSS, the printout would include the MANOVA table that you see in Exhibit 14.2. The program itself generates the difference scores and the average score, and uses the same procedures as the General Linear Model/MULTIVARIATE procedure. This means that you do not have to generate those difference scores, nor the average score, but that also means that you do not have as nice a way of looking at simple effects, which follow.

Simple Effects

In our previous analysis of these data, we looked at simple effects to try to understand better what the significant F values mean. We can do the same thing with MANOVA, and in some ways it is easier. Looking first at the interaction term, we can pull it apart in several different ways. In the first place, we could run the MANOVA analysis separately for each of the groups, checking for an Interval effect. We could do this either by feeding in separate sets of data to each analysis or telling the program to perform separate analyses on the different groups. To go even further, we could revert to the original variables (e.g., Int1) in place of the difference scores and ask for univariate analyses on them. Each of these analyses would give us meaningful information on the interaction.

To further examine Group or Individual differences, we could run multiple comparison procedures on either variable. This is particularly easy on between-subjects variables such as Group because those effects are tested as in a standard analysis of variance.

In creating the variables labeled Diff 1, Diff 2, . . . , Diff 5, I was using only one way of contrasting the levels of the original Interval variable. In effect, I was using a set of contrasts of the form

$$\begin{bmatrix} 1 & -1 & 0 & 0 & 0 & 0 \end{bmatrix}$$
$$\begin{bmatrix} 0 & 1 & -1 & 0 & 0 & 0 \end{bmatrix}$$
$$\begin{bmatrix} 0 & 0 & 1 & -1 & 0 & 0 \end{bmatrix}$$
$$\cdots$$

For example,

$$\text{Diff 1} = 1(\text{Int1}) - 1(\text{Int2}) + 0(\text{Int3}) + 0(\text{Int4}) + 0(\text{Int5}) + 0(\text{Int6})$$

$$\text{Diff 2} = 0(\text{Int1}) + 1(\text{Int2}) - 1(\text{Int3}) + 0(\text{Int4}) + 0(\text{Int5}) + 0(\text{Int6})$$

and so on. I could have broken the intervals up using a number of alternative contrast procedures. For example, I could have created new variables that would compare each interval with all of those that came after by using coefficients of the form

$$\begin{bmatrix} 1 & -1/5 & -1/5 & -1/5 & -1/5 & -1/5 \end{bmatrix}$$
$$\begin{bmatrix} 0 & 1 & -1/4 & -1/4 & -1/4 & -1/4 \end{bmatrix}$$
$$\begin{bmatrix} 0 & 0 & 1 & -1/3 & -1/3 & -1/3 \end{bmatrix}$$
$$\cdots$$

or comparing each interval with the first by using the coefficients

$$\begin{bmatrix} 1 & -1 & 0 & 0 & 0 & 0 \end{bmatrix}$$
$$\begin{bmatrix} 1 & 0 & -1 & 0 & 0 & 0 \end{bmatrix}$$
$$\begin{bmatrix} 1 & 0 & 0 & -1 & 0 & 0 \end{bmatrix}$$
$$\cdots$$

Each of these ways of forming contrasts would produce the same overall test statistics for main effects and interaction, but would speak to different questions when we look at the univariate tests.

Key Terms

Partition (Introduction)

Partialling out (Introduction)

Repeated-measures designs
  (Introduction)

$SS_{\text{between subj}}$ (SS_S) (Introduction)

$SS_{\text{within subj}}$ (Introduction)

Main diagonal (14.3)

Off-diagonal elements (14.3)

Compound symmetry (14.3)

Covariance matrix $\left(\sum \right)$ (14.3)

Sphericity (14.3)

Multivariate analysis of
  variance (MANOVA) (14.3)

Multivariate procedure (14.3)

Error$_{between}$ (14.7)

Error$_{within}$ (14.7)

Intraclass correlation (14.10)

Sequence effects (14.11)

Carryover effects (14.11)

Latin square (14.11)

Randomized blocks designs (14.11)

Matched samples (14.11)

Univariate (14.13)

Exercises

14.1 It is at least part of the folklore that repeated experience with the Graduate Record Examination (GRE) leads to better scores, even without any intervening study. We obtain eight subjects and give them the GRE verbal exam every Saturday morning for 3 weeks. The data follow:

S	First	Second	Third
1	550	570	580
2	440	440	470
3	610	630	610
4	650	670	670
5	400	460	450
6	700	680	710
7	490	510	510
8	580	550	590

a. Write the statistical model for these data.

b. Run the analysis of variance.

c. What, if anything, would you conclude about practice effects on the GRE?

14.2 Using the data from Exercise 14.1,

a. delete the data for the third session and run a (matched-sample) t test between Sessions 1 and 2.

b. Now run a repeated-measures analysis of variance on the two columns you used in part (a) and compare this F with the preceding t.

14.3 To demonstrate the practical uses of basic learning principles, a psychologist with an interest in behavior modification collected data on a study designed to teach self-care skills to severely handicapped children. An experimental group received reinforcement for activities related to self-care. A second group received an equivalent amount of attention, but no reinforcement. The children were scored (blind) by a rater on a 10-point scale of self-sufficiency. The ratings were done in a baseline session and at the end of training. The data follow:

Reinforcement		No Reinforcement	
Baseline	Training	Baseline	Training
8	9	3	5
5	7	5	5
3	2	8	10
5	7	2	5
2	9	5	3
6	7	6	10
5	8	6	9
6	5	4	5
4	7	3	7
4	9	5	5

Run the appropriate analysis and state your conclusions.

14.4 An experimenter with only a modicum of statistical training took the data in Exercise 14.3 and ran an independent-groups t test instead, using the difference scores (training minus baseline) as the raw data.

a. Run that analysis.

b. Square the value of t and compare it with the Fs you obtained in Exercise 14.3.

c. Explain why t^2 is not equal to F for Groups.

14.5 To understand just what happened in the experiment involving the training of severely handicapped children (Exercise 14.3), our original experimenter evaluated a third group at the same times as he did the first two groups, but otherwise treated them just like all other residents of the training school. In other words, these children did not receive reinforcement, or even the extra attention that the control group did. Their data follow:

Baseline:	3	5	8	5	5	6	6	6	3	4
Training:	4	5	6	6	4	7	7	3	2	2

a. Add these data to those in Exercise 14.3 and rerun the analysis.

b. Plot the results.

c. What can you conclude from the results you obtained in parts (a) and (b)?

d. Within the context of this three-group experiment, run the contrast of the two conditions that you have imported from Exercise 14.3.

e. Compute the effect size for the contrast in part (d).

14.6 For 2 years, I carried on a running argument with my daughter concerning hand calculators. She wanted one. I maintained that children who use calculators never learn to do arithmetic correctly, whereas she maintained that they do. To settle the argument, we selected five of her classmates who had calculators and five who did not, and made a totally unwarranted assumption that the presence or absence of calculators was all that distinguished these children. We then gave each child three 10-point tests (addition, subtraction, and multiplication), which they were required to do in a very short time in their heads. The scores are as follows:

	Addition	Subtraction	Multiplication
Calculator owners	8	5	3
	7	5	2
	9	7	3
	6	3	1
	8	5	1
Non-calculator owners	10	7	6
	7	6	5
	6	5	5
	9	7	8
	9	6	9

a. Run the analysis of variance.

b. Do the data suggest that I should have given in and bought my daughter a calculator? (I did anyway.)

14.7 For the data in Exercise 14.6,

a. Calculate the variance-covariance matrices.

b. Calculate $\hat{e}$ using your answers to part (a).

14.8 From the results in Exercise 14.7, do we appear to have reason to believe that we have met the assumptions required for the analysis of repeated measures?

14.9 For the data in Exercise 14.6,

a. Calculate all possible simple effects after first plotting the results.

b. Test the simple effects, calculating test terms and adjusted degrees of freedom where necessary.

14.10 In a study of the way children and adults summarize stories, we selected 10 fifth graders and 10 adults. These were further subdivided into equal groups of good and poor readers (on the hypothesis that good and poor readers may store or retrieve story information differently). All subjects read 10 short stories and were asked to summarize the story in their own words immediately after reading it. All summaries were content analyzed, and the numbers of statements related to Settings, Goals, and inferred Dispositions were recorded. The data are collapsed across the 10 stories:

Age Items	Adults			Children		
	Setting	Goal	Disp.	Setting	Goal	Disp.
Good readers	8	7	6	5	5	2
	5	6	4	7	8	4
	5	5	5	7	7	4
	7	8	6	6	4	3
	6	4	4	4	4	2
Poor readers	7	6	3	2	2	2
	5	3	1	2	0	1
	6	6	2	5	4	1
	4	4	1	4	4	2
	5	5	3	2	2	0

Run the appropriate analysis.

14.11 Refer to Exercise 14.10.

a. Calculate the simple effect of reading ability for children.

b. Calculate the simple effect of items for adult good readers.

14.12 Calculate the within-groups covariance matrices for the data in Exercise 14.10.

14.13 Suppose we had instructed our subjects to limit their summaries to 10 words. What effect might that have on the data in Exercise 14.10?

14.14 In an investigation of cigarette smoking, an experimenter decided to compare three different procedures for quitting smoking (tapering off, immediate stopping, and aversion therapy). She took five subjects in each group and asked them to rate (on a 10-point scale) their desire to smoke "right now" in two different environments (home versus work) both before and after quitting. Thus, we have one between-subjects variable (Treatment group) and two within-subjects variables (Environment and Pre/Post).

	Pre		Post	
	Home	Work	Home	Work
Taper	7	6	6	4
	5	4	5	2
	8	7	7	4
	8	8	6	5
	6	5	5	3
Immediate	8	7	7	6
	5	5	5	4
	7	6	6	5
	8	7	6	5
	7	6	5	4
Aversion	9	8	5	4
	4	4	3	2
	7	7	5	3
	7	5	5	0
	8	7	6	3

a. Run the appropriate analysis of variance.

b. Interpret the results.

14.15 Plot the results you obtained in Exercise 14.14.

14.16 Run simple effects on the data in Exercise 14.14 to clarify the results.

14.17 The abbreviated BMDP printout in Exhibit 14.3 represents the analysis of the data in Exercise 14.5.

a. Compare this printout with the results you obtained in Exercise 14.5.

b. What does a significant F for "MEAN" tell us?

c. Relate $MS_{\text{w/in cell}}$ to the table of cell standard deviations.

BMDP2V – ANALYSIS OF VARIANCE AND COVARIANCES
WITH REPEATED MEASURES.

PROGRAM CONTROL INFORMATION
/PROBLEM   TITLE IS 'BMDP2V ANALYSIS OF EXERCISE 14.5'.
/INPUT   VARIABLES ARE 3.
                  FORMAT IS '(3F2.0)'.
                  CASES ARE 30.
/VARIABLE   NAMES ARE GROUP, PRE, POST..
/DESIGN   DEPENDENT ARE 2, 3.
                  LEVELS ARE 2.
                  NAME IS TIME.
                  GROUP = 1.
/END

CELL MEANS FOR 1-ST DEPENDENT VARIABLE

				MARGINAL
GROUP =	* 1.0000	* 2.0000	* 3.0000	
TIME				
PRE 1	4.80000	4.70000	5.10000	4.86667
POST 2	7.00000	6.40000	4.60000	6.00000
MARGINAL	5.90000	5.55000	4.85000	5.43333
COUNT	10	10	10	30

STANDARD DEVIATIONS FOR 1-ST DEPENDENT VARIABLE

GROUP =	* 1.0000	* 2.000	* 3.0000
TIME			
PRE 1	1.68655	1.76698	1.52388
POST 2	2.16025	2.45855	1.89737

SOURCE	SUM OF SQUARES	DEGREES OF FREEDOM	MEAN SQUARE	F	TAIL PROBABILITY
MEAN	1771.26667	1	1771.26667	322.48	0.0000
GROUP	11.43333	2	5.71667	1.04	0.3669
1 ERROR	148.30000	27	5.49259		
TIME	19.26667	1	19.26667	9.44	0.0048
TG	20.63333	2	10.31667	5.06	0.0137
2 ERROR	55.10000	272.04074			

Exhibit 14.3

14.18 The SPSS printout in Exhibit 14.4 was obtained by treating the data in Exercise 14.10 as though all variables were between-subjects variables (i.e., as though the data represented a standard three-way factorial). Show that the error terms for the correct analysis represent a partition of the error term for the factorial analysis.

Tests of Between-Subjects Effects

Dependent Variable: DV

Source	Type III Sum of Squares	df	Mean Square	F	Sig.
Corrected Model	170.800[a]	11	15.527	9.001	.000
Intercept	1058.400	1	1058.400	613.565	.000
AGE	29.400	1	29.400	17.043	.000
READER	68.267	1	68.267	39.575	.000
PART	60.400	2	30.200	17.507	.000
AGE * READER	3.267	1	3.267	1.894	.175
AGE * PART	.000	2	.000	.000	1.000
READER * PART	.933	2	.467	.271	.764
AGE * READER * PART	8.533	2	4.267	2.473	.095
Error	82.800	48	1.725		
Total	1312.000	60			
Corrected Total	253.600	59			

[a] R Squared = .674 (Adjusted R Squared = .599)

Exhibit 14.4

14.19 Outline the summary table for an $A \times B \times C \times D$ design with repeated measures on A and B and independent measures on C and D.

14.20 Run a multivariate analysis of variance for the data in Exercise 14.10, incorporating tests on Setting versus Goal and Setting versus Disposition.

14.21 What do you gain and what do you lose by using MANOVA in Exercise 14.20 with respect to the tests you considered?

14.22 Use the results in Exhibit 14.4 to tie together the answers you received for main effects and interactions in Exercises 14.10 and 14.20. (*Hint:* Some of the names for the statistics will be different, but the numerical values will be the same.)

14.23 In the data file Stress.dat (available from the website www.uvm.edu/~dhowell/methods/), are data on the stress level reported by cancer patients and their spouses at two different times—shortly after the diagnosis and 3 months later. The data are also distinguished by the gender of the respondent. As usual, a "." indicates each missing data point. See description in Appendix Computer Data Sets, page 670.

   a. Use any statistical package to run a repeated-measures analysis of variance with Gender and Role (patient versus spouse) as between-subject variables and Time as the repeated measure.

   b. Have the program print out cell means, and plot these means as an aid in interpretation.

   c. There is a significant three-way interaction in this analysis. Interpret it along with the main effects.

14.24 Everitt reported data on a study of three treatments for anorexia in young girls. One treatment was cognitive behavior therapy, a second was a control condition with no therapy, and a third was a family therapy condition. The data follow.

   a. Run an analysis of variance on group differences in Gain scores.

Group	Pretest	Posttest	Gain	Group	Pretest	Posttest	Gain
1	80.5	82.2	1.7	2	75.1	86.7	11.6
1	84.9	85.6	.7	2	80.6	73.5	−7.1
1	81.5	81.4	−.1	2	78.4	84.6	6.2
1	82.6	81.9	−.7	2	77.6	77.4	−0.2
1	79.9	76.4	−3.5	2	88.7	79.5	−9.2
1	88.7	103.6	14.9	2	81.3	89.6	8.3
1	94.9	98.4	3.5	2	78.1	81.4	3.3
1	76.3	93.4	17.1	2	70.5	81.8	11.3
1	81.0	73.4	−7.6	2	77.3	77.3	0.0
1	80.5	82.1	1.6	2	85.2	84.2	−1.0
1	85.0	96.7	11.7	2	86.0	75.4	−10.6
1	89.2	95.3	6.1	2	84.1	79.5	−4.6
1	81.3	82.4	1.1	2	79.7	73.0	−6.7
1	76.5	72.5	−4.0	2	85.5	88.3	2.8
1	70.0	90.9	20.9	2	84.4	84.7	0.3
1	80.4	71.3	−9.1	2	79.6	81.4	1.8
1	83.3	85.4	2.1	2	77.5	81.2	3.7
1	83.0	81.6	−1.4	2	72.3	88.2	15.9
1	87.7	89.1	1.4	2	89.0	78.8	−10.2
1	84.2	83.9	−.3	3	83.8	95.2	11.4
1	86.4	82.7	−3.7	3	83.3	94.3	11.0
1	76.5	75.7	−.8	3	86.0	91.5	5.5
1	80.2	82.6	2.4	3	82.5	91.9	9.4
1	87.8	100.4	12.6	3	86.7	100.3	13.6
1	83.3	85.2	1.9	3	79.6	76.7	−2.9
1	79.7	83.6	3.9	3	76.9	76.8	−0.1
1	84.5	84.6	.1	3	94.2	101.6	7.4
1	80.8	96.2	15.4	3	73.4	94.9	21.5
1	87.4	86.7	−.7	3	80.5	75.2	−5.3
2	80.7	80.2	−.5	3	81.6	77.8	−3.8
2	89.4	80.1	−9.3	3	82.1	95.5	13.4
2	91.8	86.4	−5.4	3	77.6	90.7	13.1
2	74.0	86.3	12.3	3	83.5	92.5	9.0
2	78.1	76.1	−2.0	3	89.9	93.8	3.9
2	88.3	78.1	−10.2	3	86.0	91.7	5.7
2	87.3	75.1	−12.2	3	87.3	98.0	10.7

b. Repeat the analysis, but this time use a repeated measures design where the repeated measures are Pretest and Posttest.

c. How does the answer to part (b) relate to the answer to part (a)?

d. Plot scatterplots of the relationship between Pretest and Posttest separately for each group. What do these plots show?

e. Run a test on the null hypothesis that the Gain for the Control is 0.00. What does this analysis tell you? Are you surprised?

f. Why would significant gains in the two experimental groups not be interpretable without the control group?

Discussion Questions

14.25 In Exercise 14.23, we ignored the fact that we have pairs of subjects from the same family.

    a.   What is wrong with doing this?

    b.   Under what conditions would it be acceptable to ignore this problem?

    c.   What alternative analyses would you suggest?

14.26 In Exercise 14.23 you probably noticed that many observations at Time 2 are missing. (This is partly because for many patients it had not yet been 3 months since the diagnosis.)

    a.   Compare the means at Time 1 for those subjects who did, and who did not, have data at Time 2.

    b.   If there are differences in (a), what would this suggest to you about the data?

In a study of behavior problems in children we asked 3 "judges" to rate each of 20 children on the level of aggressive behavior. These judges were the child's Parent, the child's Teacher, and the child him or herself (Self). The data follow:

Child	1	2	3	4	5	6	7	8	9	10	11	12	13	14	15	16	17	18	19	20
Parent	10	12	14	8	16	21	10	15	18	6	22	14	19	22	11	14	18	25	22	7
Teacher	8	13	17	10	18	24	9	16	18	8	24	19	15	20	10	18	19	30	20	10
Self	12	17	16	15	24	24	13	17	21	13	29	23	16	20	15	17	21	25	25	14

These data are somewhat different from the data we saw in Section 14.10 because in that case the same people judged each child, whereas here the Parent and Self obviously change from child to child. We will ignore that for the moment and simply act as if we could somehow have the same parent and the same "self" do all the ratings.

14.27 What is the reliability of this data set in terms of the intraclass correlation coefficient?

14.28 What do your calculations tell you about the sources of variability in this data set?

14.29 Suppose that you had no concern about the fact that one source systematically rates children higher or lower than another source. How might you evaluate reliability differently?

14.30 Under what conditions might you not be interested in differences among judges?

14.31 What do you think is the importance of the fact that the "parent" who supplies the parent rating changes from child to child?

CHAPTER 15

Multiple Regression

Objectives

To show how we can examine the relationship between a criterion variable and several predictor variables simultaneously and to point out the problems inherent in this procedure.

Contents

IN CHAPTER 9, WE CONSIDERED the situation in which we have one criterion variable (Y) and one predictor variable (X) and want to predict Y on the basis of X. In this chapter, we will consider the case in which we still have only one criterion (Y) but have multiple predictors (X_1, X_2, X_3, ..., X_p), and want to predict Y on the basis of *simultaneous* knowledge of all p predictors. The situation we examined in Chapter 9 can be viewed as a special case of the one discussed in this chapter; alternatively, this chapter can be viewed as an extension of Chapter 9. We will continue to use many familiar concepts such as the correlation coefficient, the slope, the standard error of estimate, and $SS_{\text{regression}}$.

scalar algebra

A common approach to multiple regression in a book of this type uses the usual **scalar algebra** taught in high school. The main problem with that approach is that when there are more than two predictors, the arithmetic becomes appallingly laborious. Even with two predictors, the equations themselves provide little insight into what is actually going on when we solve a multiple regression problem.

matrix algebra

An alternative approach taken by more advanced texts is to cast the problem in terms of **matrix algebra** and to sneak in a "quickie" chapter on matrices. Unfortunately, even the brightest and most diligent students come away from such a chapter with only the vaguest ideas of matrix theory.

A third approach, and the one to be adopted here, is to bow to reality and to recognize that almost all multiple regression problems are solved by using readily available computer programs. What is important is not the calculations themselves but the alternative ways the problem can be approached and the interpretation of the results. Thus, what we need to concentrate on is not how the solution is actually obtained, but what the potential problems are and what interpretation can be assigned to the wealth of statistics printed out by any good regression program.

Rather than simplifying the discussion of multiple regression, the approach taken here actually complicates it. Generally, there can be little argument over formulae. On the other hand, questions about the optimal number of predictors, the use of regression diagnostics, the relative importance of various predictors, and the selection of predictors do not have universally accepted answers. Be forewarned that the opinions expressed in this chapter are only opinions and are open to dispute—but then that is part of what makes statistics interesting. Excellent and readable advanced sources for the study of multiple regression are Cohen, Cohen, West, and Aiken (2003), Darlington (1990), Draper and Smith (1981), and Stevens (1992).

15.1 Multiple Linear Regression

The problem of multiple regression is that of finding a regression equation to predict Y (sometimes denoted X_0) on the basis of p predictors (X_1, X_2, X_3, ..., X_p). Thus, we might want to predict success in graduate school (Y) on the basis of undergraduate grade point average (X_1), Graduate Record Exam (GRE) scores (X_2), number of courses taken in the major discipline (X_3), and some rating of "favorableness" of letters of recommendation (X_4). Similarly, we might want to predict the time it takes to go from one point in a city to another on the basis of number of traffic lights (X_1), posted speed limit (X_2), presence or absence of "right turn on red" signs (X_3), and traffic density (X_4). These examples are both analyzed in the same way, although in the first, we presumably care about predictions for individual applicants, whereas in the second, we might be less interested in the prediction itself and more interested in the role played by each of the predictors.

The Regression Equation

In Chapter 9, we started with the equation of a straight line ($\hat{Y} = bX + a$) and solved for the two unknowns (a and b) subject to the constraint that $\sum (Y - \hat{Y})^2$ is a minimum. In multiple regression, we are going to do the same thing, although in this case, we will solve the equation $\hat{Y} = b_0 + b_1 X_1 + b_2 X_2 + \cdots + b_p X_p$ where b_0 represents the intercept and $b_1, b_2, \ldots, b_p$ are the regression coefficients for the predictors $X_1, X_2, \ldots, X_p$, respectively. We will retain the restriction that $\sum (Y - \hat{Y})^2$ is to be minimized because it still makes sense to find predicted values that come as close as possible to the obtained values of Y.[1] As mentioned earlier, the calculations required to estimate the b_i become more cumbersome as the number of predictors increases, and we will not discuss these calculations here. Instead, we will begin with a simple example and assume that the solution was obtained by any available computer program, such as SPSS, Minitab, or SAS. (A free Java program is available on the web at www.statcrunch.com. You need to register, but it is free and painless, I would strongly recommend starting it up in a web browser and using it as you read this chapter. The main data file used here can be imported from www.uvm.edu/~dhowell/methods/ as Tab15-1.dat. This will allow you to import data into a spreadsheet and perform most things discussed in the chapter. The program is a tad slow because it runs online with Java, but what do you want for free?)

The example that we will use will provide a general overview of multiple regression. Once we have that overview, we will step back and ask specific questions about the data themselves, the choice of variables to include in the model, and ways to index the importance of variables.

A number of years ago, the student association of a large university published an evaluation of several hundred courses taught during the preceding semester. Students in each course had completed a questionnaire in which they rated a number of different aspects of the course on a 5-point scale (1 = failure, very bad ... 5 = excellent, exceptional). The data in Table 15.1 are actual data and represent mean scores on six variables for a random sample of 50 courses. These variables were (1) overall quality of lectures (Overall), (2) teaching skills of the instructor (Teach), (3) quality of the tests and exams (Exam), (4) instructor's perceived knowledge of the subject matter (Knowledge), (5) the student's expected grade in the course (Grade—F = 1, A = 5), and (6) the enrollment of the course (Enroll).

On the assumption that the best available rating of the course is the overall rating of the lectures (Overall), we will use that as the dependent variable (Y) and derive a regression equation predicting Y on the basis of the other five variables.

Before we consider the regression solution itself, we need to look at the distribution of each variable. These are shown as stem-and-leaf displays in Table 15.2 on page 497. (Notice that the display for Enroll contains a discontinuity because the few very large enrollments require the scale to move from increasing in steps of 10 to increasing in steps of 100.) From these displays, we can see that the criterion variable and four of the predictors are fairly well distributed. They are all more or less symmetric with a reasonable amount of variability. Notice that the center of each distribution is noticeably above 3.0, which was labeled on the scale as "average." Thus, there is some positive response bias in the data. This is an interesting result, but it should have no important effect on the multiple regression solution; it will merely change the intercept from what it would be if each variable had a mean of 3.0. It is particularly encouraging to notice that the distribution for scores on the instructors' Knowledge of their subject matter is biased upward. It is nice to see that students think their instructors know the material, even when they think the instructor is a poor teacher. The Teach and Exam variables each have one score that is unusually low relative to the others, but these are not so low as to bring their accuracy into question. The Enroll variable is unusually distributed; 46 of the 50 courses have enrollments below or about 100, whereas the other 4 have enrollments ranging from 220 to 800. We will consider these extreme values later.

---

[1] There are alternatives to the standard least squares criteria that often produce estimates that are in some ways superior to the estimates obtained by least squares. These procedures are less common, but many of them can be found in Rousseeuw and Leroy (1987).

Table 15.1   Course evaluation data

Overall	Teach	Exam	Knowledge	Grade	Enroll
3.4	3.8	3.8	4.5	3.5	21
2.9	2.8	3.2	3.8	3.2	50
2.6	2.2	1.9	3.9	2.8	800
3.8	3.5	3.5	4.1	3.3	221
3.0	3.2	2.8	3.5	3.2	7
2.5	2.7	3.8	4.2	3.2	108
3.9	4.1	3.8	4.5	3.6	54
4.3	4.2	4.1	4.7	4.0	99
3.8	3.7	3.6	4.1	3.0	51
3.4	3.7	3.6	4.1	3.1	47
2.8	3.3	3.5	3.9	3.0	73
2.9	3.3	3.3	3.9	3.3	25
4.1	4.1	3.6	4.0	3.2	37
2.7	3.1	3.8	4.1	3.4	83
3.9	2.9	3.8	4.5	3.7	70
4.1	4.5	4.2	4.5	3.8	16
4.2	4.3	4.1	4.5	3.8	14
3.1	3.7	4.0	4.5	3.7	12
4.1	4.2	4.3	4.7	4.2	20
3.6	4.0	4.2	4.0	3.8	18
4.3	3.7	4.0	4.5	3.3	260
4.0	4.0	4.1	4.6	3.2	100
2.1	2.9	2.7	3.7	3.1	118
3.8	4.0	4.4	4.1	3.9	35
2.7	3.3	4.4	3.6	4.3	32
4.4	4.4	4.3	4.4	2.9	25
3.1	3.4	3.6	3.3	3.2	55
3.6	3.8	4.1	3.8	3.5	28
3.9	3.7	4.2	4.2	3.3	28
2.9	3.1	3.6	3.8	3.2	27
3.7	3.8	4.4	4.0	4.1	25
2.8	3.2	3.4	3.1	3.5	50
3.3	3.5	3.2	4.4	3.6	76
3.7	3.8	3.7	4.3	3.7	28
4.2	4.4	4.3	5.0	3.3	85
2.9	3.7	4.1	4.2	3.6	75
3.9	4.0	3.7	4.5	3.5	90
3.5	3.4	4.0	4.5	3.4	94
3.8	3.2	3.6	4.7	3.0	65
4.0	3.8	4.0	4.3	3.4	100
3.1	3.7	3.7	4.0	3.7	105
4.2	4.3	4.2	4.2	3.8	70
3.0	3.4	4.2	3.8	3.7	49
4.8	4.0	4.1	4.9	3.7	64
3.0	3.1	3.2	3.7	3.3	700
4.4	4.5	4.5	4.6	4.0	27
4.4	4.8	4.3	4.3	3.6	15
3.4	3.4	3.6	3.5	3.3	40
4.0	4.2	4.0	4.4	4.1	18
3.5	3.4	3.9	4.4	3.3	90

Table 15.2   Stem-and-leaf displays for all variables for the data in Table 15.1

	Overall (Y)		Teach (X_1)		Exam (X_2)
1.		1.		1.	<u>9</u>
2*	1	2*		2*	
2t		2t	<u>2</u>	2t	
2f	5	2f		2f	
2s	677	2s	7	2s	7
2.	889999	2.	899	2.	8
3*	000111	3*	111	3*	
3t	3	3t	222333	3t	2223
3f	44455	3f	4444455	3f	455
3s	6677	3s	7777777	3s	6666666777
3.	88889999	3.	88888	3.	888889
4*	000111	4*	0000011	4*	00000111111
4t	22233	4t	22233	4t	222223333
4f	444	4f	4455	4f	4445
4s		4s			
4.	8	4.	8		
5*		5*			

	Knowledge (X_3)		Grade (X_4)		Enroll (X_5)
2*		2*		0	7
2t		2t		1	245688
2f		2f		2	0155577888
2s		2s		3	257
2.		2.	89	4	079
3*	1	3*	00011	5	00145
3t	3	3t	222222233333333	6	45
3f	55	3f	4445555	7	00356
3s	677	3s	6666777777	8	35
3.	8888999	3.	88889	9	<u>0049</u>
4*	000011111	4*	0011	1	00001
4t	2222333	4t	23	2	<u>26</u>
4f	4444555555555	4f		3	
4s	66777	4s		4	
4.	9	4.		5	
5*	0	5*		6	
				7	<u>0</u>
				8	<u>0</u>

Code: (Y to X_4)   3*|1 = 3.1   Code: X_5 9 |0 = 90
                                                      1 |0 = 100

(Underlined values would be shown as outliers in a boxplot.) Notice that the stems have been divided into five intervals per decade. The * stands for 0 and 1, t stands for 2 and 3, f stands for 4 and 5, s stands for 6 and 7, and . stands for 8 and 9.

Although each of these variables taken alone (with the possible exception of Enroll) is reasonably behaved, what is particularly noteworthy is the behavior of Course 3. This course accounts for the two extreme scores on the variables Teach and Exam, and also has the most extreme score on Enroll. It is important to keep this course in mind, as it may well contribute more than its share of influence to the final result. (A similar point was raised in Chapter 9 with respect to an example used there. Often an extreme point will pull the regression surface toward itself, especially when that point is extreme on several variables. On the other hand, such a point may fit neatly into the pattern of data points, merely representing an extreme case of a general pattern.) The influence of such scores can be examined in a number of ways, as we shall see shortly.

validities

An abbreviated printout using SPSS is presented in Exhibit 15.1, parts of which will be explained as we go along. You can see the means, standard deviations, and the number of cases (50) in the data set. The first row of the intercorrelation matrix contains the correlations of each of the predictors with the criterion. (These correlations are sometimes referred to as **validities,** because each is a measure of the degree to which the predictor [e.g., Enroll] can be used as a valid predictor of the criterion [Overall].) From the matrix, we can see that teaching skill (Teach) has the highest correlation with the rated quality of the lectures (.804). The instructor's apparent knowledge of the material and the quality of the exams come next in order of correlations with the criterion (.682 and .596, respectively), whereas the student's expected grade and the size of the class have relatively low correlations with the criterion (.301 and −.240, respectively). This latter correlation runs counter to traditional folklore among faculty, who often assume that those instructors who teach large classes will suffer poor teaching evaluations. The probability associated with a two-tailed test on the significance of the correlation between Enroll and Overall, for example, is given below the correlation and is .094.

From the correlation matrix, we also see that some of the variables have reasonably high intercorrelations with each other. The perceived quality of the exams is related to the instructor's teaching skills ($r = .720$), to the student's expected grade ($r = .610$), and to the enrollment ($r = −.558$), the last correlation probably reflecting the kinds of exams given in large lecture classes. The fact that this variable is highly correlated with several of the other

collinearity

predictors is often referred to as **collinearity,** especially if the correlations are high. In this case, the correlations suggest that Exam has much in common with several other variables and may have very little information that is unique to itself. If so, when the other variables are included as predictors, Exam will have little new to offer in the way of explaining variability in the Overall rating.

regression coefficients

The **regression coefficients** are given in the final portion of Exhibit 15.1 under the heading "B"[2]. They can be interpreted in much the same way as a regression coefficient in simple regression—with one important difference to be explained shortly. Letting b_0 represent the intercept and $b_1 − b_5$ represent the regression coefficients for the five predictors, we see from Exhibit 15.1 that

$$b_0 = −1.195$$
$$b_1 = 0.763$$
$$b_2 = 0.132$$
$$b_3 = 0.489$$
$$b_4 = −0.184$$
$$b_5 = 0.0005$$

---

[2] The top half of Exhibit 15.1 was obtained by selecting the "descriptives" option in **Regression/linear.** The lower half of the table was obtained by entering all variables at once using the Enter method.

Descriptive Statistics

	Mean	Std. Deviation	N
OVERALL	3.550	.614	50
TEACH	3.664	.532	50
EXAM	3.808	.493	50
KNOWLEDG	4.176	.408	50
GRADE	3.486	.351	50
ENROLL	88.00	145.06	50

Correlations

		OVERALL	TEACH	EXAM	KNOWLEDG	GRADE	ENROLL
OVERALL	Pearson Correlation	1.000	.804**	.596**	.682**	.301*	−.240
	Sig. (2-tailed)	.	.000	.000	.000	.034	.094
	N	50	50	50	50	50	50
TEACH	Pearson Correlation	.804**	1.000	.720**	.526**	.469**	−.451**
	Sig. (2-tailed)	.000	.	.000	.000	.001	.001
	N	50	50	50	50	50	50
EXAM	Pearson Correlation	.596**	.720**	1.000	.451**	.610**	−.558**
	Sig. (2-tailed)	.000	.000	.	.001	.000	.000
	N	50	50	50	50	50	50
KNOWLEDG	Pearson Correlation	.682**	.526**	.451**	1.000	.224	−.128
	Sig. (2-tailed)	.000	.000	.001	.	.118	.376
	N	50	50	50	50	50	50
GRADE	Pearson Correlation	.301*	.469**	.610**	.224	1.000	−.337*
	Sig. (2-tailed)	.034	.001	.000	.118	.	.017
	N	50	50	50	50	50	50
ENROLL	Pearson Correlation	−.240	−.451**	−.558**	−.128	−.337**	1.000
	Sig. (2 tailed)	.094	.001	.000	.376	.017	.
	N	50	50	50	50	50	50

**. Correlation is significant at the 0.01 level (2-tailed).
*. Correlation is significant at the 0.05 level (2-tailed).

Model Summary

Model	R	R Square	Adjusted R Square	Std. Error of the Estimate
1	.869[a]	.755	.728	.320

[a] Predictors: (Constant), ENROLL, KNOWLEDGE, GRADE, TEACH, EXAM

(continued)

Exhibit 15.1   Correlations

ANOVA[b]

Model		Sum of Squares	df	Mean Square	F	Sig.
1	Regression	13.934	5	2.787	27.184	.000[a]
	Residual	4.511	44	.103		
	Total	18.445	49			

[a] Predictors: (Constant), ENROLL, KNOWLEDG, GRADE, TEACH, EXAM
[b] Dependent Variable: OVERALL

Coefficients[a]

Model		Unstandardized Coefficients		Standardized Coefficients	t	Sig.	Collinearity Statistics	
		B	Std. Error	Beta			Tolerance	VIF
1	Constant	−1.195	.631		−1.893	.065		
	TEACH	.763	.133	.662	5.742	.000	.418	2.391
	EXAM	.132	.163	.106	.811	.422	.325	3.081
	KNOWLEDG	.489	.137	.325	3.581	.001	.675	1.482
	GRADE	−.184	.165	−.105	−1.114	.271	.620	1.614
	ENROLL	5.255E-04	.000	.124	1.347	.185	.653	1.530

[a] Dependent Variable: OVERALL

Exhibit 15.1   *(continued)*

Thus, we can write

$$\hat{Y} = -1.195 + 0.763\text{Teach} + 0.132\text{Exam} + 0.489\text{Knowledge} - 0.184\text{Grade} + 0.0005\text{Enroll}$$

This is the standard regression equation and can be used to obtain a predicted value of Y (Overall) for any specific set of values of the predictors. For example, for Course 1, substituting the data from Table 15.1, we have

$$\hat{Y} = -1.195 + 0.763(3.8) + 0.132(3.8) + 0.489(4.5) - 0.184(3.5) + 0.0005(21)$$
$$= 3.773^3$$

Because Course 1 actually obtained a mean Overall lecture rating of 3.4, our residual, or error of prediction, for that course is $e = Y - \hat{Y} = 3.4 - 3.773 = -0.373$. For Course 3, the course with the extreme observations,

$$\hat{Y} = -1.195 + 0.763(2.2) + 0.132(1.9) + 0.489(3.9) - 0.184(2.8) + 0.0005(800)$$
$$= 2.546$$

In this case, the actual Overall score was 2.6, for a residual of $2.6 - 2.546 = 0.054$. So the fit to the Course with extreme predictors was good. That is not to say, however, that we should be satisfied that extreme values are not a problem in this example. It may well be that this observation pulled the regression surface toward itself and, thus, fits well as a result.

From our regression equation, we can see that *if all other variables were held constant,* the predicted value of Y would be higher by 0.763 unit for every one-unit increase in Teach

---

[3] Predicted values have been calculated with eight-digit accuracy to increase agreement with the computer printout.

skills, and by 0.0005 for every one-unit increase in Enroll. This does not necessarily mean that if two instructors differed by one unit on the Teach variable, they would differ by 0.763 unit on the Overall course rating. The *predicted* values would differ by 0.763 unit only if the two courses had equal ratings on all other variables. Moreover, this equation cannot be interpreted to indicate that if a given instructor were to spend time improving his Exam rating by one unit, the rated Overall quality of his lectures would improve by 0.132 unit. First, such an assertion would imply a *causal* relationship, when none has necessarily been shown. Second, the pattern of interrelationships among the variables suggests that increasing the Exam rating might also be expected to alter the ratings on some of the other variables, which could in turn also affect the predicted score.

One common mistake is to treat the relative magnitudes of the b_i as an index of the relative importance of the individual predictors. By this (mistaken) logic, we might be tempted to conclude that Teach is a more important predictor than is Enroll because its coefficient is appreciably larger. Although it might actually be the case that Teach is a more important predictor, we cannot draw such a conclusion based on the regression coefficients. The relative magnitudes of the coefficients are in part a function of the standard deviations of the corresponding variables. Because the standard deviation of Enroll is much larger than the standard deviation of the other variables, its regression coefficient (b_5) is almost certain to be small regardless of the importance of that variable.

It may be easier for you to appreciate this last point if you look at the problem somewhat differently. For one instructor to have a Teach rating one point higher than another instructor would be a major accomplishment, whereas having one additional student is a trivial matter. We hardly expect on a priori grounds that these two one-point differences will lead to equal differences in the predicted Overall rating, regardless of the relative importance of the two predictors.

Standardized Regression Coefficients

importance

As we shall see later, the question of the relative importance of variables has several different answers depending on what we mean by **importance**. One measure of importance should be mentioned here, however, because it is a legitimate statistic in its own right. Suppose that before we obtained our multiple regression equation, we had standardized each of our variables. As you will recall, standardizing a variable sets its mean at 0 and its standard deviation at 1. It also expresses the result in standard deviation units. (You should recall that we standardize many of our effect size measures by dividing by the standard deviation.) Now all our variables would have equal standard deviations (1) and a one-unit difference between two courses on one variable would be comparable to a one-unit difference between those courses on any other variable. If we now solved for our regression coefficients using the standardized variables, we would obtain

$$\hat{Y}_z = 0.662Z_1 + 0.106Z_2 + 0.325Z_3 - 0.105Z_4 + 0.124Z_5$$

standardized
regression
coefficients

where Z is used to denote standardized variables. In this case, the regression coefficients are called **standardized regression coefficients**, labeled "Beta" by SPSS and denoted β_i. Thus

$$\beta_1 = 0.662$$
$$\beta_2 = 0.106$$
$$\beta_3 = 0.325$$
$$\beta_4 = -.105$$
$$\beta_5 = 0.124$$

When variables have been standardized, the intercept (β_0) is equal to 0 and is not shown.

From the preceding values of β_i, we can conclude that a one-unit difference (i.e., a difference of one standard deviation) between courses in Z_1 (the standardized Teach variable) with all other variables held constant will be associated with a difference in $\hat{Y}_Z$ of 0.662 units and therefore a difference in $\hat{Y}$ of 0.662 standard deviations. Comparable differences in Z_2 and Z_4 will be associated with differences in $\hat{Y}$ of only 0.106 and $-.105$ standard deviations, respectively. It begins to look as if Teach may be a more important predictor than either Exam or Grade. Although the relative magnitudes of the β_i are not necessarily the best indicators of "importance," they have a simple interpretation, are printed by most regression computer programs, and generally give at least a rough estimate of the relative contributions of the variables in the equation.

Lest you think you will be required to standardize the raw data to calculate the β_i, should the computer program not print them, there is an easier way. It can be shown quite easily that

$$\beta_i = \frac{b_i s_i}{s_0}$$

where s_0 is the standard deviation of the criterion. Solving for b_i instead of β_i gives

$$b_i = \frac{\beta_i s_0}{s_i}$$

For our data,

$$\beta_1 = \frac{0.763(0.532)}{0.614} = 0.662$$

and

$$\beta_2 = \frac{0.132(0.493)}{0.614} = 0.106$$

15.2 Standard Errors and Tests of Regression Coefficients

Once we have a regression coefficient, standardized or not, we normally test it for statistical significance. If the coefficient relating Teach to Overall is not statistically significantly different from 0, then Teach will serve no useful purpose in the prediction of Overall. As you might suspect, it doesn't matter whether we test the raw score regression coefficients (b_i) or the standardized coefficients (β_i). They are simply linear transformations of one another, and we would obtain the same test statistic in either case.

To test a regression coefficient (or most other statistics for that matter), we need to know the standard error of that statistic. The standard errors for the b_is are given in Exhibit 15.1 and labeled "Std. Error." For example, the standard error of b_0, the intercept, is 0.631, and the standard error for b_1 is 0.133. As with other standard errors, the standard error of the regression coefficient refers to the variability of the statistic over repeated sampling. Suppose we repeated the course-evaluation study many times on different independent samples of students. Each replication would be expected to give us a slightly different value of b_1, although each of these would be an unbiased estimate of the true coefficient in the population, which we will denote as b_1^*. The many b_1s would be normally distributed about b_1^* with a standard deviation estimated to be 0.133, the standard error of b_1.

We can use these standard errors to form a t test on the regression coefficients. Specifically,

$$t = \frac{b_j - b_j^*}{s_{b_j}}$$

on $N - p - 1$ degrees of freedom.[4]

Then to test $H_0 : b_j^* = 0$,

$$t = \frac{b_j}{s_{b_j}}$$

For a test on the regression coefficient of Teach, we have

$$t = \frac{0.763}{0.133} = 5.742$$

This is a standard Student's t on $N - p - 1 = 50 - 5 - 1 = 44$ df, and the critical value is found in Appendix t to be 2.02. Thus, we can reject H_0 and conclude that the regression coefficient in the population is not equal to 0. We don't actually need tables of t because our printout gives t and its (two-tailed) significance level. Thus, a b as large as 0.763 (for Teach) has a two-tailed probability of .000 under H_0. In other words, the predicted value of Y increases with increasing scores on Teach, and Teach thus makes a significant contribution to the prediction of Overall.

A corresponding test on the coefficient for Exam would produce

$$t = \frac{0.132}{0.163} = 0.811$$

This result is not significant ($p = .422$), meaning that given the other four predictors, Exam does not contribute significantly to the prediction of Overall. We might consider dropping this predictor from our model, but there will be more on this issue later. It is important to recognize that a test on a variable is done in the context of all other variables in the equation. A variable might have a high individual correlation with the criterion, as does Exam, with a significant Pearson r with Overall $= .596$, but have nothing useful to contribute once several other variables are included. That is the situation here.

Some computer programs prefer to print standard errors for, and test, standardized regression coefficients (β_j). It makes no difference which you do. Similarly, some programs provide an F test (on 1 and $N - p - 1$ df) instead of t. This F is simply the square of our t, so again it makes no difference which approach you take.

15.3 Residual Variance

We have just considered the standard error of the regression coefficient, recognizing that sampling error is involved in the estimation of the corresponding population regression coefficient. A somewhat different kind of error is involved in the estimation of the predicted Ys. For course evaluation, we would hope that the Overall score is, at least in part, a function of such variables as Teach, Exam, and so on. (If we didn't think that, we would not have collected data on those variables in the first place.) At the same time, we probably do not

---

[4] A number of authors (e.g., Draper and Smith, 1981; Huberty, 1989) have pointed out that in general this is not exactly distributed as Student's t. However, it is generally treated as if it were, but one should not take the associated probability too literally.

residual variance

residual error

expect that the five variables we have chosen will predict Y perfectly, even if they could be measured, and the coefficients estimated, without error. Error will still be involved in the prediction of Y after we have taken all five of our predictors into account. This error is called **residual variance** or **residual error** and is defined as

$$\frac{\sum (Y - \hat{Y})^2}{N - p - 1}$$

and is denoted as MS_{residual} or MS_{error} or $s^2_{0.12345}$. In Exhibit 15.1, it is given as the error term in the analysis of variance summary table as 0.103.

The concept of residual error is important because it is exactly the thing we hope to minimize in our study. We want our estimates of Y to be as accurate as possible. We will return to this concept later in the chapter.

The square root of MS_{residual} is called the *standard error of estimate* and has the same meaning as the standard error of estimate in Chapter 9. It is the standard deviation of the column of residual scores $(Y - \hat{Y})$. In Exhibit 15.1, it is given in the section labeled "Model Summary" before the analysis of variance summary table and denoted "Std. Error of the Estimate." In this example, that value is 0.320.

15.4 Distribution Assumptions

So far we have made no assumptions about the nature of the distributions of our variables. The statistics b_i, β_i, and R (the multiple correlation coefficient) are legitimate measures independent of any distribution assumptions. Having said that, however, it is necessary to point out that certain assumptions will be necessary if we are to use these measures in several important ways. (It may be helpful to go back to Chapter 9 and quickly reread the brief discussions in the introduction [p. 232] and in Sections 9.7 and 9.11 [pp. 249–250 and p. 263]. Those sections explained the distinction between linear-regression models and bivariate-normal models and discussed the assumptions involved.)

To provide tests on the statistics we have been discussing, we will need to make one of two different kinds of assumptions, depending on the nature of our variables. If $X_1, X_2, \ldots, X_p$ are thought of as random variables, as they are in this example because we measure the predictors as we find them rather than fixing them in advance, we will make the general assumption that

multivariate normal

the joint distribution of $Y, X_1, X_2, \ldots, X_p$ is **multivariate normal.** (This is the extension to multiple variables of the bivariate-normal distribution described in Section 9.11.) Although in theory this assumption is necessary for many of our tests, rather substantial departures from a multivariate-normal distribution are likely to be tolerable. First, our tests are reasonably robust. Second, in actual practice we are concerned not so much about whether R is significantly different from 0 as about whether R is large or small. In other words, with X_i random, we are not as interested in hypothesis testing with respect to R as we were in the analysis of variance problems. Whether $R = .10$ is statistically significant or not when it comes to prediction may be largely irrelevant because it accounts for only 1% of the variation.

If the variables $X_1, X_2, \ldots, X_p$ are fixed variables, we will simply make the assumption that the conditional distributions of Y (i.e., the distribution of Y for specific levels of X_i) are normally and independently distributed. Here again, moderate departures from normality are tolerable.

The fixed model and the corresponding assumption of normality in Y will be considered in Chapter 16. In this chapter, we generally will be concerned with random variables. The multivariate-normal assumption is more stringent than is necessary for much of what follows, but it is sufficient. For example, calculation of the standard error of b_j does not require an assumption of multivariate normality. However, a person seldom wants to find the standard error of b_j

unless he or she wants to test (or form confidence limits on) b_j, and this test requires the normality assumption. We will therefore impose this assumption on our data.

15.5   The Multiple Correlation Coefficient

multiple correlation coefficient ($R_{0.123\ldots p}$)

Exhibit 15.1 shows that the multiple correlation between Overall and the five predictors is equal to .869. The **multiple correlation coefficient** is often denoted $R_{0.123\ldots p}$. The notation denotes the fact that the criterion (Y or X_0) is predicted from predictors 1, 2, 3 ... p simultaneously. When there is no confusion about which predictors are involved, we generally drop the subscripts and use plain old R.

R is defined as the correlation between the criterion (Y) and the best linear combination of the predictors. As such, R is really nothing but $r_{Y\hat{Y}}$, where

$$\hat{Y} = b_0 + b_1 X_1 + b_2 X_2 + \cdots + b_p X_p$$

Thus, if we wanted, we could use the regression equation to generate $\hat{Y}$, and then correlate Y and $\hat{Y}$. Although no one would seriously propose calculating R in this way, it is helpful to realize that this is what the multiple correlation actually represents. In practice, R (or R^2) is printed out by every multiple regression computer program. For our data, the multiple correlation between the criterion and the five predictors taken simultaneously is .869.

The coefficient R is a regular correlation coefficient and can be treated just like any other Pearson product-moment correlation. (This is obviously true, because $R = r_{Y\hat{Y}}$.) However, in multiple correlation (as is often the case with simple correlation) we are more interested in R^2 than in R because it can be directly interpreted in terms of percentage of accountable variation. Thus, $R^2 = .869^2 = .755$, and we can say that 75.5% of the variation in the overall quality of the lectures can be predicted on the basis of the five predictors. This is about 11 percentage points more than could be predicted on the basis of Teach, our best single predictor, alone.

Unfortunately, R^2 is not an unbiased estimate of the corresponding parameter in the population ($R^{*2}_{0.123\ldots p}$). The extent of this bias depends on the relative size of N and p. When $N = p + 1$, prediction is perfect and $R = 1$, regardless of the true relationship between Y and $X_1, X_2, \ldots, X_p$ in the population. (A straight line will perfectly fit any two points; a plane, like the three legs of a milking stool will perfectly fit any three points, and so on.) A relatively unbiased estimate of R^{*2} is given by

$$\text{est } R^{*2} = 1 - \frac{(1 - R^2)(N - 1)}{N - p - 1}$$

For our data,

$$\text{est } R^{*2} = 1 - \frac{(1 - .755)(49)}{44} = .728$$

This value agrees with the "Adjusted R Square" printed by the SPSS procedure in Exhibit 15.1.

It should be apparent from the definition of R that it can take on values only between 0 and 1. This follows both from the fact that it is defined as the positive square root of R^2 and from the fact that it can be viewed as $r_{Y\hat{Y}}$—we would hardly expect $\hat{Y}$ to be negatively correlated with Y. This is an important point because if we were to predict Overall just from Enroll, the multiple correlation will be .240, whereas we know that the simple correlation was −.240. As long as you understand what is happening here, there should not be any confusion.

Because R^2_{adj} is a less biased estimate of the squared population coefficient than R^2, you might expect that people would routinely report R^2_{adj}. Actually, R^2_{adj} is seldom seen except on computer printouts. I don't know why that should be, but R or R^2 is what you would normally report.

Testing the Significance of R^2

We have seen how to ask whether each of the variables is making a significant contribution to the prediction of Y by testing its regression coefficient (b_j). But perhaps a question that should be asked first is, Does the set of variables taken together predict Y at better-than-chance levels? I suggest that this question has priority because there is little point in looking at individual variables if no overall relationship exists.

The easiest way to test the overall relationship between Y and $X_1, X_2, \ldots, X_p$ is to test the multiple correlation coefficient for statistical significance. This amounts to testing H_0: $R^* = 0$, where R^* represents the correlation coefficient in the population. By the nature of our test, it is actually easier to test R^2 than R, but that amounts to the same thing. The test on R^2 is recognizable as a simple extension of the test given in Chapter 9 when we had only one predictor. In this case, we have p predictors and

$$F = \frac{(N - p - 1)R^2}{p(1 - R^2)}$$

is distributed as the standard F distribution on p and $N - p - 1$ degrees of freedom. (With only one predictor this F statistic reduces to the familiar $(N - 2)(r^2)/(1 - r^2)$.) For our data, $N = 50$, $p = 5$, and $R^2 = .755$. Then

$$F = \frac{(50 - 5 - 1)(.755)}{5(.245)} = \frac{44(.755)}{1.225} = 27.184^5$$

This is the same F as that given in the summary table in Exhibit 15.1. An F of 27.184 on 5 and 44 df is obviously significant beyond $p = .05$, and we can therefore reject H_0: $R^* = 0$ and conclude that we can predict at better-than-chance levels. (The printout shows the probability associated with this F under H_0 to three decimal places as .000.)

Sample Sizes

As you can tell from the formula for an adjusted R^2 and from the preceding formula for F, our estimate of the correlation depends on both the size of the sample (N) and the number of predictors (p). People often assume that if there is no relation between the criterion and the predictors, R should come out near 0. In fact, the expected value of R^2 *for random data* is $p/(N - 1)$.

Thus, with 5 predictors, 50 cases, and no true relationship between the predictors and the criterion, an $R^2 = .10$ would be the expected value, not 0. So it is important that we have a relatively large sample size. A rule of thumb that has been kicking around for years is that we should have *at least* 10 observations for every predictor. Harris (1985) points out, however, that he knows of no empirical evidence supporting this rule. It certainly fails in the extreme because no one would be satisfied with 10 observations and 1 predictor. Harris advocates an alternative rule dealing not with the ratio of p to N but, rather, with their difference. His rule is that N should exceed p by at least 50. Others have suggested the slightly more liberal $N \geq p + 40$. Whereas these two rules relate directly to the reliability of a correlation coefficient, Cohen, Cohen, West, and Aiken (2003) approach the problem from the direction of statistical power. They show that in the one-predictor case, to have power $= .80$ for a population correlation of .30 would require $N = 124$. With 5 predictors, a population

---

[5] Here, as elsewhere, what you might calculate with a calculator will differ from the answers I give because of rounding error. Computer software uses far more significant digits than it prints out, and the answers are themselves more accurate. In this particular equation, using three-digit numbers would yield an answer of 27.118 when the correct answer is 27.184—quite a difference. I give the answer that agrees with the printout.

correlation of .30 would require 187 subjects for the same degree of power. As you can see, a reasonable amount of power requires fairly large samples. Perhaps Darlington's (1990) rule of thumb is the best—"more is better."

15.6 Geometric Representation of Multiple Regression

hyperspace

regression surface

Any linear multiple regression problem involving p predictors can be represented graphically in $p + 1$ dimensions. Thus, with one predictor, we can readily draw a two-dimensional scatter diagram and fit a regression line through the points. With two predictors, we can represent the data in three-dimensional space with a plane passing through the points. With more than three predictors, we would have to begin to think of **hyperspace** (multidimensional space) with the **regression surface** (the analog of the regression line or plane) fitted through the points. People have enough trouble thinking in terms of three-dimensional space, without trying to handle hyperspaces, and so we will consider here only the two-predictor case. The generalization to the case of many predictors should be apparent, even if you cannot visualize the problem.

Figure 15.1 shows a three-dimensional plot of the Overall course rating (Y) against the predictors Teach (X_1) and Knowledge (X_2). Each member of the data set is represented as the ball on top of a flagpole. The base of the flagpole is located at the point (X_1, X_2), and the height of the pole is Y.

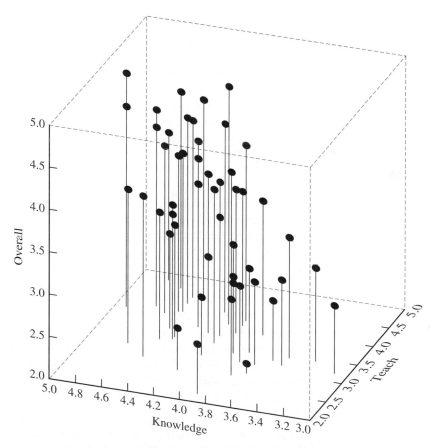

Figure 15.1   Three-dimensional representation of Y as a function of X_1 and X_2

In Figure 15.1, as you move from the lower-right front to the left back, the heights of the flagpoles (and therefore the values of Y) increase. If you had the three-dimensional model represented by this figure, you could actually pass a plane through, or near, the points to give the best possible fit. Some of the flagpoles would stick up through the plane, and some would not reach it, but the points could be fit reasonably well by this plane. The vertical distances of the points from the plane, the distances $(Y - \hat{Y})$, would be the residuals. Just as in the one-predictor case, the residuals represent the vertical distance of the points from the best-fitting line (or, in this three-dimensional case, the best-fitting plane).

We can derive one additional insight from this three-dimensional model. The plane we have been discussing forms some angle (in this case, positive) with the axis X_1 (Teach). In other words, the plane rises from right to left. The slope of that plane relative to X_1 is b_1. Similarly, the slope of the plane with respect to X_2 (Knowledge) is b_2. The height of the plane at the point $(X_1, X_2) = (0, 0)$ would be b_0.[6]

Further Interpretation of Regression Coefficients

The regression coefficients we have been discussing (both β and b) are called *partial regression coefficients*; that is, b_1 is the coefficient for the regression of Y on X_1 when we *partial out* (remove, or hold constant) the effect of $X_2, X_3, \ldots, X_p$. As such, it should actually be denoted as $b_{01.23 \ldots p}$, although we employ the shorter notation except when confusion might otherwise result. Perhaps the easiest way to see what we mean by the "partialling out of X_2" is to consider Figure 15.1. If we look at only the data for $X_2 = 4$ (for example), we see that we could fit a straight line to the regression of Y on X_1. We do not need to consider X_2 because we are considering only those cases in which $X_2 = 4 =$ a constant. Similarly, we could look at only those cases in which $X_2 = 4.2$. For any one value of X_2, we have a regression for Y on X_1. The average of the coefficients for these regressions is $b_{01.2}$. Now consider the values of Y and X_2 for only the cases in which $X_1 = 3$ (or 4, or 5, and so on). Here, we have the regression of Y on X_2, with X_1 partialled out. The average of these coefficients is $b_{02.1}$.

One common mistake is to equate $b_{01.2}$ with b_{01}—that is, with the simple regression coefficient we would obtain if we regressed Y on X_1 *without regard to X_2*. That these coefficients cannot be equated can be seen in the simple extreme example shown in Figure 15.2. The raw data used to plot this figure are

Y	2	1	4	3	6	5
X_1	1	2	3	4	5	6
X_2	2	2	4	4	6	6

Figure 15.2a represents the three-dimensional projection of Y on X_1 and X_2. Note that for any single value of X_2, the slope of the regression line of Y on X_1 is decidedly negative (in fact, $b_{01.2} = -1$). However, if you look at Figure 15.2b where Y is plotted against X_1, *ignoring* X_2, the slope is positive ($b_{01} = 0.829$). Thus, there is quite a difference between partialling out a variable and ignoring it. This is a deliberately extreme example, but it is an exaggeration of more typical cases. Only when X_1 and X_2 are independent ($r_{12} = 0$) will $b_{01.2}$ be equal to b_{01}.

There is an alternative way of viewing partial regression coefficients. It is of no practical use in calculations, but it is very useful in understanding the process of partialling. Consider the partial regression coefficient $b_{01.2}$ for the data in Figure 15.2. As mentioned earlier, $b_{01.2} = -1$. Suppose that we regress Y on X_2 and obtain the values of $Y_i - \hat{Y}_i$ for this

---

[6] Another good way to see the relationship between these variables is to use a computer program (such as www. StatCrunch.com) that will represent the data in what appears to be a three-dimensional space and then rotate the axes so that you can look at the plot from different directions. These are often called *spin plots*.

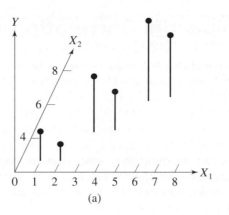
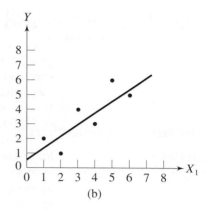

| (a) | (b) |

Figure 15.2   (a) Y as a function of X_1 and X_2 (b) Y plotted as a function of X_1 only

regression. These residual values would represent that part of Y that *cannot* be predicted by (is independent of) X_2. We will represent these residuals by the symbol Y_r. Now we regress X_1 on X_2 and generate the residuals $X_{1_i} - \hat{X}_{1_i}$. These represent the part of X_1 that is independent of X_2 and will be symbolized as X_{1r}. We now have two sets of residuals—the part of Y that is independent of X_2 and the part of X_1 that is independent of X_2. We have partialled X_2 out of Y and out of X_1. If we now regress Y_r on X_{1r}, the slope will be $b_{01.2}$. Moreover, the correlation between Y_r and X_{1r} is called the *partial correlation* of Y and X_1, with X_2 partialled out ($r_{01.2}$). Table 15.3 contains a simple illustration of what we have been discussing, using the data from Figure 15.2a.

Table 15.3   Illustrative calculation of partial regression coefficient

	Data	
Y	X_1	X_2
2	1	2
1	2	2
4	3	4
3	4	4
6	5	6
5	6	6

Y on X_2			X_1 on X_2			Y_r on X_{1r}	
$\hat{Y} = 1.0\,X_2 - 0.50$			$\hat{X}_1 = 1.0 X_2 - 0.50$				
Y	$\hat{Y}$	Y_r	X_1	$\hat{X}_1$	X_{1r}	Y_r	X_{1r}
2	1.5	0.5	1	1.5	−.5	0.5	−.5
1	1.5	−.5	2	1.5	0.5	−.5	0.5
4	3.5	0.5	3	3.5	−.5	0.5	−.5
3	3.5	−.5	4	3.5	0.5	−.5	0.5
6	5.5	0.5	5	5.5	−.5	0.5	−.5
5	5.5	−.5	6	5.5	0.5	−.5	0.5

$$b_{Y_r X_{1r}} = \frac{\text{cov}_{Y_r X_{1r}}}{s^2_{X_{1r}}} = \frac{-0.30}{0.30} = -1 = b_{01.2}$$

15.7 Partial and Semipartial Correlation

Two closely related correlation coefficients involve partialling out, or controlling for, the effects of one or more other variables. These correlations are the partial and semipartial correlation coefficients.

Partial Correlation

partial correlation $r_{01.2}$

We just saw that a **partial correlation $r_{01.2}$** is the correlation between two variables with one or more variables partialled out of *both* X and Y. More specifically, it is the correlation between the two sets of residuals formed from the prediction of the original variables by one or more other variables.

Consider an experimenter who wanted to investigate the relationship between earned income and success in college. He obtained measures for each variable and ran his correlation, which turned out to be significant. Elated with the results, he harangued his students with the admonition that if they did not do well in college they were not likely to earn large salaries. In the back of the class, however, was a bright student who realized that both variables were (presumably) related to IQ. She argued that people with high IQs tend to do well in college and also earn good salaries, and that the correlation between income and college success is an artifact of this relationship.

The simplest way to settle this argument is to calculate the partial correlation between Income and college Success with IQ partialled out of both variables. Thus, we regress Income on IQ and obtain the residuals. These residuals represent the variation in Income that cannot be attributed to IQ. You might think of this as a "purified" income measure—purified of the influence of IQ. We next regress Success on IQ and again obtain the residuals, which here represent the portion of Success that is not attributable to IQ. We can now answer the important question: Can the variation in Income not explained by (independent of) IQ be predicted by the variation in Success that is also independent of IQ? The correlation between these two variables is the partial correlation of Income and Success, partialling out IQ.

The partial correlation coefficient is represented by $r_{01.23\ldots p}$. The two subscripts to the left of the dot represent the variables being correlated, and the subscripts to the right of the dot represent those variables being partialled out of both.

A simple example illustrating the principle of partial correlation is presented in Table 15.4. Here, you can see the calculation of the residuals and the correlation between the two sets of residuals. The partial correlation ($r_{01.2} = .738$) represents the independent contribution of variable X_1 toward the prediction of Y, partialling out X_2.

Semipartial Correlation

semipartial correlation $r_{0(1.2)}$

A type of correlation that will prove exceedingly useful both here and in Chapter 16 is the **semipartial correlation $r_{0(1.2)}$** sometimes called the *part*[7] correlation. As the name suggests, a semipartial correlation is the correlation between the criterion and a partialled predictor variable. In other words, whereas the partial correlation $r_{0(1.2)}$ has variable 2 partialled out of both the criterion and predictor 1, the semipartial correlation $r_{0(1.2)}$ has variable 2 partialled out of only predictor 1. In this case, the semipartial correlation is simply the correlation between Y and the residual ($X_1 - \hat{X}_1 = X_{1r}$) of X_1 predicted on X_2. As such, it is the correlation of Y with that part of X_1 that is independent of X_2. As an illustration, the computation

---

[7] The only text that I have seen using "part correlation" was McNemar (1969) when I was just out of graduate school. But the name seems to have stuck, and you will find SPSS employing that term.

Table 15.4   Illustrative calculation of partial correlation coefficient

Data

Y	X_1	X_2
4	2	4
3	4	1
6	6	5
1	1	2
5	3	3

Y on X_2			X_1 on X_2		
$\hat{Y} = 0.90\,X_2 + 1.1$			$\hat{X}_1 = 0.50 X_2 + 1.7$		
Y	$\hat{Y}$	Y_r	X_1	$\hat{X}_1$	X_{1r}
4	4.7	−0.7	2	3.7	−1.7
3	2.0	1.0	4	2.2	1.8
6	5.6	0.4	6	4.2	1.8
1	2.9	−1.9	1	2.7	−1.7
5	3.8	1.2	3	3.2	−0.2

$$s_{X_{1r}} = 1.754$$
$$s_{Y_r} = 1.294$$
$$\text{cov}_{X_{1r}Y_r} = 1.675$$
$$r = \frac{\text{cov}_{X_{1r}Y_r}}{s_{X_{1r}}s_{Y_r}} = \frac{1.675}{(1.754)(1.294)} = .738$$

of $r_{0(1.2)}$ by way of the residuals is presented in Table 15.5 for the data in Table 15.4. A much simpler method of computation exists, however. It can be shown that

$$r^2_{0(1.2)} = R^2_{0.12} - r^2_{02}$$

Using this formula, we can verify the calculation in Table 15.5. For those data, $R^2_{0.12} = .794$, and $r^2_{02} = .547$. Thus

$$r^2_{0(1.2)} = .794 - .547 = .247$$
$$r_{0(1.2)} = \sqrt{r^2_{0(1.2)}} = \sqrt{.247} = .497$$

which agrees with the result in Table 15.5.

The preceding formula for $r_{0(1.2)}$ affords an opportunity to explore further just what multiple regression equations and correlations represent. Rearranging the formula, we have

$$R^2_{0.12} = r^2_{02} + r^2_{0(1.2)}$$

This formula illustrates that the squared multiple correlation is the sum of the squared correlation between the criterion and one of the variables plus the squared correlation between the criterion and the part of the other variable that is independent of the first. Thus, we can think of R and R^2 as being based on as much information as possible from one variable, any *additional, nonredundant* information from a second, and so on.

Table 15.5   Illustrative calculation of semipartial correlation coefficient $(\hat{X}_1 = 0.50X_2 + 1.7)$

Y	X_1	X_2	X_{1r}
4	2	4	−1.7
3	4	1	1.8
6	6	5	1.8
1	1	2	−1.7
5	3	3	−0.2

$$s_Y = 1.924$$

$$s_{X_{1r}} = 1.754$$

$$s_{YX_{1r}} = 1.675$$

$$r_{YX_{1r}} = \frac{\text{cov}_{YX_{1r}}}{s_Y s_{X_{1r}}} = \frac{1.675}{(1.924)(1.754)} = .497$$

In general

$$R^2_{0.123\ldots p} = r^2_{01} + r^2_{0(2.1)} + r^2_{0(3.12)} + \cdots + r^2_{0(p.123\ldots p-1)}$$

where $r^2_{0(3.12)}$ is the squared correlation between the criterion and variable 3, with variables 1 and 2 partialled out of 3. This way of looking at multiple regression will be particularly helpful when we consider the role of individual variables in predicting the criterion, and when we consider the least squares approach to the analysis of variance in Chapter 16. As an aside, it should be mentioned that when the predictors are independent of one another, the preceding formula reduces to

$$R^2_{0.123\ldots p} = r^2_{01} + r^2_{02} + r^2_{03} + \cdots + r^2_{0p}$$

because, if the variables are independent, there is no variance in common to be partialled out.

For the data on course evaluations, the squared semipartial correlation between Overall and Teach, partialling Exam, Knowledge, Grade, and Enroll from Teach is .183. This shows that Teach explains 18.3% of the variation in Overall *once we have accounted for the other four variables*. This value is not shown in Exhibit 15.1, but can be readily calculated using SPSS by calculating R^2 including and excluding Teach, and then obtaining the difference between the two values of R^2.

$$R^2_{0.12345} = .755 \qquad R^2_{0.2345} = .572 \,(\text{not shown})$$

$$R^2_{0.12345} - R^2_{0.2345} = .755 - .572 = .183$$

The squared *partial* correlation between Overall and Teach, partialling the other four predictors from both Overall and Teach, by the method discussed next is .428, showing that 42.8% of the variation in Overall *that could not be explained by the other predictors* can be accounted for by Teach. This point will be elaborated in the next section.

We do not need a separate significance test for semipartial or partial correlations because we already have such a test in the test on the regression coefficients. If that test is significant, then corresponding β, partial, and semipartial coefficients are also significant.[8]

---

[8] You will note that we consider both partial and semipartial correlation but only mentioned the *partial* regression coefficient (b_j). This coefficient could equally well be called the *semipartial* regression coefficient.

Therefore, from Exhibit 15.1 we also know that these coefficients for Teach are all significant. Keep in mind, however, that when we speak about the significance of a coefficient, we are speaking of it within the context of the other variables in the model. For example, Exhibit 15.1 reveals that Exam does not make a significant contribution to the model. That does not mean that it would not contribute to any other model predicting Overall. (In fact, when used as the only predictor, it predicts Overall at better-than-chance levels. See the matrix of intercorrelations.) This only means that once we have the other predictors in our model, Exam does not have any independent (or unique) contribution to make.

Alternative Interpretation of Partial and Semipartial Correlation

Venn diagrams

An alternative way of viewing the meaning of partial and semipartial correlations can be very instructive. This method is best presented in terms of what are called **Venn diagrams.**

Suppose that the box in Figure 15.3 is taken to represent all the variability in the criterion (Overall). We will set the area of the box equal to 100%—the percentage of the variation in Overall to be explained. The circle labeled Teach is taken to represent the percentage of the variation in Overall that is explained by Teach. In other words, the area of the circle is equal to $r_{01}^2 = .646$. Similarly, the area of the circle labeled Knowledge is the percentage of the variation in Overall explained by Knowledge and is equal to $r_{02}^2 = .465$. Finally, the overlap between the two circles represents the portion of Overall that both Teach and Knowledge have in common, and equals .372. The area outside of either circle but within the box is the portion of Overall that cannot be explained by either variable and is the residual variation = .261.

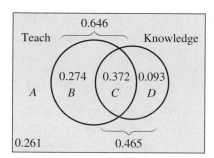

Figure 15.3   Venn diagram illustrating partial and semipartial correlation

The areas labeled B, C, and D in Figure 15.3 represent portions of the variation in Overall that can be accounted for by Teach *and/or* Knowledge. (Area A represents the portion that cannot be explained by either variable or their combination, the residual variation.) Thus, the two predictors in our example account for 73.9% of the variation of Y: $B + C + D = .274 + .372 + .093 = 0.739$. The *squared semipartial correlation* between Teach and Overall, with Knowledge partialled out of Teach, is the portion of the variation of Overall that Teach accounts for *over and above* the portion accounted for by Knowledge. As such, it is .274:

$$r_{0(1.2)}^2 = R_{0.12}^2 - r_{02}^2 = .739 - .465 = .274$$

The semipartial correlation is the square root of this quantity.

$$r_{0(1.2)} = \sqrt{.274} = .523$$

The *squared partial correlation* has a similar interpretation. Instead of being the additional percentage of Overall that Teach explains but that Knowledge does not, which is the squared *semi*partial correlation, it is the additional amount that Teach explains *relative to* the amount that Knowledge left to be explained. For example, $r_{02}^2 = .465$ and $1 - r_{02}^2 = .535$,

$$r_{01.2}^2 = \frac{r_{0(1.2)}^2}{1 - r_{02}^2}$$

$$= \frac{.274}{.535} = .512$$

$$r_{01.2} = \sqrt{.512}$$

$$= .716$$

Schematically, squared multiple, partial, and semipartial correlations can be represented as

$$r_{0(1.2)}^2 = B = \text{the squared semipartial correlation}$$

$$r_{01.2}^2 = \frac{B}{A + B} = \text{the squared partial correlation}$$

$$= \frac{r_{0(1.2)}^2}{1 - r_{02}^2}$$

In addition,

$$A = 1 - R_{0.12}^2 = \text{the residual (unexplained) variation in } Y \text{ (Overall)}$$

$$D = r_{0(2.1)}^2 = \text{the other squared semipartial correlation}$$

$$B + C + D = R_{0.12}^2 = \text{the squared multiple correlation}$$

$$B + C = r_{01}^2 = \text{the squared correlation between } Y \text{ (Overall) and } X_1 \text{ (Teach)}$$

$$C + D = r_{02}^2 = \text{the squared correlation between } Y \text{ (Overall) and } X_2$$
$$\text{(Knowledge)}$$

Why Do We Care About Partial and Semipartial Correlations?

You might ask why we bother to worry about partial and semipartial correlations. What do they add to what we already know? The answer is that they add a great deal. They allow us to control for variables that we might perceive as "nuisance" variables, and in so doing allow us to make statements of the form "The correlation between Y and A is .65, *after we control for the influence of B.*" To take an example from a study that we will discuss later in the chapter, Leerkes and Crockenberg (1999) were interested in the relationship between the maternal care a woman received when she was a child and the level of self-confidence or self-efficacy she feels toward her own mothering skills. Leerkes and Crockenberg asked whether this relationship was influenced by the fact that those who received high quality maternal care also showed high levels of self-esteem. Perhaps if we controlled for differences in self-esteem, the maternal care → self-efficacy relationship would disappear. This is a case where they are partialling out the influence of self-esteem to look at the relationship that remains. Partial and semipartial correlations are a tool to help us "get our hands around" a number of confusing relationships.

15.8 Suppressor Variables

Suppose we have a multiple regression problem in which all variables are scored so they correlate positively with the criterion. Because the scoring of variables is often arbitrary anyway, this presents no difficulty (if X is negatively related to Y, $C - X$ will be positively

suppressor variable

related to Y, where C is any constant). In such a situation, we would expect all the regression coefficients (β_i or b_i) to be positive. Occasionally, however, a regression coefficient in this situation will be *significantly* negative. Such a variable, if significant, is called a **suppressor variable.**[9]

Suppressor variables seem, at first glance, to be unreasonable. We know that the simple correlation between the criterion and the variable is positive (by our definition), yet in the resulting regression equation, an increment on this variable produces a decrement in $\hat{Y}$. Moreover, it can be shown that $R^2 = \sum \beta_i r_{0i}$. If r_{0i} is positive and β_i is negative, the product $\beta_i r_{0i}$ will be negative. Thus, by assigning β_i a negative value, the regression solution (which has the task of minimizing error) would *appear* to be reducing R^2. This does not fit with our preconceived ideas of what should be happening, yet there must be some logical explanation.

Space considerations do not allow an extensive discussion of the theory of suppressor variables, but it is important to illustrate one intuitively sensible explanation. For a more extensive discussion of suppressor variables, see Cohen and Cohen (1983) and Darlington (1968). (The discussion in Cohen and Cohen is particularly helpful.) Here, we will take an example from Darlington (1990). Suppose a speeded history examination (a long exam with a short time in which to complete it) is used as a measure of some external criterion of knowledge of history. Although knowledge of history is presumably independent of reading speed, performance on the speeded test will not be. Thus, some of the variance in test scores will reflect differences in the reading speed of the students rather than differences in their actual knowledge. What we would really like to do is penalize students who did well *only* because they read quickly and help students who did poorly *only* because they read slowly. This is precisely what is accomplished by having reading speed serve as a suppressor variable. It is suppressing some of the error in the exam scores.

As Darlington points out, a variable will serve as a suppressor variable when it correlates more highly with Y_r than with Y (where Y_r represents the residual when predicting history knowledge from history score), and will not serve as a suppressor variable when it correlates more highly with Y than Y_r. Cohen et al. (2003) point out that suppressor relationships are hard to find in psychology (at least statistically significant ones), though they are easily found in biology and economics. In those fields, suppressor relationships relate to homeostatic mechanisms, where an increase in X leads to an increase in Y, which in turn causes an increase in Z which leads back to a decrease in Y. Although these mechanisms are not as common in psychology, I am frequently asked about suppression effects—most of which turn out to be statistically nonsignificant.

15.9 Regression Diagnostics

In predicting course evaluations from variables that described the course and the instructor, we skipped an important step because of the need to first lay out some of the important concepts in multiple regression. It is now time to go back and fill that gap. Before throwing all the observations and predictors into the model and asking computer software to produce an answer to be written up and interpreted, we need to look more closely at the data. We can do this by using a variety of tools supplied by nearly all multiple regression computer programs. Once we are satisfied with the data, we can then go on and use other available tools to help us decide which variables to include in the model. Cohen et al. (2003) is a much more complete and readable treatment of the problem of regression diagnostics.

---

[9] Cohen and Cohen (1983) discuss two additional types of suppression, and their discussion is helpful when faced with results that seem contrary to intuition. That discussion has been omitted in the more recent Cohen et al. (2003), so you need to go back to the earlier edition.

The first step in examining the data has already been carried out in Table 15.2 with stem-and-leaf displays on all six variables. At that point, we noted that most of the variables were fairly nicely distributed with few outliers. Teach and Exam each had one unusually low score, whereas Enroll had four quite high scores. This should alert us to the potential influence of those cases, but aside from correcting any obvious errors that we identify, we will go ahead with our first analysis.

multivariate outliers

The fact that we don't have more outliers when we look at the variables individually does not necessarily mean that all is well. There is still the possibility of having **multivariate outliers.** A case might seem to have reasonable scores on each of the variables taken separately but have an unusual *combination* of scores on two or more variables. For example, it is not uncommon to be 6 feet tall, nor is it uncommon to weigh 125 pounds. But it clearly would be unusual to be 6 feet tall *and* weigh 125 pounds.

Having temporarily satisfied ourselves that the data set does not contain unreasonable data points and that the distributions are not seriously distorted, a useful second step is to conduct a preliminary regression analysis using all the variables, as we have done. I say "preliminary" because the point here is to use that analysis to examine the data rather than as an end in itself.

Instead of jumping directly into the course-evaluation data set, we will first investigate diagnostic tools with a smaller data set created to illustrate the use of those tools. These data are shown here and are plotted in Figure 15.4.

X:	1	1	3	3	3	4	5	5	7	6	10	13
Y:	1	2	3	5	7	6	8	10	10	5	4	14

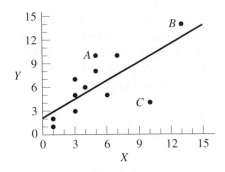

Figure 15.4 Scatterplot of *Y* on *X*

The three primary classes of diagnostic statistics, each of which is represented in Figure 15.4, are

Distance

1. **Distance,** which is useful in identifying potential outliers in the dependent variable (Y).

Leverage (h_i)

2. **Leverage (h_i),** which is useful in identifying potential outliers in the independent variables ($X_1, X_2, \ldots, X_p$).

Influence

3. **Influence,** which combines distance and leverage to identify unusually influential observations. An observation is influential if the location of the regression surface would change markedly depending on the presence or absence of that observation.

Our most common measure of distance is the residual ($Y_i - \hat{Y}_i$). It measures the vertical distance between any point and the regression line. Points *A* and *C* in Figure 15.4 have large residuals (they lie far from the regression line). Such points may represent random error, they may be data that are incorrectly recorded, or they may reflect unusual cases that don't really belong in this data set. (An example of this last point would arise if we were trying to predict physical reaction time as a function of cognitive processing features of a task, and our subjects included one individual who suffered from a neuromuscular disorder that seriously slowed his reaction time.) Residuals are a standard feature of all regression analyses, and you should routinely request and examine them in running your analyses.

Leverage (often denoted h_i, or "hat diag") measures the degree to which a case is unusual with respect to the predictor variables X_j. In the case of one predictor, leverage is simply a function of the deviation of the score on that predictor from the predictor mean. Point B in Figure 15.4 is an example of a point with high leverage because the X score for that point (13) is far from $\overline{X}$. Most programs for multiple regression compute and print the leverage of each observation if requested. Possible values on leverage range from a low of $1/N$ to a high of 1.0, with a mean of $(p + 1)/N$, where $p =$ the number of predictors. Stevens (1992) recommends looking particularly closely at those leverage values that exceed $3(p + 1)/n$.

Points that are high on either distance or leverage do not necessarily have an important influence on the regression, but they have the potential for it. For a point to be high on influence, it must have relatively high values on both distance and leverage. In Figure 15.4, Point B is very high on leverage, but it has a relatively small residual (distance). Point A, on the other hand, has a large residual but, because it is near the mean on X, has low leverage. Point C is high on leverage and has a large residual, suggesting that it is high on influence.

Cook's D

The most common measure of influence is known as **Cook's D.** It is a function of the sum of the squared *changes* in b_j that would occur if the ith observation were removed from the data and the analysis rerun.

Exhibit 15.2 contains various diagnostic statistics for the data shown in Figure 15.4. These diagnostics were produced by SAS, but similar statistics would be produced by almost any other program.

To take the diagnostic statistics in order, consider first the column headed Resid, which is a measure of distance. This column reflects what we can already see in Figure 15.4—that the 8th and 11th observations have the largest residuals. Considering that the Y values range only from 1 to 14, a residual of -5.89 seems substantial.

If the data met the underlying assumptions, we would expect the values of Y to be normally distributed about the regression line. In other words, with a very large data set, all the Y values corresponding to a specific value of X would have a normal distribution. Five percent

	OBS	X	Y	PRED	RESID	RSTUDENT	HAT DIAG H	MSE	COOK'S D
	1	1	1	3.23	−2.23	−0.87	0.20	8.22	0.10
	2	1	2	3.23	−1.22	−0.47	0.20	8.71	0.03
	3	3	3	4.71	−1.71	−0.62	0.11	8.55	0.03
	4	3	5	4.71	0.29	0.10	0.11	8.91	0.00
	5	3	7	4.71	2.29	0.85	0.11	8.26	0.05
	6	4	6	5.45	0.55	0.19	0.09	8.88	0.00
	7	5	8	6.19	1.81	0.65	0.08	8.52	0.02
"A" ->	8	5	10	6.19	3.81	1.49	0.08	7.16	0.09
	9	6	5	6.93	−1.93	−0.69	0.09	8.46	0.02
	10	7	10	7.77	2.33	0.86	0.11	8.24	0.05
"C" ->	11	10	4	9.89	−5.89	−3.54	0.26	3.73	1.01
"B" ->	12	13	14	12.11	1.89	0.98	0.54	8.06	0.55

Exhibit 15.2   Diagnostic statistics for data in Figure 15.4

Studentized residual

of these values would lie more than 1.96 adjusted standard errors from the regression line. (I use the word "adjusted" because the size of the standard error will depend partly on the degree to which X departs from the mean of X, as measured by h_i.) Within this context, it may be meaningful to ask if a point lies significantly far from the regression line. If so, we should be concerned about it. A t test on the magnitude of the residuals is given by the statistic RStudent, sometimes called the **Studentized residual.** This can be interpreted as a standard t statistic on $(N - p - 1)$ degrees of freedom. Here, we see that for case 11 RStudent $= -3.54$. This should give us pause because that is a substantial, and significant, deviation. It is often useful to think of RStudent less as a hypothesis-testing statistic and more as just an indicator of the magnitude of the residual. (Remember that here we are computing N t tests, with a resulting very large increase in the familywise error rate.) But significant or not, something that is 3.54 standard errors from the line is unusual and therefore noteworthy. We are not predicting that case well.

We now turn to leverage (h_i), shown in the column headed Hat Diag. Here, we see that most observations have leverage values that fall between about 0.00 and 0.20. The mean leverage is $(p + 1)/N = 2/12 = 0.167$, and that is about what we would expect. Notice, however, that two cases have larger leverage; namely, cases 11 and 12, which exceeds Steven's rule of thumb of $3(p + 1)/n = 3(2)/12 = .50$. We have already seen that 11 has a large residual, so its modest leverage may make it an influential point. Case 12 has a leverage value nearly twice as large. However, it falls quite close to the regression line with a fairly small residual, and it is likely to be less influential.

Cook's D, which is a measure of influence, varies as a function of distance (residual), leverage (h_i), and $MS_{residual}$. Most of the values in the last column are quite small, but cases 11 and 12 are exceptions. In particular, observation 11 has a D exceeding 1.00. The sampling distribution of Cook's D is in dispute, and there is no general rule for what constitutes a large value, but values greater than 1.00 are unusual.

We can summarize the results shown in Exhibit 15.2 by stating that each of the three points labeled in Figure 15.4 is reflected in that table. Point A has a fairly, though not significantly, large residual but has small values for both leverage and influence. Point B has a large leverage, but Cook's D is not high and its removal would not substantially reduce $MS_{residual}$. Point C has a large residual, a fairly large leverage, and a substantial Cook's D; its removal would provide a substantial reduction in $MS_{residual}$. This is the kind of observation that we should consider seriously. Although data should not be deleted merely because they are inconvenient and their removal would make the results look better, it is important to pay attention to observations such as case 11. There may be legitimate reasons to set that case aside and to treat it differently. Or it may be erroneous. Because this is not a real data set, we cannot do anything further with it.

It may seem like overkill to compute regression diagnostics simply to confirm what anyone can see simply by looking at a plot of the data. However, we have looked only at a situation with one predictor variable. With multiple predictors, there is no reasonable way to plot the data and visually identify influential points. In that situation, you should at least create univariate stem-and-leaf displays, perhaps bivariate plots of each predictor against the criterion looking for peculiar distributions of points, and compute diagnostic statistics. From those statistics, you can then target particular cases for closer study.

Returning briefly to the data on course evaluations, we can illustrate some additional points concerning diagnostic statistics. Exhibit 15.3 contains additional statistics that were not shown in Exhibit 15.1, but came from that SPSS analysis.[10] These values are obtained

---

[10] SPSS calculates leverage, and hence the Studentized Range Statistic, slightly differently than do SAS, JMP, SYSTAT, BMDP, and others. The leverage values are lower by a factor of $1/N$, but this makes no substantive difference in the interpretation (except that the mean leverage will now be p/N instead of $(p + 1)/N$).

Observation	Residual	RStudent	Cook's D	Hat Diag (h_i)
1	−0.37339	−1.19504	0.01193	0.02773
2	0.24080	0.79546	0.01252	0.08615
3	0.05357	0.25351	0.01388	0.54438
4	0.34881	1.11196	0.00862	0.02014
5	0.25758	0.90901	0.03811	0.19674
...	...	...	...	...
13	0.30483	1.00923	0.02100	0.09010
14	−0.39454	−1.27803	0.02061	0.05039
15	0.82464	2.87139	0.33384	0.17546
16	−0.20253	−0.65917	0.00623	0.05919
...	...	...	...	...
20	−0.07747	−0.24904	0.00062	0.03620
21	0.41409	1.36670	0.03635	0.08455
22	−0.11133	−0.36404	0.00212	0.06765
...	...	...	...	...
41	−0.34665	−1.10746	0.00947	0.02427
42	0.16844	0.54526	0.00368	0.04912
43	−0.15645	−0.51599	0.00511	0.08328
44	0.65304	2.14000	0.07700	0.07164
45	−0.16242	−0.69696	0.07187	0.45028
46	0.04006	0.13116	0.00028	0.06995
...	...	...	...	...

Exhibit 15.3 Diagnostic statistics

by choosing the "Save" option in the regression dialog box and then selecting the appropriate statistics.

You may recall that several observations had extreme scores on one or more variables. In particular, case 3 had unusually low ratings on Teach and Exam and a very high Enroll. It also had one of the lowest Overall ratings. Cases 4, 21, and 45 also had much larger values of Enroll than usual. If we look at these cases in the previous diagnostic statistics, we can see that none of them have very large residuals or leverage, and therefore they cannot have a high value for influence, as measured by Cook's D. (Case 3's leverage is larger than some, but is still not particularly large.) It is still possible, however, that the four courses with enrollments considerably larger than others make a disproportionate contribution to the regression. Recall that measures such as residuals or influence depend on how far that case departs from the regression surface. Cook's D in particular asks what the effect would be of removing a *single* case from the data set and rerunning the regression. When you have several cases that all depart in the same direction from the bulk of the data, removing one of them may have a minor effect because the rest are still there. If we have reason to think that large classes may differ in important ways from smaller ones and are worried about the potential effects of these outliers, it would make sense to rerun the analysis with those cases removed to see what effect this has. When you do so for the course data, the value of R^2 is virtually unaffected (it goes from .755 to .758) and none of the regression coefficients (including that for Enroll) changes substantially.

One interesting case in these data is case 15, which has a Studentized residual (RStudent) of 2.87139, although it does not have particularly unusual values on any of the variables. Recall that RStudent is actually a *t* statistic. At first glance, this would seem like a significant value of *t* (on 44 degrees of freedom). However, keep in mind that we are computing 50 different t values here, and some of them can be expected to be significant by chance. We can apply a Bonferroni correction to this test statistic by using Appendix t' with $c = 50$ and $\alpha = .05$. With interpolation, the critical value of t' would be 3.525, and our test is not significant.

15.10 Constructing a Regression Equation

A major problem for anyone who has ever attempted to write a regression equation to predict some criterion or to understand a set of relationships among variables concerns choosing the variables to be included in the model. We often suffer from having too many potential variables rather than too few. Although it would be possible to toss in all the variables to see what would happen, this solution is neither practical nor wise.

Before we look at formal stepwise methods of constructing regression equations, we should first look at two (related) statistics that deal with the relationship among the potential predictor variables. If we want to use several variables to predict Y, presumably we would like each of those variables to bring something new to the task. If variables X_1 and X_2 are themselves very highly correlated, we won't learn much more from both of them than we would from either of them alone.

tolerance

cross-correlation

Exhibit 15.1 (p. 499) contains two columns of numbers that we have not yet discussed. The first is tolerance and the second is the variance inflation factor (VIF). **Tolerance** refers to the degree to which one predictor can itself be predicted by the other predictors in the model. To be more specific, if we let R_X refer to the correlation (sometimes called the **cross-correlation**) between one predictor and all other *predictors,* tolerance is defined as $(1 - R_X^2)$. Looking at Exhibit 15.1, we can see that with Teach as the criterion and Exam, Knowledge, Grade, and Enroll as the predictors, tolerance equals .41819 (rounded). Thus,

$$\text{Tolerance} = .418 = 1 - R_X^2$$
$$R_X^2 = (1 - .418) = .582$$

The squared multiple correlation predicting the Teach predictor from the other four predictors is .58. On the other hand, if we used Enroll, Grade, or Knowledge as the criterion and the others as predictors, the tolerance would be in the .60s and R_X^2 would be in the mid .30s.

Tolerance tells us two things. First, it tells us the degree of overlap among the predictors, helping us see which predictors have information in common and which are relatively independent. (The higher the tolerance, the lower the overlap.) Just because two variables substantially overlap in their information is not reason enough to eliminate one of them, but it does alert us to the possibility that their joint contribution might be less than we would like. Note that the only other significant variable in this model using all five predictors is Knowledge, which has, as it turns out, the highest tolerance and thus the least overlap with the others.

singular

Second, the tolerance statistic alerts us to the potential problems of instability in our model. With very low levels of tolerance, the stability of the model and sometimes even the accuracy of the arithmetic can be in danger. In the extreme case where one predictor can be perfectly predicted from the others, we will have what is called a **singular** covariance (or correlation) matrix and most programs will stop without generating a model. If you see a statement in your printout that says that the matrix is singular or "not positive-definite," the most likely explanation is that one predictor has a tolerance of 0.00 and is perfectly correlated with others. In this case, you will have to drop at least one predictor to break up that relationship. Such a relationship most frequently occurs when one predictor is the simple sum or average of the others, or where all p predictors sum to a constant.

variance inflation factor (VIF)

The reciprocal of tolerance is called the **variance inflation factor (VIF),** and it refers to the degree to which the standard error of b_j is increased because X_j is correlated with the other predictors. The higher the standard error of b_j, the more that coefficient will fluctuate from sample to sample and the less confidence we can have in the particular value that we obtained. We want stable regression coefficients, and therefore we want variables with low VIFs, or high tolerances. It is often worth eliminating a redundant variable from the model to achieve that goal.

Selection Methods

There are many ways to construct some sort of "optimal" regression equation from a large set of variables. This section will briefly describe several of these approaches. But first we must raise the issue of whether this whole approach is generally appropriate. In many cases, it is not.

If we assume that you have a large set of variables and a large number of data points, and are truly interested in a question of prediction (you want to predict who will do well at some job and have no particular theoretical axe to grind), then one of these methods may be for you. However, if you are trying to test some theoretical model by looking to see if certain variables are related to some outcome (e.g., can you predict adolescents' psychological symptoms on the basis of major stressful events, daily hassles, and parental stress), then choosing a model on the basis of some criterion such as the maximum R^2 or the minimum MS_{residual} is not likely to be particularly helpful. It could even be particularly harmful by causing you to focus on statistically derived models that fit only slightly, and perhaps nonsignificantly, better than some other more logically appropriate model. Conducting a stepwise analysis, for example, to report which of two competing psychological variables is second to enter the equation often adds a spurious form of statistical elegance to a poor theory. Solid arguments against the use of stepwise regression for the purpose of ordering variables by importance have been given by Huberty (1989). Henderson and Denison (1989), in an excellent article that summarizes many of the important issues, suggest that "stepwise regression" should be called "unwise regression."

On the assumption that you still want to construct a regression model using some form of variable-selection process, we will consider three alternative approaches: all subsets regression, backward elimination, and stepwise regression. A readable and much more thorough discussion of this topic can be found in Draper and Smith (1981, Chapter 6).

All Subsets Regression

all subsets regression

The simplest of these methods at a conceptual level is called **all subsets regression** because it looks at all possible subsets of the predictor variables and chooses that set that is optimal in some way (such as maximizing R^2 or minimizing the mean square error). With three or four predictors and some patience, you could conduct such an analysis by using any standard computer package to calculate multiple analyses. However, with a large number of variables, the only way to go about this is to use a specialized program such as Minitab or SAS PROC RSQUARE. Either of these programs allows you to specify the largest and smallest number of predictors to appear in each subset and the number of subsets of each size. (For example, you can say, "Give me the eight models with the highest R^2s using five predictors.")

You can define "best" in several different ways; these ways do not always lead to the same models. You can select models on the basis of (1) the magnitude of R^2, (2) the magnitude of MS_{residual}, (3) a statistic called Mallow's C_p, and (4) a statistic called PRESS. The magnitudes of R^2 and MS_{residual} have already been discussed. We search for that combination of predictors with the highest R^2 (or better yet, adjusted R^2) or that set that minimizes error. Mallow's C_p statistic compares the relative magnitudes of the error term in any particular model with the error term in the complete model with all predictors present (see Draper & Smith, 1981, p. 299). Because the error term in the reduced model must be greater than (or equal to) the error term in the full model, we want to minimize that ratio.

PRESS (Predicted RESidual Sum of Squares) is a statistic similar to $MS_{residual}$ in that it looks at $\sum (Y_i - \hat{Y}_i)^2$, but in the case of PRESS, the predictions are made from a data set that includes all cases *except* the one to be predicted. Ordering models on the basis of PRESS would generally, though not always, be similar to ordering them on the basis of $MS_{residual}$. The advantage of PRESS is that it is more likely to focus on influential data points (see Draper & Smith, 1981, p. 325).

The major disadvantage of all subsets regression, aside from the enormous amount of computer time it could involve, is that it has a substantial potential for capitalizing on chance. By fitting all possible models to the data, or at least the best of all possible models, you run the serious risk of selecting those models that best fit the peculiar data points that are unique to your data set. The final R^2 cannot reasonably be thought of as an unbiased estimate of the corresponding population parameter.

Backward Elimination

backward elimination

The **backward elimination** procedure, as well as the stepwise regression procedure to follow, are generally lumped under the term *stepwise procedures* because they go about their task in a logical stepwise fashion. They both have the advantage of being easy to carry out interactively using standard regression procedures, although programs to carry them out automatically are readily available.

In the backward elimination procedure, we begin with a model that includes all the predictors. Having computed that model, we examine the tests on the individual regression coefficients, or look at the partial or semipartial correlations and remove the variable that contributes the least to the model (assuming that its contribution is statistically nonsignificant). We then rerun the regression without that predictor, again looking for the variable with the smallest contribution, remove that, and continue. Normally, we continue until we come to a model in which all the remaining predictors are statistically significant, although alternative stopping points are possible. For example, we could plot R^2 or $MS_{residual}$ against the number of predictors in the model and stop when that curve shows a break in direction.

Most computer programs that run backward elimination or stepwise regression use some combination of terms called "F to enter," "F to remove," "p to enter," and "p to remove." To take just one of these, consider "p to remove." If we plan to remove predictors from the model if they fail to reach significance at $\alpha = .05$, then we set "p to remove" at .05. The "F to remove" would simply be the critical value of F corresponding to that level of p.[11] (Those programs that calculate t statistics instead of F would simply make the appropriate change.) The situation is actually more complicated than I have made it seem (see Draper & Smith, 1981, p. 311), but for practical purposes, it is as I have described.

An important disadvantage of backward elimination is that it too capitalizes on chance. Because it begins with many predictors, it has the opportunity to identify and account for any suppressor relations among variables that can be found in the data. For example, if variables

---

[11] As Draper and Smith (1981) point out, when we are testing *optimal* models, the F statistics are not normal Fs and their probability values should not be interpreted as if they were. Thus, although both F and p form the basis of a legitimate ordering of potential variables, do not put too much faith in the actual probabilities. McIntyre, Montgomery, Srinwason, and Weitz (1983) address this problem directly and illustrate the liberal nature of the test. They also provide guidelines on more appropriate tests on stepwise correlation coefficients, should you want to follow this route.

7 and 8 have some sort of suppressor relationship between them, this method has a good chance of finding it and making those variables a part of the model. If that is a true relationship, then backward elimination has done what we want it to. On the other hand, if the relationship is spurious, we have just wasted extra variables explaining something that does not deserve explanation. Darlington (1990, p. 166) made this point about both backward elimination and all subsets regression. True suppressor relationships are fairly rare, but apparent ones are fairly common. Therefore, methods that systematically look for them, especially without accompanying hypothesis tests, may be misleading more often than are simpler methods that ignore them.

Stepwise Regression

stepwise regression

The **stepwise regression** method is more or less the reverse of the backward elimination method.[12] However, because at each stage we do not have all the other variables in the model and therefore immediately available to test, as we did with backward elimination, we will go about it in a slightly different way.

Stepwise regression relies on the fact that

$$R^2_{0.123\ldots p} = r^2_{01} + r^2_{0(2.1)} + r^2_{p(3.12)} + \cdots$$

If we define variable 1 as that variable with the highest validity (correlation with the criterion), then the first step in the process involves only variable 1. We then calculate all semi-partials of the form $r_{0(i.1)}, i = 2 \ldots p$. The variable (assume that it is X_2) with the highest (first-order) semipartial correlation with the criterion is the one that will produce the greatest increment in R^2. This variable is then entered, and we obtain the regression of Y on X_1 and X_2. We now test to see whether that variable contributes significantly to the model containing two variables. We could either test the regression coefficient or the semipartial correlation directly, or test to see if there was a significant increment in R^2. The result would be the same. Because the test on the increment in R^2 will prove useful later, we will do it that way here. A test on the difference between an R^2 based on f predictors and an R^2 based on r predictors (where the r predictors are a subset of the f predictors) is given by

$$F_{(f-r, N-f-1)} = \frac{(N - f - 1)\left(R^2_f - R^2_r\right)}{(f - r)\left(1 - R^2_f\right)}$$

where R^2_f is the R^2 for the full model $= R^2_{0.12}$, R^2_r is the R^2 for the reduced model $= R^2_{0.1}$, f is the number of predictors in the full model, and r is the number of predictors in the reduced model.

This process is repeated until the addition of further variables produces no significant (by whatever criterion we want to use) improvement. At each step in the process, before we add a new variable, we first ask whether a variable that was added on an earlier step should now be removed on the grounds that it is no longer making a significant contribution. If the test on a variable falls below "F to remove" (or above "p to remove"), that variable is removed before another variable is added. Procedures that do not include this step are often referred to as **forward selection** procedures.

forward selection

Of the three variable selection methods discussed here, the stepwise regression method is probably the best. Both Draper and Smith (1981) and Darlington (1990) recommend it as the best compromise between finding an "optimal" equation for predicting future randomly

---

[12] The terminology here is terrible, but you'll just have to bear with me. Backward elimination is *a* stepwise procedure, as is forward elimination, but when we refer to *the* stepwise approach, we normally mean the procedure that I'm about to discuss.

selected data sets from the same population and finding an equation that predicts the maximum variance for the specific data set under consideration. I think that the use of *any* of these methods when you are more interested in theory than in direct prediction is unwise. I use it only rarely, and sometimes wonder why I use it at all.

Cross-Validation

cross-validation

The stumbling block for most multiple regression studies is the concept of **cross-validation** of the regression equation against an independent data set. For example, we might break our data into two or more data sets and derive a regression equation for the first set. We then apply the regression coefficients obtained from that sample against the data in the other sample to obtain predicted values of Y on a cross-validation sample ($\hat{Y}_{cv}$). Our interest then focuses on the question of the relationship between Y and $\hat{Y}_{cv}$ in the new subsample. If the regression equations have any reasonable level of validity, then the cross-validation correlation (R_{cv}—the correlation between Y and $\hat{Y}_{cv}$ predicted on the *other* sample's regression equation) should be high. If they do not, our solution does not amount to much. R_{cv}^2 will in almost all cases be less than R^2, because R^2 depends on a regression equation tailored for that set of data. Essentially, we have an equation that does its best to account for every bump and wiggle (including sampling error) in the data. We should not be surprised when it does not do as well in accounting for different bumps and wiggles in a different set of data. However, substantial differences between R^2 and R_{cv}^2 are an indication that our solution lacks appreciable validity. When our regression equation for the data from Table 15.1 (using Teach, Knowledge, and Enroll as predictors) was applied to a new set of 50 courses, the correlation of Y_i and $\hat{Y}_i$ derived from the regression equation in Exhibit 15.1 was .818, representing an acceptable level of cross-validation, considering that the R for the original sample was .869.[13]

As Darlington (1990) has pointed out, our adjusted R^2, defined as

$$\text{adj } R^2 = 1 - \frac{(1 - R^2)(N - 1)}{N - p - 1}$$

is an estimate of the correlation in the population between Y and a *population* regression equation for $\hat{Y}_i$. It is not (and actually is an overestimation of) the expected correlation in a cross-validation sample. See Darlington (1990, pp. 159f) for estimates of the cross-validation correlation.

Missing Observations

Missing data are often a problem in regression analyses, and a number of alternative methods have been devised to deal with them. The most common approach is simply to delete all cases not having complete data on the variables being investigated. This is called **listwise** (or **casewise**) **deletion** because when an observation is missing we delete the whole case.

listwise deletion

casewise deletion

pairwise deletion

A second approach, which is available in SPSS but is deliberately not available in many programs, is called **pairwise deletion.** Here, we use whatever data are at hand. If the 13th subject has data on both X and Y, then that subject is included in the calculation of r_{XY}. But if subject 13 does not have a score on Z, that subject is not included in the calculation of r_{XZ} or r_{YZ}. Once the complete intercorrelation matrix has been computed using pairwise deletion, the rest of the regression solution follows directly from that matrix.

---

[13] For a more thorough discussion of cross-validation, see the entry by Stuetzle (2005).

Both of these solutions have their problems. Listwise deletion may result in relatively low sample sizes, and, if the data are not missing completely at random, in samples that are not a fair reflection of the population from which they were presumably sampled. Pairwise deletion, on the other hand, can result in an intercorrelation matrix that does not resemble the matrix that we would have if we had complete data on all cases. Pairwise deletion can result in an "impossible" intercorrelation matrix. It is well known that given r_{XY} and r_{XZ}, the correlation between Y and Z *must* fall within certain limits. But if we keep changing the data that go into the correlations, we could obtain an r_{YZ} that is inconsistent with the other two correlations. When we then try to use such an inconsistent matrix, we find ourselves in serious trouble.

A third approach has been advocated by Cohen et al. (2003). They recommend creating dummy variables having a 0 if the data for a particular variable are present, and a 1 if the data are absent. This dummy variable is then included as an independent variable in the analysis. A demonstration of this approach can be found at http://www.uvm.edu/~dhowell/StatPages/StatHomePage.html. Just click on the "Additional Material" heading.

Although there is much to be said for Cohen's approach, most researchers, especially when missing only a few observations, fall back on listwise deletion. I would certainly recommend that over the pairwise deletion approach.

15.11   The "Importance" of Individual Variables

When an investigator derives a regression equation to predict some criterion on the basis of several variables, it is logical for her to want to know which of the variables is most important in predicting Y. Unfortunately, that question has no simple answer, except in the unusual case in which the predictors are mutually independent. As we have seen, β_j (or β_j^2) is sometimes taken as a measure of importance. This is done on the grounds that β^2 can be interpreted as the *unique* contribution of each variable to the prediction of Y. Thus, X_1 has some variance in common with Y that is not shared by any of the other variables, and this variance is represented by β_1^2. The difficulty with this measure is that it has nothing to say about the portion of the variance of Y that X_1 *does* share with the other variables but that is in some sense part of the contribution of X_1 to the prediction of Y. Moreover, what does it mean to speak of the independent contribution of variables that are not independent?

Darlington (1990) has argued against using β_j as a measure of importance. β_j does represent the difference, in standard deviation units, between two cases that are equal on all other predictor variables but differ by one unit on X_j. However, this does not take into account the fact that when variables are highly correlated, such cases will rarely, if ever, exist.

multicollinearity   Basing a measure of importance on the β weights has the further serious drawback that when variables are highly correlated (a condition known as **multicollinearity**), the values of β are very unstable from sample to sample, although R^2 may change very little. Given two sets of data, it would not be particularly unusual to find

$$\hat{Y} = 0.50Z_1 + 0.25Z_2$$

in one case and

$$\hat{Y} = 0.25Z_1 + 0.50Z_2$$

in the other, with nearly equal values of R^2 associated with the two equations. If we now seek a measure of the contribution of each of the predictors in accounting for Y (rather than using regression to simply predict Y for a given set of data), we could come to quite different conclusions for the two data sets. Darlington (1968) presents an interesting discussion of this issue and concludes that β_i has only limited utility as a measure of "importance." An even stronger stand is taken by Cooley and Lohnes (1971), who point out that our estimate of β ultimately relies on our estimates of the elements of the intercorrelation matrix. Because this matrix contains $p + p(p - 1)/2$ intercorrelations that are all subject to sampling error, Cooley and Lohnes suggested that we must be exceedingly careful about attaching practical significance to the regression coefficients.

As an illustration of the variability of the regression coefficients, a second set of 50 courses, the same set as used for cross-validation, was drawn from the same source as that for the data in Table 15.1. In this case, R^2 was more or less the same as it had been for the first example ($R^2 = .710$), but the regression equation looked quite different. In terms of standardized variables,

$$Z_{\hat{Y}} = 0.371\,\text{Teach} + 0.113\,\text{Exam} + 0.567\,\text{Knowledge} - 0.27\,\text{Grade} + 0.184\,\text{Enroll}$$

If you compare this equation with the one found from Exhibit 15.1, it is clear that there are substantial differences in some of the values of β_i.

Another measure of importance, which has much to recommend it, is the squared semi-partial correlation between predictor i and the criterion (with all other predictors partialled out)—that is, $r^2_{0(i.123...p)}$. Darlington (1968) refers to this measure as the "usefulness" of a predictor. As we have already seen, this semipartial correlation squared represents the decrement in R^2 that would result from the elimination of the ith predictor from the model (or the increment that would result from its addition). When the main goal is prediction rather than explanation, this is probably the best measure of "importance." Fortunately, it is easy to obtain from most computer printouts, because

$$r^2_{0(i.123...p)} = \frac{F_i(1 - R^2_{0.123...p})}{N - p - 1}$$

where F_i is the F test on the individual β_i (or b_i) coefficients. (If your program uses t tests on the coefficient, $F = t^2$.) Because all terms except F_i are constant for $i = 1 \ldots p$, the F_is order the variables in the same way as do the squared semipartials and, thus, can be used to rank order the variables in terms of their usefulness.

Darlington (1990) has made a strong case for not squaring the semipartial correlation when speaking about the importance of variables. His case is an interesting one. However, whether or not the correlations are squared will not affect the ordering of variables. (If you want to argue persuasively about the absolute importance of a variable, you should read Darlington's argument.)

One common, but unacceptable, method of ordering the importance of variables is to rank them by the order of their inclusion in a stepwise regression solution. The problem with this approach is that it ignores the interrelationships among the variables. Thus, the first variable to be entered is entered solely on the strength of its correlation with the criterion. The second variable entered is chosen on the basis of its correlation with the criterion after partialling the first variable but ignoring all others. The third is chosen on the basis of how it correlates with the criterion after partialling the first two variables, and so on. In other words, each variable is chosen on a different basis, and it makes little sense to rank them according to order of entry. To take a simple example, assume that variables 1, 2, and 3 correlate .79, .78, and .32 with the criterion. Assume further that variables 1 and 2 are correlated .95, whereas 1 and 3 are correlated .20. They will then enter the equation in the order 1, 3, and 2, with the last entry being nonsignificant. But in what sense do we mean to say that variable 3 ranks above variable 2 in importance? I would hate to defend such a statement to a reviewer—actually, I would be hard pressed even to say what I meant by importance in this situation. A similar point has been made well by Huberty (1989). For an excellent discussion of measures of importance, see Harris (1985, pp. 79ff).

15.12   Using Approximate Regression Coefficients

I have pointed out that regression coefficients frequently show substantial fluctuations from sample to sample without producing drastic changes in R. This might lead someone to suggest that we might use rather crude approximations of these coefficients as a substitute for the more precise estimates obtained from the data. For example, suppose that a five-predictor problem produced the following regression equation:

$$\hat{Y} = 9.2 + 0.85X_1 + 2.1X_2 - 0.74X_3 + 3.6X_4 - 2.4X_5$$

We might ask how much loss we would suffer if we rounded these values to

$$\hat{Y} = 10 + 1X_1 + 2X_2 - 1X_3 + 4X_4 - 2X_5$$

The answer is that we would probably lose very little. Excellent discussions of this problem are given by Cohen et al. (2003), Dawes and Corrigan (1974), and Wainer (1976, 1978).

This method of rounding off regression coefficients is more common than you might suppose. For example, the college admissions officer who quantifies the various predictors he has available and then weights the grade point average twice as highly as the letter of recommendation is really using crude estimates of what he thinks would be the actual regression coefficients. Similarly, many scoring systems for the Minnesota Multiphasic Personality Inventory (MMPI) are based on the reduction of coefficients to convenient integers. Whether the use of these *diagnostic signs* produces results that are better than, worse than, or equivalent to the use of the usual linear regression equations is still a matter of debate. A dated but very comprehensive study of this question is presented in Goldberg (1965). Rather than undermining our confidence in multiple regression, I think the fact that rounded off coefficients do nearly as well (sometimes better if we are applying them to new data) speaks to the robustness of regression. It also suggests that you not put too much faith in small differences in coefficients.

15.13   Mediating and Moderating Relationships

One of the most frequently cited papers in the psychological literature related to multiple regression during the past 20 years has been a paper by Baron and Kenny (1986) on what they called the moderator-mediator distinction. The important point for both moderating and mediating relationships is that a third variable plays an important role in governing the relationship between two other variables.

Mediation

mediating relationship

A **mediating relationship** is what it sounds like—some variable mediates the relationship between two other variables. For example, take a situation in which high levels of care from your parents leads to feelings of competence and self-esteem on your part, which, in turn, leads to high confidence when you become a mother. Here, we would say that your feelings of competence and self-esteem *mediate* the relationship between how you were parented and how you feel about mothering your own children.

Baron and Kenny (1986) laid out several requirements that must be met before we can speak of a mediating relationship. Consider the diagram in Figure 15.5 as being representative of a mediating relationship that we want to explain.

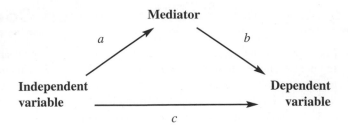

Figure 15.5 Diagram of a mediated relationship

The predominant relationship that we want to explain is labeled "c" and is the path from the independent to the dependent variable. The mediating path has two parts: "a," the path connecting the independent variable to the potential mediator, and "b," the path connecting that mediator to the dependent variable.

Baron and Kenny argued that for us to claim a mediating relationship, we need to first show that there is a significant relationship between the independent variable and the mediator. (If the mediator is not associated with the independent variable, then it couldn't mediate anything.) The next step is to show that there is a significant relationship between the mediator and the dependent variable, for reasons similar to those for the first requirement. Then we need to show that there is a significant relationship between the independent and dependent variable.

These three conditions require that the three paths (*a*, *b*, and *c*) are all individually significant. The final step consists of demonstrating that when the mediator and the independent variable are used simultaneously to predict the dependent variable, the previously significant path between the independent and dependent variables (*c*) is now greatly reduced, if not nonsignificant. Maximum evidence for mediation would occur if *c* drops to 0. I have never seen a path go away completely. Most likely to happen is that *c* becomes a weaker, though perhaps still significant, path.

Leerkes and Crockenberg (1999) were interested in studying the relationship between how children were raised by their own mothers, and their later feelings of maternal self-efficacy when they, in turn, became mothers. The sample consisted on 92 mothers of 5-month old infants. The researchers expected to find that high levels of maternal care when the mother was a child translated to high levels of self-efficacy when that child later became a mother. But Leerkes and Crockenberg went further, postulating that the mediating variable in this relationship is self-esteem. They argued that high levels of maternal care lead to high levels of self-esteem in the child and that this high self-esteem later translates into high levels of self-efficacy as a mother. This relationship is diagrammed in Figure 15.6.

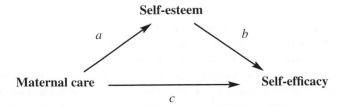

Figure 15.6 Diagram of mediation in the development of self-efficacy

The initial conditions of Baron and Kenny (1986) can be tested by looking at the simple correlations among the variables. These are shown here, as produced by SPSS.

Correlations

Pearson Correlation

	Maternal care	Self-esteem	5 month efficacy
Maternal care	1.000	.403**	.272**
Self-esteem	.403**	1.000	.380**
5 month efficacy	.272**	.380**	1.000

** Correlation is significant at the 0.01 level (2-tailed).

Exhibit 15.4a   Correlations between variables in mediation example

Here, we can see that maternal care is correlated with self-esteem and with self-efficacy, and that self-esteem is also correlated with self-efficacy. These relationships satisfy Baron and Kenny's basic prerequisites. The next step is to use both self-esteem and maternal care as predictors of self-efficacy. This is shown in Exhibit 15.4b.

Coefficients[a]

Model		Unstandardized Coefficients		Standardized Coefficients	t	Sig.	Correlations	
		B	Std. Error	Beta			Zero-order	Part
1	(Constant)	3.260	.141		23.199	.000		
	maternal care	.112	.042	.272	2.677	.009	.272	.272
2	(Constant)	2.929	.173		16.918	.000		
	maternal care	5.817E-02	.044	.142	1.334	.185	.272	.130
	self esteem	.147	.048	.323	3.041	.003	.380	.295

[a] Dependent Variable: 5 month efficacy

Exhibit 15.4b   Multiple regressions for mediation example

The first model in the previous table uses maternal care as the sole predictor. The second model has added self-esteem as a predictor. You can see that when we add self-esteem to maternal care, which was clearly significant when used alone to predict self-efficacy, maternal care is no longer significant ($t = 1.334$, $p = 0.185$). This is evidence that self-esteem is serving a mediating role between maternal care and self-efficacy. The output also shows what SPSS calls the "part correlation," but which the rest of us call the semipartial correlation. The semipartial correlation between maternal care and self-efficacy is .130, whereas the simple correlation (zero-order) between maternal care and self-efficacy was .27. It remains significant, as we can see by the t test on self-esteem, but has dropped noticeably.

These results support Leerkes and Crockenberg's hypothesis that self-esteem played a mediating role between maternal care and self-efficacy. Caring parents seem to produce children with higher levels of self-esteem, and this higher self-esteem translates into positive feelings of self-efficacy when the child, in turn, becomes a mother.

In this situation, Leerkes and Crockenberg were fortunate to have a situation in which the direct path from maternal care to self-efficacy dropped to nonsignificance when self-esteem was added. Unfortunately, that does not always happen. (Actually, it seems to happen relatively infrequently.) The more common result is that the direct path becomes less important, though it remains significant. There has been considerable discussion about what to do in this situation, but there is a relatively simple answer, developed by Sobel (1982), that was referred to by Baron and Kenny.

When we have a situation in which the direct path remains significant, though at a lower value, one way to test for a mediating relationship is to ask whether the complete mediating path from independent variable to mediator to dependent variable is significant. To do this, we need to know the regression coefficients and their standard errors for the two paths in the mediating chain. We will soon also need the regression of Self-esteem on Maternal care, so that table follows.

Coefficients[a]

Model		Unstandardized Coefficients		Standardized Coefficients	t	Sig.	Correlations		
		B	Std. Error	Beta			Zero-order	Partial	Part
1	(Constant)	2.257	.294		7.687	.000			
	pbi maternal care	.364	.087	.403	4.178	.000	.403	.403	.403

[a] Dependent Variable: self esteem

Exhibit 15.4c Regression of self-esteem on maternal care

The important statistics from the two regressions are shown in Table 15.6. Because SPSS does not report the standard error of beta, we need to calculate it. The t statistic given in these tables is either the unstandardized regression coefficient (b) divided by its standard error, or the standardized regression coefficient divided by its standard error. Thus, we can solve

$$t = \frac{\beta}{s_\beta}; \qquad s_\beta = \frac{\beta}{t} = \frac{0.403}{4.178} = 0.096$$

Similarly for the path from Self-esteem to Self-efficacy, partialling Maternal care, we have

$$t = \frac{\beta}{s_\beta}; \qquad s_\beta = \frac{\beta}{t} = \frac{0.323}{3.041} = 0.106$$

These results yield Table 15.6.

Table 15.6 Regression coefficients and standard errors for two parts of mediating path

Path a		Path b	
Maternal Care → Self-Esteem		Self-Esteem → Self-Efficacy	
β_a	0.403	β_b	.323
s_a	0.096	s_b	.106
t	4.18*	t	3.04*

Then the regression coefficient for the path from Maternal care $\rightarrow$ Self-esteem $\rightarrow$ Self-efficacy is equal to $\beta_a \times \beta_b = 0.403 \times 0.323 = 0.130$, where a and b refer to the relevant paths. (Path c is the direct path from Maternal care to Self-efficacy.) In addition, we know that the standard error of this two-part path is given by

$$s_{\beta_a \beta_b} = \sqrt{\beta_a^2 s_b^2 + \beta_b^2 s_a^2 - s_a^2 s_b^2}$$

where β_a and β_b are the paths, and s_a^2 and s_b^2 are the corresponding standard errors of the standardized regression coefficients for those paths.[14] We can calculate the standard error of the combined path as

$$s_{\beta_a \beta_b} = \sqrt{\beta_a^2 s_b^2 + \beta_b^2 s_a^2 - s_a^2 s_b^2} = \sqrt{.403^2(.106^2) + .323^2(.096^2) - (.106^2)(.096^2)}$$

$$= \sqrt{0.0027}$$

$$= 0.052$$

We now know the path coefficient ($0.403 \times 0.323 = 0.130$) and its standard error (0.052), and we can form a t ratio as

$$t = \frac{\beta_1 \beta_3}{s_{\beta_1 \beta_3}} = \frac{.130}{.052} = 2.50$$

Sobel (1982) stated that this ratio is asymptotically normally distributed, which, for large samples, would lead to rejection of the null hypothesis at $\alpha = 0.05$ when the ratio exceeds ± 1.96. It would presumably have a t distribution on $N - 3$ df for small samples. In our case, the path is clearly significant, as we would expect from the previous results. Therefore, we can conclude that we have convincing evidence of a strong mediating pathway from maternal care through self-esteem to self-efficacy. Because the regression coefficient (and semi-partial correlation) for the direct path from maternal care to self-efficacy is not significant, the main influence of maternal care is through its mediating relationship with self-esteem.

There has been considerable discussion in the literature about the best approach to testing mediation. For an online test using three alternative approaches to the standard error, go to www.unc.edu/~preacher/sobel/sobel.htm. Preacher and Hayes (2004) (available from the previous website) present SPSS and SAS macros that allow you to using bootstrapping methods (see Chapter 18) to address this question. A very well-written description of mediation has been put on the Web by Paul Jose, at the University of Wellington. It can be found at http://www.vuw.ac.nz/psyc/staff/paul-jose/files/helpcentre/help7_mediation_example.php. In additon, Jose offers a free mediation calculator, which runs under Excel, at http://www.vuw.ac.nz/psyc/staff/paul-jose/files/medgraph/medgraph.php. I have found that very useful, but be aware that there seems to be minor disagreement between the example and the results of the software. Finally, an extensive comparison of alternative approaches can be found in MacKinnon, Lockwood, Hoffman, West, and Sheets (2002).

Moderating Relationships

moderating relationships

Whereas a mediating relationship attempts to identify a variable or variables through which the independent variable acts to influence the dependent variable, **moderating relationships** refer to situations in which the relationship between the independent and dependent variables changes as a function of the level of a third variable (the moderator).

---

[14] There is some disagreement about the exact form of these equation, but the one given here is recommended by Baron and Kenny. The differences between the various equations turn out to be very minor in practice.

Wagner, Compas, and Howell (1988) hypothesized that individuals who experience more stress, as assessed by a measure of daily hassles, will exhibit higher levels of symptoms than will those who experience little stress. That is what, in analysis of variance terms, would be the main effect of hassles. However, the researchers also expected that if a person had a high level of social support to help deal with his or her stress, symptoms would increase only slowly with increases in hassles. For those who had relatively little social support, symptoms were expected to rise more quickly as hassles increased.

Wagner et al. (1988) studied students who were attending an orientation before starting their first year of collge. Students were asked to report on the number of minor stressful events (labeled hassles) that they had recently experienced and to report on their perceived level of social support. Students then completed a symptom checklist about the number of symptoms they had experienced in the past month. For this part of the study, there were complete data on 56 participants. These data are available in a file named hassles.dat (from www.uvm.edu/~dhowell/methods/).

Our first step is to look at the relationships between these variables. The correlation matrix is shown here.

Correlations

Pearson Correlation

	Hassles	Support	Symptoms
Hassles	1.000	−.167	.577**
Support	−.167	1.000	−.134
Symptoms	.577**	−.134	1.000

** Correlation is significant at the 0.01 level

As expected, there is a significant relationship between Hassles and Symptoms ($r = .577$), though Support is not related to Symptoms, or to Hassles. This does not, however, answer the question that the researchers really wanted to ask, which is whether the relationship between Hassles and Symptoms depends on the degree of social support.

If you think about this question, it starts to sound very much like the question behind an interaction in the analysis of variance. Actually, it is an interaction, and the way that we will test for that interaction is to create a variable that is the product of Hassles and Support. (This is also similar to what we will do in the general linear model approach to the analysis of variance in the next chapter.) However, if we just multiply Hassles and Support together, there will be two problems with what results. In the first place, either Hassles or Support or both will be highly correlated with their product, which will make for multicollinearity in the data. This will seriously effect the magnitude, and tests of significance, of the coefficients for the main effect of Hassles and Support. The second problem is that any effect of Hassles or Support in the regression analysis will be evaluated at a value of 0 for the other variable. In other words, the test on Hassles will be a test on whether Hassles is related to Symptoms if a participant had exactly no social support. Similarly, the test on Support would be evaluated for those participants who have exactly no hassles. Both the problem of multicollinearity and the problem of evaluating one main effect at an extreme value of the other main effect are unwelcome.

center   To circumvent these two problems, we are going to **center** our data. This means that we are going to create deviation scores by subtracting each variable's mean from the individual observations. Now a score of 0 for (centered) Hassles represents someone who has the mean level of Hassles, which seems an appropriate place to examine any effects of support, and anyone with a 0 on (centered) support represents someone with a mean level of support. This has solved one of our problems because we are now evaluating the main effects at a reasonable level of the other main effect. It has also helped to solve our other problem because if you look at the resulting correlations, multicollinearity will have been significantly reduced.

Having centered our variables, we will then form a product of our centered variables, and this will represent our interaction term. The means for hassles, support, and symptoms are 170.1964, 28.9643, and 90.4286, respectively, and the equations for creating centered variables and their interaction follow. The letter "c" at the beginning of the variable name indicates that it is centered.

$$\text{chassles} = \text{hassles} - 170.1964$$

$$\text{csupport} = \text{support} - 28.9643$$

$$\text{chassupp} = \text{chassles} \times \text{csupport}$$

The correlations among the centered (and uncentered) variables are shown in the following table. I have included the product of the uncentered variables simply to show how high the correlation between hassles and hassupp is, but we are not going to use this variable. You can see that by centering the variables we have substantially reduced the correlation between the main effects and the interactions. That was our goal. Notice that centering the variables did not change their correlations with each other—only with the interaction.

Correlations

Pearson Correlation

	Hassles	Support	Symptoms	hassupp	chassles	csupport	chassupp
Hassles	1.000	−.167	.577**	.910**	1.000**	−.167	−.297*
Support	−.167	1.000	−.134	−.510**	−.167	1.000**	.402**
Symptoms	.577**	−.134	1.000	.585**	.577**	−.134	−.391**
hassupp	.910**	−.510**	.585**	1.000	.910**	−.510**	−.576**
chassles	1.000**	−.167	.577**	.910**	1.000	−.167	−.297*
csupport	−.167	1.000**	−.134	−.510**	−.167	1.000	.402**
chassupp	−.297*	.402**	−.391**	−.576**	−.297*	.402**	1.000

  ** Correlation is significant at the 0.01 level (2-tailed).

  * Correlation is significant at the 0.05 level (2-tailed).

We can now examine the interaction of the two predictor variables by including the interaction term in the regression with the other centered predictors. The dependent variable is Symptoms. This regression is shown in Exhibit 15.5. (As long as we use the product of centered variables, it doesn't matter, except for the intercept, if we use the centered or uncentered main effects. I prefer the latter, but for no particularly good reason.)

From the printout, you can see that $R^2 = .388$, which is significant. (Without the interaction term, R^2 would have been .334 [not shown].) From the table of regression coefficients, you see that both the centered Hassles and the interaction terms are significant ($p = .000$ and .037, respectively), but the social support variable is not significant. By convention, we leave it in our regression solution because it is involved in the interaction, even though the associated t value shows that deleting that variable would not lead to a significant decrease in R^2.

Our regression equation now becomes

$$\hat{Y} = .086\text{chassles} + 0.146\text{csupport} - .005\text{chassupp} + 89.585.$$

We have answered our initial questions (social support does moderate the relationship between hassles and symptoms), but it would be helpful if we could view this graphically to

Model Summary

Model	R	R Square	Adjusted R Square	Std. Error of the Estimate
1	.623[a]	.388	.353	16.8932

[a] Predictors: (Constant), CHASSUPP, CHASSLES, CSUPPORT

ANOVA[b]

Model		Sum of Squares	df	Mean Square	F	Sig.
1	Regression	9427.898	3	3142.633	11.012	.000[a]
	Residual	14839.816	52	285.381		
	Total	24267.714	55			

[a] Predictors: (Constant), chassupp, chassles, csupport

[b] Dependent Variable: Symptoms

Coefficients[a]

Model		Unstandardized Coefficients		Standardized Coefficients	t	Sig.
		B	Std. Error	Beta		
1	(Constant)	89.585	2.292		39.094	.000
	chassles	8.594E-02	.019	.509	4.473	.000
	csupport	.146	.305	.057	.479	.634
	chassupp	−5.06E-03	.002	−.262	−2.144	.037

[a] Dependent Variable: Symptoms

Exhibit 15.5   Regression solution for moderated relationship between hassles and symptoms

interpret the meaning of the interactive effect. Excellent discussions of this approach can be found in Finney, Mitchell, Cronkite, and Moos (1984), Jaccard, Turrisi, and Wan (1990), and Aiken and West (1991). The latter is the authoritative work on moderation. Normand Péladeau has a free program called Italassi, available on the web at http://www.simstat.com/. This program will plot the interaction on your screen and provides a slider so that you can vary the level of the support variable.

The simplest solution is to look at the relationship between chassles and csymptoms for fixed levels of social support. Examination of the distribution of csupport scores shows that they range from about −21 to +19. Thus, scores of −15, 0, and +15 would represent low, neutral, and high scores on csupport. (You don't have to be satisfied with these particular values, you can use any that you like. I have picked extremes to better illustrate what is going on.

First, I will rewrite the regression equation, substituting generic labels for the regression coefficients. I will also substitute chassles × csupport for chassupp because that is the way that I calculated chssupp. Finally, I will also reorder the terms a bit just to make life easier.

$$\hat{Y} = b_1\text{chassles} + b_2\text{csupport} - b_3\text{chassupp} + b_0$$

$$\hat{Y} = b_0 + b_2\text{csupport} + b_3(\text{chassles} \times \text{csupport}) + b_1\text{chassles}$$

Collecting terms, I have

$$\hat{Y} = b_0 + b_2\text{csupport} + \text{chassles}(b_3\text{csupport} + b_1)$$

Next, I will substitute the actual regression coefficients to get

$$\hat{Y} = [89.585 + 0.146\text{csupport}] + \text{chassles}(-.005\text{csupport} + .086)$$

Notice the first term in square brackets. For any specific level of csupport (e.g., 15), this is a constant. Similarly, for the terms in parentheses after chassles, that is also a constant for a fixed level of support. To see this most easily, we can solve for $\hat{Y}$ when csupport is at 15, which is a high level of support. This gives us

$$\hat{Y} = [89.585 + 0.146 \times 15] + \text{chassles}(-.005 \times 15 + .086)$$
$$= 91.755 + 0.011 \times \text{chassles}$$

which is just a plain old linear equation. This is the equation that represents the relationship between $\hat{Y}$ and chassles when social support is high (i.e., 15).

Now we can derive two more simple linear equations, one by substituting 0 for csupport and one by substituting -15.

When csupport $= 0$,

$$\hat{Y} = 89.585 + .086 \times \text{chassles}$$

When csupport $= -15$,

$$\hat{Y} = 87.395 + .161 \times \text{chassles}$$

When I look at the frequency distribution of chassles, low, neutral, and high scores are roughly represented by -150, 0, and 150. So I will next calculate predicted values for symptoms and low, neutral, and high levels of chassles for each of low, neutral, and high levels of csupport. These are shown in Table 15.7, and they were computed using the three previous regression equations and setting chassles at -150, 0, and 150.

Table 15.7   Predicted values of symptoms at varying levels of hassles & support

		Centered Support		
		-15	0	15
Centered Hassles	-150	63.245	76.685	90.105
	0	87.395	89.585	91.755
	150	111.545	102.485	93.405

If we plot these predicted values separately for the different levels of social support, we see that with high social support increases in hassles are associated with relatively small increases in symptoms. When we move to csupport $= 0$, which puts us at the mean level of support, increasing hassles leads to a greater increase in symptoms. Finally, when we have low levels of support (csupport $= -15$), increases in hassles lead to dramatic increases in symptoms. This is shown graphically in Figure 15.7.

The use of interaction terms (e.g., $X_1 \times X_2$) in data analysis, such as the problem that we have just addressed, has become common in psychology in recent years. However, my experience and that of others has been that it is surprisingly difficult to find meaningful situations where the regression coefficient for $X_1 \times X_2$ is significant, especially in experimental settings where we deliberately vary the levels of X_1 and X_2. McClelland and Judd (1993)

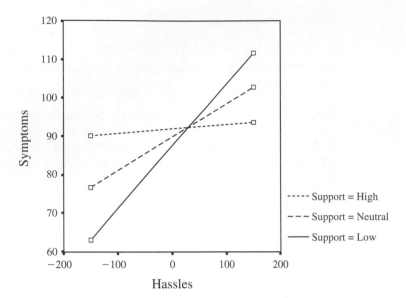

Figure 15.7   Plot of symptoms as a function of hassles for different levels of social support

have investigated this problem and have shown why our standard field study designs have so little power to detect interactions. That is an important paper for anyone investigating interaction effects in nonexperimental research.

15.14   Logistic Regression

logistic regression

In the past few years, the technique of **logistic regression** has become popular in the psychological literature. (It has been popular in the medical and epidemiological literature for much longer.) Logistic regression is a technique for fitting a regression surface to data in which the dependent variable is a dichotomy.[15] A very common situation in medicine is the case in which we want to predict response to treatment, where we might code survivors as 1 and those who don't survive as 0. In psychology, we might class clients as Improved or Not Improved, or we might rate performance as Successful or Not Successful. Whenever we have such a dichotomous outcome, we have a possible candidate for logistic regression.

But when we have a dichotomous dependent variable, we have at least two other statistical procedures as candidates for our analysis. One of them, which is not discussed in this text, is **discriminant analysis,** which is a technique for distinguishing two or more groups on the basis of a set of variables. The question is often raised about whether logistic regression is better than discriminant analysis. It isn't always clear how we might define "better," but discriminant analysis has two strikes against it that logistic regression does not. In the first place, discriminant analysis can easily produce a probability of success that lies outside the range of 0 and 1, yet we know that such probabilities are impossible. In the second place, discriminant analysis depends on certain restrictive normality assumptions on the independent variables, which are often not realistic. Logistic regression, on the other hand, does not

discriminant analysis

---

[15] Logistic regression can also be applied in situations where there are three or more levels of the dependent variable, which we refer to as a polychotomy, but we will not discuss that method here.

produce probabilities beyond 0 and 1, and requires no such restrictive assumptions on the independent variables, which can be categorical or continuous. Common practice has now moved away from discriminant analysis in favor of logistic regression.

A second alternative would be to run a standard multiple regression solution, which we have just been covering, using the dichotomous variable as our dependent variable. In many situations the results would be very similar. But there are reasons to prefer logistic regression in general, though to explain those I have to use a simple example.

We will look at actual, though slightly modified, data on variables that we hope to relate to whether or not the individual responds positively to cancer treatment. The data that we will consider were part of a study of behavioral variables and stress in people recently diagnosed with cancer. For our purposes, we will look at patients who have been in the study for at least a year, and our dependent variable (Outcome) is coded 1 for those who have improved or are in complete remission, and 0 for those who have not improved or who have died. (Any consistent method of coding, such as 1 and 2, or 5 and 8, would also work.)[16] Out of 66 cases, we have 48 patients who have improved and 18 who have not. Suppose that we start our discussion with a single predictor variable, which is the Survival rating (SurvRate) assigned by the patient's physician at the time of diagnosis. This is a number between 0 and 100 and represents the estimated probability of survival at 5 years.

One way to look at the relationship between SurvRate and Outcome would be to simply create a scatterplot of the two variables, with Outcome on the Y axis. Such a plot is given in Figure 15.8. (In this figure, I have offset overlapping points slightly so that you could see them pile up. That explains why there seems to be string of points at SurvRate = 91 and Outcome = 1, for example.) From this plot, we can see that the proportion of people who

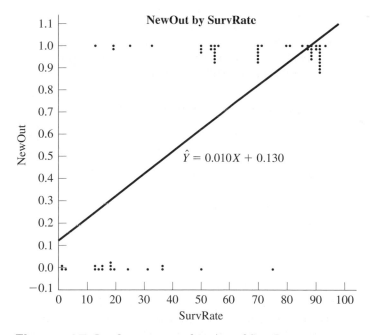

Figure 15.8 Outcome as a function of SurvRate

---

[16] You have to be careful with coding because different programs treat the same codes differently. Some will code the higher value as success and the lower as failure, and others will do the opposite. If you have a printout where the results seem exactly the opposite of what you might expect, check the manual to see how the program treats the dichotomous variable.

conditional means

improve is much higher when the survival rating is high, as we would expect. Assume for the moment that we had a great many subjects and could calculate the mean Outcome score (the mean of 0s and 1s) associated with each value of SurvRate. (These are called **conditional means** because they are conditional on the value of SurvRate.) The conditional means would be the proportion of people with that value of SurvRate who improved. If we fit a standard regression line to these data, this would be the regression line that fits the *probability* of improvement as a function of SurvRate. But as you can imagine, for many values of SurvRate, the predicted probability would be outside the bounds 0 and 1, which is impossible. That alone would make standard linear regression a poor choice. There is a second problem. If you were to calculate the *variances* of Outcome for different values of SurvRate, you would see that they are quite small for both large and small values of SurvRate (because almost everyone with low values of SurvRate has a 0 and almost everyone with high values of SurvRate has a 1). But for people with mid-level SurvRate values, there is nearly an even mix of 0s and 1s, which will produce a relatively larger variance. This will clearly violate our assumption of homogeneity of variance in arrays, to say nothing of normality. Because of these problems, standard linear regression is not a wise choice with a dichotomous dependent variable, though it would provide a pretty good estimate if the percentage of improvement scores didn't fall below 20% or above 80% across all values of SurvRate (Cox & Wermuth, 1992).

Another problem is that the true relationship is not likely to be linear. Differences in SurvRate near the center of the scale will lead to noticeably larger differences in Outcome than will comparable differences at the ends of the scale.

sigmoidal

Although a straight line won't fit the data in Figure 15.6 well, an S-shaped, or **sigmoidal** curve will. This line changes little as we move across low values of SurvRate, then changes rapidly as we move across middle values, and finally changes slowly again across high values. In no case, does it fall below 0 or above 1. This line is shown in Figure 15.9. Notice that it is quite close to the whole cluster of points in the lower left, rises rapidly for those values of SurvRate that have a roughly equal number of patients who improve and don't improve, and then comes close to the cluster of points in the upper right. When you think about how you might expect the probability of improvement to change with SurvRate, this curve makes sense.

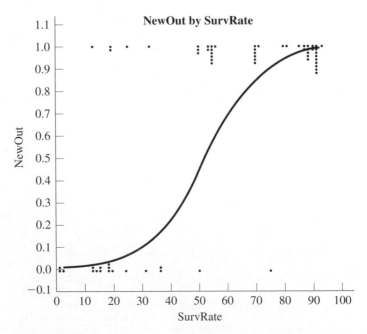

Figure 15.9   More appropriate regression line for predicting outcome

There is another way to view what is happening that provides a tie to standard linear regression. If you think back to what we have said in the past about regression, you will recall that, at least with large samples, a whole collection of *Y* values correspond to each value of *X*. You saw this diagrammatically in Figure 9.5, when I spoke about the assumptions of normality and homogeneity of variance in arrays. Rather than classifying people as improved or not improved, suppose that we could somehow measure their disease outcomes more precisely. Then for a rating of SurvRate = 20, for example, we would have a whole distribution of disease outcome scores; similarly for people with SurvRate = 30, SurvRate = 40, and so on. These distributions are shown schematically in Figure 15.10.

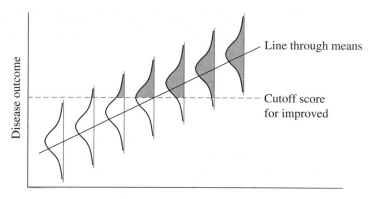

Figure 15.10   Disease outcome as a function of SurvRate

When we class someone as improved, we are simply saying that his disease outcome score is sufficiently high for us to say that he falls in that category. He may be completely cured, he may be doing quite a bit better, or he may be only slightly improved, but he at least met our criterion of "improved." Similarly, someone else may have remained constant, gotten slightly worse, or died, but in any event her outcome was below our decision point.

censored data   What we have here are called **censored data.** When I speak of censoring, I'm not talking about some nasty little man with a big black marker who blocks out things he doesn't want others to see. We are talking about a situation where something that is above a cutoff is classed as a success, and something below the cutoff is classed as a failure. It could be performance on a test, obtaining a qualifying time for the Boston Marathon, or classifying an airline flight as "on time" or "late." From this point of view, logistic regression can be thought of as applying linear regression to censored data. Because the data are censored to provide only success or failure, we have to fit our model somewhat differently.

The horizontal line across the plot in Figure 15.8 represents a critical value. Anyone scoring above that line would be classed as improved, and anyone below it would be classed as not improved. As you can see, the proportion improving, as given by the shaded area of each curve, changes slowly at first, then much more rapidly, and then slowly again as we move from left to right. This should remind you of the sigmoid curve we saw in Figure 15.9 because this is what gives rise to that curve. The regression line that you see in Figure 15.10 is the linear regression of the *continuous* measure of outcome against SurvRate, and it goes through the mean of each distribution. If we had the continuous measure, we could solve for this line. But we have censored data, containing only the dichotomous values, and for that we are much better off solving for the sigmoidal function in Figure 15.9.

We have seen that although our hypothetical continuous variable is a linear function of SurvRate, our censored dichotomous variable (or the probability of improvement) is not. But a simple transformation from p(improvement) to odds(improvement) to log odds(improvement) will give us a variable that *is* a linear function of SurvRate. Therefore, we can convert p(improvement) to log odds(improvement) and get back to a linear function.

Dabbs and Morris (1990) ran an interesting study in which they classified male military personnel as High or Normal in testosterone, and as either having, or not having, a history of delinquency. The results follow:

		Delinquent		
		Yes	No	Total
Testosterone	Normal	402	3614	4016
	High	101	345	446
		503	3959	4462

For these data, the odds of being delinquent if you are in the Normal group are (frequency delinquent)/(frequency not delinquent). (Using probabilities instead of frequencies, this comes down to $p_{\text{delinquent}}/p_{\text{not delinquent}} = p(\text{delinquent})/(1 - p(\text{delinquent})).$) For the Normal testosterone group, the odds of being delinquent are $402/3614 = .1001$. The odds of being not delinquent if you are in the Normal group is the reciprocal of this, which is $3614/402 = 8.990$. This last statistic can be read as meaning that if you are a male with normal testosterone levels, you are nearly 9 times *less likely* to be delinquent than not delinquent. If we look at the High testosterone group, however, the odds of being delinquent are $101/345 = 0.293$, and the odds of being not delinquent are $345/101 = 3.416$. Both groups of males are more likely to be not delinquent than delinquent, but that isn't saying much, because we would hope that most people are not delinquent. But notice that as you move from the Normal to the High group, your odds of being delinquent nearly triple, going from 0.111 to 0.293. If we form the ratio of these odds, we get $0.293/0.111 = 2.64$, which is the odds ratio. For these data, you are 2.64 times more likely to be delinquent if you have high testosterone levels than if you have normal levels. That is a pretty impressive statistic.

We will set aside the odds ratio for a moment and just look at odds. With our cancer data, we will focus on the odds of survival. (We can return to odds ratios any time we want simply by forming the ratio of the odds of survival for each of two different levels of SurvRate.)

For what we are doing here (predicting the odds of surviving breast cancer), we will work with the *natural logarithm*[17] of the odds, the result is called the log odds of survival. For our example, the log odds of being delinquent for a male with high testosterone,

$$\log \text{odds} = \log_e(\text{odds}) = \ln(\text{odds}) = \ln(0.293) = -0.228$$

The log odds will be positive for odds greater than $1/1$ and negative for odds less than $1/1$. (They are undefined for odds = 0.) You will sometimes see log odds referred to as the **logit** and the transformation to log odds referred to as the **logit transformation.**

logit

logit transformation

Returning to the cancer study, we will start with the simple prediction of Outcome on the basis of SurvRate. Letting $p = $ the probability of improvement and $1 - p = $ the

---

[17] The natural logarithm of X is the logarithm to the base e of X. In other words, it is the power to which e must be raised to produce X, where e is the base of the natural number system $= 2.718281$.

probability of nonimprovement, we will solve for an equation of the form:

$$\log(p/1 - p) = \log \text{odds} = b_0 + b_1 \text{ SurvRate}$$

Here b_1 will be the amount of increase in the *log odds* for a one unit increase in SurvRate. It is important to keep in mind how the data were coded. For the Outcome variable, $1 = $ no change or worse, and $2 = $ improvement. For SurvRate, a higher score represents a better prognosis. So you might expect to see that SurvRate would have a positive coefficient, being associated with a better outcome. But with SPSS that will not be the case. SPSS will transform Outcome $= 1$ and 2 to 0 and 1, and then try to predict a 0 (better). Thus, its coefficient will be negative. (SAS would try to predict a 1, and its coefficient would be positive, though of exactly the same magnitude.)

In simple linear regression, we had formulae for b_0 and b_1 and could solve the equations with pencil and paper. Things are not quite so simple in logistic regression, partly because our data consist of 0 and 1 for SurvRate, rather than the conditional proportions of improvement. For logistic regression, we are going to have to solve for our regression coefficients **iteratively.** This means that our computer program will begin with some starting values for b_0 and b_1, see how well the estimated log odds fit the data, adjust the coefficients, again examine the fit, and so on until no further adjustments in the coefficients will lead to a better fit. This is not something you would attempt by hand.

iteratively

In simple linear regression, you also had standard F and t statistics testing the significance of the relationship and the contribution of each predictor variable. We are going to have something similar in logistic regression, although here we will use χ^2 tests instead of F or t.

In Exhibit 15.6, you will see SPSS results of using SurvRate as our only predictor of Outcome. I am beginning with only one predictor just to keep the example simple. We will shortly move to the multiple predictor case, where nothing will really change except that we have more predictors to discuss. The fundamental issues are the same regardless of the number of predictors.

I will not discuss all the statistics in Exhibit 15.6 because to do so would take us away from the fundamental issues. For more extensive discussion of the various statistics, see Darlington (1990), Hosmer and Lemeshow (1989), and Lunneborg (1994). My purpose here is to explain the basic problem and approach.

The first part of the printout is analogous to the first part of a multiple regression printout, where we have a test on whether the model (all predictors taken together) predicts he dependent variable at greater than chance levels. For multiple regression, we have an F test, whereas here we have (several) χ^2 tests.

Start with the line indicating Beginning Block Number 0, and the row labeled "-2 log Likelihood." At this point there is no predictor in the model and -2 log likelihood $=$ 77.345746. The is a measure of the overall variability in the data. You might think of it as being analogous to SS_{total} in the analysis of variance. The quantity $-2 \log L$ can be interpreted as a χ^2 test on how well a model with *no* predictors would fit the data. That χ^2 is 77.3457, which is a significant departure from a good fit, as we would expect with no predictors. (χ^2 would be 0.00 if the fit were perfect.)

For the next block, SPSS adds SurvRate as the (only) predictor and produces another value of -2 log likelihood $= 37.323$. This is the amount of variability that remains after SurvRate is taken into account, and the difference ($77.345 - 37.323 = 40.022$) represents a reduction in χ^2 that can be attributed to adding the predictor. Because we have added one predictor, this is itself a χ^2 on 1 *df*, and can be evaluated as such. You can see that the significance level is given as .0000, meaning that SurvRate added significantly to our ability to predict. (You will note that there are lines labeled Model, Block, and Step, and they are all the same because we have added all of our predictors (1) at the same time.)

Number of selected cases: 66
Number rejected because of missing data: 0
Number of cases included in the analysis: 66

Dependent Variable Encoding:

Original Value	Internal Value
1.00	0
2.00	1

Dependent Variable. OUTCOME   Cancer Outcome

Beginning Block Number 0. Initial Log Likelihood Function

–2 Log Likelihood 77.345746

* Constant is included in the model.

Beginning Block Number 1. Method: Enter

Variable(s) Entered on Step Number
1..   SURVRATE   Survival Rating by Physician

–2 Log Likelihood	37.323
Goodness of Fit	57.235
Cox & Snell – R^2	.455
Nagelkerke – R^2	.659

	Chi-Square	df	Significance
Model	40.022	1	.0000
Block	40.022	1	.0000
Step	40.022	1	.0000

-- Variables in the Equation --

Variable	B	S.E.	Wald	df	Sig	R	Exp(B)
SURVRATE	–.0812	.0193	17.7558	1	.0000	–.4513	.9220
Constant	2.6836	.8113	10.9408	1	.0009		

Exhibit 15.6 Logistic analysis of cancer survival

The next section of the table contains, and tests, the individual predictors. (Here, there is only one predictor—SurvRate.) From this section, we can see that the optimal logistic regression equation is

Log odds $= -0.0812$ SurvRate $+ 2.6836$

The negative coefficient here for SurvRate indicates that the log odds go down as the physician's rating of survival increases. This reflects the fact that SPSS is trying to predict whether a patient will get worse, or even die, and we would expect that the likelihood of getting worse will decrease as the physician's rating increases.

We can also see that SurvRate is a significant predictor, as tested by Wald's $\chi^2 = 17.7558$ on 1 df, which is significant at $p = .0001$ (Wald's χ^2 is a statistic distributed approximately as the chi-square distribution). You will notice that the χ^2 test, that is, $-2 \log L$, on the whole model and the Wald χ^2 test on SurvRate disagree. Because SurvRate *is* the whole model, you might think that they should say the same thing. This is certainly the case in standard linear regression, where our F on regression is, with one predictor, just the square of our t on the regression coefficient. This disagreement stems from the fact that they are based on different estimates of χ^2. Questions have been raised about the behavior of the Wald criterion, and Hosmer and Lemeshow (1989) suggest relying on the likelihood ratio test ($-2 \log L$) instead.

Looking at the logistic regression equation we see that the coefficient for SurvRate is -0.0812, which can be interpreted to mean that a one point increase in SurvRate will decrease the log odds of getting worse by 0.0812. But you and I probably don't care about things like log odds. We probably want to at least work with odds. But that's easy—we simply exponentiate the coefficient. Don't get excited! "Exponentiate" is just an important sounding word that means "raise e to that power." If you have a calculator that cost you more than \$9.99, it probably has a button labeled e^x. Just enter -0.0812, press that button, and you'll have 0.9220. This means that if you increase SurvRate by one point you *multiply* the odds of deterioration by 0.9220. A simple example will show what this means.

Suppose we take someone with a SurvRate score of 40. That person will have a log odds of

$$\text{Log odds} = -0.0812(40) + 2.6837 = -0.5643$$

If we calculate $e^{-0.5643}$ we will get 0.569. This means that the person's odds of deteriorating are 0.569, which means that she is 0.569 times as likely to be deteriorate than improve.[18] Now suppose we take someone with SurvRate $= 41$, one point higher. That person would have predicted log odds of

$$\text{Log odds} = -0.0812(41) + 2.6837 = -0.6455$$

And $e^{-0.6455} = .524$. So this person's log odds are $-0.6455 - (-0.5643) = -.0812$ lower than the first person's, and her odds are $e^{-0.0812} = 0.9220$ times larger ($0.569 \times 0.922 = .524$). Now, 0.922 may not look like a very large number, but if you had cancer a one point higher survival rating gives you about a 7.8% lower chance of deterioration, and that's certainly not something to sneer at.

I told you that if you wanted to see the effect of SurvRate expressed in terms of odds rather than log odds you needed to take out your calculator and exponentiate. That isn't strictly true here, because SPSS does it for you. The last column in this section is labeled "Exp (B)" and contains the exponentiated value of b ($e^{-0.0812} = .9220$).

Although SurvRate is a meaningful and significant predictor of survivability of cancer, it does not explain everything. Epping-Jordan, Compas, and Howell (1994) were interested in determining whether certain behavioral variables also contribute to how a person copes with cancer. They were interested in whether people who experience a high rate of intrusive thoughts (Intrusiv) have a poorer prognosis. (People who experience intrusive thoughts are people who keep finding themselves thinking about their cancer and related events. They can't seem to put it out of their minds.) These authors were also interested in the effect of avoidant behavior (Avoid), which is exhibited by people who just don't want to think about cancer and who try to avoid dealing with the problem. [Intrusiv and Avoid are variables computed from the Impact of Events Scale (Horowitz, Wilner, and Alvarez, 1979).]

Exhibit 15.7 presents the results of using SurvRate, Intrusiv, and Avoid as predictors of Outcome. You can again see that the overall model fits at better-than-chance levels. With no predictors, $-2 \log$ likelihood $= 77.346$. Adding the three predictors to the model

---

[18] If you don't like odds, you can even turn this into a probability. Becauses odds $= p/(1 - p)$, then $p = \text{odds}/(1 + \text{odds})$.

Dependent Variable Encoding:

Original Value	Internal Value
1.00	0
2.00	1

Dependent Variable.. OUTCOME   Cancer Outcome

Beginning Block Number 0. Initial Log Likelihood Function

–2 Log Likelihood 77.345746

* Constant is included in the model.

Beginning Block Number 1. Method: Enter

Variable(s) Entered on Step Number

1..   SURVRATE   Survival Rating by Physician
             INTRUS
             AVOID

–2 Log Likelihood	31.650
Goodness of Fit	35.350
Cox & Snell – R^2	.500
Nagelkerke – R^2	.724

	Chi-Square	df	Significance
Model	45.695	3	.0000
Block	45.695	3	.0000
Step	45.695	3	.0000

-- Variables in the Equation --

Variable	B	S.E.	Wald	df	Sig	R	Exp(B)
SURVRATE	–.0817	.0211	14.9502	1	.0001	–.4092	.9215
INTRUS	–.0589	.0811	.5281	1	.4674	.0000	.9428
AVOID	.1618	.0777	4.3310	1	.0374	.1736	1.1756
Constant	1.6109	1.1780	1.8700	1	.1715		

Exhibit 15.7   Outcome as a function of Survival Rate, Intrusive thoughts, and Avoidance

reduces -2 log likelihood to 31.650, for an improvement of $77.346 - 31.650 = 45.695$. This difference is a χ^2 on 3 *df* because we have three predictors, and it is clearly significant. We would have expected a significant model because we knew that SurvRate alone was a significant predictor. From the bottom section of the table, we see that the Wald χ^2 is significant for both SurvRate and for Avoid, but not for Intrusiv. This tell us that people who exhibit a high level of avoidance behavior do not do as well as those who do less avoiding (Wald chi-square $= 4.3310$, $p = .0374$).[19] More specifically, the regression coefficient for Avoid is 0.1618. This can be interpreted to mean that a one-point increase in Avoid,

---

[19] In line with Hosmer and Lemeshow's (1989) concern with the validity of the Wald chi-square, we might treat this test with some caution. However, Wald's test tends to be conservative, so confidence in this effect is probably not misplaced. You will see some confirmation of that statement shortly.

holding the other two variables constant, increases the log odds of deterioration by 0.1618 points. Exponentiating this we obtain $e^{0.1618} = 1.1756$. Thus, a one-point increase in Avoid multiplies the odds of detereoration by 1.1756, which would increases them.

The Wald χ^2 test on Intrusiv produced a χ^2 of 0.5281, which was not even close to being significant ($p = .4674$). Thus, this variable is not contributing to our prediction. If Intrusiv is not making a significant contribution of predicting Outcome, perhaps it should be dropped from the model. There is actually a very good reason to do just that. Recall that when we had only one predictor, our overall χ^2, as given by $-2 \log L$, was 40.022. We have now added two more predictors, and our overall χ^2 has become 45.695. The nice thing about χ^2 is that a difference between two chi-squares is itself distributed as χ^2 on df equal to the difference between the df for the two models. This means that we can compare the fit of the two models by subtracting $45.695 - 40.022 = 5.673$ and testing this as a χ^2 on $3 - 1 = 2$ df. But the critical value of $\chi^2_{.05}(2) = 5.99$, which means that the degree of improvement between the two models is not significant. It is no greater than we would expect if we just added a couple of useless predictors. But we know that Avoid was significant, as well as SurvRate, so what went wrong?

Well, what went wrong is that we have taken the improvement that we gained by adding Avoid, and spread it out over the nonimprovement that we gained by adding Intrusiv, and their average is not enough to be considered significant. In other words, we have diluted the added contribution of Avoid with Intrusiv. If our goal had been to predict Outcome, rather than to test a model that includes Intrusiv, we would have been much better off if we had just stayed with Avoid. So I would suggest noting that Intrusiv does not contribute significantly and then dropping back to the two-predictor model with SurvRate and Avoid, giving us

$$\text{Log odds} = -0.0823 \text{ SurvRate} + 0.1325 \text{ Avoid} + 1.1961$$

Both of these predictors are significant, as is the degree of improvement over the one-predictor case. The fact that adding Avoid leads to a significant improvement in the model over the one-predictor case is welcome confirmation of the significant Wald chi-square for this effect.

The example that was used here included only continuous predictors because that was the nature of the data set. However, there is nothing to preclude dichotomous predictors, and they are often used. The nice thing about a dichotomous predictor is that a one-unit change in that predictor represents a shift from one category to another. For example, if we used Sex as a predictor and coded Male = 1, Female = 2, then a one-unit increase in Sex would move us from Male to Female. The exponentiated coefficient for Sex would then represent the difference in the odds between males and females. Suppose that Sex had been a predictor in the cancer study and that the coefficient was 0.40. Exponentiating this, we would have 1.49. This would mean that, holding all other variables constant, the odds of a female improving are about 1.5 times greater than the odds of a male improving. You will often see statements in the press of the form "Researchers have concluded that people who exercise regularly have a 44% lower chance of developing heart problems than those who do not." Such statements are often based on the kind of reasoning that we are discussing here.

There is much more to logistic regression than I can cover in this short introduction, but perhaps the biggest stumbling block that people experience is the movement to odds and log odds when we are used to thinking about 0 and 1 or about probabilities. My major purpose in this section was to get you past that barrier (and to supply you with arguments why you should consider logistic regression over linear regression or discriminant analysis when you have a dichotomous dependent variable). Everything else that could be said about logistic regression is mainly about the technicalities, and you can find those in a number of texts, particularly the one by Hosmer and Lemeshow (1989).

Key Terms

Scalar algebra (Introduction)

Matrix algebra (Introduction)

Validities (15.1)

Collinearity (15.1)

Regression coefficients (15.1)

Importance (15.1)

Standardized regression
 coefficients (15.1)

Residual variance (15.3)

Residual error (15.3)

Multivariate normal (15.4)

Multiple correlation coefficient
 ($R_{0.123...p}$) (15.5)

Hyperspace (15.6)

Regression surface (15.6)

Partial correlation ($r_{01.2}$) (15.7)

Semipartial correlation ($r_{0(1.2)}$) (15.7)

Venn diagrams (15.7)

Suppressor variable (15.8)

Multivariate outliers (15.9)

Distance (15.9)

Leverage (h_i) (15.9)

Influence (15.9)

Cook's D (15.9)

Studentized residual (15.9)

Tolerance (15.10)

Cross-correlation (15.10)

Singular (15.10)

Variance inflation factor (VIF) (15.10)

All subsets regression (15.10)

Backward elimination (15.10)

Stepwise regression (15.10)

Forward selection (15.10)

Cross-validation (15.10)

Listwise deletion (15.10)

Casewise deletion (15.10)

Pairwise deletion (15.10)

Multicollinearity (15.11)

Mediating relationship (15.13)

Moderating relationships (15.13)

Center (15.13)

Logistic regression (15.14)

Discriminant analysis (15.14)

Conditional means (15.14)

Sigmoidal (15.14)

Censored data (15.14)

Logit (15.14)

Logit transformation (15.14)

Iteratively (15.14)

Exercises

Note: Many of these exercises are based on a very small data set for reasons of economy of space and computational convenience. For actual applications of multiple regression, sample sizes should be appreciably larger than those used here.

15.1 A psychologist studying perceived "quality of life" in a large number of cities ($N = 150$) came up with the following equation using mean temperature (Temp), median income in $1,000 (Income), per capita expenditure on social services (Socser), and population density (Popul) as predictors.

$$[\hat{Y} = 5.37 - 0.01\,Temp + 0.05\,Income + 0.003\,Socser - 0.01\,Popul]$$

   a. Interpret the regression equation in terms of the coefficients.

   b. Assume there is a city that has a mean temperature of 55 degrees, a median income of $12,000, spends $500 per capita on social services, and has a population density of 200 people per block. What is its predicted quality of life score?

   c. What would we predict in a different city that was identical in every way except that it spent $100 per capita on social services?

15.2 Refer to Exercise 15.1. Assume that

$$\beta_0 = -0.438 \qquad \beta_1 = 0.762 \qquad \beta_2 = 0.08 \qquad \beta_3 = -0.132$$

Interpret the results.

15.3 For the values of β in Exercise 15.2, the corresponding standard errors are

$$[0.397 \qquad 0.252 \qquad 0.052 \qquad 0.025]$$

Which, if any, predictor would you be most likely to drop if you wanted to refine your regression equation?

15.4 A large corporation is interested in predicting a measure of job satisfaction among its employees. They have collected data on 15 employees who each supplied information on job satisfaction, level of responsibility, number of people supervised, rating of working environment, and years of service. The data follow:

Satisfaction:	2	2	3	3	5	5	6	6	6	7	8	8	8	9	9
Responsibility:	4	2	3	6	2	8	4	5	8	8	9	6	3	7	9
No. Supervised:	5	3	4	7	4	8	6	5	9	8	9	3	6	9	9
Environment:	1	1	7	3	5	8	5	5	6	4	7	2	8	7	9
Years of Service:	5	7	5	3	3	6	3	2	7	3	5	5	8	8	1

Exhibit 15.8 is an abbreviated form of the printout.

a. Write out the regression equation using all five predictors.

b. What are the β_is?

DEPENDENT VARIABLE . . . .   1 SATIF

TOLERANCE 0.0100

ALL DATA CONSIDERED AS A SINGLE GROUP

MULTIPLE R   0.6974   STD. ERROR OF EST.   2.0572

MULTIPLE R-SQUARE   0.4864

ANALYSIS OF VARIANCE

	SUM OF SQUARES	DF	MEAN SQUARE	F RATIO	P(TAIL)
REGRESSION	40.078	4	10.020	2.367	0.12267
RESIDUAL	42.322	10	4.232		

VARIABLE		COEFFICIENT	STD. ERROR	STD. REG COEFF	T	P(2 TAIL)	TOLERANCE
INTERCEPT		1.66926					
RESPON	2	0.60516	0.428	0.624	1.414	0.188	0.263940
NUMSUP	3	−0.33399	0.537	−0.311	−0.622	0.548	0.205947
ENVIR	4	0.48552	0.276	0.514	1.758	0.109	0.600837
YRS	5	0.07023	0.262	0.063	0.268	0.794	0.919492

Exhibit 15.8   Printout for regression analysis of data in Exercise 15.4

15.5 Refer to Exercise 15.4.

a. Which variable has the largest semipartial correlation with the criterion, partialling out the other variables?

b. The overall F in Exercise 15.4 is not significant, yet Environment correlates significantly ($r = .58$) with Y. How is this possible?

15.6 Calculate the adjusted R^2 for the data in Exercise 15.4.

15.7 All other things being equal, the ability of two variables to predict a third will increase as the correlation between them decreases. Explain this fact in terms of semipartial correlation.

15.8 All other things being equal, the stability of any given regression coefficient across different samples of data is partly a function of how that variable correlates with other predictors. Explain this fact.

15.9 What does the Tolerance column in Exhibit 15.8 contribute to the answers in Exercises 15.7 and 15.8?

15.10 Using the data in Exercise 15.4, generate $\hat{Y}$ and show that $R_{0.1234} = r_{Y\hat{Y}}$.

15.11 Use Y and $\hat{Y}$ from Exercise 15.10 to show that MS_{residual} is $\sum(Y - \hat{Y})^2/(N - p - 1)$.

15.12 Using the following (random) data, demonstrate what happens to the multiple correlation when you drop cases from the data set (e.g., use 15 cases, then 10, 6, 5, 4).

Y	5	0	5	9	4	8	3	7	0	4	7	1	4	7	9
X_1	3	8	1	5	8	2	4	7	9	1	3	5	6	8	9
X_2	7	6	4	3	1	9	7	5	3	1	8	6	0	3	7
X_3	1	7	4	1	8	8	6	8	3	6	1	9	7	7	7
X_4	3	6	0	5	1	3	5	9	1	1	7	4	2	0	9

15.13 Calculate the adjusted R^2 for the 15 cases in Exercise 15.12.

15.14 Refer to the first three variables from Exercise 15.4.

a. Use any computer program to calculate the squared semipartial correlation and the squared partial correlation for Satisfaction as the criterion and No. Supervised as the predictor, partialling out Responsibility.

b. Draw a Venn diagram to illustrate these two coefficients.

15.15 Refer to the first three variables in Exercise 15.4.

a. Draw a figure comparable to Figure 15.1.

b. Obtain the regression solution for these data and relate the solution to the figure.

15.16 The state of Vermont is divided into 10 Health Planning Districts—they correspond roughly to counties. The following data represent the percentage of live births of babies weighing under 2500 grams (Y), the fertility rate for females 17 years of age or younger (X_1), total high-risk fertility rate for females younger than 17 or older than 35 years of age (X_2), percentage of mothers with fewer than 12 years of education (X_3), percentage of births to unmarried mothers (X_4), and percentage of mothers not seeking medical care until the third trimester (X_5).

Y	X_1	X_2	X_3	X_4	X_5
6.1	22.8	43.0	23.8	9.2	6
7.1	28.7	55.3	24.8	12.0	10
7.4	29.7	48.5	23.9	10.4	5
6.3	18.3	38.8	16.6	9.8	4
6.5	21.1	46.2	19.6	9.8	5
5.7	21.2	39.9	21.4	7.7	6
6.6	22.2	43.1	20.7	10.9	7
8.1	22.3	48.5	21.8	9.5	5
6.3	21.8	40.0	20.6	11.6	7
6.9	31.2	56.7	25.2	11.6	9

A stepwise regression is shown in Exhibit 15.9. (Only the first three steps are shown to conserve space. For purposes of this exercise, we will not let the lack of statistical significance worry us.)

a. What are the values of R for the successive steps?

b. From the definition of a partial correlation (in terms of Venn diagrams), show that the R^2 at step 2 is a function of R^2 at step 1 and the partial correlation listed under step 1—"VARIABLES NOT IN EQUATION."

15.17 In Exercise 15.16, what meaning attaches to R^* as far as the Vermont Department of Health is concerned?

15.18 In Exercise 15.16, the adjusted R^2 would actually be lower for five predictors than for three predictors. Why?

15.19 In Exercise 15.16, the fifth predictor has a very low correlation with the criterion ($r = .05$) and yet plays a significant role in the regression. Why?

STEP NO. 1
VARIABLE ENTERED   3   X2
MULTIPLE R   0.6215
MULTIPLE R-SQUARE   0.3862
ADJUSTED R-SQUARE   0.3095
STD. ERROR OF EST.   0.5797

ANALYSIS OF VARIANCE

	SUM OF SQUARES	DF	MEAN SQUARE	F RATIO
REGRESSION	1.6917006	1	1.691701	5.03
RESIDUAL	2.6882995	8	0.3360374	

VARIABLES IN EQUATION							.	VARIABLES NOT IN EQUATION				
VARIABLE	COEFFICIENT	STD. ERROR OF COEFF	STD. REG COEFF	TOLERANCE	F TO REMOVE	LEVEL	. VARIABLE		PARTIAL CORR.	TOLERANCE	F TO ENTER	LEVEL
(Y-INTERCEPT	3.529)						.					
X2 3	0.069	0.031	0.621	1.00000	5.03	1	. X1	2	−0.19730	0.25831	0.28	1
							. X3	4	−0.25039	0.43280	0.47	1
							. X4	5	0.00688	0.69838	0.00	1
							. X5	6	−0.59063	0.58000	3.75	1

STEP NO. 2
VARIABLE ENTERED   6   X5
MULTIPLE R   0.7748
MULTIPLE R-SQUARE   0.6003
ADJUSTED R-SQUARE   0.4862
STD. ERROR OF EST.   0.5001

ANALYSIS OF VARIANCE

	SUM OF SQUARES	DF	MEAN SQUARE	F RATIO
REGRESSION	2.6294919	2	1.314746	5.26
RESIDUAL	1.7505082	7	0.2500726	

VARIABLES IN EQUATION							.	VARIABLES NOT IN EQUATION				
VARIABLE	COEFFICIENT	STD. ERROR OF COEFF	STD. REG COEFF	TOLERANCE	F TO REMOVE	LEVEL	. VARIABLE		PARTIAL CORR.	TOLERANCE	F TO ENTER	LEVEL
(Y-INTERCEPT	2.949)						.					
X2 3	0.113	0.035	1.015	0.58000	10.47	1	. X1	2	−0.09613	0.24739	0.06	1
X5 6	−0.223	0.115	−0.608	0.58000	3.75	1	. X3	4	−0.05399	0.37826	0.02	1
							X4	5	0.41559	0.53416	1.25	1

STEP NO. 3
VARIABLE ENTERED   5 X4
MULTIPLE R   0.8181
MULTIPLE R-SQUARE   0.6694
ADJUSTED R-SQUARE   0.5041
STD. ERROR OF EST.   0.4913

(continued)

Exhibit 15.9   Stepwise regression of data on low birth weight infants

ANALYSIS OF VARIANCE

	SUM OF SQUARES	DF	MEAN SQUARE	F RATIO
REGRESSION	2.9318295	3	0.9772765	4.05
RESIDUAL	1.4481706	6	0.2413618	

			VARIABLES IN EQUATION						VARIABLES NOT IN EQUATION			
VARIABLE	COEFFICIENT	STD. ERROR OF COEFF	STD. REG COEFF	TOLERANCE	F TO REMOVE	LEVEL	VARIABLE		PARTIAL CORR.	TOLERANCE	F TO ENTER	LEVEL
(Y-INTERCEPT	1.830)											
X2	3 0.104	0.035	0.942	0.55484	8.93	1	X1	2	−0.14937	0.24520	0.11	1
X4	5 0.190	0.170	0.359	0.53416	1.25	1	X3	4	0.14753	0.31072	0.11	1
X5	6 −0.294	0.130	−0.799	0.44362	5.14	1						

Exhibit 15.9   *(continued)*

15.20 For the data in Exercise 15.16, compute $\hat{Y} = 1X_2 + 1X_4 - 3X_5$. How well does this equation fit compared with the optimal equation? Why should this be the case?

15.21 For the data in Exercise 15.16, would it be safe to conclude that decreasing the number of mothers who fail to seek medical care before the third trimester is a good way to decrease the incidence of low-birthweight infants?

15.22 Create a set of data on 10 cases that illustrates leverage, distance, and influence. Use any standard regression program to produce statistics measuring these attributes.

15.23 Produce a set of data where the variance of *Y* values associated with large values of *X* is greater than the variance of *Y* values associated with small values of *X*. Then run the regression and plot the residuals on the ordinate against *X* on the abscissa. What pattern emerges?

Computer Exercises

15.24 Use the data set Mireault.dat (on the website www.uvm.edu/~dhowell/methods/) from Mireault (1990) to examine the relationship between current levels of depression and other variables. A reasonable model might propose that depression (DepressT) is a function of (1) the person's current perceived level of vulnerability to additional loss (PVLoss), (2) the person's level of social support (SuppTotl), and (3) the age at which the person lost a parent during childhood (AgeAtLos). Use any statistical package to evaluate the model outlined here. (Because only subjects in Group 1 lost a parent to death during their childhood, your analysis will be restricted to that group.)

15.25 A compulsive researcher who wants to cover all possibilities might throw in the total score on perceived vulnerability (PVTotal) as well as PVLoss. (The total includes vulnerability to accidents, illness, and life-style related problems.)

  a. Run this analysis adding PVTotal to the variables used in Exercise 15.24.

  b. What effect did the inclusion of PVTotal have on R^2? What effect did it have on the standard error of the regression coefficient for PVLoss? If your program will also give you tolerance and VIF, what effect does the inclusion of PVTotal have on them?

  c. What would you conclude about the addition of PVTotal to our model?

15.26 In Exercise 15.24, we posited a model in which depression was a function of perceived vulnerability, social support, and age at loss. An alternative, or additional, view might be that vulnerability itself is a function of social support and age at loss. (If you lost a parent when you were very young and you have little social support, then you might feel particularly vulnerable to future loss.)

   a. Set up the regression problem for this question and run the appropriate analysis. (Use PVLoss, SuppTotl, and AgeAtLos.)

   b. Interpret your results.

15.27 Draw one diagram to illustrate the relationships examined in Exercises 15.24 and 15.26. Use arrows to show predicted relationships, and write the standardized regression coefficients next to the arrows. (You have just run a simple path analysis.)

15.28 Notice that in the diagram in Exercise 15.27, SuppTotl has both a direct and an indirect effect on Depression. Its direct effect is the arrow that goes from SuppTotl to DepressT. The indirect effect (which here is not significant) comes from the fact that SuppTotl influences PVLoss, which in turn affects DepressT. Explain these direct and indirect effects in terms of semipartial regression coefficients.

15.29 Repeat the analysis of Exercise 15.24, requesting statistics on regression diagnostics.

   a. What, if anything, do these statistics tell you about the data set?

   b. Delete the subject with the largest measure of influence (usually indexed by Cook's D). What effect does that have for this particular data set?

15.30 It is useful to examine the effects of measurement reliability on the outcome of a regression problem. In Exercise 15.24, the variable PVLoss was actually a reasonably reliable variable. However, for purposes of illustration, we can manufacture a new, and less reliable, measure from it by adding a bit of random error to PVLoss.

   a. Create a new variable called UnrelLos with a statement *of the form*

   $$UnrelLos = PVLoss + 7.5 \times \text{"random"}$$

   Here, "random" is a random-number function available with most statistical programs. You will need to check the manual to determine the exact form of the statement. I used a multiplier of 7.5 on the assumption that the random-number function will sample from an $N(0, 1)$ population. Multiplying by 7.5 will increase the standard deviation of UnrelLos by 50% (see the variance sum law). You may want to play with other constants.

   b. Now repeat Exercise 15.24 using UnrelLos in place of PVLoss.

   c. What effect does this new variable have on the contribution of the perceived vulnerability of loss to the prediction of DepressT? How has the regression coefficient changed? How has its standard error changed? How does a test on its statistical significance change? What changes occurred for the other variables in the equation?

15.31 The data set Harass.dat (from the website www.uvm.edu/~dhowell/methods/) contains slightly modified data on 343 cases created to replicate the results of a study of sexual harassment by Brooks and Perot (1991). The dependent variable is whether or not the subjects reported incidents of sexual harassment, and the independent variables are, in order, Age, Marital Status (1 = married, 2 = single), Feminist Ideology, Frequency of the behavior, Offensiveness of the behavior, and whether or not it was reported (0 = no, 1 = yes). (For each variable, higher numbers represent more of the property.) Using any logistic regression program, examine the likelihood that a subject will report sexual harassment on the basis of the independent variables.

15.32 Repeat Exercise 15.31, but this time use just the dichotomous predictor Marital Status. Create a contingency table of Married/Unmarried by Report/No Report, calculate odds ratios, and compare those ratios with the results of the logistic regression. (The result will not be significant, but that is not important.)

15.33 I was surprised to see that frequency of the behavior was not related to the likelihood of reporting. Can you suggest reasons why this might be so?

15.34 The following question comes from a paper by Guber (1999) published in the *Journal of Statistics Education*, which is an online journal. (You can get access to the summary and the entire article by going to http://www.amstat.org/publications/jse/v7n2_abstracts.html) Guber was interested in the relationship between how much states spend on education and the performance of their students on SAT tests. That simple question can be misleading, as you will see, so she added the percentage of students taking the SAT for each state, and other variables not shown here. Her data are reproduced for some of the variables here, where SAT refers to the combined Math and Verbal score.

a. Plot a histogram of the individual variables. How would you describe these variables?

b. Create a scatterplot of the relationship between Expenditure and SAT scores.

c. Draw a scatterplot of the relationship between SAT scores and the percentage of students taking the SAT (PctSAT).

d. Run the simple regression of SAT as a function of Expenditure, and then rerun it adding PctSAT.

e. What does the answer to (c) contribute to our understanding of what is going on?

f. Rerun the multiple regression for (c), but this time calculate the residuals and plot them as a function of the predicted value. Does this leave you somewhat concerned?

g. Create a new variable that is the square root of PctSAT and use this with PctSAT and Expenditure to predict SAT. What does this show? What happens to a plot of the residuals against the predicted value? (This was a suggestion from Guber.)

State	Expend	Pct	Combined	State	Expend	Pct	Combined
Alabama	4.405	8	1029	Nebraska	5.935	9	1050
Alaska	8.963	47	934	Nevada	5.160	30	917
Arizona	4.778	27	944	New Hamp.	5.859	70	935
Arkansas	4.459	6	1005	New Jersey	9.774	70	898
Calif.	4.992	45	902	New Mexico	4.586	11	1015
Colorado	5.443	29	980	New York	9.623	74	892
Conn.	8.817	81	908	North Car.	.	60	865
Delaware	7.030	68	897	North Dak.	4.775	5	1107
Florida	5.718	48	889	Ohio	6.162	23	975
Georgia	5.193	65	854	Oklahoma	4.845	9	1027
Hawaii	6.078	57	889	Oregon	6.436	51	947
Idaho	4.210	15	979	Pennsylvania	7.109	70	880
Illinois	6.136	13	1048	Rhode Island	7.469	70	888
Indiana	5.826	58	882	South Car.	.	58	844
Iowa	5.483	5	1099	South Dak.	4.775	5	1068
Kansas	5.817	9	1060	Tennessee	4.388	12	1040
Kentucky	5.217	11	999	Texas	5.222	47	893
Louisiana	4.761	9	1021	Utah	3.656	4	1076
Maine	6.428	68	896	Vermont	6.750	68	901
Maryland	7.245	64	909	Virginia	5.327	65	896
Mass.	7.287	80	907	Wash	5.906	48	937
Michigan	6.994	11	1033	West Vir	6.107	17	932
Minn.	6.000	9	1085	Wisconsin	6.930	9	1073
Miss.	4.080	4	1036	Wyoming	6.160	10	1001
Missouri	5.383	9	1045				
Montana	5.692	21	1009				

Discussion Questions

15.35 What are some of the reasons why stepwise regression (broadly defined) would not find favor with most statisticians?

15.36 Make up a very simple example with very simple variables to illustrate how one could see the effect of an interaction between two predictors.

15.37 Using the data you created in Exercise 15.36, demonstrate the effect of "centering" your predictor variables.

15.38 As you know, the regression coefficient gives the effect of one variable holding all other variables constant. How would you view this interpretation when you have an interaction term in your model?

15.39 Paul Jose has a website referred to in the section on mediation. He discusses a problem in which he believes that stress leads to depression through a mediating path via rumination. (In other words, some stressed people ruminate, and as a consequence they become depressed.) The path diagram derived from his analysis of actual data follows. The beta given for the path from stress to depression is from the multiple regression of depression on stress and rumination. Predicting depression only from stress had a beta of 0.471. Test the decline in the coefficient for the direct path from stress to depression using Sobel's test. (You can check your work at Jose's website at http://www.vuw.ac.nz/psyc/staff/paul-jose/files/helpcentre/help7_mediation_example.php, though the answers will not be exactly equal.

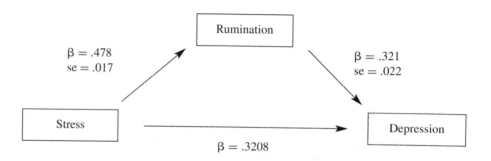

CHAPTER 16

Analyses of Variance and Covariance as General Linear Models

Objectives

To show how the analysis of variance can be viewed as a special case of multiple regression; to present procedures for the treatment of unequal sample sizes; to present the analysis of covariance.

Contents

Most people think of multiple regression and the analysis of variance as two totally separate statistical techniques that answer two entirely different sets of questions. This is not at all the case. In the first place, both techniques ask the same kind of questions, and in the second place, they return the same kind of answers, although the answers may be phrased somewhat differently. The analysis of variance tells us that three treatments (T_1, T_2, and T_3) have different means ($\overline{X}_i$). Multiple regression tells us that means ($\overline{Y}_i$) are related to treatments (T_1, T_2, and T_3), which amounts to the same thing. Furthermore, the analysis of variance produces a statistic (F) on the differences between means. The analysis of regression produces a statistic (F) on the significance of R. As we shall see shortly, these Fs are equivalent.

16.1 The General Linear Model

general linear model

Just as multiple regression and the analysis of variance are concerned with the same general type of question, so are they basically the same technique. Actually, the analysis of variance is a special case of multiple linear regression, which in turn is a special case of what is commonly referred to as the **general linear model.** The fact that the analysis of variance has its own formal set of equations can be attributed primarily to good fortune. When certain conditions are met (as they are in the analysis of variance), the somewhat cumbersome multiple-regression calculations are reduced to a few relatively simple equations. If it were not for this, there might not even be a separate set of procedures called the analysis of variance.

For the student interested solely in the application of statistical techniques, a word is in order in defense of even including a chapter on this topic. Why, you may ask, should you study what amounts to a cumbersome way of doing what you already know how to do in a simple way? Ignoring the cry of "intellectual curiosity," which is something that most people are loath to *admit* that they do not possess in abundance, there are several practical (applied) answers to such a question. First, this approach represents a relatively straightforward way of handling particular cases of unequal sample sizes, and understanding this approach helps you make intelligent decisions about various options in statistical software. Second, this approach provides us with a simple and intuitively appealing way of running, and especially of understanding, an analysis of covariance—which is a very clumsy technique when viewed from the more traditional approach. Last, and most important, this approach represents a glimpse at the direction in which statistical techniques are moving. With the greatly extended use of powerful and fast computers, many of the traditional statistical techniques are giving way to what were previously impractical procedures. The increase in the popularity of multivariate analysis of variance (with all its attendant strengths and weaknesses) is a case in point. Other examples are such techniques as structural equation modeling and that old and much-abused standby, factor analysis. Unless you understand the relationship between the analysis of variance and the general linear model (as represented by multiple linear regression), and unless you understand how the data for simple analysis of variance problems can be cast in a multiple-regression framework, you will find yourself in the near future using more and more techniques about which you know less and less. This is not to say that t, χ^2, F, and so on are likely to disappear, but only that other techniques will be added, opening up entirely new ways of looking at data. The recent rise in the use of structural equation modeling (SEM) is a case in point because much of what that entails builds on what you already know about regression, and what you will learn about underlying models of processes.

During the past 25 years, several excellent and very readable papers on this general topic have been written. The clearest presentation is still Cohen (1968). A paper by Overall and Spiegel (1969) is also worth seeing. Both of these papers appeared in the *Psychological Bulletin* and are therefore readily available. Other good discussions can be found in Overall (1972); Overall and Klett (1972); Overall, Spiegel, and Cohen (1975); Green, Marquis, Hershberger, Thompson, and McCollam (1999); and Appelbaum and Cramer (1974). Cramer and Appelbaum (1980), and Howell and McConaughy (1982) provide contrasting views on the choice of the underlying model and the procedures to be followed.

There are two different ways to read this chapter, both legitimate. The first is to look for general concepts and to go lightly over the actual techniques of calculation. That is the approach I often tell my students to follow. I want them to understand where the reasoning leads, and I want them to feel that they could carry out all the steps if they had to (with the book in front of them), but I don't ask them to commit very much of the technical material to memory. On the other hand, some instructors may want their students to grasp the material at a deeper level. There are good reasons for doing so. But I would still suggest that the first time you read the chapter, you look for general understanding. To develop greater expertise, sit down with both a computer and a calculator and work lots and lots of problems.

The Linear Model

Consider first the traditional multiple-regression problem with a criterion (Y) and three predictors (X_1, X_2, and X_3). We can write the usual model

$$Y_i = b_0 + b_1 X_{1i} + b_2 X_{2i} + b_3 X_{3i} + e_i$$

or, in terms of *vector* notation

$$\mathbf{y} = \mathbf{b_0} + b_1 \mathbf{x_1} + b_2 \mathbf{x_2} + b_3 \mathbf{x_3} + \mathbf{e}$$

where $\mathbf{y}$, $\mathbf{x_1}$, $\mathbf{x_2}$, and $\mathbf{x_3}$ are ($n \times 1$) vectors (columns) of data, $\mathbf{e}$ is a ($n \times 1$) vector of errors, and $\boldsymbol{b_0}$ is a ($n \times 1$) vector whose elements are the intercept. This equation can be further reduced to

$$\mathbf{y} + \mathbf{Xb} + \mathbf{e}$$

where $\mathbf{X}$ is a $n \times (p + 1)$ matrix of predictors, the first column of which is 1s, and $\mathbf{b}$ is a ($p + 1$) $\times 1$ vector of regression coefficients.

Now consider the traditional model for a one-way analysis of variance:

$$Y_{ij} = \mu + \tau_j + e_{ij}$$

Here the symbol τ_j is simply a shorthand way of writing $\tau_1, \tau_2, \tau_3, \ldots \tau_p$, where for any given subject we are interested in only that value of τ_j that pertains to the particular treatment in question. To see the relationship between this model and the traditional regression model, it is necessary to introduce the concept of a design matrix. Design matrices are used in a wide variety of situations, not simply the analysis of variance, so it is important to understand them.

Design Matrices

design matrix

A **design matrix** is a matrix of *coded,* or *dummy,* or *counter* variables representing group membership. The *complete* form of the design matrix (X) will have $p + 1$ columns, representing the mean (μ) and the p treatment effects. A subject is always scored 1 for μ, because μ is part of all observations. In all other columns, she is scored 1 if she is a member of the treatment associated with that column, and 0 otherwise. Thus, for three treatments with two subjects per treatment, the complete design matrix would be

$$\sum = \begin{array}{c} \\ 1 \\ 2 \\ 3 \\ 4 \\ 5 \\ 6 \end{array} \begin{array}{cccc} S & \mu & A_1 & A_2 & A_3 \\ \begin{bmatrix} 1 & 1 & 0 & 0 \\ 1 & 1 & 0 & 0 \\ 1 & 0 & 1 & 0 \\ 1 & 0 & 1 & 0 \\ 1 & 0 & 0 & 1 \\ 1 & 0 & 0 & 1 \end{bmatrix} \end{array}$$

Notice that subjects 1 and 2 (who received Treatment A_1) are scored 1 on μ and A_1, and 0 on A_2 and A_3 because they did not receive those treatments. Similarly, subjects 3 and 4 are scored 1 on μ and A_2, and 0 on A_1 and A_3.

We will now define the vector τ of treatment effects as $[\mu \quad \tau_1 \quad \tau_2 \quad \tau_3]$. Taking $\mathbf{X}$ as the design matrix, the analysis of variance model can be written in matrix terms as

$$\mathbf{y} = \mathbf{X}\tau + \mathbf{e}$$

which can be seen as being of the same form as the traditional regression equation. The elements of τ are the effects of each dummy treatment variable, just as the elements of b in the regression equation are the effects of each independent variable. Expanding, we obtain

$$\mathbf{y} = \begin{bmatrix} 1 & 1 & 0 & 0 \\ 1 & 1 & 0 & 0 \\ 1 & 0 & 1 & 0 \\ 1 & 0 & 1 & 0 \\ 1 & 0 & 0 & 1 \\ 1 & 0 & 0 & 1 \end{bmatrix} \times \begin{bmatrix} \mu \\ \tau_1 \\ \tau_2 \\ \tau_3 \end{bmatrix} + \begin{bmatrix} e_{11} \\ e_{21} \\ e_{12} \\ e_{22} \\ e_{13} \\ e_{23} \end{bmatrix}$$

$$\mathbf{y} = \qquad\qquad \mathbf{X} \qquad\qquad \times \quad \tau \quad + \quad \mathbf{e}$$

which, following the rules of matrix multiplication, produces

$$Y_{11} = \mu + \tau_1 + e_{11}$$
$$Y_{21} = \mu + \tau_1 + e_{21}$$
$$Y_{12} = \mu + \tau_2 + e_{12}$$
$$Y_{22} = \mu + \tau_2 + e_{22}$$
$$Y_{13} = \mu + \tau_3 + e_{13}$$
$$Y_{23} = \mu + \tau_3 + e_{23}$$

For each subject, we now have the model associated with her response. Thus, for the second subject in Treatment 2, $Y_{22} = \mu + \tau_2 + e_{22}$, and for the ith subject in Treatment j, we have $Y_{ij} = \mu + \tau_j + e_{ij}$, which is the usual analysis of variance model.

The point is that the design matrix allows us to view the analysis of variance in a multiple-regression framework, in that it permits us to go from

$$Y_{ij} = \mu + \tau_j + e_{ij} \quad \text{to} \quad \mathbf{y} = \mathbf{X}\mathbf{b} + \mathbf{e}$$

Moreover, the elements of $\mathbf{b}$ are the values of $\mu, \tau_1, \tau_2, \ldots, \tau_k$. In other words, these are the actual treatment effects in which we are interested.

The design matrix we have been using has certain technical limitations that must be circumvented. First, it is redundant in the sense that if we are told that a subject is not in A_1 or

A_2, we know without being told that she must be in A_3. This is another way of saying that there are only 2 *df* for treatments. For this reason, we will eliminate the column headed A_3, leaving only $a - 1$ columns for the treatment effects. A second change is necessary if we want to use any computer program that obtains a multiple-regression equation by way of first calculating the intercorrelation matrix. The column headed μ has no variance, and therefore we cannot enter it into a standard multiple-regression program—it would cause us to attempt division by 0. Thus, it too must be eliminated. This is no real loss because our ultimate solution will not be affected.

One further change will be made simply for the sake of allowing us to test the desired null hypotheses using the method to be later advocated for factorial designs. Because we have omitted a column dealing with the third (or *a*th) level of treatments, solutions given our modified design matrix would produce estimates of treatment effects in relation to $\overline{X}_3$ rather than in relation to $\overline{X}_.$. In other words, b_1 would turn out to be $(\overline{X}_1 - \overline{X}_3)$ rather than $(\overline{X}_1 - \overline{X}_.)$. This problem can be eliminated, however, by a modification of the design matrix to make the mean $(\overline{X}_i)$ of each column of $\mathbf{X}$ equal to 0. Under this new system, a subject is scored 1 in column A_i if she is a member of Treatment A_i, she is scored -1 if she is a member of the *a*th (last) treatment, and she is scored 0 if neither of these conditions apply. (This restriction corresponds to the fixed-model analysis of variance requirement that $\sum \tau_i = 0$.)

These modifications have led us from

$$\mathbf{X} = \begin{bmatrix} 1 & 1 & 0 & 0 \\ 1 & 1 & 0 & 0 \\ 1 & 0 & 1 & 0 \\ 1 & 0 & 1 & 0 \\ 1 & 0 & 0 & 1 \\ 1 & 0 & 0 & 1 \end{bmatrix} \text{ to } \begin{bmatrix} 1 & 1 & 0 \\ 1 & 1 & 0 \\ 1 & 0 & 1 \\ 1 & 0 & 1 \\ 1 & 0 & 0 \\ 1 & 0 & 0 \end{bmatrix} \text{ to } \begin{bmatrix} 1 & 1 \\ 1 & 0 \\ 0 & 1 \\ 0 & 1 \\ 0 & 0 \\ 0 & 0 \end{bmatrix} \text{ to } \begin{bmatrix} 1 & 0 \\ 1 & 0 \\ 0 & 1 \\ 0 & 1 \\ -1 & -1 \\ -1 & -1 \end{bmatrix}$$

Although these look like major changes in that the last form of $\mathbf{X}$ appears to be far removed from where we started, it actually carries all the necessary information. We have merely eliminated redundant information, removed a constant term, and then caused the treatment effects to be given as deviations from $\overline{X}_.$.

16.2 One-Way Analysis of Variance

At this point, a simple example is in order. Table 16.1 contains data for three subjects in each of four treatments. Table 16.1b shows the summary table for the corresponding analysis of variance, along with the value of η^2 (discussed in Chapter 11). Table 16.1c contains the estimated treatment effects $(\hat{\tau}_i)$ where $\hat{\tau}_i = \hat{\mu}_i - \hat{\mu}$. The fixed-model analysis of variance imposes the restriction that $\sum \tau_i = 0$, so τ_4 is automatically defined by τ_1, τ_2, and τ_3.

Now let us approach the statistical treatment of these data by means of least-squares multiple linear regression. We will take as our criterion (Y) the raw data in Table 16.1. For the predictors, we will use a design matrix *of the form*

$$\mathbf{X} = \begin{array}{c} \text{Treatment 1} \\ \text{Treatment 2} \\ \text{Treatment 3} \\ \text{Treatment 4} \end{array} \begin{array}{ccc} A_1 & A_2 & A_3 \\ \begin{bmatrix} 1 & 0 & 0 \\ 0 & 1 & 0 \\ 0 & 0 & 1 \\ -1 & -1 & -1 \end{bmatrix} \end{array}$$

Table 16.1 Illustrative calculations for simple one-way design with equal ns

(a) Data

Treatment 1	Treatment 2	Treatment 3	Treatment 4
8	5	3	6
9	7	4	4
7	3	1	9
8	5	2.667	6.333

$\overline{X}_{..} = 5.500$

(b) Summary Table

Source	df	SS	MS	F	η^2
Treatments	3	45.667	15.222	4.46	.626
Error	8	27.333	3.417		
Total	11	73.000			

(c) Estimated Treatment Effects

$\hat{\tau}_1 = \overline{X}_1 - \overline{X}_{..} = 8.0 - 5.5 = 2.5$

$\hat{\tau}_2 = \overline{X}_2 - \overline{X}_{..} = 5.0 - 5.5 = -0.5$

$\hat{\tau}_3 = \overline{X}_3 - \overline{X}_{..} = 2.67 - 5.5 = -2.83$

Here the elements of any one row are taken to apply to *all the subjects in the treatment*. The multiple-regression solution using the design matrix **X** as the matrix of predictors is presented in Exhibit 16.1. Here the dependent variable (**Y**) is the first column of the data matrix. The next three columns together form the matrix **X**. Minitab was used to generate this solution, but any standard program would be suitable. (I have made some very minor changes in the output to simplify the discussion.)

Notice the patterns of intercorrelations among the *X* variables in Exhibit 16.1. This type of pattern with constant off-diagonal correlations will occur whenever there are equal numbers of subjects in the various groups. (The fact that we don't have constant off-diagonal correlations with unequal-*n* factorial designs is what makes our life more difficult in those situations.)

Notice that the regression coefficients are written in a column. This column can be called a *vector* and is the vector **b**, or, in analysis of variance terms, the vector τ. Notice that $b_1 = 2.50$, which is the same as the estimated treatment effect of Treatment 1 shown in Table 16.1. In other words, $b_1 = \tau_1$. This also happens for b_2 and b_3. This necessarily follows from our definition of **X** and τ. Moreover, if we were to examine the significance of the b_i, given as the column of *t*-ratios, we would simultaneously have tests on the hypothesis ($H_0: \tau_j = \mu_i - \mu = 0$). Notice further that the intercept (b_0) is equal to the grand mean ($\overline{Y}$). This follows directly from the fact that we scored the *a*th treatment as -1 on all coded variables. Using the (-1) coding, the mean of every column of **X** ($\overline{X}_j$) is equal to 0 and, as a result, $\sum b_1 \overline{X}_j = 0$ and therefore $b_0 = \overline{Y} - \sum b_1 \overline{X}_j = \overline{Y} - 0 = \overline{Y}$. This situation holds only in the case of equal ns because otherwise $\overline{X}_i$ would not be 0 for all i. However, in all cases, b_0 is our best estimate of μ in a least squares sense.

Untitled - SPSS Data Editor

File   Edit   View   Data   Transform   Analyze   Graphs   Utilities   Add-ons   W

18 : Y

	Y	X1	X2	X3	var
1	8	1	0	0	
2	9	1	0	0	
3	7	1	0	0	
4	5	0	1	0	
5	7	0	1	0	
6	3	0	1	0	
7	3	0	0	1	
8	4	0	0	1	
9	1	0	0	1	
10	6	-1	-1	-1	
11	4	-1	-1	-1	
12	9	-1	-1	-1	

Correlations

		Y	X1	X2	X3
Pearson Correlation	Y	1.000	.239	−.191	−.526
	X1	.239	1.000	.500	.500
	X2	−.191	.500	1.000	.500
	X3	−.526	.500	.500	1.000

Model Summary

Model	R	R Square	Adjusted R Square	Std. Error of the Estimate
1	.791(a)	.626	.485	1.848

a Predictors: (Constant), X3, X2, X1

Coefficients(a)

Model		Unstandardized Coefficients		Standardized Coefficients	t	Sig.
		B	Std. Error	Beta		
1	(Constant)	5.500	.534		10.307	.000
	X1	2.500	.924	.717	2.705	.027
	X2	−.500	.924	−.143	−.541	.603
	X3	−2.833	.924	−.812	−3.066	.015

a Dependent Variable: Y

(continued)

Exhibit 16.1   SPSS regression analysis of data in Table 16.1

ANOVA[b]

Model		Sum of Squares	df	Mean Square	F	Sig.
1	Regression	45.667	3	15.222	4.455	.040[a]
	Residual	27.333	8	3.417		
	Total	73.000	11			

[a] Predictors: (Constant), X3, X2, X1
[b] Dependent Variable: Y

Exhibit 16.1 *(continued)*

The value of $R^2 = .626$ is equivalent to η^2 because they both estimate the percentage of variation in the dependent variable accounted for by variation among treatments.

If we test R^2 for significance, we have $F = 4.46$, $p = .040$. This is the F value we obtained in the analysis of variance, although this F can be found by the formula that we saw for testing R^2 in Chapter 15.

$$F(p, N - p - 1) = \frac{R^2(N - p - 1)}{(1 - R^2)p}$$

$$F(3, 8) = \frac{.626(8)}{.374(3)} = 4.46$$

Notice that the sums of squares for Regression, Residual, and Total in Exhibit 16.1 are exactly equivalent to the sums of squares for Treatments, Error, and Total in Table 16.1. This equality makes clear that there is complete correspondence between sums of squares in regression and the analysis of variance.

The foregoing analysis has shown the marked similarity between the analysis of variance and multiple regression. This is primarily an illustration that there is no important difference between asking whether different treatments produce different means and asking whether means are a function of treatments. We are simply looking at two sides of the same coin.

We have discussed only the most common way of forming a design matrix. This matrix could take a number of other useful forms. For a good discussion of these, see Cohen (1968).

16.3 Factorial Designs

We can readily extend the analysis of regression of two-way and higher-order factorial designs, and doing so illustrates some important features of both the analysis of variance and the analysis of regression. (A good discussion of this approach, and the decisions that need to be made, can be found in Harris (2005).) We will consider first a two-way analysis of variance with equal *n*s.

The Full Model

The most common model for a two-way analysis of variance is

$$Y_{ijk} = \mu + \alpha_i + \beta_j + \alpha\beta_{ij} + e_{ijk}$$

As we did before, we can expand the α_i and β_j terms by using a design matrix. But then how should the interaction term be handled? The answer to this question relies on the fact that an interaction represents a multiplicative effect of the component variables. Suppose we consider the simplest case of a 2×2 factorial design. *Letting the entries in each row represent the coefficients for all subjects in the corresponding cell* of the design, we can write our design matrix as

$$
\mathbf{X} = \begin{array}{c} \\ a_1b_1 \\ a_1b_2 \\ a_2b_1 \\ a_2b_2 \end{array}
\begin{array}{ccc} A_1 & B_1 & AB_{11} \\ \left[\begin{array}{ccc} 1 & 1 & 1 \\ 1 & -1 & -1 \\ -1 & 1 & -1 \\ -1 & -1 & 1 \end{array}\right] \end{array}
$$

The first column represents the main effect of A and distinguishes between those subjects who received A_1 and those who received A_2. The next column represents the main effect of B, separating B_1 subjects from B_2 subjects. The third column is the interaction of A and B. Its elements are obtained by multiplying the corresponding elements of columns 1 and 2. Thus, $1 = 1 \times 1$, $-1 = 1 \times -1$, $-1 = -1 \times 1$, and $1 = -1 \times -1$. Once again, we have as many columns per effect as we have degrees of freedom for that effect. We have no entries of 0 simply because with only two levels of each variable, a subject must either be in the first or last level.

Now consider the case of a 2×3 factorial. With two levels of A and three levels of B, we will have $df_A = 1$, $df_B = 2$, and $df_{AB} = 2$. This means that our design matrix will require one column for A and two columns each for B and AB. This leads to the following matrix:

$$
\mathbf{X} = \begin{array}{c} \\ a_1b_1 \\ a_1b_2 \\ a_1b_3 \\ a_2b_1 \\ a_2b_2 \\ a_2b_3 \end{array}
\begin{array}{ccccc} A_1 & B_1 & B_2 & AB_{11} & AB_{12} \\ \left[\begin{array}{ccccc} 1 & 1 & 0 & 1 & 0 \\ 1 & 0 & 1 & 0 & 1 \\ 1 & -1 & -1 & -1 & -1 \\ -1 & 1 & 0 & -1 & 0 \\ -1 & 0 & 1 & 0 & -1 \\ -1 & -1 & -1 & 1 & 1 \end{array}\right] \end{array}
$$

Column A_1 distinguishes between those subjects who are in treatment level A_1 and those in treatment level A_2. Column 2 distinguishes level B_1 subjects from those who are not in B_1, and Column 3 does the same for level B_2. Once again, subjects in the first $a - 1$ and first $b - 1$ treatment levels are scored 1 or 0, depending on whether or not they served in the treatment level in question. Subjects in the ath or bth treatment level are scored -1 for each column related to that treatment effect. The column labeled AB_{11} is simply the product of columns A_1 and B_1, and AB_{12} is the product of A_1 and B_2.

The analysis for a factorial design is more cumbersome than the one for a simple one-way design because we want to test two or more main effects and one or more interaction effects. If we consider the relatively simple case of a two-way factorial, however, you should have no difficulty generalizing it to more complex factorial designs. The basic principles are the same—only the arithmetic is messier.

As an illustration, we will consider a case of a 2×4 factorial with four subjects per cell. Such a design is analyzed by the conventional analysis of variance in Table 16.2, which also

Table 16.2   Sample data and summary table for 2×4 factorial design

(a) Data

	B_1	B_2	B_3	B_4	Means
A_1	5	2	8	11	
	7	5	11	15	
	9	7	12	16	
	8	3	14	10	
	7.25	4.25	11.25	13.00	8.92750
A_2	7	3	9	11	
	9	8	12	14	
	10	9	14	10	
	9	11	8	12	
	8.75	7.75	10.75	11.75	9.75000
Means	8.000	6.000	11.000	12.375	9.34375

(b) Summary Table

Source	df	SS	MS	F	η^2
A	1	5.282	5.282	<1	.014
B	3	199.344	66.448	11.452*	.537
AB	3	27.344	9.115	1.571	.074
Error	24	139.250	5.802		
Total	31	371.220			

* $p < .05$

(c) Estimated Treatment Effects

$\hat{\mu} = 9.34375$

$\hat{\alpha}_1 = \overline{A}_1 - \overline{X}_{..} = 8.9375 - 9.34375 = -0.40625$

$\hat{\beta}_1 = \overline{B}_1 - \overline{X}_{..} = 8.0000 - 9.34375 = -1.34375$

$\hat{\beta}_2 = \overline{B}_2 - \overline{X}_{..} = 6.0000 - 9.34375 = -3.34375$

$\hat{\beta}_3 = \overline{B}_3 - \overline{X}_{..} = 11.0000 - 9.34375 = 1.65625$

$\widehat{\alpha\beta} = \overline{AB}_{11} - \overline{A}_1 - \overline{B}_1 + \overline{X}_{..} = 7.2500 - 8.9375 - 8.0000 + 9.34375 = -.34375$

$\widehat{\alpha\beta}_{12} = \overline{AB}_{12} - \overline{A}_1 - \overline{B}_2 + \overline{X}_{..} = 4.2500 - 8.9375 - 6.0000 + 9.34375 = -1.34375$

$\widehat{\alpha\beta}_{13} = \overline{AB}_{13} - \overline{A}_1 - \overline{B}_3 + \overline{X}_{..} = 11.2500 - 8.9375 - 11.0000 + 9.34375 = 0.65625$

includes means, estimated effects, and values of η^2. From the summary table, it is apparent that the main effect of B is significant but that the effects of A and AB are not.

To analyze these data from the point of view of multiple regression, we begin with the following design matrix. Once again, the elements of each row apply to all subjects in the corresponding treatment combination.

$$
\mathbf{X} = \begin{array}{c}
 \\ a_1b_1 \\ a_1b_2 \\ a_1b_3 \\ a_1b_4 \\ a_2b_1 \\ a_2b_2 \\ a_2b_3 \\ a_2b_4
\end{array}
\begin{array}{ccccccc}
A_1 & B_1 & B_2 & B_3 & AB_{11} & AB_{12} & AB_{13} \\
\left[\begin{array}{ccccccc}
1 & 1 & 0 & 0 & 1 & 0 & 0 \\
1 & 0 & 1 & 0 & 0 & 1 & 0 \\
1 & 0 & 0 & 1 & 0 & 0 & 1 \\
1 & -1 & -1 & -1 & -1 & -1 & -1 \\
-1 & 1 & 0 & 0 & -1 & 0 & 0 \\
-1 & 0 & 1 & 0 & 0 & -1 & 0 \\
-1 & 0 & 0 & 1 & 0 & 0 & -1 \\
-1 & -1 & -1 & -1 & 1 & 1 & 1
\end{array}\right]
\end{array}
$$

The first step in a multiple-regression analysis is presented in Exhibit 16.2 using all seven predictors (A_1 to AB_{13}). The results were obtained using SAS PROC CORR and PROC REG, although every software package should give the same answers.

Exhibit 16.2 has several important features. First, consider the matrix of correlations among variables, often symbolized as **R**. Suppose that we simplify this matrix by defining the following sets of predictors: $A' = [A_1]$, $B' = [B_1, B_2, B_3]$, and $AB' = [AB_{11}, AB_{12}, AB_{13}]$. If we then rewrite the intercorrelation matrix, we have

$$
\begin{array}{c}
 \\ \mathbf{A'} \\ \mathbf{B'} \\ \mathbf{AB'}
\end{array}
\begin{array}{ccc}
A' & B' & AB' \\
\left[\begin{array}{ccc}
1.00 & 0.00 & 0.00 \\
0.00 & 1.00 & 0.00 \\
0.00 & 0.00 & 1.00
\end{array}\right]
\end{array}
$$

Notice that each of the effects is independent of the others. Such a pattern occurs only if there are equal (or proportional) numbers of subjects in each cell; this pattern is also what makes simplified formulae for the analysis of variance possible. The fact that this structure disappears in the case of unequal ns is what makes our life more difficult when we have missing subjects.

Next notice the vector **b**, labeled as the Parameter Estimate. The first entry (b_0) is labeled Intercep and is the grand mean of all the observations. The subsequent entries ($b_1 \ldots b_7$) are the estimates of the corresponding treatment effects. Thus, $b_1 = \alpha_1$, $b_2 = \beta_1$, $b_5 = \alpha\beta_{11}$, and so on. Tests on these regression coefficients represent tests on the corresponding treatment effects. The fact that we have only the $(a-1)(b-1) = 3$ interaction effects presents no problem because of the restrictions that these effects must sum to 0 across rows and down columns. Thus, if $\alpha\beta_{12} = -1.34$, then $\alpha\beta_{22}$ must be $+1.34$. Similarly, $\alpha\beta_{14} = 0 - \sum \alpha\beta_{1j} = -\sum \alpha\beta_{1j} = 1.03$.

The value of $R^2 = .625$ represents the percentage of variation that can be accounted for by all the variables simultaneously. With equal ns, and therefore independent effects, it is equivalent to $\eta_A^2 + \eta_B^2 + \eta_{AB}^2 = .014 + .537 + .074 = .625$. The test on R^2 produces an F of 5.711 on 7 and 24 df which, because it is significant ($p = .0006$), shows that there is a nonchance relationship between the treatment variables, considered together, and the dependent variable (Y).

Two more parallels can be drawn between Table 16.2, the analysis of variance, and Exhibit 16.2, the regression analysis. First, notice that $SS_{\text{regression}} = SS_{\text{Model}} = SS_Y(1 - R^2) = 231.969$. This is the variation that can be predicted by a linear combination of the predictors. This value is equal to $SS_A + SS_B + SS_{AB}$, although from Exhibit 16.2 we cannot yet partition the variation among the separate sources. Finally, notice that $SS_{\text{residual}} = SS_{\text{error}} = SS_Y(1 - R^2) = 139.250$, which is the error sum of squares in the analysis of variance. This makes sense when you recall that error is the variation that cannot be attributed to the separate or joint effects of the treatment variables.

Data Anova;
    infile 'Ex162.dat';
    input A1 B1 B2 B3 dv;
    AB11 = A1 * B1;
    AB12 = A1 * B2;
    AB13 = A1 * B3;
Run;

Proc Corr Data = Anova;
    Var A1 B1 B2 B3 AB11 AB12 AB13;
Run;
Proc Reg Data = Anova;
    Model dv = A1 B1 B2 B3 AB11 AB12 AB13;
Run;

Pearson Correlation Coefficients, N = 32
Prob > |r| under H0: Rho = 0

	A1	B1	B2	B3	AB11	AB12	AB13
A1	1.00000	0.00000	0.00000	0.00000	0.00000	0.00000	0.00000
		1.0000	1.0000	1.0000	1.0000	1.0000	1.0000
B1	0.00000	1.00000	0.50000	0.50000	0.00000	0.00000	0.00000
	1.0000		0.0036	0.0036	1.0000	1.0000	1.0000
B2	0.00000	0.50000	1.00000	0.50000	0.00000	0.00000	0.00000
	1.0000	0.0036		0.0036	1.0000	1.0000	1.0000
B3	0.00000	0.50000	0.50000	1.00000	0.00000	0.00000	0.00000
	1.0000	0.0036	0.0036		1.0000	1.0000	1.0000
AB11	0.00000	0.00000	0.00000	0.00000	1.00000	0.50000	0.50000
	1.0000	1.0000	1.0000	1.0000		0.0036	0.0036
AB12	0.00000	0.00000	0.00000	0.00000	0.50000	1.00000	0.50000
	1.0000	1.0000	1.0000	1.0000	0.0036		0.0036
AB13	0.00000	0.00000	0.00000	0.00000	0.50000	0.50000	1.00000
	1.0000	1.0000	1.0000	1.0000	0.0036	0.0036	

The REG Procedure
Dependent Variable: dv

Analysis of Variance

Source	DF	Sum of Squares	Mean Square	F Value	Pr > F
Model	7	231.96875	33.13839	5.71	0.0006
Error	24	139.25000	5.80208		
Corrected Total	31	371.21875			

Root MSE	2.40875	R-Square	0.6249
Dependent Mean	9.34375	Adj R-Sq	0.5155
Coeff Var	25.77928		

(continued)

Exhibit 16.2 Regression solutions using all predictors for data in Table 16.2

Parameter Estimates

Variable	DF	Parameter Estimate	Standard Error	t Value	Pr > |t|
Intercep	1	9.34375	0.42581	21.94	<.0001
A1	1	−0.40625	0.42581	−0.95	0.3496
B1	1	−1.34375	0.73753	−1.82	0.0809
B2	1	−3.34375	0.73753	−4.53	0.0001
B3	1	1.65625	0.73753	2.25	0.0342
AB11	1	−0.34375	0.73753	−0.47	0.6454
AB12	1	−1.34375	0.73753	−1.82	0.0809
AB13	1	0.65625	0.73753	0.89	0.3824

Exhibit 16.2 *(continued)*

Reduced Models

At this point, we know only the amount of variation that can be accounted for by all the predictors simultaneously. What we want to know is how this variation can be partitioned among A, B, and AB. This information can be readily obtained by computing several reduced regression equations.

In the subsequent course of the analysis, we must compute several multiple regression sums of squares relating to the different effects, so we will change our notation and use the effect labels (α, β, and $\alpha\beta$) as subscripts. For the multiple regression just computed, the model contained variables to account for α, β, and $\alpha\beta$. Thus, we will designate the sum of squares regression in that solution as $SS_{\text{regression}_{\alpha,\beta,\alpha\beta}}$. If we dropped the last three predictors (AB_{11}, AB_{12}, and AB_{13}) we would be deleting those predictors carrying information concerning the interaction but would retain those predictors concerned with α and β. Thus, we would use the designation $SS_{\text{regression}_{\alpha,\beta}}$. If we used only A, AB_{11}, AB_{12}, and AB_{13} as predictors, the model would account for only α and $\alpha\beta$, and the result would be denoted $SS_{\text{regression}_{\alpha,\alpha\beta}}$.

I have run the individual regression solutions for our example, and the results are

$$SS_{\text{regression}_{\alpha,\beta,\alpha\beta}} = 231.969$$

$$SS_{\text{regression}_{\alpha,\beta}} = 204.625$$

$$SS_{\text{regression}_{\beta,\alpha\beta}} = 226.687$$

$$SS_{\text{regression}_{\alpha,\alpha\beta}} = 32.625$$

Now this is the important part. If the interaction term accounts for any of the variation in Y, then removing the interaction predictors from the model should lead to a decrease in accountable variation. This decrease will be equal to the variation that can be attributable to the interaction. By this and similar reasoning,

$$SS_{AB} = SS_{\text{regression}_{\alpha,\beta,\alpha\beta}} - SS_{\text{regression}_{\alpha,\beta}}$$

$$SS_{A} = SS_{\text{regression}_{\alpha,\beta,\alpha\beta}} - SS_{\text{regression}_{\beta,\alpha\beta}}$$

$$SS_{B} = SS_{\text{regression}_{\alpha,\beta,\alpha\beta}} - SS_{\text{regression}_{\alpha,\alpha\beta}}$$

The relevant calculations are presented in Table 16.3. (I leave it to you to verify that these are the sums of squares for regression that result when we use the relevant predictors).

Looking first at the AB interactions, we see from Exhibit 16.2 that when the interaction terms were deleted from the model, the sum of squares that could be accounted for by the model decreased by

$$SS_{AB} = SS_{\text{regression}_{\alpha,\beta,\alpha\beta}} - SS_{\text{regression}_{\alpha,\beta}} = 231.969 - 204.625 = 27.344$$

Table 16.3 Regression solution for the data in Table 16.2

$$SS_{\text{regression}_{\alpha,\beta,\alpha\beta}} = 231.969 \qquad R^2 = .625$$

$$SS_{\text{residual}_{\alpha,\beta,\alpha\beta}} = 139.250$$

$$SS_{\text{regression}_{\alpha,\beta}} = 204.625 \qquad R^2 = .551$$

$$SS_{\text{regression}_{\beta,\alpha\beta}} = 226.687 \qquad R^2 = .611$$

$$SS_{\text{regression}_{\alpha,\alpha\beta}} = 32.625 \qquad R^2 = .088$$

$$SS_{AB} = SS_{\text{regression}_{\alpha,\beta,\alpha b}} - SS_{\text{regression}_{\alpha,\beta}} = 231.969 - 204.625 = 27.344$$

$$SS_A = SS_{\text{regression}_{\alpha,\beta,\alpha b}} - SS_{\text{regression}_{\alpha,\alpha\beta}} = 231.969 - 226.687 = 5.282$$

$$SS_B = SS_{\text{regression}_{\alpha,\beta,\alpha b}} - SS_{\text{regression}_{\alpha,\alpha\beta}} = 231.969 - 32.625 = 199.344$$

$$SS_{\text{error}} = SS_{\text{residual}_{\alpha,\beta,\alpha b}} = 139.250$$

Summary Table

Source	df	SS	MS	F
A	1	5.282	5.282	<1
B	3	199.344	66.448	11.452*
AB	3	27.344	9.115	1.571
Error	24	139.250	5.802	
	31	371.220		

* $p < .05$

This decrement can only be attributable to the predictive value of the interaction terms, and therefore

$$SS_{AB} = 27.344$$

By a similar line of reasoning, we can find the other sums of squares.

Notice that these values agree exactly with those obtained by the more traditional procedures. Notice also that the corresponding decrements in R^2 agree with the computed values of η^2.

As Overall and Spiegel (1969) pointed out, the approach we have taken in testing the effects of A, B, and AB is not the only one we could have chosen. They presented two alternative models that might have been considered in place of this one. Fortunately, however, the different models all lead to the same conclusions in the case of equal sample sizes because in this situation effects are independent of one another and therefore are additive. When we consider the case of unequal sample sizes, however, the choice of an underlying model will require careful consideration.

16.4 Analysis of Variance with Unequal Sample Sizes

The least-squares approach to the analysis of variance is particularly useful for factorial experiments with unequal sample sizes. However, special care must be used in selecting the particular restricted models that are employed in generating the various sums of squares.

Several different models could underlie an analysis of variance. Although these models all lead to the same results in the case of equal sample sizes, in the unequal n case they do not. This is because with unequal ns, the row, column, and interaction effects are no longer orthogonal and, thus, account for overlapping portions of the variance. (I would strongly recommend quickly reviewing the example given in Chapter 13, Section 13.10 [pp. 419–424].) Consider the Venn diagram in Figure 16.1. The area enclosed by the surrounding square will be taken to represent SS_{total}. Each circle represents the variation attributable to (or accounted for by) one of the effects. The area outside the circles but within the square represents SS_{error}. Finally, the total area enclosed by the circles represents $SS_{regression\,\alpha,\beta,\alpha\beta}$, which is the sum of squares for regression when all the terms are included in the model. If we had equal sample sizes, none of the circles would overlap, and each effect would be accounting for a separate, independent, portion of the variation. In that case, the decrease in $SS_{regression}$ resulting from deleting of an effect from the model would have a clear interpretation—it would be the area enclosed by the omitted circle and, thus, would be the sum of squares for the corresponding effect.

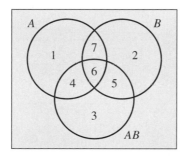

Figure 16.1   Venn diagram representing portions of overall variation

But what do we do when the circles overlap? If we were to take a model that included terms for A, B, and AB and compared it with a model containing only A and B terms, the decrement would not represent the area of the AB circle because some of that area still would be accounted for by A or B. Thus, SS_{AB}, which we calculate as $SS_{regression\,\alpha,\beta,\alpha\beta} - SS_{regression\,\alpha,\beta}$, represents only the portion of the enclosed area that is *unique* to AB—the area labeled with a "3." So far, all the models that have been seriously proposed are in agreement. SS_{AB} is that portion of the AB circle remaining after adjusting for A and B.

But now things begin to get a little sticky. Three major approaches have been put forth that differ in the way the remainder of the pie is allotted to A and B. Overall and Spiegel (1969), put forth three models for the analysis of variance, and these models continue to generate a voluminous literature debating their proper use and interpretation, even though the discussion began 35 years ago. We will refer to these models as Type I, Type II, and Type III, from the terminology used by SPSS and SAS. (Overall and Spiegel numbered them in the reverse order, just to make things more confusing.) Basically, the choice between the three models hinges on how we see the relationship between the sample size and the treatments themselves, or, more specifically, how we want to weight the various cell means to produce row and column means. Before exploring that issue, however, we must first examine the competing methods.

Method III   **Method III** (or Type III Sum of squares) is the method we used in the preceding section. In this case, each effect is adjusted for all other effects. Thus, we obtain SS_{AB} as $SS_{regression\,\alpha,\beta,\alpha\beta} - SS_{regression\,\alpha,\beta}$, SS_A as $SS_{regression\,\alpha,\beta,\alpha\beta} - SS_{regression\,\beta,\alpha\beta}$, and SS_B as $SS_{regression\,\alpha,\beta,\alpha\beta} - SS_{regression\,\alpha,\alpha\beta}$. In terms of Figure 16.1, each effect is defined as the part of the area that is unique to that effect. Thus, SS_A is represented by area "1," SS_B by area "2," and SS_{AB} by area "3."

Method II (or Type II SS) breaks up the pie differently. We continue to define SS_{AB} as area "3." But now that we have taken care of the interaction, we still have areas "1," "2," "4," "5," "6," and "7," which can be accounted for by the effects of A or B. Method II essentially redefines the full model as $SS_{\text{regression}_{\alpha,\beta}}$ and obtains $SS_A = SS_{\text{regression}_{\alpha,\beta}} - SS_{\text{regression}_\beta}$, and SS_B as $SS_{\text{regression}_{\alpha,\beta}} - SS_{\text{regression}_\alpha}$. Thus, A is allotted areas "1" and "4," whereas B is allotted areas "2" and "5." Methods II and III are summarized in Table 16.4.

Table 16.4 Alternative models for solution of nonorthogonal designs

Method III

$$Y_{ijk} = \mu + \alpha_i + \beta_j + \alpha\beta_{ij} + e_{ijk}$$

Source	df	SS	Portion of Diagram
A	$a-1$	$SS_{\text{regression}_{\alpha,\beta,\alpha\beta}} - SS_{\text{regression}_{\beta,\alpha\beta}}$	1
B	$b-1$	$SS_{\text{regression}_{\alpha,\beta,\alpha\beta}} - SS_{\text{regression}_{\alpha,\alpha\beta}}$	2
AB	$(a-1)(b-1)$	$SS_{\text{regression}_{\alpha,\beta,\alpha\beta}} - SS_{\text{regression}_{\alpha,\beta}}$	3
Error	$N-ab$	$SS_{\text{regression}_{\alpha,\beta,\alpha\beta}}$	
Total	$N-1$	SS_Y	

Method II

$$Y_{ijk} = \mu + \alpha_i + \beta_j + \alpha\beta_{ij} + e_{ijk}$$

and

$$Y_{ijk} = \mu + \alpha_i + \beta_j + e_{ijk}$$

Source	df	SS	Portion of Diagram
A	$a-1$	$SS_{\text{regression}_{\alpha,\beta}} - SS_{\text{regression}_\beta}$	$1+4$
B	$b-1$	$SS_{\text{regression}_{\alpha,\beta}} - SS_{\text{regression}_\alpha}$	$2+5$
AB	$(a-1)(b-1)$	$SS_{\text{regression}_{\alpha,\beta,\alpha\beta}} - SS_{\text{regression}_{\alpha,\beta}}$	3
Error	$N-ab$	$SS_{\text{regression}_{\alpha,\beta,\alpha\beta}}$	
Total	$N-1$	SS_Y	

Both of these methods make a certain amount of sense when looked at from the point of view of the Venn diagram in Figure 16.1. However, the diagram is only a crude approximation and we have pushed it about as far as we can go.[1]

As Carlson and Timm (1974) argued, a more appropriate way to compare the models is to examine the hypotheses they test. These authors point out that Method III represents an estimation of treatment effects when cell means are weighted equally and is particularly appropriate whenever we consider sample size to be independent of treatment conditions. A convincing demonstration of this is presented in Overall, Spiegel, and Cohen (1975). Carlson and Timm also showed that Method II produces estimates of treatment effects when row and

---

[1] From this discussion, you could easily get the impression that Method II will always account for more of the variation than Method III. This is not necessarily the case because the degree of overlap represents the correlation between effects, and suppressor relationships might appear as "black holes," canceling out accountable variation.

column means are weighted by the sample size, but only when no interaction is present. When an interaction is present, simple estimates of row and column effects cannot be made, and the null hypotheses actually tested are very bizarre indeed (see Carlson and Timm [1974] for a statement of the null hypotheses for Method II.) SPSS, which used to rely on a method similar to Method II, finally saw the light some years ago and came around to using Method III as the default. They labeled this method "Unique SS" because each effect is assigned only that portion of the variation that it uniquely explains. An excellent discussion of the hypotheses tested by different approaches is presented in Blair and Higgins (1978) and Blair (1978). As Cochran and Cox suggested, "the only complete solution of the 'missing data' problem is not to have them" (1957, p. 82).

A third method of computing sums of squares seems particularly bizarre at first. Just to make matters even more confusing than they need to be, this is the method that SPSS and SAS refer to as "Type I SS," but which I will refer to as **hierarchical sums of squares,** though it is sometimes referred to as **sequential sums of squares,** which is the term that SPSS uses. The peculiar thing about this approach is that it is dependent on the order in which you name your variables. Thus, if you tell SAS or SPSS to model (predict or account for) the dependent variable on the basis of A, B, and AB, the program will first assign $SS_A = SS_{\text{regression}_\alpha}$. Then $SS_B = SS_{\text{regression}_{\alpha,\beta}} - SS_{\text{regression}_\alpha}$, and finally $SS_{AB} = SS_{\text{regression}_{\alpha,\beta,\alpha\beta}} - SS_{\text{regression}_{\alpha,\beta}}$. In this situation, the first effect is assigned all the sums of squares it can possibly account for. The next effect is assigned all that it can account for *over and above* what was accounted for by the first one. Finally, the interaction effect is assigned only what it accounts for over and above the two main effects. But, if you ask the software to model the dependent variable on the basis of B, A, and AB, then SS_B will equal $SS_{\text{regression}_\beta}$, which is quite a different thing from $SS_{\text{regression}\,\alpha,\beta} - SS_{\text{regression}\,\alpha}$. The only time I could recommend using this approach is if you have a strong reason to want to control the variables in a particular order.[2] If you can defend the argument that Variable A is so important that it should be looked at first without controlling for any other variables, then perhaps this is a method you can use. But I have never seen a case where I would want to do that, with the possible exception of dealing with a variable as a covariate, which we will discuss shortly. The only reason that I bring the issue up here is to explain some of the choices you will have to make in using computer software.

Howell and McConaughy (1982) argued that there are very few instances in which one would want to test the peculiar null hypotheses tested by Method II. The debate over the "correct" model will probably continue for some time, mainly because no one model is universally "correct" and because there are philosophical differences in the approaches to model specification (see Howell & McConaughy [1982] and Lewis & Keren [1977] versus Appelbaum & Cramer [1974], O'Brien [1976], and Macnaughton [1998].) However, the conclusion to be drawn from the literature at present is that for the most common situations Method III is appropriate because we usually want to test unweighted means. (This is the default method employed by SPSS. Method III sum of squares are the values labeled as Type III SS in SAS, and now by more recent versions of SPSS.) It is also the method that is approximated by the *unweighted means solution* discussed in Chapter 13. (You may recall that in Chapter 13, I mentioned that the traditional label "unweighted means solution" really should be the "equally weighted means solution," if that name hadn't been appropriated in the past for a different procedure, because, using it, we are treating all means equally, regardless of the sample sizes.)

As an illustration of this method, we will take the data used in the previous example but add four scores to produce unequal cell sizes. The data are given in Table 16.5, with the unweighted and weighted row and column means and the values resulting from the various

---

[2] There is a good and honorable tradition of prioritizing variables in this way for theoretical studies using standard multiple regression with continuous variables. I have never seen a similar application in an analysis of variance framework, though I have seen a number of people talk about hypothetical examples.

Table 16.5 Illustrative calculations for nonorthogonal factorial design

	B_1	B_2	B_3	B_4	Unweighted Mean	Weighted Mean
A_1	5	2	8	11		
	7	5	11	15		
	9	7	12	16	8.975	8.944
	8	3	14	10		
		9		9		
A_2	7	3	9	11		
	9	8	12	14		
	10	9	14	10	9.625	9.778
	9	11	8	12		
			7	13		
Unweighted Means	8.000	6.475	10.625	12.1		
Weighted Mean	8.000	6.333	10.556	12.1		

Full Model

$$R^2_{\alpha,\beta,\alpha\beta} = 0.532$$
$$SS_{\text{regression}_{\alpha,\beta,\alpha\beta}} = 207.7055$$
$$SS_{\text{residual}} = 182.6001$$

Reduced Models

$$R^2_{\alpha,\beta} = 0.483$$
$$SS_{\text{regression}_{\alpha,\beta}} = 188.4301$$
$$R^2_{\beta,\alpha\beta} = 0.523$$
$$SS_{\text{regression}_{\beta,\alpha\beta}} = 203.9500$$
$$SS_{\text{regression}_{\alpha,\alpha\beta}} = 29.7499$$

regression solutions. The unweighted means are the mean of means (therefore, the mean of row_1 is the mean of the four cell means in that row). The weighted mean of row_1, for example, is just the sum of the scores in row_1 divided by the number of scores in row_1.

From Table 16.5, we see that $R^2_{\alpha,\beta,\alpha\beta} = .532$, indicating that approximately 53% of the variation can be accounted for by a linear combination of the predictor variables. We do not know, however, how this variation is to be distributed among A, B, and AB. For that, we need to form and calculate the reduced models.

Testing the Interaction Effects

First, we delete the predictors associated with the interaction term and calculate $R^2_{\alpha,\beta}$. For these data, $R^2_{\alpha,\beta} = .483$, representing a drop in R^2 of about .05. If we examine the predictable sum of squares ($SS_{\text{regression}}$), we see that eliminating the interaction terms has produced a decrement in $SS_{\text{regression}}$ of

$$SS_{\text{regression}_{\alpha,\beta,\alpha\beta}} = 207.7055$$
$$SS_{\text{regression}_{\alpha,\beta}} = \underline{188.4301}$$
$$SS_{AB} = 19.2754$$

This decrement is the sum of squares attributable to the AB interaction (SS_{AB}).

In the case of unequal ns, it is particularly important to understand what this term represents. You should recall that $SS_{\text{regression}_{\alpha,\beta,\alpha\beta}}$, for example, equals $SS_Y(R^2_{\alpha,\beta,\alpha\beta})$. Then

$$SS_{AB} = SS_Y\left(R^2_{\alpha,\beta,\alpha\beta}\right) - SS_Y\left(R^2_{\alpha,\beta}\right)$$
$$= SS_Y\left(R^2_{\alpha,\beta,\alpha\beta} - R^2_{\alpha,\beta}\right)$$
$$= SS_Y\left(R^2_{0(\alpha\beta.\alpha,\beta)}\right)$$

The final term in parentheses is the squared semipartial correlation between the criterion and the interaction effects, partialling out (adjusting for) the effects of A and B. In other words, the squared correlation between the criterion and the part of the AB interaction is orthogonal to A and B. Thus, we can think of SS_{AB} as really being $SS_{AB(\text{adj})}$, where the adjustment is for the effects of A and B. (In the equal-n case, the issue does not arise because A, B, and AB are independent, and therefore there is no overlapping variation to partial out.)[3]

Testing the Main Effects

Because we are calculating Method III SS, we will calculate the main effects of A and B in a way that is directly comparable to our estimation of the interaction effect. Here, each main effect represents the sum of squares attributable to that variable after partialling out the other main effect and the interaction.

To obtain SS_A, we will delete the predictor associated with the main effect of A and calculate $SS_{\text{regression}_{\beta,\alpha\beta}}$. For these data, $R^2_{\beta,\alpha\beta} = .523$, producing a drop in R^2 of $.532 - .523 = .009$. In terms of the predictable sum of squares ($SS_{\text{regression}}$), the elimination of α from the model produces a decrement in $SS_{\text{regression}}$ of

$$SS_{\text{regression}_{\alpha,\beta,\alpha\beta}} = 207.7055$$
$$SS_{\text{regression}_{\alpha,\alpha\beta}} = \underline{203.9500}$$
$$SS_A = 3.7555$$

This decrement is the sum of squares attributable to the main effect of A.

By the same reasoning, we can obtain SS_B by comparing $SS_{\text{regression}}$ for the full model and for a model omitting β.

$$SS_{\text{regression}_{\alpha,\beta,\alpha\beta}} = 207.7055$$
$$SS_{\text{regression}_{\alpha,\alpha\beta}} = \underline{29.7499}$$
$$SS_B = 177.9556$$

These results are summarized in Table 16.6, with the method by which they were obtained. Notice that the sums of squares do not sum to SS_{total}. This is as it should be because the overlapping portions of accountable variation (segments "4," "5," "6," and "7" of Figure 16.1) are not represented anywhere. Also notice that SS_{error} is taken as the SS_{residual} from the

---

[3] Some people have trouble understanding the concept of nonindependent treatment effects. As an aid, perhaps an extreme example will help point out how a row effect could cause an *apparent* column effect, or vice versa. Consider the following two-way table. When we look at differences among means, are we looking at a difference due to A, B, or AB? There is no way to tell.

	B_1	B_2	Means
A_1	$\overline{X} = 10$ $n = 20$	$n = 0$	10
A_2	$n = 0$	$\overline{X} = 30$ $n = 20$	30
Means	10	30	

Table 16.6 Calculation of sums of squares using Method III—the unweighted means solution

Method III (Unweighted Means)

Source	df	SS
A	$a - 1$	$SS_Y\left(R^2_{\alpha,\beta,\alpha\beta} - R^2_{\beta,\alpha\beta}\right)$
B	$b - 1$	$SS_Y\left(R^2_{\alpha,\beta,\alpha\beta} - R^2_{\alpha,\alpha\beta}\right)$
AB	$(a - 1)(b - 1)$	$SS_Y\left(R^2_{\alpha,\beta,\alpha\beta} - R^2_{\alpha,\beta}\right)$
Error	$N - ab$	$SS_Y\left(1 - R^2_{\alpha,\beta,\alpha\beta}\right)$
Total	$N - 1$	SS_Y

Summary Table for Analysis of Variance

Source	df	SS	MS	F
A	1	3.7555	3.7555	<1
B	3	177.9556	59.3185	9.10
AB	3	19.2754	6.4251	<1
Error	28	182.6001	6.5214	
Total	35	(390.3056)		

full model, just as in the case of equal sample sizes. Here again, we define SS_{error} as the portion of the total variation that cannot be explained by any one or more of the independent variables.

As I mentioned earlier, the unweighted-means solution presented in Chapter 13 is an approximation of the solution (Method III) given here. If you were to apply this approach to the data in Table 13.11, you would find that the two solutions differ only in the second or third decimal place. The two solutions are not usually that close, although in my experience they seldom differ by very much. The main reason for discussing that solution in this chapter is so that you will understand what the computer program is giving you and how it is treating the unequal sample sizes.

The very simple SAS program and its abbreviated output in Exhibit 16.3 illustrate that the Type III sums of squares from SAS PROC GLM do, in fact, produce the appropriate analysis of the data in Table 16.5.

```
Data Nonorth;
    Infile 'Table16-7.dat';
    Input A B dv;
Run;

Proc GLM Data = Nonorth;
    Class A B;
    Model dv = A B A*B;
```

Analysis of Variance

Source	DF	Type III SS	Mean Square	F Value	Pr > F
A	1	3.7555556	3.7555556	0.58	0.4543
B	3	177.9556246	59.3185415	9.10	0.0002
A*B	3	19.2755003	6.4251668	0.99	0.4139
(Error	28	182.6000000	6.5214286	)	

Exhibit 16.3 Abbreviated SAS analysis of the data in Table 16.7

16.5   The One-Way Analysis of Covariance

analysis of covariance

An extremely useful tool for analyzing experimental data is the **analysis of covariance.** As presented within the context of the analysis of variance, the analysis of covariance appears to be unpleasantly cumbersome, especially so when there is more than one covariate. Within the framework of multiple regression, however, it is remarkably simple, requiring little, if any, more work than does the analysis of variance.

Suppose we want to compare driving proficiency on three different sizes of cars to test the experimental hypothesis that small cars are easier to handle. We have available three different groups of drivers, but we are not able to match individual subjects on driving experience, which varies considerably within each group. Let us make the simplifying assumption, which will be discussed in more detail later, that the mean level of driving experience is equal across groups. Suppose further that using the number of steering errors as our dependent variable, we obtain the somewhat exaggerated data plotted in Figure 16.2. In this figure, the data have been plotted separately for each group (size of car), as a function of driving experience (the **covariate**), and the separate regression lines have been superimposed.

covariate

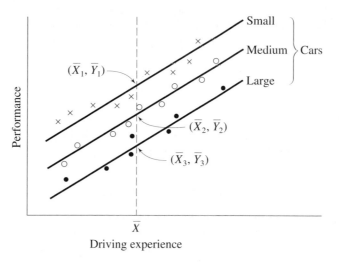

Figure 16.2   Hypothetical data illustrating error-reduction in the analysis of covariance

One of the most striking things about Figure 16.2 is the large variability in both performance and experience within each treatment. This variability is so great that an analysis of variance on performance scores would almost certainly fail to produce a significant effect. Most of the variability in performance, however, is directly attributable to differences in driving experience, which has nothing to do with what we want to study. If we could somehow remove (partial out) the variance that can be attributed to experience (the covariate), we would have a clearer test of our original hypothesis. This is exactly what the analysis of covariance is designed to do, and this is precisely the situation in which it does its job best—its job in this case being to reduce the error term.

A more controversial use of the analysis of covariance concerns situations in which the treatment groups have different covariate (driving experience) means. Such a situation (using the same hypothetical experiment) is depicted in Figure 16.3, in which two of the treatments have been displaced along the X axis. At the point at which the three regression lines intersect the vertical line $X = \overline{X}$, you can see the values $\overline{Y}_1'$, $\overline{Y}_2'$, and $\overline{Y}_3'$. These are the **adjusted Y means** and represent our best guess about what the Y means would have been *if the treatments had not differed on the covariate.* The analysis of covariance then tests

adjusted Y means

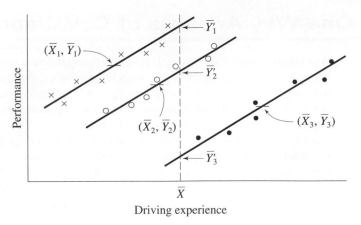

Figure 16.3 Hypothetical data illustrating mean adjustment in the analysis of covariance

whether these *adjusted* means differ significantly, again using an error term from which the variance attributable to the covariate has been partialled out. Notice that the adjusted performance means are quite different from the unadjusted means. The adjustment has increased $\overline{Y}_1$ and decreased $\overline{Y}_3$.

Although the structure and procedures of the analysis of covariance are the same regardless of whether the treatment groups differ on the covariate means, the different ways of visualizing the problem as represented in Figures 16.2 and 16.3 are instructive. In the first case, we are simply reducing the error term. In the second case, we are both reducing the error term *and* adjusting the means on the dependent variable. We will have more to say about this distinction later in the chapter.

Assumptions of the Analysis of Covariance

homogeneity of regression

Aside from the usual analysis of variance assumptions of normality and homogeneity of variance, we must add two more assumptions. First, we will assume that whatever the relationship between Y and the covariate (C), this relationship is linear.[4] Second, we will assume **homogeneity of regression**—that the regression coefficients are equal across treatments—$b_1^* = b_2^* = b_3^* = \cdots = b^*$. This is merely the assumption that the three lines in Figure 16.2 or 16.3 are parallel, and it is necessary to justify our substitution of one regression line (the pooled within-groups regression line) for the separate regression lines. As we shall see shortly, this assumption is testable. Note that no assumption has been made about the nature of the covariate; it may be either a fixed or a random variable. (It can even be a categorical variable if we create dummy variables to represent the different levels of the variable, as we did in the earlier parts of this chapter.)

Calculating the Analysis of Covariance

When viewed within the framework of multiple regression, the analysis of covariance is basically no different from the analysis of variance, except that we want to partial out the effects of the covariate. As Cohen (1968) put it, "A covariate is, after all, nothing but an independent variable which, because of the logic dictated by the substantive issues of the

---

[4] Methods for handling nonlinear relationships are available but will not be discussed here.

research, assumes priority among the set of independent variables as a basis for accounting for Y variance." (p. 439).

If we want to ask about the variation in Y after the covariate (C) has been partialled out, and if the variation in Y can be associated with only C, the treatment effect (α), and error, then $SS_{\text{regression}_{C,\alpha}}$ represents the total amount of accountable variation. If we now compare $SS_{\text{regression}_{C,\alpha}}$ with $SS_{\text{regression}_C}$, the difference will be the variation attributable to treatment effects *over and above* that attributable to the covariate.

We will take as an example a variation on the study by Conti and Musty (1984) presented in Chapter 11. As you may recall, in that study the authors were interested in examining the effects of different amounts of THC, the major active ingredient in marijuana, injected directly into the brain. The dependent variable was locomotor activity, which normally increases with the administration of THC by more traditional routes. Because of the nature of the experimental setting (all animals were observed under baseline conditions and then again after the administration of THC), activity should decrease in all animals as they become familiar and more comfortable with the apparatus. If THC has its effect through the nucleus accumbens, however, the effects of moderate doses of THC should partially compensate for this anticipated decrease, leading to relatively greater activity levels in the moderate-dose groups as compared with the low- or high-dose groups.

Conti and Musty (1984) actually analyzed postinjection activity as a percentage of preinjection activity because that is the way such data are routinely analyzed in their field. An alternative procedure would have been to run an analysis of covariance on the postinjection scores, partialling out preinjection differences. Such a procedure would adjust for the fact that much of the variability in postinjection activity could be accounted for by variability in preinjection activity. It would also control for the fact that, by chance, there were group differences in the level of preinjection activity that could contaminate postinjection scores.

As will become clear later, it is important to note here that all animals were assigned at random to groups. Therefore, we would *expect* the group means on the preinjection phase to be equal. Any differences that do appear on the preinjection phase, then, are due to chance, and, *in the absence of any treatment effect,* we would expect that postinjection means, adjusted for chance preinjection differences, would be equal. The fact that subjects were assigned at random to treatments is what allows us to expect equal adjusted group means at postinjection (if H_0 is true), and this in turn allows us to interpret group differences at postinjection to be a result of real treatment differences rather than of some artifact of subject assignment.

The data and the design matrix for the Conti and Musty (1984) study are presented in Table 16.7. The raw data have been divided by 100 simply to make the resulting sums of squares manageable.[5] In the design matrix that follows the data, only the first and last subject in each group are represented. Columns 6 through 9 of X represent the interaction of the covariate and the group variables. These columns are used to test the hypothesis of homogeneity of regression coefficients across groups:

$$H_0: b_1^* = b_2^* = b_3^* = b_4^* = b_5^*$$

The full model (including the interaction predictors) states

$$Y_{ij} = \tau_j + c + c\tau_j + e_{ij}$$

where τ_j represents the treatment effect for the jth treatment, c represents the covariate, $c\tau_j$ represents our term testing homogeneity of regression, and e_{ij} represents the error associated with the ith subject in treatment j.

---

[5] If the data had not been divided by 100, the resulting sums of squares and mean squares would be $100^2 = 10,000$ times their present size. The F and t values would be unaffected.

Table 16.7 Pre- and postinjection data from Conti and Musty (1984)

	Control		0.1 μg		0.5 μg		1 μg		2 μg	
	Pre	Post	Pre	Post	Pre	Post	Pre	Post	Pre	Post
	4.34	1.30	1.55	0.93	7.18	5.10	6.94	2.29	4.00	1.44
	3.50	0.94	10.56	4.44	8.33	4.16	6.10	4.75	4.10	1.11
	4.33	2.25	8.39	4.03	4.05	1.54	4.90	3.48	3.62	2.17
	2.76	1.05	3.70	1.92	10.78	6.36	3.69	2.76	3.92	2.00
	4.62	0.92	2.40	0.67	6.09	3.96	4.76	1.67	2.90	0.84
	5.40	1.90	1.83	1.70	7.78	4.51	4.30	1.51	2.90	0.99
	3.95	0.32	2.40	0.77	5.08	3.76	2.32	1.07	1.82	0.44
	1.55	0.64	7.67	3.53	2.86	1.92	7.35	2.35	4.94	0.84
	1.42	0.69	5.79	3.65	6.30	3.84			5.69	2.84
	1.90	0.93	9.58	4.22					5.54	2.93
Mean	3.377	1.094	5.387	2.586	6.494	3.906	5.045	2.485	3.943	1.560

Design Matrix

	Cov	T_1	T_2	T_3	T_4	CT_1	CT_2	CT_3	CT_4		
	4.34	1	0	0	0	4.34	0	0	0		1.30
	...	...	...	...	...	...	...	...	...		...
	1.90	1	0	0	0	1.90	0	0	0		0.93
	1.55	0	1	0	0	0	1.55	0	0		0.93
	...	...	...	...	...	...	...	...	...		...
	9.58	0	1	0	0	0	9.58	0	0		4.22
	7.18	0	0	1	0	0	0	7.18	0		5.10
$\mathbf{X} =$	...	...	...	...	...	...	...	...	...	$\mathbf{Y} =$	...
(47×9)	6.30	0	0	1	0	0	0	6.30	0	(47×1)	3.84
	6.94	0	0	0	1	0	0	0	6.94		2.29
	...	...	...	...	...	...	...	...	...		...
	7.35	0	0	0	1	0	0	0	7.35		2.35
	4.00	−1	−1	−1	−1	−4.00	−4.00	−4.00	−4.00		1.44
	...	...	...	...	...	...	...	...	...		...
	5.54	−1	−1	−1	−1	−5.54	−5.54	−5.54	−5.54		2.93

We can compare two models either on the basis of the change in $SS_{\text{regression}}$ between the two models (using the residual from the more complete model for our error term), or on the basis of the decrease in R^2. In this case, the latter is somewhat simpler.

The regression analysis of this model would produce

$$R^2_{\tau,c,c\tau} = .8238$$

If there is no significant difference in within-treatment regressions—that is, if the regression lines are parallel and, thus, the slopes of the regression lines that could be calculated for each group separately are homogeneous—called homogeneity of regression—the deletion of the interaction term should produce only a trivial decrement in the percentage of accountable variation. When we delete the CT terms, we have

$$R^2_{\tau,c} = .8042$$

The F test on this decrement is the usual F test on the difference between two models:

$$F(f - r, N - f - 1) = \frac{(N - f - 1)\left(R^2_{\tau,c,c\tau} - R^2_{\tau,c}\right)}{(f - r)\left(1 - R^2_{\tau,c,c\tau}\right)}$$

$$= \frac{(47 - 9 - 1)(.8238 - .8042)}{(4)(.1762)} = 1.03$$

Given an F of 1.03 on 4 and 37 degrees of freedom, we have no basis to reject the assumption of homogeneity of regression (common regression coefficients) within the five treatments. Thus, we can proceed with the analysis on the basis of the revised full model that does not include the covariate by treatment interaction:

$$Y_{ij} = \mu + \tau_j + c + e_{ij}$$

This model will serve as the basis against which we compare reduced models.

The three sets of results of the multiple-regression solutions using (1) the covariate and dummy treatment variables, (2) just the treatment variables, and then (3) just the covariates are presented in Table 16.8.

From Table 16.8, you can see that using both the covariate (Pre) and the group membership dummy variates ($T_1 \ldots T_4$), the sum of squares for regression ($SS_{\text{regression}_{\tau,c}}$) is equal to 82.6435, which is the portion of the total variation that can be accounted for by these two sets of predictors. You can also see that the residual sum of squares (SS_{residual}) is 20.1254, which is the variability that cannot be predicted. In our analysis of covariance summary table, this will become the sum of squares for error.

When we remove the dummy group membership variates from the equation and use only the covariate (Pre) as a predictor, $SS_{\text{regression}}$ drops from 82.6435 to 73.4196. The difference between $SS_{\text{regression}}$ with and without the group membership predictors must be the amount of the sum of squares that can be attributable to treatment *over and above* the amount that can be explained by the covariate. For our data, this is

$$SS_{\text{treat(adj)}} = SS_{\text{regression}_{\tau,c}} - SS_{\text{regression}_c}$$

$$= 82.6435 - 73.4196$$

$$= 9.2239$$

This last value is called the *adjusted* treatment sum of squares for the analysis of covariance because it has been adjusted for any effects of the covariate. In this case, it has been adjusted for the fact that the five groups differed on the pretest measure.

We need one additional term to form our analysis of covariance summary table, and that is the sum of squares to be attributed to the covariate. There are a number of different ways to define this term, but the most common is to define it analogously to the way the adjusted treatment effect was defined. We will attribute to the covariate that portion of the variation that cannot be defined by the treatment effect. In other words, we will take the model with both the covariate and treatment predictors and compare it with a model with only the treatment predictors. The difference in the two sums of squares due to regression will be the sum of squares that the covariate accounts for *over and above* what is accounted for by treatment effects. For our data, this is

$$SS_{\text{covariate}} = SS_{\text{regression}_{\tau,c}} - SS_{\text{regression}_\tau}$$

$$= 82.6435 - 44.3028$$

$$= 38.3407$$

Table 16.8   Regression analysis

(a) Full Model

$$\hat{Y}_{ij} = 0.4347(\text{Pre}) - 0.5922(T_1) + 0.0262(T_2) + 0.8644(T_3) + 0.0738(T_4) + 0.2183$$

$$R^2_{\tau,c} = 0.8042$$

Analysis of Variance Summary Table for Regression

Source	df	SS	MS	F
Regression	5	82.6435	16.5287	33.6726
Residual	41	20.1254	0.4909	
Total	46	102.7689		

(b) Reduced Model—Omitting Treatment Predictors

$$\hat{Y}_{ij} = 0.5311(\text{Pre}) - 0.26667$$

$$R^2_c = 0.7144$$

Analysis of Variance Summary Table for Regression

Source	df	SS	MS	F
Regression	1	73.4196	73.4196	112.5711
Residual	45	29.3493	0.6522	
Total	46	102.7689		

(c) Reduced Model—Omitting Covariate (Pre)

$$\hat{Y}_{ij} = -1.2321(T_1) + 0.2599(T_2) + 1.5794(T_3) + 0.1589(T_4) + 2.3261$$

$$R^2_\tau = .4311$$

Analysis of Variance Summary Table for Regression

Source	df	SS	MS	F
Regression	4	44.3028	11.0757	7.9564
Residual	42	58.4661	1.3921	
Total	46	102.7689		

We now have all the information necessary to construct the analysis of covariance summary table. This is presented in Table 16.9. Notice that in this table the error term is SS_{residual} from the full model and the other sums of squares are as calculated before. Notice also that there is one degree of freedom for the covariate because there is one covariate, there are $(k - 1) = (5 - 1) = 4\ df$ for the adjusted treatment effect, and there are $N - k - c = 41\ df$ for error (where k represents the number of groups and c represents the number of covariates).

Table 16.9   Summary tables for analysis of covariance

General Summary Table for One-Way Analysis of Covariance

Source	df	SS
Covariate	c	$SS_{\text{regression}_{(\tau,c)}} - SS_{\text{regression}_{(\tau)}}$
Treat (adj)	$k-1$	$SS_{\text{regression}_{(\tau,c)}} - SS_{\text{regression}_{(c)}}$
Error	$N-k-1$	$SS_{\text{residual}_{(\tau,c)}}$
Total	$N-1$	

Summary Table for Data in Table 16.7

Source	df	SS	MS	F
Covariate	1	38.3407	38.3407	78.108*
Treat (adj)	4	9.2239	2.3060	4.698*
Error	41	20.1254	0.4909	
Total	46	102.7689		

Full Model:

$$\hat{Y}_{ij} = 0.4347(Pre) - 0.5922(T_1) + 0.0262(T_2) + 0.8644(T_3) + 0.0738(T_4) + 0.2183$$

* $p < .05$

From the summary table, we see that $SS_{\text{treat(adj)}} = 9.2239$. On 4 df, this gives us $MS_{\text{treat(adj)}} = 2.3060$. Dividing that term by $MS_{\text{error}} = 0.4909$, we have $F = 4.698$ on (4,41) df, which is significant at $p < .05$. Thus, we can conclude that after we control for individual preinjection differences in activity, the treatment groups do differ on postinjection activity.

Adjusted Means

$F_{.05}(4, 41) = 2.61 < F_{\text{obt}} = 4.698$, so we have rejected H_0: $\mu_1(\text{adj}) = \mu_2(\text{adj}) = \mu_3(\text{adj})$ $\mu_4(\text{adj}) = \mu_5(\text{adj})$ and conclude that there were significant differences among the treatment means after the effect of the covariate has been partialled out of the analysis. To interpret these differences, it would be useful, if not essential, to obtain the treatment means adjusted for the effects of the covariate. We are basically asking for an estimate of what the postinjection treatment means would have been had the groups not differed on the preinjection means. The adjusted means are readily obtained from the regression solution using the covariate and treatments as predictors.

From the analysis of the revised full model, we obtained (see Table 16.8)

$$\hat{Y}_{ij} = 0.4347(\text{Pre}) - 0.5922(T_1) + 0.0262(T_2) + 0.8644(T_3) \\ + 0.0738(T_4) + 0.2183$$

Writing this in terms of means and representing adjusted means as $\overline{Y}'_j$, we have

$$\overline{Y}'_j = 0.4347(\overline{\text{Pre}}) - 0.5922(T_1) + 0.0262(T_2) + 0.8644(T_3) \\ + 0.0738(T_4) + 0.2183$$

where $\overline{\text{Pre}} = 4.8060$ (the mean preinjection score) and T_1, T_2, T_3, and T_4 are $(0, 1, -1)$ variables. (We substitute the mean Pre score for the individual Pre score because we are interested in the adjusted means for Y if all subjects had received the mean score on the covariate.) For our data, the adjusted means of the treatments are as follows:

$$\overline{Y}'_1 = 0.4347(4.8060) - 0.5922(1) + 0.0262(0) + 0.8644(0)$$
$$+ 0.0738(0) + 0.2183$$
$$= 1.7153$$

$$\overline{Y}'_2 = 0.4347(4.8060) - 0.5922(0) + 0.0262(1) + 0.8644(0)$$
$$+ 0.0738(0) + 0.2183$$
$$= 2.3336$$

$$\overline{Y}'_3 = 0.4347(4.8060) - 0.5922(0) + 0.0262(0) + 0.8644(1)$$
$$+ 0.0738(0) + 0.2183$$
$$= 3.1719$$

$$\overline{Y}'_4 = 0.4347(4.8060) - 0.5922(0) + 0.0262(0) + 0.8644(0)$$
$$+ 0.0738(1) + 0.2183$$
$$= 2.3813$$

$$\overline{Y}'_5 = 0.4347(4.8060) - 0.5922(-1) + 0.0262(-1) + 0.8644(-1)$$
$$+ 0.0738(-1) + 0.2183$$
$$= 1.9353$$

The adjusted means are plotted in Figure 16.4.

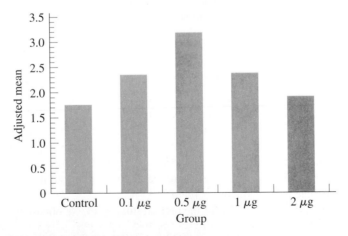

Figure 16.4 Adjusted means by group

The grand mean is

$$\overline{Y}'_. = 0.4347(4.8060) - 0.5922(0) + 0.0262(0) + 0.8644(0)$$
$$+ 0.0738(0) + 0.2183$$
$$= 2.3075$$

which is the mean of the adjusted means. (In a case in which we have equal sample sizes, the adjusted grand mean will equal the unadjusted grand mean.)[6]

Now we are about to go into deep water in terms of formulae, and I expect eyes to start glazing over. I can't imagine that anyone is going to expect you to memorize these formulae. Just try to understand what is happening and remember where to find them when you need them. Don't expect to find them printed out by most statistical software.

Any individual comparisons between treatments would now be made using these adjusted means. In this case, however, we must modify our error term from that of the overall analysis of covariance. If we let $SS_{e(c)}$ represent the error sum of squares from an analysis of variance on the *covariate*, then Huitema (1980), in an excellent and readable book on the analysis of covariance, gives as a test of the difference between two adjusted means

$$F(1, N - a - 1) = \frac{\left(\overline{Y}_j' - \overline{Y}_k'\right)^2}{MS_{error}'\left[\left(\dfrac{1}{n_j} + \dfrac{1}{n_k}\right) + \dfrac{(C_j - C_k)^2}{SS_{e(c)}}\right]}$$

MS_{error}'

where MS_{error}' is the error term from the analysis of covariance. For an excellent discussion of effective error terms and comparisons among means, see Winer (1971, pp. 771ff) and, especially, Huitema (1980).

As an example, suppose we want to compare $\overline{Y}_1'$ and $\overline{Y}_3'$, which theory had predicted would show the greatest difference. From the preceding analysis, we either know or can compute

$$MS_{error}' = 0.4909$$

$$SS_{e(c)} = 202.938 \quad \text{[calculation not shown]}$$

$$\overline{C}_1 = 3.3770 \qquad \overline{C}_3 = 6.4944$$

$$\overline{Y}_1' = 1.7153 \qquad \overline{Y}_3' = 3.1719$$

$$F(1, 41) = \frac{(1.7153 - 3.1719)^2}{0.4909\left[\left(\dfrac{1}{10} + \dfrac{1}{9}\right) + \dfrac{(3.3770 - 6.4944)^2}{202.938}\right]}$$

$$= \frac{2.1217}{0.1271} = 16.69$$

The critical value $F_{.05}(1, 41) = 4.08$. We would thus reject the null hypothesis that the adjusted means of these two conditions are equal in the population. Even after adjusting for the fact that the groups differed by chance on the pretest, we find significant postinjection differences.

Exhibit 16.4 contains SPSS output for the analysis of variance. (The pretest and posttest means were computed using the **compare means** procedure.) Notice that I requested a "spread versus level" plot from the options menu, and it reveals that there is a correlation

---

[6] An alternative approach to calculating adjusted means is to define

$$\overline{Y}_j' = \overline{Y}_j - b_w(\overline{C}_j - \overline{C}_.)$$

where $\overline{C}_j$ is the covariate mean for Group j, $\overline{C}_.$ is the covariate grand mean, and b_w is the regression coefficient for the covariate from the complete model (here $b_w = 0.4347$). This more traditional way of calculating adjusted means makes it clear that the adjusted mean is some function of how deviant that group was on the covariate. The same values for the adjusted means will result from using either approach.

Report

Treatment group		PRETEST	POSTTEST
Control	Mean	3.3770	1.0940
	N	10	10
	Std. Deviation	1.3963	.5850
0.1μg	Mean	5.3870	2.5860
	N	10	10
	Std. Deviation	3.4448	1.5332
0.5μg	Mean	6.4944	3.9056
	N	9	9
	Std. Deviation	2.3781	1.4768
1μg	Mean	5.0450	2.4850
	N	8	8
	Std. Deviation	1.6876	1.1874
2μg	Mean	3.9430	1.5600
	N	10	10
	Std. Deviation	1.2207	.8765
Total	Mean	4.8060	2.2857
	N	47	47
	Std. Deviation	2.3788	1.4947

Tests of Between-Subjects Effects

Dependent Variable: POSTTEST

Source	Type III Sum of Squares	df	Mean Square	F	Sig.	Eta Squared
Corrected Model	82.644[a]	5	16.529	33.673	.000	.804
Intercept	.347	1	.347	.707	.405	.017
PRETEST	38.341	1	38.341	78.108	.000	.656
GROUP	9.224	4	2.306	4.698	.003	.314
Error	20.125	41	.491			
Total	348.327	47				
Corrected Total	102.769	46				

[a] R Squared = .804 (Adjusted R Squared = .780)

Estimated Marginal Means

Treatment Group

Dependent Variable: POSTTEST

Treatment group	Mean	Std. Error	95% Confidence Interval	
			Lower Bound	Upper Bound
Control	1.715[a]	.232	1.246	2.185
0.1μg	2.333[a]	.223	1.882	2.785
0.5μg	3.172[a]	.248	2.671	3.672
1μg	2.381[a]	.248	1.880	2.882
2μg	1.935[a]	.226	1.480	2.391

[a] Evaluated at covariates appeared in the model: PRETEST = 4.8060.

(continued)

Exhibit 16.4 SPSS output for analysis of Conti and Musty data

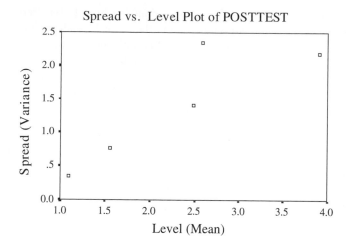

Groups: Treatment group

Exhibit 16.4   *(continued)*

between the size of the mean and the size of the variance. Notice, however, that the relationship would appear very much reduced when we plotted the relationship between the adjusted means and their standard errors.

16.6   Computing Effect Sizes in an Analysis of Covariance

As you might expect, computing effect sizes is a bit more complicated in analysis of covariance than it was in the analysis of variance. That is because we have choices to make in the means we compare and the error term we use. You may recall that with factorial designs and repeated measures designs we had a similar problem concerning the choice of the error term for the effect size.

As before, we can look at effect size with *r*-family and *d*-family measures. Normally, I would suggest *r*-family measures when looking at an omnibus *F* test and a *d*-family measure when looking at specific contrasts. We will start with a *r*-family example, and then move to the *d*-family. The example we have been using based on Conti and Musty's study produced a significant *F* on the omnibus null hypothesis. Probably the most appropriate way to talk about this particular example would make use of the fact that Group (or Dose) was a metric variable, increasing from 0 to 2 μg.[7] However, I am going to take a "second-best" approach here because most studies we run do not have the independent variable distributed as such an ordered variable.

r-Family Measure

As our *r*-family measure of association, we will use η^2, acknowledging that it is positively biased. You should recall that η^2 is defined as the treatment *SS* divided by the total *SS*. But which sums of squares for treatments should we use—the ones from an analysis of variance

---

[7] SPSS will test polynomial contrasts on the adjusted means. Just click on the CONTRAST button and ask for polynomial contrasts. For this example, there is a significant quadratic component.

on the dependent variable, or the ones from the analysis of covariance? Kline (2004) offers both of those alternatives, though he uses an adjusted SS_{total} in the second,[8] without suggesting a choice. If the covariate naturally varies in the population (as it does in this case, where we expect different animals to vary in their pretest score, then it makes the most sense to divide the SS_{treat} from the analysis of covariance by the SS_{total} (unadjusted) from that analysis. This will produce a value of η^2, which is the percentage of "normal variation" accounted for by the independent variable.[9] Then

$$\eta^2 = \frac{SS_{treat(adj)}}{SS_{total}} = \frac{9.2239}{102.7689} = .09$$

An alternative approach, which will produce the same answer, is to take η^2 as the difference between the R^2 from a model predicting the dependent variable from only the covariate (the pretest) and one predicting the dependent variable from both the covariate and the treatment. The increase in explained variation from the first of these models to the second represents what the treatment contributes after controlling for the covariate. For our example, R^2 using just the covariate is .714. (You can obtain this by an analysis of variance using the covariate as the independent variable, or by a regression of the independent variable on the covariate.) When you add in the treatment effect the R^2 is .804. These values are shown in the following table.

Step	Predictors	R^2	Change in R^2	F for change
1	Pretest	.714		
2	Pretest, Treatment	.804	.090	4.689

η^2 is the difference between these two values of R^2, which is the contribution to the explained variation of the treatment after controlling for the covariate. This is the same value we obtained by the first approach.

d-Family Measure

Measures from the d-family often are more interpretable, and they are most often used for specific contrasts between two means. The example we have been using is not a very good one for a contrast of two means because the independent variable is a continuum. But I will use the contrast between the control group and the 0.5 μg group as an example because these are the two conditions that Conti and Musty's theory would have expected to show the greatest mean difference. Because we are working with an analysis of covariance, the appropriate means to compare are the adjusted means ($\overline{Y}_i$) from that analysis. In this case, they are 3.1719 for the .5 μg condition and 1.7153 for the control condition. (You may recall that we performed a test on the difference between these adjusted means in the previous section, and it was significant.)

You should recall that we have generally used

$$d = \frac{\hat{\psi}}{\hat{\sigma}}$$

as our effect size estimate. When we are comparing two group means, $\hat{\psi}$ is simply the difference between the two means because the coefficients are $[-1\ 0\ 1\ 0\ 0]$. For our example, $\hat{\psi}$ is $3.1719 - 1.7153 = 1.4566$. But the question of most importance is what we will use for

---

[8] SPSS uses this same adjustment if you request effect sizes, and it is simply $SS_{treat} + SS_{error}$.

[9] If you were interested in the η^2 for the quadratic relationship between dose and the activity level, controlling for the pretest activity level, you could just divide the $SS_{quadratic}$ by SS_{total}.

the estimate of the standard deviation. One of several choices would be the square root of MS_{error} from an analysis of variance because this would be an estimate of the average variability within each group and would thus standardize the mean difference in the metric of the original measurements. (Recall that we used SS_{total} from the analysis of variance when we calculated η^2.) An alternative would be the square root of MS_{error} from the analysis of covariance, which would standardize the mean difference in the metric of the adjusted scores, which is a bit more difficult to understand. Cortina and Nouri (2000) have made the sensible suggestion that when the covariate normally varies in the population, as ours does, we want to include that variability in our estimate of error. This means that we would use the square root of MS_{error} from the analysis of variance on the posttest scores. In that analysis, MS_{error} is 1.392, which is simply the square root of the weighted mean of the group variances). Then

$$d = \frac{\hat{\psi}}{\hat{\sigma}} = \frac{\overline{X}_3 - \overline{X}_1}{\sqrt{MS_{error}}} = \frac{3.1719 - 1.7153}{\sqrt{1.392}} = \frac{1.4566}{1.1798} = 1.23$$

Injection of the optimal dose of THC (0.5 μg) leads to an increase of postinjection activity by nearly 1.25 standard deviations relative to the control group.

16.7 Interpreting an Analysis of Covariance

Interpreting an analysis of covariance can present certain problems, depending on the nature of the data and, more important, the design of the experiment. Huitema (1980) presents a thorough and readable discussion of most of these problems. Other important sources for consideration of these problems are Anderson (1963), Evans and Anastasio (1968), Huitema (2005), Lord (1967, 1969), Maxwell and Cramer (1975), Reichardt (1979), Smith (1957), and Weisberg (1979).

The ideal application for an analysis of covariance is an experiment in which participants are randomly assigned to treatments (or cells of a factorial design). In that situation, the *expected value* of the covariate mean for each group or cell is the same, and any differences can be attributed only to chance, assuming that the covariate was measured before the treatments were applied. In this situation, the analysis of covariance will primarily reduce the error term, but it will also, properly, remove any bias in the dependent variable means caused by chance group differences on the covariate. This was the situation in the Conti and Musty (1984) study that we have been discussing.

In a randomized experiment in which the covariate is measured *after* the treatment has been applied and has affected the covariate, interpreting the results of an analysis of covariance is difficult at best. In this situation, the expected values of the group covariate means are not equal, even though the subjects were assigned randomly. It is difficult to interpret the results of the analysis because you are asking what the groups would have been like had they not differed on the covariate, when in fact the covariate differences may be an integral part of the treatment effect. This problem is particularly severe if the covariate was measured with error (i.e., if it is not perfectly reliable). In this case, an alternative analysis, called the **true-score analysis of covariance,** may be appropriate if the other interpretive problems can be overcome. Huitema (1980, Chapter 14) discusses such an analysis.

true-score analysis of covariance

When subjects are not assigned to the treatment groups at random, interpreting the analysis of covariance can be even more troublesome. The most common example of this problem is what is called the **nonequivalent groups design.** In this design, two (or more) intact groups are chosen (e.g., schools or classrooms of children), a pretest measure is obtained from subjects in both groups, the treatment is applied to one of the groups, and the two groups

nonequivalent groups design

are then compared on some posttest measure. Because participants are not assigned to the groups at random, we have no basis for assuming that any differences that exist on the pretest are to be attributed to chance. Similarly, we have no basis for expecting the two groups to have the same mean on the posttest in the absence of a real treatment effect. Huitema (1980, pp. 149ff) gives an excellent demonstration that when the groups differ at the beginning of the experiment, the phenomenon of regression to the mean could lead to posttest differences even in the absence of a treatment effect. For alternative analyses that are useful under certain conditions, see Huitema (1980). Maris (1998) takes a different view of the issue.

The problems of interpreting results of designs in which subjects are not randomly assigned to the treatment groups are not easily overcome. This is one of the reasons why random assignment is even more important than random selection of subjects. It is difficult to overestimate the virtues of random assignment, both for interpreting data and for making causal statements about the relationship between variables. In what is probably only a slight overstatement of the issue, Lord (1967) remarked, "In the writer's opinion, the explanation is that with the data usually available for such studies, there is simply no logical or statistical procedure that can be counted on to make proper allowances for uncontrolled preexisting differences between groups" (p. 305). (Lord was *not* referring to differences that arise by chance through random assignment.) Anderson (1963) made a similar point by stating, "One may well wonder exactly what it means to ask what the data would be like if they were not what they are" (p. 170). All of this is not to say that the analysis of covariance has no place in the analysis of data in which the treatments differ on the covariate. Anyone using covariance analysis, however, must think carefully about her data and the practical validity of the conclusions she draws.

16.8 Reporting the Results of an Analysis of Covariance

The only difference between describing the results of an analysis of covariance and an analysis of variance is that we must refer to the covariate and to adjusted means. For the experiment by Conti and Musty, we could write

> Conti and Musty (1984) examined the effect of THC on locomotor activity in rats. They predicted that moderate doses of THC should show the greatest increase in activity (or the least decrease caused by adaptation). After a pretesting session, five different groups of rats were randomly assigned to receive 0, .1 μg, .5 μg, 1 μg, or 2 μg of THC. Activity level was measured in a 10-minute postinjection interval. Because there was considerable variability in pretest activity, the pretest measure was used as a covariate in the analysis.
>
> The analysis of covariance was significant ($F(4, 41) = 4.694$, $p = .003$), with intermediate doses showing greater effect. Eta-squared was .09 using a SS_{total} that has not been adjusted for the covariate. A contrast of the means of the control group and the .5 μg group revealed a significant difference ($F(1, 41) = 16.69$, $p < .05$), with a standardized effect size (d) of 1.23.

16.9 The Factorial Analysis of Covariance

The analysis of covariance applies to factorial designs just as well as it does to single-variable designs. Once again, the covariate may be treated as a variable that, because of methodological considerations, assumes priority in the analysis. In this section, we will deal only with the case of equal cell sizes, but the generalization to unequal *ns* is immediate.

The logic of the analysis is straightforward and follows that used in the previous examples. $SS_{\text{regression}_{c,\alpha,\beta,\alpha\beta}}$ is the variation attributable to a linear combination of the covariate, the main effects of A and B, and the AB interaction. Similarly, $SS_{\text{regression}_{c,\alpha,\beta}}$ is the variation attributable to the linear combination of the covariate and the main effects of A and B. The difference

$$SS_{\text{regression}_{c,\alpha,\beta,\alpha\beta}} - SS_{\text{regression}_{c,\alpha,\beta}}$$

is the variation attributable to the AB interaction, with the covariate and the main effects partialled out. With equal sample sizes, the two main effects and the interaction are orthogonal, so all that is *actually* partialled out in equal n designs is the covariate.

By the same line of reasoning,

$$SS_{\text{regression}_{c,\alpha,\beta,\alpha\beta}} - SS_{\text{regression}_{c,\alpha,\alpha\beta}}$$

represents the variation attributable to B, partialling out the covariate, and

$$SS_{\text{regression}_{c,\alpha,\beta,\alpha\beta}} - SS_{\text{regression}_{c,\beta,\alpha\beta}}$$

represents the variation attributable to the main effect of A, again partialling out the covariate.

The error term represents the variation remaining after controlling for A, B, and AB, and the covariate. As such, it is given by

$$SS_{\text{residual}_{c,\alpha,\beta,\alpha\beta}}$$

The general structure of the analysis is presented in Table 16.10. Notice that once again the error term loses a degree of freedom for each covariate. Because the independent variable and the covariate account for overlapping portions of the variation, their sums of squares will not equal SS_{total}.

Table 16.10   Structure of the analysis of covariance for factorial designs

Source	df	SS
A(adj)	$a - 1$	$SS_{\text{regression}_{c,\alpha,\beta,\alpha\beta}} - SS_{\text{regression}_{c,\beta,\alpha\beta}}$
B(adj)	$b - 1$	$SS_{\text{regression}_{c,\alpha,\beta,\alpha\beta}} - SS_{\text{regression}_{c,\alpha,\alpha\beta}}$
AB(adj)	$(a - 1)(b - 1)$	$SS_{\text{regression}_{c,\alpha,\beta,\alpha\beta}} - SS_{\text{regression}_{c,\alpha,\beta}}$
Error	$N - ab - c$	$SS_{\text{residual}_{c,\alpha,\beta,\alpha\beta}}$
Covariate	c	$SS_{\text{regression}_{c,\alpha,\beta,\alpha\beta}} - SS_{\text{regression}_{\alpha,\beta,\alpha\beta}}$
Total	$N - 1$	

As an example, consider the study by Spilich, June, and Renner (1992) that we examined in Chapter 13 on performance as a function of cigarette smoking. In that study, subjects performed either a Pattern Recognition task, a Cognitive task, or a Driving Simulation task. The subjects were divided into three groups. One group (Active Smoking) smoked during or just before the task. A second group (Delayed Smoking) were smokers who had not smoked for three hours, and a third group (NonSmoking) was composed of Nonsmokers. The dependent variable was the number of errors on the task. To make this suitable for an analysis of covariance, I have added an additional (hypothetical) variable, which is the subject's measured level of distractibility. (Higher distractibility scores indicate a greater ease at being distracted.)

The data are presented in Table 16.11 and represent a 3×3 factorial design with one covariate (Distract).

Table 16.11 Hypothetical data on smoking and performance (modeled on Spilich et al., 1992)

Pattern Recognition

NS: Errors	9	8	12	10	7	10	9	11	8	10	8	10	8	11	10
Distract	107	133	123	94	83	86	112	117	130	111	102	120	118	134	97
DS: Errors	12	7	14	4	8	11	16	17	5	6	9	6	6	7	16
Distract	101	75	138	94	138	127	126	124	100	103	120	91	138	88	118
AS: Errors	8	8	9	1	9	7	16	19	1	1	22	12	18	8	10
Distract	64	135	130	106	123	117	124	141	95	98	95	103	134	119	123

Cognitive Task

NS: Errors	27	34	19	20	56	35	23	37	4	30	4	42	34	19	49
Distract	126	154	113	87	125	130	103	139	85	131	98	107	107	96	143
DS: Errors	48	29	34	6	18	63	9	54	28	71	60	54	51	25	49
Distract	113	100	114	74	76	162	80	118	99	146	132	135	111	106	96
AS: Errors	34	65	55	33	42	54	21	44	61	38	75	61	51	32	47
Distract	108	191	112	98	128	145	76	107	128	128	142	144	131	110	132

Driving Simulation

NS: Errors	15	2	2	14	5	0	16	14	9	17	15	9	3	15	13
Distract	110	96	112	114	137	125	168	102	109	111	137	106	117	101	116
DS: Errors	7	0	6	0	12	17	1	11	4	4	3	5	16	5	11
Distract	93	102	108	100	123	131	99	116	81	103	78	103	139	101	102
AS: Errors	3	2	0	0	6	2	0	6	4	1	0	0	6	2	3
Distract	130	83	91	92	109	106	99	109	136	102	119	84	68	67	114

Table 16.12 contains an abbreviated form of the design matrix, showing only the entries for the first and last subject in each cell. Notice that the matrix contains a column for the covariate (denoted C), the usual design matrix elements for the main effects of Task (T) and Group (G), and the Task $\times$ Group interaction. I have also added columns representing the interaction of the covariate with the Task $\times$ Group interaction. The latter will be used to test the hypothesis $H_0: b_i^* = b_j^*$ for all values of i and j because the assumption of homogeneity of regression applies to any analysis of covariance.

It is important to consider just what the interactions involving the covariate represent. If I had included the terms CT_1 and CT_2 I could have used them to test the null hypothesis that the regression lines of Errors as a function of Distract have equivalent slopes in the three tasks. Similarly, if I had included CG_1 and CG_2, I could have tested homogeneity of regression in each of the three smoking groups. Because I am most interested in testing the hypothesis of homogeneity of regression in each of the nine cells, I have included only the CTG_{ij} terms.

The first regression is based on all predictors in X. From this analysis, we obtain

$$SS_{\text{regression}_{c,\alpha,\beta,\alpha\beta,c\alpha\beta}} = 36728.42272$$

$$MS_{\text{residual}_{c,\alpha,\beta,\alpha\beta,c\alpha\beta}} = 71.10333$$

Table 16.12   Design matrix for the analysis of covariance for smoking data

	C	T_1	T_2	G_1	G_2	TG_{11}	TG_{12}	TG_{21}	TG_{22}	cTG_{11}	cTG_{12}	cTG_{21}	cTG_{22}		
	107	1	0	1	0	1	0	0	0	107	0	0	0		9
	...	...	...	...	...	...	...	...	...	...	...	...	...		...
	97	1	0	1	0	1	0	0	0	97	0	0	0		10
	101	1	0	0	1	0	1	0	0	0	101	0	0		12
	...	...	...	...	...	...	...	...	...	...	...	...	...		...
	118	1	0	0	1	0	1	0	0	0	118	0	0		16
	64	1	0	−1	−1	−1	−1	0	0	−64	−64	0	0		8
	...	...	...	...	...	...	...	...	...	...	...	...	...		...
	123	1	0	−1	−1	−1	−1	0	0	−123	−123	0	0		10
	126	0	1	1	0	0	0	1	0	0	0	126	0		27
	...	...	...	...	...	...	...	...	...	...	...	...	...		...
	143	0	1	1	0	0	0	1	0	0	0	143	0		49
X =	113	0	1	0	1	0	0	0	1	0	0	0	113	**Y =**	48
	...	...	...	...	...	...	...	...	...	...	...	...	...		...
	96	0	1	0	1	0	0	0	1	0	0	0	96		49
	108	0	1	−1	−1	0	0	−1	−1	0	0	−108	−108		34
	...	...	...	...	...	...	...	...	...	...	...	...	...		...
	132	0	1	−1	−1	0	0	−1	−1	0	0	−132	−132		47
	110	−1	−1	1	0	−1	0	1	0	−110	0	−110	0		15
	...	...	...	...	...	...	...	...	...	...	...	...	...		...
	116	−1	−1	1	0	−1	0	1	0	−116	0	−116	0		13
	93	−1	−1	0	1	0	−1	0	−1	0	−93	0	−93		7
	...	...	...	...	...	...	...	...	...	...	...	...	...		...
	102	−1	−1	0	1	0	−1	0	−1	0	−102	0	−102		11
	130	−1	−1	−1	−1	1	1	1	1	130	130	130	130		3
	...	...	...	...	...	...	...	...	...	...	...	...	...		...
	114	−1	−1	−1	−1	1	1	1	1	114	114	114	114		3

If we drop the interaction terms representing the interaction of the covariate (Distract) with the Task × Group interaction, we have

$$SS_{\text{regression}_{c,\alpha,\beta,\alpha\beta}} - 36389.60175$$

The difference between these two sums of squares = 338.82097. The most complete model had 13 degrees of freedom, whereas the second had 9 *df*, meaning that the previous sum of squares is based on 13 − 9 = 4 *df*. Converting to mean squares we have

$$MS_{\text{difference}} = 338.82097/4 = 84.70524$$

We can test the difference between these two models by using MS_{residual} from the more complete model and computing

$$F = \frac{MS_{\text{difference}}}{MS_{\text{residual}}} = \frac{84.70524}{71.10333} = 1.19$$

This is an *F* on $[(f-r),(N-f-1)] = 4$ and 121 *df*. The critical value is $F_{.05}(4, 121) = 2.45$, so we will not reject the null hypothesis of homogeneity of regression. We will conclude that we have no reason to doubt that the regression lines relating Errors to Distract have the same slope in the nine cells. This will allow us to adopt a much simpler full model against which we can compare subsequent reduced models. Our revised full model is

$$\hat{Y} = b_0 + b_1 C + b_2 T_1 + b_3 T_2 + b_4 G_1 + b_5 G_2 + b_6 TG_{11} + b_7 TG_{12} + b_8 TG_{21} + b_9 TG_{22}$$

or, in more traditional analysis of variance terms,

$$Y_{ijk} = \mu + C_k + \alpha_i + \beta_j + \alpha\beta_{ij} + \varepsilon_{ijk}$$

The results of the several multiple regression solutions needed for the analysis of covariance are shown in Table 16.13. By calculating and testing the differences between full and reduced models, you will be able to compute the complete analysis of covariance.

Table 16.13 Regression results for various models for data in Table 16.11

Model	$SS_{regression}$	$MS_{residual}$	R^2
C, T, G, TG	36,389.60175	75.53859	0.8027
C, T, G	34,763.09104		0.7669
C, G, TG	12,519.11654		0.2762
C, T, TG	35,826.34433		0.7903
T, G, TG	31,744.72593		0.7003

Exhibit 16.5 contains the results of an SPSS analysis of these data. You should compare your results with the results in that exhibit.

For purposes of comparison, I have presented the analysis of variance from Exhibit 13.1. This is the analysis on the same data, but without the covariate.

Source	df	SS	MS	F
Task	2	28,661.526	14,330.763	132.895*
Group	2	354.548	177.274	1.644
Task × Group	4	2728.652	682.213	6.326*
Error	126	13,587.084	107.834	
Total	134	45,331.810		

*$p < 0.05$

Notice that in this analysis we have a significant effect due to Task, which is uninteresting because the tasks were quite different and we would expect that some tasks would lead to more errors than others. We also have a Task × Group interaction, which was what we were seeking because it tells us that smoking makes a difference in certain kinds of situations (which require a lot of cognitive processing) but not in others. Notice that we did not have an overall effect due to Group. Notice also that our MS_{error} was 107.834, whereas in the analysis of covariance, it was 71.539.

When we look at our analysis of covariance, one of the first things we see is that MS_{error} (71.539) is about one-third smaller than it was in the analysis of variance. This is because the covariate (Distract) was able to explain much of the variability in Errors that had been left unexplained in the analysis of variance.

In Exhibit 16.5, we see that we have a significant effect for Groups. This is partly a function of the smaller error term and partly a function of adjustments of group means because of small differences in mean Distract scores across groups. Unless we are willing to assume

Tests of Between-Subjects Effects

Dependent Variable: ERRORS

Source	Type III Sum of Squares	df	Mean Square	F	Sig.	Eta Squared	Noncent. Parameter	Observed Power[a]
Corrected Model	36389.602[b]	9	4043.289	56.519	.000	.803	508.671	1.000
Intercept	892.395	1	892.395	12.474	.001	.091	12.474	.939
DISTRACT	4644.876	1	4644.876	64.928	.000	.342	64.928	1.000
TASK	23870.485	2	11935.243	166.836	.000	.727	333.673	1.000
SMKGRP	563.257	2	281.629	3.937	.022	.059	7.873	.699
TASK * SMKGRP	1626.511	4	406.628	5.684	.000	.154	22.736	.977
Error	8942.324	125	71.539					
Total	90341.000	135						
Corrected Total	45331.926	134						

[a] Computed using alpha = .05

[b] R Squared = .803 (Adjusted R Squared = .789)

1. Task * Smoking Group

Dependent Variable: ERRORS

Task	Smoking Group	Mean	Std. Error	95% Confidence Interval Lower Bound	95% Confidence Interval Upper Bound
Patrecog	NonSmokers	9.805[a]	2.184	5.482	14.128
	Delayed smokers	9.732[a]	2.184	5.410	14.054
	Active Smokers	9.558[a]	2.184	5.235	13.882
Cognitive	NonSmokers	27.770[a]	2.188	23.440	32.101
	Delayed smokers	40.436[a]	2.185	36.112	44.760
	Active Smokers	43.785[a]	2.233	39.366	48.204
Driving	NonSmokers	8.505[a]	2.191	4.169	12.842
	Delayed smokers	8.921[a]	2.200	4.568	13.275
	Active Smokers	5.820[a]	2.226	1.414	10.226

[a] Evaluated at covariates appeared in the model: DISTRACT = 112.52.

2. Task

Dependent Variable: ERRORS

Task	Mean	Std. Error	95% Confidence Interval Lower Bound	95% Confidence Interval Upper Bound
Patrecog	9.699[a]	1.261	7.203	12.194
Cognitive	37.330[a]	1.274	34.810	39.851
Driving	7.749[a]	1.273	5.230	10.268

[a] Evaluated at covariates appeared in the model: DISTRACT = 112.52.

(continued)

Exhibit 16.5   SPSS analysis of covariance of Spilich data

3. Smoking Group

Dependent Variable: ERRORS

Smoking Group	Mean	Std. Error	95% Confidence Interval	
			Lower Bound	Upper Bound
NonSmokers	15.360[a]	1.264	12.859	17.862
Delayed smokers	19.696[a]	1.266	17.191	22.202
Active Smokers	19.721[a]	1.261	17.225	22.217

[a] Evaluated at covariates appeared in the model: DISTRACT = 112.52.

Exhibit 16.5 *(continued)*

that smokers in general are more distractable (and perhaps they are), then it is appropriate to adjust for random differences among groups. (An analysis of variance on the covariate [Distract] showed no significant effects.)

Notice that Exhibit 16.5 presents *partial* eta-squared for the effects. These effect-size measures can be calculated as the difference between two R^2 values, divided by $(1 - R^2_{\text{reduced}})$. For example, the model without the dummy variables for Task has an $R^2 = .2762$. This leaves $1 - .2762 = 72.38\%$ of the variation unexplained. When we add in the Task variables (going to the full model), we have $R^2 = .8027$. This is an increase of $.8027 - .2762 = .5265$, which accounts for $.5265/.7238 = 72.74\%$ of the variation *that had been left unexplained*. This is the value given in Exhibit 16.5 for Task. Similar calculations will reproduce the other values.

Adjusted Means

The method of obtaining adjusted means is simply an extension of the method employed in the Conti and Musty example. We want to know what the cell means would have been if the treatment combinations had not differed on the covariate.

From the full model, we have

$$\hat{Y} = b_0 + b_1 C + b_2 T_1 + b_3 T_2 + b_4 G_1 + b_5 G_2 + b_6 TG_{11} + b_7 TG_{12} + b_8 TG_{21} + b_9 TG_{22}$$

which equals

$$\hat{Y} = -14.654 - 8.561 T_1 + 19.071 T_2 - 2.900 G_1 + 1.437 G_2 + 3.006 TG_{11} - 1.404 TG_{12}$$
$$-6.661 TG_{21} + 1.668 TG_{22} + 0.292512 \text{Distract}$$

Because we want to know what the Y means would be if the treatments did not differ on the covariate, we will set $C = \overline{C} = 112.518$ for all treatments.

For all observations in Cell$_{11}$, the appropriate row of the design matrix, with C replaced by $\overline{C}$, is

$$1 \quad\quad 0 \quad\quad 1 \quad\quad 0 \quad\quad 1 \quad\quad 0 \quad\quad 0 \quad\quad 0 \quad\quad\quad 112.518$$

Applying the regression coefficients and taking the intercept into account, we have

$$\overline{Y}_{11} = -14.654 - 8.561\,(1) + 19.071\,(0) - 2.900\,(1) + 1.437\,(0) + 3.006\,(1)$$
$$-1.404\,(0) - 6.661\,(0) + 1.668\,(0) + 0.292512\,(112.518)$$
$$= 9.804$$

Applying this procedure to all cells, we obtain the following adjusted cell means

	Pattern Rec	Cognitive	Driving	Row Means
NonSmokers	9.805	27.770	8.505	15.360
Delayed	9.732	40.436	8.921	19.696
Active	9.558	43.785	5.820	19.721
Column Means	9.699	37.330	7.749	18.259

These are the cell means given in Exhibit 16.5, and the row and column means can be found as the mean of the cells in that row or column.

Testing Adjusted Means

The adjusted means are plotted in Figure 16.5. They illustrate the interaction and also the meaning that may be attached to the main effects. Further analyses of these data are probably unnecessary because differences caused by smoking seemed to be confined to the condition that requires high levels of cognitive processing. However, for the sake of completeness, we will assume that you want to make a comparison between the mean of the NonSmoking group and the combined means of the Active and Delayed groups. In this case, you want to compare $\overline{X}'_{1.}$ with $\overline{X}'_{2.}$ and $\overline{X}'_{3.}$ combined. This comparison requires some modification of the error term, to account for differences in the covariate. This adjustment is given by Winer (1971) as

$$MS''_{\text{error}} = MS'_{\text{error}} \left[1 + \frac{\frac{SS_{g(c)}}{g-1}}{SS_{e(c)}} \right]$$

where $SS_{g(c)}$ and $SS_{e(c)}$ represent the sum of squares attributable to Groups and Error (respectively) in an analysis of variance on the *covariate,* and MS'_{error} is the error term from the overall analysis of covariance.

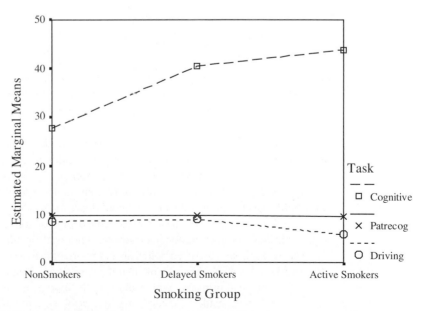

Figure 16.5   Adjusted cell means as a function of Group and Task

$$MS'_{\text{error}} = 71.538$$

$$SS_{g(c)} = 730.015$$

$$SS_{e(c)} = 54285.867$$

Thus,

$$MS''_{\text{error}} = 71.538 \left[1 + \frac{\dfrac{730.015}{2-1}}{54285.867} \right] = 72.019$$

To compare the adjusted means, we have

$$\psi = 2(15.360) - 1(19.696) - 1(19.721) = -8.697$$

$$F(1,125) = \frac{n\psi^2}{\sum a_i^2 MS''_{\text{error}}} = \frac{45(-8.697)^2}{6(72.019)} = 7.88$$

Because $F_{.05}(1,125) = 3.92$, we can reject H_0 and conclude that the Active Smoking group performs more poorly (overall) than the average of the other two groups.

Another experimenter might be interested in examining the effects of Group only for the Cognitive task. If we want to examine these simple effects, we will again need to modify our error term in some way. This is necessary because we will be looking at Groups for only some of the data, and the covariate mean of the Cognitive task subjects may differ from the covariate mean for all subjects. Probably the safest route here would be to run a separate analysis of covariance for only those subjects performing the cognitive task. Although this method has the disadvantage of costing us degrees of freedom for error, it has the advantage of simplicity and eliminates the need to make discomforting assumptions in the adjustment of our error term.

To complete our discussion of the tests that we want to conduct, consider the experimenter who wants to compare two particular adjusted cell means (whether or not they are in the same row or column). The adjusted error term for this comparison was given by Winer (1971) as

$$MS''_{\text{error}} = \frac{2MS'_{\text{error}}}{n} \left[1 + \frac{\dfrac{SS_{\text{cells}(c)}}{tg-1}}{SS_{e(c)}} \right]$$

where $SS_{\text{cells}(c)}$ is the sum of squares from an analysis of variance on the covariate.

You may wonder why we continually worry about adjusting the error term in making comparisons. The general nature of the answer is apparent when you recall what the confidence limits around the regression line looked like in Chapter 9. (They were curved—in fact, they were elliptical.) For $X = \overline{X}$, we were relatively confident about $\hat{Y}$. However, as X departed more and more from $\overline{X}$, we became less and less confident of our prediction, and consequently the confidence limits widened. If you now go back to Figure 16.3, you will see that the problem applies directly to the case of adjusted means. In that figure, $\overline{Y}'_1$ is a long way from $\overline{Y}_1$, and we would probably have relatively little confidence that we have estimated it correctly. On the other hand, we can probably have a reasonable degree of confidence in our estimate of $\overline{Y}'_2$. It is just this type of consideration that causes us constantly to adjust our error term.

The example we have used points up an important feature of the analysis of covariance—the covariate is just another variable that happens to receive priority. In designing the study, we were concerned primarily with evaluating the effects of smoking. However, we had two variables that we considered it necessary to control: type of task and distractibility. We controlled the first one (Task) by incorporating it into our design as an independent variable. We controlled the second one (Distractibility) by measuring it and treating it as a covariate.

In many respects, these are two ways of treating the same problem. Although there are obvious differences in the way these two variables are treated, there are also important similarities. In obtaining SS_{group}, we are actually partialling out *both* task and covariate. It is true that with equal ns, task is orthogonal to group, leaving nothing to partial out, but that is merely a technicality. For unequal ns, the partialling out of both variables is a very real procedure. Although it is important not to lose sight of the fact that the analysis of covariance is a unique technique with its own additional assumptions, it is equally important to keep in mind that a covariate is just another variable.

16.10 Using Multiple Covariates

We have been concerned with the use of a single covariate. There is no theoretical or practical reason, however, why we must restrict ourselves in this way. For example, a study on the effectiveness of several different teaching methods might want to treat IQ, Age, and Type of School (progressive or conservative) as covariates. When viewed from the point of view of multiple regression, this presents no particular problem, whereas when viewed within the traditional framework of the analysis of variance, the computational complexities for only a very few covariates would be overwhelming.

In the expression $R^2_{c,\alpha,\beta,\alpha\beta}$, β is really only a shorthand way of representing a set of predictors (e.g., $B_1, B_2, \ldots, B_b$). By the same token, c can be used to stand for a set of covariates ($C_1, C_2, \ldots, C_k$). Thus, in terms of the more specific notation, $R^2_{c,\alpha,\beta,\alpha\beta}$ might really represent

$$R^2_{0.IQ,Age,School,A_1,B_1,B_2,AB_{11},AB_{12}}$$

When seen this way, the use of multiple covariates is no different from that of single covariates. If C represents the covariates IQ, Age, and School, then $SS_{AB(adj)}$ remains

$$SS_{AB(adj)} = SS_{regression(IQ,Age,School,A_1,B_1,B_2,AB_{11},AB_{12})} - SS_{regression(IQ,Age,School,A_1,B_1,B_2)}$$

It should be apparent from the previous example that no restriction is placed on the nature of the covariate, other than that it is assumed to be linearly related to the criterion. The covariate can be a continuous variable, as in the case of IQ and Age, or a discrete variable, as in the dichotomous classification of Schools as progressive and conservative.

A word of warning: Just because it is possible (and in fact easy) to use multiple covariates is not a good reason for adopting this procedure. Interpreting an analysis of covariance may be difficult enough (if not impossible) with only one covariate. The problems increase rapidly with the addition of multiple covariates. Thus, it might be easy to *say*, in evaluating several methods of teaching English, that such and such a method is better if groups are equated for age, IQ, type of school, parents' occupation, and so on. But the experimenter must then ask himself if such equated groups actually exist in the population. If they do not, he has just answered a question about what would happen in groups that could never exist, and it is unlikely that he will receive much applause for his efforts. Moreover, even if it is possible to form such groups, will they behave in the expected manner? The very fact that the students are now in homogeneous classes may itself have an effect on the dependent variable that could not have been predicted.

16.11 Alternative Experimental Designs

stratification

The analysis of covariance is not the only way to handle data in which a covariate is important. Two common alternative procedures are also available: **stratification** (matched samples) and difference scores.

If we have available measures on the covariate and are free to assign subjects to treatment groups, then we can form subsets of subjects who are homogeneous with respect to the covariate, and then assign one member of each subset to a different treatment group. In the analysis of variance, we can then pull out an effect due to blocks (subsets) from the error term.

The use of matched samples and the analysis of covariance are almost equally effective when the regression of Y on C is linear. If ρ equals the correlation in the population between Y and C, and σ_e^2 represents the error variance in a straight analysis of variance on Y, then the use of matched samples reduces the error variance to

$$\sigma_e^2(1 - \rho^2)$$

The reduction due to the analysis of covariance in this situation is given by

$$\sigma_e^2(1 - \rho^2)\frac{(f_e)}{(f_e - 1)}$$

where f_e is the degrees of freedom for the error variance. For any reasonable value of f_e, the two procedures are almost equally effective, assuming linearity of regression. If the relationship between Y and C is not linear, however, matching will be more effective than covariance analysis.

difference scores   A second alternative to the analysis of covariance concerns the use of **difference scores.** If the covariate (C) represents a test score before the treatment is administered and Y a score on the same test after the administration of the treatment, the variable $C - Y$ is sometimes used as the dependent variable in an analysis of variance to control for initial differences on C. This approach will work only if C and Y are comparable measures. We could hardly justify subtracting a driving test score (Y) from an IQ score (C). If the relationship between C and Y is linear and if $b_{CY} = 1.00$, which is rarely true, the analysis of difference scores and the analysis of covariance will give the same estimates of the treatment effects. When b_{CY} is not equal to 1, the two methods will produce different results, and in this case, it is difficult to justify the use of difference scores. Actually, for the Conti and Musty (1984) data on THC, if we took the *difference* between the Pre and Post scores as our dependent variable, the results would be decidedly altered ($F_{4,42} = 0.197$). In this case, the analysis of covariance was clearly a more powerful procedure. Exercise 16.24 at the end of the chapter illustrates this view of the analysis of covariance. For a more complete treatment of this entire problem, see Harris (1963) and Huitema (1980, 2005). The thing to keep in mind here is that a slope of one on the relationship between pre- and posttest scores implies that the intervention led to a similar increase in scores, regardless of where people started. But it might be that the change is *proportional* to where people started out. Someone who is very poor in math may have much more to gain by an intervention program than will someone who was doing well, and thus, the gain score will be directly (and negatively) related to the pretest score. In the example from Conti and Musty (1984), more active animals were likely to change more than were less active animals, which may be why Conti and Musty took as their dependent variable the posttest score as a percentage of the pretest score, rather than just the difference between their two scores.

Key Terms

General linear model (16.1)	Method II (16.4)	Analysis of covariance (16.5)
Design matrix (16.1)	Hierarchical sums of squares (16.4)	Covariate (16.5)
Method III (16.4)	Sequential sums of squares (16.4)	Adjusted Y means (16.5)

Homogeneity of regression (16.5)   True-score analysis of covariance (16.7)   Stratification (16.11)

MS'_{error} (16.5)   Nonequivalent groups design (16.7)   Diference scores (16.11)

Exercises

16.1   The following hypothetical data were obtained from poor, average, and good readers on the number of eye fixations per line of text.

Poor	Average	Good
10	5	3
7	8	5
8	4	2
11	6	3
5	5	4

   a.   Construct the design matrix for these data.

   b.   Use any standard regression program to calculate a least-squares analysis of variance.

   c.   Run the analysis of variance in the traditional manner and compare your answers.

16.2   For the data in Exercise 16.1,

   a.   Calculate treatment effects and show that the regression model reproduces these treatment effects.

   b.   Demonstrate that R^2 for the regression model is equal to η^2 for the analysis of variance.

16.3   Taking the data from Exercise 16.1, add the scores 5 and 8 to the Average group and the scores 2, 3, 3, and 5 to the Good group. Rerun the analysis for Exercise 16.1 using the more complete data.

16.4   Rerun the analysis of Exercise 16.2 for the amended data from Exercise 16.3.

16.5   A psychologist was concerned with the relationship between Gender, Socioeconomic Status (SES), and perceived Locus of Control. She took eight adults (age = 25 to 30 years) in each Gender–SES combination and administered a scale dealing with Locus of Control (a high score indicates that the individual feels in control of his or her everyday life).

		SES	
	Low	Average	High
Male	10	16	18
	12	12	14
	8	19	17
	14	17	13
	10	15	19
	16	11	15
	15	14	22
	13	10	20
Female	8	14	12
	10	10	18
	7	13	14
	9	9	21
	12	17	19
	5	15	17
	8	12	13
	7	8	16

a. Run a traditional analysis of variance on these data.

b. The following sums of squares have been computed on the data using the appropriate design matrix (α = Gender, β = SES)

$$SS_Y = 777.6667 \quad SS_{reg(\alpha,\beta,\alpha\beta)} = 422.6667$$

$$SS_{reg(\alpha,\beta)} = 404.0000 \quad SS_{reg(\beta,\alpha\beta)} = 357.333$$

$$SS_{reg(\alpha,\alpha\beta)} = 84.000$$

Compute the summary table for the analysis of variance using these sums of squares.

16.6 Using the SES portion of the design matrix as our predictor, we find that $SS_{reg}(\beta) = 338.6667$.

a. Why is this value the same as SS_{SES} in the answer to Exercise 16.5?

b. Will this be the case in all analyses of variance?

16.7 When we take the data in Exercise 16.5 and delete the last two low-SES males, the last three average-SES males, and the last two high-SES females, we obtain the following sums of squares:

$$SS_Y = 750.1951 \quad SS_{reg(\alpha,\beta,\alpha\beta)} = 458.7285$$

$$SS_{reg(\alpha,\beta)} = 437.6338 \quad SS_{reg(\beta,\alpha\beta)} = 398.7135$$

$$SS_{reg(\alpha,\alpha\beta)} = 112.3392 \quad SS_{reg(\alpha)} = 95.4511$$

$$SS_{reg(\beta)} = 379.3325$$

$$SS_{reg(\alpha\beta)} = 15.8132$$

Compute the analysis of variance using these sums of squares.

16.8 Using only the SES predictors for the data in Exercise 16.7, we find $SS_{reg(\beta)} = 379.3325$. Why is this not the same as SS_{SES} in Exercise 16.7?

16.9 For the data in Exercise 16.5, the complete model is

$$1.1667A_1 - 3.1667B_1 - 0.1667B_2 + 0.8333AB_{11} - 0.1667AB_{12} + 13.4167$$

Show that this model reproduces the treatment and interaction effects as calculated by the method shown in Table 16.2.

16.10 For the data in Exercise 16.7, the complete model is

$$1.2306A_1 - 3.7167B_1 - 0.3500B_2 + 0.4778AB_{11} - 0.5444AB_{12} + 13.6750$$

Show that this model reproduces the treatment and interaction effects as calculated in Table 16.3.

16.11 Using the following data, demonstrate that Method III (the method advocated in this chapter) really deals with unweighted means.

	B_1	B_2
	5	11
	3	9
A_1		14
		6
		11
		9
	10	6
	11	2
A_2	12	
	7	

16.12 Draw a Venn diagram representing the sums of squares in Exercise 16.5.

16.13 Draw a Venn diagram representing the sums of squares in Exercise 16.7.

16.14 If you have access to SAS, use that program to analyze the data in Exercise 16.7. Add/ SS1 SS2 SS3 SS4 to the end of your Model command and show that

    a.   Type I sums of squares adjust each term in the model only for those that come earlier in the model statement.

    b.   Type II sums of squares adjust main effects only for other main effect variables, while adjusting the interaction for each of the main effects.

    c.   Type III sums of squares adjust each term for all other terms in the model.

    d.   Type IV sums of squares in this case are equal to the Type II sums of squares.

16.15 In studying the energy consumption of families, we have broken them into three groups. Group 1 consists of those who have enrolled in a time-of-day electrical-rate system (the charge per kilowatt-hour of electricity is higher during peak demand times of the day). Group 2 is made up of those who inquired into such a system but did not use it, and Group 3 represents those who have shown no interest in the system. We record the amount of the electrical bill per month for each household as our dependent variable (Y). As a covariate, we take the electrical bill for that household for the same month last year (C). The data follow:

Group 1		Group 2		Group 3	
Y	C	Y	C	Y	C
58	75	60	70	75	80
25	40	30	25	60	55
50	68	55	65	70	73
40	62	50	50	65	61
55	67	45	55	55	65

    a.   Set up the design matrix.

    b.   Run the analysis of covariance.

16.16 To refine the experiment described in Exercise 16.15, a psychologist added an additional set of households to each group. This group had a special meter installed to show them exactly how fast their electric bill was increasing. (The amount-to-date was displayed on the meter.) The data follow; the nonmetered data are the same as those in Exercise 16.15.

	Y	C	Y	C	Y	C
Nonmetered	58	75	60	70	75	80
	25	40	30	25	60	55
	50	68	55	65	70	73
	40	62	50	50	65	61
	55	67	45	55	55	65
Metered	25	42	40	55	55	56
	38	64	47	52	62	74
	46	70	56	68	57	60
	50	67	28	30	50	68
	55	75	55	72	70	76

    a.   Run the analysis of covariance on these data—after first checking the assumption of homogeneity of regression.

    b.   Draw the appropriate conclusions.

16.17 Compute the adjusted means for the data in Exercise 16.16.

16.18. Compute the energy savings per household for the data in Exercise 16.16 by subtracting this year's bill from last year's bill. Then run an analysis of variance on the savings scores and compare that to the analysis of covariance.

16.19 Use any statistical package to perform an analysis on the data from Klemchuk, Bond, and Howell (1990) in Table 13.11. Compare your answers with the answers in part (c) of that table.

Computer Exercises

16.20 Use the data set named in Epinuneq.dat (from the website www.uvm.edu/~dhowell/ methods/) to examine the results of the study by Introini-Collison and McGaugh (1986) described before Exercises 11.28–11.31. Using any statistical package, run a two-way analysis of variance with unequal sample sizes. What would you conclude from this analysis?

16.21 Use the data from Mireault (1990) in the file named Mireault.dat referred to in Exercise 7.6 to run a two-way analysis of variance on the Global Symptom Index T score (GSIT) using Gender and Group as independent variables. Plot the cell means and interpret the results.

16.22 Using the same data as in Exercise 16.21, run an analysis of covariance instead, using year in college (YearColl) as the covariate.

   a.   Why would we want to consider YearColl as a covariate?

   b.   How would you interpret the results?

16.23 In Exercise 16.22, we used YearColl as the covariate. Run an analysis of variance on YearColl, using Gender and Group as the independent variables. What does this tell us that is relevant to the preceding analysis of covariance?

16.24 Everitt reported data on a study of three treatments for anorexia in young girls. One treatment was cognitive behavior therapy, a second was a control condition with no therapy, and a third was a family therapy condition. These are the same data we examined in Chapter 14. The data follow.

Group	Pretest	Posttest	Gain	Group	Pretest	Posttest	Gain
1	80.5	82.2	1.7	1	80.2	82.6	2.4
1	84.9	85.6	.7	1	87.8	100.4	12.6
1	81.5	81.4	−.1	1	83.3	85.2	1.9
1	82.6	81.9	−.7	1	79.7	83.6	3.9
1	79.9	76.4	−3.5	1	84.5	84.6	.1
1	88.7	103.6	14.9	1	80.8	96.2	15.4
1	94.9	98.4	3.5	1	87.4	86.7	−.7
1	76.3	93.4	17.1	2	80.7	80.2	−.5
1	81.0	73.4	−7.6	2	89.4	80.1	−9.3
1	80.5	82.1	1.6	2	91.8	86.4	−5.4
1	85.0	96.7	11.7	2	74.0	86.3	12.3
1	89.2	95.3	6.1	2	78.1	76.1	−2.0
1	81.3	82.4	1.1	2	88.3	78.1	−10.2
1	76.5	72.5	−4.0	2	87.3	75.1	−12.2
1	70.0	90.9	20.9	2	75.1	86.7	11.6
1	80.4	71.3	−9.1	2	80.6	73.5	−7.1
1	83.3	85.4	2.1	2	78.4	84.6	6.2
1	83.0	81.6	−1.4	2	77.6	77.4	−0.2
1	87.7	89.1	1.4	2	88.7	79.5	−9.2
1	84.2	83.9	−.3	2	81.3	89.6	8.3
1	86.4	82.7	−3.7	2	78.1	81.4	3.3
1	76.5	75.7	−.8	2	70.5	81.8	11.3

Group	Pretest	Posttest	Gain	Group	Pretest	Posttest	Gain
2	77.3	77.3	0.0	3	82.5	91.9	9.4
2	85.2	84.2	−1.0	3	86.7	100.3	13.6
2	86.0	75.4	−10.6	3	79.6	76.7	−2.9
2	84.1	79.5	−4.6	3	76.9	76.8	−0.1
2	79.7	73.0	−6.7	3	94.2	101.6	7.4
2	85.5	88.3	2.8	3	73.4	94.9	21.5
2	84.4	84.7	0.3	3	80.5	75.2	−5.3
2	79.6	81.4	1.8	3	81.6	77.8	−3.8
2	77.5	81.2	3.7	3	82.1	95.5	13.4
2	72.3	88.2	15.9	3	77.6	90.7	13.1
2	89.0	78.8	−10.2	3	83.5	92.5	9.0
3	83.8	95.2	11.4	3	89.9	93.8	3.9
3	83.3	94.3	11.0	3	86.0	91.7	5.7
3	86.0	91.5	5.5	3	87.3	98.0	10.7

a. Run an analysis of variance on group differences in Gain scores. (You may have already done this in Chapter 14.)

b. Now run the analysis on posttest scores, ignoring pretest scores.

c. Finally, run the analysis on posttest scores using pretest as the covariate.

d. How do these three answers relate to one another, and what do they show about the differences and similarities between analysis of covariance and the treatment of gain (or change) scores?

e. Calculate η^2 on Groups for the analysis of covariance.

f. Calculate d for the contrast on the two therapy groups (ignoring the control group) using adjusted means.

Discussion Questions

16.25 I initially thought of creating an analysis of variance example from the example in Chapter 14, Section 14.7. I could have used Sex and Group as the independent variables, posttest scores as the dependent variable, and pretest scores as the covariate (ignoring FU6 and FU12 entirely). This would have made a very bad example for the analysis of covariance. Why would that be? Is there any way in which we might be able to salvage the study as an analysis of covariance example?

16.26 I said that in any experiment where we have pretest and posttest scores we could either look at the difference scores (compared across groups) or use the pretest as a covariate. These two analyses will be equivalent only when the slope relating posttest to pretest scores is 1.00. How likely do you think it is that such a condition would be met (or at least approximated)? Would does $b = 1.00$ actually imply?

16.27 Make up or find an example with respect to Exercise 16.25 where the slope is not nearly 1.0. Analyze it using both the analysis of covariance and a t test on difference scores. Do either of these analyses make sense?

CHAPTER 17

Log-Linear Analysis

Objectives

To present log-linear models as ways of exploring discrete data from experiments having multiple independent variables.

Contents

MOST OF THIS BOOK has been concerned with variables that are measured on a more-or-less continuous scale and for which the mean, or a related sample statistic, would be a typical measure of interest. However, many variables we deal with are measured categorically, such as the classic study by Geller, Witmer, and Orebaugh (1976), discussed in Chapter 6, in which a supermarket flyer on "daily specials" was categorized both by whether it contained a message about littering and where it was found at the end of the day (trash can, litter, removed from store). In that particular example, we were able to show that where a notice was left depended on whether it contained a message about littering. In other words, the two variables are not independent—they interact.

Experimenters faced with multiple categorical variables have often dealt with them two at a time, creating two-way contingency tables and computing the standard Pearson chi-square test statistic to check for independence. Recently, however, major efforts to develop procedures that deal with multiple categorical variables simultaneously have been undertaken. (I say "recently" because even though the important work in this field started with Leo Goodman at the University of Chicago in the 1960s, it generally takes at least 20 to 30 years for statistical procedures to work their way from initial development in the statistical journals, to occasional appearance in the experimental literature, to widespread accep-

Log-linear models

tance. **Log-linear models** are just beginning to make it to the latter stage.)

The presentation of log-linear models presents several challenges. In the first place, such models are much easier to understand when presented as simple contingency tables with two dimensions (variables). However, the two-dimensional case is not handled appreciably better by log-linear models than by the standard approach, and the reader can easily be left wondering "So what?" Log-linear models come into their own with three-, four-, or higher-dimensional cases, but the explanation can become unpleasantly tortuous and opaque. For this reason, we will start with the two-dimensional case, lay out most of the reasoning, and then move on to higher dimensions.

A second problem with log-linear models is that each author views them from a different perspective. If you skim several of the excellent books on such models, you might almost think that they were talking about different topics. Some authors are interested primarily in hypothesis testing, whereas others are interested primarily in model building. Some concentrate on examining individual effects, whereas others mention individual effects only in passing. Some concentrate on models in which all of the variables are treated as independent variables, whereas others focus on cases in which one or more variables are thought of as dependent variables and the others as independent variables. This chapter will try to steer a middle course, focusing on those aspects of the models that apply most directly to psychology and related disciplines. I recommend that you concentrate on the hypothesis testing aspects of log-linear models the first time through. Then go back and pay more serious attention to estimating treatment effects.

A number of excellent references on this subject are available. Some of the clearest are Agresti (1984, 1990—especially the former), Green (1988), Kennedy (1983), Marascuilo and Serlin (1990), and Wickens (1989), which is very complete and readable. An excellent presentation of the applications of standard computer software is given in Tabachnick and Fidell (1996). I have borrowed heavily from all of these sources.

My motivation in writing this chapter is a little different from the motivation for other chapters. There is a great deal of technical information here that I would not expect my students to grapple with until they had a particular need for the material. I think that this chapter is most likely to be read by someone who has found herself with a set of data on categorical variables and knows (or has been told) that log-linear models might be the way to go. This chapter was written primarily from the point of view of helping that person wade through complex and confusing material on the topic. I try to explain what all those terms are all about, and why you would care. I also try to explain what various sections of the computer printout mean. (I use SPSS GENLOG as my example program, but if you have a different program, just apply it to the example data and note the parallels in the printout. But don't be put off if your answers are slightly different. Different programs use different algorithms, which come to slightly different answers. That's not nice, but it is the way things are.)

Symmetric and Asymmetric Models

In general, log-linear analysis treats independent and dependent variables alike, ignoring the distinction between them. We as experimenters, however, build our interpretation of the data partly on whether a variable is seen, by us, as independent or dependent.

To take a simple example, suppose that we have developed a scale of myths related to rape ("If a woman is raped, she was probably partly responsible") and myths related to spouse abuse ("An abused wife is always free to leave her abuser"). Suppose further that our subjects have responded Agree, Neutral, or Disagree with the terms on both scales. If we want to look at the relation between the rape myths and spouse abuse myths, neither variable would be dependent relative to the other. This would remain so if we added yet another dimension and categorized subjects in terms of other sorts of beliefs (e.g., just-world beliefs). Relations of this sort, in which all the variables are treated alike as dependent variables, are classed as **symmetric relationships.**

symmetric relationships

Now suppose that we take another variable (Gender) and look to see whether there are differences in rape myths between males and females. Here, most people would see Gender as an independent variable and Rape myth as a dependent variable. We would account for Rape myth as a function of Gender, but would be unlikely to account for Gender on the basis of Rape myth. This is an **asymmetric relationship.**

asymmetric relationship

Log-linear models apply to both symmetric and asymmetric models. The difference comes more in the interpretation than in the mathematics. When you have an asymmetric model, you will focus more on the dependent variable and its relations with independent variables. When the model is symmetric, you will spread your interest more widely. In addition, with asymmetric models you may choose to keep certain nonsignificant variables in the model on the basis of their role in the study. With symmetric models, we are more even-handed.

17.1 Two-Way Contingency Tables

We will begin with the simplest example of a 2×2 contingency table. Although log-linear analysis does not have a great deal more to offer than the standard Pearson chi-square approach does in this situation, it will allow us to examine a number of important concepts in a simple setting. (Agresti [2002] suggests that with a single categorical response (dependent) variable, it is simpler to use logistic regression.) As we move to more complex situations, we will leave more and more of the actual calculations to computer software because such computations can become extremely cumbersome.

As an example, we will use a study by Pugh (1983) on the "blaming-the-victim" issue in prosecutions for rape. Pugh's paper is an excellent example of how to use log-linear analysis to establish a statistical model to explain experimental results. But we will, at first, simplify the underlying experiment to create an example that is more useful for our purposes. The simplification involves collapsing over, and thereby ignoring, some experimental variables. (In general, we would not collapse across variables unless we were confident that they did not play a role or we were not interested in any role they did play.)

Pugh designed a study to examine what many have seen to be the disposition of jurors to base their judgments of defendants on the alleged behavior of the victim. Defense attorneys often adopt a strategy suggesting that the victim is in some way responsible for the crime. By attacking the victim's past behavior, the victim is put on trial instead of the defendant. Pugh's study varied the gender of the juror, the level of stigma attached to the victim, and the degree to which the juror could assign fault to the victim, and then looked at the degree to which the defendants were judged guilty or not guilty. For our first example, we will collapse over two of those variables and look at the relationship between the degree to which the victim was believed to be "at fault" and the verdict. These data are shown in Table 17.1. (Expected frequencies for the standard test of the independence of these two variables are shown in parentheses.)

Table 17.1 Data from Pugh (1983) collapsed across two variables

| | | Verdict | | |
		Guilty	Not Guilty	Total
	Low	153	24	177
		(127.559)	(49.441)	
Fault	High	105	76	181
		(130.441)	(50.559)	
	Total	258	100	358

If we ran the standard Pearson chi-square test on these data, we would find, with a minor change in notation,

$$\chi_1^2 = \sum \frac{(O - E)^2}{E} = \sum \frac{(f_{ij} - F_{ij})^2}{F_{ij}} = 35.93$$

which is significant at $\alpha = .05$. The change in notation, equating f_{ij} with the observed frequency in cell$_{ij}$ and F_{ij} with the expected frequency in that cell, was instituted to bring the notation in line with the standard notation used with log-linear analysis.

If we calculate the likelihood ratio χ^2 (see Section 6.8) instead of the Pearson's chi-square, we would have

$$\chi^2 = 2 \sum f_{ij} \ln \left(\frac{f_{ij}}{F_{ij}} \right)$$

$$= 2 \left(153 \ln \frac{153}{127.559} + 24 \ln \frac{24}{49.441} + 105 \ln \frac{105}{130.441} + 76 \ln \frac{76}{50.559} \right)$$

$$= 37.3503$$

which is also approximated by the χ^2 distribution on 1 degree of freedom.[1] Again, we would reject the null hypothesis of independence of rows and columns. We would conclude that in making a judgment of guilt or innocence, the jurors base that judgment, in part, on the perceived fault of the victim.

The use of the chi-square test, whether using Pearson's statistic or the likelihood ratio statistic, focuses directly on hypothesis testing. (*From this point on, all χ^2 statistics will be likelihood ratio χ^2s unless otherwise noted.*) But we can look at these data from a different perspective—the perspective of model building. We saw the modeling approach clearly in the analysis of variance where we associated a two-way factorial design with the model

$$X_{ijk} = \mu + \alpha_i + \beta_j + \alpha\beta_{ij} + e_{ijk}$$

---

[1] In this chapter, we will frequently refer to natural logarithms. These are normally abbreviated as ln or as $\log_e$. We will use ln throughout.

In the case of the analysis of variance, we first posited this model to underlie the obtained data and then used the model and its associated error term to develop tests of the components of that model. When the data analysis was complete, we let the model stand but made statements of the form, "There is a significant effect due to variable A and the $A \times B$ interaction, but there is no significant difference due to variable B."

In Chapter 15, Section 15.10 on stepwise multiple regression, we reversed the process. We used the data themselves to create a model rather than using an a priori model as in the analysis of variance. Using the backward solution, which is most relevant here, we continued to remove variables from our model as long as their removal did not produce a significant decrement in R^2 (or until we met some similar criterion). When we were done, we were left with a model all of whose components contributed significantly to the prediction of Y.

For log-linear models, we generally fall somewhere between these two approaches. We use a model-building approach, as in the regression situation, but the resultant model may, as in the analysis of variance, contain nonsignificant terms.

Consider Pugh's data and a variety of different models that *might* be posited to account for those data. I don't remotely believe that the first few models are likely to be true, but they are possible models, and they are models that you must understand. Moreover, they are models that might be included in a complete analysis, if only as a basis for comparison of alternative models.

Equiprobability Model

At the simplest level, we might hypothesize that respondents distribute themselves among the four cells at random. In others words, $p(\text{Low, Guilty}) = p(\text{High, Guilty}) = p(\text{Low, Not Guilty}) = p(\text{High, Not Guilty}) = .25$. This model basically says that nothing interesting is going on in this study and one-quarter of the subjects ($.25 \times 358 = 89.5$) would be expected to fall in each cell.

Using the likelihood ratio χ^2 to test this model, we have

Observed:	153	24	105	76
Expected:	89.5	89.5	89.5	89.5

$$
\begin{aligned}
\chi^2 &= 2 \sum f_{ij} \ln \left(\frac{f_{ij}}{F_{ij}} \right) \\
&= 2 \left(153 \ln \frac{153}{89.5} + 24 \ln \frac{24}{89.5} + 105 \ln \frac{105}{89.5} + 76 \ln \frac{76}{89.5} \right) \\
&= 109.5889
\end{aligned}
$$

This can be evaluated as a χ^2 on $4 - 1 = 3$ *df* (we lose one degree of freedom due to the restriction that the cell totals must sum to N), and from Appendix χ^2 we find that $\chi^2_{.05}(3) = 7.82$. We can reject H_0 and conclude that this model does not fit the data. In other words, the individual cell frequencies cannot be fit by a model in which all cells are considered equally probable. Notice that rejection of H_0 is equivalent to rejection of the underlying model. This is an important point and comes up whenever we are trying to decide on a suitable model.

Conditional Equiprobability Model

Our first model really had no variable contributing to the observed frequency (not differences due to Fault, not differences due to Verdict, and not differences due to the interaction of those variables). A second model, however, might hold that the individual cell frequencies represent differences due to assignment of Verdict because noticeably more people were found guilty than were found innocent. (Notice that Verdict is likely to be an important variable not because it has any theoretical significance, but because there were more Guilty verdicts than NonGuilty ones, and we have to take that into account.)[2] By this model, $258/358 = 72.1\%$ of the observations fall in column 1 and 27.9% fall in column 2. Beyond that, however, observations are assumed to be equally likely to fall in rows 1 and 2. In other words, the null hypothesis states that once we have conditioned on the judgment of guilt or innocence) (i.e., adjusted for the fact that more people were judged guilty than not guilty), assignment to Fault levels is equally probable. By this model, we would have the expected frequencies (shown in parentheses) contained in Table 17.2. (The expected frequencies in this model came from assuming that half of the column 1 total would fall in row 1 and half in row 2; similarly for column 2.)

Table 17.2   Observed and expected frequencies for the first conditional equiprobability model

		Verdict		Total
		Guilty	Not Guilty	
	Low	153	24	177
		(129)	(50)	
Fault	High	105	76	181
		(129)	(50)	
	Total	258	100	358

$$\chi^2 = 2 \sum f_{ij} \ln\left(\frac{f_{ij}}{F_{ij}}\right)$$
$$= 2\left(153 \ln\frac{153}{129} + 24 \ln\frac{24}{50} + 105 \ln\frac{105}{129} + 76 \ln\frac{76}{50}\right)$$
$$= 37.3960$$

This model has $4 - 2 = 2$ degrees of freedom because we have imposed two restrictions—the cell frequencies in each column must sum to the expected frequency for that column. Because $\chi^2_{.05}(2) = 5.99$, we will again reject H_0 and conclude that the model does not fit the observed data either.

A second conditional equiprobability model could be created by assuming that cell frequencies are affected only by differences in levels of Fault. In this case, probabilities are equal within each Fault condition but different between them. The expected frequencies in this case are given in Table 17.3.

This χ^2 has $4 - 2 = 2$ degrees of freedom for the same reason that the model in Table 17.2 did, and again the significant χ^2 shows that this model is an inadequate fit to the data. Thus, we have so far concluded that the data *cannot* be explained by assuming that observations fall in the four cells at random. Nor can they be explained by positing differences due simply to an unequal distribution across either Verdict or Fault. More would appear to be happening in the data. The next step would be to propose a model involving both Verdict and

---

[2] To emphasize this point, assume that I wanted to look at Gender and Age in studying Intelligence, and suppose that I had 150 women and 10 men in my sample. (Then Gender will almost certainly have to be in the model even if it has absolutely nothing to do with intelligence—how else would we explain the radical difference in the number of males and females in the cells?)

Table 17.3   Observed and expected frequencies for the second conditional equiprobability model

		Verdict		
		Guilty	Not Guilty	Total
	Low	153	24	177
		(88.5)	(88.5)	
Fault	High	105	76	181
		(90.5)	(90.5)	
	Total	258	100	358

$$\chi^2 = 2 \sum f_{ij} \ln\left(\frac{f_{ij}}{F_{ij}}\right)$$

$$= 2\left(153 \ln \frac{153}{88.5} + 24 \ln \frac{24}{88.5} + 105 \ln \frac{105}{90.5} + 76 \ln \frac{76}{90.5}\right)$$

$$= 109.544$$

Fault *operating independently of one another.* This is the standard null model routinely tested by a chi-square test on a contingency table. It is so standard that we often lose sight of the fact that it is the model we usually test (and hope to reject).

Mutual Dependence Model

We are now testing a model that assumes that two factors operate jointly, *but independently,* to produce expected cell frequencies. If the two variables are independent, then

$$F_{ij} = \frac{RT \times CT}{GT} = \frac{f_{i\cdot} \times f_{\cdot j}}{f_{\cdot\cdot}}$$

where RT stands for the row total, CT for the column total, GT for the grand total, and the "dot notation" is used to show that we have collapsed across that dimension. This is the same formula for expected frequencies that we saw in Chapter 6.

We began this chapter by testing this hypothesis of independence. The expected frequencies and the likelihood ratio χ^2 are given on page 608. From those calculations, we found that

$$\chi^2 = 37.3503$$

which is significant on 1 *df.* Thus, we can further conclude, and importantly so in this case, that a model that posits an independence between Fault and Verdict also does not fit the data. The only conclusion remaining is that the likelihood of a jury convicting a defendant of rape depends on an interaction between Fault and Verdict. Perceived guilt is, in part, a function of the blame that is attributed to the victim.

If we use the model that includes both Verdict and Fault, as well as their interaction, we **saturated model** will have what is called a **saturated model.** This model has as many parameters (a mean, a row effect, a column effect, and an interaction effect) as it has cells, and it is guaranteed to fit perfectly with $\chi^2 = 0$.

The models that we have examined are listed in Table 17.4—for the moment you can ignore the second column. You can see that all but the last one fail to fit the data (i.e., have a significant χ^2. That means that unless you allow for an interaction of the variables, you will not be able to fit the data adequately. Thus, Verdict and Fault interact.)

Table 17.4 Five possible models for data in Table 17.1

Model	Representation	χ^2	df	p
1. Equiprobability	$\ln(F_{ij}) = \lambda$	109.5889	3	<0.05
2. Conditional Equiprobability	$\ln(F_{ij}) = \lambda + \lambda_i^V$	37.3960	2	<0.05
3. Conditional Equiprobability	$\ln(F_{ij}) = \lambda + \lambda_j^F$	109.5442	2	<0.05
4. Independence	$\ln(F_{ij}) = \lambda + \lambda_i^V + \lambda_j^F$	37.3503	1	<0.05
5. Saturated	$\ln(F_{ij}) = \lambda + \lambda_i^V + \lambda_j^F + \lambda_{ij}^{VF}$	0.00	0	—

17.2 Model Specification

The models we have been discussing can be represented algebraically as well as descriptively. The algebraic notation can seem awkward, but it allows us to learn a great deal more about the data. It is somewhat confusing because we start out with one set of parameters, represented as τ (tau), usually with a superscript, and then shortly convert to the natural logarithm of τ, represented as λ (lambda), with superscripts. Both of these statistics strongly resemble the grand mean (μ) and treatment effects (α, β, and $\alpha\beta$) that we saw in the analysis of variance. (You might think that we would be satisfied with one or the other, but both have their uses.) I would urge you to read the next two sections fairly quickly just to see where we are heading, and then come back to it after you see how such parameter estimates are used in more complex models.

The following gets a bit confusing at first, but it's not really that hard. Remember in the analysis of variance that we had models like

$$X_{ijk} = \mu + \alpha_i + \beta_j + \alpha\beta_{ij} + e_{ijk}$$

All that I'm going to do is derive some terms that are parallel to these. First, you'll see τ. Think of it as μ. Then you'll see τ_i^F and τ_j^V. Think of these as α_i and β_j. I'll make mention of τ_{ij}^{FV}—you can guess what that is like. And finally, I'll take logs of all this stuff. That's just so that I can add them up the same way we added μ, α_i, β_j, and $\alpha\beta_{ij}$, to get an expected frequency.

geometric mean

In the simplest equiprobability model, all cell frequencies are explained by a single parameter τ, where τ is estimated by the **geometric mean** of the expected cell frequencies given by the model. In other words,

$$F_{ij} = \hat{\tau}$$

(This model corresponds to the equiprobability model discussed in the previous section.) A geometric mean is the nth root of the product of n terms, so in this case, the geometric mean of the four expected frequencies is

$$\sqrt[4]{(89.5)(89.5)(89.5)(89.5)} = 89.5$$

which is not a very exciting result.

For the first conditional equiprobable models, we have to go further. We again define $\hat{\tau}$ as the geometric mean of the expected cell frequencies in that model, but here those expected frequencies are different from the equiprobability model because they take differences caused by Verdict into account.

$$\hat{\tau} = \sqrt[4]{(129)(129)(50)(50)} = 80.3119$$

We also define $\hat{\tau}_1^V$ (where the superscript "V" stands for "Verdict") as the ratio of the geometric mean of the expected frequencies for the first (Guilty) column to the geometric mean of the expected frequencies of all the cells (the grand mean) ($\hat{\tau}$). Then

$$\hat{\tau}_1^V = \frac{\sqrt{(129)(129)}}{\hat{\tau}} = \frac{129}{80.3119} = 1.6062$$

You can think of $\hat{\tau}_j^V$ very much the way you thought of the treatment effect (β_j) in the analysis of variance. It is the contribution of column$_j$. But for the analysis of variance, β_j was the amount that was *added* to the grand mean to obtain the column mean. Here, on the other hand, $\hat{\tau}_j^V$ is the amount by which we *multiply* $\hat{\tau}$ to obtain the column's expected frequency. $\hat{\tau}_j^V = 1.6062$ says that the column one expected frequency is 1.6062 times larger than the overall mean—or 160.62% of it. For the Not Guilty column,

$$\hat{\tau}_2^V = \sqrt{\frac{(50)(50)}{\hat{\tau}}} = \frac{50}{80.3119} = 0.6226$$

Then, we can show that for this model

$$F_{ij} = \hat{\tau}\hat{\tau}_j^V$$

For cell 11, we would have $80.3119 \times 1.6062 = 129$, which has reproduced the expected frequency that we used in Table 17.2.

We have a similar model when we consider just the Fault variable instead of just the Verdict variable. Here we have

$$F_{ij} = \hat{\tau}\hat{\tau}_i^F$$

To go one step further, we can consider the independence model (Table 17.1), which contained both Fault and Verdict effects but not their interaction. Here, we will need both τ_j^V and τ_i^F to account for both Verdict and Fault. Working with the expected frequencies from the independence model we have the following:

$$\hat{\tau} = \sqrt[4]{(127.559)(49.441)(130.441)(50.559)} = 80.3069$$

$$\hat{\tau}_1^V = \frac{\sqrt{(127.559)(130.441)}}{\hat{\tau}} = \frac{128,9920}{80.3069} = 1.6062$$

$$\hat{\tau}_2^V = \frac{\sqrt{(49.441)(50.559)}}{\hat{\tau}} = \frac{49.9969}{80.3069} = 0.6226$$

$$\hat{\tau}_1^F = \frac{\sqrt{(127.559)(49.441)}}{\hat{\tau}} = \frac{79.4144}{80.3069} = 0.9889$$

$$\hat{\tau}_2^F = \frac{\sqrt{(130.441)(50.559)}}{\hat{\tau}} = \frac{81.2094}{80.3069} = 1.0112$$

Then, for example,

$$F_{11} = \hat{\tau}\hat{\tau}_1^V\hat{\tau}_1^F = 80.3069 \times 1.6062 \times 0.9889 = 127.557$$

which agrees, within rounding error, with the actual expected value for the independence model. You should verify for yourself that in the general case, for the independence model, the expected frequency for cell$_{ij}$ is

$$F_{ij} = \hat{\tau}\hat{\tau}_i^V\hat{\tau}_j^F$$

I have led you through the last few paragraphs to make a simple but very important point. In the analysis of variance, we wrote an *additive* linear model for observations in each cell as

$$X_{ijk} = \mu + \alpha_i + \beta_j + \alpha\beta_{ij} + e_{ijk}$$

With log-linear models of categorical data, we have seen that we can write the *multiplicative* independence model for expected cell *frequencies* as

$$F_{ij} = \hat{\tau}\hat{\tau}_i^V\hat{\tau}_j^F$$

This model is multiplicative rather than additive and doesn't look much like the analysis of variance model.[3] But if you recall your high-school algebra, you will remember that products become sums when you take logs. Thus, we can convert the preceding expression to

$$\ln(F_{ij}) = \ln(\hat{\tau}) + \ln(\hat{\tau}_i^V) + \ln(\hat{\tau}_j^F)$$

and we have something that very closely resembles the analysis of variance model. We can then confuse everyone a little more by substituting the symbol λ (**lambda**) to represent the natural log of τ and have

λ (lambda)

$$\ln(F_{ij}) = \lambda + \lambda_i^V + \lambda_j^F$$

which is an additive linear expression directly analogous to the model we had for the analysis of variance. This model is linear in the logs, hence the name *log-linear models*.

To summarize, in the analysis of variance, we modeled expected cell means as the sum of the grand mean and row and column treatment effects. In log-linear models, we model the log of expected cell frequencies as the sum of the *logs* of the overall geometric mean and the row and column effects. The arithmetic is slightly different and we are modeling different things, but the logic is the same.

Given my new notation, I can now go back and characterize the separate models by their underlying equations. The models are numbered in the order of their presentation and were shown in column two of Table 17.4.

1. Equiprobability model: $\ln(F_{ij}) = \lambda$
2. Conditional equiprobability model 1: $\ln(F_{ij}) = \lambda + \lambda_i^V$
3. Conditional equiprobability model 2: $\ln(F_{ij}) = \lambda + \lambda_j^F$
4. Mutual independence model: $\ln(F_{ij}) = \lambda + \lambda_i^V + \lambda_j^F$
5. Saturated model: $\ln(F_{ij}) = \lambda + \lambda_i^V + \lambda_j^F + \lambda_{ij}^{VF}$

The interaction term (λ_{ij}^{VF}) in the saturated model is defined as what is left unexplained when we fit model 4 earlier. Thus,

$$\lambda_{ij}^{VF} = \ln(f_{ij}) - \lambda - \lambda_i^V - \lambda_j^F = \ln(f_{ij}) - \ln(F_{ij})$$

In this model, every expected frequency is forced to be exactly equal to every obtained frequency, and χ^2 will be exactly 0.00. A saturated model *always* fits the data perfectly.

Whereas in the analysis of variance we usually set up the complete model and *test* for interaction, the highest-order interaction in the log-linear analysis is not tested directly. The interaction model in the $R \times C$ case is basically the model that we adopt if the simpler mutual independence model (also called the **additive model**) does not fit.

additive model

17.3   Testing Models

The central issue in log-linear analysis is the issue of choosing an optimal model to fit the data. In a normal chi-square test on a two-dimensional contingency table, we just jump in, posit what I have called the additive model, and, if we reject it, conclude that an interaction

---

[3] In the analysis of variance, we have variation within cells and can thus calculate an error term. In log-linear models, we are working with cell frequencies and will not have an error term. Therefore, our models will have nothing comparable to e_{ijk}.

term is necessary because the variables are not independent. I have done some of that here, but I have shown you five possible models instead of one or two. That would certainly be unnecessary if we were just interested in the two-variable case where we have only one serious alternative to the saturated model. But the approach will be very useful when we come to more complex designs.

From Table 17.4 (on page 612), we see that the first four models all have significant χ^2 values. This means that for each of these models, there is a significant difference between observed and expected values; *none of them fits the obtained data.* From such results, we must conclude that only a model that incorporates the interaction term can account for the results. Thus, as we have previously concluded, Fault and Verdict interact and, within the context of Pugh's experiment, we cannot model the data without considering this interaction. Because, for Pugh, Verdict is a dependent variable, we conclude that decisions about guilt or innocence are dependent on perceptions about Fault. (This is one of the few places in inferential statistics where we actually seek nonsignificant results.)

From the point of view of fitting models, these results suggest that we should conclude that

$$\ln(F_{ij}) = \lambda + \lambda_i^V + \lambda_j^F + \lambda_{ij}^{VF}$$

But when viewed from the perspective of the analysis of variance, something is missing in such a conclusion. In the analysis of variance, we start out with (and generally retain) a model such as this, but we also test the individual elements of the model. In other words, we asked, "Within the complete model are there significant effects due to V, to F, and to their interaction?" That is a question we haven't really asked here. When we tested the model $\ln(F_{ij}) = \lambda + \lambda_j^F$, we were asking whether such a model fit the data, but we were not asking the equally important question, When we *adjust for other effects,* is there a difference attributable to Fault?

There are two ways of asking these questions using log-linear models—the easy way and the harder way, paralleling what we did in the equal-n case of the analysis of variance in Chapter 16. The advantage of the more complicated way is that it generalizes to the process we will use on interactions in more complex designs.

Let's start with the easy way because it supplies a frame of reference. If you want to know whether there is a difference in the data attributable to Verdict (i.e., are there significantly more decisions of Guilty than Not Guilty), why not just ask that question directly by looking at the marginal totals? In other words, just run a one-dimensional likelihood ratio χ^2, as shown in Table 17.5. The $\chi^2 = 72.1929$ is a significant result on 1 *df*, and we would conclude that there is a difference in the number of cases judged guilty and not guilty.

Table 17.5   Test on differences due to Verdict

	Verdict	
	Guilty	Not Guilty
f_{ij}	258	100
F_{ij}	179	179

$$\chi^2 = 2 \sum f_{ij} \ln\left(\frac{f_{ij}}{F_{ij}}\right)$$
$$= 2\left(258 \ln \frac{258}{179} + 100 \ln \frac{100}{179}\right)$$
$$= 72.1929$$

Now let's ask the same question about low and high levels of Fault (see Table 17.6). This effect (0.0447) is clearly not significant—nor would Pugh have expected it to be given the design of the experiment, which deliberately placed nearly equal numbers of observations in the two levels of Fault.

Table 17.6 Test on differences due to Fault

| | Fault | |
	Low	High
f_{ij}	177	181
F_{ij}	179	179

$$\chi^2 = 2 \sum f_{ij} \ln \left(\frac{f_{ij}}{F_{ij}} \right)$$

$$= 2 \left(177 \ln \frac{177}{179} + 181 \ln \frac{181}{179} \right)$$

$$= 0.0447$$

The interaction itself we have already tested at the beginning of the chapter. There we found that $\chi^2 = 37.3503$, and we concluded that Fault and Verdict were not independent.

We have run each of these tests separately. Now let's see how we can derive them from the log-linear models that we have already created. (In higher-order designs we can still test the effect of single variables [what the analysis of variance labels as main effects], but not interaction, in the way we just did. However, the model-comparison approach to be adopted generalizes to interaction effects as well.)

We have found that the simplest model $[\ln(F_{ij}) = \lambda]$ produces a $\chi^2 = 109.5889$. When we added λ^V to this model, χ^2 dropped to 37.3960, reflecting the variation in cell frequencies attributable to Verdict. This drop $(109.5889 - 37.3960 = 72.1929)$ is the χ^2 for Verdict, and its degrees of freedom equal the difference between the degrees of freedom in the two models $(3 - 2 = 1)$. This is exactly the same value we obtained in Table 17.5 when we compared the marginal frequencies. In other words, adjusting $\lambda + \lambda_i^V$ for λ yields the same result as basing our results on the marginals.

By a similar line of reasoning, we can note that taking Fault into account and going from $\ln(F_{ij}) = \lambda$ to $\ln(F_{ij}) = \lambda + \lambda_j^F$ reduces χ^2 from 109.5889 to 109.5442, for a decrease of 0.0447. This is the same as the marginal χ^2 on Fault that we obtained in Table 17.6.

Finally, we should note that when we go from a model of $\ln(F_{ij}) = \lambda + \lambda_i^V + \lambda_j^F$ to $\ln(F_{ij}) = \lambda + \lambda_i^V + \lambda_j^F + \lambda_{ij}^{VF}$, χ^2 drops from 37.3503 to 0.00. This drop (37.3503) is the same as the χ^2 for the interaction based on marginal frequencies. This equality will not generally hold for more complex designs unless we are looking at the highest-order interaction.

One other feature of log-linear models should be mentioned. The minimal model $[\ln(F_{ij}) = \lambda]$ produced $\chi^2 = 109.5889$. The individual components of the saturated model had χ^2 values of 72.1929, 0.0447, and 37.3503. These sum to 109.5889. In other words, these likelihood ratio χ^2 values are additive. This would not have been the case had we computed the Pearson chi-square statistic instead, which is one very good reason to concentrate on likelihood ratio χ^2.

At this point, you should have an overview of parameter estimates. It would be smart to go back to the beginning of Section 17.2 and reread that section. (I would if I were you, and I wrote the chapter.)

Differences Between Log-Linear Models and the Analysis of Variance

Although I have frequently compared the analysis of variance and log-linear models and pointed to the many real similarities between the two techniques, this comparison may at times lead to confusion. The purpose behind the models is not quite the same in the two situations. The analysis of variance models cell means, whereas log-linear analysis models cell frequencies.

To take a simple example, assume that we have an experiment looking at the effects of Previous artistic experience and Gender (two independent variables) on the quality of a written Composition (the dependent variable). First, suppose that Composition is measured on a continuous scale, that Artistic experience and Gender are dichotomies, and that we have 20 male and 40 female subjects. Note those cell sizes; they are important! Further assume that Gender has absolutely nothing to do with Composition. Then in an analysis of variance framework with Gender (β_j) included, our model would be

$$X_{ijk} = \mu + \alpha_i + \beta_j + \alpha\beta_{ij} + e_{ijk}$$

Here we would expect the main effect of Gender to be 0.00 because we have assumed the condition that Gender does not influence Composition. On the other hand, if differences did exist between the quality of Composition for males and females, a significant main effect would appear. The presence or absence of an effect due to Gender relates to whether or not male and female subjects differ on the scores on Composition.

Now assume the same experiment, again with 20 males and 40 females, but this time record Composition scores as high, medium, and low, and include Composition as a categorical variable in our log-linear model. We fit a log-linear model to these data. This time, even if there are no differences in Composition between males and females, we will still need to include Gender in our model, *and its effects will be significant.* The reason is quite simple. With our log-linear model, we are *not* trying to model mean Composition; we are trying to model cell frequencies. We are not trying to ask whether males have better composition scores than females. We are trying to explain why there are more scores in some cells than in others. Those cells dealing with female subjects will have relatively larger frequencies than do those cells with male subjects (all other things equal) because there are more female subjects. Similarly, if we had equal numbers of male and female subjects, even with huge differences in quality of Composition between the two sexes, the effect of Gender would be 0.00.

I point this out, and will come back to it again, because it is too easy and seductive to see Gender playing the same role in the two kinds of experiments. In fact, in asymmetric log-linear models, the main effects associated with our independent variables (and their interactions with each other) are often of no interest whatsoever. They may merely reflect our sampling plan. They need to be included to model the data properly, but they do not have a substantive role. In such models, it is the *interaction* of these variables that is of interest (and that parallels main effects in the analysis of variance.)

17.4 Odds and Odds Ratios

Before we move to complex designs, two other basic concepts are more easily explained with simple tables than with higher-order tables. These concepts were discussed in Chapters 6 and 15, but deserve review.

conditional odds Looking at our original data in Table 17.1, we note that in the low fault condition 153 people were found guilty and 24 were found not guilty. Thus, the **conditional odds** of being judged guilty *given* that the victim was seen as low on Fault is $153/24 = 6.3750$. (This can be read to mean that in the low fault group, the odds in favor of being found guilty are 6.3750:1.) For every person in that group who is found not guilty, 6.375 are found guilty. These are equivalent ways of saying the same thing. The conditional odds of being found guilty given that the victim is seen as having a *high* degree of fault are only $105/76 = 1.3816$.

If there had been no interaction between Fault and Verdict, the odds of being found guilty would have been the same in the two Fault conditions. Therefore, the *ratio* of the two odds would have been approximately 1.00. Instead, the ratio of the two conditional odds, the

odds ratio (Ω)

odds ratio (Ω), is $6.3750/1.3816 = 4.6142$. The odds that a defendant will be found guilty in the low fault condition are about 4.6 times that in the high fault condition. The "blame the victim" strategy, whether fair or not, seems to work.

An important feature of the odds ratio is that it is independent of the size of the sample, whereas χ^2 is not. A second advantage is that within the context of a 2×2 table, a test on the odds ratio would be equivalent to a likelihood ratio χ^2 test of independence. A third advantage of Ω is that its magnitude will not be artificially affected by the presence of unequal marginal distributions. In other words, if we doubled the number of cases in the high fault condition (but still held other things constant), χ^2 (either Pearson's or likelihood ratio) and phi would change. The odds ratio (Ω), however, would not be affected.

17.5 Treatment Effects (Lambda)

As we have already seen, log-linear models have a nice parallel with the analysis of variance, and that parallelism extends to the treatment effects. In the analysis of variance, treatment effects are denoted by terms like μ, α_i, β_j, and $\alpha\beta_{ij}$, whereas in log-linear models, we denote these effects by λ, λ_i^V, λ_j^F, and λ_{ij}^{VF}. As you know, log-linear models work with the natural logs of frequencies rather than with the frequencies themselves.

In Section 17.3, we saw that the independence model (the model without the interaction term) did not fit the data from Pugh's study ($\chi^2 = 37.35$). To model the data adequately, we are going to have to use a model that contains the interaction term $\ln(F_{ij}) = \lambda + \lambda_i^V + \lambda_j^F + \lambda_{ij}^{VF}$. Remember that for the fully saturated model, the observed and expected frequencies are the same. Thus, we will start with the logs of these frequencies as the raw data, as shown in Table 17.7. Notice that the table also contains the row and column marginal means and the grand mean.

Table 17.7 Natural logs of cell frequencies

		Verdict		
		Guilty	Not Guilty	Marginals
Fault	High	5.03043	3.17805	4.10424
	Low	4.65396	4.33073	4.49235
	Marginals	4.84220	3.75439	4.29830

Now recall that when we calculated treatment effects in the analysis of variance, we took deviations of means around the grand mean. We will do something similar here. Thus,

ANOVA Effects	Log-linear Model Effects
$\hat{\mu} = \overline{X}_{..}$	$\lambda = 4.29830$
$\hat{\alpha} = \overline{X}_{i.} - \overline{X}_{..}$	$\lambda_1^F = 4.10424 - 4.29830 = -0.19406$
$\hat{\beta}_j = \overline{X}_{.j} - \overline{X}_{..}$	$\lambda_1^V = 4.84220 - 4.29830 = 0.54390$
$\hat{\alpha}\hat{\beta}_{ij} = \overline{X}_{ij} - \overline{X}_{i.} - \overline{X}_{.j} + \overline{X}_{..}$	$\lambda_{11}^{FV} = 5.03043 - 4.10424 - 4.84220 + 4.29830$
	$= 0.38229$

Note the parallelism. Further, $\sum \lambda_i^F = \sum \lambda_j^V = \sum_i \lambda_{ij}^{FV} = \sum_j \lambda_{ij}^{FV} = 0.00$. Thus, we can calculate all the rest of the effects directly.

$$\lambda_2^F = 0.19406$$
$$\lambda_2^V = -0.54390$$
$$\lambda_{12}^{FV} = -0.38229$$
$$\lambda_{21}^{FV} = -0.38229$$
$$\lambda_{22}^{FV} = 0.38229$$

17.6 Three-Way Tables

We now have all the concepts that are necessary to move to more complex designs. Log-linear models come into their own once we move to contingency tables of more than two dimensions. These are the situations in which standard chi-square analyses are not able to reveal a full understanding of the data. In this section, we will concentrate on three-way tables because they illustrate all the essential points. Extrapolation to tables of higher dimensionality is direct. Good examples of the analysis and interpretation of four- and five-way tables can be found in Pugh (1983) and Tabachnick and Fidell (1996), respectively. (If you try to duplicate the results in Pugh's paper, instruct your program to add 0.5 to the cell frequencies in the four-way table before running any analyses. This is normally done anyway on a temporary basis when the program deals with the highest interaction to avoid problems of cell frequencies of zero; ln(0) is undefined. Pugh instructed BMDP4F to leave the 0.5 in while computing all tables.)

When we move beyond a simple $R \times C$ table, the calculations of expected frequencies, especially for interactions involving subsets of variables, become appreciably more complex. Such calculations are usually carried out by an iterative process in which initial estimates are continually refined until the result meets some specified criterion. Most analyses at this level are solved by computers, and that is the approach adopted here. This chapter will focus on analyses computed by SPSS GENLOG, though SPSS has two other procedures for log-linear analysis—SPSS LOGLINEAR and SPSS HILOGLINEAR. SYSTAT, or PROC CATMOD in SAS, are also possibilities. Results of different programs sometimes vary because they use different algorithms for their solutions.

Assumptions

One of the pleasant things about log-linear models is the relative absence of assumptions. Like the more traditional chi-square test, log-linear analysis does not make assumptions about population distributions, although it does assume, as does Pearson's chi-square, that observations are independent. You may apply log-linear analysis in a wide variety of circumstances, including even the analysis of badly distributed (ill-behaved) continuous variables that have been classified into discrete categories.

The major problem with log-linear analysis is the same problem that we encountered with traditional chi-square: The expected frequencies have to be sufficiently large to allow the assumption that frequencies in each cell would be normally distributed over repeated sampling. In the case of chi-square, we set the rule that all (or at least most) of the expected frequencies should be at least 5. We also saw that serious departures from this rule were probably acceptable, as long as all expected frequencies exceeded 1 and 80% were greater than 5. In such cases, however, we would have unacceptably low power. We have a similar situation with log-linear analysis. Once again, we require that at least all cells have expected frequencies greater than 1 and that no more than 20% of the cells have expected frequencies less than 5. The biggest problem comes with what are called

sparse matrices

sparse matrices, which are contingency tables with a large number of empty cells. In these cases, you may want to combine categories on the basis of some theoretical rationale, increase sample sizes, collapse across variables, or do whatever you can to increase the expected frequencies. Regardless of the effects such small cells have on the level of Type I errors, you are virtually certain to have very low levels of power.

Hierarchical and Nonhierarchical Models

hierarchical model

Most, but not all, analyses of log-linear models involve what are called hierarchical models. You can think of a **hierarchical model** as one for which the presence of an interaction term *requires* the presence of all lower-order interactions and main effects involving the components of that higher-order interaction. For example, suppose that we had four variables, *A*, *B*, *C*, and *D*. If you include in the model the three-way interaction *ACD*, a hierarchical model would also have to include *A*, *C*, *D*, *AC*, *AD*, and *CD*, because each of these terms is a subset of *ACD*. Similarly, if your model included *ABC* and *ABD*, the model would actually include *A*, *B*, *C*, *D*, *AB*, *AC*, *BC*, *AD*, and *BD*. It need not include *CD*, *ACD*, *BCD*, or *ABCD* because those are not components of either of the three-way interactions.

Hierarchical models are in many ways parallel to models used in the analysis of variance. If you turn to any of the models in Chapters 13, 14, and 16, you will note that they are all hierarchical—for a three-way analysis of variance, all main effects and two-way interactions are included, along with the three-way interaction. Just as in the analysis of variance, the presence of a term in log-linear models does not necessarily mean that it will make a significant contribution. (If we design a study having exactly as many males as females, the contribution of Gender to a log-linear model will be precisely 0. We still usually include it in the model because of its influence on other expected frequencies.) SPSS HILOGLINEAR, and SYSTAT TABLES handle only hierarchical models. On the other hand, SPSS GENLOG, SPSS LOGLINEAR, SAS PROC CATMOD, and SYSTAT LOGIT are capable of analyzing nonhierarchical models. We will deal only with hierarchical models in this chapter. Agresti (2002, p. 317) states, "Nonhierarchical models are rarely sensible in practice. Using them is analogous to using ANOVA or regression models with interaction terms but without the corresponding main effects."

One of the convenient things about hierarchical designs is that they allow us to specify models very clearly and simply. Assume that we have four variables (*A*, *B*, *C*, and *D*). The notation *ABC* specifies a model that includes the *ABC* interaction, and, because we are speaking about hierarchical models, also includes *A*, *B*, *C*, *AB*, *AC*, and *BC*. We do not have to write out the latter to specify the model (well, you do in SPSS GENLOG)—*ABC* will suffice. Similarly, the label *AB* stands for a model that includes *A*, *B*, and *AB*, but not *C* or any interactions involving *C*. Finally, a model written as *AB*, *ACD* is really the model that involves *A*, *B*, *C*, *D*, *AB*, *AC*, *AD*, *CD*, and *ACD*, but not *BC*, *BD*, *ABC*, *ABD*, or *BCD*. In much of what follows, we will characterize models by the interactions that define them (some-

defining set

generating class

times called their **defining set,** or **generating class**). Because the program that we will use for the following examples (SPSS GENLOG) is not restricted to hierarchical models, if we want to tell it to use a hierarchical model for *AB*, *AC*, we need to explicitly specify the model as *A*, *B*, *C*, *AB*, *AC*. With a program such as SPSS HILOGLINEAR, the same model would be specified as *AB*, *AC* because the rest would be assumed. I chose GENLOG because its printout most nearly fits the material that I want to present.

A Three-Way Example

In the previous section, we examined the relationship between Fault and Verdict in the study of rape by Pugh (1983). Pugh also attempted to manipulate a third variable (Moral) by varying the trial transcript to present the *victim* as someone with "high moral character," "low moral character," or "neutral" on this dimension. We now have three variables by which to categorize the data: Fault (F), Moral (M), and Verdict (V). Fault and Moral refer to characteristics attributed to the victim, whereas Verdict represents a judgment on the defendant. Fault and Verdict each have two levels, whereas Moral has three levels. Pugh's data collapsed across a fourth variable (Gender of juror) are given in Table 17.8.

Table 17.8 Pugh's data collapsed across Gender

Verdict	Fault	Moral			Total
		High	Neutral	Low	
Guilty	Low	42	79	32	153
	High	23	65	17	105
	Total	65	144	49	258
Not Guilty	Low	4	12	8	24
	High	11	41	24	76
	Total	15	53	32	100
	Column total	80	197	81	358

Possible Models

Our task is to try to explain the pattern of obtained cell frequencies in Table 17.8. We could ask a variety of possible questions in seeking an explanation, including the following:

1. Can the pattern of cell frequencies be explained (solely) by differences in the number of participants in the three Moral conditions?

2. Can the pattern of cell frequencies be explained by differences in the number of people judged Guilty and Not Guilty?

3. Can the pattern be explained by a combination of the number of participants in the three Moral conditions and a higher incidence of Guilty over Not Guilty?

4. Can the pattern be explained by an interaction of Moral and Verdict—for example, are there more judgments of Guilty when the victim is seen as being of "high moral character" and fewer when she is seen as being of "low moral character" or "neutral moral character"?

5. Can the pattern be explained by both the Moral × Verdict interaction *and* the difference in the number of cases where the victim was seen as high or low in Fault?

6. Can the pattern be explained by both a Moral × Verdict interaction *and* a Moral × Fault interaction?

7. Can the pattern be explained by a three-way interaction involving Fault, Moral, and Verdict?

Each of these possibilities—and there are a total of 18 if you count the hypothesis that the cell frequencies are random (equiprobable)—represents a possible underlying model. Our task will be to decide which of these models both fits the data and is parsimonious. (I already know that the saturated model, which by definition involves the highest-order interaction, will fit the data perfectly—but it is certainly not parsimonious.)

This list of questions corresponds directly to a list of different models. Letting F, M, and V stand for Fault, Moral, and Verdict, we can associate the first question with a model specified as *M*. To be more precise, our underlying structural model, which is almost certainly much too simple, would be

$$\ln(F_{ijk}) = \lambda + \lambda^M$$

In the same way, we can write out the other models, as shown in Table 17.9.

Table 17.9 Some possible models for data in Table 17.8

Question	Model	Specification
1	$\ln(F_{ij}) = \lambda + \lambda^M$	*M*
2	$\ln(F_{ij}) = \lambda + \lambda^V$	*V*
3	$\ln(F_{ij}) = \lambda + \lambda^M + \lambda^V$	*M, V*
4	$\ln(F_{ij}) = \lambda + \lambda^M + \lambda^V + \lambda^{MV}$	*MV*
5	$\ln(F_{ij}) = \lambda + \lambda^F + \lambda^M + \lambda^V + \lambda^{MV}$	*MV, F*
6	$\ln(F_{ij}) = \lambda + \lambda^F + \lambda^M + \lambda^V + \lambda^{MV} + \lambda^{FM}$	*FM, MV*
7	$\ln(F_{ij}) = \lambda + \lambda^F + \lambda^M + \lambda^V + \lambda^{MV} + \lambda^{FM} + \lambda^{FV} + \lambda^{FMV}$	*FMV*

Notice once again that this is *not* an analysis of variance—that is, we are not trying to explain variability in a single dependent variable (Verdict) on the basis of two independent variables (Fault and Moral). It is easy to keep falling into that trap. We are trying to explain a pattern of observed cell frequencies, and the explanation may involve any or all the variables (dependent or independent) and their interactions. Even where you have one clearly defined dependent variable and two clearly defined independent variables, part of the variability may involve just the independent variables—for example, higher frequencies in the Group 1 cells may be due to the often inconsequential fact that you assigned more participants to Group 1.

Examining the Saturated Model

In considering two-way tables, we defined a saturated model as one that includes all possible effects. The same holds for three-way and higher-order tables. Consider the model that can be designated as *FMV* or written as

$$\ln(F_{ijk}) = \lambda + \lambda^F + \lambda^M + \lambda^V + \lambda^{MV} + \lambda^{FM} + \lambda^{FV} + \lambda^{FMV}$$

This is the saturated model for our data. It includes all possible effects and exhausts the degrees of freedom available in the data. (One degree of freedom goes to estimating λ, one each to estimating *F*, *V*, and *FV*, and two each to estimating *M*, *FM*, *MV*, and *FMV*; *M* has three levels and thus two degrees of freedom for it and its interactions.) These sum to 12, and because we have 12 cells, there isn't anything left over.) If we knew the values of the various lambdas, and eventually we will, the resultant expected frequencies would exactly equal the observed frequencies, leaving nothing else to be explained. For this reason, we know without even looking at the data that the likelihood ratio χ^2 for this model will be exactly 0.00. We should not be any happier with this perfect fit than we are when we draw a straight line to fit perfectly any two points, and for the same reason—the model exhausts the degrees of freedom.

We do not fit a saturated model to data just because we hope that it will fit—we know that before we start. We usually fit it hoping that it will help us identify simpler models by revealing nonsignificant effects. If we could show, for example, that we could do about as well by eliminating the three-way interaction and two of the two-way interactions, we would be well on our way to representing the data by a relatively simple model.

In Exhibit 17.1, you will see part of the printout from the SPSS GENLOG analysis of the saturated model. You can either run GENLOG from syntax or from drop-down menus. The syntax is given first, and the only line that will change in further analyses is the /Design statement. You can see from the output that chi-square is precisely 0.000, as it should be,

GENLOG

    Verdict Fault Moral

    /PRINT = FREQ RESID ADJRESID ZRESID DEV ESTIM CORR COV

    /CRITERIA = CIN(95) ITERATE(20) CONVERGE(.001) DELTA(.5)

    /DESIGN Fault Moral Verdict Fault*Moral Fault*Verdict Moral*Verdict Fault*Moral*Verdict

    .

Goodness-of-Fit Tests[a,b]

	Value	df	Sig.
Likelihood Ratio	.000	0	.
Pearson Chi-Square	.000	0	.

[a] Model: Poisson

[b] Design: Constant + Fault + Moral + Verdict + Fault * Moral + Verdict * Fault + Verdict * Moral + Verdict * Fault * Moral

Cell Counts and Residuals[b]

Verdict	Fault	Moral	Observed Count	Observed %	Expected Count	Expected %	Residual	Standardized Residual	Adjusted Residual	Deviance
1	1	1	42.500	11.7%	42.500	11.7%	.000	.000		.000
		2	79.500	21.8%	79.500	21.8%	.000	.000	.000	.000
		3	32.500	8.9%	32.500	8.9%	.000	.000	.000	.000
	2	1	23.500	6.5%	23.500	6.5%	.000	.000		.000
		2	65.500	18.0%	65.500	18.0%	.000	.000	.000	.000
		3	17.500	4.8%	17.500	4.8%	.000	.000	.000	.000
2	1	1	4.500	1.2%	4.500	1.2%	.000	.000	.000	.000
		2	12.500	3.4%	12.500	3.4%	.000	.000	.000	.000
		3	8.500	2.3%	8.500	2.3%	.000	.000	.000	.000
	2	1	11.500	3.2%	11.500	3.2%	.000	.000	.000	.000
		2	41.500	11.4%	41.500	11.4%	.000	.000	.000	.000
		3	24.500	6.7%	24.500	6.7%	.000	.000	.000	.000

[a] Model: Poisson

[b] Design: Constant + Fault + Moral + Verdict + Fault * Moral + Verdict * Fault + Verdict * Moral + Verdict * Fault * Moral

Exhibit 17.1   Saturated model applied to Pugh's data on three variables

and that the expected frequencies exactly match the obtained frequencies. What we would like to do is to find a model that fits nearly as well but has fewer components.

17.7 Deriving Models

The saturated model is certainly not the only possible model that would fit these data, and we are going to search for a model that will fit nearly as well and have a simpler structure. I am not going to show you a printout for all possible models, but for an example, suppose that we start with a model that tries to explain the cell frequencies on the basis of Verdict, Fault, Moral, and the Verdict × Moral interaction. (I chose this model for an example almost at random.) The generating class for this model would be $F, M \times V$, but GENLOG requires that we specify it explicitly as "/Design Verdict Fault Moral Verdict * Moral." The result follows in Exhibit 17.2.

Notice that the likelihood ratio chi-square for this model is 40.163 on 5 degrees of freedom, which is statistically significant at $p = .000$. Thus, this model does not present an adequate fit

Goodness-of-Fit Tests[a,b]

	Value	df	Sig.
Likelihood Ratio	40.163	5	.000
Pearson Chi-Square	38.602	5	.000

[a] Model: Poisson

[b] Design: Constant + Fault + Moral + Verdict + Verdict * Moral

Cell Counts and Residuals[b]

Verdict	Fault	Moral	Observed Count	Observed %	Expected Count	Expected %	Residual	Standardized Residual	Adjusted Residual	Deviance
1	1	1	42	11.7%	32.137	9.0%	9.863	1.740	2.705	1.661
		2	79	22.1%	71.196	19.9%	7.804	.925	1.682	.909
		3	32	8.9%	24.226	6.8%	7.774	1.579	2.391	1.505
	2	1	23	6.4%	32.863	9.2%	−9.863	−1.721	−2.705	−1.820
		2	65	18.2%	72.804	20.3%	−7.804	−.915	−1.682	−.932
		3	17	4.7%	24.774	6.9%	−7.774	−1.562	−2.391	−1.657
2	1	1	4	1.1%	7.416	2.1%	−3.416	−1.254	−1.802	−1.376
		2	12	3.4%	26.204	7.3%	−14.204	−2.775	−4.228	−3.109
		3	8	2.2%	15.821	4.4%	−7.821	−1.966	−2.898	−2.175
	2	1	11	3.1%	7.584	2.1%	3.416	1.241	1.802	1.161
		2	41	11.5%	26.796	7.5%	14.204	2.744	4.228	2.543
		3	24	6.7%	16.179	4.5%	7.821	1.944	2.898	1.813

[a] Model: Poisson

[b] Design: Constant + Fault + Moral + Verdict + Verdict * Moral

Exhibit 17.2   Test of simplified model F, $M \times V$

to the data. Notice that this conclusion is bolstered by the substantial differences between the observed and expected cell counts. One valuable thing about hierarchical models is that they allow us to compare individual models by subtracting their corresponding likelihood ratio chi-squares. For model F, $M \times V$ in Exhibit 17.2, chi-square $= 40.163$ on 5 df. The saturated model had a chi-square $= 0.00$ on 0 df. We can ask the question "Does F, $M \times V$ represent a significantly worse fit than $F \times M \times V$?" by taking the difference between the two values of chi-square and treating that as a chi-square on the difference in the degrees of freedom. Here

$$\chi^2 = 40.163 - 0 = 40.163$$

on $5 - 0 = 5$ df. The critical value for 5 df is 11.07, which means that the new model fits significantly worse than the saturated one.

Partly for completeness and partly to help in arriving at an optimal model, I have run the syntax for each of the 17 possible models. (The 18th would be the model with no predictors.) Some programs will generate all the possible models on command, but GENLOG will not. The results of these analyses are presented in Exhibit 17.3. The models are again specified by their "generating class" or "defining set," meaning that if there are interactions, the main effects are assumed and not listed. Thus, the generating class MV, FM implies that M, V, and F are part of the model because they are part of the interaction. If the generating class were MF, the model would contain M, F, and MF, but not V or any of its interactions. (As I stated before, you cannot use generating classes in SPSS GENLOG, but have to list each main effect and interaction that you want. This is not true with SPSS HILOGLINEAR or SAS PROC CATMOD, which automatically create hierarchical models.)

MODEL	D.F.	LIKELIHOOD-RATIO CHISQ	PROB.	PEARSON CHISQ	PROB.
M	9	121.17	.0000	109.51	.0000
F	10	191.88	.0000	201.66	.0000
V	10	119.73	.0000	125.38	.0000
M, F	8	121.12	.0000	110.43	.0000
F, V	9	119.68	.0000	125.03	.0000
V, M	8	48.98	.0000	49.13	.0000
M, F, V	7	48.93	.0000	49.02	.0000
MF	6	118.21	.0000	105.99	.0000
MV	6	40.21	.0000	38.64	.0000
FV	8	82.33	.0000	84.99	.0000
M, FV	6	11.58	.0720	11.63	.0709
F, MV	5	40.16	.0000	38.60	.0000
V, MF	5	46.01	.0000	45.02	.0000
MF, MV	3	37.25	.0000	35.94	.0000
MV, FV	4	2.81	.5898	2.80	.5921
FV, MF	4	8.66	.0701	8.74	.0680
MF, MV, FV	2	.26	.8801	.26	.8802
MVF	0	0.00	1.000	0.00	1.000

Exhibit 17.3   Test of all possible models

From this table, you can see that four models are nonsignificant at $\alpha = .05$, and are thus possibilities. (The corresponding rows are shaded.) These are (*MF, MV, FV*), (*FV, MF*), (*MV, FV*), and (*M, FV*). I am ignoring the saturated model because we know that it fits perfectly. Because our models are hierarchical, the difference in the log-likelihood chi-square values attached to the models is itself a test on whether we would lose a significant amount of predictability by going to the simpler model. The difference between the chi-square for the model containing (*MF, MV, FV*) and the complete model is $0.26 - 0.00 = 0.26$, which is a chi-square statistic on $2 - 0 = 2$ *df*, which is a nonsignificant decrease. So we are as well off with the *MF, MV, FV* model as we were with the saturated model. Now we can move up the table and compare the *MF, MV, FV* model with each of the models that include two-way interactions. If we compare chi-square for the *MF, MV, FV* model with chi-square for the *FV, MF* model, the difference in chi-square values is $8.66 - 0.26 = 8.40$, which is a chi-square on $4 - 2 = 2$ *df*. This decrement is statistically significant, indicating that we have lost real interpretive power by dropping *MV* from the model. So we don't want to do that. But if we now compare the *MF, MV, FV* model with the *MV, FV* model we have $\chi^2 = 2.81 - 0.26 = 2.55$ on $4 - 2$ *df*, which is not statistically significant. This suggests that we do not need *MF* in our model. Moving up one additional row, we see that dropping *FV* from our model would lead to a significant decrement. This leaves us with *MV, FV* as our best model to date. If you compare the likelihood ratio chi-square for that model with the likelihood ratio chi-square for any of the models above it, you see that dropping any other components of the model would lead to a statistically significant decrement. For example, although the model *M, FV* is not statistically significant ($p = .0720$), and therefore fits the data at least adequately, it is significantly different from *MV, FV* ($11.58 - 2.81 = 8.77$ on $6 - 4 = 2$ *df*. (One very good reason for using hierarchical models is that they allow us to test differences between models in this way. If we don't have hierarchical models, we cannot always test the decrement in chi-square resulting from omitting a term from the model.) As a result of these tests, we are left with the model (*MV, FV*).

Stepwise Procedures

Just as with multiple regression, there are stepwise procedures for model building. SPSS HILOGLINEAR includes just such a procedure, which starts with the saturated model and shows what would happen if various parts of the model were eliminated. An example of such an approach can be seen in Exhibit 17.4.

********H I E R A R C H I C A L   L O G   L I N E A R********

DESIGN 1 has generating class

    Verdict*Fault*Moral

Note: For saturated models .500 has been added to all observed cells. This value may be changed by using the CRITERIA = DELTA subcommand.

Backward Elimination (p = .050) for DESIGN 1 with generating class

    Verdict*Fault*Moral

Likelihood ratio chi square = .00000 DF = 0 P = .

- -

If Deleted Simple Effect is	DF	L.R. Chisq Change	Prob	Iter
Verdict*Fault*Moral	2	.255	.8801	3

Step 1

  The best model has generating class

    Verdict*Fault
    Verdict*Moral
    Fault*Moral

Likelihood ratio chi square =   .25546   DF = 2   P =   .880

- -

If Deleted Simple Effect is	DF	L.R. Chisq Change	Prob	Iter
Verdict*Fault	1	36.990	.0000	2
Verdict*Moral	2	8.406	.0149	2
Fault*Moral	2	2.556	.2786	2

Step 2

  The best model has generating class

    Verdict*Fault
    Verdict*Moral

Likelihood ratio chi square =   2.81175   DF = 4   P =   .590

- -

If Deleted Simple Effect is	DF	L.R. Chisq Change	Prob	Iter
Verdict*Fault	1	37.351	.0000	2
Verdict*Moral	2	8.768	.0125	2

********H I E R A R C H I C A L   L O G   L I N E A R********

The final model has generating class

    Verdict*Fault
    Verdict*Moral

Exhibit 17.4   Stepwise solution

From Exhibit 17.4, you can see that the program begins with the saturated model. It then considers what would happen if each of the two-variable interactions were removed. We see that if the Verdict × Fault interaction were removed, the change in the log likelihood chi-square would be 36.990 on 1 *df*. That would be a significant decrement in the fit of the model, so we won't want to drop that. Similarly, dropping Verdict × Moral would also lead to a significant decrement. However, dropping Fault × Moral would only produce a change in chi-square of 2.556 on 2 *df*, which is not significant. This leaves us with the more optimal model of (Verdict × Fault Verdict × Moral). The program looks to see if either of the remaining two-way interactions can be deleted and finds that both would result in a significant decrement.

You might expect the program to then see what happens if we were to drop one of the main effects, but that won't do. Remember that for a hierarchical model, any effect that appears in an interaction must also appear as a main effect. Thus, if our model contains Verdict × Fault, both Verdict and Fault most also appear. And if the model also contains Verdict × Moral, Moral must also appear. Therefore, the simplest possible model is (Verdict × Fault Verdict × Moral), which is the same conclusion we came to earlier.

17.8   Treatment Effects

Now that we have chosen a model, we can return to the treatment effect statistics that were discussed in conjunction with two-dimensional tables. Here, we can see how they add to our understanding of the data. We can ask SPSS GENLOG to fit our model, produce observed and expected frequencies, and calculate treatment effects (lambdas). Exhibit 17.5 contains this information for the model

$$\ln(F_{ijk}) = \lambda + \lambda^F + \lambda^M + \lambda^V + \lambda^{FV} + \lambda^{MV}$$

It is important to understand that estimates of λ depend on the way the program you are using codes the data internally. For example, I entered 1s, 2s, and 3s as the values for Moral. SPSS GENLOG takes my codes and converts them to dummy variables, where $Moral_2$ is coded 1 if the observation came from the second level of Moral, and 0 otherwise. Because there are only 2 *df* for Moral, there is no dummy variable corresponding to the last level of Moral. (Remember, all this is done internally, and you won't see the recoding. You have to read the manual to see what the program does.) Other programs, however, use a 1, 0, −1 type of coding, which we saw when we discussed the analysis of variance. The net result is that SPSS GENLOG forces the coefficient for the highest level of a variable to be 0, whereas other programs force the sum of the coefficients for that variable to be 0, making the last one equal to −1 times the sum of the others. For this reason, you may see very drastic differences between the parameter estimates produce by different programs. The result in terms of expected values will be the same, but the solutions may look very different.

From Exhibit 17.5, we see that our model can be written as

$$\begin{aligned}
\ln(F_{ijk}) &= \lambda + \lambda^F + \lambda^M + \lambda^V + \lambda^{FV} + \lambda^{MV} \\
&= 3.191 - 1.153F_1 - 0.758M_1 + 0.505M_2 - 0.198V_1 \\
&\quad + 1.529VF_{11} + 1.040VM_{11} + 0.573VM_{12}
\end{aligned}$$

Parameter Estimates[c,d]

Parameter	Estimate	Std. Error	Z	Sig.	95% Confidence Interval Lower Bound	95% Confidence Interval Upper Bound
Constant	3.191[a]					
[Fault = 1]	−1.153	.234	−4.923	.000	−1.612	−.694
[Fault = 2]	0[b]	.	.	.	.	.
[Moral = 1]	−.758	.313	−2.422	.015	−1.371	−.145
[Moral = 2]	.505	.224	2.254	.024	.066	.943
[Moral = 3]	0[b]	.	.	.	.	.
[Verdict = 1]	−.198	.246	−.807	.420	−.680	.283
[Verdict = 2]	0[b]	.	.	.	.	.
[Verdict = 1] * [Fault = 1]	1.529	.266	5.744	.000	1.007	2.051
[Verdict = 1] * [Fault = 2]	0[b]	.	.	.	.	.
[Verdict = 2] * [Fault = 1]	0[b]	.	.	.	.	.
[Verdict = 2] * [Fault = 2]	0[b]	.	.	.	.	.
[Verdict = 1] * [Moral = 1]	1.040	.366	2.845	.004	.324	1.757
[Verdict = 1] * [Moral = 2]	.573	.278	2.060	.039	.028	1.119
[Verdict = 1] * [Moral = 3]	0[b]	.	.	.	.	.
[Verdict = 2] * [Moral = 1]	0[b]	.	.	.	.	.
[Verdict = 2] * [Moral = 2]	0[b]	.	.	.	.	.
[Verdict = 2] * [Moral = 3]	0[b]	.	.	.	.	.

[a] Constants are not parameters under the multinomial assumption. Therefore, their standard errors are not calculated.

[b] This parameter is set to zero because it is redundant.

[c] Model: Multinomial

[d] Design: Constant + Fault + Moral + Verdict + Verdict * Fault + Verdict * Moral

Exhibit 17.5   Parameter estimates for the model $V \times FV \times M$

Because SPSS codes an observation as 1 if it is the member of a particular treatment or interaction level, and 0 if it is not, we can calculate expected frequencies by substituting 1s or 0s in the model and solving for $\ln(F_{ij})$. We then exponentiate the result to obtain the expected frequency. The easiest case is an observation in the (Not Guilty, High Fault, Low Moral) cell because it would be coded 0 on everything. That would lead us to $\ln(F_{113}) = 3.191 + 0 + 0 + 0 + \cdots + 0 = 3.191$. Then $e^{3.191} = 24.31$, which, as we will see in Exhibit 17.6, is the expected value for that cell.

Expected Cell Frequencies

Finally, let us look at how well our model predicts the observed cell frequencies. The results are shown in Exhibit 17.6.

In this exhibit, you see the observed and expected cell frequencies followed by a statistical test on the residuals (deviates)—the difference between observed and expected frequencies. This test is easy to compute because the standard error of a residual is simply the square root of the expected frequency. Thus,

$$z = \frac{\text{Observed} - \text{Expected}}{\sqrt{\text{Expected}}}$$

is conservatively a standard normal deviate (Agresti, 1990). Standardized deviates (z) in excess of ±1.96 should give cause for concern (the model did not fit that cell well), but because you are running a large number of such tests, a Bonferroni correction would be in order. To do this, treat the deviates as though they were t values on an infinite number of

Cell Counts and Residuals[b]

Verdict	Fault	Moral	Observed		Expected		Residual	Standardized Residual	Adjusted Residual	Deviance
			Count	%	Count	%				
1	1	1	42	11.7%	38.547	10.8%	3.453	.556	1.008	.548
		2	79	22.1%	85.395	23.9%	−6.395	−.692	−1.632	−.701
		3	32	8.9%	29.058	8.1%	2.942	.546	.950	.537
	2	1	23	6.4%	26.453	7.4%	−3.453	−.671	−1.008	−.687
		2	65	18.2%	58.605	16.4%	6.395	.835	1.632	.821
		3	17	4.7%	19.942	5.6%	−2.942	−.659	−.950	−.676
2	1	1	4	1.1%	3.600	1.0%	.400	.211	.262	.207
		2	12	3.4%	12.720	3.6%	−.720	−.202	−.338	−.204
		3	8	2.2%	7.680	2.1%	.320	.115	.161	.115
	2	1	11	3.1%	11.400	3.2%	−.400	−.118	−.262	−.119
		2	41	11.5%	40.280	11.3%	.720	.113	.338	.113
		3	24	6.7%	24.320	6.8%	−.320	−.065	−.161	−.065

[a] Model: Poisson
[b] Design: Constant + Fault + Moral + Verdict + Verdict * Fault + Verdict * Moral

Exhibit 17.6   Estimated cell frequencies for optimal model

degrees of freedom and use Appendix t' to adjust for the number of independent tests. For our example, we have no significant deviations.

Interpreting the Model

From the analysis we have just gone through, we can say quite a bit about our data. Generally statements about main effects are less interesting than statements about interactions, but I will discuss both. In the first place, the frequencies were a function of the level of the Moral variable, but because these frequencies were largely fixed by the experimenter, they are of no great interest. Similarly, the data reflect small, but significant, differences in the attribution of Fault to the victim, with slightly more subjects seeing the victim as high in fault. This again was partly attributable to the experimenter's sampling plan. What was not under the direct control of the experimenter, and is of more interest, is a significantly higher number of defendants judged guilty than judged not guilty. Collapsing across the other dimensions, the odds in favor of a guilty judgment are $258/100 = 2.58$.

When we look at the interactions, we see that there is an interaction between Moral and Verdict. A guilty verdict is more likely when the victim is seen as of high moral character than when she is seen as of low moral character. Put another way, the odds in favor of a guilty verdict for the High, Neutral, and Low Moral conditions are $65/15 = 4.33$, $144/53 = 2.72$, and $49/32 = 1.53$, respectively. Whether a defendant is seen as guilty appears to depend on events beyond the alleged crime itself.

Finally, there is an interaction between Fault and Verdict. When the victim is seen as low in fault, the odds in favor of a guilty verdict are $153/24 = 6.38$. In the high fault condition, those same odds are $105/76 = 1.38$. (Thus, the odds ratio is $6.38/1.38 = 4.62$). A judgment of guilty depends on the degree to which the victim is seen as being at fault. These data shed light on the tendency of defense attorneys to try to put the blame on the victim, in that they show that juries' judgments of guilt or innocence are influenced by attributions of fault and low moral character to the victim.

Ordinal Variables

In this chapter, we have treated our variables as if they were measured on a nominal scale, although Moral did have an ordinal scale of Low, Neutral, and High. If variables are measured on an ordinal scale, standard log-linear analysis, though legitimate, does not use that information. Scrambling the levels of each variable would lead to the same statistical results.

Recently, attention has focused on alternative treatments that allow us to use ordinal scaling of variables where it is available. Discussions of log-linear models with ordinal variables can be found in Green (1988) and Agresti (1984, 1990). SPSS can accommodate such analyses.

Key Terms

Log-linear models (Introduction)

Symmetric relationships (Introduction)

Asymmetric relationship (Introduction)

Saturated model (17.1)

Geometric mean (17.2)

λ (lambda) (17.2)

Additive Model (17.2)

Conditional odds (17.4)

Odds ratio (Ω) (17.4)

Sparse matrices (17.6)

Hierarchical model (17.6)

Defining set (17.6)

Generating class (17.6)

Exercises

All the problems in this chapter will require solution by one or more computer programs. My answers in the back of the book are based on SPSS GENLOG, and they may differ from answers you receive if you use a different program.

Bell, Buerkel-Rothfuss, and Gore (1987) examined the relationships among aspects of idiomatic communication in 100 romantically involved heterosexual couples. Couples provided information on "any words, phrases, or nonverbal signs that they had created that had meaning to their relationship." These idioms were classified along several dimensions, but we will focus only on who invented the idiom and the function the idiom serves (confrontations, affection, labels for outsiders, nicknames, requests, sexual invitations, sexual references, and teasing insults.) (The title of the paper was "Did you bring the yarmulke for the Cabbage Patch kid?" I leave speculation about its meaning to your imagination.)

The data follow:

Function	Female Partner	Male Partner	Others/Unknown
Confrontation	14	10	6
Affection	27	41	11
Labels outsiders	23	43	11
Nickname—female	6	57	16
Nickname—male	68	11	10
Requests	12	21	3
Sexual invitation	26	41	13
Sexual reference	27	38	5
Teasing insults	35	62	10

This is only a two-dimensional table, but it allows you the opportunity to specify the model, test the fit of that model, and calculate treatment effects.

17.1 What are the possible models that could be hypothesized to underlie the data matrix?

17.2 Test each of the alternative models using the likelihood ratio χ^2.

17.3 Compute and interpret λ coefficients for the complete model.

Hansen and Swanson (1983) collected data on the Strong-Campbell Interest Inventory for college majors. Marascuilo and Busk (1987) discuss this example at length. In this study, subjects were classified as female or male (Gender), and as satisfied or unsatisfied with their major (Satisfaction). In addition, a subject's major was classified as a direct or an indirect hit (Hit), depending on whether that major was clearly represented by an Occupational Scale of the Strong-Campbell. Finally, the Validity of the Strong-Campbell prediction was classified as excellent, moderate, or poor, depending on how well it fit the student's major. The latter two variables (Hit and Validity) can be thought of as dependent variables, whereas the first two (Gender and Satisfaction) are thought of as independent variables (see Marascuilo & Busk, 1987). (There is some basis for arguing that Satisfaction can serve as a dependent variable, but the major focus is on the other dependent variables as a function of Satisfaction.) The data follow:

		Hit	
	Group	Direct	Indirect
Female	Satisfied		
	Excellent	35	8
	Moderate	14	0
	Poor	22	2
	Unsatisfied		
	Excellent	55	11
	Moderate	16	5
	Poor	68	9
Male	Satisfied		
	Excellent	16	8
	Moderate	8	4
	Poor	11	2
	Unsatisfied		
	Excellent	31	15
	Moderate	16	7
	Poor	53	12

17.4 Collapse the Hansen and Swanson data into a simple two-dimensional table of Gender × Hit and compute the likelihood ratio χ^2. Interpret the results.

17.5 If Hit is taken as a dependent variable, does the Strong-Campbell do a better job for Males than Females? Answer this question with respect to the odds and odds ratios for both genders.

17.6 Now collapse the same data into a Gender × Satisfaction table and compute a likelihood ratio χ^2. What would you conclude from this analysis?

17.7 Ignore the Hit dimension in Hansen and Swanson's data and compute a log-linear analysis on the three-dimensional table that results. Interpret your results.

17.8 Use all four dimensions of the Hansen and Swanson data and compute the complete log-linear analysis. Interpret your results.

17.9 Compare the statistics you obtained in each of the analyses on this data set. How do they change, or not change, as additional variables are entered into the analysis?

17.10 What difference does it make in Exercises 17.4–17.9 if you treat some of the variables as dependent and others as independent, rather than regarding all variables alike?

Dabbs and Morris (1990) investigated the effects of elevated testosterone levels in a representative sample of U.S. adult males (mean age 37). Subjects were classified as High (upper 10%) or Normal on testosterone, as high or low on socioeconomic status (SES), and as engaging (or not engaging) in adult delinquency. Their data follow.

| | Low SES | | High SES | |
| | Normal | High | Normal | High |
Delinquent	Testosterone	Testosterone	Testosterone	Testosterone
Yes	190	62	53	3
No	1104	140	1114	70

17.11 Calculate the odds of being classed as an adult delinquent for each of the categories in the preceding table.

17.12 What are the odds ratios of delinquency for the four SES/Testosterone groups in the preceding table?

17.13 Apply a log-linear model to the data from Dabbs and Morris and interpret the results.

17.14 Dabbs and Morris collected data on a number of other variables, including childhood delinquency, hard-drug use, and many sex partners. Why would it be inappropriate to create a dimension labeled Behavior (adult delinquency, childhood delinquency, hard-drug use, many sex partners) and use that as an additional variable in the analysis? In other words, what is wrong with analyzing SES × Testosterone × Behavior?

This chapter was based heavily on a study by Pugh (1983) on the blaming-the-victim phenomenon in rape cases. The complete data from Pugh follow, adding the Gender of the judge as the final variable.

| | Stigma (S) | | Verdict (V) | |
Gender (G)	(Moral Character)	Fault (F)	Guilty	Not Guilty
Male	High	Low	17	4
		High	11	7
Male	Neutral	Low	36	4
		High	23	18
Male	Low	Low	10	6
		High	4	18
Female	High	Low	25	0
		High	12	4
Female	Neutral	Low	43	8
		High	42	23
Female	Low	Low	22	2
		High	13	6

17.15 Run the complete analysis on these data. What effect does adding Gender to the analysis produce? You can compare your conclusions against the results given in Pugh's paper.

17.16 Agresti (1990) presents data on the relationship of the assignment of the death penalty, the defendant's race, and the victim's race. The data follow.

RaceDefendant	RaceVictim	DeathPenalty	Frequency
1	1	1	19
1	1	2	132
1	2	1	0
1	2	2	9
2	1	1	11
2	1	2	52
2	2	1	6
2	2	2	97

a. Fit an appropriate model.

b. Summarize the conclusions you would draw.

Discussion Questions

17.17 If you search on PsychINFO under "log-linear," you will find many studies that used such models. Find an interesting study that contains the necessary cell frequencies and write a short example that can be used to illustrate the material covered in this chapter.

17.18 Apply the same data to two or more software packages and note the similarities and differences in the output. How, if at all, can you resolve the discrepancies?

CHAPTER 18

Resampling and Nonparametric Approaches to Data

Objectives

To present resampling and nonparametric (distribution-free) procedures that can be used for testing hypotheses but that rely on less restrictive assumptions about populations than do previously discussed tests.

Contents

MOST OF THE STATISTICAL PROCEDURES we have discussed throughout this book have involved estimating one or more parameters of the distribution of scores in the population(s) from which the data were sampled and assumptions concerning the shape of that distribution. For example, the t test uses the sample variance (s^2) as an estimate of the population variance (σ^2) and also requires the assumption that the population from which the sample was drawn is normal. Tests such as the t test, which involve either assumptions about specific parameters or their estimation, are referred to as **parametric tests.**

There is a class of tests, however, that does not rely on parameter estimation or distribution assumptions. Such tests are usually referred to as **nonparametric tests** or **distribution-free tests.** By and large, if a test is nonparametric, it is also distribution-free, and the distribution-free nature of the test is most valuable to us. Although the two names are often used interchangeably, the tests will be referred to here as nonparametric tests because that term is somewhat more common.

Another approach to statistical analysis, which is predominantly nonparametric in nature, has become considerably more popular in recent years because of the increased computing power we now enjoy. These are called **resampling procedures**. I will discuss several of these that do not require strict parametric assumptions. These techniques are useful either when we are uncomfortable with the assumptions that a parametric test, such as t, would require, or when we just don't have good parametric procedures to do what we want—such as forming a confidence interval on a median when we doubt that the distribution is normally distributed. I will discuss these procedures first because I believe that in a short time,[1] they will overtake what are now the more common nonparametric tests, and may eventually overtake the traditional parametric tests.[2]

The major advantage generally attributed to nonparametric tests is also the most obvious—they do not rely on any very seriously restrictive assumptions concerning the shape of the sampled population(s). This is not to say that nonparametric tests do not make any distribution assumptions, but only that the assumptions they do require are far more general than those required for the parametric tests. The exact null hypothesis being tested may depend, for example, on whether two populations are symmetric or have a similar shape. None of these tests, however, makes an a priori assumption about the specific shape of the distribution; that is, the validity of the test is not affected by whether the variable is normally distributed in the population. A parametric test, on the other hand, usually includes some type of normality assumption, and, if that assumption is false, the conclusions drawn from that test may be inaccurate. To take the standard t test as an example, you can *compute* a t no matter what the parent populations look like. But the t that you compute is compared with values from Student's t distribution and that distribution was derived from normal populations. So if your samples came from very nonnormal populations, you are really comparing apples and oranges. In addition, some violations of parametric test assumptions may cause that test to be less powerful for a specific set of data than is the corresponding nonparametric test. Perhaps the most articulate spokesperson for nonparametric/distribution free tests has been Bradley (1968), who still has one of the clearest descriptions of the underlying assumptions and their role.

Another characteristic of nonparametric tests that often acts as an advantage is that many of them, especially the ones discussed in this chapter, are more sensitive to medians than to means. Thus, if the nature of your data is such that you are interested primarily in medians, the tests presented here may be particularly useful to you.

---

[1] "Short" is a relative term, and in the field of statistics things change very slowly. But they do change, and permutation and bootstrapping procedures will take over—the only question is when.

[2] Long and his colleagues (personal communication) have shown that randomization procedures applied to the data in Table 15.1 produce quite different test results on the regression coefficients. This may lead to an important discussion of how best to approach data sets that have reasonable-looking distributions but that do not meet the standard normality assumptions of least squares regression.

Those who argue in favor of using parametric tests in almost every case do not deny that nonparametric tests are more liberal in the assumptions they require. Proponents of parametric tests argue, however, that the assumptions normally cited as being required of parametric tests are overly restrictive in practice and that the parametric tests are remarkably unaffected by violations of distribution assumptions. See Rasmussen (1987) for an example where parametric tests win out even with their assumptions violated.

The major disadvantage generally attributed to nonparametric tests is their (reputed) lower power relative to the corresponding parametric test. In general, when the assumptions of the parametric test are met, the nonparametric test requires somewhat more observations than does the comparable parametric test for the same level of power. Thus, for a given set of data, the parametric test is more likely to lead to rejection of a false null hypothesis than is the corresponding nonparametric test. Moreover, even when the distribution assumptions are violated to a moderate degree, the parametric tests are thought to maintain their advantage. A number of studies, however, have shown that for perfectly reasonable data sets, nonparametric tests may have greater power than the corresponding parametric tests do. The problem is that we generally do not know when the nonparametric test will be more powerful.

Some nonparametric tests have an additional advantage. Because many of them rank the raw scores and operate on those ranks, they offer a test of differences in central tendency that are not affected by one or a few very extreme scores (outliers). An extreme score in a set of data actually can make the parametric test less powerful, because it inflates the variance, and hence the error term, as well as biasing the mean by shifting it toward the outlier (the latter may increase or decrease the mean difference).

Nonparametric tests can be divided into several different approaches. One group of tests, which we will discuss in the second half of the chapter, depend on ranking the data and carrying out the statistical test on the ranks. These are the most commonly known nonparametric procedures and are particularly useful when the ranking procedure reduces problems with outliers. A second group of tests are broadly known under the title of "resampling statistics," and these tests rely on drawing repeated samples from some population and evaluating the distribution of the resulting test statistic. Within the resampling statistics, the bootstrapping procedures, to be discussed next, rely on random **sampling, *with replacement,*** from a population whose characteristics reflect the characteristics of the sample. **Bootstrapping procedures** are particularly important in those situations where we are interested in statistics, such as the median, whose sampling distribution and standard error cannot be derived analytically (i.e., from a standard formula, such as the formula for the standard error of the mean) unless we are willing to assume a normally distributed population.[3] The next section will be an introduction to bootstrapping.

sampling, *with replacement*

bootstrapping procedures

permutation tests

randomization tests

sampling without replacement

After looking at the bootstrap, we will move on to other resampling procedures that do not rely on drawing repeated samples, with replacement, from some population. Instead, we will consider all possible permutations, or rearrangements, of the data. These are often called **permutation** or **randomization tests,** and they are covered in sections 18.2–18.4. Whereas bootstrapping involves sampling with replacement, permutation tests involve **sampling without replacement**.

18.1 **Bootstrapping as a General Approach**

Think for the moment about the standard *t* test on the difference between two population means. (Everything that I am about to say would apply, with only the obvious changes, if I had chosen any other parametric test, but the *t* test is a good example.) To carry out our

---

[3] If the population is normally distributed, the standard error of the median is approximately 1.25 times the standard error of the mean. If the distribution is skewed, however, the standard error of the median cannot easily be calculated.

t test, we first assumed that we drew our samples from two normal populations and that the populations had the same variance (σ^2). We then assume that the null hypothesis was true, and ask what kinds of differences between means (or what values of *t*) we would expect if we drew an infinite number of pairs of samples from these normal populations, calculated the means, and then took their differences. Notice in all of this that we ask about sampling from normal populations with equal variances. To go one step further, if we actually computed all these samples from the specified population, the resulting sampling distribution of *t* would be the same as the tabled sampling distribution that we normally use to compute the probability of *t* under the null hypothesis.

But suppose that we are not willing to assume that our data came from normal populations, or that we are not willing to assume that these populations had equal variances. Perhaps if we knew enough statistics, which neither you nor I do, and we were willing to assume that the populations have some other specified distribution (e.g., an exponential distribution), we could derive something comparable to our *t* test, and use that for our purposes. Of course that test, if we could derive one, would still only apply when data come from that particular kind of distribution. But suppose that we think that our populations are not distributed according to any of the common distributions. Then what do we do? Bootstrapping gives us a way to solve this problem. Before I talk about how we would perform a bootstrapped hypothesis test, however, let's look at another problem that we can deal with using the bootstrap.

If I asked you to calculate a confidence interval on a mean, and I told you that the population from which the data came was normal, you could solve the problem. In particular, you know that the standard error of the mean is equal to the population standard deviation (perhaps estimated by the sample standard deviation) divided by the square root of *n*. You could then measure the appropriate number of standard errors from the mean using the normal (or *t*) distribution, and you would have your answer. But, suppose that I asked you for the confidence limit on the median instead of the mean. Now you are stuck, because you don't have a nice simple formula to calculate the standard error of the median. So what do you do? Again, you use the bootstrap.

Macauley (1999, personal communication) collected mental status information on older adults. One of her dependent variables was a memory score on the Neurobehavioral Cognitive Status Examination for the 20 participants who were 80 to 84 years old. As you might expect, these data were negatively skewed because some, but certainly not all, of her participants had lost some cognitive functioning. Her actual data are shown in Figure 18.1. Macauley wanted to establish confidence limits on the population median for this age group. Here she was faced with both problems described earlier. It does not seem reasonable to base that confidence interval on the assumption that the population is normally distributed (it most clearly is not), and we want confidence limits on a median, but don't have a convenient formula for the standard error of the median. What's a body to do?

What we will do is to assume that the population is distributed *exactly as our sample*. In other words, we will assume that the shape of the parent population is as shown in Figure 18.1.

It might seem like a substantial undertaking to create an infinitely large population of numbers such as that seen in Figure 18.1, but it is trivially easy. All that we have to do is to take the sample on which it is based, as represented in Figure 18.1, and draw as many observations as we need, *with replacement,* from that sample. This is the way that all bootstrapping programs work, as you will see. In other words, 20 individual observations from an infinite population shaped as in Figure 18.1 is exactly the same as 20 individual observations drawn *with replacement* from the sample distribution. In the future when I speak of a population created to exactly mirror the shape of the sample data, I will refer to this as a pseudo-population.

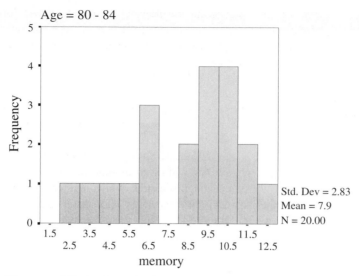

Figure 18.1   Sample distribution of memory scores for participants 80–84 years of age

18.2 Bootstrapping with One Sample

Macauley (1999, personal communication) was interested in defining a 95% confidence interval on the median of memory scores of older participants. She had reason to doubt that the population of scores was normally distributed, and there is no general formula defining the standard error of the median. But neither of those considerations interferes with computing the confidence interval she sought. All that she had to do was to assume that the shape of the population was accurately reflected in the distribution of her sample, then draw a large number of new samples (each of $n = 20$) from that population. For each of these samples, she computed the median, and when she was through, she examined the distribution of these medians. She could then empirically determine those values that encompassed 95% of the sample medians.

It is quite easy to solve Macauley's problem using a program named *Resampling Stats* by Simon and Bruce (1999). The syntax and the results are shown in Figure 18.2, and a histogram of the results is presented in Figure 18.3. There is no particular reason for you to learn the sequence of commands that are required for *Resampling Stats*, but a cursory look at the program is enlightening. The first two lines of the program describe the problem and set aside sufficient space to store 10,000 sample medians. Then the data are read in to create a pseudo-population from which we can sample with replacement. The next two lines calculate and print the median of the original sample. At this point, the program goes into a loop that repeats 10,000 times, each time drawing a sample of 20 observations from our pseudo-population, computing its median, and labeling that median as "bmedian." After 10,000 medians have been drawn and stored in an array called "medians," the program prints a frequency distribution and histogram of the results, calculates the standard deviation of these medians, which is the standard error of the median, and prints that. The amazing thing is that it probably took me 5 minutes to compose, type, and revise this paragraph, but it only took the program 7.8 *seconds* to draw those 10,000 samples and print the results.

The results in Figures 18.2 and 18.3 are interesting for several reasons. In the first place, they show you what happens when you try to calculate medians of a large number of relatively small samples. The distribution in Figure 18.3 is quite discrete, because the median is going to be the middle value in a limited set of numbers. You couldn't get a median of 9.63,

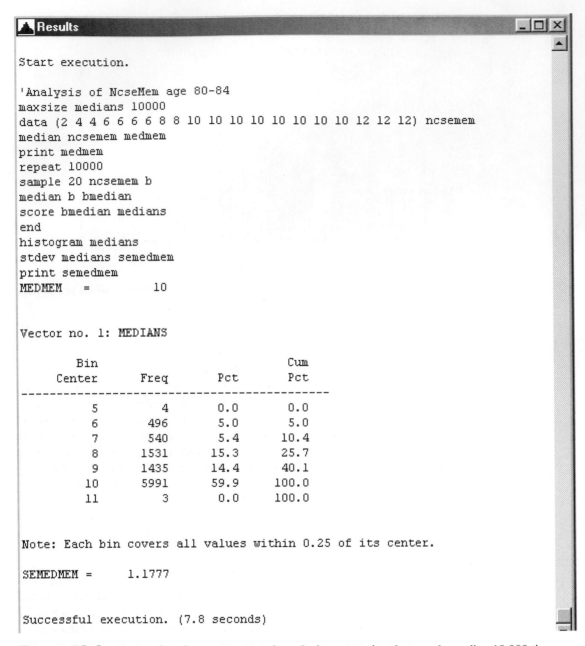

```
Results                                                    _ □ ✕

Start execution.

'Analysis of NcseMem age 80-84
maxsize medians 10000
data (2 4 4 6 6 6 6 8 8 10 10 10 10 10 10 10 10 12 12 12) ncsemem
median ncsemem medmem
print medmem
repeat 10000
sample 20 ncsemem b
median b bmedian
score bmedian medians
end
histogram medians
stdev medians semedmem
print semedmem
MEDMEM    =           10


Vector no. 1: MEDIANS

          Bin                       Cum
       Center     Freq      Pct     Pct
       ------------------------------------------
            5        4      0.0     0.0
            6      496      5.0     5.0
            7      540      5.4    10.4
            8     1531     15.3    25.7
            9     1435     14.4    40.1
           10     5991     59.9   100.0
           11        3      0.0   100.0


Note: Each bin covers all values within 0.25 of its center.

SEMEDMEM =    1.1777


Successful execution. (7.8 seconds)
```

Figure 18.2   *Resampling Stats* program and results bootstrapping the sample median 10,000 times

for example, no matter how many samples you drew. For this particular population, the medians must be an integer (or the average of two integers in the ordered array) between 5 and 11. There are no other possibilities.

Ideally, to calculate a 95% confidence interval, we would like to find those outcomes that cut off 2.5% of the observations at each end of the distribution.[4] With the very discrete

---

[4] This is the simplest approach to obtaining confidence limits and relies on the 2.5 and 97.5 percentiles of the sampling distribution of the median. There are a number of more sophisticated estimators, but the one given here best illustrates the approach.

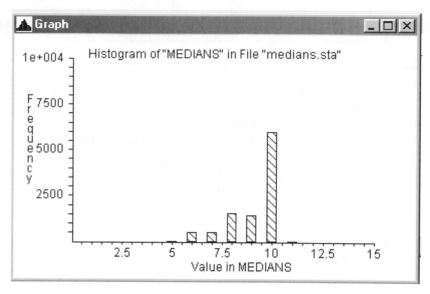

Figure 18.3   Histogram of the results displayed in Figure 18.2

distribution we have with medians, there is no point that cuts off the lowest 2.5% of the distribution. At the extreme, 4/10,000 = .04% lie at or below a median of 5, and (496 + 4)/ 10,000 = 5.00% lie at or below a median of 6. At the other end of the distribution, 4006/10,000 = 40.6% lie at or below 9, and 9997/10,000 = 99.97% lie at or below 10. To be conservative, we would choose the extremes of each of these sets and put the confidence interval at 5–10, which includes virtually all the distribution. We really have a 99.97% confidence interval, which is probably close enough for any purpose to which we would be likely to put these data. If we were willing to let the lower bound represent the 5% point, we would have an interval at 6–10. What is important here, and the reason why Macauley wanted these limits in the first place, is that for this memory test, the lower bound of what is classed as "normal functioning" is a score of 10. The confidence interval does include 10 as its upper limit, and so we cannot reject the null hypothesis that people in this age group, *on average*, fall in the normal range. An examination of the sampling distribution reinforces this view and perhaps gives us a more complete understanding of the performance of this age cohort. That there are a number of individuals whose scores are well below 10 might lead us to seek a different confidence interval, that being limits on the *proportion* of people in that age group who fall below 10. Although that would be a perfectly legitimate use of bootstrapping for these data, we will not pursue that question here.

This may not seem like the most inspiring example of bootstrapping because it makes bootstrapping look rather imprecise. It is a good example, nonetheless, because it reflects the sometimes awkward nature of real data. As we will see, however, not all data lead to such discrete distributions. In addition, the discreteness of the result is inherent in the data, not merely in the process itself. If we drew 10,000 samples from this population and calculated *t* values, the resulting *t* distribution would be almost as discrete. The problem comes from drawing samples from a distribution with a limited number of different values, instead of modeling the results of drawing from continuous (e.g., normal) distributions. If it is not reasonable to assume normality, it is not reasonable to draw from normal distributions just to get a prettier graph.

18.3 Resampling with Two Paired Samples

We will now move from the bootstrap, where we drew large numbers of samples from a pseudo-population using sampling with replacement, to procedures that involve taking the full set of observations, randomly shuffling them, and assigning them to conditions randomly. These are called randomization tests or permutation tests, and we will start with one of the simplest types.

Hoaglin, Mosteller, and Tukey (1983) looked at the role of beta-endorphins in response to stress. The researchers were interested in testing whether beta-endorphin levels rose in stressful situations. They recorded beta-endorphin levels in 19 patients 12 hours before surgery and again, for the same patients, 10 minutes before surgery. The data[5] follow in fmol/ml.

12 hours	10.0	6.5	8.0	12.0	5.0	11.5	5.0	3.5	7.5	5.8	4.7
10 min.	20.0	14.0	13.5	18.0	14.5	9.0	18.0	6.5	7.4	6.0	25.0
Difference	10.0	7.5	5.5	6.0	9.5	−2.5	13.0	3.0	−0.1	0.2	20.3

12 hours	8.0	7.0	17.0	8.8	17.0	15.0	4.4	2.0
10 min.	12.0	15.0	42.0	16.0	52.0	11.5	2.5	2.1
Difference	4.0	8.0	25.0	7.2	35.0	−3.5	−1.9	0.1

Because these are paired scores, we are primarily interested in the difference scores. We want to test the null hypothesis that the average difference score was 0.0, which would indicate that there was no change in beta-endorphin levels on average. The difference scores are shown in the bottom line of the table, where it is clear that most differences are positive, and those that are negative are relatively small. If you were to plot the differences in this example, you would find that they are very positively skewed, which might discourage us from using a standard parametric t test. Moreover, if we were particularly interested in the median of the differences, a t test would not be appropriate. We will solve our problem by drawing on resampling statistics.

Our resampling procedure is based on the idea that if the null hypothesis is true, a patient's 10-minute score was just as likely to be larger than his 12-hour score as it was to be smaller. If a patient has scores of 8.0 and 13.5, and if the null hypothesis is true, the 13.5 could just as likely come from the 12-hour measurement as from the 10-minute measurement. Under H_0, each difference had an equal chance of being positive or negative. This tells us how to model what the data would look like under H_0. We will simply draw a very large number of samples of 19 difference scores each, in such a way that the difference score has a 50:50 chance of being positive or negative. For each sample, we will calculate the median of the differences, and then plot the sampling distribution of these differences. Remember, this is the sampling distribution of the differences when H_0 is true. We can compare our obtained median difference against this distribution to test H_0.

The way that we will conduct this test using Simon and Bruce's *Resampling Stats* is to take all 19 difference scores and randomly attach the sign of the difference. (*Assigning the sign at random is exactly equivalent to randomly assigning one score to the 12-hour condition and the other to the 10-minute condition.*) We will then calculate the median difference and store that. This procedure will be repeated many times (in this case, 10,000 times). The program and results are shown in Figure 18.4, with the resulting histogram in Figure 18.5.

---

[5] I have made two very trivial changes to avoid difference scores of 0.0, just to make the explanation easier. With differences of zero, we normally simply remove those cases from the data.

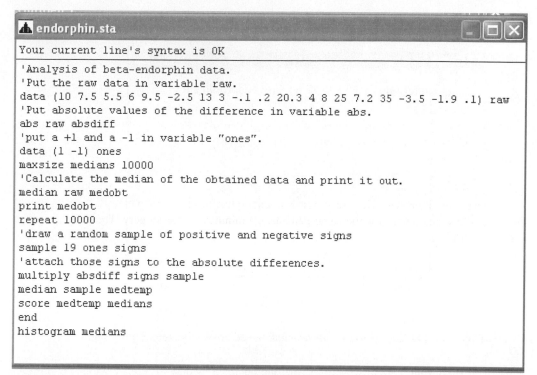

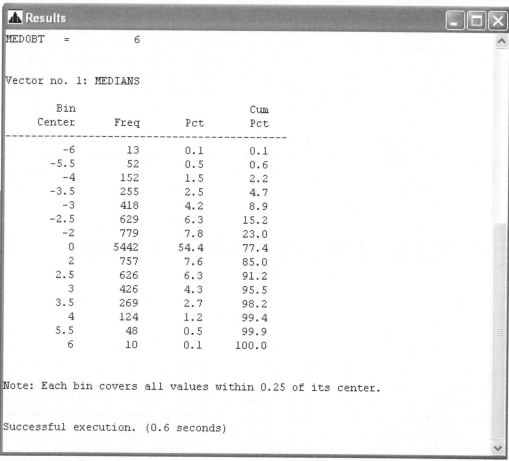

Figure 18.4 Resampling program and results for beta-endorphin data

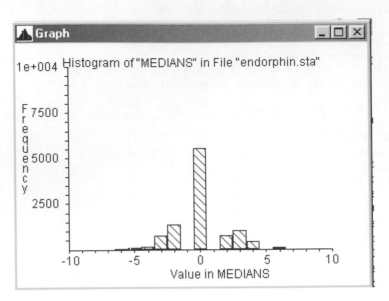

Figure 18.5 Histogram of resampled medians for beta-endorphin study

From Figure 18.4, we can see that the obtained median difference score was 6. From either the frequency distribution in Figure 18.4 or the histogram in Figure 18.5, we see the results of drawing 10,000 samples from a model in which the null hypothesis is true. Figure 18.5 is reassuring because it shows us that when the null is true, the resampled medians are distributed symmetrically about 0, which is what we would expect. From that figure, we can also see that our obtained median of 6 is certainly extreme under H_0. Going back to Figure 18.4, we see that there were 10 resampled medians as large as 6, and 13 resampled values as low as −6. If we want a two-tailed test, the probability of a median as extreme as the one we obtained is $(10 + 13)/10,000 = .0023$, which is certainly a small probability. These results, thus, tell us that if we were sampling from a model where H_0 is true, the probability is very small that we would obtain a sample median as extreme as the one we obtained. Therefore, we will reject the null hypothesis and conclude that beta-endorphin levels do increase as the time for surgery approaches. This is really a very good thing, because beta-endorphins act as the body's pain pills.

18.4 Resampling with Two Independent Samples

Now we will move on to the resampling equivalent of the *t* test for two independent samples. The example we will use involves data collected by Epping-Jordon, Compas, and Howell (1994) on the effect of avoidance on women's recovery from breast cancer. Epping-Jordon and colleagues were interested in examining the question of whether people who try to actively avoid thinking about their cancer have a poorer prognosis over a 1-year period than those who do not report high levels of avoidance behavior. She collected data on the incidence of avoidance shortly after patients had been diagnosed with breast cancer. At the end of one year, she sorted patients into those who were in remission (49 cases) and those who were no better or who had died (28 cases). These groups were labeled Success and Fail, respectively. The data are shown in Table 18.1. Epping-Jordon then compared the earlier reported level of avoidance for the two groups.

For this example, we will compare the medians of the two groups, although we could just as easily compare their means.

Table 18.1 Data on avoidance from Epping-Jordon et al. (1994)

	Success				Fail			
	19	14	17	10	18	17	17	21
	23	12	10	14	17	15	8	12
	20	21	8	12	16	11	27	18
	8	11	13	23	13	22	18	18
	11	9	8	20	22	16		
	13	15	18	15				
	13	8	16	15				
	16	14	11	19				
	10	12	12	15				
	8	12	12	17				
	20	18	25	12				
	9	23	11	21				
	13							
Median		14				17		
n		49				18		

   If the null hypothesis is true in Epping-Jordon's case, the two samples (Success and Fail) can be thought of as having been drawn from one population. Any particular Avoidance score would be as likely to be a member of the Success group as to be a member of the Fail group. We could model this null situation by assigning a random sample of 49 of the scores to the Success group and the remaining 18 scores to the Fail group. (Notice here that we are sampling without replacement.) The difference between those two groups' medians would be an example of a median difference that we might reasonably obtain under H_0. We could repeat this procedure (randomly assigning 49 scores to the Success group and 18 scores to the Fail group) many times, and look at the median differences we obtain. Finally, we could compare the difference we actually found with those we obtained when we modeled the null hypothesis.

   This procedure is quite easy to do because we simply shuffle the complete data set, split the result into the first 49 cases and the last 18 cases, compute and record the medians and the median differences, shuffle the data again, and repeat this process 10,000 times. The result of such a procedure is shown in Figure 18.6 and 18.7. I have omitted the program syntax because it would not add to the presentation.

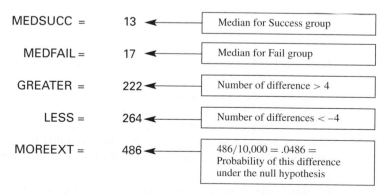

Figure 18.6 Summary results of resampling from Epping-Jordon et al. data

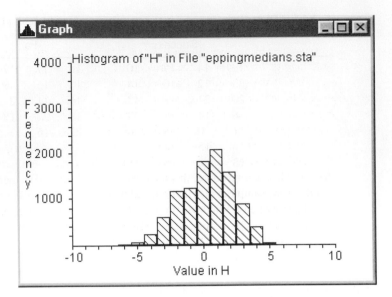

Figure 18.7   Frequency distribution of median differences from Epping-Jordan et al. (1994)

From Figure 18.6, we can see that the median Avoidance score for the Success group was 13, and the median for the Fail group was 17. The group who failed to improve exhibited more avoidance behavior early in treatment. The difference in median avoidance is –4. From the output, you can also see that when we model the null hypothesis, 222 of the resamples were greater than a difference of 4; 264 results were less than a difference of –4; and 486 results were greater than ±4. Of 10,000 samples, this represents $486/10,000 = 4.86\%$ of the cases. Thus, only 4.9% of the resampling statistics were more extreme than our result, and we can reject the null hypothesis at $\alpha = .05$. We can conclude that those in the Fail group experienced significantly more avoidance behavior early in treatment than did those who later were classed as successes.[6]

18.5   Bootstrapping Confidence Limits on a Correlation Coefficient

The standard approach to correlation problems is to calculate a correlation coefficient and then to apply a hypothesis test with the hope of showing that the correlation is significantly different from 0.00. However, many significant correlations are so low that they are not particularly important, even if they are significantly different from 0.00. Along with the recent emphasis on effect size measures should be an increase in the use of confidence limits.

As we saw in Chapter 9, Fisher's arcsine transformation

$$r' = (0.5) \log_e \left| \frac{1 + r}{1 - r} \right|$$

provides one way to adjust for the skewed sampling distribution of r when $\rho \neq 0$. An attractive alternative is to draw bootstrapped samples on the assumption that the bivariate

---

[6] If we had run a standard t test on the means of these data, that probability would have been .0397.

data reflect the relationship in the population and then to obtain confidence limits simply by taking the cutoffs for the $\alpha/2$ percent of each end of the distribution.

As an example, we can look at the data from Macauley on the mental status scores of older adults. Macauley's data included 123 adults between the ages of 60 and 97, and we can look at the relationship between memory performance and age. We would probably expect to see a negative correlation between the two variables, but the significance of the correlation is not as useful as confidence limits on this correlation, which give us a better sense of how strong the relationship really is.

The bootstrap approach to obtaining these confidence limits would involve sampling 123 cases, with replacement, from the XY pairs in the sample, computing the correlation between the variables, and repeating this a large number of times. We then find the 2.5 and 97.5 percentile of the sampling distribution, and that gives us our 95% confidence limits.

I have written a Windows program (available at www.uvm.edu/~dhowell/methods/) that will carry out this procedure. (It will also calculate a number of other resampling procedures.) The results of drawing 2,000 resamples with replacement from the pseudo-population of pairs of scores are shown in Figure 18.8.

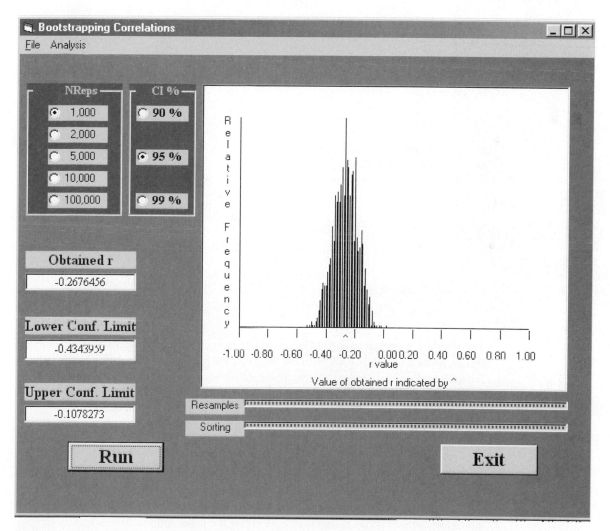

Figure 18.8   Sampling distribution and confidence limits on correlation between age and memory performance in older adults

In the center of this figure, you can see the sampling distribution of r. To the left is the obtained correlation ($-.268$) and upper and lower confidence limits. These are $-.43$ and $-.11$. Because they are both on the same side of 0.00, we also know that our correlation is significant. The confidence interval may strike you as surprisingly wide, but confidence intervals on correlation coefficients often are.

The example from Macauley involved a fairly low correlation coefficient that, because it was only $-.268$, was nearly symmetrically distributed around 0.00. If we run the same analysis on the beta-endorphin data that we used earlier, we can easily see the skewed nature of the sampling distribution for large correlations. This result is shown in Figure 18.9.

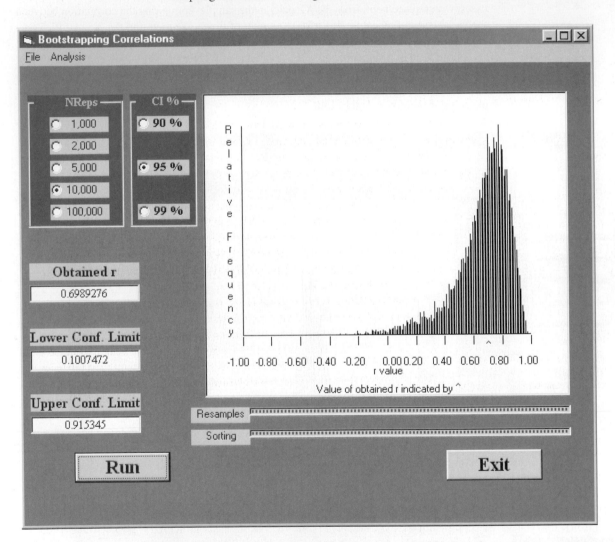

Figure 18.9   Sampling distribution of r for beta-endorphin data for 10,000 resamples

Figure 18.9 presents two interesting results. In the first place, notice that, because the correlation is fairly large ($r = .699$), the sampling distribution is very negatively skewed. In addition, notice how asymmetrical the confidence limits are. The upper limit is .91, which is a bit more than 20 points higher than r. However the lower limit is .101, which is approximately 59 points lower. Whenever we have large correlations, the sampling distribution will be skewed and our confidence limits will be asymmetrical.

An excellent discussion of bootstrapped estimates of confidence limits can be found in Mooney and Duval (1993). They discuss corrections for bias that are relatively easy to apply. Excellent sources on both bootstrapping and randomization tests can be found in Edgington (1995), Manly (1997), and Efron and Tibshirani (1993). Efron has probably been the most influential developer of the bootstrap approach, and his book with Tibshirani is an important source. Good (2000) has an excellent presentation of permutation tests, and Lunnenborg (2000) addresses resampling methods at a sophisticated level.

Additional information on resampling and bootstrapping is available from the website that I maintain at http://www.uvm.edu/~dhowell/StatPages/StatHomePage.html. These particular pages are being updated on a regular basis, and the material there goes well beyond what is available in this chapter. This is a rapidly expanding field, and new results are being published on a regular basis.

18.6 Wilcoxon's Rank-Sum Test

Wilcoxon rank-sum test

We will now move away from bootstrapping and randomization to the more traditional nonparametric tests. One of the most common and best known of these tests is the **Wilcoxon rank-sum test** for two independent samples. This test is often thought of as the nonparametric analogue of the t test for two independent samples, although it tests a slightly different, and broader, null hypothesis. Its null hypothesis is the hypothesis that the two samples were drawn at random from identical populations (not just populations with the same mean), but it is especially sensitive to population differences in central tendency. Thus, rejection of H_0 is generally interpreted to mean that the two distributions had different central tendencies, but it is possible that rejection actually resulted from some other difference between the populations. Notice that when we gain one thing (freedom from assumptions), we pay for it with something else (loss of specificity).

The logical basis of Wilcoxon's rank-sum test is particularly easy to understand. Assume that we have two independent treatment groups, with n_1 observations in group 1 and n_2 observations in group 2. Further assume that the null hypothesis is *false* to a very substantial degree and that the population from which group 1 scores have been sampled contains values generally lower than the population from which group 2 scores were drawn. Then, if we were to rank all $n_1 + n_2 = N$ scores from lowest to highest without regard to group membership, we would expect that the lower ranks generally would fall to group 1 scores and the higher ranks to group 2 scores. Going one step further, if we were to sum the ranks assigned to each group, the sum of the ranks in group 1 would be expected to be appreciably smaller than the sum of the ranks in group 2.

Now consider the opposite case, in which the null hypothesis is *true* and the scores for the two groups were sampled from identical populations. In this situation, if we were to rank all N scores without regard to group membership, we would expect some low ranks and some high ranks in each group, and the sum of the ranks assigned to group 1 would be roughly equal to the sum of the ranks assigned to group 2. These situations are illustrated in Table 18.2.

Wilcoxon based his test on the logic just described, using the sum of the ranks in one of the groups as his test statistic. If that sum is too small relative to the other sum, we will reject the null hypothesis. More specifically, we will take as our test statistic the sum of the ranks assigned to the *smaller* group, or, if $n_1 = n_2$, the *smaller* of the two sums. Given this value, we can use tables of the Wilcoxon statistic (W_S) to test the null hypothesis.

To take a specific example, consider the following hypothetical data on the number of recent stressful life events reported by a group of Cardiac Patients in a local hospital and a control group of Orthopedic Patients in the same hospital. It is well known that stressful life events (marriage, new job, death of spouse, and so on) are associated with illness, and it is

Table 18.2   Illustration of typical results to be expected under H_0 false and H_0 true

H_0 False

Raw Data	10	12	17	13	19	20	30	26	25	33	18	27
Ranks (R_i)	1	2	4	3	6	7	11	9	8	12	5	10
$\sum(R_i)$			23						55			

H_0 True

Raw Data	22	28	32	19	24	33	18	25	29	20	23	34
Ranks (R_i)	4	8	10	2	6	11	1	7	9	3	5	12
$\sum(R_i)$			41						37			

reasonable to expect that, on average, many cardiac patients would have experienced more recent stressful events than would orthopedic patients (who just happened to break an ankle while tearing down a building or a leg while playing touch football). It would appear from the data that this expectation is borne out. Because we have some reason to suspect that life stress scores probably are not symmetrically distributed in the population (especially for cardiac patients, if our research hypothesis is true), we will choose to use a nonparametric test. In this case, we will use the Wilcoxon rank-sum test because we have two independent groups.

	Cardiac Patients						Orthopedic Patients				
Raw Data	32	8	7	29	5	0	1	2	2	3	6
Ranks	11	9	8	10	6	1	2	3.5	3.5	5	7

To apply Wilcoxon's test, we first rank all 11 scores from lowest to highest, assigning tied ranks to tied scores (see the discussion on ranking in Chapter 10). The orthopedic group is the smaller of the two and, if those patients generally have had fewer recent stressful life events, then the sum of the ranks assigned to that group should be relatively low. Letting W_S stand for the sum of the ranks in the smaller group (the orthopedic group), we find

$$W_S = 2 + 3.5 + 3.5 + 5 + 7 = 21$$

We can evaluate the obtained value of W_S by using Wilcoxon's table (Appendix W_S), which gives the *smallest* value of W_S that we would expect to obtain by chance if the null hypothesis were true. From Appendix W_S, we find that for $n_1 = 5$ subjects in the smaller group and $n_2 = 6$ subjects in the larger group (n_1 is *always* the number of subjects in the smaller group if group sizes are unequal), the entry for $\alpha = .025$ (one-tailed) is 18. This means that for a difference between groups to be significant at the one-tailed .025 level, or the two-tailed .05 level, W_S must be less than or equal to 18. We found W_S to be equal to 21, so we cannot reject H_0. (By way of comparison, if we ran a t test on these data, ignoring the fact that one sample variance is almost 50 times the other and that the data suggest that our prediction of the shape of the distribution of cardiac scores may be correct, t would be 1.92 on 9 df, a nonsignificant result with $p = .110$. Using a resampling program on the means of the raw data, the probability of an outcome this extreme would be .059. A similar test on medians would yield $p = .059$.)

The entries in Appendix W_S are for a one-tailed test and will lead to rejection of the null hypothesis only if the sum of the ranks for the smaller group is sufficiently *small*. It is possible, however, that the larger ranks could be congregated in the smaller group, in which case if H_0 is false, the sum of the ranks would be larger than chance expectation rather than smaller. One rather awkward way around this problem would be to rank the data all over again, this time ranking from high to low. If we did this, then the smaller ranks would now appear in the smaller group and we could proceed as before. We do not have to go through the process of reranking data, however. We can accomplish the same thing by using the symmetric properties of the distribution of the rank sum by calculating a statistic called W_S'. The statistic W_S' is the sum of the ranks for the smaller group that we would have found if we had reversed our ranking and ranked from highest to lowest:

$$W_S' = 2\overline{W} - W_S$$

where $2\overline{W} = n_1(n_1 + n_2 + 1)$ and is shown in the table in Appendix W_S. We can then evaluate W_S' against the tabled value and have a one-tailed test on the *upper* tail of the distribution. For a two-tailed test of H_0 (which is what we normally want), we calculate W_S and W_S', enter the table with whichever is smaller, and double the listed value of α.

To illustrate W_S and W_S', consider the two sets of data in Table 18.3. Notice that the two data sets exhibit the same degree of *extremeness,* in the sense that for the first set four of the five lowest ranks are in group 1, and in the second set four of the five highest ranks are in group 1. Moreover, W_S for set 1 is equal to W_S' for set 2, and vice versa. Thus, if we establish the rule that we will calculate both W_S and W_S' for the *smaller* group and refer the *smaller* of W_S and W_S' to the tables, we will come to the same conclusion with respect to the two data sets.

Table 18.3   Sample data for Wilcoxon's rank-sum test

Set 1		Group 1					Group 2		
X	2	15	16	19	18	23	25	37	82
Ranks	1	2	3	5	4	6	7	8	9
$W_S = 11$									
$W_S' = 29$									

Set 2		Group 1					Group 2		
X	60	40	24	21	23	18	15	14	4
Ranks	9	8	7	5	6	4	3	2	1
$W_S = 29$									
$W_S' = 11$									

The Normal Approximation

Appendix W_S is suitable for all cases in which n_1 and n_2 are less than or equal to 25. For larger values of n_1 or n_2, we can make use of the fact that the distribution of W_S approaches a normal distribution as sample sizes increase. This distribution has

$$\text{Mean} = \frac{n_1(n_1 + n_2 + 1)}{2}$$

and

$$\text{Standard error} = \sqrt{\frac{n_1 n_2 (n_1 + n_2 + 1)}{12}}$$

Because the distribution is normal, and we know its mean and standard deviation (the standard error), we can calculate

$$z = \frac{\text{Statistic} - \text{Mean}}{\text{Standard deviation}} = \frac{W_S - \dfrac{n_1(n_1 + n_2 + 1)}{2}}{\sqrt{\dfrac{n_1 n_2 (n_1 + n_2 + 1)}{12}}}$$

and obtain from the tables of the normal distribution an approximation of the true probability of a value of W_S at least as low as the one obtained. (It is immaterial whether we use W_S or W_S' in this situation because they will produce equal values of z, differing only in sign.)

To illustrate the computations for the case in which the larger ranks fall in the smaller groups and to illustrate the use of the normal approximation (although we do not really need to use an approximation for such small sample sizes), consider the data in Table 18.4. These data are hypothetical (but not particularly unreasonable) data on the birthweight (in grams) of children born to mothers who did not seek prenatal care until the third trimester and those born to mothers who received prenatal care starting in the first trimester.

Table 18.4 Hypothetical data on birthweight of infants born to mothers with different levels of prenatal care

Beginning of Care			
Third Trimester		First Trimester	
Birthweight	Rank	Birthweight	Rank
1680	2	2940	10
3830	17	3380	16
3110	14	4900	18
2760	5	2810	9
1700	3	2800	8
2790	7	3210	15
3050	12	3080	13
2660	4	2950	11
1400	1		
2775	6		

$$W_S = \sum(\text{Ranks in Group 2}) = 100$$

$$W_S' = 2\overline{W} - W_S = 152 - 100 = 52$$

$$z = \frac{W_S - \dfrac{n_1(n_1 + n_2 + 1)}{2}}{\sqrt{\dfrac{n_1 n_2(n_1 + n_2 + 1)}{12}}}$$

$$= \frac{100 - \dfrac{8(8 + 10 + 1)}{2}}{\sqrt{\dfrac{8(10)(8 + 10 + 1)}{12}}}$$

$$= \frac{100 - 76}{\sqrt{126.6667}} = 2.13$$

For the data in Table 18.4, the sum of the ranks in the smaller group equals 100. From Appendix W_S, we find $2\overline{W} = 152$, and thus, $W_S' = 2\overline{W} - W_S = 52$. Because 52 is smaller than 100, we enter Appendix W_S with $W_S' = 52$, $n_1 = 8$, and $n_2 = 10$. (n_1 is defined as the smaller sample size.) We want a two-tailed test, so we will double the tabled value of α. The critical value of W_S (or W_S') for a two-tailed test at $\alpha = .05$ is 53, meaning that only 5% of the time would we expect a value of W_S or W_S' less than or equal to 53 if H_0 is true. Our obtained value of W_S' is 52, which thus falls in the rejection region, and we will reject H_0. We will conclude that mothers who do not receive prenatal care until the third trimester tend to give birth to smaller babies. This probably does not mean that not having care until the third trimester *causes* smaller babies, but only that variables associated with delayed care (e.g., young mothers, poor nutrition, or poverty) are also associated with lower birthweight.

The use of the normal approximation for evaluating W_S is illustrated in the bottom part of Table 18.3. Here we find that $z = 2.13$. From Appendix z, we find that the probability of W_S as large as 100 or as small as 52 (a z as extreme as ± 2.13) is $2(.0166) = .033$. Because this value is smaller than our traditional cutoff of $\alpha = .05$, we will reject H_0 and again conclude that there is sufficient evidence to say that failing to seek early prenatal care is related to lower birthweight. Note that both the exact solution and the normal approximation lead to the same conclusion with respect to H_0. However, a resampling test on the means using randomization would yield $p = .059$ (two-tailed). (It would be instructive for you to calculate t for the same set of data.)

The Treatment of Ties

When the data contain tied scores, any test that relies on ranks is likely to be somewhat distorted. Ties can be dealt with in several different ways. You can assign tied ranks to tied scores (as we have been doing), you can flip a coin and assign consecutive ranks to tied scores, or you can assign untied ranks in whatever way will make it hardest to reject H_0. In actual practice, most people simply assign tied ranks. Although that may not be the best way to proceed statistically, it is clearly the most common and is the method that we will use here.

The Null Hypothesis

Wilcoxon's rank-sum test evaluates the null hypothesis that the two sets of scores were sampled from identical populations. This is broader than the null hypothesis tested by the corresponding t test, which dealt specifically with means (primarily as a result of the underlying assumptions that ruled out other sources of difference). If the two populations are assumed to have the same shape and dispersion, then the null hypothesis tested by the rank-sum test will actually deal with the central tendency (in this case the medians) of the two populations, and if the populations are also symmetric, the test will be a test of means. In any event, the rank-sum test is particularly sensitive to differences in central tendency.

Wilcoxon's Test and Resampling Procedures

rank-randomization tests

An interesting feature of Wilcoxon's test is that it is actually not anything you haven't seen before. Wilcoxon derived his test as a permutation test on ranked data, and such tests are often referred to as **rank-randomization tests.** In other words, if you took the data we had earlier, converted them to ranks, and ran a standard permutation test (which is really a randomization test where we draw every possible permutation once and only once), you would obtain the same result that Wilcoxon's test produces. The reason that Wilcoxon was able to derive his test many years before computers could reasonably do the calculations, and why he could create tables for it, is that he uses ranks. We know a good many things about ranks, such as their sum and mean, without having to do the calculations. If we have five numbers, we know that their ranks will be the numbers $1-5$ regardless of what their individual values are. This allowed Wilcoxon to derive the resulting sampling distributions once, and only once, and thus create his tables.

The Mann–Whitney *U* Statistic

Mann–Whitney *U* test

A common competitor to the Wilcoxon rank-sum test is the **Mann–Whitney *U* test.** We do not need to discuss the Mann–Whitney test at any length, however, because the two are equivalent tests, and there is a perfect linear relationship between W_S and U. The only reason for its inclusion here is that you may run across a reference to U, and therefore you should know what it is. Very simply,

$$U = \frac{n_1(n_1 + 2n_2 + 1)}{2} - W_S$$

From this formula, we can see that for any given set of sample sizes, U and W_S differ by only a constant (as do their critical values). Because we have this relationship between the two statistics, we can always convert U to W_S and evaluate W_S using Appendix W_S.

18.7   Wilcoxon's Matched-Pairs Signed-Ranks Test

Wilcoxon matched-pairs signed-ranks test

Wilcoxon is credited with developing not only the most popular nonparametric test for independent groups, but also the most popular test for matched groups (or paired scores). This test is the nonparametric analogue of the t test for related samples, and it tests the null hypothesis that two related (matched) samples were drawn either from identical populations or from symmetric populations with the same mean. More specifically, it tests the null hypothesis that the distribution of difference scores (in the population) is symmetric about zero. This is the same hypothesis tested by the corresponding t test when that test's normality assumption is met.

The development of the logic behind the **Wilcoxon matched-pairs signed-ranks test** is as straightforward as it was for his rank-sum test and can be illustrated with a simple example. Assume that we want to test the often-stated hypothesis that a long-range program of running will reduce blood pressure. To test this hypothesis, we measure the blood pressure of a number of participants, ask them to engage in a systematic program of running for 6 months, and again test their blood pressure at the end of that period. Our dependent variable will be the change in blood pressure over the 6-month interval. If running does reduce blood pressure, we would expect most of the participants to show a lower reading the second time, and thus, a positive pre–post difference. We also would expect that those whose blood pressure actually went up (and thus have a negative pre–post difference) would be only *slightly* higher. On the other hand, if running is worthless as a method of controlling blood pressure, then about one-half of the difference scores will be positive and one-half will be negative, and the positive differences will be about as large as the negative ones. In other words, if H_0 is really true, we would no longer expect most changes to be in the predicted direction with only small changes in the unpredicted direction. Notice that we have two expectations here: (1) Most of the changes will be in the same direction; (2) those that are in the opposite direction will be small ones. We will relax that second expectation when we shortly come to the sign test, but with a concomitant loss in power.

As is illustrated in the following numerical example, in carrying out the Wilcoxon matched-pairs signed ranks test, we first calculate the difference score for each pair of measurements. We then rank all difference scores *without* regard to the sign of the difference, then assign the algebraic sign of the differences to the ranks themselves, and finally sum the positive and negative ranks separately. The test statistic (T) is taken as the smaller of the absolute values (i.e., ignoring the sign) of the two sums and is evaluated against the tabled entries in Appendix T. (It is important to note that in calculating T, we attach algebraic signs to the ranks only for convenience. We could just as easily, for example, circle those ranks that went with improvement and underline those that went with deterioration. We are merely trying to differentiate between the two cases.)

Assume that the study previously described produced the following data on systolic blood pressure before and after the 6-month training session:

Before:	130	170	125	170	130	130	145	160
After:	120	163	120	135	143	136	144	120
Difference $(B - A)$:	10	7	5	35	−13	−6	1	40
Rank of Difference:	5	4	2	7	6	3	1	8
Signed Rank:	5	4	2	7	−6	−3	1	8

$$T_+ = \sum (\text{positive ranks}) = 27$$

$$T_- = \sum (\text{negative ranks}) = -9$$

The first two rows contain the participants' blood pressures as measured before and after a 6-month program of running. The third row contains the difference scores, obtained by subtracting the "after" score from the "before." Notice that only two participants showed a negative change—increased blood pressure. These difference scores do not appear to reflect a population distribution that is anywhere near normal, so we have chosen to use a nonparametric test. In the fourth row, all the difference scores have been ranked without regard to the direction of the change; in the fifth row, the appropriate sign has been appended to the ranks to discriminate those participants whose blood pressure decreased from those whose blood pressure increased. At the bottom of the table we see the sum of the positive and negative ranks (T_+ and T_-). T is defined as the smaller absolute value of T_+ and T_-, thus $T = 9$.

To evaluate T, we refer to Appendix T, a portion of which is shown in Table 18.5. This table has a format somewhat different from that of the other tables we have seen. The easiest way to understand what the entries in the table represent is by way of an analogy. Suppose that to test the fairness of a coin, you were going to flip it eight times and reject the null hypothesis, at $\alpha = .05$ (one-tailed), if there were too few heads. Out of eight flips of a coin, there is no set of outcomes that has a probability of *exactly* .05 under H_0. The probability of one or fewer heads is .0352, and the probability of two or fewer heads is .1445. Thus, if we want to work at $\alpha = .05$, we can either reject for one or fewer heads, in which case the probability of a Type I error is actually .0352 (less than .05), or we can reject for two or fewer heads, in which case the probability of a Type I error is actually .1445 (very much greater than .05). The same kind of problem arises with T because it, like the binomial distribution that gave us the probabilities of heads and tails, is a discrete distribution.[7]

In Appendix T, we find that for a one-tailed test at $\alpha = .025$ (or a two-tailed test at $\alpha = .05$) with $n = 8$, the entries are 3(0.0195) and 4(0.0273). This tells us that if we want to work at a (one-tailed) $\alpha = .025$, which is the equivalent of a two-tailed test at $\alpha = .05$, we can either reject H_0 for $T \leq 3$ (in which case α actually equals .0195) or we can reject for $T \leq 4$ (in which case, the true value of α is .0273). We want a two-tailed test, so the probabilities should be doubled to 3(0.0390) and 4(0.0546). Because we obtained a T value of 9, we would not reject H_0, whichever cutoff we chose. We will conclude, therefore, that we have no reason to doubt that blood pressure is unaffected by a short (6-month) period of daily running. It is going to take a lot more than 6 months to make up for a lifetime of dissipated habits.

---

[7]A similar situation arises for the Wilcoxon rank-sum test, but the standard tables for that test give only the conservative cutoff.

Table 18.5 Critical lower-tail values of T and their associated probabilities (abbreviated version of Appendix T)

	Nominal α (One-Tailed)											
	0.05			0.025			0.01			0.005		
N	T	α		T	α		T	α		T	α
5	0	0.0313									
	1	0.0625									
6	2	0.0469		0	0.0156						
	3	0.0781		1	0.0313						
7	3	0.0391		2	0.0234		0	0.0078			
	4	0.0547		3	0.0391		1	0.0156			
8	5	0.0391		3	0.0195		1	0.0078		0	0.0039
	6	0.0547		4	0.0273		2	0.0117		1	0.0078
9	8	0.0488		5	0.0195		3	0.0098		1	0.0039
	9	0.0645		6	0.0273		4	0.0137		2	0.0059
10	10	0.0420		8	0.0244		5	0.0098		3	0.0049
	11	0.0527		9	0.0322		6	0.0137		4	0.0068
11	13	0.0415		10	0.0210		7	0.0093		5	0.0049
	14	0.0508		11	0.0269		8	0.0122		6	0.0068
...	...	...		...	...		...	...		...	...

Ties

Ties can occur in the data in two different ways. One way would be for a participant to have the same before and after scores, leading to a difference score of 0, which has no sign. In this case, we normally eliminate that participant from consideration and reduce the sample size accordingly, although this leads to some bias in the data.

In addition, we could have tied difference scores that lead to tied rankings. If both the tied scores are of the same sign, we can break the ties in any way we want (or assign tied ranks) without affecting the final outcome. If the scores are of opposite signs, we normally assign tied ranks and proceed as usual.

The Normal Approximation

When the sample size is larger than 50, which is the limit for Appendix T, a normal approximation is available to evaluate T. For larger sample sizes, we know that the sampling distribution of T is approximately normally distributed with

$$\text{Mean} = \frac{n(n+1)}{4} \quad \text{and} \quad \text{Standard error} = \sqrt{\frac{n(n+1)(2n+1)}{24}}$$

Thus, we can calculate

$$z = \frac{T - \dfrac{n(n+1)}{4}}{\sqrt{\dfrac{n(n+1)(2n+1)}{24}}}$$

and evaluate z using Appendix z. The procedure is directly analogous to that used with the rank-sum test and will not be repeated here.

Another interesting example of the use of Wilcoxon's signed-ranks matched-pairs test is found in a study by Manning, Hall, and Gold (1990). These investigators were interested in studying the role of glucose in memory, in particular its effects on performance of memory tasks for elderly people. There has been considerable suggestion in the literature that participants with poor glucose regulation show poor memory and decreased performance on other kinds of neuropsychological tests.

Manning et al. asked 17 elderly volunteers to perform a battery of tests early in the morning after having drunk an 8-ounce lemon-flavored drink sweetened with either glucose or saccharin. Participants performed these tasks under both conditions, so we have matched sets of data. On one of these tasks, for which they had data on only 16 people, participants were read a narrative passage and were asked for recall of that passage 5 minutes later. The dependent variable was not explicitly defined, but we will assume that it was the number of specific propositions recalled from the passage.

The data given in Table 18.6 were generated to produce roughly the same means, standard deviations, and test results as the data found by Manning et al. From Appendix T with $N = 16$ and a two-tailed test at $\alpha = .05$, we find that the critical value of T is 35 or 36, depending on whether you prefer to err on the liberal or conservative side. Our value of $T_{\text{obt}} = 14.5$ is less than either and is therefore significant. This is the same conclusion that Manning et al. came to when they reported improved recall in the Glucose condition.

Table 18.6   Recall scores for elderly subjects after drinking a glucose or saccharin solution

Subject	1	2	3	4	5	6	7	8	9	10	11	12	13	14	15	16
Glucose	0	10	9	4	8	6	9	3	12	10	15	9	5	6	10	6
Saccharin	1	9	6	2	5	5	7	2	8	8	11	3	6	8	8	4
Difference	−1	1	3	2	3	1	2	1	4	2	4	6	−1	−2	2	2
Positive ranks		3	12.5	8.5	12.5	3	8.5	3	14.5	8.5	14.5	16			8.5	8.5
Negative ranks	−3												−3	−8.5		

$$T_+ = \sum (\text{positive ranks}) = 121.5$$

$$T_- = \sum (\text{negative ranks}) = 14.5$$

As an example of using the normal approximation, we can solve for the normal variate (z score) associated with a T of 14.5 for $N = 16$. In this case,

$$z = \frac{T - \dfrac{n(n+1)}{4}}{\sqrt{\dfrac{n(n+1)(2n+1)}{24}}} = \frac{14.5 - \dfrac{(16)(17)}{4}}{\sqrt{\dfrac{(16)(17)(33)}{24}}} = -2.77$$

which has a two-tailed probability under H_0 of .0056. A resampling procedure on the means would produce $p = .002$ (two-tailed).

18.8 The Sign Test

sign test

The Wilcoxon matched-pairs signed-ranks test is an excellent distribution-free test for differences with matched samples. Unlike Student's t test, it makes less than maximum use of the data, in that it substitutes ranks for raw score differences, thus losing some of the subtle differences among the data points. When the assumptions of Student's t hold, that test also has somewhat less power. When those assumptions do not hold, however, it may have greater power. A test that goes even further in the direction of gaining freedom from assumptions at the cost of power is the **sign test.** This test loses even more information by ignoring the values altogether and looking only at the sign of the differences. As a result, it loses even more power.

We can use the example from Manning et al. (1990) in the preceding section. It might be argued that this is a good candidate for such a test because the Wilcoxon was forced to rely on a large number of tied ranks. This argument is not all that persuasive because the results would have been the same no matter how you had broken the tied ranks, but it would be comforting to know that Manning et al.'s results are sufficiently solid that a sign test would also reveal their statistical significance.

The data from Manning et al. are repeated in Table 18.7. From these data, you can see that 13 out of 16 participants showed higher recall under the Glucose condition, whereas only 3 of the 16 showed higher recall under the Saccharin condition. The sign test consists simply of asking the question of whether a 3-to-13 split would be likely to occur if recall under the two conditions were equally good.

Table 18.7 Data from Manning et al. (1990)

Participant	1	2	3	4	5	6	7	8	9	10	11	12	13	14	15	16
Glucose	0	10	9	4	8	6	9	3	12	10	15	9	5	6	10	6
Saccharin	1	9	6	2	5	5	7	2	8	8	11	3	6	8	8	4
Difference	−1	1	3	2	3	1	2	1	4	2	4	6	−1	−2	2	2
Sign	−	+	+	+	+	+	+	+	+	+	+	+	−	−	+	+

This test could be set up in several ways. We could solve for the binomial probability of 13 or more successes out of 16 trials given $p = .50$. From standard tables, or the binomial formula, we would find

$$p(13) = .0085$$
$$p(14) = .0018$$
$$p(15) = .0002$$
$$p(16) = \underline{.0000}$$

Sum　　.0105

The binomial distribution is symmetric for $p = .50$, so we would then double this probability to obtain the two-tailed probability, which in this case is .021. Because this probability is less than .05, we would reject the null hypothesis and conclude that recall is greater in the Glucose condition.

Yet another, which is logically equivalent to the first, is to use a goodness of fit χ^2 test. In this case, we would take 8 as our expected frequency for each cell because if the two conditions lead to equal recall, we would expect half of our 16 participants to do better by chance under each condition. We would then set up the table

	Glucose	Saccharin
Observed	13	3
Expected	8	8

$$\chi^2 = \sum \frac{(O - E)^2}{E} = \frac{(13 - 8)^2}{8} + \frac{(3 - 8)^2}{8} = 6.25$$

The critical value of χ^2 on 1 df is 3.84, so we can reject H_0 and again conclude that the difference is significant. (The probability of $\chi^2 \geq 6.25$ is .0124, which agrees well enough, given the small sample size, with the exact binomial probability.) Both of these tests are more or less equivalent, and you can use whichever is most convenient.

18.9 Kruskal–Wallis One-Way Analysis of Variance

Kruskal–Wallis one-way analysis of variance

The **Kruskal–Wallis one-way analysis of variance** is a direct generalization of the Wilcoxon rank-sum test to the case in which we have three or more independent groups. As such, it is the nonparametric analogue of the one-way analysis of variance discussed in Chapter 11. It tests the hypothesis that all samples were drawn from identical populations and is particularly sensitive to differences in central tendency.

To perform the Kruskal–Wallis test, we simply rank all scores without regard to group membership and then compute the sum of the ranks for each group. The sums are denoted by R_i. If the null hypothesis is true, we would expect the R_is to be more or less equal (aside from difference caused by the size of the samples). A measure of the degree to which the R_i differ from one another is provided by

$$H = \frac{12}{N(N + 1)} \sum_{i=1}^{k} \frac{R_i^2}{n_i} - 3(N + 1)$$

where

$k = $ the number of groups

$n_i = $ the number of observations in group$_i$

$R_i = $ the sum of the ranks in group$_i$

$N = \sum n_i = $ total sample size

H is then evaluated against the χ^2 distribution $k - 1$ df.

As an example, assume that the data in Table 18.8 represent the number of simple arithmetic problems (out of 85) solved (correctly or incorrectly) in 1 hour by participants given a depressant drug, a stimulant drug, or a placebo. Notice that in the Depressant group, three of the participants were too depressed to do much of anything, and in the Stimulant group, three of the participants ran up against the limit of 85 available problems. These data are decidedly nonnormal, and we will use the Kruskal–Wallis test. The calculations are shown in the lower part of the table. The obtained value of H is 10.36, which can be treated as χ^2 on $3 - 1 = 2$ df. The critical value of $\chi^2_{.05}(2)$ is found in Appendix χ^2 to be 5.99. Because $10.36 > 5.99$, we can reject H_0 and conclude that the three drugs lead to different rates of performance.

Table 18.8   Kruskal–Wallis test applied to data on problem solving

Depressant		Stimulant		Placebo	
Score	Rank	Score	Rank	Score	Rank
55	9	73	15	61	11
0	1.5	85	18	54	8
1	3	51	7	80	16
0	1.5	63	12	47	5
50	6	85	18		
60	10	85	18		
44	4	66	13		
		69	14		
R_i	35		115		40

$$H = \frac{12}{N(N+1)} \sum_{i=1}^{k} \frac{R_i^2}{n_i} - 3(N+1)$$

$$= \frac{12}{19(20)} \left(\frac{35^2}{7} + \frac{115^2}{8} + \frac{40^2}{4} \right) - 3(19+1)$$

$$= \frac{12}{380} (2228.125) - 60$$

$$= 70.36 - 60$$

$$= 10.36$$

$$\chi_{0.05}^2(2) = 5.99$$

18.10 Friedman's Rank Test for k Correlated Samples

Friedman's rank test for k correlated samples

The last test to be discussed in this chapter is the nonparametric analogue of the one-way repeated-measures analysis of variance, **Friedman's rank test for k correlated samples.** It was developed by the well-known economist Milton Friedman—in the days before he was a well-known economist. This test is closely related to a standard repeated-measures analysis of variance applied to ranks instead of raw scores. It is a test on the null hypothesis that the scores for each treatment were drawn from identical populations, and it is especially sensitive to population differences in central tendency.

Assume that we want to test the hypothesis that the judged quality of a lecture is related to the number of visual aids used. The experimenter obtains 17 people who frequently give lectures to local business groups on a variety of topics. Each lecturer delivers the same lecture to three different, but equivalent, audiences—once with no visual aids, once with a few transparencies to illustrate major points, and once with transparencies and flip charts to illustrate every point made. At the end of each lecture, the audience is asked to rate the lecture on a 75-point scale, and the mean rating across all members of the audience is taken as the dependent variable. The same lecturers serve under all three conditions, so we would expect the data to be correlated. Terrible lecturers are terrible no matter how many visual aids they use. Hypothetical data are presented in Table 18.9, in which a higher score represents a more favorable rating. The ranking of the raw scores *within each participant* are shown in parentheses.

Table 18.9   Hypothetical data on rated quality of lectures

	Number of Visual Aids		
Lecturer	None	Few	Many
1	50 (1)	58 (3)	54 (2)
2	32 (2)	37 (3)	25 (1)
3	60 (1)	70 (3)	63 (2)
4	58 (2)	60 (3)	55 (1)
5	41 (1)	66 (3)	59 (2)
6	36 (2)	40 (3)	28 (1)
7	26 (3)	25 (2)	20 (1)
8	49 (1)	60 (3)	50 (2)
9	72 (1)	73 (2)	75 (3)
10	49 (2)	54 (3)	42 (1)
11	52 (2)	57 (3)	47 (1)
12	36 (2)	42 (3)	29 (1)
13	37 (3)	34 (2)	31 (1)
14	58 (3)	50 (1)	56 (2)
15	39 (1)	48 (3)	44 (2)
16	25 (2)	29 (3)	18 (1)
17	51 (1)	63 (2)	68 (3)
	30	45	27

$$\chi_F^2 = \frac{12}{Nk(k+1)} \sum_{i=1}^{k} R_i^2 - 3N(k+1)$$

$$= \frac{12}{17(3)(4)}(30^2 + 45^2 + 27^2) - 3(17)(4)$$

$$= \frac{12}{204}(3654) - 204$$

$$= 10.94$$

If the null hypothesis is true, we would expect the rankings to be randomly distributed within each lecturer. Thus, one lecturer might do best with no visual aids, another might do best with many aids, and so on. If this were the case, the sum of the rankings in each condition (column) would be approximately equal. On the other hand, if a few visual aids were to lead to the most popular lecture, then most lecturers would have their highest rating under that condition, and the sum of the rankings for the three conditions would be decidedly unequal.

To apply Friedman's test, we rank the raw scores for each lecturer separately and then sum the rankings for each condition. We then evaluate the variability of the sums by computing

$$\chi_F^2 = \frac{12}{Nk(k+1)} \sum_{i=1}^{k} R_i^2 - 3N(k+1)$$

where

R_i = the sum of the ranks for the ith condition

N = the number of subjects (lecturers)

k = the number of conditions

This value of χ_F^2 can be evaluated with respect to the standard χ^2 distribution on $k-1$ df.

For the data in Table 18.9, $\chi_F^2 = 10.94$ on 2 df. Because $\chi_{.05}^2(2) = 5.99$, we will reject H_0 and conclude that the judged quality of a lecture differs as a function of the degree to which visual aids are included. The data suggest that some visual aids are helpful, but that too many of them can detract from what the lecturer is saying. (*Note:* The null hypothesis we have just tested says nothing about differences among participants [lecturers], and participant differences are completely eliminated by the ranking procedure.)

Key Terms

Parametric tests (Introduction)

Nonparametric tests (Introduction)

Distribution-free tests (Introduction)

Resampling procedures (Introduction)

Sampling with replacement (Introduction)

Bootstrapping procedures (Introduction)

Permutation tests (Introduction)

Randomization tests (Introduction)

Sampling without replacement (Introduction)

Wilcoxon rank-sum test (18.6)

Rank-randomization tests (18.6)

Mann–Whitney U test (18.6)

Wilcoxon matched-pairs signed-ranks test (18.7)

Sign test (18.8)

Kruskal–Wallis one-way analysis of variance (18.9)

Friedman's rank test for k correlated samples (18.10)

Exercises

18.1 McConaughy (1980) has argued that younger children organize stories in terms of simple descriptive ("and then . . .") models, whereas older children incorporate causal statements and social inferences. Suppose that we asked two groups of children differing in age to summarize a story they just read. We then counted the number of statements in the summary that can be classed as inferences. The data follow:

Younger Children:	0	1	0	3	2	5	2
Older Children:	4	7	6	4	8	7	

    a.   Analyze these data using the two-tailed rank-sum test.

    b.   What can you conclude?

    c.   How would you go about analyzing these data if you had access to a program that would do resampling for you?

18.2 Kapp, Frysinger, Gallagher, and Hazelton (1979) have demonstrated that lesions in the amygdala can reduce certain responses commonly associated with fear (e.g., decreases in heart rate). If fear is really reduced, then it should be more difficult to train an avoidance response in lesioned animals because the aversiveness of the stimulus will be reduced. Assume two groups of rabbits: One group has lesions in the amygdala, and the other is an untreated control group. The following data represent the number of trials to learn an avoidance response for each animal:

Group with Lesions:	15	14	15	8	7	22	36	19	14	18	17
Control Group:	9	4	9	10	6	6	4	5	9		

Analyze the data using the Wilcoxon rank-sum test (two-tailed).

What can you conclude?

18.3 Repeat the analysis in Exercise 18.2 using the normal approximation.

18.4 Repeat the analysis in Exercise 18.2 using the appropriate one-tailed test.

18.5 Nurcombe and Fitzhenry-Coor (1979) have argued that training in diagnostic techniques should lead a clinician to generate (and test) more hypotheses in coming to a decision about a case. Suppose we take 10 psychiatric residents who are just beginning their residency and ask them to watch a videotape of an interview and to record their thoughts on the case every few minutes. We then count the number of hypotheses each resident includes in his or her written remarks. The experiment is repeated at the end of the residency with a comparable videotape. The data follow:

Subject:	1	2	3	4	5	6	7	8	9	10
Before:	8	4	2	2	4	8	3	1	3	9
After:	7	9	3	6	3	10	6	7	8	7

    a.   Analyze the data using Wilcoxon's matched-pairs signed-ranks test.

    b.   What can you conclude?

18.6 Refer to Exercise 18.5.

    a.   Repeat the analysis using the normal approximation.

    b.   How well do the two answers (18.5a and 18.6a) agree? Why do they not agree exactly?

18.7 How would you go about applying a resampling procedure to test the difference between Before and After scores in Exercise 18.6?

18.8 It has been argued that first-born children tend to be more independent than later-born children. Suppose we develop a 25-point scale of independence and rate each of 20 first-born children and their second-born siblings using our scale. We do this when both siblings are adults, thus eliminating obvious age effects. The data on independence are as follows (a higher score means that the person is more independent):

Sibling Pair:	1	2	3	4	5	6	7	8	9	10
First Born:	12	18	13	17	8	15	16	5	8	12
Second Born:	10	12	15	13	9	12	13	8	10	8

Sibling Pair:	11	12	13	14	15	16	17	18	19	20
First Born:	13	5	14	20	19	17	2	5	15	18
Second Born:	8	9	8	10	14	11	7	7	13	12

    a.   Analyze the data using Wilcoxon's matched-pairs signed-ranks test.

    b.   What can you conclude?

18.9 Rerun the analysis in Exercise 18.8 using the normal approximation.

18.10 How would we run a standard resampling test for the data in Exercise 18.8?

18.11 The results in Exercise 18.8 are not quite as clear-cut as we might like. Plot the differences as a function of the first-born's score. What does this figure suggest?

18.12 What is the difference between the null hypothesis tested by Wilcoxon's rank-sum test and the corresponding t test?

18.13 What is the difference between the null hypothesis tested by Wilcoxon's matched-pairs signed-ranks test and the corresponding t test?

18.14 One of the arguments put forth in favor of nonparametric tests is that they are more appropriate for ordinal-scale data. This issue was addressed earlier in the book in a different context. Give a reason why this argument is not a good one.

18.15 Why is rejection of the null hypothesis using a t test a more specific statement than is rejection of the null hypothesis using the appropriate nonparametric test?

18.16 Three rival professors teaching English I all claim the honor of having the best students. To settle the issue, eight students are randomly drawn from each class and are given the same exam, which is graded by a neutral professor who does not know from which class the students came.

The data follow:

Professor A:	82	71	56	58	63	64	62	53
Professor B:	55	88	85	83	71	70	68	72
Professor C:	65	54	66	68	72	78	65	73

Run the appropriate test and draw the appropriate conclusions.

18.17 A psychologist operating a group home for delinquent adolescents needs to show that it is successful at reducing delinquency. He samples nine adolescents living in their parents' home that the police have identified as having problems, nine similar adolescents living in foster homes, and nine adolescents living in the group home. As an indicator variable, he uses truancy (number of days truant in the past semester), which is readily obtained from school records. On the basis of the following data, draw the appropriate conclusions.

Natural Home:	15	18	19	14	5	8	12	13	7
Foster Home:	16	14	20	22	19	5	17	18	12
Group Home:	10	13	14	11	7	3	4	18	2

18.18 As an alternative method of evaluating a group home, suppose that we take 12 adolescents who have been declared delinquent. We take the number of days truant (1) during the month before they are placed in the home, (2) during the month they live in the home, and (3) during the month after they leave the home.

The data follow:

Adolescent:	1	2	3	4	5	6	7	8	9	10	11	12
Before:	10	12	12	19	5	13	20	8	12	10	8	18
During:	5	8	13	10	10	8	16	4	14	3	3	16
After:	8	7	10	12	8	7	12	5	9	5	3	2

Apply Friedman's test. What do you conclude?

18.19 I did not discuss randomization tests on the evaluation of data that are laid out like a one-way analysis of variance (as in Exercise 18.17), but you should be able to suggest an analysis that would be appropriate if we had the software to carry out the calculations. How would you outline that test?

18.20 The test referred to in Exercise 18.19 is on the website for this book (www.uvm.edu/~dhowell/methods/). Run that program on the data for Exercise 18.18 and report the results. (The "read-me" file will tell you how to run the resampling program.

18.21 What advantage does the study described in Exercise 18.18 have over the study described in Exercise 18.17?

18.22 It would be possible to apply Friedman's test to the data in Exercise 18.5. What would we lose if we did?

18.23 For the data in Exercise 18.5, we could say that 3 out of 10 residents used fewer hypotheses the second time and 7 used more. We could test this with χ^2. How would this differ from Friedman's test applied to those data?

18.24 The history of statistical hypothesis testing really began with a tea-tasting experiment (Fisher, 1935), so it seems fitting for this book to end with one. The owner of a small tearoom does not think that people really can tell the difference between the first cup made with a given tea bag and the second and third cups made with the same bag (perhaps that is why it is still a small tearoom). He chooses eight different brands of tea bags, makes three cups of tea with each, reusing the same tea bag, and then has a group of customers rate each cup on a 20-point scale (without knowing which cup is which). The data are shown here, with higher ratings indicating better tea.

Tea Brands	First Cup	Second Cup	Third Cup
1	8	3	2
2	15	14	4
3	16	17	12
4	7	5	4
5	9	3	6
6	8	9	4
7	10	3	4
8	12	10	2

Using Friedman's test, draw the appropriate conclusions.

Appendices

Appendix: Data Set

Howell and Huessy (1985) reported on a study of 386 children who had, and had not, exhibited symptoms of attention deficit disorder (ADD)—previously known as hyperkinesis or minimal brain dysfunction—during childhood. In 1965, teachers of all second-grade school children in a number of schools in northwestern Vermont were asked to complete a questionnaire for each of their students dealing with behaviors commonly associated with ADD. Questionnaires on these same children were again completed when the children were in the fourth and fifth grades and, for purposes of this data set only, those three scores were averaged to produce a score labeled ADDSC. The higher the score, the more ADD-like behaviors the child exhibited. At the end of ninth grade and again at the end of twelfth grade, information on the performances of these children was obtained from school records. These data offer the opportunity to examine questions about whether later behavior can be predicted from earlier behavior and to examine academically related variables and their interrelationships. The data are referred to in many of the exercises at the end of each chapter. A description of each variable follows.

ADDSC	Average of the three ADD-like behavior scores obtained in elementary school
GENDER	1 = male; 2 = female
REPEAT	1 = repeated at least one grade; 0 = did not repeat a grade
IQ	IQ obtained from a group-administered IQ test
ENGL	Level of English in ninth grade: 1 = college prep; 2 = general; 3 = remedial
ENGG	Grade in English in ninth grade: 4 = A; 3 = B; and so on
GPA	Grade point average in ninth grade
SOCPROB	Social problems in ninth grade: 1 = yes; 0 = no
DROPOUT	1 = dropped out before completing high school; 0 = did not drop out

Appendix: Computer Data Sets

The website (www.uvm.edu/~dhowell/methods/) contains many data sets. The data sets represent a combination of data from actual studies, data that have been created to mimic the data from actual studies, data from all examples and exercises at the end of each chapter. It also contains two sets of random numbers that have been generated to illustrate certain points.

All of these data sets are standard ASCII files, meaning that they can be read by virtually all computer programs and can be edited if necessary with standard editors available on any computer system (for example, Microsoft Wordpad). In addition, they can be edited by any word processor that can produce an ASCII file (sometimes referred to as a text file or a DOS file).

The following, unusually complex, data sets are the focus of a number of homework exercises in many different chapters. The descriptions that follow are intended to explain the study from which the data were drawn and to describe how the data are arranged in the data set. You should refer to these descriptions when working with these data sets. The data sets drawn directly from tables and exercises are much simpler, and their structure can be inferred from the text.

In addition, this disk contains copies of data from most of the examples and exercises in the book. Those data sets are described in a file on the disk, and will not be described further here.

Add.dat

The data in this file come from a study by Howell and Huessy (1985). The data are described above.

Variable Name	Columns	Description
ID	1–3	Subject identification number
ADDSC	5–6	ADD score averaged over 3 years
GENDER	8	1 = male; 2 = female
REPEAT	10	1 = repeated a grade, 0 = did not repeat
IQ	12–14	IQ obtained from group-administered IQ test
ENGL	16	Level of English: 1 = college prep; 2 = general; 3 = remedial
ENGG	18	Grade in English: 4 = A, 3 = B, and so on
GPA	20–23	Grade point average in ninth grade
SOCPROB	25	Social problems: 0 = no, 1 = yes
DROPOUT	27	1 = Dropped out of school before finishing 0 = Did not drop out

The first four lines of data are shown here:

1	45	1	0	111	2	3	2.60	0	0
2	50	1	0	102	2	3	2.75	0	0
3	49	1	0	108	2	4	4.00	0	0
4	55	1	0	109	2	2	2.25	0	0

Badcancr.dat

For a description of both the study behind these data and the data set, see the following section on Cancer.dat. The data in this file differ from those in Cancer.dat only by the inclusion of deliberate errors.

These data have been deliberately changed for purposes of an assignment. Errors have been added, and at least one variable has been distorted. The correct data are in Cancer.dat, which should be used for all *future* analyses. Virtually any program is likely to fail at first until errors are found and corrected, and even when it runs, impossible values will remain. The quickest way to find many of the errors is to print out the file and scan the columns.

Cancer.dat

The data in this file come from a study by Compas (1990, personal communication) on the effects of stress in cancer patients and their families. Only a small portion of the data that were collected are shown here, primarily data related to behavior problems in children and psychological symptoms in the patient and her or his spouse. The file contains data on 89 families, and many of the data points are missing because of the time in the study at which these data were selected. This example does, however, offer a good opportunity to see preliminary data on important psychological variables.

The codebook (the listing of variables, descriptions, location, and legitimate values) for the data in Cancer.dat is shown following the sample data.

Missing observations are represented with a period. The first four lines of data are shown here as an example.

101	2	62	50	52	39	52	1	42	44	41	40	42	.	.	.	.	.	.	.
104	1	56	65	55	40	57	2	53	73	68	67	71	1	11	12	28	58	57	60
105	1	56	57	67	65	61	2	41	67	63	66	65	2	7	7	15	47	48	45
106	2	41	61	64	53	57	1	60	60	59	67	62	1	6	10	15	49	52	48

Variable	Description	Columns	Legal Values
FamNum	Family ID number	1–3	100–400
	GSI Variables		
	Patient Variables		
SexP	Gender of patient	5	1 = male; 2 = female
SomTP	Somaticism T score	8–9	41–80
DepTP	Depression T score	12–13	42–80
AnxTP	Anxiety T score	16–17	38–80
HosTP	Hostility T score	20–21	39–80
GSITP	Global Symptom Index T score	24–25	33–80
	Spouse Variables		
SexS	Gender of spouse	27	1 = male; 2 = female
SomTS	Somaticism T score	30–31	41–80
DepTS	Depression T score	34–35	42–80
AnxTS	Anxiety T score	38–39	38–80
HosTS	Hostility T score	42–43	39–80
GSITS	GSI T score	46–47	33–80
	Child Behavior Checklist Variables		
SexChild	Gender of child	49	1 = male; 2 = female
Intern	Internalizing subscale	51–52	0–98
Extern	Externalizing subscale	54–55	0–102
TotBP	Total behavior problems	57–58	0–240
InternT	Internalizing T score	60–61	33–100
ExternT	Externalizing T score	63–64	30–100
TotBPT	Total behavior problem T score	66–67	30–100

Epineq.dat, Epinuneq.dat

Introini-Collison and McGaugh (1986) examined the hypothesis that hormones normally produced in the body can play a role in memory. Specifically, they looked at the effect of post-training injections of epinephrine on retention of a previously learned discrimination. They first trained mice to escape mild shock by choosing the left arm of a Y-maze. Immediately after training, the researchers injected the mice with either 0.0, 0.3, or 1.0 mg/kg of epinephrine. They predicted that low doses of epinephrine would facilitate retention, whereas high doses would inhibit it.

Either 1 day, 1 week, or 1 month after original training, each mouse was again placed in the Y-maze. But this time, running to the right arm of the maze led to escape from shock. Presumably, the stronger the memory of the original training, the more it would interfere with the learning of this new task and the more errors the subjects would make.

This experiment has two data sets, named Epineq.dat and Epinuneq.dat. The original study used 18 animals in the three dosage groups tested after 1 day, and 12 animals in each group tested after intervals of 1 week and 1 month. Hypothetical data that closely reproduce the original results are contained in Epinuneq.dat, although five subjects having a 1-month recall interval have been deleted from the 1.0 mg/kg condition. A second data set was created with 12 observations in each of the 9 cells, and is called Epineq.dat. In both cases, the need to create data that were integers led to results that are slightly conservative relative to the actual data. But the conclusions with respect to H_0 are the same.

For both data sets, there is a three-digit ID; dosage is coded (1, 2, or 3) in column 5; the retention interval is coded (1, 2, or 3) in column 7; and the number of errors in learning the second discrimination is coded in column 9. The first four lines of data follow:

001	1	1	0
002	1	1	3
003	1	1	4
004	1	1	2

Mireault.dat

Mireault (1990) collected data from 381 college students, some of whom had lost a parent by death during their childhood. She had three groups of students. Group 1 was composed of subjects who had lost a parent. Group 2 was composed of subjects whose parents were still alive and married to each other. Group 3 consisted of students whose parents were divorced.

Mireault was interested in observing the effects of parental loss on the person's current level of symptomatology (as measured by the Brief Symptom Inventory, Derogatis, 1983) and on the individual's self-perceived vulnerability to future loss. In the interest of space, the data set includes only the total vulnerability measure, and not the subscales. There is also a single measure for social support. For all measures, a higher score represents more of the concept being measured.

The variables, and their location in the file, are listed following the sample data.

Missing data are represented by a period. The first three lines of data are shown below as an example.

```
002 2 1 1 4 2 .    42 53 59 57 49 57 47 51 46 51 112 24 66
007 1 2 1 2 . 1 18 65 80 64 71 72 73 63 67 67 72 100 23 73
008 2 2 1 1 4 . .  52 67 60 62 65 78 60 65 58 65 118 28 64
```

Variable Name	Columns	Description
ID	1–3	Subject identification number
Group	5	1 = loss; 2 = married; 3 = divorced
Gender	7	1 = male; 2 = female
YearColl	9	1 = first year; 2 = sophomore; and so on
College	11	1 = arts and sciences; 2 = health;
		3 = engineering; 4 = business; 5 = agriculture
GPA	13	4 = A; 3 = B; 2 = C; 1 = D; 0 = F
LostPGen	15	Gender of lost parent
AgeAtLos	17–18	Age at parent's death
SomT	20–21	Somatization T score
ObsessT	23–24	Obsessive-compulsive T score
SensitT	26–27	Interpersonal sensitivity T score
DepressT	29–30	Depression T score
AnxT	32–33	Anxiety T score

(continued)

Variable Name	Columns	Description
HostT	35–36	Hostility T score
PhobT	38–39	Phobic anxiety T score
ParT	41–42	Paranoid ideation T score
PsyT	44–45	Psychoticism T score
GSIT	47–48	Global symptom index T score
PVTotal	50–52	Perceived vulnerability total score
PVLoss	54–56	Perceived vulnerability to loss
SuppTotl	58–60	Social support score

Stress.dat

The data in this file are a subset of data being collected by Compas and his colleagues on stress and coping in cancer patients. The file contains the family number, the gender of the respondent (1 = Male; 2 = Female), the role of the respondent (1 = Patient; 2 = Spouse), and two stress measures (one obtained shortly after diagnosis and one 3 months later). The variables are in the following order: FamNum, Gender, Role, Time1, Time2. The first six cases follow:

```
101   2   1   2   .
101   1   2   2   .
104   1   1   4   .
104   2   2   5   .
105   1   1   3   4
105   2   2   5   4
```

Appendix χ²: Upper Percentage Points of the χ² Distribution

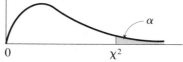

df	0.995	0.990	0.975	0.950	0.900	0.750	0.500	0.250	0.100	0.050	0.025	0.010	0.005
1	0.00	0.00	0.00	0.00	0.02	0.10	0.45	1.32	2.71	3.84	5.02	6.63	7.88
2	0.01	0.02	0.05	0.10	0.21	0.58	1.39	2.77	4.61	5.99	7.38	9.21	10.60
3	0.07	0.11	0.22	0.35	0.58	1.21	2.37	4.11	6.25	7.82	9.35	11.35	12.84
4	0.21	0.30	0.48	0.71	1.06	1.92	3.36	5.39	7.78	9.49	11.14	13.28	14.86
5	0.41	0.55	0.83	1.15	1.61	2.67	4.35	6.63	9.24	11.07	12.83	15.09	16.75
6	0.68	0.87	1.24	1.64	2.20	3.45	5.35	7.84	10.64	12.59	14.45	16.81	18.55
7	0.99	1.24	1.69	2.17	2.83	4.25	6.35	9.04	12.02	14.07	16.01	18.48	20.28
8	1.34	1.65	2.18	2.73	3.49	5.07	7.34	10.22	13.36	15.51	17.54	20.09	21.96
9	1.73	2.09	2.70	3.33	4.17	5.90	8.34	11.39	14.68	16.92	19.02	21.66	23.59
10	2.15	2.56	3.25	3.94	4.87	6.74	9.34	12.55	15.99	18.31	20.48	23.21	25.19
11	2.60	3.05	3.82	4.57	5.58	7.58	10.34	13.70	17.28	19.68	21.92	24.72	26.75
12	3.07	3.57	4.40	5.23	6.30	8.44	11.34	14.85	18.55	21.03	23.34	26.21	28.30
13	3.56	4.11	5.01	5.89	7.04	9.30	12.34	15.98	19.81	22.36	24.74	27.69	29.82
14	4.07	4.66	5.63	6.57	7.79	10.17	13.34	17.12	21.06	23.69	26.12	29.14	31.31
15	4.60	5.23	6.26	7.26	8.55	11.04	14.34	18.25	22.31	25.00	27.49	30.58	32.80
16	5.14	5.81	6.91	7.96	9.31	11.91	15.34	19.37	23.54	26.30	28.85	32.00	34.27
17	5.70	6.41	7.56	8.67	10.09	12.79	16.34	20.49	24.77	27.59	30.19	33.41	35.72
18	6.26	7.01	8.23	9.39	10.86	13.68	17.34	21.60	25.99	28.87	31.53	34.81	37.15
19	6.84	7.63	8.91	10.12	11.65	14.56	18.34	22.72	27.20	30.14	32.85	36.19	38.58
20	7.43	8.26	9.59	10.85	12.44	15.45	19.34	23.83	28.41	31.41	34.17	37.56	40.00
21	8.03	8.90	10.28	11.59	13.24	16.34	20.34	24.93	29.62	32.67	35.48	38.93	41.40
22	8.64	9.54	10.98	12.34	14.04	17.24	21.34	26.04	30.81	33.93	36.78	40.29	42.80
23	9.26	10.19	11.69	13.09	14.85	18.14	22.34	27.14	32.01	35.17	38.08	41.64	44.18
24	9.88	10.86	12.40	13.85	15.66	19.04	23.34	28.24	33.20	36.42	39.37	42.98	45.56
25	10.52	11.52	13.12	14.61	16.47	19.94	24.34	29.34	34.38	37.65	40.65	44.32	46.93
26	11.16	12.20	13.84	15.38	17.29	20.84	25.34	30.43	35.56	38.89	41.92	45.64	48.29
27	11.80	12.88	14.57	16.15	18.11	21.75	26.34	31.53	36.74	40.11	43.20	46.96	49.64
28	12.46	13.56	15.31	16.93	18.94	22.66	27.34	32.62	37.92	41.34	44.46	48.28	50.99
29	13.12	14.26	16.05	17.71	19.77	23.57	28.34	33.71	39.09	42.56	45.72	49.59	52.34
30	13.78	14.95	16.79	18.49	20.60	24.48	29.34	34.80	40.26	43.77	46.98	50.89	53.67
40	20.67	22.14	24.42	26.51	29.06	33.67	39.34	45.61	51.80	55.75	59.34	63.71	66.80
50	27.96	29.68	32.35	34.76	37.69	42.95	49.34	56.33	63.16	67.50	71.42	76.17	79.52
60	35.50	37.46	40.47	43.19	46.46	52.30	59.34	66.98	74.39	79.08	83.30	88.40	91.98
70	43.25	45.42	48.75	51.74	55.33	61.70	69.34	77.57	85.52	90.53	95.03	100.44	104.24
80	51.14	53.52	57.15	60.39	64.28	71.15	79.34	88.13	96.57	101.88	106.63	112.34	116.35
90	59.17	61.74	65.64	69.13	73.29	80.63	89.33	98.65	107.56	113.14	118.14	124.13	128.32
100	67.30	70.05	74.22	77.93	82.36	90.14	99.33	109.14	118.49	124.34	129.56	135.82	140.19

Source: The entries in this table were computed by the author.

Appendix *F*: Critical Values of the *F* Distribution

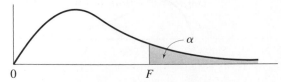

0   F

Table 1 $\alpha = 0.05$

Degrees of Freedom for Numerator

	1	2	3	4	5	6	7	8	9	10	15	20	25	30	40	50
1	161.4	199.5	215.8	224.8	230.0	233.8	236.5	238.6	240.1	242.1	245.2	248.4	248.9	250.5	250.8	252.6
2	18.51	19.00	19.16	19.25	19.30	19.33	19.35	19.37	19.38	19.40	19.43	19.44	19.46	19.47	19.48	19.48
3	10.13	9.55	9.28	9.12	9.01	8.94	8.89	8.85	8.81	8.79	8.70	8.66	8.63	8.62	8.59	8.58
4	7.71	6.94	6.59	6.39	6.26	6.16	6.09	6.04	6.00	5.96	5.86	5.80	5.77	5.75	5.72	5.70
5	6.61	5.79	5.41	5.19	5.05	4.95	4.88	4.82	4.77	4.74	4.62	4.56	4.52	4.50	4.46	4.44
6	5.99	5.14	4.76	4.53	4.39	4.28	4.21	4.15	4.10	4.06	3.94	3.87	3.83	3.81	3.77	3.75
7	5.59	4.74	4.35	4.12	3.97	3.87	3.79	3.73	3.68	3.64	3.51	3.44	3.40	3.38	3.34	3.32
8	5.32	4.46	4.07	3.84	3.69	3.58	3.50	3.44	3.39	3.35	3.22	3.15	3.11	3.08	3.04	3.02
9	5.12	4.26	3.86	3.63	3.48	3.37	3.29	3.23	3.18	3.14	3.01	2.94	2.89	2.86	2.83	2.80
10	4.96	4.10	3.71	3.48	3.33	3.22	3.14	3.07	3.02	2.98	2.85	2.77	2.73	2.70	2.66	2.64
11	4.84	3.98	3.59	3.36	3.20	3.09	3.01	2.95	2.90	2.85	2.72	2.65	2.60	2.57	2.53	2.51
12	4.75	3.89	3.49	3.26	3.11	3.00	2.91	2.85	2.80	2.75	2.62	2.54	2.50	2.47	2.43	2.40
13	4.67	3.81	3.41	3.18	3.03	2.92	2.83	2.77	2.71	2.67	2.53	2.46	2.41	2.38	2.34	2.31
14	4.60	3.74	3.34	3.11	2.96	2.85	2.76	2.70	2.65	2.60	2.46	2.39	2.34	2.31	2.27	2.24
15	4.54	3.68	3.29	3.06	2.90	2.79	2.71	2.64	2.59	2.54	2.40	2.33	2.28	2.25	2.20	2.18
16	4.49	3.63	3.24	3.01	2.85	2.74	2.66	2.59	2.54	2.49	2.35	2.28	2.23	2.19	2.15	2.12
17	4.45	3.59	3.20	2.96	2.81	2.70	2.61	2.55	2.49	2.45	2.31	2.23	2.18	2.15	2.10	2.08
18	4.41	3.55	3.16	2.93	2.77	2.66	2.58	2.51	2.46	2.41	2.27	2.19	2.14	2.11	2.06	2.04
19	4.38	3.52	3.13	2.90	2.74	2.63	2.54	2.48	2.42	2.38	2.23	2.16	2.11	2.07	2.03	2.00
20	4.35	3.49	3.10	2.87	2.71	2.60	2.51	2.45	2.39	2.35	2.20	2.12	2.07	2.04	1.99	1.97
22	4.30	3.44	3.05	2.82	2.66	2.55	2.46	2.40	2.34	2.30	2.15	2.07	2.02	1.98	1.94	1.91
24	4.26	3.40	3.01	2.78	2.62	2.51	2.42	2.36	2.30	2.25	2.11	2.03	1.97	1.94	1.89	1.86
26	4.23	3.37	2.98	2.74	2.59	2.47	2.39	2.32	2.27	2.22	2.07	1.99	1.94	1.90	1.85	1.82
28	4.20	3.34	2.95	2.71	2.56	2.45	2.36	2.29	2.24	2.19	2.04	1.96	1.91	1.87	1.82	1.79
30	4.17	3.32	2.92	2.69	2.53	2.42	2.33	2.27	2.21	2.16	2.01	1.93	1.88	1.84	1.79	1.76
40	4.08	3.23	2.84	2.61	2.45	2.34	2.25	2.18	2.12	2.08	1.92	1.84	1.78	1.74	1.69	1.66
50	4.03	3.18	2.79	2.56	2.40	2.29	2.20	2.13	2.07	2.03	1.87	1.78	1.73	1.69	1.63	1.60
60	4.00	3.15	2.76	2.53	2.37	2.25	2.17	2.10	2.04	1.99	1.84	1.75	1.69	1.65	1.59	1.56
120	3.92	3.07	2.68	2.45	2.29	2.18	2.09	2.02	1.96	1.91	1.75	1.66	1.60	1.55	1.50	1.46
200	3.89	3.04	2.65	2.42	2.26	2.14	2.06	1.98	1.93	1.88	1.72	1.62	1.56	1.52	1.46	1.41
500	3.86	3.01	2.62	2.39	2.23	2.12	2.03	1.96	1.90	1.85	1.69	1.59	1.53	1.48	1.42	1.38
1000	3.85	3.01	2.61	2.38	2.22	2.11	2.02	1.95	1.89	1.84	1.68	1.58	1.52	1.47	1.41	1.36

Degrees of Freedom for Denominator

Source: The entries in this table were computed by the author.

Table 2 $\alpha = 0.025$

Degrees of Freedom for Numerator

	1	2	3	4	5	6	7	8	9	10	15	20	25	30	40	50
1	647.8	799.5	864.2	899.6	921.8	937.1	948.2	956.7	963.3	968.6	984.9	993.1	998.1	1001	1006	1008
2	38.51	39.00	39.17	39.25	39.30	39.33	39.36	39.37	39.39	39.40	39.43	39.45	39.46	39.46	39.47	39.48
3	17.44	16.04	15.44	15.10	14.89	14.73	14.62	14.54	14.47	14.42	14.25	14.17	14.12	14.08	14.04	14.01
4	12.22	10.65	9.98	9.60	9.36	9.20	9.07	8.98	8.90	8.84	8.66	8.56	8.50	8.46	8.41	8.38
5	10.01	8.43	7.76	7.39	7.15	6.98	6.85	6.76	6.68	6.62	6.43	6.33	6.27	6.23	6.18	6.14
6	8.81	7.26	6.60	6.23	5.99	5.82	5.70	5.60	5.52	5.46	5.27	5.17	5.11	5.07	5.01	4.98
7	8.07	6.54	5.89	5.52	5.29	5.12	4.99	4.90	4.82	4.76	4.57	4.47	4.40	4.36	4.31	4.28
8	7.57	6.06	5.42	5.05	4.82	4.65	4.53	4.43	4.36	4.30	4.10	4.00	3.94	3.89	3.84	3.81
9	7.21	5.71	5.08	4.72	4.48	4.32	4.20	4.10	4.03	3.96	3.77	3.67	3.60	3.56	3.51	3.47
10	6.94	5.46	4.83	4.47	4.24	4.07	3.95	3.85	3.78	3.72	3.52	3.42	3.35	3.31	3.26	3.22
11	6.72	5.26	4.63	4.28	4.04	3.88	3.76	3.66	3.59	3.53	3.33	3.23	3.16	3.12	3.06	3.03
12	6.55	5.10	4.47	4.12	3.89	3.73	3.61	3.51	3.44	3.37	3.18	3.07	3.01	2.96	2.91	2.87
13	6.41	4.97	4.35	4.00	3.77	3.60	3.48	3.39	3.31	3.25	3.05	2.95	2.88	2.84	2.78	2.74
14	6.30	4.86	4.24	3.89	3.66	3.50	3.38	3.29	3.21	3.15	2.95	2.84	2.78	2.73	2.67	2.64
15	6.20	4.77	4.15	3.80	3.58	3.41	3.29	3.20	3.12	3.06	2.86	2.76	2.69	2.64	2.59	2.55
16	6.12	4.69	4.08	3.73	3.50	3.34	3.22	3.12	3.05	2.99	2.79	2.68	2.61	2.57	2.51	2.47
17	6.04	4.62	4.01	3.66	3.44	3.28	3.16	3.06	2.98	2.92	2.72	2.62	2.55	2.50	2.44	2.41
18	5.98	4.56	3.95	3.61	3.38	3.22	3.10	3.01	2.93	2.87	2.67	2.56	2.49	2.44	2.38	2.35
19	5.92	4.51	3.90	3.56	3.33	3.17	3.05	2.96	2.88	2.82	2.62	2.51	2.44	2.39	2.33	2.30
20	5.87	4.46	3.86	3.51	3.29	3.13	3.01	2.91	2.84	2.77	2.57	2.46	2.40	2.35	2.29	2.25
22	5.79	4.38	3.78	3.44	3.22	3.05	2.93	2.84	2.76	2.70	2.50	2.39	2.32	2.27	2.21	2.17
24	5.72	4.32	3.72	3.38	3.15	2.99	2.87	2.78	2.70	2.64	2.44	2.33	2.26	2.21	2.15	2.11
26	5.66	4.27	3.67	3.33	3.10	2.94	2.82	2.73	2.65	2.59	2.39	2.28	2.21	2.16	2.09	2.05
28	5.61	4.22	3.63	3.29	3.06	2.90	2.78	2.69	2.61	2.55	2.34	2.23	2.16	2.11	2.05	2.01
30	5.57	4.18	3.59	3.25	3.03	2.87	2.75	2.65	2.57	2.51	2.31	2.20	2.12	2.07	2.01	1.97
40	5.42	4.05	3.46	3.13	2.90	2.74	2.62	2.53	2.45	2.39	2.18	2.07	1.99	1.94	1.88	1.83
50	5.34	3.97	3.39	3.05	2.83	2.67	2.55	2.46	2.38	2.32	2.11	1.99	1.92	1.87	1.80	1.75
60	5.29	3.93	3.34	3.01	2.79	2.63	2.51	2.41	2.33	2.27	2.06	1.94	1.87	1.82	1.74	1.70
120	5.15	3.80	3.23	2.89	2.67	2.52	2.39	2.30	2.22	2.16	1.94	1.82	1.75	1.69	1.61	1.56
200	5.10	3.76	3.18	2.85	2.63	2.47	2.35	2.26	2.18	2.11	1.90	1.78	1.70	1.64	1.56	1.51
500	5.05	3.72	3.14	2.81	2.59	2.43	2.31	2.22	2.14	2.07	1.86	1.74	1.65	1.60	1.52	1.46
1000	5.04	3.70	3.13	2.80	2.58	2.42	2.30	2.20	2.13	2.06	1.85	1.72	1.64	1.58	1.50	1.45

Degrees of Freedom for Denominator

Source: The entries in this table were computed by the author.

Table 3 $\alpha = 0.01$

Degrees of Freedom for Numerator

	1	2	3	4	5	6	7	8	9	10	15	20	25	30	40	50
1	4048	4993	5377	5577	5668	5924	5992	6096	6132	6168	6079	6168	6214	6355	6168	6213
2	98.50	99.01	99.15	99.23	99.30	99.33	99.35	99.39	99.40	99.43	99.38	99.48	99.43	99.37	99.44	99.59
3	34.12	30.82	29.46	28.71	28.24	27.91	27.67	27.49	27.34	27.23	26.87	26.69	26.58	26.51	26.41	26.36
4	21.20	18.00	16.69	15.98	15.52	15.21	14.98	14.80	14.66	14.55	14.20	14.02	13.91	13.84	13.75	13.69
5	16.26	13.27	12.06	11.39	10.97	10.67	10.46	10.29	10.16	10.05	9.72	9.55	9.45	9.38	9.29	9.24
6	13.75	10.92	9.78	9.15	8.75	8.47	8.26	8.10	7.98	7.87	7.56	7.40	7.30	7.23	7.14	7.09
7	12.25	9.55	8.45	7.85	7.46	7.19	6.99	6.84	6.72	6.62	6.31	6.16	6.06	5.99	5.91	5.86
8	11.26	8.65	7.59	7.01	6.63	6.37	6.18	6.03	5.91	5.81	5.52	5.36	5.26	5.20	5.12	5.07
9	10.56	8.02	6.99	6.42	6.06	5.80	5.61	5.47	5.35	5.26	4.96	4.81	4.71	4.65	4.57	4.52
10	10.04	7.56	6.55	5.99	5.64	5.39	5.20	5.06	4.94	4.85	4.56	4.41	4.31	4.25	4.17	4.12
11	9.65	7.21	6.22	5.67	5.32	5.07	4.89	4.74	4.63	4.54	4.25	4.10	4.01	3.94	3.86	3.81
12	9.33	6.93	5.95	5.41	5.06	4.82	4.64	4.50	4.39	4.30	4.01	3.86	3.76	3.70	3.62	3.57
13	9.07	6.70	5.74	5.21	4.86	4.62	4.44	4.30	4.19	4.10	3.82	3.66	3.57	3.51	3.43	3.38
14	8.86	6.51	5.56	5.04	4.69	4.46	4.28	4.14	4.03	3.94	3.66	3.51	3.41	3.35	3.27	3.22
15	8.68	6.36	5.42	4.89	4.56	4.32	4.14	4.00	3.89	3.80	3.52	3.37	3.28	3.21	3.13	3.08
16	8.53	6.23	5.29	4.77	4.44	4.20	4.03	3.89	3.78	3.69	3.41	3.26	3.16	3.10	3.02	2.97
17	8.40	6.11	5.18	4.67	4.34	4.10	3.93	3.79	3.68	3.59	3.31	3.16	3.07	3.00	2.92	2.87
18	8.29	6.01	5.09	4.58	4.25	4.01	3.84	3.71	3.60	3.51	3.23	3.08	2.98	2.92	2.84	2.78
19	8.18	5.93	5.01	4.50	4.17	3.94	3.77	3.63	3.52	3.43	3.15	3.00	2.91	2.84	2.76	2.71
20	8.10	5.85	4.94	4.43	4.10	3.87	3.70	3.56	3.46	3.37	3.09	2.94	2.84	2.78	2.69	2.64
22	7.95	5.72	4.82	4.31	3.99	3.76	3.59	3.45	3.35	3.26	2.98	2.83	2.73	2.67	2.58	2.53
24	7.82	5.61	4.72	4.22	3.90	3.67	3.50	3.36	3.26	3.17	2.89	2.74	2.64	2.58	2.49	2.44
26	7.72	5.53	4.64	4.14	3.82	3.59	3.42	3.29	3.18	3.09	2.81	2.66	2.57	2.50	2.42	2.36
28	7.64	5.45	4.57	4.07	3.75	3.53	3.36	3.23	3.12	3.03	2.75	2.60	2.51	2.44	2.35	2.30
30	7.56	5.39	4.51	4.02	3.70	3.47	3.30	3.17	3.07	2.98	2.70	2.55	2.45	2.39	2.30	2.25
40	7.31	5.18	4.31	3.83	3.51	3.29	3.12	2.99	2.89	2.80	2.52	2.37	2.27	2.20	2.11	2.06
50	7.17	5.06	4.20	3.72	3.41	3.19	3.02	2.89	2.78	2.70	2.42	2.27	2.17	2.10	2.01	1.95
60	7.08	4.98	4.13	3.65	3.34	3.12	2.95	2.82	2.72	2.63	2.35	2.20	2.10	2.03	1.94	1.88
120	6.85	4.79	3.95	3.48	3.17	2.96	2.79	2.66	2.56	2.47	2.19	2.03	1.93	1.86	1.76	1.70
200	6.76	4.71	3.88	3.41	3.11	2.89	2.73	2.60	2.50	2.41	2.13	1.97	1.87	1.79	1.69	1.63
500	6.69	4.65	3.82	3.36	3.05	2.84	2.68	2.55	2.44	2.36	2.07	1.92	1.81	1.74	1.63	1.57
1000	6.67	4.63	3.80	3.34	3.04	2.82	2.66	2.53	2.43	2.34	2.06	1.90	1.79	1.72	1.61	1.54

Degrees of Freedom for Denominator (vertical axis label)

Source: The entries in this table were computed by the author.

Appendix *ncF*: Critical Values of the Noncentral *F* Distribution

Power = 1 – (Table Entry)

						ϕ					
		0.50	1.0	1.2	1.4	1.6	1.8	2.0	2.2	2.6	3.0
df_e	α						$df_t = 1$				
2	0.05	0.93	0.86	0.83	0.78	0.74	0.69	0.64	0.59	0.49	0.40
	0.01	0.99	0.97	0.96	0.95	0.94	0.93	0.91	0.90	0.87	0.83
4	0.05	0.91	0.80	0.74	0.67	0.59	0.51	0.43	0.35	0.22	0.12
	0.01	0.98	0.95	0.93	0.90	0.87	0.83	0.78	0.73	0.62	0.50
6	0.05	0.91	0.78	0.70	0.62	0.52	0.43	0.34	0.26	0.14	0.06
	0.01	0.98	0.93	0.90	0.86	0.81	0.75	0.69	0.61	0.46	0.31
8	0.05	0.90	0.76	0.68	0.59	0.49	0.39	0.30	0.22	0.11	0.04
	0.01	0.98	0.92	0.89	0.84	0.78	0.70	0.62	0.54	0.37	0.22
10	0.05	0.90	0.75	0.66	0.57	0.47	0.37	0.28	0.20	0.09	0.03
	0.01	0.98	0.92	0.87	0.82	0.75	0.67	0.58	0.49	0.31	0.17
12	0.05	0.90	0.74	0.65	0.56	0.45	0.35	0.26	0.19	0.08	0.03
	0.01	0.97	0.91	0.87	0.81	0.73	0.65	0.55	0.46	0.28	0.14
16	0.05	0.90	0.74	0.64	0.54	0.43	0.33	0.24	0.17	0.07	0.02
	0.01	0.97	0.90	0.85	0.79	0.71	0.61	0.52	0.42	0.24	0.11
20	0.05	0.90	0.73	0.63	0.53	0.42	0.32	0.23	0.16	0.06	0.02
	0.01	0.97	0.90	0.85	0.78	0.69	0.59	0.49	0.39	0.21	0.10
30	0.05	0.89	0.72	0.62	0.52	0.40	0.31	0.22	0.15	0.06	0.02
	0.01	0.97	0.89	0.83	0.76	0.67	0.57	0.46	0.36	0.19	0.08
∞	0.05	0.89	0.71	0.60	0.49	0.38	0.28	0.19	0.12	0.04	0.01
	0.01	0.97	0.88	0.81	0.72	0.62	0.51	0.40	0.30	0.14	0.05
df_e	α						$df_t = 2$				
2	0.05	0.93	0.88	0.85	0.82	0.78	0.75	0.70	0.66	0.56	0.48
	0.01	0.99	0.98	0.97	0.96	0.95	0.94	0.93	0.92	0.89	0.86
4	0.05	0.92	0.82	0.77	0.70	0.62	0.54	0.46	0.38	0.24	0.14
	0.01	0.98	0.96	0.94	0.92	0.89	0.85	0.81	0.76	0.66	0.54
6	0.05	0.91	0.79	0.71	0.63	0.53	0.43	0.34	0.26	0.13	0.05
	0.01	0.98	0.94	0.91	0.87	0.82	0.76	0.70	0.62	0.46	0.31
8	0.05	0.91	0.77	0.68	0.58	0.48	0.37	0.28	0.20	0.08	0.03
	0.01	0.98	0.93	0.89	0.84	0.78	0.70	0.61	0.52	0.34	0.19
10	0.05	0.91	0.75	0.66	0.55	0.44	0.34	0.24	0.16	0.06	0.02
	0.01	0.98	0.92	0.88	0.82	0.74	0.65	0.55	0.45	0.26	0.13
12	0.05	0.90	0.74	0.64	0.53	0.42	0.31	0.22	0.14	0.05	0.01
	0.01	0.98	0.91	0.86	0.80	0.71	0.61	0.51	0.40	0.22	0.09
16	0.05	0.90	0.73	0.62	0.51	0.39	0.28	0.19	0.12	0.04	0.01
	0.01	0.97	0.90	0.84	0.77	0.67	0.57	0.45	0.34	0.16	0.06
20	0.05	0.90	0.72	0.61	0.49	0.36	0.26	0.17	0.11	0.03	0.01
	0.01	0.97	0.90	0.83	0.75	0.65	0.53	0.42	0.31	0.14	0.04
30	0.05	0.90	0.71	0.59	0.47	0.35	0.24	0.15	0.09	0.02	0.00
	0.01	0.97	0.88	0.82	0.72	0.61	0.49	0.37	0.26	0.10	0.03
∞	0.05	0.89	0.68	0.56	0.43	0.30	0.20	0.12	0.06	0.01	0.00
	0.01	0.97	0.86	0.77	0.66	0.53	0.40	0.28	0.18	0.05	0.01

(continued)

Appendix *ncF* (continued)

		φ									
		0.50	1.0	1.2	1.4	1.6	1.8	2.0	2.2	2.6	3.0
df_e	α	$df_t = 3$									
2	0.05	0.93	0.89	0.86	0.83	0.80	0.76	0.73	0.69	0.60	0.52
	0.01	0.99	0.98	0.97	0.96	0.96	0.95	0.94	0.93	0.90	0.88
4	0.05	0.92	0.83	0.77	0.71	0.63	0.55	0.47	0.39	0.25	0.14
	0.01	0.98	0.96	0.94	0.92	0.89	0.86	0.82	0.77	0.67	0.55
6	0.05	0.91	0.79	0.71	0.62	0.52	0.42	0.33	0.24	0.11	0.04
	0.01	0.98	0.94	0.91	0.87	0.82	0.76	0.69	0.61	0.44	0.29
8	0.05	0.91	0.76	0.67	0.57	0.46	0.35	0.25	0.17	0.06	0.02
	0.01	0.98	0.93	0.89	0.84	0.77	0.68	0.59	0.49	0.30	0.16
10	0.05	0.91	0.75	0.65	0.53	0.41	0.30	0.21	0.13	0.04	0.01
	0.01	0.98	0.92	0.87	0.80	0.72	0.62	0.52	0.41	0.22	0.09
12	0.05	0.90	0.73	0.62	0.50	0.38	0.27	0.18	0.11	0.03	0.01
	0.01	0.98	0.91	0.85	0.78	0.69	0.58	0.46	0.35	0.17	0.06
16	0.05	0.90	0.71	0.60	0.47	0.34	0.23	0.14	0.08	0.02	0.00
	0.01	0.97	0.90	0.83	0.74	0.64	0.51	0.39	0.28	0.11	0.03
20	0.05	0.90	0.70	0.58	0.45	0.32	0.21	0.13	0.07	0.01	0.00
	0.01	0.97	0.89	0.82	0.72	0.60	0.47	0.35	0.24	0.08	0.02
30	0.05	0.89	0.68	0.55	0.42	0.29	0.18	0.10	0.05	0.01	0.00
	0.01	0.97	0.87	0.79	0.68	0.55	0.42	0.29	0.18	0.05	0.01
∞	0.05	0.88	0.64	0.50	0.36	0.23	0.13	0.07	0.03	0.00	0.00
	0.01	0.97	0.84	0.73	0.59	0.44	0.30	0.18	0.10	0.02	0.00
df_e	α	$df_t = 4$									
2	0.05	0.94	0.89	0.87	0.84	0.81	0.77	0.74	0.70	0.62	0.54
	0.01	0.99	0.98	0.97	0.97	0.96	0.95	0.94	0.93	0.91	0.88
4	0.05	0.92	0.83	0.78	0.71	0.64	0.55	0.47	0.39	0.25	0.14
	0.01	0.98	0.96	0.94	0.92	0.89	0.86	0.82	0.78	0.67	0.56
6	0.05	0.92	0.79	0.71	0.62	0.52	0.41	0.31	0.23	0.10	0.04
	0.01	0.98	0.94	0.91	0.87	0.82	0.76	0.68	0.60	0.43	0.28
8	0.05	0.91	0.76	0.66	0.55	0.44	0.33	0.23	0.15	0.05	0.01
	0.01	0.98	0.93	0.89	0.83	0.76	0.67	0.57	0.47	0.28	0.14
10	0.05	0.91	0.74	0.63	0.51	0.39	0.27	0.18	0.11	0.03	0.01
	0.01	0.98	0.92	0.86	0.79	0.70	0.60	0.49	0.37	0.19	0.07
12	0.05	0.90	0.72	0.61	0.48	0.35	0.24	0.15	0.08	0.02	0.00
	0.01	0.98	0.91	0.85	0.76	0.66	0.55	0.42	0.31	0.13	0.04
16	0.05	0.90	0.70	0.57	0.44	0.31	0.19	0.11	0.06	0.01	0.00
	0.01	0.97	0.89	0.82	0.72	0.60	0.47	0.34	0.23	0.08	0.02
20	0.05	0.89	0.68	0.55	0.41	0.28	0.17	0.09	0.04	0.01	0.00
	0.01	0.97	0.88	0.80	0.69	0.56	0.42	0.29	0.18	0.05	0.01
30	0.05	0.89	0.66	0.52	0.37	0.24	0.14	0.07	0.03	0.00	0.00
	0.01	0.97	0.86	0.77	0.64	0.50	0.35	0.22	0.13	0.03	0.00
∞	0.05	0.88	0.60	0.45	0.29	0.17	0.08	0.04	0.01	0.00	0.00
	0.01	0.96	0.81	0.68	0.53	0.36	0.22	0.11	0.05	0.01	0.00

Source: Abridged from M. L. Tiku (1967), Tables of the power of the *F* test, *Journal of the American Statistical Association, 62,* 525–539, with the permission of the author and the editors.

Appendix Polynomial: Orthogonal Polynomial Coefficients

k	Polynomial	$X = 1$	2	3	4	5	6	7	8	9	10	$\sum a_i^2$
3	Linear	−1	0	1								2
	Quadratic	1	−2	1								6
4	Linear	−3	−1	1	3							20
	Quadratic	1	−1	−1	1							4
	Cubic	−1	3	−3	1							20
5	Linear	−2	−1	0	1	2						10
	Quadratic	2	−1	−2	−1	2						14
	Cubic	−1	2	0	−2	1						10
	Quartic	1	−4	6	−4	1						70
6	Linear	−5	−3	−1	1	3	5					70
	Quadratic	5	−1	−4	−4	−1	5					84
	Cubic	−5	7	4	−4	−7	5					180
	Quartic	1	−3	2	2	−3	1					28
7	Linear	−3	−2	−1	0	1	2	3				28
	Quadratic	5	0	−3	−4	−3	0	5				84
	Cubic	−1	1	1	0	−1	−1	1				6
	Quartic	3	−7	1	6	1	−7	3				154
8	Linear	−7	−5	−3	−1	1	3	5	7			168
	Quadratic	7	1	−3	−5	−5	−3	1	7			168
	Cubic	−7	5	7	3	−3	−7	−5	7			264
	Quartic	7	−13	−3	9	9	−3	−13	7			616
	Quintic	−7	23	−17	−15	15	17	−23	7			2184
9	Linear	−4	−3	−2	−1	0	1	2	3	4		60
	Quadratic	28	7	−8	−17	−20	−17	−8	7	28		2772
	Cubic	−14	7	13	9	0	−9	−13	−7	14		990
	Quartic	14	−21	−11	9	18	9	−11	−21	14		2002
	Quintic	−4	11	−4	−9	0	9	4	−11	4		468
10	Linear	−9	−7	−5	−3	−1	1	3	5	7	9	330
	Quadratic	6	2	−1	−3	−4	−4	−3	−1	2	6	132
	Cubic	−42	14	35	31	12	−12	−31	−35	−14	42	8580
	Quartic	18	−22	−17	3	18	18	3	−17	−22	18	2860
	Quintic	−6	14	−1	−11	−6	6	11	1	−14	6	780

Source: The entries in this table were computed by the author.

Appendix Power: Power as a Function of δ and Significance Level (α)

	α for Two-Tailed Test			
δ	*0.10*	*0.05*	*0.02*	*0.01*
1.00	0.26	0.17	0.09	0.06
1.10	0.29	0.20	0.11	0.07
1.20	0.33	0.22	0.13	0.08
1.30	0.37	0.26	0.15	0.10
1.40	0.40	0.29	0.18	0.12
1.50	0.44	0.32	0.20	0.14
1.60	0.48	0.36	0.23	0.17
1.70	0.52	0.40	0.27	0.19
1.80	0.56	0.44	0.30	0.22
1.90	0.60	0.48	0.34	0.25
2.00	0.64	0.52	0.37	0.28
2.10	0.68	0.56	0.41	0.32
2.20	0.71	0.60	0.45	0.35
2.30	0.74	0.63	0.49	0.39
2.40	0.78	0.67	0.53	0.43
2.50	0.80	0.71	0.57	0.47
2.60	0.83	0.74	0.61	0.51
2.70	0.85	0.77	0.65	0.55
2.80	0.88	0.80	0.68	0.59
2.90	0.90	0.83	0.72	0.63
3.00	0.91	0.85	0.75	0.66
3.10	0.93	0.87	0.78	0.70
3.20	0.94	0.89	0.81	0.73
3.30	0.95	0.91	0.84	0.77
3.40	0.96	0.93	0.86	0.80
3.50	0.97	0.94	0.88	0.82
3.60	0.98	0.95	0.90	0.85
3.70	0.98	0.96	0.92	0.87
3.80	0.98	0.97	0.93	0.89
3.90	0.99	0.97	0.94	0.91
4.00	0.99	0.98	0.95	0.92
4.10	0.99	0.98	0.96	0.94
4.20	—	0.99	0.97	0.95
4.30	—	0.99	0.98	0.96
4.40	—	0.99	0.98	0.97
4.50	—	0.99	0.99	0.97
4.60	—	—	0.99	0.98
4.70	—	—	0.99	0.98
4.80	—	—	0.99	0.99
4.90	—	—	—	0.99
5.00	—	—	—	0.99

Source: The entries in this table were computed by the author.

Appendix q: Critical Values of the Studentized Range Statistic (q)

Table 1 $\alpha = 0.05$

r = Number of Steps Between Ordered Means

Error df	2	3	4	5	6	7	8	9	10	11	12	13	14	15
1	17.97	26.98	32.82	37.08	40.41	43.12	45.40	47.36	49.07	50.59	51.96	53.20	54.33	55.36
2	6.08	8.33	9.80	10.83	11.74	12.44	13.03	13.54	13.99	14.39	14.75	15.08	15.38	15.65
3	4.50	5.91	6.82	7.50	8.04	8.48	8.85	9.18	9.46	9.72	9.95	10.15	10.35	10.53
4	3.93	5.04	5.76	6.29	6.71	7.05	7.35	7.60	7.83	8.03	8.21	8.37	8.52	8.66
5	3.64	4.60	5.22	5.67	6.03	6.33	6.58	6.80	7.00	7.17	7.32	7.47	7.60	7.72
6	3.46	4.34	4.90	5.30	5.63	5.90	6.12	6.32	6.49	6.65	6.79	6.92	7.03	7.14
7	3.34	4.16	4.68	5.06	5.36	5.61	5.82	6.00	6.16	6.30	6.43	6.55	6.66	6.76
8	3.26	4.04	4.53	4.89	5.17	5.40	5.60	5.77	5.92	6.05	6.18	6.29	6.39	6.48
9	3.20	3.95	4.42	4.76	5.02	5.24	5.43	5.60	5.74	5.87	5.98	6.09	6.19	6.28
10	3.15	3.88	4.33	4.65	4.91	5.12	5.30	5.46	5.60	5.72	5.83	5.94	6.03	6.11
11	3.11	3.82	4.26	4.57	4.82	5.03	5.20	5.35	5.49	5.60	5.71	5.81	5.90	5.98
12	3.08	3.77	4.20	4.51	4.75	4.95	5.12	5.26	5.40	5.51	5.62	5.71	5.79	5.88
13	3.06	3.74	4.15	4.45	4.69	4.88	5.05	5.19	5.32	5.43	5.53	5.63	5.71	5.79
14	3.03	3.70	4.11	4.41	4.64	4.83	4.99	5.13	5.25	5.36	5.46	5.55	5.64	5.71
15	3.01	3.67	4.08	4.37	4.60	4.78	4.94	5.08	5.20	5.31	5.40	5.49	5.57	5.65
16	3.00	3.65	4.05	4.33	4.56	4.74	4.90	5.03	5.15	5.26	5.35	5.44	5.52	5.59
17	2.98	3.63	4.02	4.30	4.52	4.70	4.86	4.99	5.11	5.21	5.31	5.39	5.47	5.54
18	2.97	3.61	4.00	4.28	4.50	4.67	4.82	4.96	5.07	5.17	5.27	5.35	5.43	5.50
19	2.96	3.59	3.98	4.25	4.47	4.64	4.79	4.92	5.04	5.14	5.23	5.32	5.39	5.46
20	2.95	3.58	3.96	4.23	4.45	4.62	4.77	4.90	5.01	5.11	5.20	5.28	5.36	5.43
24	2.92	3.53	3.90	4.17	4.37	4.54	4.68	4.81	4.92	5.01	5.10	5.18	5.25	5.32
30	2.89	3.49	3.84	4.10	4.30	4.46	4.60	4.72	4.82	4.92	5.00	5.08	5.15	5.21
40	2.86	3.44	3.79	4.04	4.23	4.39	4.52	4.64	4.74	4.82	4.90	4.98	5.04	5.11
60	2.83	3.40	3.74	3.98	4.16	4.31	4.44	4.55	4.65	4.73	4.81	4.88	4.94	5.00
120	2.80	3.36	3.69	3.92	4.10	4.24	4.36	4.47	4.56	4.64	4.71	4.78	4.84	4.90
∞	2.77	3.31	3.63	3.86	4.03	4.17	4.29	4.39	4.47	4.55	4.62	4.68	4.74	4.80

Source: Abridged from H. L. Harter (1960), Tables cf range and Studentized range, *Annals of Mathematical Statistics, 31,* 1122–1147, with permission of the author and the publisher.

Table 2 α = 0.01

r = Number of Steps Between Ordered Means

Error df	2	3	4	5	6	7	8	9	10	11	12	13	14	15
1	90.03	135.0	164.3	185.6	202.2	215.8	227.2	237.0	245.6	253.2	260.0	266.2	271.8	277.0
2	14.04	19.02	22.29	24.72	26.63	28.20	29.53	30.68	31.69	32.59	33.40	34.13	34.81	35.43
3	8.26	10.62	12.17	13.33	14.24	15.00	15.64	16.20	16.69	17.13	17.53	17.89	18.22	18.52
4	6.51	8.12	9.17	9.96	10.58	11.10	11.55	11.93	12.27	12.57	12.84	13.09	13.32	13.53
5	5.70	6.98	7.80	8.42	8.91	9.32	9.67	9.97	10.24	10.48	10.70	10.89	11.08	11.24
6	5.24	6.33	7.03	7.56	7.97	8.32	8.62	8.87	9.10	9.30	9.48	9.65	9.81	9.95
7	4.95	5.92	6.54	7.00	7.37	7.68	7.94	8.17	8.37	8.55	8.71	8.86	9.00	9.12
8	4.75	5.64	6.20	6.62	6.96	7.24	7.47	7.68	7.86	8.03	8.18	8.31	8.44	8.55
9	4.60	5.43	5.96	6.35	6.66	6.92	7.13	7.32	7.50	7.65	7.78	7.91	8.02	8.13
10	4.48	5.27	5.77	6.14	6.43	6.67	6.88	7.06	7.21	7.36	7.48	7.60	7.71	7.81
11	4.39	5.15	5.62	5.97	6.25	6.48	6.67	6.84	6.99	7.13	7.25	7.36	7.46	7.56
12	4.32	5.05	5.50	5.84	6.10	6.32	6.51	6.67	6.81	6.94	7.06	7.17	7.26	7.36
13	4.26	4.96	5.40	5.73	5.98	6.19	6.37	6.53	6.67	6.79	6.90	7.01	7.10	7.19
14	4.21	4.90	5.32	5.63	5.88	6.08	6.26	6.41	6.54	6.66	6.77	6.87	6.96	7.05
15	4.17	4.84	5.25	5.56	5.80	5.99	6.16	6.31	6.44	6.56	6.66	6.76	6.84	6.93
16	4.13	4.79	5.19	5.49	5.72	5.92	6.08	6.22	6.35	6.46	6.56	6.66	6.74	6.82
17	4.10	4.74	5.14	5.43	5.66	5.85	6.01	6.15	6.27	6.38	6.48	6.57	6.66	6.73
18	4.07	4.70	5.09	5.38	5.60	5.79	5.94	6.08	6.20	6.31	6.41	6.50	6.58	6.66
19	4.05	4.67	5.05	5.33	5.55	5.74	5.89	6.02	6.14	6.25	6.34	6.43	6.51	6.58
20	4.02	4.64	5.02	5.29	5.51	5.69	5.84	5.97	6.09	6.19	6.28	6.37	6.45	6.52
24	3.96	4.55	4.91	5.17	5.37	5.54	5.69	5.81	5.92	6.02	6.11	6.19	6.26	6.33
30	3.89	4.46	4.80	5.05	5.24	5.40	5.54	5.65	5.76	5.85	5.93	6.01	6.08	6.14
40	3.82	4.37	4.70	4.93	5.11	5.26	5.39	5.50	5.60	5.69	5.76	5.84	5.90	5.96
60	3.76	4.28	4.60	4.82	4.99	5.13	5.25	5.36	5.45	5.53	5.60	5.67	5.73	5.78
120	3.70	4.20	4.50	4.71	4.87	5.01	5.12	5.21	5.30	5.38	5.44	5.51	5.56	5.61
∞	3.64	4.12	4.40	4.60	4.76	4.88	4.99	5.08	5.16	5.23	5.29	5.35	5.40	5.45

Source: Abridged from H. L. Harter (1960), Tables of range and Studentized range, *Annals of Mathematical Statistics, 31,* 1122–1147, with permission of the author and the publisher.

Appendix *r'*: Table of Fisher's Transformation of *r* to *r'*

r	r'	r	r'	r	r'	r	r'	r	r'
0.000	0.000	0.200	0.203	0.400	0.424	0.600	0.693	0.800	1.099
0.005	0.005	0.205	0.208	0.405	0.430	0.605	0.701	0.805	1.113
0.010	0.010	0.210	0.213	0.410	0.436	0.610	0.709	0.810	1.127
0.015	0.015	0.215	0.218	0.415	0.442	0.615	0.717	0.815	1.142
0.020	0.020	0.220	0.224	0.420	0.448	0.620	0.725	0.820	1.157
0.025	0.025	0.225	0.229	0.425	0.454	0.625	0.733	0.825	1.172
0.030	0.030	0.230	0.234	0.430	0.460	0.630	0.741	0.830	1.188
0.035	0.035	0.235	0.239	0.435	0.466	0.635	0.750	0.835	1.204
0.040	0.040	0.240	0.245	0.440	0.472	0.640	0.758	0.840	1.221
0.045	0.045	0.245	0.250	0.445	0.478	0.645	0.767	0.845	1.238
0.050	0.050	0.250	0.255	0.450	0.485	0.650	0.775	0.850	1.256
0.055	0.055	0.255	0.261	0.455	0.491	0.655	0.784	0.855	1.274
0.060	0.060	0.260	0.266	0.460	0.497	0.660	0.793	0.860	1.293
0.065	0.065	0.265	0.271	0.465	0.504	0.665	0.802	0.865	1.313
0.070	0.070	0.270	0.277	0.470	0.510	0.670	0.811	0.870	1.333
0.075	0.075	0.275	0.282	0.475	0.517	0.675	0.820	0.875	1.354
0.080	0.080	0.280	0.288	0.480	0.523	0.680	0.829	0.880	1.376
0.085	0.085	0.285	0.293	0.485	0.530	0.685	0.838	0.885	1.398
0.090	0.090	0.290	0.299	0.490	0.536	0.690	0.848	0.890	1.422
0.095	0.095	0.295	0.304	0.495	0.543	0.695	0.858	0.895	1.447
0.100	0.100	0.300	0.310	0.500	0.549	0.700	0.867	0.900	1.472
0.105	0.105	0.305	0.315	0.505	0.556	0.705	0.877	0.905	1.499
0.110	0.110	0.310	0.321	0.510	0.563	0.710	0.887	0.910	1.528
0.115	0.116	0.315	0.326	0.515	0.570	0.715	0.897	0.915	1.557
0.120	0.121	0.320	0.332	0.520	0.576	0.720	0.908	0.920	1.589
0.125	0.126	0.325	0.337	0.525	0.583	0.725	0.918	0.925	1.623
0.130	0.131	0.330	0.343	0.530	0.590	0.730	0.929	0.930	1.658
0.135	0.136	0.335	0.348	0.535	0.597	0.735	0.940	0.935	1.697
0.140	0.141	0.340	0.354	0.540	0.604	0.740	0.950	0.940	1.738
0.145	0.146	0.345	0.360	0.545	0.611	0.745	0.962	0.945	1.783
0.150	0.151	0.350	0.365	0.550	0.618	0.750	0.973	0.950	1.832
0.155	0.156	0.355	0.371	0.555	0.626	0.755	0.984	0.955	1.886
0.160	0.161	0.360	0.377	0.560	0.633	0.760	0.996	0.960	1.946
0.165	0.167	0.365	0.383	0.565	0.640	0.765	1.008	0.965	2.014
0.170	0.172	0.370	0.388	0.570	0.648	0.770	1.020	0.970	2.092
0.175	0.177	0.375	0.394	0.575	0.655	0.775	1.033	0.975	2.185
0.180	0.182	0.380	0.400	0.580	0.662	0.780	1.045	0.980	2.298
0.185	0.187	0.385	0.406	0.585	0.670	0.785	1.058	0.985	2.443
0.190	0.192	0.390	0.412	0.590	0.678	0.790	1.071	0.990	2.647
0.195	0.198	0.395	0.418	0.595	0.685	0.795	1.085	0.995	2.994

Source: The entries in this table were computed by the author.

Appendix *t*: Percentage Points of the *t* Distribution

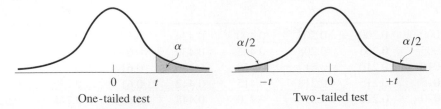

	One-tailed test				Two-tailed test			

Level of Significance for One-Tailed Test

0.25	0.20	0.15	0.10	0.05	0.025	0.01	0.005	0.0005

Level of Significance for Two-Tailed Test

df	0.50	0.40	0.30	0.20	0.10	0.05	0.02	0.01	0.001
1	1.000	1.376	1.963	3.078	6.314	12.706	31.821	63.657	636.620
2	0.816	1.061	1.386	1.886	2.920	4.303	6.965	9.925	31.599
3	0.765	0.978	1.250	1.638	2.353	3.182	4.541	5.841	12.924
4	0.741	0.941	1.190	1.533	2.132	2.776	3.747	4.604	8.610
5	0.727	0.920	1.156	1.476	2.015	2.571	3.365	4.032	6.869
6	0.718	0.906	1.134	1.440	1.943	2.447	3.143	3.707	5.959
7	0.711	0.896	1.119	1.415	1.895	2.365	2.998	3.499	5.408
8	0.706	0.889	1.108	1.397	1.860	2.306	2.896	3.355	5.041
9	0.703	0.883	1.100	1.383	1.833	2.262	2.821	3.250	4.781
10	0.700	0.879	1.093	1.372	1.812	2.228	2.764	3.169	4.587
11	0.697	0.876	1.088	1.363	1.796	2.201	2.718	3.106	4.437
12	0.695	0.873	1.083	1.356	1.782	2.179	2.681	3.055	4.318
13	0.694	0.870	1.079	1.350	1.771	2.160	2.650	3.012	4.221
14	0.692	0.868	1.076	1.345	1.761	2.145	2.624	2.977	4.140
15	0.691	0.866	1.074	1.341	1.753	2.131	2.602	2.947	4.073
16	0.690	0.865	1.071	1.337	1.746	2.120	2.583	2.921	4.015
17	0.689	0.863	1.069	1.333	1.740	2.110	2.567	2.898	3.965
18	0.688	0.862	1.067	1.330	1.734	2.101	2.552	2.878	3.922
19	0.688	0.861	1.066	1.328	1.729	2.093	2.539	2.861	3.883
20	0.687	0.860	1.064	1.325	1.725	2.086	2.528	2.845	3.850
21	0.686	0.859	1.063	1.323	1.721	2.080	2.518	2.831	3.819
22	0.686	0.858	1.061	1.321	1.717	2.074	2.508	2.819	3.792
23	0.685	0.858	1.060	1.319	1.714	2.069	2.500	2.807	3.768
24	0.685	0.857	1.059	1.318	1.711	2.064	2.492	2.797	3.745
25	0.684	0.856	1.058	1.316	1.708	2.060	2.485	2.787	3.725
26	0.684	0.856	1.058	1.315	1.706	2.056	2.479	2.779	3.707
27	0.684	0.855	1.057	1.314	1.703	2.052	2.473	2.771	3.690
28	0.683	0.855	1.056	1.313	1.701	2.048	2.467	2.763	3.674
29	0.683	0.854	1.055	1.311	1.699	2.045	2.462	2.756	3.659
30	0.683	0.854	1.055	1.310	1.697	2.042	2.457	2.750	3.646
40	0.681	0.851	1.050	1.303	1.684	2.021	2.423	2.704	3.551
50	0.679	0.849	1.047	1.299	1.676	2.009	2.403	2.678	3.496
100	0.677	0.845	1.042	1.290	1.660	1.984	2.364	2.626	3.390
∞	0.674	0.842	1.036	1.282	1.645	1.960	2.326	2.576	3.291

Source: The entries in this table were computed by the author.

Appendix *T*: Critical Lower-Tail Values of *T* (and Their Associated Probabilities) for Wilcoxon's Matched-Pairs Signed-Ranks Test

	Nominal α (One-Tailed)										
		0.05			0.025			0.01			0.005	
N	*T*	α		*T*	α		*T*	α		*T*	α
5	0	0.0313									
	1	0.0625									
6	2	0.0469		0	0.0156						
	3	0.0781		1	0.0313						
7	3	0.0391		2	0.0234		0	0.0078			
	4	0.0547		3	0.0391		1	0.0156			
8	5	0.0391		3	0.0195		1	0.0078		0	0.0039
	6	0.0547		4	0.0273		2	0.0117		1	0.0078
9	8	0.0488		5	0.0195		3	0.0098		1	0.0039
	9	0.0645		6	0.0273		4	0.0137		2	0.0059
10	10	0.0420		8	0.0244		5	0.0098		3	0.0049
	11	0.0527		9	0.0322		6	0.0137		4	0.0068
11	13	0.0415		10	0.0210		7	0.0093		5	0.0049
	14	0.0508		11	0.0269		8	0.0122		6	0.0068
12	17	0.0461		13	0.0212		9	0.0081		7	0.0046
	18	0.0549		14	0.0261		10	0.0105		8	0.0061
13	21	0.0471		17	0.0239		12	0.0085		9	0.0040
	22	0.0549		18	0.0287		13	0.0107		10	0.0052
14	25	0.0453		21	0.0247		15	0.0083		12	0.0043
	26	0.0520		22	0.0290		16	0.0101		13	0.0054
15	30	0.0473		25	0.0240		19	0.0090		15	0.0042
	31	0.0535		26	0.0277		20	0.0108		16	0.0051
16	35	0.0467		29	0.0222		23	0.0091		19	0.0046
	36	0.0523		30	0.0253		24	0.0107		20	0.0055
17	41	0.0492		34	0.0224		27	0.0087		23	0.0047
	42	0.0544		35	0.0253		28	0.0101		24	0.0055
18	47	0.0494		40	0.0241		32	0.0091		27	0.0045
	48	0.0542		41	0.0269		33	0.0104		28	0.0052
19	53	0.0478		46	0.0247		37	0.0090		32	0.0047
	54	0.0521		47	0.0273		38	0.0102		33	0.0054
20	60	0.0487		52	0.0242		43	0.0096		37	0.0047
	61	0.0527		53	0.0266		44	0.0107		38	0.0053
21	67	0.0479		58	0.0230		49	0.0097		42	0.0045
	68	0.0516		59	0.0251		50	0.0108		43	0.0051

(continued)

Appendix *T* *(continued)*

Nominal α (One-Tailed)

N	0.05 T	0.05 α	0.025 T	0.025 α	0.01 T	0.01 α	0.005 T	0.005 α
22	75	0.0492	65	0.0231	55	0.0095	48	0.0046
	76	0.0527	66	0.0250	56	0.0104	49	0.0052
23	83	0.0490	73	0.0242	62	0.0098	54	0.0046
	84	0.0523	74	0.0261	63	0.0107	55	0.0051
24	91	0.0475	81	0.0245	69	0.0097	61	0.0048
	92	0.0505	82	0.0263	70	0.0106	62	0.0053
25	100	0.0479	89	0.0241	76	0.0094	68	0.0048
	101	0.0507	90	0.0258	77	0.0101	69	0.0053
26	110	0.0497	98	0.0247	84	0.0095	75	0.0047
	111	0.0524	99	0.0263	85	0.0102	76	0.0051
27	119	0.0477	107	0.0246	92	0.0093	83	0.0048
	120	0.0502	108	0.0260	93	0.0100	84	0.0052
28	130	0.0496	116	0.0239	101	0.0096	91	0.0048
	131	0.0521	117	0.0252	102	0.0102	92	0.0051
29	140	0.0482	126	0.0240	110	0.0095	100	0.0049
	141	0.0504	127	0.0253	111	0.0101	101	0.0053
30	151	0.0481	137	0.0249	120	0.0098	109	0.0050
	152	0.0502	138	0.0261	121	0.0104	110	0.0053
31	163	0.0491	147	0.0239	130	0.0099	118	0.0049
	164	0.0512	148	0.0251	131	0.0105	119	0.0052
32	175	0.0492	159	0.0249	140	0.0097	128	0.0050
	176	0.0512	160	0.0260	141	0.0103	129	0.0053
33	187	0.0485	170	0.0242	151	0.0099	138	0.0049
	188	0.0503	171	0.0253	152	0.0104	139	0.0052
34	200	0.0488	182	0.0242	162	0.0098	148	0.0048
	201	0.0506	183	0.0252	163	0.0103	149	0.0051
35	213	0.0484	195	0.0247	173	0.0096	159	0.0048
	214	0.0501	196	0.0257	174	0.0100	160	0.0051
36	227	0.0489	208	0.0248	185	0.0096	171	0.0050
	228	0.0505	209	0.0258	186	0.0100	172	0.0052
37	241	0.0487	221	0.0245	198	0.0099	182	0.0048
	242	0.0503	222	0.0254	199	0.0103	183	0.0050

Source: The entries in this table were computed by the author.

Appendix *T*   *(continued)*

				Nominal α (One-Tailed)					
	0.05		0.025		0.01		0.005		
N	*T*	α	*T*	α	*T*	α	*T*	α
38	256	0.0493	235	0.0247	211	0.0099	194	0.0048
	257	0.0509	236	0.0256	212	0.0104	195	0.0050
39	271	0.0492	249	0.0246	224	0.0099	207	0.0049
	272	0.0507	250	0.0254	225	0.0103	208	0.0051
40	286	0.0486	264	0.0249	238	0.0100	220	0.0049
	287	0.0500	265	0.0257	239	0.0104	221	0.0051
41	302	0.0488	279	0.0248	252	0.0100	233	0.0048
	303	0.0501	280	0.0256	253	0.0103	234	0.0050
42	319	0.0496	294	0.0245	266	0.0098	247	0.0049
	320	0.0509	295	0.0252	267	0.0102	248	0.0051
43	336	0.0498	310	0.0245	281	0.0098	261	0.0048
	337	0.0511	311	0.0252	282	0.0102	262	0.0050
44	353	0.0495	327	0.0250	296	0.0097	276	0.0049
	354	0.0507	328	0.0257	297	0.0101	277	0.0051
45	371	0.0498	343	0.0244	312	0.0098	291	0.0049
	372	0.0510	344	0.0251	313	0.0101	292	0.0051
46	389	0.0497	361	0.0249	328	0.0098	307	0.0050
	390	0.0508	362	0.0256	329	0.0101	308	0.0052
47	407	0.0490	378	0.0245	345	0.0099	322	0.0048
	408	0.0501	379	0.0251	346	0.0102	323	0.0050
48	426	0.0490	396	0.0244	362	0.0099	339	0.0050
	427	0.0500	397	0.0251	363	0.0102	340	0.0051
49	446	0.0495	415	0.0247	379	0.0098	355	0.0049
	447	0.0505	416	0.0253	380	0.0100	356	0.0050
50	466	0.0495	434	0.0247	397	0.0098	373	0.0050
	467	0.0506	435	0.0253	398	0.0101	374	0.0051

Source: The entries in this table were computed by the author.

Appendix t' : Critical Values of Bonferroni Multiple Comparison Test

Table 1 $\alpha = 0.05$

Number of Comparisons

df	2	3	4	5	6	7	8	9	10	15	20	25	30	35	40	45	50	55
5	3.16	3.53	3.81	4.03	4.22	4.38	4.53	4.66	4.77	5.25	5.60	5.89	6.14	6.35	6.54	6.71	6.87	7.01
6	2.97	3.29	3.52	3.71	3.86	4.00	4.12	4.22	4.32	4.70	4.98	5.21	5.40	5.56	5.71	5.84	5.96	6.07
7	2.84	3.13	3.34	3.50	3.64	3.75	3.86	3.95	4.03	4.36	4.59	4.79	4.94	5.08	5.20	5.31	5.41	5.50
8	2.75	3.02	3.21	3.36	3.48	3.58	3.68	3.76	3.83	4.12	4.33	4.50	4.64	4.76	4.86	4.96	5.04	5.12
9	2.69	2.93	3.11	3.25	3.36	3.46	3.55	3.62	3.69	3.95	4.15	4.30	4.42	4.53	4.62	4.71	4.78	4.85
10	2.63	2.87	3.04	3.17	3.28	3.37	3.45	3.52	3.58	3.83	4.00	4.14	4.26	4.36	4.44	4.52	4.59	4.65
11	2.59	2.82	2.98	3.11	3.21	3.29	3.37	3.44	3.50	3.73	3.89	4.02	4.13	4.22	4.30	4.37	4.44	4.49
12	2.56	2.78	2.93	3.05	3.15	3.24	3.31	3.37	3.43	3.65	3.81	3.93	4.03	4.12	4.19	4.26	4.32	4.37
13	2.53	2.75	2.90	3.01	3.11	3.19	3.26	3.32	3.37	3.58	3.73	3.85	3.95	4.03	4.10	4.16	4.22	4.27
14	2.51	2.72	2.86	2.98	3.07	3.15	3.21	3.27	3.33	3.53	3.67	3.79	3.88	3.96	4.03	4.09	4.14	4.19
15	2.49	2.69	2.84	2.95	3.04	3.11	3.18	3.23	3.29	3.48	3.62	3.73	3.82	3.90	3.96	4.02	4.07	4.12
16	2.47	2.67	2.81	2.92	3.01	3.08	3.15	3.20	3.25	3.44	3.58	3.69	3.77	3.85	3.91	3.96	4.01	4.06
17	2.46	2.65	2.79	2.90	2.98	3.06	3.12	3.17	3.22	3.41	3.54	3.65	3.73	3.80	3.86	3.92	3.97	4.01
18	2.45	2.64	2.77	2.88	2.96	3.03	3.09	3.15	3.20	3.38	3.51	3.61	3.69	3.76	3.82	3.87	3.92	3.96
19	2.43	2.63	2.76	2.86	2.94	3.01	3.07	3.13	3.17	3.35	3.48	3.58	3.66	3.73	3.79	3.84	3.88	3.93
20	2.42	2.61	2.74	2.85	2.93	3.00	3.06	3.11	3.15	3.33	3.46	3.55	3.63	3.70	3.75	3.80	3.85	3.89
21	2.41	2.60	2.73	2.83	2.91	2.98	3.04	3.09	3.14	3.31	3.43	3.53	3.60	3.67	3.73	3.78	3.82	3.86
22	2.41	2.59	2.72	2.82	2.90	2.97	3.02	3.07	3.12	3.29	3.41	3.50	3.58	3.64	3.70	3.75	3.79	3.83
23	2.40	2.58	2.71	2.81	2.89	2.95	3.01	3.06	3.10	3.27	3.39	3.48	3.56	3.62	3.68	3.72	3.77	3.81
24	2.39	2.57	2.70	2.80	2.88	2.94	3.00	3.05	3.09	3.26	3.38	3.47	3.54	3.60	3.66	3.70	3.75	3.78
25	2.38	2.57	2.69	2.79	2.86	2.93	2.99	3.03	3.08	3.24	3.36	3.45	3.52	3.58	3.64	3.68	3.73	3.76
30	2.36	2.54	2.66	2.75	2.82	2.89	2.94	2.99	3.03	3.19	3.30	3.39	3.45	3.51	3.56	3.61	3.65	3.68
40	2.33	2.50	2.62	2.70	2.78	2.84	2.89	2.93	2.97	3.12	3.23	3.31	3.37	3.43	3.47	3.51	3.55	3.58
50	2.31	2.48	2.59	2.68	2.75	2.81	2.85	2.90	2.94	3.08	3.18	3.26	3.32	3.38	3.42	3.46	3.50	3.53
75	2.29	2.45	2.56	2.64	2.71	2.77	2.81	2.86	2.89	3.03	3.13	3.20	3.26	3.31	3.35	3.39	3.43	3.45
100	2.28	2.43	2.54	2.63	2.69	2.75	2.79	2.83	2.87	3.01	3.10	3.17	3.23	3.28	3.32	3.36	3.39	3.42
∞	2.24	2.39	2.50	2.58	2.64	2.69	2.73	2.77	2.81	2.94	3.02	3.09	3.14	3.19	3.23	3.26	3.29	3.32

Source: The entries in this table were computed by the author.

Table 2 α = 0.01

Number of Comparisons

df	2	3	4	5	6	7	8	9	10	15	20	25	30	35	40	45	50	55
5	4.77	5.25	5.60	5.89	6.14	6.35	6.54	6.71	6.87	7.50	7.98	8.36	8.69	8.98	9.24	9.47	9.68	9.87
6	4.32	4.70	4.98	5.21	5.40	5.56	5.71	5.84	5.96	6.43	6.79	7.07	7.31	7.52	7.71	7.87	8.02	8.16
7	4.03	4.36	4.59	4.79	4.94	5.08	5.20	5.31	5.41	5.80	6.08	6.31	6.50	6.67	6.81	6.94	7.06	7.17
8	3.83	4.12	4.33	4.50	4.64	4.76	4.86	4.96	5.04	5.37	5.62	5.81	5.97	6.11	6.23	6.34	6.44	6.53
9	3.69	3.95	4.15	4.30	4.42	4.53	4.62	4.71	4.78	5.08	5.29	5.46	5.60	5.72	5.83	5.92	6.01	6.09
10	3.58	3.83	4.00	4.14	4.26	4.36	4.44	4.52	4.59	4.85	5.05	5.20	5.33	5.44	5.53	5.62	5.69	5.76
11	3.50	3.73	3.89	4.02	4.13	4.22	4.30	4.37	4.44	4.68	4.86	5.00	5.12	5.22	5.31	5.38	5.45	5.52
12	3.43	3.65	3.81	3.93	4.03	4.12	4.19	4.26	4.32	4.55	4.72	4.85	4.96	5.05	5.13	5.20	5.26	5.32
13	3.37	3.58	3.73	3.85	3.95	4.03	4.10	4.16	4.22	4.44	4.60	4.72	4.82	4.91	4.98	5.05	5.11	5.17
14	3.33	3.53	3.67	3.79	3.88	3.96	4.03	4.09	4.14	4.35	4.50	4.62	4.71	4.79	4.87	4.93	4.99	5.04
15	3.29	3.48	3.62	3.73	3.82	3.90	3.96	4.02	4.07	4.27	4.42	4.53	4.62	4.70	4.77	4.83	4.88	4.93
16	3.25	3.44	3.58	3.69	3.77	3.85	3.91	3.96	4.01	4.21	4.35	4.45	4.54	4.62	4.68	4.74	4.79	4.84
17	3.22	3.41	3.54	3.65	3.73	3.80	3.86	3.92	3.97	4.15	4.29	4.39	4.47	4.55	4.61	4.66	4.71	4.76
18	3.20	3.38	3.51	3.61	3.69	3.76	3.82	3.87	3.92	4.10	4.23	4.33	4.42	4.49	4.55	4.60	4.65	4.69
19	3.17	3.35	3.48	3.58	3.66	3.73	3.79	3.84	3.88	4.06	4.19	4.28	4.36	4.43	4.49	4.54	4.59	4.63
20	3.15	3.33	3.46	3.55	3.63	3.70	3.75	3.80	3.85	4.02	4.15	4.24	4.32	4.39	4.44	4.49	4.54	4.58
21	3.14	3.31	3.43	3.53	3.60	3.67	3.73	3.78	3.82	3.99	4.11	4.20	4.28	4.34	4.40	4.45	4.49	4.53
22	3.12	3.29	3.41	3.50	3.58	3.64	3.70	3.75	3.79	3.96	4.08	4.17	4.24	4.31	4.36	4.41	4.45	4.49
23	3.10	3.27	3.39	3.48	3.56	3.62	3.68	3.72	3.77	3.93	4.05	4.14	4.21	4.27	4.33	4.37	4.42	4.45
24	3.09	3.26	3.38	3.47	3.54	3.60	3.66	3.70	3.75	3.91	4.02	4.11	4.18	4.24	4.29	4.34	4.38	4.42
25	3.08	3.24	3.36	3.45	3.52	3.58	3.64	3.68	3.73	3.88	4.00	4.08	4.15	4.21	4.27	4.31	4.35	4.39
30	3.03	3.19	3.30	3.39	3.45	3.51	3.56	3.61	3.65	3.80	3.90	3.98	4.05	4.11	4.15	4.20	4.23	4.27
40	2.97	3.12	3.23	3.31	3.37	3.43	3.47	3.51	3.55	3.69	3.79	3.86	3.92	3.98	4.02	4.06	4.09	4.13
50	2.94	3.08	3.18	3.26	3.32	3.38	3.42	3.46	3.50	3.63	3.72	3.79	3.85	3.90	3.94	3.98	4.01	4.04
75	2.89	3.03	3.13	3.20	3.26	3.31	3.35	3.39	3.43	3.55	3.64	3.71	3.76	3.81	3.85	3.88	3.91	3.94
100	2.87	3.01	3.10	3.17	3.23	3.28	3.32	3.36	3.39	3.51	3.60	3.66	3.72	3.76	3.80	3.83	3.86	3.89
∞	2.81	2.94	3.02	3.09	3.14	3.19	3.23	3.26	3.29	3.40	3.48	3.54	3.59	3.63	3.66	3.69	3.72	3.74

Source: The entries in this table were computed by the author.

Appendix t_d: Critical Values of Dunnett's t Statistic (t_d)

Two-Tailed Comparisons
k = Number of Treatment Means, Including Control

Error df	α	2	3	4	5	6	7	8	9	10
5	0.05	2.57	3.03	3.29	3.48	3.62	3.73	3.82	3.90	3.97
	0.01	4.03	4.63	4.98	5.22	5.41	5.56	5.69	5.80	5.89
6	0.05	2.45	2.86	3.10	3.26	3.39	3.49	3.57	3.64	3.71
	0.01	3.71	4.21	4.51	4.71	4.87	5.00	5.10	5.20	5.28
7	0.05	2.36	2.75	2.97	3.12	3.24	3.33	3.41	3.47	3.53
	0.01	3.50	3.95	4.21	4.39	4.53	4.64	4.74	4.82	4.89
8	0.05	2.31	2.67	2.88	3.02	3.13	3.22	3.29	3.35	3.41
	0.01	3.36	3.77	4.00	4.17	4.29	4.40	4.48	4.56	4.62
9	0.05	2.26	2.61	2.81	2.95	3.05	3.14	3.20	3.26	3.32
	0.01	3.25	3.63	3.85	4.01	4.12	4.22	4.30	4.37	4.43
10	0.05	2.23	2.57	2.76	2.89	2.99	3.07	3.14	3.19	3.24
	0.01	3.17	3.53	3.74	3.88	3.99	4.08	4.16	4.22	4.28
11	0.05	2.20	2.53	2.72	2.84	2.94	3.02	3.08	3.14	3.19
	0.01	3.11	3.45	3.65	3.79	3.89	3.98	4.05	4.11	4.16
12	0.05	2.18	2.50	2.68	2.81	2.90	2.98	3.04	3.09	3.14
	0.01	3.05	3.39	3.58	3.71	3.81	3.89	3.96	4.02	4.07
13	0.05	2.16	2.48	2.65	2.78	2.87	2.94	3.00	3.06	3.10
	0.01	3.01	3.33	3.52	3.65	3.74	3.82	3.89	3.94	3.99
14	0.05	2.14	2.46	2.63	2.75	2.84	2.91	2.97	3.02	3.07
	0.01	2.98	3.29	3.47	3.59	3.69	3.76	3.83	3.88	3.93
15	0.05	2.13	2.44	2.61	2.73	2.82	2.89	2.95	3.00	3.04
	0.01	2.95	3.25	3.43	3.55	3.64	3.71	3.78	3.83	3.88
16	0.05	2.12	2.42	2.59	2.71	2.80	2.87	2.92	2.97	3.02
	0.01	2.92	3.22	3.39	3.51	3.60	3.67	3.73	3.78	3.83
17	0.05	2.11	2.41	2.58	2.69	2.78	2.85	2.90	2.95	3.00
	0.01	2.90	3.19	3.36	3.47	3.56	3.63	3.69	3.74	3.79
18	0.05	2.10	2.40	2.56	2.68	2.76	2.83	2.89	2.94	2.98
	0.01	2.88	3.17	3.33	3.44	3.53	3.60	3.66	3.71	3.75
19	0.05	2.09	2.39	2.55	2.66	2.75	2.81	2.87	2.92	2.96
	0.01	2.86	3.15	3.31	3.42	3.50	3.57	3.63	3.68	3.72
20	0.05	2.09	2.38	2.54	2.65	2.73	2.80	2.86	2.90	2.95
	0.01	2.85	3.13	3.29	3.40	3.48	3.55	3.60	3.65	3.69
24	0.05	2.06	2.35	2.51	2.61	2.70	2.76	2.81	2.86	2.90
	0.01	2.80	3.07	3.22	3.32	3.40	3.47	3.52	3.57	3.61
30	0.05	2.04	2.32	2.47	2.58	2.66	2.72	2.77	2.82	2.86
	0.01	2.75	3.01	3.15	3.25	3.33	3.39	3.44	3.49	3.52
40	0.05	2.02	2.29	2.44	2.54	2.62	2.68	2.73	2.77	2.81
	0.01	2.70	2.95	3.09	3.19	3.26	3.32	3.37	3.41	3.44
60	0.05	2.00	2.27	2.41	2.51	2.58	2.64	2.69	2.73	2.77
	0.01	2.66	2.90	3.03	3.12	3.19	3.25	3.29	3.33	3.37
120	0.05	1.98	2.24	2.38	2.47	2.55	2.60	2.65	2.69	2.73
	0.01	2.62	2.85	2.97	3.06	3.12	3.18	3.22	3.26	3.29
∞	0.05	1.96	2.21	2.35	2.44	2.51	2.57	2.61	2.65	2.69
	0.01	2.58	2.79	2.92	3.00	3.06	3.11	3.15	3.19	3.22

Source: Reproduced from C. W. Dunnett (1964), New tables for multiple comparisons with a control, *Biometrics 20*, 482–491. With permission of The Biometric Society.

Appendix W_S: Critical Lower-Tail Values of W_S for Rank-Sum Test for Two Independent Samples ($N_1 \leq N_2$)

$N_1 = 1$

N_2	0.001	0.005	0.010	0.025	0.05	0.10	2$\overline{W}$
2							4
3							5
4							6
5							7
6							8
7							9
8						—	10
9						1	11
10						1	12
11						1	13
12						1	14
13						1	15
14						1	16
15						1	17
16						1	18
17						1	19
18					—	1	20
19					1	2	21
20					1	2	22
21					1	2	23
22					1	2	24
23					1	2	25
24					1	2	26
25	—	—	—	—	1	2	27

$N_1 = 2$

0.001	0.005	0.010	0.025	0.05	0.10	2$\overline{W}$	N_2
					—	10	2
					3	12	3
			—		3	14	4
			3		4	16	5
			3		4	18	6
			—	3	4	20	7
		3	4	5		22	8
		3	4	5		24	9
		3	4	6		26	10
		3	4	6		28	11
		—	4	5	7	30	12
	3	4	5	7		32	13
	3	4	6	8		34	14
	3	4	6	8		36	15
	3	4	6	8		38	16
	3	5	6	9		40	17
	—	3	5	7	9	42	18
3	4	5	7	10		44	19
3	4	5	7	10		46	20
3	4	6	8	11		48	21
3	4	6	8	11		50	22
3	4	6	8	12		52	23
3	4	6	9	12		54	24
—	3	4	6	9	12	56	25

$N_1 = 3$

N_2	0.001	0.005	0.010	0.025	0.05	0.10	2$\overline{W}$
3					6	7	21
4				—	6	7	24
5			6	7	8		27
6			—	7	8	9	30
7		6	7	8	10		33
8		—	6	8	9	11	36
9	6	7	8	10	11		39
10	6	7	9	10	12		42
11	6	7	9	11	13		45
12	7	8	10	11	14		48
13	7	8	10	12	15		51
14	7	8	11	13	16		54
15	8	9	11	13	16		57

$N_1 = 4$

0.001	0.005	0.010	0.025	0.05	0.10	2$\overline{W}$	N_2
		—	10	11	13	36	4
	—	10	11	12	14	40	5
	10	11	12	13	15	44	6
	10	11	13	14	16	48	7
	11	12	14	15	17	52	8
—	11	13	14	16	19	56	9
10	12	13	15	17	20	60	10
10	12	14	16	18	21	64	11
10	13	15	17	19	22	68	12
11	13	15	18	20	23	72	13
11	14	16	19	21	25	76	14
11	15	17	20	22	26	80	15

(continued)

Appendix W_S *(continued)*

			$N_1 = 3$								$N_1 = 4$					
N_2	0.001	0.005	0.010	0.025	0.05	0.10	$2\overline{W}$	0.001	0.005	0.010	0.025	0.05	0.10	$2\overline{W}$	N_2
16	—	8	9	12	14	17	60	12	15	17	21	24	27	84	16
17	6	8	10	12	15	18	63	12	16	18	21	25	28	88	17
18	6	8	10	13	15	19	66	13	16	19	22	26	30	92	18
19	6	9	10	13	16	20	69	13	17	19	23	27	31	96	19
20	6	9	11	14	17	21	72	13	18	20	24	28	32	100	20
21	7	9	11	14	17	21	75	14	18	21	25	29	33	104	21
22	7	10	12	15	18	22	78	14	19	21	26	30	35	108	22
23	7	10	12	15	19	23	81	14	19	22	27	31	36	112	23
24	7	10	12	16	19	24	84	15	20	23	27	32	38	116	24
25	7	11	13	16	20	25	87	15	20	23	28	33	38	120	25

			$N_1 = 5$								$N_1 = 6$					
N_2	0.001	0.005	0.010	0.025	0.05	0.10	$2\overline{W}$	0.001	0.005	0.010	0.025	0.05	0.10	$2\overline{W}$	N_2
5		15	16	17	19	20	55								
6		16	17	18	20	22	60	—	23	24	26	28	30	78	6
7	—	16	18	20	21	23	65	21	24	25	27	29	32	84	7
8	15	17	19	21	23	25	70	22	25	27	29	31	34	90	8
9	16	18	20	22	24	27	75	23	26	28	31	33	36	96	9
10	16	19	21	23	26	28	80	24	27	29	32	35	38	102	10
11	17	20	22	24	27	30	85	25	28	30	34	37	40	108	11
12	17	21	23	26	28	32	90	25	30	32	35	38	42	114	12
13	18	22	24	27	30	33	95	26	31	33	37	40	44	120	13
14	18	22	25	28	31	35	100	27	32	34	38	42	46	126	14
15	19	23	26	29	33	37	105	28	33	36	40	44	48	132	15
16	20	24	27	30	34	38	110	29	34	37	42	46	50	138	16
17	20	25	28	32	35	40	115	30	36	39	43	47	52	144	17
18	21	26	29	33	37	42	120	31	37	40	45	49	55	150	18
19	22	27	30	34	38	43	125	32	38	41	46	51	57	156	19
20	22	28	31	35	40	45	130	33	39	43	48	53	59	162	20
21	23	29	32	37	41	47	135	33	40	44	50	55	61	168	21
22	23	29	33	38	43	48	140	34	42	45	51	57	63	174	22
23	24	30	34	39	44	50	145	35	43	47	53	58	65	180	23
24	25	31	35	40	45	51	150	36	44	48	54	60	67	186	24
25	25	32	36	42	47	53	155	37	45	50	56	62	69	192	25

Appendix W_S *(continued)*

			$N_1 = 7$								$N_1 = 8$					
N_2	0.001	0.005	0.010	0.025	0.05	0.10	$2\overline{W}$	0.001	0.005	0.010	0.025	0.05	0.10	$2\overline{W}$	N_2
7	29	32	34	36	39	41	105								
8	30	34	35	38	41	44	112	40	43	45	49	51	55	136	8
9	31	35	37	40	43	46	119	41	45	47	51	54	58	144	9
10	33	37	39	42	45	49	126	42	47	49	53	56	60	152	10
11	34	38	40	44	47	51	133	44	49	51	55	59	63	160	11
12	35	40	42	46	49	54	140	45	51	53	58	62	66	168	12
13	36	41	44	48	52	56	147	47	53	56	60	64	69	176	13
14	37	43	45	50	54	59	154	48	54	58	62	67	72	184	14
15	38	44	47	52	56	61	161	50	56	60	65	69	75	192	15
16	39	46	49	54	58	64	168	51	58	62	67	72	78	200	16
17	41	47	51	56	61	66	175	53	60	64	70	75	81	208	17
18	42	49	52	58	63	69	182	54	62	66	72	77	84	216	18
19	43	50	54	60	65	71	189	56	64	68	74	80	87	224	19
20	44	52	56	62	67	74	196	57	66	70	77	83	90	232	20
21	46	53	58	64	69	76	203	59	68	72	79	85	92	240	21
22	47	55	59	66	72	79	210	60	70	74	81	88	95	248	22
23	48	57	61	68	74	81	217	62	71	76	84	90	98	256	23
24	49	58	63	70	76	84	224	64	73	78	86	93	101	264	24
25	50	60	64	72	78	86	231	65	75	81	89	96	104	272	25

			$N_1 = 9$								$N_1 = 10$					
N_2	0.001	0.005	0.010	0.025	0.05	0.10	$2\overline{W}$	0.001	0.005	0.010	0.025	0.05	0.10	$2\overline{W}$	N_2
9	52	56	59	62	66	70	171								
10	53	58	61	65	69	73	180	65	71	74	78	82	87	210	10
11	55	61	63	68	72	76	189	67	73	77	81	86	91	220	11
12	57	63	66	71	75	80	198	69	76	79	84	89	94	230	12
13	59	65	68	73	78	83	207	72	79	82	88	92	98	240	13
14	60	67	71	76	81	86	216	74	81	85	91	96	102	250	14
15	62	69	73	79	84	90	225	76	84	88	94	99	106	260	15
16	64	72	76	82	87	93	234	78	86	91	97	103	109	270	16
17	66	74	78	84	90	97	243	80	89	93	100	106	113	280	17
18	68	76	81	8'7	93	100	252	82	92	96	103	110	117	290	18
19	70	78	83	90	96	103	261	84	94	99	107	113	121	300	19
20	71	81	85	93	99	107	270	87	97	102	110	117	125	310	20
21	73	83	88	95	102	110	279	89	99	105	113	120	128	320	21
22	75	85	90	98	105	113	288	91	102	108	116	123	132	330	22
23	77	88	93	101	108	117	297	93	105	110	119	127	136	340	23
24	79	90	95	104	111	120	306	95	107	113	122	130	140	350	24
25	81	92	98	107	114	123	315	98	110	116	126	134	144	360	25

(continued)

Appendix W_S *(continued)*

N_2	$N_1 = 11$								$N_1 = 12$							
	0.001	0.005	0.010	0.025	0.05	0.10	$2\overline{W}$		0.001	0.005	0.010	0.025	0.05	0.10	$2\overline{W}$	N_2
11	81	87	91	96	100	106	253									
12	83	90	94	99	104	110	264		98	105	109	115	120	127	300	12
13	86	93	97	103	108	114	275		101	109	113	119	125	131	312	13
14	88	96	100	106	112	118	286		103	112	116	123	129	136	324	14
15	90	99	103	110	116	123	297		106	115	120	127	133	141	336	15
16	93	102	107	113	120	127	308		109	119	124	131	138	145	348	16
17	95	105	110	117	123	131	319		112	122	127	135	142	150	360	17
18	98	108	113	121	127	135	330		115	125	131	139	146	155	372	18
19	100	111	116	124	131	139	341		118	129	134	143	150	159	384	19
20	103	114	119	128	135	144	352		120	132	138	147	155	164	396	20
21	106	117	123	131	139	148	363		123	136	142	151	159	169	408	21
22	108	120	126	135	143	152	374		126	139	145	155	163	173	420	22
23	111	123	129	139	147	156	385		129	142	149	159	168	178	432	23
24	113	126	132	142	151	161	396		132	146	153	163	172	183	444	24
25	116	129	136	146	155	165	407		135	149	156	167	176	187	456	25

N_2	$N_1 = 13$								$N_1 = 14$							
	0.001	0.005	0.010	0.025	0.05	0.10	$2\overline{W}$		0.001	0.005	0.010	0.025	0.05	0.10	$2\overline{W}$	N_2
13	117	125	130	136	142	149	351									
14	120	129	134	141	147	154	364		137	147	152	160	166	174	406	14
15	123	133	138	145	152	159	377		141	151	156	164	171	179	420	15
16	126	136	142	150	156	165	390		144	155	161	169	176	185	434	16
17	129	140	146	154	161	170	403		148	159	165	174	182	190	448	17
18	133	144	150	158	166	175	416		151	163	170	179	187	196	462	18
19	136	148	154	163	171	180	429		155	168	174	183	192	202	476	19
20	139	151	158	167	175	185	442		159	172	178	188	197	207	490	20
21	142	155	162	171	180	190	455		162	176	183	193	202	213	504	21
22	145	159	166	176	185	195	468		166	180	187	198	207	218	518	22
23	149	163	170	180	189	200	481		169	184	192	203	212	224	532	23
24	152	166	174	185	194	205	494		173	188	196	207	218	229	546	24
25	155	170	178	189	199	211	507		177	192	200	212	223	235	560	25

N_2	$N_1 = 15$								$N_1 = 16$							
	0.001	0.005	0.010	0.025	0.05	0.10	$2\overline{W}$		0.001	0.005	0.010	0.025	0.05	0.10	$2\overline{W}$	N_2
15	160	171	176	184	192	200	465									
16	163	175	181	190	197	206	480		184	196	202	211	219	229	528	16
17	167	180	186	195	203	212	495		188	201	207	217	225	235	544	17
18	171	184	190	200	208	218	510		192	206	212	222	231	242	560	18
19	175	189	195	205	214	224	525		196	210	218	228	237	248	576	19
20	179	193	200	210	220	230	540		201	215	223	234	243	255	592	20
21	183	198	205	216	225	236	555		205	220	228	239	249	261	608	21
22	187	202	210	221	231	242	570		209	225	233	245	255	267	624	22
23	191	207	214	226	236	248	585		214	230	238	251	261	274	640	23
24	195	211	219	231	242	254	600		218	235	244	256	267	280	656	24
25	199	216	224	237	248	260	615		222	240	249	262	273	287	672	25

Appendix W_S (continued)

N_2				$N_1 = 17$								$N_1 = 18$				
	0.001	0.005	0.010	0.025	0.05	0.10	$2\overline{W}$	0.001	0.005	0.010	0.025	0.05	0.10	$2\overline{W}$	N_2
17	210	223	230	240	249	259	595								
18	214	228	235	246	255	266	612	237	252	259	270	280	291	666	18
19	219	234	241	252	262	273	629	242	258	265	277	287	299	684	19
20	223	239	246	258	268	280	646	247	263	271	283	294	306	702	20
21	228	244	252	264	274	287	663	252	269	277	290	301	313	720	21
22	233	249	258	270	281	294	680	257	275	283	296	307	321	738	22
23	238	255	263	276	287	300	697	262	280	289	303	314	328	756	23
24	242	260	269	282	294	307	714	267	286	295	309	321	335	774	24
25	247	265	275	288	300	314	731	273	292	301	316	328	343	792	25

N_2				$N_1 = 19$								$N_1 = 20$				
	0.001	0.005	0.010	0.025	0.05	0.10	$2\overline{W}$	0.001	0.005	0.010	0.025	0.05	0.10	$2\overline{W}$	N_2
19	267	283	291	303	313	325	741								
20	272	289	297	309	320	333	760	298	315	324	337	348	361	820	20
21	277	295	303	316	328	341	779	304	322	331	344	356	370	840	21
22	283	301	310	323	335	349	798	309	328	337	351	364	378	860	22
23	288	307	316	330	342	357	817	315	335	344	359	371	386	880	23
24	294	313	323	337	350	364	836	321	341	351	366	379	394	900	24
25	299	319	329	344	357	372	855	327	348	358	373	387	403	920	25

N_2				$N_1 = 21$								$N_1 = 22$				
	0.001	0.005	0.010	0.025	0.05	0.10	$2\overline{W}$	0.001	0.005	0.010	0.025	0.05	0.10	$2\overline{W}$	N_2
21	331	349	359	373	385	399	903								
22	337	356	366	381	393	408	924	365	386	396	411	424	439	990	22
23	343	363	373	388	401	417	945	372	393	403	419	432	448	1012	23
24	349	370	381	396	410	425	966	379	400	411	427	441	457	1034	24
25	356	377	388	404	418	434	987	385	408	419	435	450	467	1056	25

N_2				$N_1 = 23$								$N_1 = 24$				
	0.001	0.005	0.010	0.025	0.05	0.10	$2\overline{W}$	0.001	0.005	0.010	0.025	0.05	0.10	$2\overline{W}$	N_2
23	402	424	434	451	465	481	1081								
24	409	431	443	459	474	491	1104	440	464	475	492	507	525	1176	24
25	416	439	451	468	483	500	1127	448	472	484	501	517	535	1200	25

N_2				$N_1 = 25$			
	0.001	0.005	0.010	0.025	0.05	0.10	$2\overline{W}$
25	480	505	517	536	552	570	1275

Source: Table 1 in L. R. Verdooren (1963), Extended tables of critical values for Wilcoxon's test statistic, *Biometrika, 50,* 177–186, with permission of the author and the editor.

Appendix z: The Normal Distribution (z)

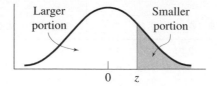

z	Mean to z	Larger Portion	Smaller Portion	y	z	Mean to z	Larger Portion	Smaller Portion	y
0.00	0.0000	0.5000	0.5000	0.3989	0.36	0.1406	0.6406	0.3594	0.3739
0.01	0.0040	0.5040	0.4960	0.3989	0.37	0.1443	0.6443	0.3557	0.3725
0.02	0.0080	0.5080	0.4920	0.3989	0.38	0.1480	0.6480	0.3520	0.3712
0.03	0.0120	0.5120	0.4880	0.3988	0.39	0.1517	0.6517	0.3483	0.3697
0.04	0.0160	0.5160	0.4840	0.3986	0.40	0.1554	0.6554	0.3446	0.3683
0.05	0.0199	0.5199	0.4801	0.3984	0.41	0.1591	0.6591	0.3409	0.3668
0.06	0.0239	0.5239	0.4761	0.3982	0.42	0.1628	0.6628	0.3372	0.3653
0.07	0.0279	0.5279	0.4721	0.3980	0.43	0.1664	0.6664	0.3336	0.3637
0.08	0.0319	0.5319	0.4681	0.3977	0.44	0.1700	0.6700	0.3300	0.3621
0.09	0.0359	0.5359	0.4641	0.3973	0.45	0.1736	0.6736	0.3264	0.3605
0.10	0.0398	0.5398	0.4602	0.3970	0.46	0.1772	0.6772	0.3228	0.3589
0.11	0.0438	0.5438	0.4562	0.3965	0.47	0.1808	0.6808	0.3192	0.3572
0.12	0.0478	0.5478	0.4522	0.3961	0.48	0.1844	0.6844	0.3156	0.3555
0.13	0.0517	0.5517	0.4483	0.3956	0.49	0.1879	0.6879	0.3121	0.3538
0.14	0.0557	0.5557	0.4443	0.3951	0.50	0.1915	0.6915	0.3085	0.3521
0.15	0.0596	0.5596	0.4404	0.3945	0.51	0.1950	0.6950	0.3050	0.3503
0.16	0.0636	0.5636	0.4364	0.3939	0.52	0.1985	0.6985	0.3015	0.3485
0.17	0.0675	0.5675	0.4325	0.3932	0.53	0.2019	0.7019	0.2981	0.3467
0.18	0.0714	0.5714	0.4286	0.3925	0.54	0.2054	0.7054	0.2946	0.3448
0.19	0.0753	0.5753	0.4247	0.3918	0.55	0.2088	0.7088	0.2912	0.3429
0.20	0.0793	0.5793	0.4207	0.3910	0.56	0.2123	0.7123	0.2877	0.3410
0.21	0.0832	0.5832	0.4168	0.3902	0.57	0.2157	0.7157	0.2843	0.3391
0.22	0.0871	0.5871	0.4129	0.3894	0.58	0.2190	0.7190	0.2810	0.3372
0.23	0.0910	0.5910	0.4090	0.3885	0.59	0.2224	0.7224	0.2776	0.3352
0.24	0.0948	0.5948	0.4052	0.3876	0.60	0.2257	0.7257	0.2743	0.3332
0.25	0.0987	0.5987	0.4013	0.3867	0.61	0.2291	0.7291	0.2709	0.3312
0.26	0.1026	0.6026	0.3974	0.3857	0.62	0.2324	0.7324	0.2676	0.3292
0.27	0.1064	0.6064	0.3936	0.3847	0.63	0.2357	0.7357	0.2643	0.3271
0.28	0.1103	0.6103	0.3897	0.3836	0.64	0.2389	0.7389	0.2611	0.3251
0.29	0.1141	0.6141	0.3859	0.3825	0.65	0.2422	0.7422	0.2578	0.3230
0.30	0.1179	0.6179	0.3821	0.3814	0.66	0.2454	0.7454	0.2546	0.3209
0.31	0.1217	0.6217	0.3783	0.3802	0.67	0.2486	0.7486	0.2514	0.3187
0.32	0.1255	0.6255	0.3745	0.3790	0.68	0.2517	0.7517	0.2483	0.3166
0.33	0.1293	0.6293	0.3707	0.3778	0.69	0.2549	0.7549	0.2451	0.3144
0.34	0.1331	0.6331	0.3669	0.3765	0.70	0.2580	0.7580	0.2420	0.3123
0.35	0.1368	0.6368	0.3632	0.3752	0.71	0.2611	0.7611	0.2389	0.3101

Appendix z   *(continued)*

z	Mean to z	Larger Portion	Smaller Portion	y	z	Mean to z	Larger Portion	Smaller Portion	y
0.72	0.2642	0.7642	0.2358	0.3079	1.14	0.3729	0.8729	0.1271	0.2083
0.73	0.2673	0.7673	0.2327	0.3056	1.15	0.3749	0.8749	0.1251	0.2059
0.74	0.2704	0.7704	0.2296	0.3034	1.16	0.3770	0.8770	0.1230	0.2036
0.75	0.2734	0.7734	0.2266	0.3011	1.17	0.3790	0.8790	0.1210	0.2012
0.76	0.2764	0.7764	0.2236	0.2989	1.18	0.3810	0.8810	0.1190	0.1989
0.77	0.2794	0.7794	0.2206	0.2966	1.19	0.3830	0.8830	0.1170	0.1965
0.78	0.2823	0.7823	0.2177	0.2943	1.20	0.3849	0.8849	0.1151	0.1942
0.79	0.2852	0.7852	0.2148	0.2920	1.21	0.3869	0.8869	0.1131	0.1919
0.80	0.2881	0.7881	0.2119	0.2897	1.22	0.3888	0.8888	0.1112	0.1895
0.81	0.2910	0.7910	0.2090	0.2874	1.23	0.3907	0.8907	0.1093	0.1872
0.82	0.2939	0.7939	0.2061	0.2850	1.24	0.3925	0.8925	0.1075	0.1849
0.83	0.2967	0.7967	0.2033	0.2827	1.25	0.3944	0.8944	0.1056	0.1826
0.84	0.2995	0.7995	0.2005	0.2803	1.26	0.3962	0.8962	0.1038	0.1804
0.85	0.3023	0.8023	0.1977	0.2780	1.27	0.3980	0.8980	0.1020	0.1781
0.86	0.3051	0.8051	0.1949	0.2756	1.28	0.3997	0.8997	0.1003	0.1758
0.87	0.3078	0.8078	0.1922	0.2732	1.29	0.4015	0.9015	0.0985	0.1736
0.88	0.3106	0.8106	0.1894	0.2709	1.30	0.4032	0.9032	0.0968	0.1714
0.89	0.3133	0.8133	0.1867	0.2685	1.31	0.4049	0.9049	0.0951	0.1691
0.90	0.3159	0.8159	0.1841	0.2661	1.32	0.4066	0.9066	0.0934	0.1669
0.91	0.3186	0.8186	0.1814	0.2637	1.33	0.4082	0.9082	0.0918	0.1647
0.92	0.3212	0.8212	0.1788	0.2613	1.34	0.4099	0.9099	0.0901	0.1626
0.93	0.3238	0.8238	0.1762	0.2589	1.35	0.4115	0.9115	0.0885	0.1604
0.94	0.3264	0.8264	0.1736	0.2565	1.36	0.4131	0.9131	0.0869	0.1582
0.95	0.3289	0.8289	0.1711	0.2541	1.37	0.4147	0.9147	0.0853	0.1561
0.96	0.3315	0.8315	0.1685	0.2516	1.38	0.4162	0.9162	0.0838	0.1539
0.97	0.3340	0.8340	0.1660	0.2492	1.39	0.4177	0.9177	0.0823	0.1518
0.98	0.3365	0.8365	0.1635	0.2468	1.40	0.4192	0.9192	0.0808	0.1497
0.99	0.3389	0.8389	0.1611	0.2444	1.41	0.4207	0.9207	0.0793	0.1476
1.00	0.3413	0.8413	0.1587	0.2420	1.42	0.4222	0.9222	0.0778	0.1456
1.01	0.3438	0.8438	0.1562	0.2396	1.43	0.4236	0.9236	0.0764	0.1435
1.02	0.3461	0.8461	0.1539	0.2371	1.44	0.4251	0.9251	0.0749	0.1415
1.03	0.3485	0.8485	0.1515	0.2347	1.45	0.4265	0.9265	0.0735	0.1394
1.04	0.3508	0.8508	0.1492	0.2323	1.46	0.4279	0.9279	0.0721	0.1374
1.05	0.3531	0.8531	0.1469	0.2299	1.47	0.4292	0.9292	0.0708	0.1354
1.06	0.3554	0.8554	0.1446	0.2275	1.48	0.4306	0.9306	0.0694	0.1334
1.07	0.3577	0.8577	0.1423	0.2251	1.49	0.4319	0.9319	0.0681	0.1315
1.08	0.3599	0.8599	0.1401	0.2227	1.50	0.4332	0.9332	0.0668	0.1295
1.09	0.3621	0.8621	0.1379	0.2203	1.51	0.4345	0.9345	0.0655	0.1276
1.10	0.3643	0.8643	0.1357	0.2179	1.52	0.4357	0.9357	0.0643	0.1257
1.11	0.3665	0.8665	0.1335	0.2155	1.53	0.4370	0.9370	0.0630	0.1238
1.12	0.3686	0.8686	0.1314	0.2131	1.54	0.4382	0.9382	0.0618	0.1219
1.13	0.3708	0.8708	0.1292	0.2107	1.55	0.4394	0.9394	0.0606	0.1200

(continued)

Appendix z *(continued)*

z	Mean to z	Larger Portion	Smaller Portion	y	z	Mean to z	Larger Portion	Smaller Portion	y
1.56	0.4406	0.9406	0.0594	0.1182	1.98	0.4761	0.9761	0.0239	0.0562
1.57	0.4418	0.9418	0.0582	0.1163	1.99	0.4767	0.9767	0.0233	0.0551
1.58	0.4429	0.9429	0.0571	0.1145	2.00	0.4772	0.9772	0.0228	0.0540
1.59	0.4441	0.9441	0.0559	0.1127	2.01	0.4778	0.9778	0.0222	0.0529
1.60	0.4452	0.9452	0.0548	0.1109	2.02	0.4783	0.9783	0.0217	0.0519
1.61	0.4463	0.9463	0.0537	0.1092	2.03	0.4788	0.9788	0.0212	0.0508
1.62	0.4474	0.9474	0.0526	0.1074	2.04	0.4793	0.9793	0.0207	0.0498
1.63	0.4484	0.9484	0.0516	0.1057	2.05	0.4798	0.9798	0.0202	0.0488
1.64	0.4495	0.9495	0.0505	0.1040	2.06	0.4803	0.9803	0.0197	0.0478
1.65	0.4505	0.9505	0.0495	0.1023	2.07	0.4808	0.9808	0.0192	0.0468
1.66	0.4515	0.9515	0.0485	0.1006	2.08	0.4812	0.9812	0.0188	0.0459
1.67	0.4525	0.9525	0.0475	0.0989	2.09	0.4817	0.9817	0.0183	0.0449
1.68	0.4535	0.9535	0.0465	0.0973	2.10	0.4821	0.9821	0.0179	0.0440
1.69	0.4545	0.9545	0.0455	0.0957	2.11	0.4826	0.9826	0.0174	0.0431
1.70	0.4554	0.9554	0.0446	0.0940	2.12	0.4830	0.9830	0.0170	0.0422
1.71	0.4564	0.9564	0.0436	0.0925	2.13	0.4834	0.9834	0.0166	0.0413
1.72	0.4573	0.9573	0.0427	0.0909	2.14	0.4838	0.9838	0.0162	0.0404
1.73	0.4582	0.9582	0.0418	0.0893	2.15	0.4842	0.9842	0.0158	0.0396
1.74	0.4591	0.9591	0.0409	0.0878	2.16	0.4846	0.9846	0.0154	0.0387
1.75	0.4599	0.9599	0.0401	0.0863	2.17	0.4850	0.9850	0.0150	0.0379
1.76	0.4608	0.9608	0.0392	0.0848	2.18	0.4854	0.9854	0.0146	0.0371
1.77	0.4616	0.9616	0.0384	0.0833	2.19	0.4857	0.9857	0.0143	0.0363
1.78	0.4625	0.9625	0.0375	0.0818	2.20	0.4861	0.9861	0.0139	0.0355
1.79	0.4633	0.9633	0.0367	0.0804	2.21	0.4864	0.9864	0.0136	0.0347
1.80	0.4641	0.9641	0.0359	0.0790	2.22	0.4868	0.9868	0.0132	0.0339
1.81	0.4649	0.9649	0.0351	0.0775	2.23	0.4871	0.9871	0.0129	0.0332
1.82	0.4656	0.9656	0.0344	0.0761	2.24	0.4875	0.9875	0.0125	0.0325
1.83	0.4664	0.9664	0.0336	0.0748	2.25	0.4878	0.9878	0.0122	0.0317
1.84	0.4671	0.9671	0.0329	0.0734	2.26	0.4881	0.9881	0.0119	0.0310
1.85	0.4678	0.9678	0.0322	0.0721	2.27	0.4884	0.9884	0.0116	0.0303
1.86	0.4686	0.9686	0.0314	0.0707	2.28	0.4887	0.9887	0.0113	0.0297
1.87	0.4693	0.9693	0.0307	0.0694	2.29	0.4890	0.9890	0.0110	0.0290
1.88	0.4699	0.9699	0.0301	0.0681	2.30	0.4893	0.9893	0.0107	0.0283
1.89	0.4706	0.9706	0.0294	0.0669	2.31	0.4896	0.9896	0.0104	0.0277
1.90	0.4713	0.9713	0.0287	0.0656	2.32	0.4898	0.9898	0.0102	0.0270
1.91	0.4719	0.9719	0.0281	0.0644	2.33	0.4901	0.9901	0.0099	0.0264
1.92	0.4726	0.9726	0.0274	0.0632	2.34	0.4904	0.9904	0.0096	0.0258
1.93	0.4732	0.9732	0.0268	0.0620	2.35	0.4906	0.9906	0.0094	0.0252
1.94	0.4738	0.9738	0.0262	0.0608	2.36	0.4909	0.9909	0.0091	0.0246
1.95	0.4744	0.9744	0.0256	0.0596	2.37	0.4911	0.9911	0.0089	0.0241
1.96	0.4750	0.9750	0.0250	0.0584	2.38	0.4913	0.9913	0.0087	0.0235
1.97	0.4756	0.9756	0.0244	0.0573	2.39	0.4916	0.9916	0.0084	0.0229

Appendix z   *(continued)*

z	Mean to z	Larger Portion	Smaller Portion	y	z	Mean to z	Larger Portion	Smaller Portion	y
2.40	0.4918	0.9918	0.0082	0.0224	2.75	0.4970	0.9970	0.0030	0.0091
2.41	0.4920	0.9920	0.0080	0.0219	2.76	0.4971	0.9971	0.0029	0.0088
2.42	0.4922	0.9922	0.0078	0.0213	2.77	0.4972	0.9972	0.0028	0.0086
2.43	0.4925	0.9925	0.0075	0.0208	2.78	0.4973	0.9973	0.0027	0.0084
2.44	0.4927	0.9927	0.0073	0.0203	2.79	0.4974	0.9974	0.0026	0.0081
2.45	0.4929	0.9929	0.0071	0.0198	2.80	0.4974	0.9974	0.0026	0.0079
2.46	0.4931	0.9931	0.0069	0.0194	2.81	0.4975	0.9975	0.0025	0.0077
2.47	0.4932	0.9932	0.0068	0.0189	2.82	0.4976	0.9976	0.0024	0.0075
2.48	0.4934	0.9934	0.0066	0.0184	2.83	0.4977	0.9977	0.0023	0.0073
2.49	0.4936	0.9936	0.0064	0.0180	2.84	0.4977	0.9977	0.0023	0.0071
2.50	0.4938	0.9938	0.0062	0.0175	2.85	0.4978	0.9978	0.0022	0.0069
2.51	0.4940	0.9940	0.0060	0.0171	2.86	0.4979	0.9979	0.0021	0.0067
2.52	0.4941	0.9941	0.0059	0.0167	2.87	0.4979	0.9979	0.0021	0.0065
2.53	0.4943	0.9943	0.0057	0.0163	2.88	0.4980	0.9980	0.0020	0.0063
2.54	0.4945	0.9945	0.0055	0.0158	2.89	0.4981	0.9981	0.0019	0.0061
2.55	0.4946	0.9946	0.0054	0.0154	2.90	0.4981	0.9981	0.0019	0.0060
2.56	0.4948	0.9948	0.0052	0.0151	2.91	0.4982	0.9982	0.0018	0.0058
2.57	0.4949	0.9949	0.0051	0.0147	2.92	0.4982	0.9982	0.0018	0.0056
2.58	0.4951	0.9951	0.0049	0.0143	2.93	0.4983	0.9983	0.0017	0.0055
2.59	0.4952	0.9952	0.0048	0.0139	2.94	0.4984	0.9984	0.0016	0.0053
2.60	0.4953	0.9953	0.0047	0.0136	2.95	0.4984	0.9984	0.0016	0.0051
2.61	0.4955	0.9955	0.0045	0.0132	2.96	0.4985	0.9985	0.0015	0.0050
2.62	0.4956	0.9956	0.0044	0.0129	2.97	0.4985	0.9985	0.0015	0.0048
2.63	0.4957	0.9957	0.0043	0.0126	2.98	0.4986	0.9986	0.0014	0.0047
2.64	0.4959	0.9959	0.0041	0.0122	2.99	0.4986	0.9986	0.0014	0.0046
2.65	0.4960	0.9960	0.0040	0.0119	3.00	0.4987	0.9987	0.0013	0.0044
2.66	0.4961	0.9961	0.0039	0.0116	. . .	. . .	. . .	. . .	. . .
2.67	0.4962	0.9962	0.0038	0.0113	3.25	0.4994	0.9994	0.0006	0.0020
2.68	0.4963	0.9963	0.0037	0.0110	. . .	. . .	. . .	. . .	. . .
2.69	0.4964	0.9964	0.0036	0.0107	3.50	0.4998	0.9998	0.0002	0.0009
2.70	0.4965	0.9965	0.0035	0.0104	. . .	. . .	. . .	. . .	. . .
2.71	0.4966	0.9966	0.0034	0.0101	3.75	0.4999	0.9999	0.0001	0.0004
2.72	0.4967	0.9967	0.0033	0.0099	. . .	. . .	. . .	. . .	. . .
2.73	0.4968	0.9968	0.0032	0.0096	4.00	0.5000	1.0000	0.0000	0.0001
2.74	0.4969	0.9969	0.0031	0.0093					

Source: The entries in this table were computed by the author.

References

Achenbach, T. M. (1991a). *Manual for the Child Behavior Checklist/4-18 and 1991 profile.* Burlington, VT: University of Vermont Department of Psychiatry.

Achenbach, T. M. (1991b). *Manual for the Youth Self-Report and 1991 profile.* Burlington, VT: University of Vermont Department of Psychiatry.

Adams, H. E., Wright, L. W., Jr., & Lohr, B. A. (1996). Is homophobia associated with homosexual arousal? *Journal of Abnormal Psychology, 105,* 440–445.

Agresti, A. (1984). *Analysis of ordinal categorical data.* New York: Wiley.

Agresti, A. (1990). *Categorical data analysis.* New York: Wiley.

Agresti, A. (1996). *An introduction to categorical data analysis.* New York: Wiley.

Agresti, A. (2002). *Categorical data analysis.* New York: Wiley.

Aiken, L. S., & West, S. G. (1991). *Multiple regression: Testing and interpreting interactions.* Newbury Park, CA: Sage.

Anderson. N. H. (1963). Comparison of different populations: Resistance to extinction and transfer. *Psychological Review, 70,* 162–179.

Appelbaum, M. I., & Cramer, E. M. (1974). Some problems in the nonorthogonal analysis of variance. *Psychological Bulletin, 81,* 335–343.

Aronson, J., Lustina, M. J., Good, C., Keough, K., Steele, C. M., & Brown, J. (1998). When white men can't do math: Necessary and sufficient factors in stereotype threat. *Journal of Experimental Social Psychology, 35,* 29–46.

Baker, S. P., & Lew, R. (1987). A Monte Carlo comparison of Tukey's and Bonferroni's methods with an optimized Bonferroni multiple comparison procedure in repeated measures experiments. Paper presented at the annual meeting of the American Statistical Association. San Francisco, CA.

Baron, R. M., & Kenny, D. A. (1986). The moderator-mediator variable distinction in social psychological research: Conceptual, strategic, and statistical considerations. *Journal of Personality and Social Psychology, 51,* 1173–1182.

Bell, R. A., Buerkel-Rothfuss, N. L., & Gore, K. E. (1987). "Did you bring the yarmulke for the Cabbage Patch kid?": The idiomatic communication among young lovers. *Human Communication Research, 14,* 47–67.

Berger, V. I. (2005). Mid-p values. In Everitt, B. S., and Howell, D. C. *Encyclopedia of statistics in behavioral sciences.* Chichester, England: Wiley.

Blair, R. C. (1978). I've been testing some statistical hypotheses. . . . Can you guess what they are? *Journal of Educational Research, 72,* 116–118.

Blair, R. C., & Higgins, J. J. (1978). Tests of hypotheses for unbalanced factorial designs under various regression/coding method combinations. *Educational and Psychological Measurement, 38,* 621–631.

Blanchard, E. B., Theobald, D. E., Williamson, D. A., Silver, B. V., & Brown, D. A. (1978). Temperature biofeedback in the treatment of migraine headaches. *Archives of General Psychiatry, 35,* 581–588.

Bohj, D. S. (1978). Testing equality of means of correlated variates with missing data on both responses. *Biometrika, 65,* 225–228.

Boik, R. J. (1981). A priori tests in repeated measures designs: Effects of nonsphericity. *Psychometrika, 46,* 241–255.

Boneau, C. A. (1960). The effects of violations of assumptions underlying the *t* test. *Psychological Bulletin, 57,* 49–64.

Borenstein, M., & Cohen, J. (1988). *Statistical power analysis: A computer program.* Hillsdale, NJ: Erlbaum.

Bouton, M., & Swartzentruber, D. (1985). Unpublished raw data. University of Vermont.

Box, G. E. P. (1953). Non-normality and tests on variance. *Biometrika, 40,* 318–335.

Box, G. E. P. (1954a). Some theorems on quadratic forms applied in the study of analysis of variance problems: I. Effect of inequality of variance in the one-way classification. *Annals of Mathematical Statistics, 25,* 290–302.

Box, G. E. P. (1954b). Some theorems on quadratic forms applied in the study of analysis of variance problems: II. Effect of inequality of variance and of correlation of errors in the two-way classification. *Annals of Mathematical Statistics, 25,* 484–498.

Bradley, D. R., Bradley, T. D., McGrath, S. G., & Cutcomb, S. D. (1979). Type I error rate of the chi-square test of independence in $R \times C$ tables that have small expected frequencies. *Psychological Bulletin, 86,* 1290–1297.

Bradley, D. R. (1988). *DATASIM.* Lewiston, ME: Desktop Press.

Bradley, D. R., Russell, R. L., & Reeve, C. P. (1996). Statistical power in complex experimental designs. *Behavioral Research Methods, Instruments, and Computers, 28,* 319–326.

Bradley, J. V. (1963). *Studies in research methodology: IV. A sampling study of the central limit theorem and the robustness of one sample parametric tests* (AMRL Tech. Rep. No. 63-29). Aerospace Medical Research Laboratories, Wright-Patterson Air Force Base, OH.

Bradley, J. V. (1964). *Studies in research methodology: VI. The central limit effect for a variety of populations and the robustness of z, t, and F* (AMRL Tech. Rep. No. 64-123). Aerospace Medical Research Laboratories, Wright-Patterson Air Force Base, OH.

Bradley, J. V. (1968). *Distribution free statistical tests.* Englewood Cliffs, NJ: Prentice-Hall.

Brooks, L., & Perot, A. R. (1991). Reporting sexual harassment. *Psychology of Women Quarterly, 15,* 31–47.

Brown, M. B., & Forsythe, A. B. (1974). The ANOVA and multiple comparisons for data with heterogeneous variances. *Biometrics, 30,* 719–724.

Camilli, G., & Hopkins, K. D. (1979). Testing for association in 2×2 contingency tables with very small sample sizes. *Psychological Bulletin, 86,* 1011–1014.

Campbell, A., Converse, P. E., & Rodgers, W. L. (1976). *The quality of American life.* New York: Russell Sage Foundation.

Carlson, J. E., & Timm, N. H. (1974). Analysis of nonorthogonal fixed-effects designs. *Psychological Bulletin, 81,* 563–570.

Carmer, S. G., & Swanson, M. R. (1973). An evaluation of ten multiple comparison procedures by Monte Carlo methods. *Journal of the American Statistical Association, 68,* 66–74.

Clark, K. B., & Clark, M. K. (1939). The development of consciousness of self in the emergence of racial identification in Negro pre-school children. *Journal of Social Psychology, 10,* 591–599.

Cochran, W. G., & Cox, G. M. (1957). *Experimental designs* (2nd ed.). New York: Wiley.

Cochrane, A. I., St. Leger, A. S., & Moore, F. (1978). Health service "input" and mortality "output" in developed countries. *Journal of Epidemiology and Community Health, 32,* 200–205.

Cohen, J. (1960). A coefficient of agreement for nominal scales. *Educational and Psychological Measurement, 20,* 37–46.

Cohen, J. (1962). The statistical power of abnormal-social psychological research: A review. *Journal of Abnormal and Social Psychology, 65,* 145–153.

Cohen, J. (1965). Some statistical issues in psychological research. In B. B. Wolman (Ed.), *Handbook of clinical psychology.* New York: McGraw-Hill.

Cohen, J. (1968). Multiple regression as a general data-analytic system. *Psychological Bulletin, 70,* 426–443.

Cohen, J. (1969). *Statistical power analysis for the behavioral sciences* (1st ed.) Hillsdale, NJ: Erlbaum.

Cohen, J. (1973). Eta-squared and partial eta-squared in fixed factor ANOVA designs. *Educational and Psychological Measurement, 33,* 107–112.

Cohen, J. (1988). *Statistical power analysis for the behavioral sciences* (2nd ed.) New York: Academic Press.

Cohen, J. (1992a). A power primer. *Psychological Bulletin, 112,* 155–159.

Cohen, J. (1992b). Statistical power analysis. *Current Directions in Psychological Science, 1,* 98–101.

Cohen, J., & Cohen, P. (1983). *Applied multiple regression/Correlation analysis for the behavioral sciences,* 2nd ed. Hillsdale, NJ: Erlbaum.

Cohen, J., Cohen, P., West, S. G., & Aiken, L. S. (2003). *Applied multiple regression/correlation analysis for the behavioral sciences.* (3rd ed.) Mahwah, NJ: Lawrence Erlbaum Associates.

Cohen, S., Kaplan, J. R., Cunnick, J. E., Manuck, S. B., & Rabin, B. S. (1992). Chronic social stress, affiliation, and cellular immune response in nonhuman primates. *Psychological Science, 3,* 301–304.

Collier, R. O., Jr., Baker, F. B., & Mandeville, G. K. (1967). Tests of hypotheses in a repeated measures design from a permutation viewpoint. *Psychometrika, 32,* 15–24.

Collier, R. O., Jr., Baker, F. B., Mandeville, G. K., & Hayes, T. F. (1967). Estimates of test size for several test procedures based on conventional variance ratios in the repeated measures design. *Psychometrika, 32,* 339–353.

Compas, B. E., Howell, D. C., Phares, V., Williams, R. A., & Giunta, C. (1989). Risk factors for emotional/behavioral problems in young adolescents: A prospective analysis of adolescent and parental stress symptoms. *Journal of Consulting and Clinical Psychology, 57,* 732–740.

Compas, B. E., Worsham, N. S., Grant, K., Mireault, G., Howell, D. C., & Malcarne, V. L. (1994). When mom or dad has cancer: I. Symptoms of depression and anxiety in cancer patients, spouses, and children. *Health Psychology, 13,* 507–515.

Conti, L., & Musty, R. E. (1984). The effects of delta-9tetrahydrocannabinol injections to the nucleus accumbens on the locomotor activity of rats. In S. Aquell et al. (Eds.), *The cannabinoids: Chemical, pharmacologic, and therapeutic aspects.* New York: Academic Press.

Cooley, W. W., & Lohnes, P. R. (1971). *Multivariate data analysis.* New York: Wiley.

Cortina, J. M., & Nouri, H. (2000). *Effect size for ANOVA designs.* Thousand Oaks, CA: Sage.

Cox, D. R., & Wermuth, N. (1992). A comment on the coefficient of determination for binary responses. *American Statistician, 46,* 1–4.

Crawford, J. R., Garthwaite, P. H., Howell, D. C., & Venneri, A. (2003). Intra-individual measures of association in neuropsychology: Inferential methods for comparing a single case with a control or normative sample. *Journal of International Neuropsychological Society, 9,* 989–1000.

Craik, F. I. M., & Lockhart, R. S. (1972). Levels of processing: A framework for memory research. *Journal of Verbal Learning and Verbal Behavior, 11,* 671–684.

Cramer, E. M., & Appelbaum, M. I. (1980). Nonorthogonal analysis of variance—Once again. *Psychological Bulletin, 87,* 51–57.

Cramér, H. (1946). *Mathematical methods of statistics.* Princeton, NJ: Princeton University Press.

Cumming, G., & Finch, S. (2001). A primer on the understanding, use, and calculation of confidence intervals that are based on central and noncentral distribution. *Educational & Psychological Measurement, 61,* 532–574.

Czitrom, V. (1999). One-factor-at-a-time versus designed experiments, *American Statistician, 53,* 126–131.

Dabbs, J. M., Jr., & Morris, R. (1990). Testosterone, social class, and antisocial behavior in a sample of 4462 men. *Psychological Science, 1,* 209–211.

Darley, J. M., & Latané B. (1968). Bystander intervention in emergencies: Diffusion of responsibility. *Journal of Personality and Social Psychology, 8,* 377–383.

Darlington, R. B. (1968). Multiple regression in psychological research and practice. *Psychological Bulletin, 69,* 161–182.

Darlington, R. B. (1990). *Regression and linear models.* New York: McGraw-Hill.

Davey, G. C. L., Startup, H. M., Zara, A., MacDonald, C. B., & Field, A. P. (2003). Perseveration of checking thoughts and mood-as-input hypothesis. *Journal of Behavior Therapy & Experimental Psychiatry, 34,* 141–160.

Dawes, R. M., & Corrigan, B. (1974). Linear models in decision making. *Psychological Bulletin, 81,* 95–106.

Delucchi, K. L. (1983). The use and misuse of chi-square: Lewis and Burke revisited. *Psychological Bulletin, 94,* 166–176.

Derogatis, L. R. (1983). *SCL90-R: Administration, scoring, and procedures manual* (vol. 1.). Towson, MD: Clinical Psychometric Research.

Dodd, D. H., & Schultz, R. F., Jr. (1973). Computational procedures for estimating magnitude of effect for some analysis of variance designs. *Psychological Bulletin, 79,* 391–395.

Doob, A. N., & Gross, A. E. (1968). Status of frustrator as an inhibitor of horn-honking responses. *Journal of Social Psychology, 76,* 213–218.

Draper, N. R., & Smith, H. (1981). *Applied regression analysis* (2nd ed.). New York: Wiley.

Dunn, O. J. (1961). Multiple comparisons among means, *Journal of the American Statistical Association, 56,* 52–64.

Dunnett, C. W. (1955). A multiple comparison procedure for comparing several treatments with a control. *Journal of the American Statistical Association, 50,* 1096–1121.

Dunnett, C. W. (1964). New tables for multiple comparisons with a control. *Biometrics, 20,* 482–491.

Edgington, E. S. (1995). *Randomization tests.* New York: Marcel Dekker.

Edwards, A. L. (1985). *Experimental design in psychological research* (5th ed.). New York: Harper & Row.

Efron, B., & Tibshirani, R. (1993). *An introduction to the bootstrap.* New York: Chapman and Hall.

Einot, I., & Gabriel, K. R. (1975). A study of the powers of several methods of multiple comparisons. *Journal of the American Statistical Association, 70,* 574–583.

Epping-Jordan, J. E., Compas, B. E., & Howell, D. C. (1994). Predictors of cancer progression in young adult men and women: Avoidance, intrusive thoughts, and psychological symptoms. *Health Psychology, 13,* 539–547.

Eron, L. D., Huesmann, L. R., Lefkowitz, M. M., & Walden, L. O. (1972). Does television violence cause aggression? *American Psychologist, 27,* 253–263.

Evans, S. H., & Anastasio, E. J. (1968). Misuse of analysis of covariance when treatment effect and covariate are confounded. *Psychological Bulletin, 69,* 225–234.

Everitt, B. (1994). Cited in Hand, D. J., Daly, F., Lunn, A. D., McConway, K. J., & Ostrowski, E. *Handbook of small data sets.* London: Chapman & Hall. p. 229.

Eysenck, M. W. (1974). Age differences in incidental learning. *Developmental Psychology, 10,* 936–941.

Federer, W. T. (1955). *Experimental design: Theory and application.* New York: Macmillan.

Finney, J. W., Mitchell, R. E., Cronkite, R. C., & Moos, R. H. (1984). Methodological issues in estimating main and interactive effects: Examples from coping/social support and stress field. *Journal of Health and Social Behavior, 25,* 85–98.

Fisher, R. A. (1921). On the probable error of a coefficient of correlation deduced from a small sample. *Metron, 1,* 3–32.

Fisher, R. A. (1935). *The design of experiments.* Edinburgh: Oliver & Boyd.

Fisher, R. A., & Yates, F. (1953). *Statistical tables for biological, agricultural, and medical research* (4th ed.). Edinburgh, Scotland: Oliver & Boyd.

Fleiss, J. L. (1969). Estimating the magnitude of experimental effects. *Psychological Bulletin, 72,* 273–276.

Foa, E. B., Rothbaum, B. O., Riggs, D. S., & Murdock, T. B. (1991). Treatment of posttraumatic stress disorder in rape victims: A comparison between cognitive-behavioral procedures and counseling. *Journal of Consulting and Clinical Psychology, 59,* 715–723.

Fowler, R. L. (1985). Point estimates and confidence intervals in measures of association. *Psychological Bulletin, 98,* 160–165.

Frigge, M., Hoaglin, D. C., & Iglewicz B. (1989). Some implementations of the boxplot. *American Statistician, 43,* 50–54.

Galton, F. (1886). Regression towards mediocrity in hereditary stature. *Journal of the Anthropological Institute, 15,* 246–263.

Games, P. A. (1978a). A three-factor model encompassing many possible statistical tests on independent groups. *Psychological Bulletin, 85,* 168–182.

Games, P. A. (1978b). A four-factor structure for parametric tests on independent groups. *Psychological Bulletin, 85,* 661–672.

Games, P. A., & Howell, J. F. (1976). Pairwise multiple comparison procedures with unequal *n's* and/or variances: A Monte Carlo study. *Journal of Educational Statistics, 1,* 113–125.

Games, P. A., Keselman, H. J., & Rogan, J. C. (1981). Simultaneous pairwise multiple comparison procedures for means when sample sizes are unequal. *Psychological Bulletin, 90,* 594–598.

Geller, E. S., Witmer, J. F., & Orebaugh, A. L. (1976). Instructions as a determinant of paper disposal behaviors. *Environment and Behavior, 8,* 417–439.

Gibson, L., & Leitenberg, H. (2000). Child sexual abuse prevention programs: Do they decrease the occurrence of child sexual abuse? *Child Abuse & Neglect, 24,* 1115–1125.

Gigerenzer, G., Swijtink, Z., Porter, T., Daston, L., Beatty, J., and Krüger, L. (1989). *The empire of chance.* Cambridge: Cambridge University Press.

Glass, G. V. (1976). Primary, secondary, and meta-analysis of research. *Educational Researcher, 5,* 3–8.

Glass, G. V., McGaw, B., & Smith, M. L. (1981). *Meta-analysis in social research.* Newbury Park, CA: Sage.

Goldberg, L. R. (1965). Diagnosticians versus diagnostic signs: The diagnosis of psychosis versus neurosis from the MMPI. *Psychological Monographs, 79* (9, Whole No. 602).

Goldstein, R. (1989). Power and sample size via MS/PC-DOS computers. *American Statistician, 43,* 253–260.

Good, P. I. (1999). *Resampling methods: A practical guide to data analysis.* Boston: Birkhäuser.

Good, P. (2000). *Permutation tests: A practical guide to resampling methods for testing hypotheses* (2nd ed.). New York: Springer-Verlag.

Green, J. A. (1988). Loglinear analysis of cross-classified ordinal data: Applications in developmental research. *Child Development, 59,* 1–25.

Green, S. B., Marquis, J. G., Hershberger, S. L., Thompson, M. S., & McCollam, K. M. (1999). The overparameterized analysis of variance model. *Psychological Methods, 4,* 214–233.

Greenhouse, S. W., & Geisser, S. (1959). On methods in the analysis of profile data. *Psychometrika, 24,* 95–112.

Grissom, R. J. (2000). Heterogeneity of variance in clinical data. *Journal of Consulting and Clinical Psychology, 68,* 155–165.

Gross, J. S. (1985). Weight modification and eating disorders in adolescent boys and girls. Unpublished doctoral dissertation, University of Vermont.

Guber, D. L. (1999). Getting what you pay for: The debate over equity in public school expenditures. *Journal of Statistics Education, 7,* (2).

Hand, D. J., Daly, F., Lunn, A. D., McConway, K. J., & Ostrowski, E. (1994). *Handbook of small data sets.* London: Chapman & Hall.

Hansen, J. C., & Swanson, J. L. (1983). Stability of interests and the predictive and concurrent validity of the 1981 Strong-Campbell Interest Inventory for college majors. *Journal of Counseling Psychology, 30,* 194–201.

Harris, C. W. (Ed.). (1963). *Problems in measuring change.* Madison: University of Wisconsin Press.

Harris, R. J. (1985). *A primer of multivariate statistics* (2nd ed.). New York: Academic Press.

Harris, R. J. (2005). Classical statistical inference: Practice versus presentation. In Everitt, B. S., & Howell, D. C. *Encyclopedia of Statistics in Behavioral Science* (268–278). Chichester: Wiley.

Harter, H. L. (1960). Tables of range and Studentized range. *Annals of Mathematical Statistics, 31,* 1122–1147.

Hays, W. L. (1981). *Statistics* (3rd ed.). New York: Holt, Rinehart & Winston.

Hays, W. L. (1994). *Statistics* (5th ed.). New York: Harcourt, Brace.

Hedges, L. V. (1981). Distribution theory for Glass's estimator of effect size and related estimators. *Journal of Educational Statistics, 6,* 107–128.

Hedges, L. V. (1982). Estimation of effect size from a series of independent experiments. *Psychological Bulletin, 92,* 490–499.

Henderson, D. A., & Denison, D. R. (1989). Stepwise regression in social and psychological research. *Psychological Reports, 64,* 251–257.

Hindley, C. B., Filliozat, A. M., Klackenberg, G., Nicolet-Meister, D., & Sand, E. A. (1966). Differences in age of walking for five European longitudinal samples. *Human Biology, 38,* 364–379.

Hoaglin, D. C., Mosteller, F., & Tukey, J. W. (1983). *Understanding robust and exploratory data analysis.* New York: Wiley.

Hochberg, Y., & Tamhane, A. C. (1987). *Multiple comparison procedures.* New York: Wiley.

Hoenig, J. M., & Heisey, D. M. (2001). The abuse of power: The pervasive fallacy of power calculations for data analysis. *American Statistician, 55,* 19–24.

Holm, S. (1979). A simple sequentially rejective multiple test procedure. *Scandinavian Journal of Statistics, 6,* 65–70.

Holmes, T. H., & Rahe, R. H. (1967). The social readjustment rating scale. *Journal of Psychosomatic Research, 11,* 213.

Holway, A. H., & Boring, E. G. (1940). The moon illusion and the angle of regard. *American Journal of Psychology, 53,* 509–516.

Horowitz, M. J., Wilner, N., & Alvarez, W. (1979). Impact of event scale: A measure of subjective stress. *Psychosomatic Medicine, 41,* 209–218.

Hosmer, D. W., & Lemeshow, S. (1989). *Applied logistic regression.* New York: Wiley.

Hotelling, H. (1931). The generalization of Student's ratio. *Annals of Mathematical Statistics, 2,* 360–378.

Hout, M., Duncan, O. D., & Sobel, M. E. (1987). Association and heterogeneity: Structural models of similarities and differences. In Clegg, C. C., *Sociological methodology, 17,* 145ff.

Howell, D. C. (1987). *Statistical methods for psychology* (2nd ed.). Boston: PWS-KENT.

Howell, D. C. (1997). *Statistical methods for psychology* (4th ed.). Pacific Grove, CA: Duxbury.

Howell, D. C. (2004). *Fundamental statistics for the behavioral sciences* (5th ed.). Pacific Grove, CA: Duxbury.

Howell, D. C., & Gordon, L. R. (1976). Computing the exact probability of an R × C contingency table with fixed marginal totals. *Behavior Research Methods and Instrumentation, 8,* 317.

Howell, D. C., & Huessy, H. R. (1981). Hyperkinetic behavior followed from 7 to 21 years of age. In M. Gittelman (Ed.), *Intervention strategies with hyperactive children* (pp. 201–214). White Plains, NY: M. E. Sharpe.

Howell, D. C., & Huessy, H. R. (1985). A fifteen-year follow-up of a behavioral history of Attention Deficit Disorder (ADD). *Pediatrics, 76,* 185–190.

Howell, D. C., & McConaughy, S. H. (1982). Nonorthogonal analysis of variance: Putting the question before the answer. *Educational and Psychological Measurement, 42,* 9–24.

Hraba, J., & Grant, G. (1970). Black is beautiful: A reexamination of racial preference and identification. *Journal of Personality and Social Psychology, 16,* 398–402.

Huberty, C. J. (1989). Problems with stepwise methods–better alternatives. In B. Thompson (Ed.), *Advances in social science methodology* (Vol. 1) (43–70). Greenwich, CT: JAI Press.

Huitema, B. E. (1980). *The analysis of covariance and alternatives.* New York: Wiley.

Huitema, B. E. (2005). Analysis of covariance. In Everitt, B. S., & Howell, D. C. *Encyclopedia of Statistics in Behavioral Science.* Chichester, England: Wiley.

Hunter, J. E. (1997). Needed: A ban on the significance test. *Psychological Science, 8,* 3–7.

Huynh, H., & Feldt, L. S. (1970). Conditions under which mean square ratios in repeated measurement designs have exact F distributions. *Journal of the American Statistical Association, 65,* 1582–1589.

Huynh, H., & Feldt, L. S. (1976). Estimation of the Box correction for degrees of freedom from sample data in the randomized block and split plot designs. *Journal of Educational Statistics, 1,* 69–82.

Huynh, H., & Mandeville, G. K. (1979). Validity conditions in repeated measures designs. *Psychological Bulletin, 86,* 964–973.

Introini-Collison, I., & McGaugh, J. L. (1986). Epinephrine modulates long-term retention of an aversively-motivated discrimination task. *Behavioral and Neural Biology, 45,* 358–365.

Jaccard, J., Turrisi, R., & Wan, C. K. (1990). *Interaction effects in multiple regression.* Newbury Park, CA: Sage.

Jones, L. V., & Tukey, J. W. (2000). A sensible formulation of the significance test. *Psychological Methods, 5,* 411–414.

Judd, C. M., & McClelland, G. H. (1989). *Data analysis: a model comparison approach.* San Diego, CA: Harcourt Brace Jovanovich.

Judd, C. M., McClelland, G. H., & Culhane, S. E. (1995). Data analysis: Continuing issues in the everyday analysis of psychological data. *Annual Review of Psychology, 46,* 433–435.

Kapp, B., Frysinger, R., Gallagher, M., & Hazelton, J. (1979). Amygdala central nucleus lesions: Effects on heart rate conditioning in the rabbit. *Physiology and Behavior, 23,* 1109–1117.

Katz, S., Lautenschlager, G. J., Blackburn, A. B., & Harris, F. H. (1990). Answering reading comprehension items without passages on the SAT. *Psychological Science, 1,* 122–127.

Kaufman, L., & Rock, I. (1962). The moon illusion, I. *Science, 136,* 953–961.

Kaufman, L., & Kaufman, J. (2000). Explaining the Moon illusion. *Proceedings of the National Academy of Sciences, 97,* 500–504.

Kendall, M. G. (1948). *Rank correlation methods.* London: Griffin.

Kennedy, J. J. (1983). *Analyzing qualitative data: Introductory loglinear analysis for behavioral research.* New York: Praeger.

Kenny, D. A., & Judd, C. M. (1986). Consequences of violating the independence assumption in analysis of variance. *Psychological Bulletin, 99,* 422–431.

Keppel, G. (1973). *Design and analysis: A researcher's handbook.* Englewood Cliffs, NJ: Prentice-Hall.

Keselman, H. J., Games, P. A., & Rogan, J. C. (1979). Protecting the overall rate of Type I errors of pairwise comparisons with an omnibus test statistic. *Psychological Bulletin, 86,* 884–888.

Keselman, H. J., Holland, B., & Cribbie, R. A. (2005). Multiple comparison procedures. In Everitt, B. S., & Howell, D. C. *Encyclopedia of Statistics in Behavioral Sciences.* Chichester, England: Wiley.

Keselman, H. J., & Keselman, J. C. (1988). Repeated measures multiple comparison procedures: Effects of violating multisample sphericity in unbalanced designs. *Journal of Educational Statistics, 13,* 215–226.

Keselman, H. J., & Rogan, J. C. (1977). The Tukey multiple comparison test: 1953–1976. *Psychological Bulletin, 84,* 1050–1056.

Keselman, H. J., Rogan, J. C., Mendoza, J. L., & Breen, L. J. (1980). Testing the validity conditions of repeated measures F tests. *Psychological Bulletin, 87,* 479–481.

King, D. A. (1986). Associative control of tolerance to the sedative effects of a short-acting benzodiazepine. Unpublished doctoral dissertation, University of Vermont.

Kirk, R. E. (1968). *Experimental design: Procedures for the behavioral sciences.* Belmont, CA: Brooks/Cole.

Klemchuk, H. P., Bond, L. A., & Howell, D. C. (1990). Coherence and correlates of level 1 perspective taking in young children. *Merrill-Palmer Quarterly, 36,* 369–387.

Kline, R. B. (2004). *Beyond significance testing.* Washington, DC: American Psychological Association.

Koele, P. (1982). Calculating power in analysis of variance. *Psychological Bulletin, 92,* 513–516.

Kohr, R. L., & Games, P. A. (1974). Robustness of the analysis of variance, the Welch procedure, and a Box procedure to heterogeneous variances. *Journal of Experimental Education, 43,* 61–69.

Landwehr, J. M., & Watkins, A. E. (1987). *Exploring data.* Palo Alto, CA: Dale Seymour Publications.

Lane, D. M., & Dunlap, W. P. (1978). Estimating effect size: Bias resulting from the significance criterion in editorial decisions. *British Journal of Mathematical and Statistical Psychology, 31,* 107–112.

Langlois, J. H., & Roggman, L. A. (1990). Attractive faces are only average. *Psychological Science, 1,* 115–121.

Larzelere, R. E., & Mulaik, S. A. (1977). Single-sample tests for many correlations. *Psychological Bulletin, 84,* 557–569.

Latané, B., & Dabbs, J. M., Jr. (1975). Sex, group size, and helping in three cities. *Sociometry, 38,* 180–194.

Leerkes, E., & Crockenberg, S. (1999). The development of maternal self-efficacy and its impact on maternal behavior. Poster presentation at the Biennial Meetings of the Society for Research in Child Development, Albuquerque, NM, April.

Lehmann, E. L. (1993). The Fisher, Neyman-Pearson theories of testing hypotheses: One theory or two? *Journal of the American Statistical Association, 88,* 1242–1249.

Lenth, R. V. (2001). Some practical guidelines for effective sample size determination. *American Statistician, 55,* 187–193.

Levene, H. (1960). Robust tests for the equality of variance. In I. Olkin (Ed.), *Contributions to probability and statistics.* Palo Alto, CA: Stanford University Press.

Lewis, C., & Keren, G. (1977). You can't have your cake and eat it too: Some considerations of the error term. *Psychological Bulletin, 84,* 1150–1154.

Lewis, D., & Burke, C. J. (1949). The use and misuse of the chi square test. *Psychological Bulletin, 46,* 433–489.

Little, R. J. A., & Rubin, D. B. (1987). *Statistical analysis with missing data.* New York: Wiley.

Lord, F. M. (1953). On the statistical treatment of football numbers. *American Psychologist, 8,* 750–751.

Lord, F. M. (1967). A paradox in the interpretation of group comparisons. *Psychological Bulletin, 68,* 304–305.

Lord, F. M. (1969). Statistical adjustments when comparing preexisting groups. *Psychological Bulletin, 72,* 336–337.

Lunneborg, C. E. (1994). *Modeling experimental and observational data.* Belmont, CA: Duxbury.

Lunneborg, C. E. (2000). *Data analysis by resampling: Concepts and applications.* Pacific Grove, CA: Duxbury.

Macnaughton, D. B. (1998). Which sums of squares are best in unbalanced analysis of variance. Unpublished paper available at http://www.matstat.com/ss/

MacKinnon, D. P., Lockwood, C. M., Hoffman, J. M., West, S. G., & Sheets, V. (2002). A comparison of methods to test mediation and other intervening variable effects. *Psychological Methods, 7,* 83–104.

Manly, B. F. J. (1997). *Randomization, bootstrap, and Monte Carlo methods in biology* (2nd ed.). London: Chapman & Hall.

Manning, C. A., Hall, J. L., & Gold, P. E. (1990). Glucose effects on memory and other neuropsychological tests in elderly humans. *Psychological Science, 1,* 307–311.

Marascuilo, L. A., & Busk, P. L. (1987). Loglinear models: A way to study main effects and interactions for multi-dimensional contingency tables with categorical data. *Journal of Counseling Psychology, 34,* 443–455.

Marascuilo, L. A., & Serlin, R. C. (1990). *Statistical methods for the social and behavioral sciences.* New York: Freeman.

Maris, E. (1998). Covariance adjustment versus gain scores— revisited. *Psychological Methods, 3,* 309–327.

Mauchly, J. W. (1940). Significance test for sphericity of a normal n-variate distribution. *Annals of Mathematical Statistics, 11,* 204–209.

Maxwell, A. E. (1961). *Analyzing Quantitative Data.* London: Methuen.

Maxwell, S. E. (1980). Pairwise multiple comparisons in repeated measures designs. *Journal of Educational Statistics, 5,* 269–287.

Maxwell, S. E., & Cramer, E. M. (1975). A note on analysis of covariance. *Psychological Bulletin, 82,* 187–190.

Maxwell, S. E., & Delaney, H. D. (1990). *Designing experiments and analyzing data: A model comparison approach.* Belmont, CA: Wadsworth.

McClelland, G. H. (1997). Optimal design in psychological research. *Psychological Methods, 2,* 3–19.

McClelland, G. H., & Judd, C. M. (1993). Statistical difficulties of detecting interactions and moderator effects. *Psychological Bulletin, 114,* 376–390.

McConaughy, S. H. (1980). Cognitive structures for reading comprehension: Judging the relative importance of ideas in short stories. Unpublished doctoral dissertation, University of Vermont.

McIntyre, S. H., Montgomery, D. B., Srinwason, V., & Weitz, B. A. (1983). Evaluating the statistical significance of models developed by stepwise regression. *Journal of Marketing Research, 10,* 1–11.

McNemar, Q. (1969). *Psychological statistics* (4th ed.). New York: Wiley.

Miller, R. G., Jr. (1981). *Simultaneous statistical inference* (2nd ed.). New York: McGraw-Hill.

Mireault, G. C. (1990). Parent death in childhood, perceived vulnerability, and adult depression and anxiety. Unpublished M.A. thesis, University of Vermont.

Mood, A. M. (1950). *Introduction to the theory of statistics.* New York: McGraw-Hill.

Mood, A. M., & Graybill, F. A. (1963). *Introduction to the theory of statistics* (2nd ed.). New York: McGraw-Hill.

Mooney, C. Z., & Duval, R. D. (1993). *Bootstrapping: A nonparametric approach to statistical inference.* Newbury Park, CA: Sage.

Moore, D. S., & McCabe, G. P. (1989). *Introduction to the practice of statistics.* New York: Freeman.

Murphy, K. R., & Myors, B. (1999). *Statistical power analysis: A simple and general model for traditional and modern hypothesis tests.* Mahwah, NJ: Lawrence Erlbaum.

Myers, J. L. (1979). *Fundamentals of experimental design* (3rd ed.). Boston: Allyn & Bacon.

Neter, J., & Wasserman, W. (1974). *Applied linear statistical models.* Homewood, IL: Richard D. Irwin.

Newsweek: Crowley, G. (1991). Can sunshine save your life? December 30, p. 56.

Neyman, J., & Pearson, E. S. (1933). On the problem of the most efficient tests of statistical hypotheses. *Philosophic Transactions of the Royal Society of London (Series A), 231,* 289–337.

Nickerson, R. S. (2000). Null hypothesis significance testing: A review of an old and continuing controversy. *Psychological Methods, 5,* 241–301.

Nie, N. H., Hull, C. H., Jenkins, J. G., Steinbrenner, K., & Bent, D. H. (1975). *SPSS Statistical Package for the Social Sciences.* New York: McGraw-Hill.

Norton, D. W. (1953). Study reported in E. F. Lindquist, *Design and analysis of experiments in psychology and education.* New York: Houghton Mifflin.

Norusis, M. J. (1985). SPSS^X *advanced statistics guide.* New York: McGraw-Hill.

Nurcombe, B., & Fitzhenry-Coor, I. (1979). Decision making in the mental health interview: I. An introduction to an education and research problem. Paper delivered at the Conference on Problem Solving in Medicine, Smuggler's Notch, Vermont.

Nurcombe, B., Howell, D. C., Rauh, V. A., Teti, D. M., Ruoff, P., & Brennan, J. (1984). An intervention program for mothers of low birthweight infants: Preliminary results. *Journal of the American Academy of Child Psychiatry, 23,* 319–325.

Oakes, M. (1990). *Statistical inference.* Chestnut Hill, MA: Epidemiology Resources.

O'Brien, R. G. (1976). Comment on "Some problems in the nonorthogonal analysis of variance." *Psychological Bulletin, 83,* 72–74.

O'Brien, R. G. (1981). A simple test for variance effects in experimental designs. *Psychological Bulletin, 89,* 570–574.

O'Brien, R. G., & Kaiser, M. K. (1985). MANOVA method for analyzing repeated measures designs: An extensive primer. *Psychological Bulletin, 97,* 316–333.

O'Grady, K. E. (1982). Measures of explained variation: Cautions and limitations. *Psychological Bulletin, 92,* 766–777.

O'Neil, R., & Wetherill, G. B. (1971). The present state of multiple comparison methods. *Journal of the Royal Statistical Society (Series B), 33,* 218–250.

Overall, J. E. (1972). Computers in behavioral science: Multiple covariance analysis by the general least squares regression method. *Behavioral Science, 17,* 313–320.

Overall, J. E. (1980). Power of chi-square tests for 2×2 contingency tables with small expected frequencies. *Psychological Bulletin, 87,* 132–135.

Overall, J. E., & Klett, C. J. (1972). *Applied multivariate analysis.* New York: McGraw-Hill.

Overall, J. E., & Spiegel, D. K. (1969). Concerning least squares analysis of experimental data. *Psychological Bulletin, 72,* 311–322.

Overall, J. E., Spiegel, D. K., & Cohen, J. (1975). Equivalence of orthogonal and nonorthogonal analysis of variance. *Psychological Bulletin, 82,* 182–186.

Ozer, D. J. (1985). Correlation and the coefficient of determination. *Psychological Bulletin, 97,* 307–315.

Pearson, K. (1900). On a criterion that a given system of deviations from the probable in the case of a correlated system of variables is such that it can reasonably be supposed to have arisen in random sampling. *Philosophical Magazine, 50,* 157–175.

Preacher, K. J., & Hayes, A. F. (2004). SPSS and SAS procedures for estimating indirect effects in simple mediation models. *Behavior research methods, instruments, and computers, 36,* 717–737.

Prentice, D. A., & Miller, D. T. (1992). When small effects are impressive. *Psychological Bulletin, 112,* 160–164.

Pugh, M. D. (1983). Contributory fault and rape convictions: Loglinear models for blaming the victim. *Social Psychology Quarterly, 46,* 233–242.

Raghunathan, T. E., Rosenthal, R., & Rubin, D. B. (1996). Comparing correlated but non-overlapping correlations, *Psychological Methods, 1,* 178–183.

Rasmussen, J. L. (1987). Estimating correlation coefficients: Bootstrap and parametric approaches. *Psychological Bulletin, 101,* 136–139.

Reichardt, C. S. (1979). The statistical analysis of data from nonequivalent control group designs. In T. D. Cook & D. T. Campbell (Eds.), *Quasi-experimentation: Design and analysis issues for field settings* (pp. 147–205). Boston: Houghton Mifflin.

Reilly, T. P., Drudge, O. W., Rosen, J. C., Loew, D. E., & Fischer, M. (1985). Concurrent and predictive validity of the WISC-R, McCarthy Scales, Woodcock-Johnson, and academic achievement. *Psychology in the Schools, 22,* 380–382.

Reynolds, C. R., & Richmond, B. O. (1978). What I think and feel: A revised measure of children's manifest anxiety. *Journal of Abnormal Child Psychology, 6,* 271–280.

Robson, D. S. (1959). A simple method for constructing orthogonal polynomials when the independent variable is unequally spaced. *Biometrics, 15,* 187–191.

Rogers, R. W., & Prentice-Dunn, S. (1981). Deindividuation and anger-mediated aggression: Unmasking regressive racism. *Journal of Personality and Social Psychology, 41,* 63–73.

Rosenthal, R. (1990). How are we doing in soft psychology? Comment in *American Psychologist, 45,* 775–777.

Rosenthal, R. (1994). Parametric measures of effect size. In H. Cooper, & L. V. Hedges (Eds), *The handbook of research synthesis.* New York: Russell Sage Foundation.

Rosenthal, R., & Rubin, D. B. (1982). A simple, general purpose display of magnitude of experimental effect. *Journal of Educational Psychology, 74,* 166–169.

Rosenthal, R., & Rubin, D. B. (1984). Multiple contrasts and ordered Bonferroni procedures. *Journal of Educational Psychology, 76,* 1028–1034.

Rosnow, R. L., & Rosenthal, R. (1989). Definition and interpretation of interaction effects. *Psychological Bulletin, 105,* 143–146.

Rousseeuw, P. J., & Leroy, A. M. (1987). *Robust regression and outlier detection.* New York: Wiley.

Ruback, R. B., & Juieng, D. (1997). Territorial defense in parking lots: Retaliation against waiting drivers. *Journal of Applied Social Psychology, 27,* 821–834.

Ryan, B. F., Joiner, B. L., & Ryan, T. A. (1985). *Minitab handbook* (2nd ed.). Boston: Duxbury Press.

Ryan, T. A. (1959). Multiple comparisons in psychological research. *Psychological Bulletin, 56,* 26–47.

Ryan, T. A. (1960). Significance tests for multiple comparisons of proportions, variances, and other statistics. *Psychological Bulletin, 57,* 318–328.

Saint-Exupery, A. de (1943). *The little prince.* Tr. by Woods, K. New York: Harcourt Brace Jovanovich.

Satterthwaite, F. E. (1946). An approximate distribution of estimates of variance components. *Biometrics Bulletin, 2,* 110–114.

Scheffé, H. A. (1953). A method for judging all possible contrasts in the analysis of variance. *Biometrika, 40,* 87–104.

Scheffé, H. A. (1959). *The analysis of variance.* New York: Wiley.

Sedlmeier, P., & Gigerenzer, G. (1989). Do studies of statistical power have an effect on the power of studies? *Psychological Bulletin, 105,* 309–316.

Seligman, M. E. P., Nolen-Hoeksema, S., Thornton, N., & Thornton, K. M. (1990). Explanatory style as a mechanism of disappointing athletic performance. *Psychological Science, 1,* 143–146.

Sgro, J. A., & Weinstock, S. (1963). Effects of delay on subsequent running under immediate reinforcement. *Journal of Experimental Psychology, 66,* 260–263.

Shrout, P. E., & Fleiss, J. L. (1979). Intraclass correlations: Uses in assessing rater reliability. *Psychological Bulletin, 86,* 420–428.

Sĭdák, Z. (1967). Rectangular confidence regions for the means of multivariate normal distributions. *Journal of the American Statistical Association, 62,* 623–633.

Siegel, S. (1975). Evidence from rats that morphine tolerance is a learned response. *Journal of Comparative and Physiological Psychology, 80,* 498–506.

Simon, J. L., & Bruce, P. *Resampling Stats.* Software available at http://www.resample.com.

Smith, H. F. (1957). Interpretation of adjusted treatment means and regression in analysis of covariance. *Biometrics, 13,* 282–308.

Smith, M. L., & Glass, G. V. (1977). Meta-analysis of psychotherapy outcome studies. *American Psychologist, 32,* 752–760.

Sobel, M. E. (1982). Asymptotic confidence intervals for indirect effects in structural equation models. In S. Leinhart (Ed.), *Sociological methodology.* pp. 290–312. San Francisco: Jossey-Bass.

Spilich, G. J., June, L., & Renner, J. (1992). Cigarette smoking and cognitive performance. *British Journal of Addiction 87,* 1313–1326.

St. Lawrence, J. S., Brasfield, T. L., Shirley, A., Jefferson, K. W., Alleyne, E., & O'Bannon, R. E. III. (1995). Cognitive-behavioral intervention to reduce African American adolescents' risk for HIV infection. *Journal of Consulting and Clinical Psychology, 63,* 221–237.

St. Leger, A. S., Cochrane, A. L., & Moore, F. (1978). The anomaly that wouldn't go away. *Lancet, ii,* 1153.

Steiger, J. H. (1980). Tests for comparing elements of a correlation matrix. *Psychological Bulletin, 87,* 245–251.

Sternberg, S. (1966). High speed scanning in human memory. *Science, 153,* 652–654.

Stevens, J. (1992). *Applied multivariate statistics for the social sciences.* Hillsdale, NJ: Erlbaum.

Stevens, S. S. (1951). Mathematics, measurement, and psychophysics. In S. S. Stevens (Ed.), *Handbook of experimental psychology* (pp. 1–49). New York: Wiley.

Stone, W. S., Rudd, R. J., Ragozzino, M. E., & Gold, P. E. (1992). Glucose attenuation of deficits in memory retrieval in altered light/dark cycles. *Psychobiology, 20,* 47–50.

Stuetzle, W. (2005). Cross-validation. In Everitt, B. S., & Howell, D. C. *Encyclopedia of Statistics in Behavioral Sciences.* Chichester, England: Wiley.

Tabachnick, B. G., & Fidell, L. S. (1996). *Using multivariate statistics* (3rd ed.). New York: Harper & Row.

Tiku, M. L. (1967). Tables of the power of the *F* test. *Journal of the American Statistical Association, 62,* 525–539.

Tolman, E. C., Ritchie, B. F., & Kalish, D. (1946). Studies in spatial learning: I. Orientation and the short cut. *Journal of Experimental Psychology, 36,* 13–24.

Tomarken, A. J., & Serlin, R. C. (1986). Comparison of ANOVA alternatives under variance heterogeneity and specific noncentrality structures. *Psychological Bulletin, 99,* 90–99.

Toothaker, L. (1991). *Multiple comparisons for researchers.* Newbury Park, CA: Sage.

Tufte, E. R. (1983). *The visual display of quantitative information.* Cheshire, CT: Graphics Press.

Tukey, J. W. (1949). One degree of freedom for nonadditivity. *Biometrics, 5,* 232–242.

Tukey, J. W. (1953). The problem of multiple comparisons. Unpublished manuscript, Princeton University.

Tukey, J. W. (1977). *Exploratory data analysis.* Reading, MA: Addison-Wesley.

Tversky, A., & Kahneman, C. (1980). Causal schemata in judgments under uncertainty. In Fishbein, M., (Ed.), *Progress in Social Psychology,* Vol. 1. Hillsdale, NJ: Erlbaum.

U.S. Department of Commerce. (1977). *Social indicators, 1976.* (Document #C3.2: S01/2/976). Washington, DC: U.S. Government Printing Office.

U.S. Department of Justice, Bureau of Justice Statistics. (1983). *Prisoners in 1982,* Bulletin NCJ-87933. Washington, DC: U.S. Government Printing Office.

Vaughan, G. M., & Corballis, M. C. (1969). Beyond tests of significance: Estimating strength of effects in selected ANOVA designs. *Psychological Bulletin, 72,* 204–223.

Verdooren, L. R. (1963). Extended tables of critical values for Wilcoxon's test statistic. *Biometrika, 50,* 177–186.

Vermont Department of Health (1982). *1981 annual report of vital statistics in Vermont.* Burlington, VT.

Visintainer, M. A., Volpicelli, J. R., & Seligman, M. E. P. (1982). Tumor rejection in rats after inescapable or escapable shock. *Science, 216,* 437–439.

Von Eye, A., Bogat, G. A., & Von Weber, S. F. (2005). Goodness of fit for categorical variables. In B. S. Everitt & D. C. Howell, *Encyclopedia of Statistics in Behavioral Sciences.* Chichester, England: Wiley.

Wagner, B. M., Compas, B. E., & Howell, D. C. (1988). Daily and major life events: A test of an integrative model of psychosocial stress. *American Journal of Community Psychology, 61,* 189–205.

Wainer, H. (1976). Estimating coefficients in linear models: It don't make no nevermind. *Psychological Bulletin, 83,* 213–217.

Wainer, H. (1978). On the sensitivity of regression and regressors. *Psychological Bulletin, 85,* 267–273.

Wainer, H. (1984). How to display data badly. *American Statistician, 38,* 137–147.

Watkins, A. E. (1995). The law of averages. *Chance, 8,* 28–32.

Weaver, K. A. (1999). The statistically marvelous medical growth chart: A tool for teaching variability. *Teaching of Psychology, 26,* 284–286.

Weinberg, C. R., & Gladen, B. C. (1986). The beta-geometric distribution applied to comparative fecundability studies. *Biometrics, 42,* 547–560.

Weisberg, H. I. (1979). Statistical adjustments and uncontrolled studies. *Psychological Bulletin, 86,* 1149–1164.

Welch, B. L. (1938). The significance of the difference between two means when the population variances are unequal. *Biometrika, 29,* 350–362.

Welch, B. L. (1947). The generalization of Student's problem when several difference population variances are involved. *Biometrika, 34,* 29–35.

Welch, B. L. (1951). On the comparison of several mean values: An alternative approach. *Biometrika, 38,* 330–336.

Welkowitz, J., Ewen, R. B., & Cohen, J. (2000). *Introductory statistics for the behavioral sciences* (5th ed.). New York: Harcourt/Academic Press.

Welsch, R. E. (1977). Stepwise multiple comparison procedures. *Journal of the American Statistical Association, 72,* 566–575.

Werner, M., Stabenau, J. B., & Pollin, W. (1970). TAT methods for the differentiation of families of schizophrenics, delinquents, and normals. *Journal of Abnormal Psychology, 75,* 139–145.

Wickens, T. D. (1989). *Multiway contingency table analysis for the social sciences.* Hillsdale, NJ: Erlbaum.

Wilcox, R. R. (1986). Critical values for the correlated *t*-test when there are missing observations. *Communications in Statistics, Simulation and Computation. 15,* 709–714.

Wilcox, R. R. (1987a). New designs in analysis of variance. *Annual Review of Psychology, 38,* 29–60.

Wilcox, R. R. (1987b). *New statistical procedures for the social sciences.* Hillsdale, NJ: Erlbaum.

Wilcox, R. R. (1992). Why can methods for comparing means have relatively low power, and what can you do to correct the problem? *Current Directions in Psychological Science, 1,* 101–105.

Wilcox, R. R. (1993). Analyzing repeated measures or randomized block designs using trimmed means. *British Journal of Mathematical and Statistical Psychology, 46,* 63–76.

Wilcox, R. R. (1995). ANOVA: The practical importance of heteroscedastic methods, using trimmed means versus means, and designing simulation studies. *British Journal of Mathematical and Statistical Psychology, 48,* 99–114.

Wilcox, R. R. (2005). Trimmed means. In B. S. Everitt, & D. E. Howell, *Encyclopedia of Statistics in Behavioral Sciences.* Chichester, UK: Wiley.

Wilkinson, L. (1999). Statistical methods in psychology journals: Guidelines and explanations. *American Psychologist, 54,* 594–604.

Williams, E. J. (1959). The comparison of regression variables. *Journal of the Royal Statistical Society (Series B), 21,* 396–399.

Winer, B. J. (1962). *Statistical principles in experimental design.* New York: McGraw-Hill.

Winer, B. J. (1971). *Statistical principles in experimental design* (2nd ed.). New York: McGraw-Hill.

Winer, B. J., Brown, D. R., & Michels, K. M. (1991). *Statistical principles in experimental design* (3rd ed.). New York: McGraw-Hill.

Yates, F. (1934). Contingency tables involving small numbers and the χ^2 test. Supplement. *Journal of the Royal Statistical Society (Series B), 1,* 217–235.

Yuen, K. K., & Dixon, W. J. (1973). The approximate behavior and performance of the two-sample trimmed *t. Biometrika, 60,* 369–374.

Zuckerman, M., Hodgins, H. S., Zuckerman, A., & Rosenthal, R. (1993). Contemporary issues in the analysis of data. *Psychological Science, 4,* 49–53.

Zumbo, B. D., & Zimmerman, D. W. (2000). Scales of measurement and the relation between parametric and nonparametric statistical tests. In Thompson, B. (Ed.) *Advances in social science methodology, 6.* Greenwich, CT: JAI Press.

Answers to Exercises

I am supplying the answers to most of the odd-numbered exercises. Some have been omitted because they ask that you draw a figure or compare computer output to the results of hand calculation. Others have been omitted when the question simply asks you to apply computer software to a set of data. Either you will be able to do that, and will almost certainly get the correct answer, or you won't be able to set up the problem in the first place (in which case, the numerical answer is of no help). You will sometimes be frustrated because I have often omitted answers to the Discussion Questions. Very often there is no simple answer to these questions. On other occasions, there is a straightforward answer, but I want you to think about the problem a while and see what you come up with. Frequently, you will find much more of interest than the simple answer that I might give. I recognize that it is frustrating when you can't figure out what the exercise is getting at; I, too, hate those situations. But that's the nature of discussion questions. Students sometimes write and ask for answers to even-numbered items. I cannot supply those because many instructors do not want those answers available to their students.

Chapter 1

1.1 The entire student body of your college or university would be considered a population under any circumstances in which you want to generalize only to the student body of your school.

1.3 The students of your college or university are a nonrandom sample of U.S. students, for example, because not all U.S. students have an equal chance of being included in the sample.

1.5 Independent variables: first-grade students who attended kindergarten versus those who did not; seniors, masters, submasters, and juniors as categories of marathon runners. Dependent variables: social-adjustment scores assigned by first-grade teachers; time to run 26.2 miles.

1.7 Continuous variables: length of gestation; typing speed in words/minute; number of books in the library collection.

1.9 The planners of a marathon race would like to know the average times of senior, master, submaster, and junior runners so they can plan accordingly.

1.11 Categorical data examples: The number of Brown University students in an October 1984 referendum voting for and the number voting against the university's stockpiling suicide pills in case of nuclear disaster; the number of students in a small Midwestern college who are white, African American, Hispanic American, Asian, or other; one year after an experimental program to treat alcoholism, the number of participants who are "still on the wagon," "drinking without having sought treatment," or "again under treatment."

1.13 Children's scores in an elementary school could be reported numerically (a measurement variable), or the students could be categorized as Bluebirds (rating > 90), Robins (rating = 70 − 90), or Cardinals (rating < 70).

1.15 For adults of a given height and gender, weight is a ratio scale of body weight, but it is at best an ordinal scale of strength.

1.17 Speed is probably a much better index of motivation than of learning.

1.19 Examples: The final grade point average for low-achieving students taking courses that interest them could be compared with the averages of low-achieving students taking courses that do not interest them; the frequency of sexual intercourse could be compared for happily versus unhappily married couples.

1.21 An interesting study of the health effects of smoking in China can be found at http://www.berkeley.edu/news/media/releases/2005/09/04 smoking.shtml.

Chapter 2

2.1 (b) Unimodal and positively skewed.

2.3 The problem with making a stem-and-leaf display of the data in Exercise 2.1 is that almost all of the values fall on only two leaves if we use the usual tens' digits for stems. And things are not much better even if we double the number of stems.

2.11 The first quartile for males is approximately 77, whereas for females it is about 80. The third quartiles are nearly equal for males and females, with a value of 87.

2.13 The shape of the distribution of number of movies attended per month for the next 200 people you met would be positively skewed with a peak at 0 movies per month and a sharp dropoff to essentially the baseline by about 5 movies per month.

2.17 (a) 9, 10, 8 (b) 77 (c) $\sum_{i=1}^{10} X_i$

2.19 (a) $\left(\sum X\right)^2 = 5929; \sum X^2 = 657$

(b) 7.7 (c) the average, or the mean

2.21 (a) $\sum XY = (10)(9) + (8)(9) + \cdots + (7)(2)$
$= 460$

(b) $\sum X \sum Y = (77)(57) = 4389$

(c) $\dfrac{\sum XY - \dfrac{\sum X \sum Y}{N}}{N-1} = \dfrac{460 - \dfrac{4389}{10}}{9} = 2.344$

2.25 You could compare the reaction times for those cases in which the correct response was "Yes" and those cases in which it was "No." If we process information sequentially, the reaction times, on average, should be longer for the "No" condition than for the "Yes" condition because we would have to make comparisons against all stimuli in the comparison set. In the "Yes" condition we could stop as soon as we found a match.

2.27 For animals raised in a stable environment, there is little or no difference in immunity depending on Affiliation. However, for animals raised in an unstable environment, High Affiliation subjects showed much greater immunity than Low Affiliation subjects. Stability seems to protect against the negative effects of low affiliation.

2.29 There are any number of ways that these data could be plotted. Perhaps the simplest is to look at the change in the *percentages* of each ethnic group's representation from 1982 to 1991.

2.31 One way to look at these data is to plot the percentage of households headed by women and the family size separately against years. Notice that there is an uneven sampling of years.

2.33 The mean falls above the median.

2.35 Mean = 21.33; median = 21

2.41 Range = 30; variance = 20.214; standard deviation = 4.496

2.43 The two standard deviations are roughly the same, although the range for the children is about twice the range for the adults.

2.45 $\overline{X} \pm 2s_X^2 = 10.2 \pm 2(3.405) = 10.2 \pm 6.81 = 3.39$ to 17.01. This interval includes 96% of the scores.

2.49 -0.893, 0.535, -1.846, 0.535, -0.417, 1.012, 1.012, 0.059

2.53   For Exercise 2.1

$$\text{cv} = s/\overline{X} = 4.496/18.9 = 0.238$$

For Exercise 2.4

$$\text{cv} = s/\overline{X} = 3.405/10.2 = 0.334$$

The adult sample shows somewhat greater variability when its smaller mean is taken into account.

2.55   The answers are listed in a file on the book's website (www.uvm.edu/~dhowell/methods/) named BadCancr.txt.

2.57   10% Winsorized standard deviation of data in Table 2.6:

3.13 3.17 3.19 319 3.20 3.20 3.22 3.23 3.25 3.26
3.27 3.29 3.29 3.30 3.31 3.31 3.34 3.34 3.36 3.38

Ten percent Windsorizing would replace the two lowest observations with 3.19 and the two highest observations with 3.34. This leaves

3.19 3.19 3.19 3.19 3.20 3.20 3.22 3.23 3.25 3.26
3.27 3.29 3.29 3.30 3.31 3.31 3.34 3.34 3.34 3.34

The standard deviation of the Winsorized sample is 0.058, whereas the standard deviation of the original sample was 0.069.

2.59   A transformation will alter the shape of a distribution when it is a nonlinear transformation.

Chapter 3

3.1   (b) $-3, -2, -2, -1, -1, -1, 0, 0, 0, 0, 1, 1, 1, 2, 2, 3$
   (c) $-1.84, -1.23, -1.23, -0.61, -0.61, -0.61, 0, 0,$
         $0, 0, 0.61, 0.61, 0.61, 1.23, 1.23, 1.84$

3.3   (a)   68%   (b)   50%   (c)   84%

3.5 $z = (950 - 975)/15 = -1.67$; only 4.75% of the time would we expect a count as low as 950, given what we know about the distribution. The two-tailed probability would be .095.

3.7   The answers to parts (b) and (c) of Exercise 3.6 will be equal when the two distributions have the same standard deviation.

3.9   (a)   \$2,512.68   (b)   \$1,342.00

3.11   Multiply the raw scores by 10/7 to raise the standard deviation to 10, and then add 11.43 points to each new score to bring the mean up to 80.

3.13 $z = (600 - 489)/126 = 0.88$. Therefore 81% of the scores fall below this, so 600 represents the 81st percentile.

3.15 $z = 0.79$, $p = .7852$; $X = 586.591$

For seniors and nonenrolled college graduates, a GRE score of 600 is at the 79th percentile, and a score of 587 would correspond to the 75th percentile.

3.17   The 75th percentile for GPA is 3.04.

3.19   The differences are caused by random sampling error because we are dealing with a sample and not a population.

3.21   I would expect that you would do reasonably well if you treated these as normally distributed, especially if you calculated a trimmed mean and a Winsorized standard deviation. The extreme salaries probably come from people who have either stayed at the rank of assistant professor for many years, possibly because they don't have the highest degree in their field, or those who have come to the university with considerable nonacademic experience.

Chapter 4

4.1   (a)   I set up the null hypothesis that last night's game was actually an NHL hockey game.
   (b)   On the basis of that hypothesis, I expected that each team would earn somewhere between 0 and 6 points. I then looked at the actual points and concluded that they were way out of line with what I would expect if this were an NHL hockey game. I therefore rejected the null hypothesis.

4.3   Concluding that I had been shortchanged when in fact I had not.

4.5   The critical value would be that amount of change below which I would decide that I had been shortchanged. The rejection region would be all amounts of change less than the critical value—that is, all amounts that would lead to rejection of H_0.

4.7 $z = (490 - 650)/50 = -3.2$. The probability that a student drawn at random from those properly admitted would have a GRE score as low as 490 is .0007. I suspect that the fact that his mother was a member of the board of trustees played a role in his admission.

4.9   The distribution would drop away smoothly to the right for the same reason that it always does—there are few high-scoring people. It would drop away steeply to the left because fewer of the borderline students would be admitted (no matter how high the borderline is set).

4.11 M is called a test statistic.

4.13   The alternative hypothesis is that this student was sampled from a population of students whose mean is not equal to 650.

4.15   The word "distribution" refers to the set of values obtained for any set of observations. The phrase "sampling distribution" is reserved for the distribution of outcomes (either theoretical or empirical) of a sample statistic.

4.17   (a)   *Research hypothesis:* Children who attend kindergarten adjust to first grade faster than those who do not. *Null hypothesis:* First-grade adjustment rates are equal for children who did and did not attend kindergarten.
   (b)   *Research hypothesis:* Sex education in junior high school decreases the rate of pregnancies among unmarried mothers in high school. *Null hypothesis:* The rate of pregnancies among unmarried mothers in high school is the same regardless of the presence or absence of sex education in junior high school.

4.19 For $\alpha = .01$, z must be -2.327. The cutoff score is therefore approximately 53.46. The corresponding value for z when a cutoff score of 53.46 is applied to the curve for H_1, is $z = -1.33$. From Appendix Power we find $\beta = .9082$.

4.21 To determine whether there is a true relationship between grades and course evaluations, I would find a statistic that reflected the degree of relationship between two variables. (You will see such a statistic [r] in Chapter 9.) I would then calculate the sampling distribution of that statistic in a situation in which there is no relationship between two variables. Finally, I would calculate the statistic for a representative set of students and classes, and compare my sample value with the sampling distribution of that statistic.

4.23 (a) You could draw a large sample of boys and a large sample of girls in the class and calculate the mean allowance for each group. The null hypothesis would be the hypothesis that the mean allowance, in the population, for boys is the same as the mean allowance, in the population, for girls.

(b) I would use a two-tailed test because I want to be able to reject the null hypothesis whether girls receive significantly more allowance or significantly less allowance than boys.

(c) I would reject the null hypothesis if the difference between the two sample means were greater than I could expect to find due to chance. Otherwise, I would not reject the null.

(d) The most important thing to do would be to have some outside corroboration for the amount of allowance reported by the children.

4.25 In the parking lot example, the traditional approach to hypothesis testing would test the null hypothesis that the mean time to leave a space is the same whether someone is waiting or not. If the statistical test failed to reject the null hypothesis the researchers would simply fail to reject the null hypothesis, and would do so at a two-tailed level of $\alpha = .05$. Jones and Tukey, on the other hand, would not consider that the null hypothesis of equal population means could possibly be true. They would focus on making a conclusion about which population mean is higher. A "nonsignificant result" would only mean that they didn't have enough data to draw any conclusion. Jones and Tukey would also be likely to work with a one-tailed $\alpha = .025$, but be actually making a two-tailed test because they would not have to specify a hypothesized direction of difference.

Chapter 5

5.1 (a) *Analytic:* If two tennis players are exactly evenly skillful—so that the outcome of their match is random, the probability is .50 that Player A will win his upcoming match.

(b) *Relative frequency:* If in past matches Player A has beaten Player B on 13 of the 17 occasions on which they played, then Player A has a probability of $13/17 = .76$ of winning their upcoming match.

(c) *Subjective:* Player A's coach feels that he has a probability of .90 of winning his upcoming match against Player B.

5.3 (a) $p = 1/9 = .111$ that you will win second prize given that you do not win first prize.

(b) $p = (2/10)(1/9) = (.20)(.111) = .022$ that he will win first and you second.

(c) $p = (1/10)(2/9) = (.111)(.20) = .022$ that you will win first and he second.

(d) p (you are first and he is second [$= .022$]) $+ p$(he is the first and you second [$= .022$]) $= p$(you and he will be first and second) $= .044$.

5.5 Conditional probabilities were involved in Exercise 5.3a.

5.7 *Conditional probabilities:* What is the probability that skiing conditions will be good on Wednesday, *given* that they are good today?

5.9 $p = (2/13)(3/13) = (.154)(.231) = .036$.

5.11 A continuous distribution for which we care about the probability of an observation's falling within some specified interval is exemplified by the probability that your baby will be born on its due date.

5.13 Two examples of discrete variables: variety of meat served at dinner tonight; brand of desktop computer owned.

5.15 (a) 20%, or 60 applicants, will fall at or above the 80th percentile and 10 of these will be chosen. Therefore p(that an applicant with the highest rating will be admitted) $= 10/60 = .167$.

(b) No one below the 80th percentile will be admitted, therefore p(that an applicant with the lowest rating will be admitted) $= .00$.

5.17 (a) $z = -.33$; p(larger portion) $= .6293$

(b) $29/55 = 53\% > 50$; $32/55 = 58\% \geq 50$.

5.19 Compare the probability of dropping out of school, ignoring the ADDSC score, with the conditional probability of dropping out given that ADDSC in elementary school exceeded some value (e.g., 66).

5.21 Probabilities of correct choices on trial 1 of a 5-choice task:

$p(0) = .1074$	$p(6) = .0055$
$p(1) = .2684$	$p(7) = .0008$
$p(2) = .3020$	$p(8) = .0001$
$p(3) = .2013$	$p(9) = .0000$
$p(4) = .0881$	$p(10) = .0000$
$p(5) = .0264$	

5.23 At $\alpha = .05$, as many as 4 correct choices indicate chance performance, but 5 or more correct choices would lead me to conclude that they are no longer performing at chance levels.

5.25 If there is no housing discrimination, then a person's race and whether or not they are offered a particular unit of housing are independent events. We could calculate the probability that a particular unit (or a unit in a particular section of the city) will be offered to anyone in a specific income group. We can also calculate the probability that the customer is a member of an ethnic minority. We can then calculate the probability of that person being shown the unit *assuming independence,* and compare that answer against the actual proportion of times a member of an ethnic minority was offered such a unit.

5.27 The number of subjects needed in the Exercise 5.26 verbal learning experiment if each subject can see only two of the four classes of words is the number of permutations of 4 things taken 2 at a time = $4!/2! = 12$.

5.29 The total number of ways of making ice cream cones = 63. (You can't have an ice cream cone without ice cream, so exclude the combination of *6* things taken 0 at a time.)

5.31 Because the probability of 11 correct by chance is .16, the probability of 11 *or more* correct must be greater than .16. Therefore, we cannot reject the hypothesis that $p = .50$ (the student is guessing) at $\alpha = .05$.

5.33 Driving test passed by 22 out of 30 drivers when 60% expected to pass:
$$z = \frac{22 - 30(.60)}{\sqrt{30(.60)(.40)}} = 1.49;$$ we cannot reject H_0 at $\alpha = .05$.

5.35 Students should come to understand that nature does not have a responsibility to make things come out even in the end, and that it has a terrible memory of what has happened in the past. Any "law of averages" refers to the results of a long-term series of events, and it describes what we would expect to see. It does not have any self-correcting mechanism built into it.

5.37 It is low because the probability of breast cancer is itself very low. But don't be too discouraged. Having collected some data (a positive mammography), the probability is 7.8 times higher than it would otherwise have been. (And if you are a woman, please don't stop having mammographies.)

Chapter 6

6.1 $\chi^2 = 11.33$ on 2 *df*; reject H_0 and conclude that students do not enroll at random.

6.3 $\chi^2 = 2.4$ on 4 *df*; do not reject H_0 that the child's sorting behavior is in line with the theory.

6.5 $\chi^2 = 29.35$ on 1 *df*; reject H_0 and conclude that the children did not choose dolls at random (at least with respect to color).

6.7 $\chi^2 = 34.184$ on 1 *df*; reject H_0 and conclude that the distribution of choices between black and white dolls

was different in the two studies. Choice is *not* independent of Study. We are no longer asking whether one color of doll is preferred over the other color, but whether the *pattern* of preference is constant across studies.

6.9 (a) Take a group of subjects at random and sort them by gender and by life style (categorized across three levels).
     (b) Deliberately take an equal number of males and females and ask them to specify a preference between three types of life styles.
     (c) Deliberately take 10 males and 10 females and have them divide themselves into two teams of 10 players each.

6.11 (a) $\chi^2 = 10.306$
      (b) This demonstration shows that the obtained value of χ^2 is exactly doubled, whereas the critical value remains the same. Thus, the sample size plays a very important role, with larger samples being more likely to produce significant results—as is also true with other tests.

6.13 $\chi^2 = 5.50$. Reject H_0 and conclude that women voted differently from men. Women were much more likely to vote for civil unions—the odds ratio is $(35/9)/(60/41) = 3.89/1.46 = 2.66$. The odds for women are 2.66 times those for men. That is a substantial difference and likely reflects fundamental differences in attitude.

6.15 $\chi^2 = 7.908$ on 2 *df*; reject H_0 and conclude that the number of bystanders influences whether or not subjects seek help.

6.17 (a) $\chi^2 = 37.141$ on 2 *df*; reject H_0 and conclude that adolescent girls' preferred weight varies with race.
      (b) The number of girls desiring to lose weight was far in excess of the number of girls who were really overweight.

6.19 Likelihood ratio $\chi^2 = 12.753$ on 7 *df*; do not reject H_0.

6.21 As the data are originally presented, chi-square would not be appropriate because the observations are not independent. The same subjects contributed twice to the data matrix.

6.23 (b) Row percents take entries as a percentage of row totals, whereas column percents take entries as a percentage of column totals.
      (c) These are the probabilities (to four decimal places) of a $\chi^2 \geq \chi^2_{obs}$, under H_0.
      (d) The correlation between the variables is approximately .25.

6.25 (a) Cramér's $V = \sqrt{26.903/22{,}071} = 0.0349$.
      (b) Odds Fatal | Placebo $= 18/10{,}845 = .00166$. Odds Fatal | Aspirin $= 5/10{,}933 = .000453$. Odds Ratio $= .00166/.000453 = 3.66$. The odds that you will die from a myocardial infarction if you do not take aspirin are 3.66 greater than if you do.

6.27 For Table 6.4, the odds ratios for smoking, as a function of gender $= (150/350)/(100/400) = 1.7$. The odds of smoking for men are 1.7 times greater than for women. This is a substantial difference. The odds of being the primary shopper as a function of gender $= (15/4)/(4/15) = 14.06$. The odds of women being the primary shopper are 14.06 times the odds for men. This gender difference is much more extreme than it was for smoking.

6.29 For the Dabbs and Morris (1990) study, $\chi^2 = 64.08$ on 1 *df*. We can reject H_0 and conclude that antisocial behavior in males is linked to testosterone levels.

6.31 (a) $\chi^2 = 15.57$ on 1 *df*. Reject H_0.
  (b) There is a significant relationship between high levels of testosterone in adult men and a history of delinquent behaviors during childhood.
  (c) This result shows that we can tie the two variables (delinquency and testosterone) together historically.

6.33 $\chi^2 = 9.79$. Reject H_0.
  (b) Odds ratio $= (43/457)/(50/268) = 0.094/.186 = .505$. Those who receive the program have about half the odds of subsequently suffering abuse.

6.35 (a) $\chi^2 = 0.232$, $p = .630$
  (b) There is no relationship between the gender of the parent and the gender of the child.
  (c) We would be unable to separate effects due to parent's gender from effects due to the child's gender. They would be completely confounded.

6.37 We could ask a series of similar questions, evenly split between "right" and "wrong" answers. We could then sort the replies into positive and negative categories and ask whether faculty were more likely than students to give negative responses.

Chapter 7

7.3 The mean and standard deviation of the sample are 4.10 and 2.82. The mean is surprisingly low, but the standard deviation is close when compared with the parameters of the population from which the sample was drawn (4.5 and 2.6, respectively). The mean of the distribution of means is 4.47, which is close to the population mean, and the standard deviation is 1.23.
  (a) The central limit theorem would predict a sampling distribution of the mean with a mean of 4.5 and a standard deviation of $2.6\sqrt{5} = 1.16$.
  (b) These values are close to what we would expect.

7.5 If you had drawn 50 samples of size 15, the mean of the sampling distribution should still approximate the mean of the population, but the standard error of that distribution would now be only $2.6\sqrt{15} = 0.67$.

7.7 First, these students scored better than we might have predicted, not worse. Second, these students are certainly not a random sample of high school students. Finally, there is no definition of what is meant by "a terrible state," nor any idea of whether or not the SAT measures such a concept.

7.9 This answer differs substantially from Exercise 7.8 because the sample sizes are so different. I deliberately sought examples where the means were nearly the same, but with that large difference in sample size, so the resulting z values, and associated probabilities, are very different.

7.11 Mean gain $= 3.01$, standard deviation $= 7.3$. $t = 2.22$. With 28 *df*, the critical value $= 2.048$, so we will reject the null hypothesis and conclude that the girls gained at better than chance levels.

7.13 (a) $t = 20.70$ on 27 *df*. We can reject the null hypothesis.
  (b) This does not address the question of whether the SAT is a valid measure, but it does show that people who do well at guessing at answers also do well on the SAT. This is not very surprising.

7.15 $CI_{.95} = 3.51 \le \mu \le 5.27$. An interval formed as this one was has a probability of .95 of encompassing the mean of the population. Because this interval includes the hypothesized population mean, it is consistent with the results in Exercise 7.14.

7.17 We used a matched-sample t test in Exercise 7.16 because the data were paired in the sense of coming from the same subject. Some subjects showed generally more beta-endorphins at both times than others, and we wanted to eliminate this subject-to-subject variability that had nothing to do with stress. Actually, there isn't much of a relationship between the two measures, but we can't fairly ignore it after the fact.

7.19 The means for males and females were 2.73 and 2.79 respectively, producing a $t = -0.485$ on 90 *df*. We cannot reject the null hypothesis that males and females are equally satisfied. A matched-sample t is appropriate because it would not seem reasonable to assume that the sexual satisfaction of a husband is independent of that of his wife.

7.21 The correlation between the scores of husbands and wives was .334, which is significant, and which confirms the assumption that the scores would be related.

7.23 The important question is what would the sampling distribution of the mean (or differences between means) look like, and with 91 pairs of scores, that sampling distribution would be substantially continuous with a normal distribution of means.

7.25 (a) We want to test the null hypothesis that the mean weight was the same before and after treatment.
  (b) $t = -4.185$ on 16 *df*, which tells us that there was a significant gain in weight over the course of therapy.

7.27 $CI_{.95} = -10.94 \le \mu \le -3.58$

7.29 (a) *Null hypothesis:* There is no significant difference in test scores between those who have read the passage and those who have not.

(b) *Alternative hypothesis:* There is a significant difference between the two conditions.

(c) $t = 8.89$ on 43 *df* if we pool the variances. This difference is significant.

(d) We can conclude that students do better on this test if they read the passage on which they are going to answer questions.

7.31 Girls in the Control group lost an average of 0.45 pounds, while girls in the Family Therapy group gained 7.26 pounds. A *t* on two independent groups $= -3.223$ on 41 *df*, which is significant. Family therapy led to significantly greater weight gain. (Variances were homogeneous.)

7.33 If those means had actually come from independent samples, we could not remove differences due to couples, and the resulting *t* would have been somewhat smaller.

7.35 The correlation was fairly low.

7.37 (a) I would assume that the experimenters expected that there would be more stories exhibiting positive parent-child relationships among the mothers or children in the Normal group.

(b) The means were 3.5 and 2.1 for the normal and schizophrenic groups, respectively, with $t = 2.66$ on 38 *df*, which is significant. The experimental hypothesis in (a) was supported.

7.39 It is just as likely that having a schizophrenic child might lead to deterioration in parent-child relationships. Because we can't assign children to groups at random, we cannot speak confidently about causation.

7.41 Confidence limits on Exercise 7.40: CI$_{.95} = 1.153 \pm (2.131)(1.965 = -3.03 \leq \mu \leq 5.34$). Because the confidence limits include 0, these results are in agreement with the previous nonsignificant result.

7.43 $t = 2.134$. Because the variances are very dissimilar, we could run a conservative test by using the smaller of $n_1 - 1$ and $n_2 - 1$ *df*, which would produce a nonsignificant result. We would have a nonsignificant result even if we used the full $n_1 + n_2 - 2$ *df*.

7.45 If you take the absolute differences between the observations and their group means and run a *t* test comparing the two groups on the absolute differences, you obtain $t = 0.625$. Squaring this, you have $F = 0.391$, which makes it clear that Levene's test in SPSS is operating on the absolute differences. (The *t* for squared differences would equal 0.213, which would give an *F* of 0.045.)

7.47 Women show higher levels of depression, but not of anxiety.

7.49 The effect size (d) was 0.62 using the standard deviation of weights before therapy. This indicates a gain of roughly 2/3 of a standard deviation over the course of therapy.

7.51 (a) The scale of measurement is important because if we rescaled the categories as 1, 2, 4, and 6, for example, we would have quite different answers.

(b) The first exercise asks if there is a relationship between the satisfaction of husbands and wives. The second simply asks if males (husbands) are more satisfied, on average, than females (wives).

(c) You could adapt the suggestion made in the text about combining the *t* on independent groups and the *t* on matched groups.

(d) I'm really not very comfortable with the *t* test because I am not pleased with the scale of measurement. An alternative would be a ranked test, but the number of ties is huge, and that probably worries me even more.

Chapter 8

8.1 (a) 0.250 (b) 2.50 (c) .71

8.3 $n = 99, 126$, and 169 (I have rounded up because *N is* always an integer.)

8.7 (a) For power $= .50$, $\delta = 1.95$ and $n = 15.21 \approx 16$
(b) For power $= .80$, $\delta = 2.80$ and $n = 31.36 \approx 32$

8.9 $d = .50$, $\delta = 1.46$, power $= .31$

8.11 *t* is numerically equal to δ, although *t* is calculated from statistics and δ is calculated from parameters. In other words, δ equals the *t* that you would get if the data exactly match what you think are the values of the parameters.

8.15 He should use the Dropout group. Assuming equal standard deviations, the H.S. dropout group of 25 would result in a higher value of δ and therefore higher power. (You can let σ be any value as long as it is the same for both calculations. Then calculate δ for each situation.)

8.17

Effect size	*d*	One-sample *t*	Two-sample *t*
Small	0.20	121	484
Medium	0.50	20	78
Large	0.80	8	32

8.19 Power will equal the probability of a Type II error when the mean of the distribution under H_1 lies at the critical value under H_0.

8.21 (a) I would not be assigning subjects to groups at random, and there might be differences between labs that would confound the results.

(b) I should pool my subjects and randomly assign them to conditions.

(c) Sex differences, if they exist, would confound the results. We would need to use a procedure (see Chapter 13) that separates any sex differences and looks for different patterns of results in males and females.

8.23 Both of these questions point to the need to design studies carefully so that the results are clear and interpretable.

Chapter 9

9.3 $r = .35$

9.7 $r = .99, .71,$ and $-.99$. Three arrangements will work: 2 8 6 4 or 6 4 2 8 or 6 2 8 4; $r = .14$ for each

9.9   (a) $d = .20, \delta = .98,$ power $= .17$
       (b) $N = 197$

9.11 $s_{Y.X} = 0.580$

9.13   If the high-risk fertility rate jumped to 70, we would predict 8.36% of infants would be LBW.

9.15   The predicted number of symptoms $= 80.156$.

9.17 $s'_{Y.X} = 17.563\sqrt{1 + \dfrac{1}{107} + \dfrac{(X_i - \overline{X})^2}{(N-1)s_X^2}}$

       $t_{\alpha/2} = 1.984$

       $CI(Y) = \hat{Y} \pm (t_{\alpha/2})s'_{Y.X}$

       You can calculate $\hat{Y}$ and $s'_{Y.X}$ for several different values of X, and then plot the results.

9.19   When the data are standardized, the slope equals r. Therefore, the slope will be less than 1 for all but the most trivial problems, and predicted deviations from the mean will be less than actual parental deviations.

9.21   For power $= .80, \delta = 2.80$. Therefore, $N = 50$.

9.23   (a) $z = 0.797$. The correlations are not significantly different.
       (b)   We do not have reason to argue that the relationship between performance and prior test scores is affected by whether or not the student read the passage.

9.25   It is difficult to tell whether the significant difference is to be attributable to the larger sample sizes or the higher (and thus more different) values of r'. It is likely to be the former.

9.27   (a) $r = .224, p = .509$. Do not reject H_0.
       (b)   The Irish are heavy smokers, but they certainly are not heavy drinkers compared with other regions.
       (c)   The inclusion of Northern Ireland distorts the data. If we leave them out, $r = .784, p = .007$, and there is a strong relationship between smoking and drinking.

9.29   (a)   See table below.

       (b)   All of these correlations are significant, showing that the symptoms are correlated with one another.

9.31   (b)   For a one-inch gain in height, we would expect a 4.356 pound gain in weight. Someone who is 0 inches tall would be expected to weight -149.934 pounds. The unreasonable answer reflects curvilinearity of the relationship at the extremes.
       (c)   The correlation is .604.
       (d)   Both the slope and the correlation are significant.

9.33   As a 5'8" male, my predicted weight is $\hat{Y} = 4.356(\text{Height}) - 149.934 = 4.356 \times 68 - 149.934 = 146.27$ pounds.
       (a)   I weigh 146 pounds. (Well, I did a few years ago.) Therefore, the residual in the prediction is $Y - \hat{Y} = 146 - 146.27 = -0.27$.
       (b)   If the students on which this equation is based under- or over-estimated their own height or weight, the prediction for my weight will be based on invalid data and will be systematically in error.

9.35   The male would be predicted to weigh 137.562 pounds, whereas the female would be predicted to weigh 125.354 pounds. The predicted difference between them would be 12.712 pounds.

9.37   Although the regression line has a slight positive slope, the slope is not significantly different from zero. The equation for the regression line is $\hat{Y} = 0.429X + 221.843$.

Chapter 10

10.1   (b) $r_{pb} = -.540; t = -2.72$
       (c)   Performance in the morning is significantly related to people's perception of their peak periods.

10.3   It looks as though morning people vary their performance across time, but evening people are uniformly poor performers.

10.5 $t = 2.725$. This is equal to the t test on r_{pb}.

10.7 $\hat{Y} = 0.202X + 0.093$; when $X = \overline{X} = 2.903$, $\hat{Y} = 0.68 = \overline{Y}$.

10.9   (b) $\phi = .256$ (c) $t = 1.27$, not significant.

10.11 (a) $\phi = .628$ (b) $\chi^2 = 12.62, p < .05$

	SomT	ObsessT	SensitT	DepressT	AnxT	HostT	PhobT	ParT	PsyT
ObsessT	0.482								
SensitT	0.377	0.539							
DepressT	0.400	0.599	0.654						
AnxT	0.569	0.621	0.550	0.590					
HostT	0.420	0.470	0.451	0.508	0.475				
PhobT	0.466	0.509	0.613	0.568	0.528	0.411			
ParT	0.400	0.524	0.677	0.621	0.547	0.494	0.540		
PsyT	0.334	0.503	0.625	0.725	0.509	0.404	0.529	0.651	
GSIT	0.646	0.791	0.770	0.820	0.786	0.633	0.679	0.766	0.741

10.13 (a) $\tau = .886$ (b) $z = 4.60$, $p < .05$.

10.15 $\tau = .733$

10.17 (b) An $r^2 = .0512$ would correspond to $\chi^2 = 10.24$. The closest you can come to this result is if the subjects were split 61/39 in the first condition and 39/61 in the second (rounding to integers.)

10.19 (a) $\chi^2 = 2.815$ ($p = .245$); $\phi_c = .087$.
    (b) The difference in Exercise 7.48 was not significant either. There the probability was .79.
    (c) This approach would be preferred to the approach used in Chapter 7 if you had reason to believe that differences in depression scores below the clinical cutoff were of no importance and should be ignored.

10.21 (a) If a statistic is not significant, that means that we have no reason to believe that it is reliably different from 0 (or whatever the parameter under H_0). Here we have no reason to believe that there is a relationship between the variables. Therefore, it cannot be important.
    (b) With the exception of issues of power, sample size will not make an effect more important than it is. It will simply increase the level of significance.

Chapter 11

11.1

Source	df	SS	MS	F
Group	2	2100.00	1050.000	40.127*
Error	15	392.50	26.167	
Total	17	2492.50		

*$p < .05$ $[F_{.05}(2, 15) = 3.68]$

11.3 (a)

Source	df	SS	MS	F
Group	3	1059.80	353.267	53.301*
Error	36	238.60	6.628	
Total	39	1298.40		

*$p < .05$ $[F_{.05}(3, 36) = 2.89]$

(b)

Source	df	SS	MS	F
Group	1	792.10	792.10	59.451*
Error	38	506.30	13.324	
Total	39	1298.40		

*$p < .05$ $[F_{.05}(1, 38) = 4.10]$

(c) The results are difficult to interpret because the error term now includes variance between younger and older participants. Moreover, we don't know if the levels of processing effect applies to both age groups.

11.5 (a)

Source	df	SS	MS	F
Group	1	224.583	224.583	18.8*
Error	20	238.917	11.946	
Total	21	463.500		

*$p < .05$ $[F_{.05}(1, 20) = 4.35]$

(b) t without pooled variance $= 4.27$ $t^2 = 18.2$.
(c) t with pooled variance $= 4.34$ $t^2 = 18.8$.
(d) The t with pooled variances is equivalent to the F in (a).

11.7 $\eta^2 = .816$; $\omega^2 = .796$

11.9 $\eta^2 = .182$; $\omega^2 = .120$. Eta-squared tells us that approximately 18% of the variability in the severity of symptoms can be accounted for by differences in treatment, whereas omega-squared tells us that a less biased estimate would be 12%. The F was significant, so both of these estimates are at better than chance levels.

11.11 The results are basically the same as ours, although we are presented with confidence limits on group means and r^2 (which is really η^2).

11.13 $X_{ij} = \mu + \tau_j + e_{ij}$ where μ is the grand mean, τ_j is the effect for the jth treatment, and e_{ij} is the unit of error for the ith subject in treatment j.

11.15 $X_{ij} = \mu + \tau_j + e_{ij}$ where μ is the grand mean, τ_j is the effect for the jth treatment, and e_{ij} is the unit of error for the ith subject in treatment j.

11.17

Source	df	SS	MS	F
Group	7	44.557	6.365	7.27*
Error	264	231.282	0.876	
Total	271	275.839		

*$p < .05$ $[F_{.05}(7, 264) = 2.06]$

11.21 $\eta^2 = .16$; $\omega^2 = .14$

11.23 Transforming time to speed involves a reciprocal transformation. The effect of the transformation is to decrease the relative distance between large values.

11.25 Parts of speech are a fixed variable because we deliberately chose which parts of speech to use. Words within the noun category are most likely random because we would probably choose our nouns at random (within certain constraints, such as the number of letters in the word). We would choose nouns at random because we care how people respond to nouns in general, not specifically to "house," "car," "tree," and so forth.

11.27 The $F = 4.48$, and we can reject the null hypothesis. This question addresses differences among all three groups, rather than simply pairwise differences, but I would be most interested in negative versus positive.

11.29–11.33 Computer and discussion questions.

Chapter 12

12.1 (a)

Source	df	SS	MS	F
Treatments	4	816.00	204.00	36.43**
1,2 vs 3,4,5	1	682.67	682.67	121.90**
1 vs 2	1	90.00	90.00	16.07**
3,4 vs 5	1	3.33	3.33	<1
3 vs 4	1	40.00	40.00	7.14*
Error	20	112.00	5.60	
Total	24	928.00		

$*p < .05$ $[F_{.05}(1, 20) = 4.35; F_{.05}(4, 20) = 2.87]$

$**p < .01$ $[F_{.01}(1, 20) = 8.10; F_{.01}(4, 20) = 4.43]$

(b) Orthogonality of contrasts:

Cross-products of coefficients:

$$\sum a_j b_j = (.5)(1) + (.5)(-1) + (.333)(0)$$
$$+ (.333)(0) + (.333)(0)$$
$$= 0$$

$$\sum a_j c_j = (.5)(0) + (.5)(0) + (.333)(.5)$$
$$+ (.333)(.5) + (.333)(-1)$$
$$= 0$$

$$\sum a_j d_j = (.5)(0) + (.5)(0) + (.333)(1)$$
$$+ (.333)(-1) + (.333)(0)$$
$$= 0$$

$$\sum b_j c_j = (1)(0) + (-1)(0) + (0)(.5)$$
$$+ (0)(.5) + (0)(-1)$$
$$= 0$$

$$\sum c_j d_j = (0)(0) + (0)(0) + (.5)(1)$$
$$+ (.5)(-1) + (1)(0)$$
$$= 0$$

(c) $682.67 + 90.00 + 3.33 + 40.00 = 816.00$

12.3 for $\alpha - .05$; PC $= \alpha$, FW $= 1 - (1 - \alpha)^2 = .0975$

12.5 $q = 7.101; t\sqrt{2} = 7.101$

12.7 $t'_1 = -5.861$ reject H_0. $t'_2 = -6.77$; reject H_0

12.9

Source	F	t	c	$t'_{.05}(20, c)$	Signif
1,2 vs 3,4,5	121.905	11.04	4	2.74	*
1 vs 2	16.071	4.01	3	2.61	*
3 vs 4	7.143	2.67	2	2.42	*
3,4 vs 5	<1	<1	1	2.09	

12.11 Tukey: $W_5 = W_4 = W_3 = W_2 = 3.973$. For this analysis, we have the same pattern of differences we had in Exercise 12.10.

12.13 Group 1 is different from all other groups. Groups 2, 3, and 4 are different from Group 5.

12.15 The variances are approximately equal, and so are the sample sizes, so we will use the harmonic mean of the n, which is 9.3264.

$$W_r = q_r \sqrt{\frac{MS_{error}}{\overline{n}_h}} = q_r \sqrt{\frac{240.35}{9.3264}} = q_r(5.0765)$$

The 0.5 μg group is different from the control and the 2 μg group. No other differences are significant. The maximum familywise error rate is .05.

12.17 Simply run the tests as standard t tests. Almost all software will give you the actual p value. Reject H_0 for each p value less than α/c.

12.19 $SS_{linear} = 0.0088; F = 0.145$; no significant linear trend. $SS_{quad} = 1.149; F = 18.99$; significant quadratic trend.

12.23 Trend analysis for Epineq.dat separately at each interval.

One Day: $F_{Linear} = 9.44$ ($p = .0042$); $F_{Quad} = 20.43$ ($p = .0001$)

One Week: $F_{Linear} = 4.33$ ($p = .0453$); $F_{Quad} = 13.23$ ($p = .0009$)

One Month: $F_{Linear} = 6.91$ ($p = .0129$); $F_{Quad} = 8.60$ ($p = .0061$)

12.25 (b) Using actual dose, $F_{Linear} = 0.548$;

(c) Using 1, 2, ... 6, $F_{Linear} = 11.03$.

(d) When we use the group number coding in our trend analysis we find a significant linear trend. As the dose of sucrose increases, memory increases accordingly.

Results for Exercise 12.15

	Control	2 μg	1 μg	0.1 μg	0.5 μg			
	34.00	38.10	48.50	50.80	60.33	r	q_r	W_r
34.00	⋯	4.1	14.5	16.8	26.33*	5	4.04	20.51
38.10		⋯	10.4	12.7	22.23*	4	4.04	20.51
48.50			⋯	2.3	11.83	3	4.04	20.51
50.80				⋯	9.53	2	4.04	20.51

(e) The choice of coding system in not always obvious. Using 1, 2, . . . , 6 actually ranks the dose levels and ignores the fact that dose increases in an extreme way. (In other words, the difference between the first 2 doses is 1 mg/kg, whereas the difference between the last two doses is 250 mg/kg. Using 1, 2, . . . , 6 deliberately ignores this relationship. Apparently the human body responds in a nonlinear way to the increase in actual dose levels.

12.27 Effect sizes for Exercise 12.1

$\psi_1 = 10.667 \quad \psi_2 = -6 \quad \psi_3 = -1 \quad \psi_4 = -4$

$MS_{error} = 5.60$

12.29 The contrast between the Positive and Negative mood conditions was significant ($t(27) = 3.045, p < .05$). This leads to an effect size of $d = \psi/\sqrt{MS_{error}} = 5.6/\sqrt{16.907} = 5.6/4.11 = 1.36$. The two groups differ by over 1 1/3 standard deviations. It is evident that inducing a negative mood leads to more checking behavior than introducing a positive mood. (If we had compared the Positive and No mood conditions, the difference would not have been significant. However, I had not planned to make that comparision.)

Chapter 13

13.1

Source	df	SS	MS	F
Parity	1	13.067	13.067	3.354
Size/Age	2	97.733	48.867	12.541*
$P \times S$	2	17.733	8.867	2.276
Error	54	210.400	3.896	
Total	59	338.933		

$*p < .05 \quad [F_{.05}(2, 54) = 3.17]$

13.3 The mean for these primiparous mothers would not be expected to be a good estimate of the mean for the population of all primiparous mothers because the sample is not representative of the population. For example, 50% of the population of primiparous mothers would not be expected to give birth to LBW infants.

13.5 Exercise 11.3 as a factorial

Source	df	SS	MS	F
Delay	2	188.578	94.289	3.22
Area	2	356.044	178.022	6.07*
$D \times A$	4	371.956	92.989	3.17*
Error	36	1055.200	29.311	
Total	44	1971.778		

$*p < .05 \quad [F_{.05}(2, 36) = 3.27; F_{.05}(4, 36) = 2.64]$

13.7 In Exercise 13.5, if A refers to Area:

$\hat{\alpha}_1 =$ the treatment effect for the Neutral site

$\overline{X}_{.1} - \overline{X}_{..} = 24.222 - 28.2 = 3.978$

13.9 Group N vs Group A: $t = 3.03$. Group N vs Group B: $t = 3.00$. With two t tests, each on 36 df (for MS_{error}) with $\alpha = .05/2 = .025$ (two-tailed), the critical value is ± 2.339. We would reject H_0 in each case.

13.11

Source	df	SS	MS	F
Age	1	115.60	115.60	17.44*
Level	1	792.10	792.10	119.51*
Age × Level	1	152.10	152.10	22.95*
Error	36	238.60	6.6278	
Total	39	1298.40		

$*p < .05 \quad [F_{.05}(1, 36) = 4.12]$

13.13 Made-up data with main effects but no interaction:

Cell means	8	12
	4	6

13.15 The interaction was of primary interest in an experiment by Nisbett in which he showed that obese people varied the amount of food they consumed depending on whether a lot or a little food was visible, while normal weight subjects ate approximately the same amount under the two conditions.

13.17

Source	df	SS	MS	F
Hospital	1	350.117	350.117	15.57*
Therapy	1	350.117	350.117	15.57*
$H \times T$	1	107.195	107.195	4.77
Error	11	247.417	22.492	
Total	14			

$*p < .05 \quad [F_{.05}(1, 11) = 4.84]$

13.19

$$\eta_P^2 = \frac{SS_{parity}}{SS_{total}} = \frac{13.067}{338.933} = .04$$

$$\eta_S^2 = \frac{SS_{size}}{SS_{total}} = \frac{97.733}{338.933} = .29$$

$$\eta_{PS}^2 = \frac{SS_{PS}}{SS_{total}} = \frac{17.733}{338.933} = .05$$

$$\omega_P^2 = \frac{SS_{parity} - (p-1)MS_{error}}{SS_{total} + MS_{error}}$$

$$= \frac{13.067 - (1)(3.896)}{338.933 + 3.896} = .03$$

$$\omega_S^2 = \frac{SS_{size} - (s-1)MS_{error}}{SS_{total} + MS_{error}}$$

$$= \frac{97.733 - (2)(3.896)}{338.933 + 3.896} = .26$$

$$\omega_{PS}^2 = \frac{SS_{PS} - (p-1)(s-1)MS_{error}}{SS_{total} + MS_{error}}$$

$$= \frac{17.733 - (1)(2)(3.896)}{338.933 + 3.896} = .03$$

13.21 Magnitude of effect for avoidance learning data in Exercise 13.5:

$$\eta_D^2 = \frac{SS_{delay}}{SS_{total}} = \frac{188.578}{1971.778} = .10$$

$$\eta_A^2 = \frac{SS_{Area}}{SS_{total}} = \frac{356.044}{1971.778} = .18$$

$$\eta_{DA}^2 = \frac{SS_{DA}}{SS_{total}} = \frac{371.956}{1971.778} = .19$$

$$\omega_D^2 = \frac{SS_{delay} - (d-1)MS_{error}}{SS_{total} + MS_{error}}$$

$$= \frac{188.578 - (2)(29.311)}{1971.778 + 29.311} = .06$$

$$\omega_A^2 = \frac{SS_{area} - (a-1)MS_{error}}{SS_{total} + MS_{error}}$$

$$= \frac{356.044 - (2)(29.311)}{1971.778 + 29.311} = .15$$

$$\omega_{DA}^2 = \frac{SS_{DA} - (d-1)(a-1)MS_{error}}{SS_{total} + MS_{error}}$$

$$= \frac{371.956 - (2)(2)(29.311)}{1971.778 + 29.311} = .13$$

13.23

Source	df	SS	MS	F
Experience	3	2931.667	977.222	3.544*
Intensity	2	2326.250	1163.125	4.218*
Cond Stim	1	4563.333	4563.333	16.550*
$E \times I$	6	67.083	11.181	<1
$E \times C$	3	4615.000	1538.333	5.579*
$I \times C$	2	55.417	27.708	<1
$E \times I \times C$	6	121.250	20.208	<1
Error	96	26,471.000	275.740	
Total	119	41,151.000		

*$p < .05$ [$F_{.05}(1, 96) = 3.94$; $F_{.05}(2, 96) = 3.09$;
$F_{.05}(3, 96) = 2.70$; $F_{.05}(6, 96) = 2.19$]

There are significant main effects for all variables with a significant Experience × Conditioned Stimulus interaction.

13.25 Analysis of Epineq.dat: see table at bottom of page.

13.27 Tukey on Dosage data from Exercise 13.25

Multiple Comparisons

Dependent Variable: Trials to reversal
Tukey HSD

(I) dosage of epinephrine	(J) dosage of epinephrine	Mean Difference (I − J)	Std. Error	Sig.
0.0 mg/kg	0.3 mg/kg	−1.67*	.35	.000
	1.0 mg/kg	1.03*	.35	.010
0.3 mg/kg	0.0 mg/kg	1.67*	.35	.000
	1.0 mg/kg	2.69*	.35	.000
1.0 mg/kg	0.0 mg/kg	−1.03*	.35	.010
	0.3 mg/kg	−2.69*	.35	.000

Based on observed means.

* The mean difference is significant at the .05 level.

All of these groups differed from each other at $p \leq .05$.

13.29 Simple effects on data in Exercise 13.28

Source	df	SS	MS	F
Condition	1	918.750	918.75	34.42*
Cond @ Inexp.	1	1014.00	1014.00	37.99*
Cond @ Exp.	1	121.50	121.50	4.55*
Cond × Exper	1	216.750	216.75	8.12*
Other Effects	9	2631.417		
Error	36	961.000	26.694	
Total	47	4727.917		

*$p < .05$ [$F_{.05}(1, 36) = 4.12$]

Results for Exercise 13.25

Tests of Between-Subjects Effects

Dependent Variable: Trials to reversal

Source	Type III Sum of Squares	df	Mean Square	F	Sig.
Corrected Model	141.130[a]	8	17.641	8.158	.000
Intercept	1153.787	1	1153.787	533.554	.000
DOSE	133.130	2	66.565	30.782	.000
DELAY	2.296	2	1.148	.531	.590
DOSE × DELAY	5.704	4	1.426	.659	.622
Error	214.083	99	2.162		
Total	1509.000	108			
Corrected Total	355.213	107			

[a] R Squared = .397 (Adjusted R Squared = .349)

Chapter 14

14.1 (a)

$$X_{ij} = \mu + \pi_i + \tau_j + \pi\tau_{ij} + e_{ij} \ \text{ or } \ X_{ij} = \mu + \pi_i + \tau_j + e_{ij}$$

(b)

Source	df	SS	MS	F
Subjects	7	189666.66		
Within subj	16	5266.67		
Test session	2	1808.33	904.17	3.66 ns
Error	14	3458.33	247.02	
Total	23	194,933.33		

$F_{.05}(2, 14) = 3.74$

(c) There is no significant difference among the session means—scores don't increase as a function of experience.

14.3

Source	df	SS	MS	F
Between subj	19	106.475		
Groups	1	1.125	1.125	<1
Ss w/in Grps	18	105.250	5.847	
Within subj	20	83.500		
Phase	1	38.025	38.025	15.26*
$P \times G$	1	0.625	0.625	<1
$P \times Ss$ w/in Grps	18	44.850	2.492	
Total	39	189.975		

$*p < .05 \quad [F_{.05}(1, 18) = 4.41]$

There is a significant change from baseline to training, but it does not occur differentially between the two groups, and there are no overall differences between the groups.

14.5 (a)

Source	df	SS	MS	F
Between subj	29	159.7333		
Groups	2	11.4333	5.7166	1.04
Ss w/in Grps	27	148.3000	5.4926	
Within subj	30	95.0000		
Phase	1	19.2667	19.2667	9.44*
$P \times G$	2	20.6333	10.3165	5.06*
$P \times Ss$ w/in Grps	27	55.1000	2.0407	
Total	59	254.733		

(c) Reinforcement or attention is sufficient to produce a change in baseline performance, but the additional control group demonstrates that this change is real and not one that would have happened anyway.

14.7 (b) $\hat{e} = .771$

14.9 (a) and (b)

$SS_{\text{group at add}} = 0.9$	$F < 1$
$SS_{\text{group at subt}} = 3.6$	$F = 1.96$
$SS_{\text{group at mult}} = 52.9$	$F = 28.86^*$
$SS_{\text{prob at calc}} = 78.534$	$F = 112.19^*$
$SS_{\text{prob at noncalc}} = 11.20$	$F = 5.51^*$

$*p < .05 \quad [F_{.05}(2, 16) = 3.63]$

14.11 (a) $SS_{\text{reading at child}} = 50.7$; $F = 11.655^*$

(b) $SS_{\text{items at adult good}} = 4.133$; $F = 4.133^*$

14.13 There would be a very decided lack of independence among items because an increase in one category would necessitate a decrease in another—that is, the subject would have less opportunity to draw from all categories.

14.17 (b) The F for MEAN is a test on H_0: $\mu = 0$.

(c) $MS_{\text{within cell}}$ is the average of the cell variances.

14.19 Source column of summary table for a four-way ANOVA with repeated measures on A and B

Source
Between Ss
C
D
CD
Ss w/in groups
Within Ss
A
AC
AD
ACD
$A \times Ss$ w/in groups
B
BC
BD
BCD
$B \times Ss$ w/in groups
AB
ABC
ABD
$ABCD$
$AB \times Ss$ w/in groups
Total

14.21 By using MANOVA in Exercise 14.20, we have gained freedom from the sphericity assumption, but at the potential loss of a small amount of power.

14.23

Source	df	SS	MS	F	Pillai F	Prob
Between subj.	97	137.683				
Gender	1	7.296	7.296	5.64*		
Role	1	8.402	8.402	6.49*		
$G \times R$	1	0.298	0.298	<1		
Ss w/in Grps	94	121.687	1.294			
Within subj.	97	87.390				
Time	1	1.064	1.064	1.23*	1.23	0.2700
$T \times G$	1	0.451	0.451	<1	0.52	0.4720
$T \times R$	1	0.001	0.001	<1	0.00	0.9708
$T \times G \times R$	1	4.652	4.652	5.38*	5.38	0.0225
T Ss w/in grps	94	81.222	0.864			
Total	194	225.073				

(c) The univariate and multivariate F values agree because we have only two levels of each independent variable.

14.25 (a) The problem with ignoring the fact that people are from the same family is that the responses are no longer independent.

(b) This would be less of a problem if the correlations between family members were very low.

(c) An alternative would be to restrict the analysis to only one respondent per family, or to treat family members as another repeated measure.

14.27 Intraclass correlation

$$IC = \frac{MS_{\text{subjects}} - MS_{J \times S}}{\left\{ \begin{array}{c} MS_{\text{subjects}} + (j-1)MS_{J \times S} \\ + j(MS_{\text{Judge}} - MS_{J \times S})/n \end{array} \right\}}$$

$$= \frac{82.57 - 4.08}{82.57 + 2(4.08) + 3(70.12 - 4.08)/20}$$

$$= \frac{78.49}{82.57 + 8.16 + 6.30} = .85$$

14.29 I would leave the variability due to Judge out of my calculations entirely.

14.31 The fact that the "parent" who supplies the data changes from case to case simply adds additional variability to our data, and this variability is confounded with differences between children.

Chapter 15

15.1 Predicting Quality of Life:

(a) All other variables held constant, a difference of +1 degree in Temperature is associated with a difference of −.01 in perceived Quality of life. A difference of $1000 in median Income, again all other variables held constant, is associated with a +.05 difference in perceived Quality of Life. A similar interpretation applies to b_3 and b_4. Values of

0.00 cannot reasonably occur for all predictors, so the intercept has no meaningful interpretation.

(b) $\hat{Y} = 5.37 - .01(55) + .05(12)$
$+ .003(500) - .01(200)$
$= 4.92$

(c) $\hat{Y} = 5.37 - .01(55) + .05(12)$
$+ .003(100) - .01(200)$
$= 3.72$

15.3 I would thus delete Temperature because it has the smallest $t (t = -1.104)$, and, therefore, the smallest semipartial correlation with the dependent variable.

15.5 (a) Environment has the largest semipartial correlation with the criterion because it has the largest value of t.

(b) The gain in prediction (from $r = .58$ to $R = .697$), which we obtain by using all the predictors is more than offset by the loss of power we sustain as p became large relative to N.

15.7 As the correlation between two variables decreases, the amount of variance in a third variable that they share decreases. Thus, the higher will be the possible squared semipartial correlation of each variable with the criterion. They each can account for more previously unexplained variation.

15.9 Numsup and Respon are fairly well correlated with the other predictors, whereas YRS is nearly independent of them.

15.13 $R^2_{\text{adj}} = \text{est } R^{2*} = -.158$. Because a squared value cannot be negative, we will declare it undefined. This is all the more reasonable given that we cannot reject $H_0: R^* = 0$.

15.17 It has no meaning because we have the data for the population of interest (the 10 districts).

15.19 It plays an important role through its correlation with the residual components of the other variables.

15.21 Within the context of a multiple-regression equation, we cannot look at one variable alone. The slope for one variable is only the slope for that variable when all other variables are held constant.

15.25 (b) The value of R^2 was virtually unaffected. However, the standard error of the regression coefficient for PVLoss increased from 0.105 to 0.178. Tolerance for PVLoss decreased from .981 to .345, whereas VIF increased from 1.019 to 2.900.

(c) PVTotal should not be included in the model because it is redundant with the other variables.

15.27

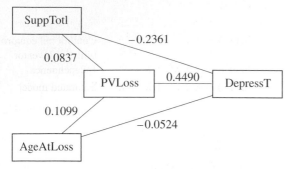

15.29 (a) Case #104 has the largest value of Cook's D (.137) but not a very large Studentized residual ($t = -1.88$).

(b) When we delete this case, the squared multiple correlation is increased slightly. More importantly, the standard error of regression and the standard error of one of the predictors (PVLoss) also decrease slightly. This case is not sufficiently extreme to have a major impact on the data.

15.31 The only predictor that contributes significantly is the Offensiveness of the behavior, which has a Wald χ^2 of 26.43. The exponentiation of the regression coefficient yields 0.9547. This would suggest that as the offensiveness of the behavior increases, the likelihood of reporting *decreases*. That's an odd result. But remember that we have all variables in the model. If we were simply predicting reporting by using Offensiveness, exp(B) = 1.65, which mean that a 1-point increase in Offensiveness multiplies the odds of reporting by 1.65. We have some work to do to make sense of these data. I leave that to you.

15.33 It may well be that the frequency of the behavior is tied in with its offensiveness, which is related to the likelihood of reporting. In fact, the correlation between those two variables is .20, which is significant at $p < .000$. (I think my explanation would be more convincing if Frequency were a significant predictor when used on its own.)

15.35 Most statisticians dislike stepwise regression because it is very dependent on chance results. A different set of data on the same study could easily produce a quite different solution.

15.37 You should see no change in the interaction term when you center the data, but you should see important differ-

ences in the "main effects" themselves. Look at the matrix of intercorrelations of the predictors.

15.39 $z = 12.47$, which would lead us to conclude that there is a significant mediating relationship in Jose's data. Rumination does mediate the relationship between stress and depression.

Chapter 16

16.1 (b)

Source	df	SS	MS	F
Group	2	57.733	28.867	9.312*
Error	12	37.200	3.100	
Total	14	94.933		

*$p < .05$ [$F_{.05}(2, 12) = 3.89$]

16.3 (a)

Source	df	SS	MS	F
Group	2	79.0095	39.5048	14.92*
Error	18	47.6571	2.6476	
Total	20	126.6666		

*$p < .05$ [$F_{.05}(2, 18) = 3.55$]

16.5

Source	df	SS	MS	F
Gender	1	65.333	65.333	7.730*
SES	2	338.667	169.333	20.034*
$G \times S$	2	18.667	9.333	1.104
Error	42	355.000	8.452	
Total	47	777.667		

*$p < .05$ [$F_{.05}(1, 42) = 4.08$; $F_{.05}(2, 42) = 3.23$]

16.7

Source	df	SS	MS	F
Gender	1	60.015	60.015	7.21*
SES	2	346.389	173.195	20.80*
$G \times S$	2	21.095	10.547	1.27
Error	35	291.467	8.328	
Total	40			

*$p < .05$ [$F_{.05}(1, 35) = 4.12$; $F_{.05}(2, 35) = 3.27$]

16.9 $\hat{\mu} = 13.4167$; $\alpha_1 = 1.167$; $\beta_1 = -3.167$; $\beta_2 = -0.167$; $\alpha\beta_{11} = 0.833$; $\alpha\beta_{12} = -0.167$

16.11 If we are actually dealing with unweighted means, SS_A and SS_B will be 0 because means of means are 7 for all rows and columns.

16.13

SS(total)

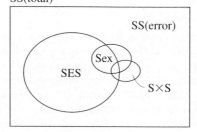

16.15 (a) Design matrix using only the first entry in each group for illustration purposes:

$$X = \begin{bmatrix} 1 & 0 & 58 & 75 \\ \cdots & \cdots & \cdots & \cdots \\ 0 & 1 & 60 & 70 \\ \cdots & \cdots & \cdots & \cdots \\ -1 & -1 & 75 & 80 \end{bmatrix}$$

(b)

Source	df	SS	MS	F
Covariate	1	1250.6779	1250.6779	55.81*
A (Group)	2	652.9228	326.4614	14.57*
Error	11	246.5221	22.4111	
Total	14	2615.733		

* $p < .05$ [$F_{.05}(1, 11) = 4.84$; $F_{.05}(2, 11) = 3.98$]

16.17 $\hat{Y} = -7.9099A_1 + 0.8786A_2 - 2.4022B$
$\qquad + 0.5667AB_{11} + 0.1311AB_{21} + 0.7260C$
$\qquad + 6.3740$

16.21 Analysis of GSIT in Mireault.dat: see table below.

16.23 If we assume that college represents a stressful environment that influences the general level of symptoms exhibited by students, we would want to eliminate that source of variability from our data. The significant F for YearColl indicates that, in fact, the Global Symptom Index varies as a function of year in college. The significant effect that we found for Gender, with males scoring higher than females, is no longer significant once we take YearColl into account.

16.25 One of the most obvious problems is that males and females differed substantially on condom use at pretest.

Adjusting for pretest differences would indirectly control for Gender, which is not something we are likely to want to do.

Chapter 17

17.1 Possible models for data on idiomatic communication.

$\ln(F_{ij}) = \lambda$	Equiprobability
$\ln(F_{ij}) = \lambda + \lambda^F$	Conditional equiprobability on Function
$\ln(F_{ij}) = \lambda + \lambda^I$	Conditional equiprobability on Inventor
$\ln(F_{ij}) = \lambda + \lambda^F + \lambda^I$	Independence
$\ln(F_{ij}) = \lambda + \lambda^F + \lambda^I + \lambda^{FI}$	Saturated model

17.3 **Lambda** values for the complete (saturated) model.

(a) $\lambda = 2.8761 = $ mean of $\ln(\text{cell}_{ij})$

(b) $\lambda^{\text{Inventor}} = .199 \;\; .540 \; -.739$

The effect for "female partner" is .199, indicating that the ln(frequencies in row 1) are slightly above average.

(c) $\lambda^{\text{Function}} = -.632 \;\; .260 \;\; .222 \; -.007 \;\; .097 \; -.667$
$\qquad\qquad\quad .303 \; -.029 \;\; .452$

The effect of Confrontation is $-.632$. Confrontation contributes somewhat less than its share of idiosyncratic expressions.

(d)

$$\lambda^{\text{Inventor} \times \text{Function}} = \begin{matrix} .196 & -.039 & \cdots & .249 & .028 \\ -.481 & .038 & \cdots & .250 & .259 \\ .286 & .001 & \cdots & -.500 & -.287 \end{matrix}$$

The unique effect of cell$_{11}$ is .196. It contributes slightly more than would be predicted from the row and column totals above.

Results for Exercise 16.21

Tests of Between-Subjects Effects

Dependent Variable: GSIT

Source	Type III Sum of Squares	df	Mean Square	F	Sig.
Corrected Model	1216.924[a]	5	243.385	2.923	.013
Intercept	1094707.516	1	1094707.516	13146.193	.000
GENDER	652.727	1	652.727	7.839	.005
GROUP	98.343	2	49.172	.590	.555
GENDER × GROUP	419.722	2	209.861	2.520	.082
Error	30727.305	369	83.272		
Total	1475553.000	375			
Corrected Total	31944.229	374			

[a] R Squared = .038 (Adjusted R Squared = .025)

17.5 For females, the odds in favor of a direct hit are 6.00, whereas for males, they are only 2.8125. This leaves an odds ratio of $6.00/2.8125 = 2.133$. A female odds of a direct hit are 2.133 times greater than the odds for a male.

17.7 Letting S represent Satisfaction, G represent Gender, and V represent Validity, and with 0.50 added to all cells because of small frequencies, the optimal model is

$$\ln(F_{ij}) = \lambda + \lambda^G + \lambda^S + \lambda^V + \lambda^{SV}$$

The χ^2 for this model is 4.53 on 5 df, for $p = .4763$.

17.9 You should examine the pattern of changes in the alternative designs. Although the marginal frequencies stay constant from design to design, the chi-square tests on those effects, the values of λ, and the tests on λ change as variables are added. This differs from what we see in the analysis of variance, where sums of squares remain unchanged as we look at additional independent variables (all other things being equal).

17.11 Odds delinquent:

Normal testosterone, low SES $= 190/1104 = .1721$

High testosterone, low SES $= 62/140 = .4429$

Normal testosterone, high SES $= 53/1114 = .0476$

High testosterone, high SES $= 3/70 = .0429$

17.13 The optimal model that results is one including all main effects and first-order interactions, but not the three-way interactions. The value of χ^2 for this model is 3.52 on 1 df, for $p = .0607$. If any main effect or interaction were dropped from the model, the χ^2 would be significant.

17.15 When you add Gender to the analysis of Pugh's data, it contributes significantly to the fit of the model. Pugh forced the Gender $\times$ Stigma interaction into the model, even though its removal did not significantly affect the fit of the model because the sampling plan for the study was based on those variables. Other models fit nearly as well, and for the specific arguments in favor of this model, you should see Pugh's paper.

Chapter 18

18.1 (a) $W_s = 23$; $W_{.025} = 27$.

(b) Reject H_0 and conclude that older children include more inferences in their summaries.

18.3 $z = -3.15$; reject H_0.

18.5 (a) $T = 8.5$; $T_{.025} = 8$; do not reject H_0.

(b) We cannot conclude that we have evidence supporting the hypothesis that there is a reliable increase in hypothesis generation and testing over time. (This is a case in which alternative methods of breaking ties could lead to different conclusions.)

18.7 I would randomly assign the order within each pair of Before and After scores, and for each set of assignments I would calculate a statistic. (That statistic could be the mean of the difference scores, or a t test on the difference scores.) I would then calculate the number of times I came out with a result as extreme as the one I actually obtained, and that, divided by the number of resamples, would give me the probability under the null.

18.9 $z = -2.20$; $p(z \geq \pm 2.20 = .0278)$. Again reject H_0, which agrees with our earlier conclusion.

18.11 The scatter plot shows that the difference between the pairs is heavily dependent upon the score for the first born.

18.13 The Wilcoxon matched-pairs signed-ranks test tests the null hypothesis that paired scores were drawn from identical populations or from symmetric populations with the same mean (and median). The corresponding t test tests the null hypothesis that the paired scores were drawn from populations with the same mean and assumes normality.

18.15 Rejection of the H_0 by a t test is a more specific statement than rejection using the appropriate distribution free test because, by making assumptions about normality and homogeneity of variance, the t test refers specifically to population means.

18.17 $H = 6.757$; reject H_0.

18.19 Take the data for all N subjects and shuffle them to random order. Then take the first n_1 observations and assign them to Treatment 1, the next n_2 observations and assign them to Treatment 2, and so on. Then calculate an F statistic on that set of resampled data and record the F. Repeat this a large number of times (e.g., 1,000) and look at the sampling distribution of F. The proportion of F values that are equal to, or greater than the F obtained on the original data will give you the probability under the null.

18.21 The study in Exercise 18.18 has an advantage over the one in Exercise 18.17 in that it eliminates the influence of individual differences (differences in overall level of truancy from one person to another).

18.23 These are equivalent tests in this case.

Index

TO THE OWNER OF THIS BOOK:

I hope that you have found *Statistical Methods for Psychology*, Sixth Edition useful. So that this book can be improved in a future edition, would you take the time to complete this sheet and return it? Thank you.

School and address:_____

Department:_____

Instructor's name:_____

1. What I like most about this book is:_____

_____

_____

2. What I like least about this book is:

_____

_____

3. My general reaction to this book is:

_____

_____

4. The name of the course in which I used this book is:

_____

5. Were all of the chapters of the book assigned for you to read?_____

   If not, which ones weren't?_____

6. In the space below, or on a separate sheet of paper, please write specific suggestions for
   improving this book and anything else you'd care to share about your experience
   in using this book.

_____

_____

_____

_____

FOLD HERE

THOMSON

WADSWORTH

NO POSTAGE
NECESSARY
IF MAILED
IN THE
UNITED STATES

BUSINESS REPLY MAIL
FIRST-CLASS MAIL   PERMIT NO. 102   MONTEREY CA

POSTAGE WILL BE PAID BY ADDRESSEE

Attn: Vicki Knight, Psychology Publisher

Wadsworth/Thomson Learning
60 Garden Ct Ste 205
Monterey CA 93940-9967

FOLD HERE

OPTIONAL:

Your name:_____ Date: _____

May we quote you, either in promotion for *Statistical Methods for Psychology*, Sixth Edition, or in future publishing ventures?

Yes: _____ No: _____

Sincerely yours,

David C. Howell